Introduction to
CRIMINOLOGY

edition 7

Titles of Related Interest

Introduction to
CRIMINOLOGY
Theories, Methods, and Criminal Behavior

edition 7

FRANK E. HAGAN
Mercyhurst College

Los Angeles | London | New Delhi
Singapore | Washington DC

For information:

SAGE Publications, Inc.
2455 Teller Road
Thousand Oaks, California 91320
E-mail: order@sagepub.com

SAGE Publications Ltd.
1 Oliver's Yard
55 City Road
London EC1Y 1SP
United Kingdom

SAGE Publications India Pvt. Ltd.
B 1/I 1 Mohan Cooperative Industrial Area
Mathura Road, New Delhi 110 044
India

SAGE Publications Asia-Pacific Pte. Ltd.
33 Pekin Street #02-01
Far East Square
Singapore 048763

Printed in Canada

Library of Congress Cataloging-in-Publication Data

Hagan, Frank E.
Introduction to criminology: Theories, methods, and criminal behavior/Frank E. Hagan.—7th ed.
       p. cm.
Includes bibliographical references and index.
ISBN 978-1-4129-7971-9 (pbk.)
    1. Criminology. 2. Criminology—United States. I. Title.

HV6025.H26 2011
364—dc22   2009043891

Printed on acid-free paper.

10   11   12   13   14   10   9   8   7   6   5   4   3   2

Acquiring Editor:	Jerry Westby
Assistant Editor:	Leah Mori
Editorial Assistant:	Nichole O'Grady
Production Editor:	Sarah K. Quesenberry
Typesetter:	C&M Digitals (P) Ltd.
Proofreader:	Jenifer Kooiman
Cover Designer:	Bryan Fishman
Cover Artist:	Olga Storm

Brief Contents

Preface   xvii

Acknowledgments   xix

1   Introduction   1

2   Research Methods in Criminology   21

3   General Characteristics of Crime and Criminals   55

4   Early and Classical Criminological Theories   91

5   Biological and Psychological Theories   115

6   Sociological Mainstream Theories   145

7   Sociological Critical Theories and Integrated Theories   175

8   Violent Crimes   207

9   Property Crime: Occasional, Conventional, and Professional   247

10   White Collar Crime: Occupational and Corporate   295

11   Political Crime and Terrorism   343

12   Organized Crime   381

13   Public Order Crime   425

14   Computer Crime   453

15   Epilogue: The Future of Crime   471

Glossary   481

References   491

Photo Credits   539

Index   541

About the Author   553

Contents

Preface   xvii

Acknowledgments   xix

1 Introduction   1

Criminology   1
Fads and Fashions in Crime   2
The Emergence of Criminology   3

Crime and Deviance   3
■ Crime File 1.1 The FBI's Ten Most Wanted Fugitives   4
Sumner's Types of Norms   6
Mala in Se *and* Mala Prohibita   6

Social Change and the Emergence of Law   7
Consensus Versus Conflict Versus Interactionist Model of Law   8
■ Crime File 1.2 Crimes of the Twentieth Century   9

Crime and Criminal Law   10
Who Defines Crime? Criminological Definitions   11
■ Crime File 1.3 What Is Crime?   12

The Crime Problem   13
The Cost of Crime   14

Summary   16
■ Key Concepts   17 ■ Review Questions   17 ■ Web Sources   18
■ Web Exercises   18
■ Selected Readings   18

2 Research Methods in Criminology   21

Theory and Methodology   21

The Research Enterprise of Criminology   22
Objectivity   22
Ethics in Criminological Research   22

Who Is Criminal?   24

Official Police Statistics—The Uniform Crime Reports (UCR)   25
Sources of Crime Statistics   25
The Crime Indexes: Violent and Property Crime   27
Issues and Cautions in Studying UCR Data   29
■ Crime File 2.1 The Crime Dip   32

Alternative Data-Gathering Strategies   33

Experiments in Criminology   34
Some Examples of Experiments in Criminology   35
Evidence-Based Research   37

Surveys   38
Victim Surveys   38
National Crime Victimization Surveys (NCVS)   38

■ Crime File 2.2 Were You a Victim of Crime? 39
Issues and Cautions in Studying Victim Data 40
Self-Report Measures of Crime 41
■ Crime File 2.3 Self-Reported Delinquency Items 41

Participant Observation 42
Participant Observation of Criminals 43
Evaluation of the Method of Participant Observation 43

Life History and Case Studies 44
■ Crime File 2.4 Confessions of a Dying Thief 45

Unobtrusive Measures 45
■ Crime File 2.5 Useful Sources for Criminological Research 46

Validity, Reliability, and Triangulation 47
■ Crime File 2.6 The FBI Reading Room 48

Summary 50
▪ Key Concepts 51
▪ Review Questions 51
▪ Web Sources 52 ▪ Web Exercises 52
▪ Selected Readings 52

3 General Characteristics of Crime and Criminals 55

Caution in Interpreting Crime Data 55

International Variations in Crime 56
The Prevalence of Crime 59

Trends in Crime 60
■ Crime File 3.1 American Crime Problems From a Global Perspective 62
Age and Crime 65
■ Crime File 3.2 What Is the Relationship Between Age and Crime? 66
Gender Differences in Criminality 68
Social Class and Crime 70
Race and Crime 71
■ Crime File 3.3 Racial Profiling 72
■ Crime File 3.4 Native Americans and Crime 74
Regional Variation in Crime 77
Urban/Rural Differences in Crime 77

Institutions and Crime 78
The Family and Crime 78
Education and Crime 80
Religion and Crime 81
War and Crime 82
Economy and Crime 82
Mass Media and Crime 83

Summary 85
▪ Key Concepts 87 ▪ Review Questions 87 ▪ Web Sources 88
▪ Web Exercises 88 ▪ Selected Readings 88

4 Early and Classical Criminological Theories 91

Theory 92
■ Crime File 4.1 The Nacirema Undergraduate as Criminal: A Criminological "Why do it?" 93

Major Theoretical Approaches 95
Demonological Theory 96
Classical Theory 97
Neoclassical Theory 100
■ Crime File 4.2 "Designing Out" Gang Homicides and Street Assaults: Situational Crime Prevention 102
■ Crime File 4.3 Justifications for Punishment 103
Ecological Theory 104
Forerunners of Modern Criminological Thought 108
Economic Theory 108

The Theory–Policy Connection 110

Summary 111
▪ Key Concepts 112 ▪ Review
Questions 112 ▪ Web Sources 112
▪ Web Exercises 112 ▪ Selected
Readings 113

5 Biological and Psychological Theories 115

Positivist Theory 115
Precursors of Positivism 116

Biological Theories 117
Cesare Lombroso 117
Charles Goring 119
The Jukes and Kallikaks 119
Earnest Hooton 120
Body Types 120
A Critique of Early Biological Theories 121

More Recent Biological Theories 122
Brain Disorders 123
Twin Studies 123
Adoption Studies 124
Problems With Twin/Adoption Studies 124
XYY Syndrome 124
Other Biological Factors 125
A Critique of Neobiological Theories 129

Psychological Theories 130
Freudian Theory 130
Psychometry 131
Hans Eysenck 131
▪ Crime File 5.1 Crime Profiling 132
B. F. Skinner 133
Albert Bandura 133
Samuel Yochelson and Stanton Samenow 133
Intelligence and Crime 134
▪ Crime File 5.2 The Insanity Defense 136

The Theory–Policy Connection 139

Summary 140

▪ Key Concepts 142 ▪ Review
Questions 142 ▪ Web Sources 142
▪ Web Exercises 143 ▪ Selected
Readings 143

6 Sociological Mainstream Theories 145

Major Sociological Theoretical Approaches in Criminology 146

Anomie Theories 146
Émile Durkheim and Anomie 146
Merton's Theory of Anomie 147
Robert Agnew's General Strain Theory (GST) 150

7 Sociological Critical Theories and Integrated Theories   175

Mainstream Versus Critical Criminology   175

Critical Criminology   176

Labeling Theory   177
Lemert's "Secondary Deviance"   178
A Critique of Labeling Theory   179
John Braithwaite's Shaming Theory   180

Conflict Criminology   181
Austin Turk   181
*William Chambliss and Richard Quinney—
    Conflict Theory*   182
W. E. B. Du Bois   182
Jeffrey Reiman   183

Feminist Criminology   183

New Critical Criminology   185
Left Realism   185
Peacemaking   186
■ Crime File 7.1 Incorporating Restorative
    and Community Justice Into American
    Sentencing and Corrections   187
Postmodernism   190

Radical "Marxist" Criminology   190
Richard Quinney—Radical Criminology   190
William Chambliss   191

Conflict Versus Marxist Criminology   192

**Critiques of Conflict and Radical
    Criminology**   192
Klockars's Critique   193

Integrated Theories of Crime   193

Subcultural Theories   151
Cohen's Lower-Class Reaction Theory   151
*Cloward and Ohlin's Differential Opportunity
    Theory*   152

Social Process Theories   153
The Chicago School   153
*Shaw and McKay's Social Disorganization
    Theory*   154
*Sutherland's Theory of Differential
    Association*   157
■ Crime File 6.1 Designing Out Crime   158
Miller's Focal Concerns Theory   160
Matza's Theory of Delinquency and Drift   162

Social Control Theories   163
Reckless's Containment Theory   164
Hirschi's Social Bond Theory   165
*Gottfredson and Hirschi's General Theory of
    Crime*   166
John Hagan's Power-Control Theory   166

**Developmental and Life Course (DLC)
Theories**   167
*Farrington's Antisocial Potential (AP)
    Theory*   168
*Sampson and Laub's Life Course
    Criminality*   168

The Theory–Policy Connection   169

Summary   170
■ Key Concepts   171 ■ Review
Questions   172 ■ Web Sources   172
■ Web Exercises   172 ■ Selected
Readings   173

Delbert Elliott's Integrative Theory 194
Terence Thornberry's Interactional Theory 195

Criminal Typologies 195
A Critique of Typologies 196
A Defense of Typologies 196
Criminal Behavior Systems 196
■ Crime File 7.2 Some Sociological Typologies
of Criminal Behavior 197

Theoretical Range and Criminological
Explanation 198
The Global Fallacy 199

The Theory–Policy Connection 200

Summary 203
■ Key Concepts 204 ■ Review
Questions 204 ■ Web Sources 204
■ Web Exercises 205 ■ Selected
Readings 205

8 Violent Crimes 207

History of Violence in the United States 207
Murder and Mayhem 208
Types of Multiple Murders: Multicide 209
■ Crime File 8.1 International Violent
Crime 210
■ Crime File 8.2 The Virginia Tech
Massacre 212
Victim Precipitation 213
■ Crime File 8.3 The D.C. Snipers, the BTK
Killer, and the Red Lake Massacre 214
Typology of Violent Offenders 216

Legal Aspects 216

Homicide and Assault Statistics 216

Patterns and Trends in Violent Crime 219
Workplace Violence 220
■ Crime File 8.4 Workplace Violence: Issues
in Response 221
School Violence 222
Guns 222
■ Crime File 8.5 Deadly Lessons: The Secret
Service Study of School Shooters 223

Sexual Assault 225
Acquaintance Rape 227
■ Crime File 8.6 The Problem of
Acquaintance Rape of College
Students 228

Amir Versus Brownmiller 229
Rape as a Violent Act 229

Robbery 231
■ Crime File 8.7 Name That Bank
Robber 232
Conklin's Typology of Robbers 233

Domestic Violence 234
Child Abuse 235
Spouse Abuse 236
Elder Abuse 238
Kidnapping 238

Criminal Careers of Violent Offenders 238
Culture of Violence 238
Subculture of Violence 239
Career Criminals/Violent Offenders 240

Societal Reaction 240

Theory and Crime 241

Summary 242
■ Key Concepts 243 ■ Review
Questions 244 ■ Web Sources 244
■ Web Exercises ■ Selected
Readings 245

9 Property Crime: Occasional, Conventional, and Professional 247

Occasional Property Crimes 248
Shoplifting 248
Vandalism 250
■ Crime File 9.1 Graffiti 252

Motor Vehicle Theft 253
Check Forgery 253

Conventional Property Crimes 255
Burglary 256
Fencing Operations 257
Stings 258
Larceny-Theft 258

Arson: A Special-Category Offense 259

Criminal Careers of Occasional and Conventional Property Criminals 260

Societal Reaction 261

Professional Crime 262
The Concept of Professional Crime 263

Characteristics of Professional Crime 264
Argot 264
A Model of Professional Crime 265
Edelhertz's Typology 266

Scams 267

Big Cons 269
Maurer's The Big Con 269
Identity Theft 269
Ponzi Schemes 269
■ Crime File 9.2 Identity Theft 270
Pyramid Schemes 272
■ Crime File 9.3 The Bernie Madoff Affair: "One Big Lie" 273
Religious Cons 273
The PTL Scandal 274
■ Crime File 9.4 Emerging Patterns of Professional Crime 275
■ Crime File 9.5 Nigerian Letter Scams 276

Boosters 276

Cannons 277
■ Crime File 9.6 Shoplifting 278

Professional Burglars 279
The Box Man 280

The Professional Fence 280

Paper Hangers 281
■ Crime File 9.7 Intellectual Property Theft 282
■ Crime File 9.8 Busting the Biggest Band of Cable Pirates in U.S. History 283

Professional Robbers 284

Professional Arsonists 285

Professional Auto Theft Rings 285

Professional Killers 286
■ Crime File 9.9 Car Cloning: A New Twist on an Old Crime 287

Criminal Careers of Professionals 287

Societal Reaction 289

Theory and Crime 290

Summary 290
■ Key Concepts 292 ■ Review Questions 292 ■ Web Sources 293 ■ Web Exercises 293 ■ Selected Readings 293

10 White Collar Crime: Occupational and Corporate 295

White Collar Crime—The Classic Statement 295
Related Concepts 296

The Measurement and Cost of Occupational and Corporate Crime 297

The History of Corporate, Organizational, and Occupational Crime 299

Legal Regulation 302
Occupations and the Law 302
Organizations and the Law 302

Occupational Crime 305
Crimes by Employees 305
Crimes by Employees Against Individuals (the Public) 305

11 Political Crime and Terrorism 343

Ideology 343

Political Crime: A Definition 344
- ■ Crime File 11.1 September 11, 2001 345

Legal Aspects 346
The Nuremberg Principle 346
*The Universal Declaration of Human
  Rights* 347
International Law 348

Crimes by Government 348
Secret Police 348
Human Rights Violations 349
Patriarchal Crime 350
Genocide 351
Crimes by Police 352
*Illegal Surveillance, Disruption, and
  Experiments* 352
Scandal 354
- ■ Crime File 11.2 White House Crime
  and Scandal: From Washington to
  George W. Bush 356

Crimes Against Government 358
Protest and Dissent 358
Social Movements 359
Assassination 360
Espionage 361
- ■ Crime File 11.3 The Robert Hanssen Spy
  Case 363
Political "Whistleblowing" 365
Terrorism 365
- ■ Crime File 11.4 The Turner Diaries,
  ZOG, and the Silent Brotherhood—The
  Order 370

Crimes by Employees Against Employees 310
*Crimes by Employees Against
  Organizations* 311
*Crimes by Individuals (or Members of
  Occupations)* 312

Corporate Crime 316
*Crimes by Organizations/Corporations Against
  Individuals (the Public)* 316
- ■ Crime File 10.1 Financial Crimes: FBI
  Releases Annual Report to the Public 320
- ■ Crime File 10.2 The Great Savings and Loan
  Scandal: The Biggest White Collar Crime in
  U.S. History 322
- ■ Crime File 10.3 The Donora Fluoride Death
  Fog: A Secret History of America's Worst Air
  Pollution Disaster 326
*Crimes by Organizations Against
  Employees* 328
*Crimes by Organizations (Corporations) Against
  Organizations* 331
- ■ Crime File 10.4 Pirates of the Internet:
  Criminal Warez Groups 333

Criminal Careers of Occupational and
Organizational Offenders 334
Corporate Environment and Crime 334
Corporate Concentration 334
Rationalizations 335

Societal Reaction 335
Why the Leniency in Punishment? 336

Theory and Crime 337

Summary 338
- ■ Key Concepts 339 ■ Review
  Questions 340 ■ Web Sources 340
- ■ Web Exercises 340 ■ Selected
  Readings 341

■ Crime File 11.5 Patterns of Global
Terrorism 372

Criminal Careers of Political Criminals 373
The Doctrine of Raison d'Etat **374**
Terrorism and Social Policy 375

Societal Reaction 375

Theory and Crime 376

Summary 376
■ Key Concepts 378 ■ Review
Questions 378 ■ Web Sources 378
■ Web Exercises 379 ■ Selected
Readings 379

12 Organized Crime 381

**Organized Crime: A Problematic
Definition 381**

**Sources of Information on Organized
Crime 382**

**Types of Organized Crime (Generic
Definitions) 383**

The Organized Crime Continuum 385

Street Gangs 386

International Organized Crime 388
Yakuza 388
Chinese Triad Societies 389
Russian Organized Crime 391

The Nature of Organized Crime 393
Ethnicity and Organized Crime 393

Money Laundering 394

Drug Trafficking 396

Colombian Cartels 397
The Underground Empire 397
Mexico's Drug War 397
■ Crime File 12.1 Amado Carrillo Fuentes
and Operation Casablanca 398

**Theories of the Nature of Syndicate Crime in
the United States 399**
*The Cosa Nostra Theory (The Cressey
Model)* 399
■ Crime File 12.2 The Origin of the
Mafia 400
The Patron Theory (The Albini Model) 402
The Italian American Syndicate (IAS) 402

The Classic Pattern of Organized Crime 403
Strategic and Tactical Crimes 404
Illegal Businesses and Activities 404
■ Crime File 12.3 "Snakeheads" and Software
Mobsters 406
■ Crime File 12.4 Mobsters, Unions, and the
Feds 407
Big Business and Government 409

**A Brief History of Organized Crime in the
United States 410**
Before 1930 410
The Luciano Period 412
The Genovese Period 413
The Apalachin Meetings 413
The Gambino Period 414
The Commission Trials 414

Criminal Careers of Organized Criminals 416

Public and Legal Reaction 417
Drug Control Strategies 417
Investigative Procedures 417
Laws and Organized Crime 418

Theory and Crime 419

Summary 420
■ Key Concepts 422 ■ Review
Questions 422 ■ Web Sources 422
■ Web Exercises 423 ■ Selected
Readings 423

13 Public Order Crime 425

Nuts, Guts, Sluts, and "Preverts" 426

Broken Windows 426

Prostitution 427

Types of Prostitution 428
Massage Parlors 430
Johns 430
Underaged Prostitutes 431

Homosexual Behavior 431
■ Crime File 13.1 Laud Humphreys's
Tearoom Trade 433

Sexual Offenses 433
Paraphilia 434
Nonvictimless Sexual Offenses 435
Sexual Predators 435
■ Crime File 13.2 Child Sexual Abuse by
Catholic Priests 436
Incest 437
Characteristics of Sex Offenders 438

Drug Abuse 438
Drugs and History 439
■ Crime File 13.3 Moral Panics and the
Strange Career of Captain Richmond
Hobson—Moral Entrepreneur 440
Drug Use in the United States: The Drug Dip? 441
Drug Abuse and Crime 441
Drunkenness 443

Special Populations 444

Societal Reaction 444
Overcriminalization 445
Decriminalization 446

Theory and Crime 447

Summary 448
Key Concepts 449 ■ Review
Questions 449 ■ Web Sources 450
■ Web Exercises 450 ■ Selected
Readings 450

14 Computer Crime 453

Computer Crime 453
Types of Computer Crime 454
Types of Attacks on Computer Systems 456
Argot of Computer Crime 457
■ Crime File 14.1 Operation Bot Roast: Bot-
Herders Charged as Part of Initiative 458
■ Crime File 14.2 Cracking Down on Sexual
Predators on the Internet 460
Online Predators 460
■ Crime File 14.3 The Bogeyman: Online
Sexual Predators 461
■ Crime File 14.4 Protecting Children
in Cyberspace: The ICAC Task Force
Program 462

Cyberterrorism 464

Public and Legal Reaction 464

Theory and Crime 464
■ Crime File 14.5 Cyberspace Security:
Breaking Ground in the New Frontier 465
■ Crime File 14.6 A Fine Point: Mapping Intel
Sources 466

Summary 467
■ Key Concepts 467 ■ Review
Questions 468 ■ Web Sources 468
■ Web Exercises 468 ■ Selected
Readings 469

Crimewarps 473
The Future of Digital Crime 473
Other Predictions 474
British Home Office Predictions 474
■ Crime File 15.1 Anticipating Future Trends in Crime and Disorder Audits 474
■ Crime File 15.2 Hot Products: Understanding, Anticipating, and Reducing Demand for Stolen Goods 477

Summary 478
  ▪ Key Concepts 478 ▪ Review Questions 479 ▪ Web Sources 479 ▪ Web Exercises 479 ▪ Selected Readings 479

15 Epilogue: The Future of Crime 471

The Future of Crime 471
  Predicting the Future of Crime: Methods 471
  Other Crime Predictions 472

Glossary 481

References 491

Photo Credits 539

Index 541

About the Author 553

Preface

The seventh edition maintains the purpose of the original text: to serve the needs of instructors in criminology who wish to avoid the excessively legal and crime-control orientation of many recent textbooks. Certainly, some familiarity with the legal and crime-control orientation is both necessary and desirable, but in emphasizing these elements, some introductory texts give short shrift to the real and vital core of criminology—theory, method, and criminal behavior. To overstress detailed analyses of social-control agencies while neglecting to provide adequate descriptions of criminal activity produces a text that would more accurately be called an introduction to criminal justice systems. An introduction to criminology, by contrast, should offer thorough descriptions and explanations of criminal behavior, because that is the basis on which effective social policy and social agencies must be developed. Many recent texts have also become increasingly encyclopedic, attempting to cover everything ever written in the field in one introductory class. This text views itself as an introductory one that will hopefully whet students' appetite for the field without overwhelming them.

This book is intended for the introductory criminology class typically offered in the sophomore or junior year. It is written for both the college-university as well as community college markets. Professors are welcome to alter the order in which they present the chapters in their classes. Chapter 1 offers a general introduction to the field, while Chapter 2 examines the area of research methods. General patterns and variations in crime are the focus of Chapter 3 while Chapters 4–7 explore the subject of theory beginning with early and classical theories (Chapter 4), biological and psychological theories (Chapter 5), sociological mainstream theories (Chapter 6) and ending with critical and integrated theories (Chapter 7). Chapters 8–13 examine specific types of criminal behavior. Violent crime in Chapter 8 is followed by property crime in Chapter 9.

Chapter 10 details the world of white collar crime, Chapter 11 undertakes to explain the world of political crime, and organized crime is analyzed in Chapter 12. Finally, Chapter 13 discusses public order crime. New chapters on computer crime (Chapter 14) and the future of crime (Chapter 15) have been added to this edition.

Many new topics have been added to this edition including: "the Lucifer effect in research," "Confessions of a Dying Thief," new "theory and crime" and the "theory–policy connection" sections , new Web sources and Web exercises at the end of each chapter, "Acquaintance Rape of College Students," "Name That Bank Robber," and "The Bogeyman: Online Sexual Predators."

New Crime File selections include "Graffiti," "Shoplifting," and "Car Cloning." The Bernie Madoff Ponzi scheme is discussed. New trends in espionage and the Mexican drug wars are also examined. "Operation Bot Roast," "Protecting Children in Cyberspace," and predicting the future of crime are also featured.

A password-protected instructor resources site is available to qualified instructors at www.sage pub.com/haganintrocrim7e. Materials on the site include PowerPoint presentations, class activities, and Web resources. An electronic test bank CD is also available by request. To acquire a copy of the CD, please contact SAGE Customer Care at 1-800-818-7243.

An open-access companion student study site is also available with the text at www.sagepub.com/haganintrocrim7e. The site features resources such as e-flashcards, chapter quizzes, SAGE journal articles, author-created podcasts, and much more.

Acknowledgments

Seven editions of a book require much help along the way. I would like to thank the many people who assisted in this endeavor. For their help with earlier editions, I again thank Jonathan Turner (University of California, Riverside), Lawrence Travis III (University of Cincinnati), George Evans (William Rainey Harper College), John Burian (Moraine Valley Community College), E. Ernest Wood (Edinboro University of Pennsylvania), and Sylvia Hill (University of the District of Columbia). I would also like to thank Stan Shernock (Norwich University), Peter Kratcowski (Kent State University), Susan Williams (Kansas State University), Nanette Davis (Western Washington University), Frank Taylor (Rhode Island Community College), and Barbara Perry (Northern Arizona University). Other reviewers include: Raymond Michalowski (Northern Arizona University), Dennis Brewster (Oklahoma State University), Jonathan Cella (Central Texas College), and Pamela Wilcox (University of Cincinnati). The reviewers for this edition were: Brian Baker (Penn State), David Baker (University of Toledo), Allan Barnes (University of Alaska, Anchorage), Alison Burke (Southern Oregon University), Cory Colyer (West Virginia University), Carol Erbes (Old Dominion University), Joshua Guetzkow (University of Arizona), Alexander Gerould (San Francisco State University), Helen Lim (California Lutheran University), and Emmanuel Onyeozili (University of Maryland, Eastern Shore).

I would also like to acknowledge my debt to my first criminology professor, the late Dan Koenig (University of Victoria). My appreciation is also extended to the Sage Publications team led by Senior Editor Jerry Westby. Others included Associate Editors Elise Smith and Leah Mori, Editorial Assistant Eve Oettinger, and Production Editor Supervisor Laureen Gleason. Special thanks to Production Editor Sarah Quesenberry, who managed to keep this project on track despite my apparent efforts to sabotage it.

Finally, I would like to dedicate this book to my granddaughter, Lily Alise Glennon.

Criminology
  Fads and Fashions in Crime
  The Emergence of Criminology
Crime and Deviance
  Crime File 1.1 The FBI's Ten Most Wanted Fugitives
  Sumner's Types of Norms
  Mala in Se and *Mala Prohibita*
Social Change and the Emergence of Law
  Consensus Versus Conflict Versus Interactionist Model of Law
  Crime File 1.2 Crimes of the Twentieth Century
Crime and Criminal Law
  Who Defines Crime? Criminological Definitions
  Crime File 1.3 What Is Crime?
The Crime Problem
  The Cost of Crime
Summary
Key Concepts
Review Questions
Web Sources
Web Exercises
Selected Readings

chapter 1

Introduction

Imagine a society of saints, a perfect cloister of exemplary individuals. Crimes, properly so-called, will there be unknown; but faults which appear venial to the layman will create there the same scandal that the ordinary offense does in ordinary consciousness.

—Émile Durkheim (1895/1950, pp. 68–69)

Crime is a sociopolitical artifact, not a natural phenomenon. . . . We can have as much or as little crime as we please, depending on what we choose to count as criminal.

—Herbert Packer (1968, p. 364)

Criminology

Remorseless suicidal terrorists hijack four airplanes and, with all passengers aboard, are successful in crashing two of these into the World Trade Center and one into the Pentagon, murdering nearly 3,000 people in the worst terrorist attack in history. A disturbed student at Virginia Tech University kills 32 in the worst mass murder in U.S. history. Major corporations and their accounting firms conspire and cause a major stock market plunge, losing stockholders billions of dollars.

What all of these events have in common is that they refer to various forms of criminal behavior; as we have just begun the twenty-first century, we can only guess what new, unforeseen horrors await us. The field that addresses this issue of crime and criminal behavior and attempts to define, explain, and predict it is criminology.

Criminology is generally defined as the science or discipline that studies crime and criminal behavior. Specifically, the field of criminology concentrates on forms of criminal behavior, the causes of crime, the definition of criminality, and the societal reaction to criminal activity; related areas of inquiry may include juvenile delinquency and victimology (the study of victims). While there is considerable overlap between criminology and criminal justice, criminology shows a greater interest in the causal explanations of crime, whereas criminal justice is more occupied with practical, applied concerns, such as technical aspects of policing and corrections. In reality, the fields are highly complementary and interrelated, as indicated by overlapping membership in the two professional organizations representative of the fields: the American Society of Criminology and the Academy of Criminal Justice Sciences.

If you tell your friends that you are taking a course in criminology, many will assume that you are a budding Sherlock Holmes, on your way to becoming a master detective trained in investigating crime scenes. That describes the field of *criminalistics* (the scientific evaluation of physical evidence), which is sometimes confused in the media and public mind with criminology. Criminology is more concerned with analyzing the phenomena of crime and criminality, in performing scientifically accurate studies, and in developing sound theoretical explanations of crime and criminal behavior. It is hoped that such criminological knowledge and scientific research can inform and direct public policies to solve some crime problems. The major concentration in this text will be on the central areas of criminal behavior, research methodology, and criminological theory. Of particular interest will be the exploration of crime typologies, the attempt to classify various criminal activity and criminals by type.

Fads and Fashions in Crime

Video Link 1.1
View a video on Bernie Madoff.

A variety of crimes were of major concern in the past but appear in modern societies only in old movies on the late show. Train robbery, piracy, stagecoach robbery, cattle rustling, gunfights such as that at the OK Corral, and grave robbery have some modern remnants, but for the most part have disappeared. Some of these practices have reappeared in different forms. In the seventies, South Vietnamese "boat people" attempting to escape from their homeland were robbed, raped, and murdered by Thai pirates. In the early twenty-first century, pirates operating off the coast of Somalia took over ships and held their occupants for ransom. Brinks trucks have replaced stagecoaches, and semi-trailer trucks full of prepared beef are hijacked instead of herds of live cattle. Post–Civil War gangs of Wild West robbers such as those of Doc Holliday, Jesse James, the Daltons, Black Bart, the Younger brothers, and Butch Cassidy disappeared with the settlement of the frontier only to reappear on wheels during the Depression of the thirties in the persons of such infamous characters as John Dillinger, "Pretty Boy" Floyd, the Barrows, Bonnie Parker, and the Ma Barker gang. Mobile, organized gangs of bank robbers have largely faded into a quaint, unsavory history; they are now replaced by cybercriminals who can commit global electronic robbery.

Photo 1.1
Group portrait of a police department liquor squad posing with cases of confiscated alcohol and distilling equipment during Prohibition.

Skyjacking, a major problem in the sixties, was virtually eliminated as a result of better security measures, only to reappear in the United States in the early eighties as an attempt by Cuban refugees to escape or by suicidal terrorists to wreak mass destruction. The skyjacking of four jumbo jets with the intention of using them as weapons of international terrorism represented the horrific events of 9/11. Kidnapping, a major concern in the United States in the thirties (as illustrated by the famous Lindbergh case), is less of a concern today despite the rash of child kidnappings by noncustodial parents. On the other hand, since the seventies, kidnapping has become a major crime in Italy, as best illustrated by the highly publicized kidnapping of billionaire J. Paul Getty's grandson; the kidnappers mailed one of the young man's ears to a daily newspaper to impress upon the family the seriousness of their intentions. In 1995 in Colombia, a kidnapping was reported every 6 hours. This was believed to have been precipitated by huge income disparities and inefficient police. The United States, by contrast,

has experienced fewer than 12 kidnappings for ransom every year (Brooke, 1995, p. A7). Slavery continues to be practiced in the form of human trafficking. Nostalgic views of the past tend to romanticize bygone violence or suppress its memory. Most apt to be forgotten are conditions of the past that more than match any chronicle of horrors of the present.

Crime File 1.1 examines the FBI's "Ten Most Wanted Fugitives" list. It features photographs of the most wanted criminals, including one of Osama bin Laden. Consult http://www.fbi.gov for the most recent list.

The Emergence of Criminology

French sociologist Auguste Comte (1798–1857) viewed the **progression of knowledge** as consisting of three stages, from the predominantly *theological* explanations to *metaphysical* (philosophical) approaches to *scientific* explanations (Comte, 1851/1877). Prior to the emergence of modern criminal law in the eighteenth century, religion was the primary basis of social control beyond kinship organization. Theological explanations used supernatural or otherworldly bases for understanding reality. Recall, for instance, the papal condemnation of Galileo for heretically questioning biblical descriptions of the earth and of astronomy. In the metaphysical stage, philosophy sought secular (worldly) events to provide understanding through a new spirit of inquiry—rationality and logical argument. The two features of the scientific stage combined this rational spirit of investigation with the scientific method, emphasizing empiricism or experimentation. The scientific orientation emphasized measurement, observation, proof, replication (repetition of observation), and verification (analyzing the validity of observations).

Systematic application of the scientific method enabled humankind to unlock many of the mysteries of the ages. At first, breakthroughs in knowledge took place in the physical sciences; more recently, changes have also begun to occur in the social sciences, such as sociology and criminology. Since the scientific method provided major understanding and ability to predict and control physical reality, the hope is that these same methods are applicable to and will prove useful in the social sciences. While many view criminology as a science, others, such as Sutherland and Cressey (1974), view it as an art similar to medicine, a field based on many sciences and disciplines.

Criminology as a field of inquiry had its beginnings in Europe in the late 1700s in the writings of various philosophers, physicians, physical scientists, sociologists, and social scientists. Much of the early theory was heavily couched in biological frameworks that have largely been abandoned by modern American criminology (Gibbons, 1982). Criminology emerged along with eighteenth-century criminal law. In fact, it was the early writings of Cesare Beccaria (1738–1794), especially his famous essay *On Crimes and Punishments* (1963), which was first published in 1764, that led to the reform of criminal law in Western Europe.

Despite its European roots, most of the major developments in modern criminology took place in the United States. Criminology was closely linked with the development of sociology, gaining its place on the U.S. academic scene between 1920 and 1940. Criminology had been largely a subdiscipline of sociology; even though criminology is interdisciplinary in focus, sociologists have devoted the most attention to the issue of criminality. Since the 1960s, criminology has emerged as a discipline in its own right. The earliest U.S. textbooks in the field were by Maurice Parmelee, John Gillin, Philip Parsons, and Fred Hayes, but it was the text and later writings of Edwin H. Sutherland, the acknowledged "dean of criminology," that received the most deserved recognition.

Handbook Article Link 1.1
Read an article on criminology as social science.

Crime and Deviance

Deviance or *deviant behavior* may refer to a broad range of activities that the majority in society may view as eccentric, dangerous, annoying, bizarre, outlandish, gross, abhorrent, and the like. It refers to behavior that is outside the range of normal societal toleration.

Video Link 1.2
View a video on the pornography industry.

Crime File 1.1

The FBI's Ten Most Wanted Fugitives

In 1950, a news reporter asked the FBI for the 10 worst "tough guys" that they were hunting. The resulting publicity was so great that the list became an official FBI program. It satisfied the public's hunger for details about notorious criminals and served as a means of exposing fugitives and encouraging citizen participation.

Edward Eugene Harper

Usama Bin Laden

Jason Derek Brown

James J. Bulger

Emigdio Preciado Jr

Robert William Fisher

Victor Manuel Gerena

Glen Stewart Godwin

Jorge Alberto Lopez-Orozco

Alexis Flores

The FBI claims that since the program's initiation, 134 of the "Ten Most Wanted Fugitives" have been apprehended as a result of citizen recognition. Perhaps the most memorable case was the arrest of bank robber Willie Sutton when a clothing salesman recognized him on the New York City subway. After the citizen's story was run in *The New York Times*, mobster Albert Anastasia had the salesman killed because, as he stated, "I hate squealers."

The list has reflected very well the social climate of various time periods in the United States. The 1950s list consisted primarily of bank robbers, burglars, and car thieves, while the 1960s version featured revolutionaries and radicals. The 1970s list was dominated by organized criminals and terrorists, and while this emphasis continues, serial murderers and drug-related offenders abound in later lists. A recent "Ten Most Wanted Fugitives" list (shown on previous page) features the following:

Edward Harper—conspiracy to commit sexual battery, child fondling, sexual battery

Jason Brown—first degree murder, armed robbery

Jorge Lopez-Orozco—unlawful flight to avoid prosecution, murder (murdered three people: a woman and two young sons, aged 2 and 4)

Emigdio Preciado, Jr.—unlawful flight to avoid prosecution, attempted murder of a police officer

Victor Manuel Gerena—bank robbery, unlawful flight to avoid prosecution, armed robbery, theft from interstate shipment

Usama Bin Laden—murders of U.S. nationals, attack on federal facility resulting in death

James J. Bulger—racketeering influenced corrupt organizations (RICO), murder (18 counts), conspiracy to commit murder, extortion, narcotics distribution, money laundering

Robert W. Fisher—unlawful flight to avoid prosecution, first degree murder (3 counts), arson

Glen S. Godwin—unlawful flight to avoid confinement, murder, escape

Alexis Flores—unlawful flight to avoid prosecution, kidnapping, murder

Sources: "In Demand for 50 Years: The FBI's 'Most Wanted' List: Good Publicity, and a History of Success," by J. Glasser, March 20, 2000, *U.S. News and World Report*, p. 60.

Web Research Project
Visit the FBI Web site and examine the latest "Ten Most Wanted Fugitives" list. Are any of the people the same as in our "Crime File" section? If yes, do they differ in any way from the "Crime File" above? Are any of them women or white collar criminals?

Visit http://www.fbi.gov and report on any changes in the FBI's "Ten Most Wanted Fugitives" list from the ones discussed in this Crime File. Also examine its "Most Wanted Terrorists" list. While visiting the FBI site, look up "Headline Archive: Top Ten Quiz on the Top Ten Program" and see how many questions you can answer correctly.

Definitions of deviance are relative to the time, the place, and the person(s) making the evaluation, and some acts are more universally defined than others. For instance, in the mid-nineteenth century in the United States, bathing in a tub was considered immoral as well as unhealthy.

All societies have *cultural values*—practices and beliefs that are prized by or believed to be of benefit to the group. For instance, despite cultural relativity in defining deviance, anthropologists have identified a number of cultural universals—practices or customs that in general form exist in all known cultures. All cultures that have been studied look dimly on indiscriminate lying, cheating, stealing, and killing. Societies protect their values by creating norms, which are basically rules or prescribed modes of conduct.

Sumner's Types of Norms

Audio Link 1.1
Listen to a discussion of norms of civil behavior.

Early American sociologist William Graham Sumner, in his classic work *Folkways* (1906), identifies three types of **norms:** folkways, mores, and laws. These norms reflect the values of a given culture; some norms are regarded by its members as more important than others. **Folkways** are the least serious norms and refer to usages, traditions, customs, or niceties that are preferred, but are not subject to serious sanctions: manners, etiquette, and dress styles, for example. The character Reb Tevye in the musical *Fiddler on the Roof,* when learning that his daughter has rejected the marriage mate chosen by the matchmaker, wails, "Tradition—without our traditions, life would be as precarious as a fiddler on the roof." Recognizing changing times or folkways, however, he ultimately accepts his daughter's decision to choose her own mate. **Mores** refer to more serious customs that involve moral judgments as well as sanctions (rewards or punishments). The mores cover prohibitions against behaviors that are felt to be seriously threatening to a group's way of life. Our previous examples of lying, cheating, stealing, and killing are most certainly included in the mores. Both folkways and mores are examples of informal modes of social control and are characteristic of small, homogeneous cultures that feature simple technology and wide-scale consensus.

Laws represent formal modes of control, codified rules of behavior. If one accepts the consensus model of law (to be discussed shortly), laws represent an institutionalization or "crystallization" of the mores.

Mala in Se and *Mala Prohibita*

We have already identified deviant acts as those that violate group expectations and crime as any act that violates criminal law. Crime and its definition are social products. Society (human groups) decides what is a crime and what is not.

Criminologists make the distinction between acts *mala prohibita* and acts *mala in se.* Acts that are defined as *mala prohibita* refer to those that are "bad because they have been prohibited." That is, such acts are not viewed as bad in themselves but are violations because the law defines them as such. Traffic violations, gambling, and infractions of various municipal ordinances might serve as examples.

Photo 1.2

Drug abuse is an example of overcriminalization.

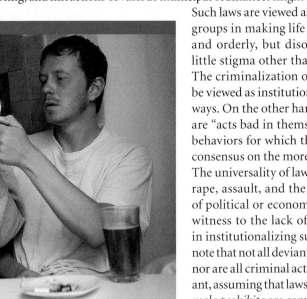

Such laws are viewed as assisting human groups in making life more predictable and orderly, but disobedience carries little stigma other than (usually) fines. The criminalization of such acts might be viewed as institutionalization of folkways. On the other hand, acts *mala in se* are "acts bad in themselves," forbidden behaviors for which there is wide-scale consensus on the mores for prohibition. The universality of laws against murder, rape, assault, and the like, irrespective of political or economic systems, bears witness to the lack of societal conflict in institutionalizing such laws. One can note that not all deviant acts are criminal, nor are all criminal acts necessarily deviant, assuming that laws against many acts *mala prohibita* are commonly violated.

Definitions of criminal activity may exhibit both undercriminalization and overcriminalization. **Undercriminalization** refers to the fact that the criminal law fails to prohibit acts that many feel are *mala in se*. Elements of corporate violence, racism, structured inequality, and systematic wrongdoing by political officials are examples. **Overcriminalization** involves the overextension of criminal law to cover acts that are inappropriately or not responsibly enforced by such measures. Examples are the legislation of morality and attempts to regulate personal conduct that does not involve a clear victim (drug abuse, sexual conduct, and the like). Morris and Hawkins (1970) claim that the United States has one of the most moralistic systems of criminal law in history, although one might suspect that ecclesiastical regimes such as Iran would more than give it a run for the money.

Social Change and the Emergence of Law

Western societies have undergone a long-term evolutionary development from sacred or Gemeinschaft societies to secular or Gesellschaft societies (Becker, 1950; Toennies, 1957). **Gemeinschaft** societies are simple, communal, relatively homogeneous societies that lack an extensive division of labor and are also characterized by normative consensus. Social control is assured by the family, extended kinship groups, and the community through informal modes of control: the folkways and mores. Such societies lack and do not need formally codified laws since sacred tradition, the lack of change, and cultural similarity and isolation assure a degree of understanding and control. **Gesellschaft** societies are complex, associational, more individualistic, and heterogeneous (pluralistic); they are characterized by secularity, an extensive division of labor, and (in free societies) a variety of moral views and political pressure groups. Social control is attempted by means of formal controls—codified laws administered by bureaucratic agencies of the state. Complex societies must rely more and more on such formal controls. As the mores or informal modes of control become weaker, the need for laws becomes greater. For example, as the family as an agent of social control becomes weaker, much of its responsibility is passed on to the state.

Sumner (1906) suggested a general maxim: If laws do not have the support of, or are not in agreement with, the mores of a particular culture, they will be ineffective. The introduction of changes or new laws in society can be explored using Merton's (1961) concepts of manifest and latent functions. The classic example is what has been described as "the noble experiment," the Prohibition Era in the United States. **Manifest functions** are intended, planned, or anticipated consequences of introduced changes or of existing social arrangements. In perhaps the last gasp of rural Protestant religious power in the United States, one group managed to pressure Congress into passing the Eighteenth Amendment prohibiting alcohol in 1919. Alcohol abuse was (and still is) a major problem, and the well-intended goal was for it to be stamped out by totally forbidding alcohol consumption by law. **Latent functions** entail unintended or unanticipated consequences, ones that may have either positive or negative outcomes. The latent functions of Prohibition included increased corruption, disobedience, and public disrespect for the law. By eliminating legitimate suppliers of a commodity in high public demand, the state in effect created a monopoly for illegitimate entrepreneurs. It was Prohibition that converted small, localized gangs into large, powerful, and wealthy regional and even national organized criminal syndicates.

Laws are by no means the most efficient means of social control; the passage of more and more laws may indicate that social solidarity and informal modes of control in the society are weakening. The police and the criminal justice system become the agents or agencies of last resort. Many people view crime as an evil intrusion into an otherwise healthy society, whereas increased crime levels may be latent functions of increased freedom, affluence, competition, and other desirable manifest functions in society. Sociologist Durkheim (1950) suggested that crime may be a normality, a positive product, a functional necessity in a healthy society. As reflected in the quotation with which we began this chapter, Durkheim's theory of the **functional necessity of crime** proposes that wrongdoing or

Photo 1.3

The Criminal Hall of Fame showcases wax replicas of notorious criminals throughout history.

crime serves to force societal members to react, condemn, and thus establish the borders of and reconfirm societal values. It is this organized resentment that upholds social solidarity.

The phrase "crime of the century" seems to be perennially used to refer to the latest dramatic crime. Crime File 1.2 explores crimes to which this label was attached over this past century.

Consensus Versus Conflict Versus Interactionist Model of Law

The **consensus model** of the origin of criminal law envisions it as arising from agreement among the members of a society as to what constitutes wrongdoing. Reflecting the "social contract theory" of Locke, Hobbes, and Rousseau, criminal law is viewed, as in our previous discussion of Sumner, as a "crystallization of the mores," reflecting social values that are commonly held within the society. The **conflict model,** on the other hand, sees the criminal law as originating in the conflict of interests of different groups. In this view, the definition of crime is assumed to reflect the wishes of the most powerful interest groups, who gain the assistance of the state in opposing rival groups. The criminal law, then, is used primarily to control the behavior of the "defective, dependent, and delinquent," the dangerous classes (Skolnick & Currie, 1988, p. 2); the crimes of the wealthy are very often not even covered. While the consensus model views criminal law as a mechanism of social control, the conflict approach sees the law as a means of preserving the status quo on behalf of the powerful.

A third model of law is the *interactionist* approach, which takes its name from the symbolic interactionist school of criminology. This school of thought views humans as responding to abstract meanings and symbols as well as to concrete meanings. According to George Herbert Mead (1934), even the mind and self-consciousness are social creations. Reflected in labeling theory (see Chapter 7), criminality is viewed as a label or stigma attached by a societal reaction that is subject to shifting standards. Laws are viewed as reflecting moral entrepreneurship on the part of labelers.

Crime File 1.2

Crimes of the Twentieth Century

Every year, it seems some particularly notorious or atrocious crime occurs that is described by the media as "the crime of the century." Now that the twentieth century is over, we might take stock of some that have been candidates. APBnews.com (http://www.apbnews.com), an Internet service specializing in crime news, chose the "Ten Crimes of the Century" based on input from its editors, historians, criminal justice experts, and users who voted in its poll, as well as those answering its telephone survey. The "Ten Crimes of the Century" in the APBnews.com survey, listed chronologically, were:

President McKinley's assassination

The St. Valentine's Day massacre

The Lindbergh baby kidnapping

The Rosenberg spy trial

President Kennedy's assassination

Martin Luther King, Jr.'s assassination

Watergate break-in

Ted Bundy serial killings

O. J. Simpson trial

Oklahoma City bombing

The assassination of President William McKinley in 1901 by Leon Czolgosz was a political crime in support of a hoped-for class revolt, while the St. Valentine's Day massacre by the Capone mob in the 1920s illustrated the ascendancy of ruthless organized crime groups during Prohibition. The tragic kidnapping and murder of the Lindbergh baby led to legislation designating kidnapping as a federal offense. The trial and subsequent execution of Julius and Ethel Rosenberg, native-born Americans who betrayed their country by giving America's atomic secrets to the Soviets, solidified the Cold War. The assassinations of President John F. Kennedy in 1963 and 5 years later of civil rights leader the Rev. Dr. Martin Luther King, Jr., gave rise to numerous conspiracy theories that secret, sinister forces were responsible.

The Watergate affair in the 1970s would lead to the first forced resignation of an elected president in disgrace in American history, and it remains the benchmark against which all political scandals are compared. Ted Bundy, the serial killer, represents just one of a number of bizarre multiple killers who seemed to proliferate in post–World War II America. The O. J. Simpson murder trial, in which a former National Football League star was found not guilty of murdering his ex-wife and her friend, despite considerable evidence to the contrary, exemplifies the numbers of celebrity cases that have attracted public attention over the years. Finally, the Oklahoma City terrorist bombing (and the 1993 World Trade Center bombing, which is not listed) demonstrated the growing vulnerability to terrorism in modern society. The 1995 Oklahoma City bombing represented the worst terrorist attack, in terms of casualties, on American soil up to that time. It also punctuated for a complacent America the fact that "it can happen here."

In its "Crime Stories of the Century," *U.S. News and World Report* included

Murder of Stanford White

Execution of IWW leader Joe Hill

St. Valentine's Day massacre

Lindbergh kidnapping

Rosenberg spy case

Lynching of Emmett Till

Charles Manson murders

Audio Link 1.2
Listen to a court reporter discuss memorable trials.

Video Link 1.3
View a video on the O. J. Simpson trial.

(Continued)

(Continued)

"Son of Sam" killings

Jeffrey Dahmer, cannibal

While the *U.S. News* list includes many of the same entries as APBnews.com, it also includes lesser-known events such as the high-society murder of Stanford White, a prominent architect, due to a romantic dispute. The execution of labor leader Joe Hill, of the radical union International Workers of the World, for allegedly killing company owners exemplifies the labor unrest in the early twentieth century. Other additions are more serial murderers: Manson; "Son of Sam" Berkowitz; and Jeffrey Dahmer, the personification of our worst nightmares. Many of these acts changed the country, inspired new laws, mesmerized a nation waiting for a verdict, or tore at the American collective conscience. While we might not agree with the specific selection of "crimes of the century," most candidates share a celebrity quality, bizarre violent characteristics, or political implications. In fact, of those listed on the APB list, 6 of the 10 involved political crime, that is, crime for ideological purposes by those supporting a cause. The remainder illustrated organized crime, celebrity involvement (Lindbergh and Simpson), or bizarre violence (Bundy). Bundy seems to be a stand-in for any number of monsters of multiple murder in the twentieth century. Note also that the list is of crimes in the United States and does not include crimes such as Hitler's Holocaust, for example.

While the fascinating and mesmerizing nature of these crimes gives them a timeless quality that still enthralls the public—a dance macabre that appalls yet entices—it is their rare, atypical quality that gives them notoriety. The typical picture of crime in most societies is far less dramatic, but often just as deadly, traumatic, or fear-inspiring. Domestic violence, rape, robbery, murder, burglary, and theft bring crime up close and personal to its victims and will be more the subject of this text.

Source: List of "crime stories" taken from "Crime Stories of the Century," by A. Cannon, December 6, 1999, *U.S. News and World Report.*

Web Research Project

What do you think was the "Crime of the Twentieth Century"? Visit the Web and see if you can find other nominees for a "Top Ten Crimes of the Century" list.

Visit http://www.fbi.gov and find an interesting investigation that they have posted on their site.

Crime and Criminal Law

A *purist legal view* of crime would define it as violation of criminal law. No matter how morally outrageous or unacceptable an act, it is not a crime unless defined as such by criminal law. Vernon Fox (1985) indicates, "Crime is a sociopolitical event rather than a clinical condition. . . . It is not a clinical or medical condition which can be diagnosed and specifically treated" (p. 28). In this view, which is technically correct, unless an act is specifically prohibited by criminal law, it is not a crime. There are four **characteristics of criminal law:**

1. It is assumed by political authority. The state assumes the role of plaintiff, or the party bringing charges. Murder, for example, is no longer just an offense against a person, but is also a crime against the state. In fact, the state prohibits individual revenge in such matters; perpetrators must pay their debt to society, not to the individual wronged.

2. It must be specific, defining both the offense and the prescribed punishment.

3. The law is uniformly applied. That is, equal punishment and fairness for all, irrespective of social position, are intended.

4. The law contains penal sanctions enforced by punishments administered by the state (Sutherland & Cressey, 1974, pp. 4–7).

Criminal law has very specific criteria: "Crime is an intentional act or omission in violation of criminal law (statutory and case law), committed without defense or justification, and sanctioned by the state as a felony or misdemeanor" (Tappan, 1960, p. 10). *Felonies* generally refer to offenses punishable by a year or more in a state or federal prison, whereas *misdemeanors* are less serious offenses punished by less than a year in jail. Some specific criteria that must be met in the U.S. criminal law in order for an act to be considered a crime include the following:

1. The act is *prohibited by law* and contains legally prescribed punishments. *Nullum crimen sine lege* (no crime without law) is the Latin expression, which can be expanded to include the notion that *ex post facto* (after-the-fact) laws are inappropriate. The act must be forbidden by law in advance of the act.

2. A criminal act, *actus reus* (the act itself, or the physical element), must have taken place.

3. Social harm of a conscious, voluntary nature is required. There must be injury to the state or to people.

4. The act is performed intentionally (although cases of negligence and omission may be exceptions). *Mens rea* (criminal intent or "guilty mind") is important in establishing guilt. A person who may have committed a criminal act (for example, John Hinckley, who shot former President Reagan) may be found not guilty under certain conditions, such as insanity or a history of mental disturbance.

5. The voluntary misconduct must be causally related to the harm. It must be shown that the decision or act directly or indirectly caused harm.

Crimes were originally considered to be private matters: The offended party had to seek private compensation or revenge. Later, only offenses committed against the king and, still later, the king's subjects were considered crimes. When compensation developed, fines were levied on behalf of the king (the state), thus making the state the wronged party. In addition to being defined by legislative statute (statutory law), criminality may also be interpreted by means of case law (common law). In contrast to laws enacted by legislatures, common law is based on judicial decision, with its roots in precedence, or previous decisions. In addition, administrative law, as enforced by federal regulatory agencies, may carry criminal penalties for offenders. Thus, criminal law provisions may be contained in statutory law, common law, and administrative law. Crime File 1.3 describes some typical legal definitions of crime in the United States.

Who Defines Crime? Criminological Definitions

Since crime was previously defined as any violation of criminal law, should criminologists restrict their inquiry solely to acts so defined? Should the subject matter of criminology be decided by lawyers and politicians? This would relegate the field of criminology to a position as status quo handmaiden of political systems. Hitler's genocide or Stalin's purges were accepted conduct within their political ideological systems. Criminologists must study the deviants—the criminals—as well as the social structural contexts that define them. Skolnick and Currie (1988), in examining the analysis of social problems, state,

> In spite of its claim to political neutrality, the social science of the 1960s typically focused on the symptoms of social ills, rather than their sources: criminals, rather than the laws; the mentally ill, rather than the quality of life; the culture of the poor, rather than the decisions of the rich; the "pathology" of the ghetto, rather than problems of the economy. (p. 11)

Crime File 1.3

Audio Link 1.3
Listen to a discussion on legal questions raised by the underground economy.

What Is Crime?

Crimes are defined by law.

In this report we define crime as all behaviors and acts for which a society provides formally sanctioned punishment. In the United States, what is criminal is specified in the written law, primarily state statutes. What is included in the definition of crime varies among federal, state, and local jurisdictions.

Criminologists devote a great deal of attention to defining crime in both general and specific terms. This definitional process is the first step toward the goal of obtaining accurate crime statistics.

To provide additional perspectives on crime it is sometimes viewed in ways other than those suggested by the standard legal definitions. Such alternatives define crime in terms of the type of victim (child abuse), the type of offender (white collar crime), the object of the crime (property crime), or the method of criminal activity (organized crime). Such definitions usually cover one or more of the standard legal definitions. For example, organized crime may include fraud, extortion, assault, or homicide.

What is considered criminal by society changes over time.

Some types of events, such as murder, robbery, and burglary, have been defined as crimes for centuries. Such crimes are part of the common-law definition of crime. Other types of conduct traditionally have not been viewed as crimes. As social values and mores change, society has codified some conduct as criminal while decriminalizing other conduct. The recent movement toward increased "criminalization" of drunk driving is an example of such change.

New technology also results in new types of conduct not anticipated by the law. Changes in the law may be needed to define and sanction these types of conduct. For example, the introduction of computers has added to the criminal codes in many states so that acts such as the destruction of programs or data could be defined as crimes.

What are some other common crimes in the United States?

Drug abuse violations. Offenses relating to growing, manufacturing, making, possessing, using, selling, or distributing narcotic and dangerous nonnarcotic drugs. A distinction is made between possession and sale/manufacturing.

Sex offenses. In current statistical usage, the name of a broad category of varying content, usually consisting of all offenses having a sexual element except for forcible rape and commercial sex offenses, which are defined separately.

Fraud offenses. The crime type comprising offenses sharing the elements of practice of deceit or intentional misrepresentation of fact, with the intent of unlawfully depriving a person of his or her property or legal rights.

Drunkenness. Public intoxication, except "driving under the influence."

Disturbing the peace. Unlawful interruption of the peace, quiet, or order of a community, including offenses called "disorderly conduct," "vagrancy," "loitering," "unlawful assembly," and "riot."

Driving under the influence. Driving or operating any vehicle or common carrier while drunk or under the influence of liquor or drugs.

Liquor law offenses. State or local liquor law violations, except drunkenness and driving under the influence. Federal violations are excluded.

Gambling. Unlawful staking or wagering of money or other thing of value on a game of chance or on an uncertain event.

Kidnapping. Transportation or confinement of a person without authority of law and without his or her consent, or of a minor without the consent of his or her guardian.

Vandalism. Destroying or damaging, or attempting to destroy or damage, the property of another without his or her consent, or public property—except by burning, which is arson.

Public order offenses. Violations of the peace or order of the community or threats to the public health through unacceptable public conduct, interference with governmental authority, or violation of civil rights or liberties. Weapons offenses, bribery, escape, and tax law violations, for example, are included in this category.

How do violent crimes differ from property crimes?

The outcome of a criminal event determines whether it is a property crime or a violent crime. Violent crime refers to events such as homicide, rape, and assault that may result in injury to a person. Robbery is also considered a violent crime because it involves the use or threat of force against a person.

Property crimes are unlawful acts with the intent of gaining property not involving the use or threat of force against an individual. Larceny and motor vehicle theft are examples of property crimes.

In the National Crime Survey (NCS) a distinction is also made between crimes against persons (violent crimes and personal larceny) and crimes against households (property crimes, including household larceny).

How do felonies differ from misdemeanors?

Criminal offenses are also classified according to how they are handled by the criminal justice system. Most jurisdictions recognize two classes of offenses: felonies and misdemeanors.

Felonies are not distinguished from misdemeanors in the same way in all jurisdictions, but most states define felonies as offenses punishable by a year or more in a state prison. The most serious crimes are never misdemeanors and the most minor offenses are never felonies.

Sources: Bureau of Justice Statistics, *BJS Dictionary of Criminal Justice Data Terminology,* 2nd ed. (Washington, DC: Government Printing Office, 1981); Bureau of Justice Statistics, *BJS Criminal Victimization in the U.S.* (Washington, DC: Government Printing Office,1985); FBI, *Crime in the United States 1985* (Washington, DC: Government Printing Office, 1985); Bureau of Justice Statistics, *Report to the Nation on Crime and Justice,* 2nd ed. (Washington, DC: Government Printing Office, March 1988), pp. 2–3.

Web Research Project

Using EBSCO, search laws, regulations, and criminal law. Read and discuss any article related to the Imperial Criminal Code, criminal law reform, self-defense, or *mens rea.*

Visit http://www.uscourts.gov, http://www.ljx.com/courthouse/staterules.html, and http://www.atlanet.org. What perspective on the criminal law have you gained by visiting these sites?

A *sociological view* of crime does not restrict its concept of criminality to those convicted of crime in a legal sense.

Were we to restrict analysis of crime solely to the legal definition in most countries, we would discuss primarily "crime in the streets" and ignore "crime in the suites." We would study the poor, dumb, slow criminal and conclude that low IQs and inferior genetics cause crime; we would ignore the fast, smart, slick violator and the possibility that maybe Ivy League educations and working on Wall Street or for the defense industry also cause crime. Hyperbole (exaggeration) is useful at times for effect, and obviously we must not loosely throw around the label *criminal,* but neither should we ignore dangerous acts that do great harm, simply because the criminal justice system chooses to ignore them.

Audio Link 1.4
Listen to a discussion of a cultural defense for murder.

The Crime Problem

Radzinowicz and King (1977), in commenting on the relentless international upsurge in crime in the latter decades of the twentieth century, indicate, "No national characteristics, no political regime, no system of law, police punishment, treatment, or even terror, has rendered a country exempt from crime. . . . What is indisputable is that new and much higher levels of crime become established as a reflex of affluence" (pp. 3–5). Despite rival explanations such as problems with statistics, there has been an obvious increase in crime internationally since World War II.

Photo 1.4

Young men in gangs embody the public perception of crime in the streets. In this photo, Suffolk County, New York, police officers of the antigang unit question two suspects on Long Island in October 2005. The suspects are members of the "Bloods."

Video Link 1.4
View a video on the cost of crime.

Journal Article Link 1.1
Examine literature regarding the costs of sexual violence.

The Cost of Crime

It is difficult, if not impossible, to measure the economic costs of crime. Estimates of the actual financial operation take us into the "megabucks" range where notions such as "give or take a few billion dollars" stagger the imagination and numb us to the reality of the amounts we are really talking about.

In 1992 in Los Angeles, riots broke out in response to a jury verdict of not guilty for police officers accused of the brutal videotaped beating of Rodney King. Those riots resulted in one of the bloodiest occurrences of civil unrest in the United States in recent history. Fifty-three people were killed, more than 2,000 were injured, 15,000 were arrested, and property damage was estimated at nearly $800 million. Although far less dramatic, losses at the nation's savings and loan companies in the eighties and early nineties are estimated to have cost the American taxpayer $500 billion, or 625 Los Angeles riots.

Perhaps the most ambitious comprehensive attempt to assess the total cost of crime was by David Anderson (1999) in an article titled "The Aggregate Burden of Crime." Anderson includes the cost of the legal system and criminal justice agencies as well as opportunity costs of victims', criminals', and prisoners' time; the fear of crime; and cost of private policing. His basic theme is "How much could the United States save if we had a crime-free environment?" His answer: $1.7 trillion. The aggregate burden of crime for 1998 consisted of

- Crime-induced production = $397 billion
- Opportunity (time) costs = $130 billion
- Risks to life and health = $574 billion
- Transfers = $603 billion
- Aggregate burden = $1.7 trillion
- Net transfers = $1.1 trillion

Photo 1.5
A man passes in front of a line of policemen during the 1992 L.A. riots that broke out in the aftermath of the Rodney King beating. Losses at the nation's savings and loan institutions cost the American taxpayers an estimated $500 billion, or the equivalent of 625 L.A. riots.

Crime-induced production refers to costs in resources to fight crime. This includes police ($47.1 billion), corrections ($35.9 billion), locks and safes ($4 billion), surveillance cameras ($1.4 billion), computer security ($8 billion), and federal agencies to fight crime ($23 billion) and drug trafficking (the highest cost at $160 billion).

Opportunity costs refer to lost time by potential victims and perpetrators, who could have spent their time doing something more productive. David Anderson (1999) estimates these opportunity costs as follows:

- Time spent securing assets = $89.6 billion
- Lost work days in prison and planning crime = $40 billion
- Victim lost work days = $0.8 billion
- Neighborhood Watch time = $0.7 billion
- Approximate total = $130 billion (p. 44)

For life and health costs, Anderson estimates roughly 72,000 crime-related deaths per year and 2.5 million crime-related injuries per year at $6.1 million per death and $52,637 per injury for a total of $574 billion. While these figures strike the author as somewhat high, Anderson is using accepted court, medical, and insurance estimates.

Transfers refer to money obtained through fraud and theft. It is called a transfer since the money (or property) is transferred from one person to the next. Estimated at $603.1 billion, this figure includes items such as workplace fraud ($203 billion), unpaid taxes ($123 billion), health insurance fraud ($108 billion), telemarketing fraud ($16.8 billion), and motor vehicle theft ($8.9 billion). The aggregate burden of crime in 1999 using Anderson's figures was $4,118 per person living in the United States.

While recent estimates rank the sale of illegal narcotics as the criminal world's greatest source of income, there is a problem with such assessments. These estimates do not even begin to measure the full impact of corporate price-fixing and other criminal activities. Added to these costs are economic costs incurred by victims of crime and the costs of running the criminal justice system. Not considered at all in these economic estimates are the social and psychological costs to society and to crime victims. Fear, mistrust, a curtailing of public activity, and a decline in the quality of life are but a few of the inestimable impacts of crime on society. Horror stories abound of the impact of crime on the forgotten figure in the criminal justice equation—the crime victim. As stated earlier, the costly Los Angeles riots of 1992 were dwarfed by the cost of the collapse of the nation's savings and loans.

Summary

Criminology is the science or discipline that studies crime and criminal behavior. Major areas of investigation include criminal behavior, etiology (theories of crime causation), and the sociology of law and societal reaction; related areas include juvenile delinquency and victimology. Criminology also shares with the field of criminal justice the areas of policing, the courts, and corrections.

Knowledge is defined as one's understanding of reality. This understanding is made possible through the creation of symbols or abstractions. Comte identified three stages in the progression of knowledge: the theological, metaphysical (philosophical), and scientific. *Science* combines the spirit of rationality of philosophy with the scientific method, which is characterized by the search for empirical proof. Criminology and sociology are more recent applicants for the scientific credentials already enjoyed by the physical sciences. Having its origins in the eighteenth century in Europe, particularly in the writing of Beccaria, who was influential in codifying modern continental law, criminology has largely become a twentieth-century U.S. discipline. This is particularly reflected in the work of Sutherland, who has been identified as "the dean of criminology."

Deviant behavior refers to activities that fall outside the range of normal societal toleration. Definitions of such activities are relative to time, place, and persons. *Values* are practices or beliefs that are prized in society and that are protected by *norms,* which are rules or prescribed modes of conduct. Sumner in his classic work *Folkways* identifies three types of norms: folkways, mores, and laws. While *folkways* are less serious customs or traditions, *mores* are serious norms that contain moral evaluations as well as penal sanctions. Both folkways and mores are examples of informal modes of control. *Laws*—codified rules of behavior—represent formal methods of attempting to assure social control.

Acts *mala in se* refer to acts that are "bad in themselves," such as murder, rape, and the like; acts *mala prohibita* are ones that are "bad because they are prohibited," such as vagrancy and gambling. While not all criminal acts are viewed as deviant, neither are all deviant acts criminal. *Undercriminalization* involves the failure of the law to cover acts *mala in se,* while *overcriminalization* entails overextension of the law to cover acts that may more effectively be enforced through the mores. As societies undergo transition from *Gemeinschaft* (communal, sacred societies) to *Gesellschaft* (associational, secular societies), they must rely more on formal agencies of control. In order to be effective, laws require the support of the mores.

Manifest functions are intended or planned consequences of social arrangements, whereas *latent functions* refer to unintended or unanticipated consequences. While the manifest function of Prohibition was to eliminate alcohol abuse, its latent functions were to encourage corruption, organized crime, and public disrespect. Durkheim viewed crime as a normal condition in society that served a positive function by the reactions it developed to encourage reaffirmation of values. *Crime,* a violation of criminal law, is characterized by politicality, specificity, uniformity, and sanctions.

In explaining the origin of criminal law, the *consensus model* views it as reflecting agreement or public will, while the *conflict model* claims that it represents the interest of the most powerful group(s) in society. In reality, criminal law reflects elements of both models.

For official purposes, crimes are identified as felonies, misdemeanors, and (in some states) summary offenses. Although there is variation by state in the actual assignment to categories, a *felony* refers to more serious crime that bears a penalty of at least 1 year in a state prison, whereas a *misdemeanor* is a less serious offense subject to a small fine or short imprisonment.

The issue of "Who defines crime?" should not be answered simply by accepting current definitions, since to do so would permit others to define criminology's subject matter. The crime problem is a growing international concern; the costs of crime are economic (which can only be estimated), psychological, and social in nature. The full social costs are inestimable.

> **Criminology on the Web**
>
> Log on to the Web-based student study site at http://www.sagepub.com/haganintrocrim7e/ for author-created podcasts, e-flashcards, quizzes, and more.

KEY CONCEPTS

Characteristics of Criminal Law

Consensus Versus Conflict Model of Law

Costs of Crime

Crime

Criminal Law

Criminology

Cultural Values

Deviance

Durkheim's "Crime as Functional Necessity"

Felony

Folkways

Gemeinschaft

Gesellschaft

Latent Functions

Laws

Mala in Se

Mala Prohibita

Manifest Functions

Misdemeanor

Mores

Norms

Overcriminalization

Stages of Progression of Knowledge

Undercriminalization

REVIEW QUESTIONS

1. What are some crimes that were not much regarded as problems in the past but are currently? Conversely, what are some crimes that were problems in the past and no longer loom as major concerns? Do you have any predictions of emerging, future crimes?

2. Besides Prohibition, what are some other social policies that have contained latent functions?

3. Do you think the American criminal justice system reflects a *consensus* or *conflict* model of law? Explain and defend your judgments.

4. Why don't criminologists simply use the legal classifications of criminals in their studies of crime and criminal behavior?

5. What are the differences between criminal law, statutory law, case law, civil law, and administrative law?

WEB SOURCES

Academy of Criminal Justice Sciences
http://www.acjs.org

American Society of Criminology
http://www.asc41.com

Bureau of Justice Statistics
http://www.ojp.usdoj.gov/bjs

CIA
http://www.cia.gov

Federal Bureau of Investigation
http://www.fbi.gov

National Criminal Justice Reference Service
http://www.ncjrs.org

National Institute of Justice
http://www.ojp.usdoj.gov/nij

Office of Juvenile Justice and Delinquency Prevention
http://www.ojjdp.ncjrs.org

truTV Online (formerly Court TV)
http://www.trutv.com

World Factbook of the Criminal Justice System
http://www.ojp.usdoj.gov/bjs/abstracts.wfci.htm

WEB EXERCISES

Using this chapter's recommended Web sites, explore the field of criminology.

1. What are the largest professional associations in the field, and what did you find out about them?

2. What types of information are available on government sites such as the Bureau of Justice Statistics, CIA, National Institute of Justice, and the Office of Juvenile Justice and Delinquency Prevention?

3. Of what use is the National Criminal Justice Reference Service (NCJRS)?

4. What type of information did you find on truTV Online?

5. What information does the *World Factbook on Criminal Justice Systems* include on countries throughout the world?

6. Using your Web browser search NCJRS for "FBI's Most Wanted" and "Crimes of the Century." Did you turn up anything new?

SELECTED READINGS

Robert Bohm. 1987. "The Myths About Criminology and Criminal Justice: A Review," *Justice Quarterly, 4,* 631–642.

This classic article reviews the many works on, and claims about, myths in criminology and criminal justice. A review such as this of misconceptions in criminology is an excellent means by which to assess the state of current knowledge in the field.

Elliott Currie. 1985. *Confronting Crime: Why There Is So Much Crime in America and What We Can Do About It.* New York: Pantheon.

Elliott Currie has remained a consistent voice of liberal thought in the area of crime policy. He challenges conservative beliefs of crime, which tend to be defeatist and tend to ignore the role of structural conditions such as inequality and discrimination in crime causation.

John R. Fuller and Eric W. Hickey, editors. 1999. *Controversial Issues in Criminology.* Boston: Allyn & Bacon.

Fuller and Hickey provide an excellent collection of 14 articles, each featuring a debate between two criminologists.

Don Gibbons. 1979. *The Criminological Enterprise: Theories and Perspectives.* Englewood Cliffs, NJ: Prentice Hall.

In this slim volume, Don Gibbons provides a very readable history of criminological thought that provides undergraduates with an excellent overview of the history of the development of criminology and criminological theory.

Philip Jenkins. 1984. *Crime and Justice: Issues and Ideas.* Monterey, CA: Brooks/Cole.

This book by one of the most lucid writers in criminology provides an engaging, original, and scholarly discussion of a number of criminological controversies such as serial murders and moral panics.

Steven F. Messner and Richard Rosenfeld. 1994. *Crime and the American Dream.* Belmont, CA: Wadsworth.

In this small volume, the authors build on Robert Merton's concept of anomie or strain theory by speaking to the development of an institutionalization of deviant means in American society. According to this theory, legitimate institutions begin to adopt illegal means of achieving their competitive objectives.

Carl E. Pope, Rick Lovell, and Steven G. Brandl, editors. 2001. *Voices From the Field: Readings in Criminal Justice Research.* Belmont, CA: Wadsworth.

This collection features 18 articles on research on a variety of criminological issues. It underlines the point that research in criminology stresses scientific analysis over mere opinion.

Jeffrey Reiman. 1995. *The Rich Get Rich and the Poor Get Prison* (6th ed.). Boston: Allyn & Bacon.

Reiman's work is an acknowledged classic in the field that documents the continuing inequality in the U.S. criminal justice system. His analysis serves as an excellent illustration of the conflict model in criminology.

Frank R. Scarpitti and Amie Nielson, editors. 1998. *Crime and Criminals: Contemporary and Classic Readings in Criminology.* Los Angeles: Roxbury.

Surprisingly, there is a shortage of good general readers in criminology. Scarpitti and Nielson more than fill the gap with a fine selection of 41 readings that serves to complement texts in criminology, including this one.

Sam Walker. 2005. *Sense and Nonsense About Crime* (3rd ed.). Belmont, CA: Wadsworth.

In this short book, Sam Walker provides a succinct and readable coverage of critical issues in criminology using as an organizational gimmick the theme of myths in criminology.

Theory and Methodology

The Research Enterprise of Criminology

   Objectivity

   Ethics in Criminological Research

Who Is Criminal?

Official Police Statistics—The Uniform Crime Reports (UCR)

   Sources of Crime Statistics

   The Crime Indexes: Violent and Property Crime

   Issues and Cautions in Studying UCR Data

   Crime File 2.1 The Crime Dip

Alternative Data-Gathering Strategies

Experiments in Criminology

   Some Examples of Experiments in Criminology

   Evidence-Based Research

Surveys

   Victim Surveys

   National Crime Victimization Surveys (NCVS)

   Crime File 2.2 Were You a Victim of Crime?

   Issues and Cautions in Studying Victim Data

   Self-Report Measures of Crime

   Crime File 2.3 Self-Reported Delinquency Items

Participant Observation

   Participant Observation of Criminals

   Evaluation of the Method of Participant Observation

Life History and Case Studies

   Crime File 2.4 Confessions of a Dying Thief

Unobtrusive Measures

   Crime File 2.5 Useful Sources for Criminological Research

Validity, Reliability, and Triangulation

   Crime File 2.6 The FBI Reading Room

Summary

Key Concepts

Review Questions

Web Sources

Web Exercises

Selected Readings

chapter 2

Research Methods in Criminology

We measure the extent of crime with elastic rulers whose units of measurement are not defined.

—*Edwin H. Sutherland and Donald Cressey, 1978,*
Principles of Criminology *(p. 17)*

Theory and Methodology

Two critical features of any discipline are its theory and its methodology. **Theory**, which will be the subject of Chapters 4–7, addresses the questions of "Why?" and "How?" **Methodology (methods)**, on the other hand, is concerned with "What is?"

Theories involve attempts to develop reasonable explanations of reality. They are efforts to structure, summarize, or explain the essential elements of the subject in question. What causes crime? Why do some individuals become criminals? Why are some nations or areas more criminogenic than others? Theories represent the intellectual leaps of faith that provide fundamental insights into how things operate; they attempt to illuminate or shed light upon the darkness of reality. Without the generation of useful theoretical explanations, a field is intellectually bankrupt; it becomes merely a collection of "war stories" and carefully documented encyclopedic accounts. It fails to explain, summarize, or capture the essential nature of its subject matter. Studying a field devoid of theory would be akin to a mystery novel in which the author told us neither "whodunit" nor how and why they did it.

Methodology involves the collection and analysis of accurate data or facts. With respect to criminology, this would comprise information regarding: How much crime is there? Who commits crime? How do commissions of crime or definitions of crime vary? and the like. If the facts regarding crime are provided by defective models they will be in error, and then theories or attempted explanations of this incorrectly described reality will most certainly be misdirected.

In the social sciences there at times exists a chasm between those who are primarily interested in theory or broad conceptual analysis, analogous to philosophy, and those who are methodologists. Theory devoid of method, explanation without accurate supportive data, is just as much a dead end as method devoid of interpretive theory. The former resembles armchair theorizing, the latter a fruitless bookkeeping operation. In reality, in order to realize mature development, criminology needs both incisive theory and sound, accurate methodology. The purpose of this chapter on methodology is to identify the research base on which the findings presented in this book rest, and to point out their relative strengths and shortcomings.

The Research Enterprise of Criminology

Objectivity

A basic canon of scientific research is that researchers attempt to maintain **objectivity**. This requires that the investigators strive to be "value free" in their inquiry (Weber, 1949) and, in a sense, to permit the findings to speak for themselves. A researcher may occasionally find the attitudes, behavior, or beliefs of a group he or she is studying repugnant or immoral; however, the researcher is trained not to judge but rather to objectively record and to determine what meaning these findings have for the field of criminology and to the development of its knowledge base.

Ethics in Criminological Research

Handbook Article Link 2.1
Read an article about criminal justice ethics.

Because it is part of the social sciences, the subject matter of criminology is different in kind from that of the physical sciences. While the latter concentrates on physical facts, criminology's subject matter—crime, criminal behavior, victims, and the criminal justice system—is concerned with human behavior, attitudes, groups, and organizations. Like physical science investigations, criminological inquiry must be concerned with its potentially adverse impacts on human subjects.

Ultimately, **ethical conduct in research** is an individual responsibility tied into deep moral judgments; a blind adherence to any checklist grossly oversimplifies a very complex decision. Until recently the fields of criminology and criminal justice relied on the codes of ethics of parent fields such as sociology or psychology for guidance. Beginning in 1998, however, both the Academy of Criminal Justice Sciences and the American Society of Criminology began compiling and later adopting **codes of ethics**. While space does not permit full discussion of each, some of the guidelines of both of these codes of ethics include (ACJS, 1998; ASC, 1998):

- Researchers should strive for the highest technical standards in research.
- Acknowledge limitations of research.
- Fully report findings.
- Disclose financial support and other sponsorship.
- Honor commitments.
- Make data available to future researchers.
- Not misuse their positions as fraudulent pretext for gathering intelligence.
- Human subjects have the right to full disclosure of the purposes of the research.
- Subjects have the right to **confidentiality.** This requires the researcher to protect the identity of his or her subject(s).
- Research should not expose subjects to more than minimal risk. If risks are greater than the risks of everyday life, then informed consent must be obtained.

- Avoid privacy invasion and protect vulnerable populations.
- All research should meet with human subject protection requirements imposed by educational institutions and funding sources.
- Researchers should properly acknowledge the work of others.
- Criminologists have an obligation not to create social injustice such as discrimination, oppression, or harassment in their work.

Photo 2.1

Philip Zimbardo, best known for his 1971 prison experiment, gives his last lecture in 2007.

Ethical horror stories in criminology and the social sciences include both biomedical and social science examples (Hagan, 2010). During World War II, Nazi doctors tortured, maimed, and murdered innocent captive subjects in the name of research. In the famous Tuskegee syphilis study, the U.S. Public Health Service withheld penicillin, a known cure for syphilis, from 425 uneducated black male sharecroppers who suffered and most of whom eventually died of untreated syphilis. During the Cold War, U.S. intelligence agencies, with the cooperation of the scientific community, performed bizarre and dangerous experiments on subjects without their permission. While most of these examples were biomedical in nature, social and behavioral research can likewise put subjects at risk. The three most cited social science examples are Stanley Milgram's *Obedience to Authority* (1974), Philip Zimbardo's simulated prison study (1972, 1973, 1974), and Laud Humphreys' *Tearoom Trade* (1970).

In his *Obedience to Authority* study, Stanley Milgram wanted to discover how "normal" people come to commit monstrous acts. Volunteers were recruited and paid to act as "teachers" while "confederates" (fake subjects) acted as "learners." The teachers were deceived into believing that each time they threw a lever on a shock apparatus they were administering higher levels of shock to the pupils. The teachers were willing to administer what they believed were painful shocks despite cries to stop from the subjects, when assured by the presence of scientific authorities. Do experimenters have the ethical right to deceive and put subjects in a position of emotional stress in the name of science?

In Zimbardo's simulated prison study, male undergraduate, paid participants played the roles of guard or prisoner in a mock prison setting set up in the basement of a Stanford University building. The experiment was cancelled after 6 days (of a planned 14) when participants became carried away with their roles. In *The Lucifer Effect: Understanding How Good People Turn Evil* (2007a), Zimbardo coined the term "Lucifer effect" to describe a transformation of human character that may cause good people to commit evil actions. This could include sexual degradation and torture as described at Abu Ghraib prison in Iraq. One of Zimbardo's associates, after observing a humiliating experiment called the "humping experiment," in which the prisoners simulated sodomy, berated Zimbardo for contributing to the suffering of human beings. This snapped Zimbardo back to his senses and led him to cancel the experiment (Zimbardo, 2007b).

Laud Humphreys' *Tearoom Trade* (1970) involved studying secret male homosexual activities in public restrooms. Acting as a voyeur (or watch queen), Humphreys served as a lookout, but also without the permission of his subjects, as a hidden observer. He copied down their license plate numbers and traced the participants back to their homes and showed up under the guise of being a mental health researcher. All three of these examples raised highly controversial ethical questions and most likely would not be approved today by codes of research ethics or institutional review boards.

In an incredibly insensitive experiment later dubbed "the Monster study," for 4 months during the Depression, researcher and graduate student Mary Tudor and her professor Wendell Johnson taught

Video Link 2.1
View a video of the Stanford Prison Experiment.

Photo 2.2

"Deep Throat" was the alias for W. Mark Felt, the anonymous source who leaked secrets about President Nixon's Watergate cover-up to *The Washington Post*.

children at an orphanage in Iowa a "lesson they would never forget"—how to stutter ("Lessons", 2001). While the experiment helped thousands of children overcome speech difficulties, this took place at the expense of some of the children unnecessarily being subject to lives as outcasts and misfits. Thirteen of the subjects who were still alive learned of the experiment in 2001, when it was reported in the *San Jose Mercury News*. The now stuttering children had been divided into two groups of 11, one labeled normal speakers and given positive speech therapy and the other group taught to stutter. Eight members of the treatment group are permanent stutterers. While Tudor felt remorse and returned to the orphanage a number of times in attempts to reverse the damage, Johnson did nothing and became famous in the field of speech pathology due to the study. Tudor describes how during the experiment trusting orphans greeted her, running to her car and carrying materials for the experiment. In 2007, the state of Iowa agreed to pay $925,000 to six subjects of the study who had been harmed by the University of Iowa researchers. The 1939 study became known as the "Monster study" because of the methods used by the researchers. Mary Tudor had been instrumental in breaking the story (Associated Press, 2007).

In the name of research, criminologists should have no interest in behaving as "mad scientists" who inhumanely pursue science for its own sake. In most research, informed consent of participants based on knowledge of the experiment is essential. If some form of deception is necessary, it is even more incumbent on the researcher to prevent harm and, where possible, to debrief, reassure, and explain the purposes of the project afterward. Obviously, criminology cannot afford to limit its inquiry to volunteers. **Reciprocity** involves a system of mutual trust and obligation between the researcher and subject. Subjects are asked to share themselves in the belief that this baring of information will not be used in an inappropriate, harmful, or embarrassing manner. A basic tenet of any scholarly research is the dictum that the investigator maintain objectivity and professional integrity in both the performance and the reporting of research. The researcher, first and foremost, is an investigator and not a hustler, huckster, salesperson, or politician. Researchers should avoid purposely choosing and reporting only those techniques that tend to shed the best light on their data, or "lying with statistics" (Huff, 1966). Related to these issues is the fact that the researcher should take steps to protect the confidentiality and privacy of respondents. One procedure for attempting to protect the identity of subjects, organizations, or communities is the use of pseudonyms, aliases, or false names. Names such as "Doc," "Chic," "The Lupollo Family," "Vince Swaggi," "Deep Throat," and "Wincanton," to mention just a few, have become legend in criminology.

Who Is Criminal?

To illustrate the importance of methodological precision, let us examine the basic but deceptively complex questions, "Who is criminal?" and "How much crime is there?" While an initial response to these questions might be, "Why, of course, we know," the answers are not as obvious as they seem.

Taking what would appear to be the easiest question—"Who is criminal?"—most would agree that long-term recidivists who have repeatedly been found guilty are criminals. Yet some ideologues (those committed to a strict adherence to a distinctive political belief system) might even on this point maintain that some of these "career criminals" are in fact not criminals, but are, from the conflict perspective, political prisoners. They are viewed as victims of an unfair class system or of a politically oppressive system (Turk, 1982). Additionally, not all apprehended individuals or persons accused of crime are guilty; and what about those who commit crimes but are not arrested?

It becomes apparent that the manner in which the variable "criminal" is operationalized will have a major influence on the definition of the concept of criminal. A **variable** is a concept that has been operationalized or measured in a specific manner and that can vary or take on different values, usually of a quantitative nature. **Operationalization** involves the process of defining concepts by describing how they are being measured; the notion of operationalization can be practically explained by completing the statement "I measured it by _____." In Chapters 4–7 we will describe many theories that assume excess criminality among lower-class groups based on official statistics; however, what methodological problems and biases in addressing this issue are introduced by relying solely on one measure of crime?

Official Police Statistics— The Uniform Crime Reports (UCR)

Internationally, until relatively recently the major source of information regarding crime statistics was official police statistics. Gathered for government administrative purposes with only secondary attention paid to their usefulness for social science research, these data tended to be uneven in quality and were not gathered or recorded in any systematic manner. Basically, criminologists had no efficient statistics to consult in order to answer even basic questions such as whether crime was increasing or decreasing.

Since 1930 the U.S. Department of Justice has compiled national crime statistics, the **Uniform Crime Reports (UCR),** with the Federal Bureau of Investigation (FBI) assuming responsibility as the clearinghouse and publisher. Although participation in the UCR program by local police departments is purely voluntary, the number of departments reporting and the comprehensiveness of the information have steadily improved over the years, with police departments from large metropolitan areas historically the most reliable participants.

Handbook Article Link 2.2
Read an article on crime reports and statistics.

Sources of Crime Statistics

Returning to our question "How much crime is there?" an examination of the UCR and its relationship to sources of data on crime and criminals is useful. Figure 2.1 illustrates the relationship between crime committed and the **sources of crime statistics,** including the UCR. It is unclear whether an accurate estimate of the amount of crime committed is possible, for several reasons. Not all crimes that are committed are discovered. Some crimes may be known only to the perpetrators, in which case the victim is unaware of loss. Perhaps there is no identifiable victim, as in the case of a gambling violation. The further a source of statistics is from the "crimes committed" category, the less useful it is as a measure of the extent of crime. Not all crimes that are discovered are reported to the police; similarly, not all reported crimes are recorded by police. See Figure 2.1.

Audio Link 2.1
Listen to a discussion of violent crime in major cities.

In addition, some law enforcement agencies may purposely conceal recorded crimes; a number of purported crimes may be **unfounded** or defined by investigating officers as not constituting a criminal matter. For instance, when a complainant reports an attempted burglary, investigating officers may conclude that there is not enough evidence to support that a crime took place.

Despite this problematic relationship between crimes recorded and crimes committed, the UCR until recently represented the best statistics available on crime commission and, as will be discussed later in this chapter, still represents one of the best sources. Again, as shown in Figure 2.1, once we move

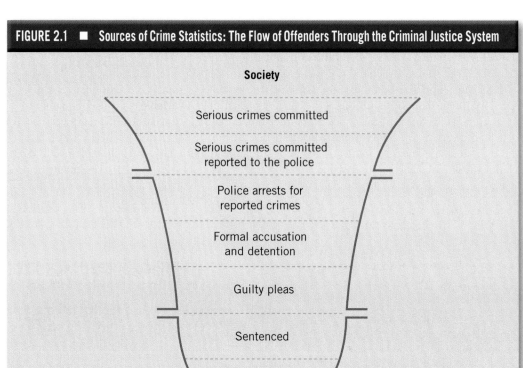

FIGURE 2.1 ■ Sources of Crime Statistics: The Flow of Offenders Through the Criminal Justice System

Society

Serious crimes committed

Serious crimes committed
reported to the police

Police arrests for
reported crimes

Formal accusation
and detention

Guilty pleas

Sentenced

Probation

Jail

Prison

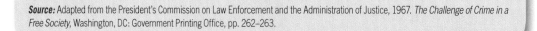

Source: Adapted from the President's Commission on Law Enforcement and the Administration of Justice, 1967. *The Challenge of Crime in a Free Society,* Washington, DC: Government Printing Office, pp. 262–263.

beyond crimes recorded as a measure of crime commission, we are getting further removed from the accurate measurement of crime commission. Thus arrest statistics, indictments, convictions, incarcerations, and other dispositions such as probation and parole are not as useful. Such statistics have much more to do with police efficiency or allocations to the criminal justice system and general societal policies toward crime control policy than they do with measuring the extent of the crime problem.

Most media accounts of changes in the crime rate are based on the annual summary presented in the UCR. While the UCR contains many qualifying remarks regarding the meaning of these statistics, in most instances the press tends to report these data uncritically and often in an alarmist manner. Obviously, the researcher who chooses to utilize UCR data must become as familiar as possible with any shortcomings or sources of bias in these statistics. The FBI receives its information for the UCR from local police departments. Considerable variation exists in state penal codes regarding criminal offenses and their definitions, although participating departments receive instruction in uniform crime recording in order to standardize their reports for use in compiling nationwide figures. In the majority of states, UCR systems require that all local departments report their statistics to the state. These data are then shared with the FBI. The Census Bureau estimates that about 97 percent of the total national population is covered by the report.

The Crime Indexes: Violent and Property Crime

Historically, the UCR has been divided into two parts: **Part I crimes** consist of the **index crimes,** major felonies that are believed to be serious, to occur frequently, and to have a greater likelihood of being reported to the police. The index offenses are:

1. Murder and non-negligent manslaughter

2. Forcible rape

3. Robbery

4. Aggravated assault

5. Burglary

6. Larceny-theft

7. Motor vehicle theft

8. Arson

The original index and the one used for historical comparison consist of the first seven offenses. Arson was added as a result of a law passed by the U.S. Congress in October 1978. As we will see shortly, the crime rate is calculated with the index offenses. Figure 2.2 defines the various offenses in uniform crime reporting. In 2004, the FBI decided to stop reporting the **crime index** and to report a violent crime index and property crime index instead. An advisory board had felt that the crime index had been distorted by including the category of larceny-theft.

The violent crime index consists of:

1. Murder and non-negligent manslaughter

2. Forcible rape

3. Robbery

4. Aggravated assault

Photo 2.3

The use of technology by police has been credited in part for crime reduction in the 1990s.

The property crime index consists of:

1. Burglary

2. Larceny-theft

3. Motor vehicle theft
*Arson listed, but not calculated

Part II crimes are nonindex offenses and are not used in the calculation of the crime rate. These include:

- Simple assault
- Forgery and counterfeiting
- Fraud
- Embezzlement
- Receiving stolen property
- Vandalism

FIGURE 2.2 ■ **Offenses in Uniform Crime Reporting**

Offenses in Uniform Crime Reporting are divided into two groupings, Part I and Part II. Information on the volume of Part I offenses known to law enforcement, those cleared by arrest or exceptional means, and the number of persons arrested is reported monthly. Only arrest data are reported for Part II offenses.

The Part I offenses are:

Criminal homicide: Murder and non-negligent manslaughter: the willful (non-negligent) killing of one human being by another. Deaths caused by negligence, attempts to kill, assaults to kill, suicides, accidental deaths, and justifiable homicides are excluded. Justifiable homicides are limited to: (1) the killing of a felon by a law enforcement officer in the line of duty; (2) the killing of a felon, during the commission of a felony, by a private citizen; (3) manslaughter by negligence: the killing of another person through gross negligence. Traffic fatalities are excluded. While manslaughter by negligence is a Part I crime, it is not included in the Crime Index.

Forcible rape: The carnal knowledge of a female forcibly and against her will. Included are rapes by force and attempts or assaults to rape. Statutory offenses (no force used—victim under age of consent) are excluded.

Robbery: The taking or attempt to take anything of value from the care, custody, or control of a person or persons by force or threat of force or violence and/or by putting the victim in fear.

Aggravated assault: An unlawful attack by one person upon another for the purpose of inflicting severe or aggravated bodily injury. This type of assault is usually accompanied by the use of a weapon or by a means likely to produce death or great bodily harm. Simple assaults are excluded.

Burglary/breaking or entering: The unlawful entry of a structure to commit a felony or theft. Attempted forcible entry is included.

Larceny-theft (except motor vehicle theft): The unlawful taking, carrying, leading, or riding away of property from the possession or constructive possession of another. Examples are thefts of bicycles or automobile accessories, shoplifting, pocket-picking, or the stealing of any property or article which is not taken by force and violence or by fraud. Attempted larcenies are included. Embezzlement, "con" games, forgery, worthless checks, etc., are excluded.

Motor vehicle theft: The theft or attempted theft of a motor vehicle. A motor vehicle is self-propelled and runs on the surface and not on rails. Specifically excluded from this category are motorboats, construction equipment, airplanes, and farming equipment.

Arson: Any willful or malicious burning or attempt to burn, with or without intent to defraud, a dwelling house, public building, motor vehicle or aircraft, personal property of another, etc.

The Part II offenses are:

Other assaults (simple): Assaults and attempted assaults where no weapon is used and which do not result in serious or aggravated injury to the victim.

Forgery and counterfeiting: Making, altering, uttering, or possessing, with intent to defraud, anything false in the semblance of that which is true. Attempts are included.

Fraud: Fraudulent conversion and obtaining money or property by false pretenses. Included are confidence games and bad checks, except forgeries and counterfeiting.

Embezzlement: Misappropriation or misapplication of money or property entrusted to one's care, custody, or control.

Stolen property; buying, receiving, possessing: Buying, receiving, and possessing stolen property, including attempts.

Vandalism: Willful or malicious destruction, injury, disfigurement, or defacement of any public or private property, real or personal, without consent of the owner or persons having custody or control.

Weapons-carrying, possessing, etc.: All violations of regulations or statutes controlling the carrying, using, possessing, furnishing, and manufacturing of deadly weapons or silencers. Attempts are included.

Prostitution and commercialized vice: Sex offenses of a commercialized nature, such as prostitution, keeping a bawdy house, procuring, or transporting women for immoral purposes. Attempts are included.

Sex offenses (except forcible rape, prostitution, and commercialized vice): Statutory rape and offenses against chastity, common decency, morals, and the like. Attempts are included.

Drug abuse violations: State and/or local offenses relating to the unlawful possession, sale, use, growing, and manufacturing of narcotic drugs. The following drug categories are specified: opium or cocaine and their derivatives (morphine, heroin, codeine); marijuana; synthetic narcotics—manufactured narcotics that can cause true addiction (Demerol, methadone); and dangerous non-narcotic drugs (barbiturates, benzedrine).

Gambling: Promoting, permitting, or engaging in illegal gambling.

Offenses against the family and children: Nonsupport, neglect, desertion, or abuse of family and children.

Driving under the influence: Driving or operating any vehicle or common carrier while drunk or under the influence of liquor or narcotics.

Liquor law violations: State and/or local liquor law violations, except drunkenness and driving under the influence. Federal violations are excluded.

Drunkenness: Offenses relating to drunkenness or intoxication. Driving under the influence is not included.

Disorderly conduct: Breaking of the peace.

Vagrancy: Vagabondage, begging, loitering, etc.

All other offenses: All violations of state and/or local laws, except those listed above and traffic offenses.

Suspicion: No specific offense; suspect released without formal charges being placed.

Curfew and loitering laws (persons under age 18): Offenses relating to violations of local curfew or loitering ordinances where such laws exist.

Runaways (persons under age 18): Limited to juveniles taken into protective custody under provisions of local statutes.

Source: Federal Bureau of Investigation. 1995. *Crime in the United States, 1994, Uniform Crime Reports.* Washington, DC: Government Printing Office, pp. 383–384.

Web Research Project
Visit http://www.fbi.gov and examine any of its latest reports on crime such as burglary or bank robbery.

- Illegal carrying of weapons
- Prostitution and related offenses
- Sex offenses (statutory rape, etc.)
- Drug law violations
- Liquor law violations
- Public drunkenness
- Disorderly conduct
- Vagrancy
- Curfew violations/loitering
- Runaways
- All other violations of state and local laws (except traffic violations)

Issues and Cautions in Studying UCR Data

An extensive literature has accumulated regarding shortcomings of UCR statistics. While the UCR has steadily improved and been refined since its inception in 1930, researchers utilizing these data should exercise caution and be aware of certain limitations. Some primary shortcomings of the UCR include the following:

1. The recorded statistics represent only a portion of the true crime rate of a community. Victim surveys suggest that there is possibly twice as much crime committed as appears in official statistics.

2. The big increase in the crime rate beginning in the mid-1960s may be explained in part by better communications, more professional and more efficient police departments, and better recording and reporting of crime. Larger, improved, and professionalized police departments appear to be positively related to rising crime rates. This was particularly the case in larger urban areas. Photo 2.3 (on p. 27) shows technology in the patrol car.

3. Increased citizen concern and awareness of crime, higher standards of expected public morality, and greater reporting of and response to ghetto crime may all have had impacts on increasing the recorded crime rate.

4. Most federal offenses, "victimless" crimes, and white collar crimes do not appear in the UCR.

5. Changes in record-keeping procedures (such as computerization), transition in police administrations, and political shenanigans can have a major impact on crime recording. The FBI attempts to monitor and control abuses. In 1999 the Philadelphia Police Department's Sex Crimes Unit was found to have dismissed as noncrimes several thousand reports of crime. In order to attract and host the 1996 Olympics, Atlanta was accused of undercounting crime to the tune of 22,000.

6. In interpreting UCR statistics, keep in mind what arrest statistics do and do not mean:
   a. Arrests do not equal crimes solved or suspects found guilty.
   b. Many reported crimes are declared unfounded by police.
   c. In the situation involving multiple offenses, only the most serious offense is recorded for UCR purposes.
   d. The majority of crimes committed are not index offenses.

7. The crime index is made up primarily of property crimes.

8. The crime index is an unweighted index; it is a simple summated scale in which a murder counts the same as a bicycle theft. Surprisingly, most bodily injury crimes are "nonindex" offenses (Savitz, 1978).

9. The existence of the crime index may encourage concentration by police agencies on these offenses at the expense of others.

10. The crime rate is calculated on the basis of decennial census population figures. Rapidly growing cities of the Southwest would, under this system, appear to have worse rates since, for example, 1979 crimes would be divided by a 1970 population base.

11. Demographic shifts may provide partial explanations for changing crime rates. Some criminologists had prophesied the crime dip (a decline in the crime rate trend) in the 1980s based on a general aging of the baby boom generation (children born in the post–World War II era, from 1946 through the mid-1950s). This larger-than-normal population cohort overwhelmed hospital nursery wards, elementary and secondary schools, and later colleges. These establishments now have extra space. Similarly, the criminal justice system was overwhelmed by a larger-than-normal proportion in the maximal crime-committing ages (15–24), as the job market and housing industry inherit this now "middle-age boom."

The Crime Rate

The **crime rate** is a calculation that expresses the total number of index crimes per 100,000 population:

$$\text{Index Crimes/Population} \times 100,000 = \text{Crime Rate}$$

As previously indicated, in 2004 the FBI decided to drop the additional calculation of the crime index rate. The purpose of an index (like the Dow Jones Industrial Average or the Consumer Price Index) is to provide a composite measure, one that does not rely too heavily on any one factor. An index also allows controlling for population size, thus permitting fair comparisons of different-sized units. As previously indicated, it is this UCR crime rate that one reads about in the newspaper, with accounts of crime either rising or falling by a given percentage. A principal difficulty with the UCR crime rate as an index of crime in the United States is that it is an unweighted index. That is, each crime, whether murder or bicycle theft, is added into the total index with no weight given to the relative seriousness of the offense. Thus, no monetary or psychological value is assigned. For instance, a city with 100 burglaries per 100,000 population and one with 100 homicides per 100,000 population would have the same crime rate.

One alternative that has been proposed is the calculation of a weighted index using crime-seriousness scales (Rossi, Waite, Bose, & Berk, 1974; Sellin & Wolfgang, 1964). In a weighted crime index, criminal incidents are assigned weights on the basis of variables such as amount stolen, method of intimidation, degree of harm inflicted, and similar salient factors.

Redesign of the UCR Program: NIBRS

The redesigned UCR program is called **NIBRS (National Incident-Based Reporting System)**. In 1982, in response to the criticisms and limitations of the UCR program, the Bureau of Justice Statistics and the FBI formed a joint task force and contracted with a private research firm (Abt Associates, Inc.) to undertake revisions in the UCR program. This was the first in the program's then more than 50 years of existence (Poggio et al., 1985; Rovetch, Poggio & Rossman, 1984). On the basis of recommendations of a steering committee made up of police practitioners, academicians, and the media, the NIBRS suggestions for changes in the UCR included:

- A new two-level reporting system in which most agencies continue to report basic offense and arrest data much as they do at present (Level I), while a small sample of agencies report more extensive information (Level II).
- The entire UCR system is to be converted into unit-record reporting in which police agencies report on the characteristics of each criminal incident (for example, location, time, presence of weapon) and on the characteristics of each individual arrest.
- Distinguish attempted from completed offenses.
- Distinguish crimes against businesses, individuals, or households from crimes against other entities.
- Institute ongoing audits of samples of participating UCR agencies to check for errors in the new program.
- Support better user services, particularly in making databases more available to outside researchers.
- NIBRS will collect data on each single incident and arrest in 22 crime categories (U.S. Department of Justice, 1988, p. 82).

It is believed that these revisions in the program, which are taking longer to implement than anticipated, will overcome a number of past criticisms as well as provide a database that will be more useful for both researchers and policymakers.

Crime File 2.1

The Crime Dip

From the first compilation of crime statistics by the Federal Bureau of Investigation in the early 1930s until the early 1960s, the crime rate in the United States had been declining. Some experts had even unwisely predicted that, given existing trends and growing affluence, crime might become a rarity by the twenty-first century. By the mid-1960s, however, recorded crime made a reversal and rose to unprecedented levels, producing in its wake yet more predictions of unrepentant explosions in the crime rate. A brief leveling off in the early 1980s was followed by an epidemic of youth violence beginning in the mid-1980s with the advent of crack cocaine and widespread use of weapons to defend disputed drug trafficking turf. By the 1990s, an assumed inevitability of rising crime rates was greeted by unexpected declines, beginning in large cities such as New York. Between 1993 and 2000, index crimes had declined over 30 percent.

The causes of this crime dip are a subject of dispute. Factors associated with the crime dip that began in the 1990s include:

- A healthy economy
- Crime prevention programs
- Decline in domestic violence
- An incarceration binge
- CompStat and community policing
- A decline in the crack cocaine epidemic
- Legalized abortion

The most prosperous American economy in over 30 years highlighted by low unemployment and low inflation may be the major reason for falling crime rates. Such an explanation may not be the case, however. During the 1960s crime rates rose sharply at a time of low unemployment. More recently, Sun Belt cities with low unemployment have had higher crime rates than older cities with high unemployment. New York City's murder rate in the 1990s fell over 66 percent despite high unemployment (Witkin, 1998, p. 30).

Crime prevention, which shows much promise for early prevention programs with high-risk juveniles, has shown only modest impacts on crime rates.

Domestic murders (among intimates) demonstrated a 40 percent decline between 1976 and 1996. Part of the explanation for this was a decline in marriages among 20–24-year-olds, as well as greater opportunities for abused women to escape bad relationships.

America's incarceration binge has been phenomenal, increasing from 744,000 inmates in 1985 to approximately 1.8 million in 1998. This is the largest imprisoned population of any country in the world outside Russia. While locking up an extra million prisoners must have some impact, New York City showed the most dramatic drop in crime, while the state of New York (with 70 percent of its prison population from New York City) increased its prison population by only 8 percent between 1993 and 1996. Utah, on the other hand, raised its incarceration rate by 19 percent between 1993 and 1996, but its violent crime rate went up (Witkin, 1998, p. 31). By 2002, 6.6 million Americans were behind bars or on probation or parole. This represented 1 of every 32 adults.

Another candidate for explanation has been better and more effective policing. "CompStat" (computer statistics) was used to computer map and identify "hot spots" (high-crime areas) by the New York City police to assign target patrols. Kelling and Wilson's "Broken Windows" (1999) theory emphasized focusing on small, nuisance crimes under the assumption that such crimes left unpunished breed more serious crimes. The fact that many cities that did not employ community policing strategies also experienced major declines in recorded crime—and some innovative departments experienced increases—leaves the more effective policing explanation in question.

A rival explanation is that the police departments are manipulating statistics to show lower crime rates. While this may occur in individual cases, such a mass conspiracy by most departments seems unlikely. In 1998 the Philadelphia Police Department was accused of systematically underreporting crime for years. The *Philadelphia Inquirer* reported routine downgrading of the seriousness of crimes in which stabbings and beatings were redefined as "hospital cases" and burglaries became "lost property" ("Philadelphia Crime Statistics Questioned," 1998).

Blumstein and Rosenfeld (1998) point out that all of the increase in homicide in the late 1980s to early 1990s was among younger people (under 21), and this was primarily due to a crack cocaine epidemic in American cities beginning in 1986 that peaked in 1993. This epidemic was accompanied by a great increase in the carrying of firearms to settle turf wars.

A final intriguing explanation in an article by Levitt and Donohue (1999) argues that legalized abortion is responsible for falling crime rates. They claim that half of the drop in crime since 1991 might reflect the Supreme Court's 1973 *Roe v. Wade* decision legalizing abortion. Some unwanted, potential criminals were not born because their potential mothers had abortions. The decline in crime began in 1992 just when those youths who would have been born in the mid-1970s would have hit their peak crime years (18–24). Even Levitt and Donohue admit, however, that other factors may be more explanatory of the crime dip than abortion. Just as criminologists debated the causes of the rise of crime, there is no consensus regarding explanations for the decline in crime or even prognostications as to when crime might rise again.

Sources: G. A. Kelling and J. Q. Wilson (1999, October). *Broken Windows and Police Discretion.* Washington, DC: U.S. Department of Justice. NCJ 178259; Philadelphia Crime Statistics Questioned. (1998, November 2). Associated Press; Alfred Blumstein and Richard Rosenfeld (1998). Assessing the Recent Ups and Downs in U.S. Homicide Rates. *National Institute of Justice Journal,* October, 9–11; Steven Levitt and John Donohue (1999, August 8). Legalized Abortion and Crime. *Chicago Tribune;* Gordon Witkin (1998, May 25). The Crime Bust: What's Behind the Dramatic Drug Bust? *U.S. News and World Report,* pp. 28–37.

Web Research Project
Using a Web browser, locate articles on the "crime dip." What explanations do they provide?

The nineteenth-century British Prime Minister Benjamin Disraeli has often been cited as having remarked, "There are three types of lies: lies, damn lies and statistics." Obviously, caution must be exercised in examining graphic devices and statistical reports (Huff, 1966; Zeisel, 1957). In the 1980s and early 1990s, rising juvenile violent crime led conservative commentators such as Robert Bennett and John DiIulio to forecast grim prophecies of exploding juvenile crime among violent criminal predators raised in mean minority ghettos and in maternal, single-parent households—a foreboding inevitability born of moral rot. In the 1990s these "hopeless areas" showed the greatest decline in crime, one which few had predicted. Crime File 2.1 assesses this "Crime Dip."

Alternative Data-Gathering Strategies

Official crime statistics published by national governments have their uses; however, criminologists would be remiss in their duty as scholars and scientists if they were to restrict their inquiries and sources of statistics to data gathered for administrative purposes by government bodies. In some totalitarian regimes, for instance, there would be nothing to study, since the official government ideology might simply hold that there is no crime in the people's paradise. Even in open societies, official statistics seldom cover crimes of the elite. Fortunately, criminologists have at their disposal a veritable arsenal of techniques whose application is limited only by the researcher's imagination and skill.

Figure 2.3 offers a model or paradigm (schema) with which to consider and compare the alternative data-gathering strategies that can be employed in criminal-justice and criminological research. As an illustrative device, Figure 2.3 is an attempt to broadly describe the relative advantages and disadvantages of the different data-gathering strategies. The model suggests that, as we move up the list of techniques or vertical arrows to experiments, we tend to obtain quantitative measurement (which lends itself to sophisticated statistical treatment), greater control over other factors that may interfere with one's findings, and increased internal validity (or accuracy in being certain that the variable[s] assumed to be responsible for one's findings are indeed the causal agent[s])—but at the expense of artificiality. The latter point suggests that, as a result of controlling for error, the researcher may have created an antiseptic or atypical group or situation that no longer resembles the "real world" that one is attempting to describe.

Generally, as one proceeds down the vertical arrows or list of techniques, the methodology employed becomes more qualitative. Qualitative techniques involve less commitment to quantitative

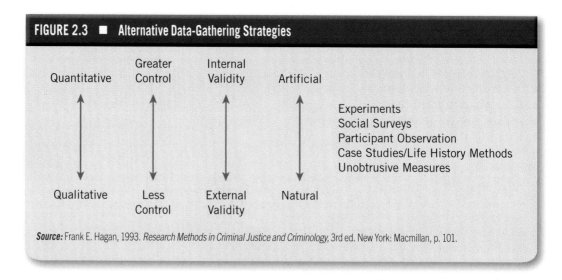

FIGURE 2.3 ■ Alternative Data-Gathering Strategies

Quantitative — Qualitative

Greater Control — Less Control

Internal Validity — External Validity

Artificial — Natural

Experiments
Social Surveys
Participant Observation
Case Studies/Life History Methods
Unobtrusive Measures

Source: Frank E. Hagan, 1993. *Research Methods in Criminal Justice and Criminology,* 3rd ed. New York: Macmillan, p. 101.

measurement on the part of the researcher, more engagement with field and observational strategies, and less direct means of obtaining information. Generally, as one moves down the list, one has less control over manipulating the research setting and rival causal factors. Such procedures, however, increase external validity (the ability to generalize to larger populations) as well as present the opportunity to study subjects in more natural settings. Criminologists, like other researchers, tend to favor their own particular methods of data gathering; this is to be expected. At times, however, academic battles break out among those who claim that their preferred method contains some inherent superiority over other procedures. Such **methodological narcissism** (or methodologism) is a fanatical adherence to a particular research method, often at the expense of a concern for substance (Bayley, 1978; "Martinson Attacks His Own Earlier Work," 1978, p. 4; Martinson, 1979). This "methods for methods' sake" orientation ignores the fact that methodology is not an end in itself but a means to an end—the development of criminological knowledge. It is more useful to permit the subject to dictate the proper methodology than to assert that, unless a subject lends itself to deployment of one's favorite method, it is not worthy of study.

Experiments in Criminology

The **experiment** is the lodestone or benchmark for comparison with all other research methods. It is the most effective means of controlling for error or rival factors before the fact through the very design of the study (Campbell & Stanley, 1963). While there are myriad variations of the experiment, the point of departure or prototype is the classic experimental design. The **classic experimental design** contains three key elements:

- Equivalence
- Pretests and posttests
- Experimental and control groups

Basically, equivalence means the assignment of subjects to experimental and control groups in such a manner that they are assumed to be alike in all major respects. This can be done either through random assignment (where each subject has an equal probability of appearing in either group) or through matching (a procedure in which subjects with similar age, sex, and other characteristics exhibited by the experimental group are recruited for the control group). The *experimental group* is to receive the treatment (*X*), while the *control group* will receive no treatment but will be observed in order to compare it with the experimental group. Both groups are given pretests (preobservations in order to note conditions that exist prior to treatment) designated as 01, or observation time one, and posttests, or observations after the experimental treatment (*X*) has taken place. The logic of the experiment assumes that, since both groups were equivalent in the pretest period, any differences in the posttest observation must be due to the fact that one group received a particular treatment and the other did not. Increasingly, such experiments are being utilized in order to inform public policy decision making. Such experiments are seen as giving answers that enable fairly clear policy direction (Garner & Visher, 1988; Kelling, 1988b).

Photo 2.4

"Scared Straight" programs were designed to expose delinquents to "heart-to-heart" talks with inmates with the aim of literally scaring them into becoming straight, or nondelinquent.

Some Examples of Experiments in Criminology

Candid Camera

Before the era of having video cameras in nearly every business, an experiment with cameras showed great promise. In an attempt to increase both the apprehension and the conviction rates of robbers of commercial establishments, the Seattle Police Department created a field experiment using high-risk establishments, some of which were designated as the experimental group, others as the control group. The treatment for the experimental group involved installation of special hidden cameras that could be triggered during a holdup by a clerk's pulling a "trip" bill from the cash drawer; prints of the photograph of the robber would be made available immediately. A posttest of the two types of sites found 55 percent of robberies in the experimental group cleared by arrest compared to 25 percent for control locations. (*Clearance* indicates that suspects have been arrested, charged, and turned over to the court for prosecution or that the police consider further investigation unnecessary.) While 48 percent of the robbers at camera sites were convicted, only 19 percent of the control group brigands were found guilty ("Hidden Cameras Project," 1978).

Scared Straight

Much fanfare was raised in the United States in the late 1970s over a novel program intended to deter wayward juveniles from progression to more serious criminal activity by means of blunt, "heart-to-heart" talks in prison with specially selected inmates (see Photo 2.4). Portrayed in a film, *Scared Straight,* the initial Rahway, New Jersey, prison project was intended to counteract the glamorized image associated with criminal life. Although many jurisdictions rushed to imitate what appeared to be the latest panacea in corrections, further research suggested that this optimism was premature. Yarborough evaluated the JOLT (Juvenile Offenders Learn Truth) program at the Jackson State Prison, Michigan, by randomly assigning youths to experimental and control groups. He then measured their delinquency rates 3 and 6 months afterward and found no significant differences between those who had attended the JOLT sessions (experimentals) and those who had not (controls; "Scared Straight Found Ineffective Again," 1979).

FIGURE 2.4 ■ Preventing Crime: What Works, What Doesn't, What's Promising

In 1996 Congress required that the Attorney General and the National Institute of Justice evaluate the effectiveness of 500 funded programs in a manner that would be "independent in nature" and "employ rigorous and scientifically recognized standards and methodologies." The Institute on Criminology and Criminal Justice at the University of Maryland was contacted to undertake this task and to serve as a clearinghouse. It issued its report titled "Preventing Crime: What Works, What Doesn't, What's Promising." These evaluations are regularly updated; full reports or research in brief summaries can be downloaded from http://www.preventingcrime.org. They can also be obtained from the Bureau of Justice Statistics Web site (http://www.ojp.usdoj.gov/bjs/). A few of the programs included in the list are:

What Doesn't Work:

- Gun "buyback " programs
- Drug Abuse Resistance Education (D.A.R.E)
- Arrest of unemployed suspects for domestic assault
- Storefront police offices
- Correctional boot camps using traditional military basic training
- "Scared Straight" programs whereby minor juvenile offenders visit adult prisons
- Shock probation, shock parole
- Home detention with electronic monitoring
- Intensive supervision on parole or probation
- Residential programs for juvenile offenders using challenging experiences in rural settings

What Works:

- For infants—frequent home visits by nurses and other professionals
- For delinquents and at-risk preadolescents—family therapy and parent training
- For schools:
  - organizational development for innovation
  - communication and reinforcement of consistent norms
  - teaching of social competency skills
  - coaching in "thinking skills" for high-risk youth
- For older male ex-offenders—vocational training
- Extra police patrols for high-crime "hot spots"
- For high-risk offenders:
  - monitoring by specialized police units
  - incarceration
- For employed, domestic abusers—arrest
- For convicted offenders—rehabilitation programs with risk-focused treatments
- For drug-using offenders in prison—therapeutic community treatment programs

What's Promising:

- Proactive drunk-driving arrests with breath testing
- Police showing greater respect to arrested offenders (may reduce repeat offending)
- Higher number of police officers in cities (may reduce crime generally)
- Gang monitoring by community workers and probation and police officers
- Community monitoring by Big Brothers/Big Sisters of America (may prevent drug abuse)
- Community-based after-school recreation programs
- Battered women's shelters
- Job Corps residential training for at-risk youth
- Prison-based vocational education programs
- Two clerks on duty in already robbed convenience stores
- Metal detectors

- Proactive arrest for carrying concealed weapons (may reduce gun crime)
- Drug courts
- Drug treatment in jails followed by urine testing
- Intensive supervision and aftercare of juvenile offenders

None of these evaluations as "working" or "not working" is final; constant replication (repeated experiments) and reevaluation is required, but a persistent, independent, scientific program of evaluation will go a long way in replacing what we think works or what doesn't with what actually does work.

Sources: Irvin Waller and Brandon Welsh (1998). Reducing Crime in Harnessing International Best Practice. *NIJ Journal,* October, pp. 26–32; and Lawrence Sherman et al. (1998). Preventing Crime: What Works, What Doesn't, What's Promising? *NIJ Research in Brief,* July. See also: Anthony J. Petrosino et al. (2003, June). Toward Evidence-Based Criminology and Criminal Justice: Systematic Reviews and the Campbell Collaboration, and the Crime and Justice Group. *International Journal of Comparative Criminology, 3,* 142–161; and Sharon Mihalic et al. (2004). Blueprints for Violence Prevention Report. *Office of Juvenile Justice and Delinquency Prevention.* NCJ 204274, July.

Web Sources: National Institute of Justice: http://www.ojp.gov/nij; Justice Information Center: http://www.ncjrs.org; University of Maryland preventing crime project: http://www.preventingcrime.org; International data on what works: http://www.crime-prevention-intl.org.

Web Research Project
Using one of the titles of the programs described above (e.g., boot camps or drug courts) find an article that describes one of these programs and whether the program worked or not.

Visit http://www.preventingcrime.org and examine a specific program. Does this program work, show promise of working, or not work?

Evidence-Based Research

Those who are impatient with or question the need for research in criminology or criminal justice often raise the questions of "So what?" or "Of what practical use are all of these research projects?" Perhaps in answer to such questions, in 1996 the U.S. Congress required the Attorney General to provide a "comprehensive evaluation of the effectiveness" of over $3 billion spent annually in Department of Justice grants that had been designed to assist state and local law enforcement and communities in preventing crime (see Figure 2.4).

Evidence-based research is an attempt to base knowledge and practice on well-researched evidence. The "what works" in criminology and criminal justice approach utilized by the Department of Justice is based on the assumption that it makes little sense to continue to invest in programs that do not work. Why not find out which programs do work or are promising and put our scarce funding into those programs? This evidence-based research employs a problem-solving approach using local, national, and international evidence on what works (http://www.crimereduction.gov.uk/tool kits/p010403.htm).

The most ambitious effort in this regard is the **Campbell Collaboration** (C2; http://www.campbell collaboration.org). Named in honor of the late Donald Campbell, a pioneer in research design, the purpose of the organization is to facilitate the preparation, maintenance, and accessibility of systematic program reviews. In support of this the group keeps a register of systematic studies. C2 was based on the highly successful Cochrane Collaboration in health care that attempted to address the lack of evidence guiding medical and health care practices. Chaired by David Farrington at Cambridge University, during one year C2 solicited program reviews in 25 areas including boot camps, street lighting, restorative justice, child skills training, and hot spots policing.

The Campbell Collaboration intends to produce the best evidence on what works to inform decision makers, researchers, and the general public. "Best evidence" means systematic reviews that are rigorous, updated in light of new studies and criticisms, relevant and accessible to end users, cover studies published worldwide, and are open to criticism and comment (Petrosino et al., 2003). Another

Audio Link 2.2
Listen to a discussion of crime prevention strategies.

Audio Link 2.3
Listen to a discussion of forensics.

example of a comprehensive effort to evaluate successful programs is the "Blueprints for Violence Prevention" program at the University of Colorado (Mihalic, Fagan, Irwin, Ballard, & Elliott, 2004).

Surveys

Most readers are familiar with the use of surveys in public opinion polls, voting-prediction studies, and marketing research. Surveys are also used in criminology, particularly in analyzing victimization, self-reported crime, public ratings of crime seriousness, measurements of fear of crime, and attitudes toward the police and the criminal justice system. The principal methods employed in gathering data for **surveys** are variations of questionnaires, interviews, or telephone surveys. Just as experiments control for error and rival causal factors before the fact by the very design of the study, survey researchers attempt to control for these factors after the fact through the use of statistical procedures.

Victim Surveys

One of the major shortcomings of such official police statistics as the UCR is that they fail to account for undiscovered or unreported crime; the **"dark figure of crime"** is the phrase early European criminologists used to refer to offenses that escape official notice. The assumption was that for every crime that came to the attention of authorities, there were an unspecified number of undiscovered crimes—"the dark figure."

Audio Link 2.4
Listen to a discussion of crime reporting by victims.

Victim surveys are specifically designed to record an estimate of claimed victimizations by a representative sample of the population. One major finding, beginning with the U.S. surveys of the late 1960s, was that overall about twice as much crime was reported to interviewers as appeared in official police records (Biderman et al., 1967; Ennis, 1967; Reiss, 1967).

National Crime Victimization Surveys (NCVS)

Beginning in 1972, the National Crime Surveys were conducted. The NCS (now called the **National Crime Victimization Survey,** or NCVS) consisted of the Central City Surveys and the National Crime Panel Surveys.

The Central City Surveys were essentially cross-sectional studies of households and commercial establishments in selected cities. Initially, probability samples of approximately 10,000 households and 1,000 to 5,000 commercial establishments were surveyed in 26 central cities. The great expense of such surveys in each city led to their discontinuance (U.S. Department of Justice, 1974, 1975a, 1975b, 1976, 1978). The National Crime Panels employed a sophisticated probability sample of housing units and businesses throughout the United States. In contrast to the Central City Surveys, which were cross-sectional or studies of one time only, the panels were longitudinal in nature, that is, studies over time of a particular group. This enabled bounding of victim reports or the use of pretests in order to have a reference point for the survey reporting period. The initial interview acted as a boundary or time period benchmark with which to compare future reported victimizations. Consisting of about 50,000 households to be interviewed every 6 months and 15,000 (later upped to 50,000) businesses, the national panels repeated the interviews twice a year in order to achieve the bounding feature previously described. Each housing unit remained in the sample for 3 years, while every 6 months a subsample of 10,000 was rotated out of the sample and replaced by a new group. While the initial findings were heralded at the time as the first accurate statistics on crime, further analysis suggests that this conclusion may have been prematurely optimistic. Just as the UCR was found to have shortcomings, so any measure of crime, including victim surveys, can be found wanting in some respects. Crime File 2.2 provides examples of the types of questions asked in the NCVS.

Crime File 2.2

Were You a Victim of Crime?

Household Screen Questions

38. Now I'd like to ask some questions about crime. They refer only to the last 6 months: between ___, 20___ and ___, 20 ___. During the last 6 months, did anyone break into or somehow illegally get into your (apartment/home), garage, or another building on your property?

Yes: How many times? _____

No

39. Other than the incident(s) just mentioned, did you find a door jimmied, a lock forced, or any other signs of an ATTEMPTED break-in?

Yes: How many times? _____

No

40. Was anything at all stolen that is kept outside your home or happened to be left out, such as a bicycle, a garden hose, or lawn furniture (other than any incidents already mentioned)?

Yes: How many times? _____

No

41. Did anyone take something belonging to you or to any member of this household, from a place where you or they were temporarily staying, such as a friend's or relative's home, a hotel or motel, or a vacation home?

Yes: How many times? _____

No

42. How many DIFFERENT motor vehicles (cars, trucks, motorcycles, etc.) were owned by you or any other member of this household during the last 6 months?

None: Skip to 45

1

2

3

4 or more

43. Did anyone steal, TRY to steal, or use (it/any of them) without permission?

Yes: How many times? _____

No

Did anyone steal or TRY to steal parts attached to (it/any of them), such as a battery, hubcaps, tape-deck, etc.?

Yes: How many times? _____

No

Source: National Crime Victimization Survey screening instrument. Bureau of Justice Statistics, 1987.

Web Research Project

Using a Web browser, search the term "victims." What issues exist in the current literature regarding victims of crime?

Issues and Cautions in Studying Victim Data

Some possible problems in victim surveys include, but are not limited to: the expense of compiling large samples, false or mistaken reports, memory failure or decay, telescoping of events, sampling bias, over- or underreporting, interviewer effects, and coding and mechanical errors.

1. While large-scale public opinion polls such as those by Gallup or Roper can be conducted with sample sizes of less than 1,000, the rarity of some types of victimization, such as rape, requires large samples in order to turn up a few victims. Hundreds may need to be surveyed in order to find one victim (Glaser, 1978, p. 63).

2. A parallel could be drawn with attempting to survey lottery winners on the basis of a sample of the general population. Many would have to be canvassed before turning up only a few winners. If the chances of winning the lottery were one in a million, in order to discover one winner by chance the researcher would have to interview 1 million players.

3. False or mistaken reports can result in error. Levine, for example, found inaccuracies in respondent reports regarding their voting behavior, finances, academic performance, business practices, and even sexual activity (Levine, 1976, p. 307). Should we assume greater precision in victim reports?

4. Memory failure or decay tends to increase with the distance between the actual time of the event and the interview concerning the event (Gottfredson & Hindelang, 1977; Panel for the Evaluation of Crime Surveys, 1976, p. 21).

5. Telescoping of events, a type of memory misfire, involves the moving of events that took place in a different time period (for example, before the reference period) into the time studied. A victimization of two years ago is mistakenly assumed to have occurred this past year. Subjects may even unconsciously telescope events in order to please interviewers (Biderman et al., 1967). Such demand characteristics or overagreeability on the part of respondents can certainly bias victim studies.

6. Sampling bias may produce an undernumeration of the young, males, and minorities. These very groups that tend to be undercounted by the U.S. Census are also more heavily victimized.

7. Overreporting in victim surveys generally involves subjects' reporting incidents to interviewers that they normally would view as too trivial or unimportant to call for police involvement. Much of the dark figure of crime consists of minor property crime, much of which could be considered unfounded by police (Black, 1970).

Controlling for Error in Victim Surveys

Some ways of controlling for error in victim surveys include, but are not limited to: the use of panels and bounding of target groups, evaluations of coding and other sources of human or mechanical error in data processing, reverse record checks of known groups, reinterviews of the same group, and interviews with significant others. Panels (longitudinal studies of the same group) were discussed previously as a means of bounding (establishing the time period during which events were recalled as having taken place), thus controlling for forward telescoping (the tendency to move prior incidents into the time frame being studied). Reinterviews of the same group in the National Crime Panel enables a tracking of reported crime incidents and the checking of responses with significant others (those who know the respondent well). The primary benefit of victim surveys is that they provide us with another independent measure of crime, separate from official statistics. Neither official statistics nor victim surveys begin to tap the extent of occupational, corporate, and public order crime; in that regard, both measures seriously underestimate the extent of crime.

Redesign of the National Crime Victimization Survey

Criticisms of the NCVS, particularly of its inability to gather accurate information regarding sexual assaults and domestic violence, prompted development of improved methodology that enhanced the ability of respondents to recall events. The survey changes increased the number of rapes and aggravated and simple assaults reported. The redesigned instrument also gathered information on other victimizations, such as nonrape sexual assault and unwanted or coerced sexual contact, for the first time. Improvements in technology and survey methodology were incorporated in the new design (Bureau of Justice Statistics, 1994).

An analysis of available data indicates that we have only a limited idea of the proportion of crime that is committed by any category of individuals or groups in a particular society. This is certainly the case if we rely entirely on official statistics for our discussions.

Self-Report Measures of Crime

As with victim surveys, **self-report** measures attempt to provide an alternative to official statistics in measuring the extent of crime in a society (Menard, 1987). Criminologists ask individuals, as in the illustration in Crime File 2.3, to admit to various crimes and/or delinquent acts. This may be achieved

Journal Article Link 2.1
Examine literature regarding self-reported copycat crime.

Crime File 2.3

Self-Reported Delinquency Items

Please indicate if you have ever done the following:

1. Stolen items of little value (less than $50).
2. Stolen items of great value ($50 or more).
3. Destroyed the property of others.
4. Used someone's vehicle without his or her permission.
5. Hit or physically attacked someone.
6. Been truant from school.
7. Consumed alcoholic beverages.
8. Used illegal drugs such as marijuana, heroin, or cocaine.
9. Indecently sexually exposed yourself in public.
10. Been paid for having sexual relations.

Web Research Project
Read Thornberry and Krohn's assessment of "The Self-Report Method for Measuring Delinquency and Crime" at http://www.ncjrs.gov/criminal_justice2000/volume_4/04b.pdf. What is their assessment of the method?

through anonymous questionnaires or surveys in which the respondent is identifiable that can be validated by later interviews or police records. Additionally, signed instruments that can be checked against official records, validation through later interviews or threats of polygraph (lie-detector) test, and interviews alone, as well as interviews that are then checked against official records, may be used (Nettler, 1978, pp. 97–113).

Most self-report surveys that have been conducted in the United States have been of "captive audiences," such as school or college populations (Glaser, 1978, p. 72; Hood & Sparks, 1971, p. 19). Few studies have been done of the adult population. One of the earliest, by Wallerstein and Wyle (1947), found that 99 percent of their adult sample had committed at least one offense. Some of the percentages of admission for males and females, respectively, were: larceny—89 and 83 percent; indecency—77 and 74 percent; assault—49 and 5 percent; grand larceny (except auto)—13 and 11 percent; and tax evasion—57 and 40 percent. These figures suggest a remarkable level of criminality on the part of an assumed noncriminal population.

Controlling for Error in Self-Report Surveys

Reliance on self-reported data as a measure of crime commission poses a major question with respect to the relationship between claimed behavior and actual behavior. Nettler states that "asking people questions about their behavior is a poor way of observing it" (Nettler, 1978, pp. 97–113). If people are inaccurate in reporting other aspects of their behavior, such as voting, medical treatment, and the like, it may be questionable to assume any greater accuracy in admitting deviant behavior. Some problems with self-report studies include: possibly inaccurate reports, the use of poor or inconsistent instruments, deficient research design, and poor choice of subjects. While mistaken or inaccurate reports may impinge on such surveys, Hood and Sparks (1971) question the number of trivial offenses that are labeled delinquent in the United States and are included in such studies. They point out that in Europe delinquency is a synonym for crime committed by the young. While small and unrepresentative samples are problematic, self-report surveys are also affected by possible lying, poor memory, and telescoping (Elliott & Ageton, 1980).

A particularly innovative program for checking self-reports was ADAM (Arrestee Drug Abuse Monitoring program), formerly the Drug Use Forecasting (DUF) program sponsored by the National Institute of Justice. Groups of arrestees are asked questions regarding their drug-use behavior and then are asked to voluntarily provide urine specimens that can be tested for drug use. Besides providing an ingenious way of estimating drug use among criminal populations, the program provided a barometer on the impact of various policies on drug usage. ADAM provided state and local drug policymakers, courts, law enforcement agencies, treatment providers, and prevention specialists with information that can be used to conduct local research and evaluation and to inform local policy decisions (National Institute of Justice, 2003). In 1998, NIJ launched International ADAM, which involved a partnership among criminal justice agencies in many countries, providing a global assessment of drug use. In conclusion, while self-report surveys have certain problems, they—like victim studies—provided us with an independent measure of crime commission. Unfortunately, the program was discontinued by the George W. Bush administration due to budget cuts in 2004.

Participant Observation

Participant observation involves a variety of strategies in which the researcher studies or observes a group through varying degrees of participation in the activities of that group. Ned Polsky's classic *Hustlers, Beats, and Others* (1967) presents both a moving statement on the need for deployment of this strategy and sound advice in this regard.

Participant Observation of Criminals

Contrary to the advice given at one time in most criminology textbooks (Sutherland & Cressey, 1960), uncaught criminals can be studied in the field. Biologists have long noted that gorillas in a zoo act differently than gorillas in their natural habitat. It is imperative that criminologists break their habit of studying the confined, slower, less intelligent, lower-class criminal. Polsky (1967), in advocating field studies of criminals, states:

> Until the criminologist learns to suspend his personal distaste for the values and life-styles of the untamed savages, until he goes out in the field to the cannibals and headhunters and observes them without trying either to civilize them or turn them over to colonial officials, he will be only a veranda anthropologist. That is, he will be only a jailhouse or courthouse sociologist, unable to produce anything like a genuinely scientific picture of crime. (pp. 117–149)

One of the reasons often given for discouraging such research is the belief that the researcher must pretend to be part of the criminal world. In fact, such a strategy would be highly inadvisable, not to mention unworkable and dangerous. Polsky suggests that the distance between criminal and conventional types is not as wide as many would suggest and that the difficulty in gaining access to such subjects is highly exaggerated.

There are, of course, problems in studying criminals *au naturel*. The researcher must realize that he or she is more of an intruder than would be the case in a prison setting. Unincarcerated criminals have more to lose than those already in jail do. And on their own turf, criminals are freer to put the researcher down or to refuse to be observed. Having successfully employed participant observation in studying uncaught pool hustlers, organized criminals, and drug addicts, Polsky (1967, pp. 117–149) offers some sage advice regarding procedures to employ in studying criminals in the field:

- Avoid using gadgets such as tape recorders, questionnaires, and the like. Construct field notes later, after leaving the scene for the day.
- Keep your eyes and ears open, but keep your mouth shut.
- Learn the argot, the specialized language or jargon of a group, but don't overuse it.
- You can often gain entry into the setting through common recreational interests, for example, card games, the track, or poolrooms.
- Do not pretend to be one of them. As soon as practicable, make them aware of your purposes.

Finally, Polsky raises a number of related issues to be considered in field studies of criminals. In some ways, researchers may be breaking the law or be considered accessories to the fact. Honoring reciprocity with respondents, observers must be prepared to be "stand-up guys" under police questioning. Although their actual legal status is unclear, social researchers in many cases have no guaranteed right to confidentiality or privileged information and are vulnerable to subpoena.

Evaluation of the Method of Participant Observation

Participant observation is an excellent procedure for studying little-understood groups. Some examples of participant observation studies with criminological ramifications have been Whyte's *Street Corner Society* (1955); Polsky's *Hustlers, Beats, and Others* (1967); Yablonsky's *Synanon* (1965) and *The Violent Gang* (1962); Ianni's *A Family Business* (1972); Albini's (1986) study of the Guardian Angels; and Humphreys's *Tearoom Trade* (1970). Eleanor Miller (1986) did field research interviewing 64 prostitutes in Milwaukee; Marquart (1986) worked as a prison guard; Hopper (1991) studied outlaw motorcycle gangs; and Sanchez-Jankowski (1991) spent 10 years living with and studying street gangs in Los Angeles, Boston, and New York.

Journal Article Link 2.2
Examine literature regarding ethnographic methodology.

The usefulness of such field studies in exploring settings that would not readily lend themselves to quantitative analysis is illustrated by some recent studies. Philipe Bourgois, author of *In Search of Respect: Selling Crack in El Barrio* (1995), spent the 5 years from 1985 to 1990 in East Harlem studying young Puerto Rican men on street corners and in crack houses, bars, and homes. Elijah Anderson's *A Place on the Corner* (1981) took place in the 1970s and reported on Chicago ghetto life from Jelly's, a bar and liquor store that he studied for over 3 years. Anderson's *Streetwise* (1990) describes two other Philadelphia neighborhoods. Mark Hamm's *American Skinheads* (1993) reports on his field study of neo-Nazi hate groups, which included communications with skinheads via the WAR (White Aryan Resistance) Web site. Jim Aho in *This Thing of Darkness* (1994) conducted a participant observation study of Idaho Christian Patriots until he defined such involvement as increasingly too dangerous. Miller and Tewksbury in *Extreme Methods: Innovative Approaches to Social Science Research* (2000) and Ferrell, Hamm, and Adler in *Ethnography at the Edge: Crime, Deviance, and Field Research* (1998) provide very interesting collections of articles on difficult-to-access deviant groups that require more innovative, and sometimes controversial, means of investigation.

The major advantages of participant observation relate to the qualitative detail that it can produce. Using this sensitizing or *verstehen* strategy, the researcher is less influenced by prejudgments. The technique is very flexible and less artificial, and enables the investigator to observe subjects in their natural environment. Such ethnographic methods provide insider accounts and acquaint students with the perspectives of the subjects (Cromwell, 1996). This technique has produced some of the most exciting and enthralling literature in the field, rivaling even some of the best of modern fiction. Examples from this genre will be presented in subsequent chapters. Some potential disadvantages of participant observation include the extremely time-consuming nature of the technique; it may exact high demands on the personal life of the observer (Carey, 1972). The observer faces the dual dangers of overidentification with, or aversion to, the group being studied, often testing to the limits the researcher's commitment to objectivity. In addition to possible observer bias and the challenge of making sense of a mass of nonquantitative data, participant observation can pose major ethical dilemmas.

Life History and Case Studies

A classic illustration of the use of **case study** and **life history** in criminology was Edwin Sutherland's *The Professional Thief* (1937), based on his interviews with an incarcerated professional thief given the pseudonym "Chic" Conwell. Like participant observation, case studies/life histories represent an interest in an in-depth close-up of only one or a few subjects in order to obtain a greater understanding or *verstehen* (Weber, 1949) that a more aggregate analysis might obscure. This method may employ diaries, letters, biographies, and autobiographies in order to attempt to capture a detailed view of either a unique or a representative subject. Some more recent examples of the life history approach have been Chambliss's *Box Man* (1975a); Klockars's *The Professional Fence* (1974); Steffensmeier's *The Fence* (1986); Shaw's *The Jackroller* (1930); and Snodgrass's *The Jack-Roller at Seventy* (1982).

Darrell Steffensmeier and Jeffrey Ulmer (2006) updated Steffensmeier's classic *The Fence: In the Shadow of Two Worlds* (1986) by presenting three decades in the life of Sam Goodman (pseudonym), a professional thief and fence. Their work was titled *Confessions of a Dying Thief: Understanding Criminal Careers and Illegal Enterprises* (Steffensmeier & Ulmer, 2006). The close relationship that developed between Steffensmeier and a dying Sam Goodman underlines the fact that research subjects and researchers become more than just observers and subjects.

Crime File 2.4

Confessions of a Dying Thief

Confessions of a Dying Thief: Understanding Criminal Careers and Illegal Enterprises by Darrell Steffensmeier and Jeffrey Ulmer (2006) is in part a 20-year follow-up to Steffensmeier's *The Fence: In the Shadow of Two Worlds* (1986), but it is more than this. It uses Sam Goodman's ethnography to address important methodological and theoretical ideas. Goodman (a pseudonym) was a professional thief and the authors attempt to use his life as a way of examining important issues in criminology.

His life illustrates a subculture that is often ignored by contemporary criminologists and sociologists. Persistent criminals are not deviant in all aspects of their lives. Goodman's biography does not support the life course/developmental theory of crime that does not account for the rewards and motives of criminal entrepreneurship. Goodman's life challenges simplistic views of criminal opportunity. *Confessions* provides an in-depth life history and picture of the criminal underworld, as well as a look at criminal entrepreneurship more generally.

The book is more than a case study, but a theory and methods book illustrated by a longitudinal case study. Sam's story was constantly checked against theory and methods. It is reminiscent of earlier longitudinal, qualitative case studies such as Snodgrass' *The Jackroller at Seventy* (1982), a follow-up to Shaw's *The Jackroller* (1930), a study of Stanley (a mugger), and Gans's *The Urban Villagers* (1962), a follow-up of Whyte's *Street Corner Society* (1943). Steffensmeier and Ullmer attempt to correct for the fact that the prison samples used in most existing studies fail to represent successful offenders for whom crime is very rewarding. They feel that the field of criminology has become dominated by theories on petty criminals and that the "life course" perspective is not only not a new perspective, but one that ignores a portion of chronic serious offending. They criticize writers such as Moffit (1999), who they claim inaccurately sees the cause of persistent criminality as biological inferiority and inherited differences. The relationship between Steffensmeier and Goodman obviously was far more than one of researcher and subject and provides a vivid picture of the world of professional crime.

Sources: Much of this crime file is drawn from Frank Hagan's remarks at "The Author Meets the Critics" session of the Academy of Criminal Justice Sciences meetings in Los Angeles, California, March 2006, on the occasion of *Confessions of a Dying Thief* receiving the Hindelang Award for Best Book of the Year; Steffensmeier, Darrell. *The Fence: In the Shadow of Two Worlds,* Totowa, NJ: Rowman and Littlefield, 1986; Steffensmeier, Darrell, and Ulmer, Jeffrey. *Confessions of a Dying Thief,* Totowa, NJ: Rowman and Littlefield, 2007; Snodgrass, Jon. *The Jackroller at Seventy: A Fifty Year Follow-up,* Lexington, MA: Heath, 1982; Shaw, Clifford, *The Jackroller,* Chicago: University of Chicago Press, 1930; Gans, Herbert, *The Urban Villagers,* New York: The Free Press, 1962; Whyte, William Foote, Jr., *Street Corner Society,* Chicago: University of Chicago Press, 1943; Moffit, Terry, "Adolescent Limited and Life Course Persistent Anti-Social Behavior: A Developmental Theory," in *Crime and Criminality,* Frank Scarpitti and Anne Nielsen, eds. Los Angeles: Roxbury, 1999, pp. 206–231.

Unobtrusive Measures

Unobtrusive measures are clandestine, secretive, or nonreactive methods of gathering data (Webb, Campbell, & Schwartz, 1981). Such techniques attempt to avoid reactivity, the tendency of subjects to behave differently when they are aware that they are being studied. This certainly has been a problem in much prison research, where the question might be asked whether research volunteers are indeed volunteers. Major types of unobtrusive methods include: physical trace analysis; the use of existing records like archives, available data, and autobiographies; and simple and disguised observation, as well as simulation.

Physical trace analysis involves studying deposits, accretion of matter, and other remains of human activity, while archival and existing records contain information that may be useful in providing historical overviews of criminological issues.

The uses of *available data* include procedures such as content analysis and secondary analysis. Content analysis refers to the systematic classification and study of the content of mass media, for

Crime File 2.5

Useful Sources for Criminological Research

Selected Journals

American Journal of Criminal Justice
American Journal of Sociology
American Sociological Review
Crime and Delinquency
Criminal Justice Policy Review
Criminal Justice Review
Criminology
Federal Probation
Journal of Criminal Justice

Journal of Research in Crime and Delinquency
Justice Quarterly
Law and Society Review
NIJ Reports
Social Forces
Social Problems
Sociology and Social Research
Victimology

Abstracts/Indexes

Crime and Delinquency Abstracts
C J Abstracts (online)
EBSCO Select College Edition (online)
National Criminal Justice Reference Service
New York Times Index

Police Science Abstracts
Psychological Abstracts
Reader's Guide to Periodical Literature
Social Science Index
Sociological Abstracts

This is only a small selection of available sources. Check the periodicals and reference sections of your college library for more.

Web Research Project

Using one of the abstract services above [hint: some are online], search for an article under the title "crime research." What methods were used and what were some of the findings of this article?

---

example, newspapers, magazines, and the like. Secondary analysis consists of the reanalysis of data that was previously gathered for other purposes. The use of all of these types of data-gathering procedures is an excellent, cost-effective means of obtaining data, particularly in a period of growing respondent hostility to studies. In an interesting example of the imaginative use of existing data, criminologist John Laub discovered more than 60 boxes of dusty files in the subbasement of the Harvard Law School Library (Associated Press, 1994). These turned out to be the research files of Eleanor Glueck and Sheldon Glueck, who had been at Harvard from the 1920s to the 1970s. They had conducted one of the first longitudinal studies in criminology in which male juveniles were followed from age 14 until age 32, attempting to predict the cause of criminal behavior. In an example of secondary analysis, Laub computerized their data and analyzed it. Crime File 2.5 presents lists of useful sources for criminological research.

Observation requires the researcher to keep participation with subjects to a minimum while carefully recording their activities; in disguised observation the investigator secretly studies groups by temporarily deceiving them as to his or her real purpose. For example, in order to study difficult subjects in the field, researchers have posed as "thieves and victims" (Stewart & Cannon, 1977), a "watch queen" (Humphreys, 1970), a "mental patient" (Caudill, 1958), "Black Panther supporters" (Heussenstamm, 1971), "a naive international tourist" (Feldman, 1968) and "a caretaker" (Sherif & Sherif, 1966), among other roles.

Simulation entails research strategies that attempt to mimic or imitate a more complex social reality. For example, since actual research into jury deliberations is prohibited, researchers may set

Photo 2.5
Researchers may set up simulated juries such as this to investigate the decision-making process.

up simulated juries by reenacting actual trial conditions in order to investigate the decision-making process (see Photo 2.5).

While the obvious advantage of unobtrusive measures is that they are nonreactive—that is, they prevent subject awareness of being observed and ideally escape reactivity—such techniques also have the strength of being more natural and of evading the overreliance upon attitudinal data. By making use of data that have already been gathered, researchers are able to exercise great economies of time and expense. Too many researchers assume that doing a study must necessarily involve the expense and time of gathering new data when, in fact, vast storehouses of potential information exist right under their noses, as close as the nearest library and scattered throughout the records of public and private organizations. On the debit side of the ledger, unobtrusive methods raise potential problems of privacy invasion. Does a researcher have the right to observe the private behavior of individuals without their permission? Compounding this ethical issue is that criminological researchers have no state-recognized right to confidentiality or claim to privileged communication comparable to that in a doctor-patient relationship. In addition, nonreactive measures may yield atypical subjects, be time consuming, and be prone to observer bias. Crime File 2.6 describes the FBI Reading Room, where one may electronically peruse a variety of files made available through the Freedom of Information Act.

Validity, Reliability, and Triangulation

In the past a number of researchers have been critical of the accuracy of much criminological research. Bailey (1971), in a review of 100 correctional research studies, pointed out that much of the research was invalid, unreliable, and based on poor research design. In an analysis of the quality of publications in criminology, Wolfgang, Figlio, and Thornberry (1978) judged that the methodological sophistication was very poor and that a greater display of concern was needed for adequate research design and execution. Although later modifying his view and admitting **methodological narcissism,** Martinson (1974; "Martinson", 1978) blasted correctional research, claiming that in his review

Crime File 2.6

The FBI Reading Room

What do Jackie Kennedy, the Beatles, Albert Einstein, Gracie Allen, Thurgood Marshall, and Walter Winchell have in common?

Give up?

They are all part of historical FBI records . . . though for a variety of reasons. One had a background investigation for government service. One received extortionate threats. One had open communist affiliations. One needed security for a family trip abroad. One actively helped FBI investigations. One tried to smuggle jewelry into the U.S. And not necessarily in that order.

Interested in all the details? Just go to the *Electronic Reading Room* at the FBI's Freedom of Information Act (FOIA) Web site. These files include some of the 50+ new additions to the site, posted there for researchers interested in federal records on everything from Alcoholics Anonymous to UFOs.

Why are we releasing all these records? It's the law. Following passage of the Freedom of Information Act, the Privacy Act, and some amendments to them, the FBI (with every other federal agency) began disclosing its records, upon written request, on a case by case basis . . . only blacking out information cited in the laws' nine exemptions and three exclusions, which are largely designed to protect national and economic security and to protect the privacy of persons who appear in FBI records.

How many requests are we talking about? Hundreds of thousands . . . and still counting.

How many pages of records are we talking about? Don't be shocked: millions . . . and still counting. After all, information is the business of law enforcement—writing down all those interviews and recording all that crime scene evidence.

Why a Reading Room? It turned out that so many people were interested in the same files that it just made sense to put them in a physical library at Headquarters for researchers to visit and use freely. But it was tough on researchers, who had to travel all the way to Washington and compete with others for the few chairs in what was generally regarded as pretty cramped space. When the Web evolved, we couldn't wait to begin digitizing documents to create the current Virtual Reading Room, for all the world to access. Good thing too, as that also became law. Now we continue to expand it as resources allow.

So pull up a chair . . . and decide where you want to start. *Spies? Celebrities? Gangsters? Violent criminals? Historical figures, issues, and events?* Or *"Unusual phenomena"?*

We recently posted FBI records online in our *virtual reading room* that concern 66 different people and organizations—an incredible assortment that paints a vivid picture of American history.

Some were helping with investigations . . . some were under investigation . . . some were just interested citizens writing in for one reason or another.

There's a one-page letter to J. Edgar Hoover about the Lizzie Borden murder case . . . there's 492 pages of documents on President Carter's brother Billy . . . and everything in between.

Just in case you're interested, here's the line-up:

Edward Abbey	Marian Anderson
Bud Abbott	Elizabeth Arden
Jane Addams	Louis Armstrong
Spiro T. Agnew	Desi Arnaz
Joseph Aiuppa	Aryan Circle
Alcoholics Anonymous	Aryan Nations
All American Anti-Imperialist League	Arthur Ashe
Gracie Allen	Vincent Astor
Lillie Belle Allen	Gene Autry
Louis Allen	Arthur Barker
Steve Allen	Fred Barker
American Civil Liberties Union (ACLU)	Herman Barker
American Committee of Jewish Writers, Artists and Scientists	Lloyd Barker
American Deserters Committee	"Ma" Barker

Bernard Baruch	Truman Capote
The Beatles	Stokely Carmichael
Melvin Belli	Rudolph Carnap
Jack Benny	Rachel Carson
Black September	Billy Carter
Sonny Bono	Center for Constitutional Rights
Lizzie Borden	Wilt Chamberlain
Werner von Braun	John Chancellor
William J. Brennan, Jr.	Chappaquiddick (Mary Jo Kopechne)
"Pat" Brown	Caryl Chessman
Lennie Bruce	American GI Forum
"Lepke" Buchalter	American Nazi Party
Pearl S. Buck	American Negro Labor Congress
George Burns	Albert Anastasia
James Cagney	Jane Anderson
Taylor Caldwell	George Jackson Brigade
Cab Calloway	Chinese Workers Mutual Aid Association
Eddie Cantor	Christian Crusade
Frank Capone	Christian Identity Movement

Source: http://www.fbi.gov.

Web Research Project

Search the FBI "Freedom of Information Act" files (FOIA) and locate a group or individual that was the subject of investigation. What do you conclude as a result of your perusal of these files?

of the evidence of programs in corrections and their impact on recidivism, he found that "nothing works." Methodological narcissism refers to the belief that one's favorite method is the only way to do research and all other methods are inferior. What is to be said of this sad state of affairs? If the data regarding "what is?" with respect to crime are defective, then what might we expect of the theories that are based upon these data? Fortunately, criminologists have plenty of methodological company with economists, psychiatrists, and meteorologists, to mention just a few. The problem of imprecise measurement is not unique to the field of criminology and, furthermore, is not an insoluble one.

Validity concerns accuracy of measurement. It asks the questions, "Does my measuring instrument in fact measure what it claims to measure?" or "Is it a true and accurate measure of the subject in question?" **Reliability,** on the other hand, involves the consistency and/or stability of measurement. If repeated measures were made of the same entity, would stable and uniform measures ensue? Obviously, validity is a more crucial issue than reliability; if a measurement is inaccurate, the consistency of inaccuracy becomes a moot question.

The problem of inadequate methods in criminology arises not because of the inherent shortcomings of any particular method, but because a given method is used alone. It is foolhardy to concentrate on the insufficiencies, the reliability, and/or the validity of any one concept, measured at one time using one measure. **Triangulation** involves the use of multiple methods in measuring the same entity. It is similar to the notion of corroborating evidence in law; if different measures of the same concept produce convergence or similar results, then we have greater confidence in the validity of an observation or finding.

Sanders in *The Sociologist as Detective* (1976) makes clever use of Arthur Conan Doyle's fictional sleuth Sherlock Holmes as a means of illustrating the notion of triangulation. Holmes, in attempting to answer the question "Whodunit?" employed multiple methods (triangulation) like those a social scientist might employ. In attempting to discover "who killed the lord of the manor," Holmes observed carefully, attempted reenactment of the crime (simulation), questioned suspects and witnesses, and carefully collected and evaluated the physical evidence at the crime scene. He collected some data through direct questioning, other data through astute observation. "Did the family dog bark the evening of the suspected murder?" If not, perhaps the murderer was a family member or friend. "Did any of the questioned suspects develop a nervous tic?" "Were there footprints or clues?" By combining these various methods, Holmes was able to make a reasonable guess as to which hypotheses to reject or accept (see also Truzzi, 1976).

This chapter has exposed the reader to a variety of methods that criminologists use in obtaining information on the nature of crime and criminals. The outcomes or findings that result from the application of these methods will be presented in forthcoming chapters. It is hoped that the reader has been alerted to viewing this material with a critical methodological eye, carefully weighing the sources of evidence for the materials presented. For more detail on research methods, see Hagan (2010).

Summary

Theory and methodology are the two critical features of any discipline, including criminology. Theory is an attempt to provide plausible explanations of reality and addresses the question "Why?" Method (methodology) involves procedures for the collection and analysis of accurate data or facts and is concerned with the issue "What is?"

The research enterprise of criminology involves certain basic procedures. Objectivity, a commitment to a "value-free," nonbiased approach to the subject matter, is an essential canon of research. Despite conflicting roles, the criminologist's primary role is that of scientist. Some general principles of ethical conduct in criminology include that the researcher should avoid harmful procedures, honor commitments and reciprocity, exercise objectivity and integrity, and protect the privacy of subjects, as well as maintain confidentiality.

The process of methodological thinking was illustrated by means of the research question "Who is criminal?" Until recently the primary source of information regarding crime statistics has been official police statistics, which represent crimes recorded by police. The Uniform Crime Reports (UCR) presents such statistics for the United States. Such statistics fail to account for unrecorded crime, "the dark figure of crime."

The UCR "crime index" from which the crime rate is calculated consists of Part I crimes: murder and non-negligent manslaughter, forcible rape, robbery, aggravated assault, burglary, larceny-theft, motor vehicle theft, and arson. Researchers should be cognizant of shortcomings of official data such as the UCR. The redesigned UCR (NIBRS, National Incident-Based Reporting System) is an attempt to improve the system.

Other alternative measures of crime and criminal activity include crime seriousness measures, which attempt to provide a weighted index of crime. Alternative data-gathering strategies include experiments, social surveys, participant observation, case studies/life history methods, and unobtrusive methods. Each possesses relative strengths and weaknesses vis-à-vis the others with respect to quantitative/qualitative control, internal/external validity, and degrees of artificiality/naturalness.

A key point is that, contrary to methodological narcissism (fanatical adherence to one's favorite method), no one method has any inherent superiority over any other. Methodology is a tool and not an end in itself. For each method, the text provides descriptions as well as examples of the method's application in criminological research. For instance, victim surveys are a critical alternative measure of criminality. Similarly, self-report surveys are a useful means of tapping hidden criminality. The basic strategy of participant observation (field studies), life histories, and case studies in criminology is delineated. A particularly moving pitch for the need for such studies emerges from Ned Polsky's research. Unobtrusive (nonreactive) methods are a cost-effective and neglected means of obtaining data. These include techniques such as physical trace analysis, use of archives/existing data (including content and secondary analysis), and autobiographies. Other procedures include simple and disguised observation and simulation.

Much of the criticism of criminological research centers on the validity (accuracy) and reliability (consistency/stability) of the methodology that has been employed. Triangulation (the use of multiple methods) is proposed as the logical path to resolve this issue.

Criminology on the Web

Log on to the Web-based student study site at http://www.sagepub.com/haganintrocrim7e/ for author-created podcasts, e-flashcards, quizzes, and more.

KEY CONCEPTS

Campbell Collaboration (C2)

Cautions in the UCR

Classic Experimental Design

Code of Ethics for Research

Confidentiality

Crime Index

Crime Rate

Dark Figure of Crime

Ethics in Research

Evidence-Based Research

Experiments

Index Crimes

Life History/Case Study

Methodological Narcissism

Methodology (Methods)

National Crime Victimization Survey (NCVS)

National Incident-Based Recording System (NIBRS)

Objectivity

Operationalization

Part I Crimes

Participant Observation

Reciprocity

Reliability

Sampling

Self-Reports of Crime

Simulation

Sources of Crime Statistics

Surveys

Theory

Triangulation

Unfounded Crimes

Uniform Crime Reports (UCR)

Unobtrusive Measures

Validity

Variables

Victim Surveys

REVIEW QUESTIONS

1. Reviewing Crime File 2.1, The Crime Dip, which factor(s) do you find to be most plausible in explaining the crime dip? Using these same factors, do you predict that crime will continue to decrease or do you foresee an increase in the near future? Explain your reasoning.

2. Examining the codes of ethics of the Academy of Criminal Justice Sciences and the American Society of Criminology, what stipulations do you regard as most important and which are of least importance? Are you familiar with any additional studies that have raised ethical concerns? Search the Web, InfoTrac® College Edition, C J Abstracts, and NCJRS under titles such as research ethics or codes of ethics and see if you can turn up any recent controversies.

3. What are some sources of information used by criminologists to examine the extent of crime in the United States?

4. Compare the UCR with the NCVS. Which of these is the better measure of crime?

5. How does the FBI compile and calculate the crime rate? What types of crime does this include?

6. What are some problems or shortcomings of the UCR?

7. What are some other ways of gathering data in criminology besides reliance on official police statistics? Give an example of each.

8. How accurate is the NCVS? Is it a better measure of crime than the UCR?

9. What is ADAM and what does it measure? Is there any way of checking on its accuracy?

WEB SOURCES

Bureau of Justice Statistics
http://www.ojp.usdoj.gov/bjs

Central Intelligence Agency
http://www.cia.gov/index.html

Department of Justice Career Opportunities
http://www.usdoj.gov/careers.html

Federal Bureau of Investigation
http://www.fbi.gov

General Accounting Office
http://www.gao.gov

Justinfo Online
http://www.ncjrs.org/justinfo/dates.html

Library of Congress
http://www.loc.gov

National Criminal Justice Reference Service
http://www.ncjrs.gov

National Institute of Justice
http://www.ojp.usdoj.gov/nij

U.S. Department of Justice
http://www.usdoj.gov

This is only a small selection of available sources. Check the periodicals and reference sections of your college library for more.

WEB EXERCISES

Using this chapter's recommended Web sites, examine the various sources available in research methods.

1. What can be learned by examining Department of Justice sites such as the National Institute of Justice, Bureau of Justice Statistics, National Criminal Justice Reference Service, and Justinfo Online?

2. Specifically, what types of careers are listed on the Department of Justice Career Opportunities site?

3. What types of studies are available on the General Accounting Office site?

4. Were you impressed by the Library of Congress site? Explain.

5. Using your Web browser, search the terms "Uniform Crime Reports" and "National Crime Victimization Survey" for recent crime statistics.

SELECTED READINGS

Jim Aho. (1994). *This Thing of Darkness: The Sociology of the Enemy.* Seattle: University of Washington Press.

    The author conducts a participant observational study of right-wing Christian Patriot groups in Idaho.

Bruce Berg. (2004). *Qualitative Research Methods in the Social Sciences* (5th ed.). Boston: Allyn & Bacon.

    This is a text devoted exclusively to qualitative methods in the social sciences, a subject which is sometimes neglected in standard methods texts.

Paul Cromwell, editor. (2003). *In Their Own Words: Field Research on Crime and Criminals—An Anthology* (3rd ed.) Los Angeles: Roxbury.

    This is an excellent anthology of articles involving field research of uncaught criminals on their own turf.

Jeffery Ferrell and Mark S. Hamm, editors, (1998). *Ethnography at the Edge: Crime and Deviance in Field Research.* Boston: Northeastern University Press.

    The authors describe their articles as unorthodox invitations to heresy. Research among uncaught criminals often put researchers in compromising positions on the edge of illegality.

Frank E. Hagan. (2006). *Research Methods in Criminal Justice and Criminology* (7th ed.) Boston: Allyn & Bacon.

    This is the author's own text on criminological and criminal justice research and features detailed coverage of issues only introduced in our account in this chapter.

John M. Hagedorn. (1994, May). Homeboys, Dope Fiends, Legits, and New Jacks. *Criminology, 32,* 197–219.

This field study of gang turf and drug use utilizes people from the neighborhood as coresearchers.

Mark S. Hamm. (1993). *American Skinheads: The Criminology and Control of Hate Crime.* Westport, CT: Praeger.
    Author Mark Hamm uses imaginative, triangulated methods in order to study skinheads.

Martin Sanchez-Jankowski. (1991). *Islands in the Streets: Gangs and American Urban Society.* Berkeley: University of California Press.
    Forty-four youth gangs are studied in the field over a 10-year period by Sanchez-Jankowski.

Pamela Tontodonato and Frank E. Hagan, editors. (1998). *The Language of Research in Criminal Justice.* Boston: Allyn & Bacon.
    Thirty-four articles are featured, illustrating research in criminology and criminal justice. These are articles that are often cited in texts.

Richard T. Wright and Scott Decker. (1996). *Burglars on the Job: Street Life and Residential Break-Ins.* Boston: Northeastern University Press.
    The authors interview uncaught burglars in the field asking questions such as how they choose targets and avoid being arrested.

Caution in Interpreting Crime Data

International Variations in Crime

The Prevalence of Crime

Trends in Crime

Crime File 3.1 American Crime Problems From a Global Perspective

Age and Crime

Crime File 3.2 What Is the Relationship Between Age and Crime?

Gender Differences in Criminality

Social Class and Crime

Race and Crime

Crime File 3.3 Racial Profiling

Crime File 3.4 Native Americans and Crime

Regional Variation in Crime

Urban/Rural Differences in Crime

Institutions and Crime

The Family and Crime

Education and Crime

Religion and Crime

War and Crime

Economy and Crime

Mass Media and Crime

Summary

Key Concepts

Review Questions

Web Sources

Web Exercises

Selected Readngs

chapter 3

General Characteristics of Crime and Criminals

This is our basic conclusion: Our nation is moving toward two societies, one black, one white—separate and unequal. . . . White racism is essentially responsible for the explosive mixture which has been accumulating in our cities since the end of World War II.

—*U.S. Riot Commission Report (Kerner, 1968, pp. 1, 10)*

Caution in Interpreting Crime Data

In Chapter 1 we discussed at length the necessity of carefully examining the database or other sources of criminological research findings and conclusions. This advice is especially applicable to the material presented in this chapter. Descriptions of characteristics of crime and criminals can vary immensely, depending on the sources of information—for example, official statistics, victim surveys, self-reports—as well as on the type of crime or criminality that is being addressed, whether traditional crime or crime by the elite. The particular method chosen for analysis provides data that flavor the types of theories developed; likewise, the theoretical framework for analysis may subjectively influence the methods of analysis. While the process of inquiry is seldom entirely value-free, tri-angulation assists in providing multiple assessments of the subject matter. As previously indicated, statistics regarding crime and delinquency are not easily measured. Realizing the limitations of these statistics, we will attempt to avoid misleading and incorrect inferences.

International Variations in Crime

International or cross-cultural comparisons of crime statistics are hazardous given the different definitions of criminal activity, the quality of data, ideological considerations, and the sheer logistical problems of compilation (see Dammer & Fairchild, 2005; Nordstrom, 2007; Rounds, 2000; Terrill, 2007; Van Dijk, 2008). Analysis of cross-cultural crime rates can produce some interesting conclusions. For instance, the inexorable rise in crime in the United States and other industrialized countries in the sixties was contradicted by a declining crime rate in Japan, which discredited the assumption that modernization inevitably produces increased criminality. Freda Adler's *Nations Not Obsessed by Crime* (1983) and Clinard's *Cities With Little Crime* (1978) also indicate that crime is not a major concern in such countries as Switzerland. In a contrary view, others indicate that the Swiss police omit crime statistics and the media ignore criminality, not to mention that Switzerland is a haven for white collar crime (Balrig, 1988; Gerber, 1991). The low crime rate in Japan is achieved by a strong *Gemeinschaft* (communal) orientation and group conformity, a high level of unchallenged police power, and a tendency to ignore violations of human and individual rights that would be found unacceptable in the Western democracies (J. Williams, 1991). Utilizing data from the International Police Organization (Interpol) and the World Health Organization (WHO), Brantingham and Brantingham (1984) point out,

> At the world level of resolution, clearly different patterns emerge for crimes of violence against the person and for crimes against property. The highest overall crime rates were experienced by the nations of the Caribbean region during the mid-1970s, followed by the nations of Western Europe, North America, and Oceania. The highest levels of violent crimes against the person were experienced in the Caribbean, in North Africa and the Middle East, in sub-Saharan Africa, and in Latin America. Property crimes were highest in Western Europe, North America, and Oceania. Crime patterns appear to be closely associated with high economic development and with income inequality; and high levels of violent crimes against the person are associated with lack of economic development and with high income inequality. Modernization and urbanization are both associated with higher levels of property crime and lower levels of violent crime. (p. 295)

Similar patterns with respect to the impact of income inequality and the lack of economic development and high crime rates have been noted by others (e.g., Clinard & Abbott, 1973; Krahn, Hartnagel, & Gartrell, 1986). At the beginning of the twenty-first century, we find that the breakdown of political order, lack of police training, and growing urbanization—particularly in developing countries—produce higher global crime. The Overseas Security Advisory Council of the U.S. State Department issues timely travel advisories to tourists warning of special dangers. Nonviolent theft and pickpocketing are the most common crimes that business travelers endure. In 1999, attacks in unregulated taxicabs in Mexico City were of concern, while pickpockets and thieves were more prevalent in Eastern Europe. Particularly dangerous were capital cities of West and East Africa, where the breakdown of tribal authority and poor police training cultivated high crime. The former Soviet Union in general had experienced high levels of violence, organized crime, and corruption (Nicolova, 1999). In Thailand, a favorite tourist destination, economic collapse and an increase in drug-related crime had increased violence against foreigners, and weak law enforcement had attracted foreign criminal gangs (Cheesman, 1999).

Particularly interesting is concern among Americans about being potential victims in some foreign lands, when in fact the murder and rape rates in general are higher in the United States than in other developed countries. By the year 2000, however, 7 years of declining official crime rates in the United States indicated that for assault, burglary, robbery, and motor vehicle thefts, the U.S. rates were actually lower than rising rates in some other developed countries such as England and Wales (Langan & Farrington, 1998).

Cross-national crime statistics have steadily improved over the years with Interpol, the United Nations, and WHO publishing such data. In addition, the United Nations has sponsored the International Crime Victimization Survey (ICVS), as well as the United Nations Surveys on Crime Trends and Operations and International ADAM (Arrestee Drug Abuse Monitoring), a self-report survey of arrests modeled after the U.S. ADAM.

Van Dijk and Kangaspunta (2000) indicate the following difficulties in analyzing crime data across countries:

- Varying definitions—legal codes define crimes in different ways.
- Recording practices—different police departments record things differently (e.g., bicycle thefts are vehicle thefts in some countries).
- Operating practices—in some countries, only crimes reaching court are recorded.
- Factual inequalities—hidden factors may affect crime rates, such as age, urbanism, and the like.
- Problems especially associated with recorded crime—governments may regard such statistics as indicators of criminal justice system workloads rather than accurate indicators of crime prevalence.

Table 3.1 presents Interpol crime rates for recent years. Given our previous discussion, the reader should interpret these official police statistics with caution. After 2004, Interpol decided to no longer make its statistics available except to law enforcement agencies. Data presented in this book are the latest available from this organization.

TABLE 3.1 ■ **Crime Rate in Selected Countries: Interpol Data (rate per 100,000 population)**

Country	Homicide	Rape	Robbery	Burglary	Auto Theft
USA (2004)	5.5	32.2	136.7	729.9	421.3
Australia (2000)	3.6		121.7	2280.8	726.2
Bahamas (1999)	27.1	55.9	457.2	1560.3	
Brazil (2001)	23	8.5			88.5
Canada (2001)	4.1		88.2	908.9	547.6
Denmark (2003)	4.1	8.7	59.1	1876	440.5
France (2002)	3.6	17.5	210	683.6	388.5
Ireland (2002)	1.6	13.7	260.2	685.5	3.8
Jamaica (2001)	43.7	35	80.9	83.8	3.7
Japan (2002)	1.1	1.9	5.5	265.5	49.2
Mongolia (2001)	15.2		26.5	464.7	0.4
Russia (2002)	22.4	5.6	148.9	425.6	33
Saudi Arabia (2002)	1	0.3	0.1		89.8

U.S. statistics were not included in the Interpol report; data are from the Uniform Crime Reports for 2004. Blank fields were unreported.

Source: International Crime Statistics, 2003. Interpol, 2004. Lyons, France: Interpol Secretariat.

A comparison of U.S. crime rates with those of other countries using Interpol data found the following:

- For 1998, the U.S. homicide rate was 6.3 using Uniform Crime Reports (UCR) data; the rate in Europe was generally between 1 and 2 using Interpol data. Colombia (56.3), Jamaica (37.2), Russia (18), Mongolia (24.6), and other, mostly Latin American and African countries had higher rates than the United States. Not shown in the table is South Africa's murder rate of 56.9.
- The U.S. rape rate was 34.4 using UCR data. This rate was virtually tied with Jamaica, Namibia, and Zimbabwe, although rape statistics tend to be notoriously underestimated in many countries. Rates in most developed countries were under 10.
- The U.S. robbery rate was 165.2. Others with high rates were Australia (127.6), France (144.1), Guyana (168.8), Jamaica (116.9), Russia (106.1), Spain (162.9), and Trinidad (193.8). Japan was lowest with 2.1.
- The U.S. burglary rate of 862 was exceeded by other developed countries. Australia (2,338.4), Belgium (2,031.3), Denmark (1,925.2), Finland (1,757), Germany (1,183.1), Israel (1,122.3), and Switzerland (1,172.5) were all higher.
- The U.S. rate of 459 for auto theft was also exceeded by Belgium (530.2), Canada (1,155.7), Australia (706.2), Denmark (709.5), Finland (482.2), Israel (694.1), Norway (487.5), and Switzerland (1,129.9).

In addition to traditional criminal activity, the multinational nature of organized, white collar, and more sophisticated crimes is a growing phenomenon. In November 1995, a 138-nation United Nations–sponsored conference on international crime was held in Naples, Italy. The end of the Cold War has opened borders and provided an opportunity for collaboration among organized crime groups. Transnational crime poisons business climates, corrupts political leaders, undermines banking and finance, and represents a global challenge of immense proportions.

As an illustration of ideological impacts on crime statistics, one need only look at Russia in the late eighties and the influence of Gorbachev's *glasnost* (openness). In the first publication of crime statistics in more than half a century, the Soviet Interior Ministry reported a sharp rise in crime of 18 percent for 1988 over 1987 ("Soviet Crime Rate Up," 1989) and 32 percent in 1989 (Bogert, 1990). In 1991, the number of robberies, muggings, burglaries, and thefts had jumped by 90 percent over 1990 (M. Shapiro, 1992). Instrumental in this increase was the collapse of an authoritarian regime (whose police were omnipresent) combined with declining economic conditions and general political confusion.

Organized crime groups often filled the power void. Having previously controlled the black market, such groups were well trained for the new, legal capitalism. Such increased levels of crime have also been experienced throughout former Soviet satellite countries and throughout the world—in the Baltic states, the People's Republic of China, South Africa, and even Vietnam (Larimer, 1996, p. A7).

There have been a number of International Crime Victimization Surveys sponsored by the United Nations, with studies begun in 1989, 1992, 1996, and 2000. The 2000 study involved 92 surveys in 56 countries and was expanding to over 100 surveys, involving cities as well as countries. The last survey found that the United States, Canada, and the Czech

Republic rank among the highest for burglary, motor vehicle theft, and petty crime. Other countries with higher levels of these crimes were Bulgaria, Estonia, and Slovakia. Low property crime victimization rates were found in Belarus, Norway, Switzerland, and Macedonia. High violence rates were found in countries of the former Soviet Union such as Estonia, Kazakhstan, Kyrgyzstan, and Russia. The United States has high violence scores compared to levels in Canada and Western European nations. Particularly low violence levels were to be found in Western Europe, Hungary, and Macedonia (Van Dijk & Kangaspunta, 2000, p. 39). For more information on international crime and surveys, see G. Newman (1999) and the Web site http://www.vn.fi/etusivu/en.jsp.

The reader is reminded that, due to unreliability in such statistics, international comparisons are risky; however, R. R. Bennett and Lynch (1990) found similar results when they compared four widely used data sets.

The Prevalence of Crime

Estimates of the extent of crime commission depend upon how far or wide one may wish to cast the net. As one moves from official statistics (UCR) to victim surveys (NCVS) to self-report surveys, the rates increase. Inclusion of other forms of nontraditional criminality, such as corporate crime or tax avoidance, would make crime seem even more pervasive.

Table 3.2 presents the UCR index of serious crimes in the United States, sometimes referred to as Part I offenses. Examination of crimes known to police in Table 3.2 underlines the point that the bulk of offenses are made up of property crimes.

In June 2004, the Criminal Justice Information System Committee (CJIS) of the International Association of Chiefs of Police Advisory Policy Board (APB), functioning in an advisory capacity concerning the UCR, decided to discontinue the use of the crime index in the UCR program. They directed the FBI to instead publish a violent crime total and a property crime total until a new index

Audio Link 3.1
Listen to a discussion of violent crime surges.

TABLE 3.2 ■ Index of Crime in the United States

Offense	Number	Rate per 100,000
Violent Crime	1,417,745	473.5
Property Crime	9,983,368	3,334.5
Murder	17,034	5.7
Forcible Rape	92,455	30.9
Robbery	447,403	149.4
Aggravated Assault	860,853	287.5
Burglary	2,183,746	729.4
Larceny-Theft	6,607,013	2,206.8
Motor Vehicle Theft	1,192,809	398.4
Arson*		

* Sufficient data are not available to estimate this offense.

Source: Modified from *Crime in the United States, 2007,* Federal Bureau of Investigation, 2007, Washington, DC: Government Printing Office.

TABLE 3.3 ▪ Number of Persons Arrested for the Ten Most Frequent Offenses

Offense	Number Arrested
1. Drug Abuse Violations	1,889,810
2. Driving Under the Influence	1,460,498
3. Other Assaults	1,305,757
4. Larceny-Theft*	1,081,157
5. Disorderly Conduct	703,504
6. Liquor Laws	645,734
7. Drunkenness	553,188
8. Aggravated Assault*	447,948
9. Burglary*	304,801
10. Fraud	280,693

*Denotes index crimes.

Source: Modified from *Crime in the United States, 2006,* Federal Bureau of Investigation, 2007, Washington, DC: Government Printing Office.

could be developed. They felt that the index was inflated by including larceny-theft, which was the category that had the highest crime rate.

Table 3.3 reports the crimes that most frequently result in arrest. These data refer to persons arrested and not, as in the case of the index offenses, simply crimes known to police. Examination of these primarily Part II offenses indicates the extent to which policing is occupied with drunk driving, drunkenness, and disorderly conduct.

Trends in Crime

Video Link 3.1
View a video of
police tracking
crime trends.

As discussed in the previous chapter, official crime statistics represented by the UCR have risen dramatically since the first study in the early 1930s. Figures 3.1 and 3.2 present trends in these index crimes.

Although, as we learned in Chapter 1, caution should be exercised in interpreting these statistics, these trend lines certainly dramatically depict an inexorable rise in official recorded crime since the mid-sixties, declining in the nineties. Despite this rise in official rates, victim surveys since the seventies reported relatively stable or falling rates, perhaps reinforcing the point that better recording and reporting may have in part accounted for some of the rise in official statistics.

The public alarm concerning the rapid rise in UCR crime statistics beginning in the mid-sixties was abetted by the fact that the decades of the 1940s and 1950s with their postwar prosperity demonstrated relative stability in many categories of crime. The new "crime wave" appeared particularly out of place. Historians of crime and violence in the United States remind us of our myopia in this regard and that waves of crime and violence, however difficult to measure, were characteristic of this land since colonial times, particularly in the post–Civil War era. (This subject will be given greater scope in Chapter 8.)

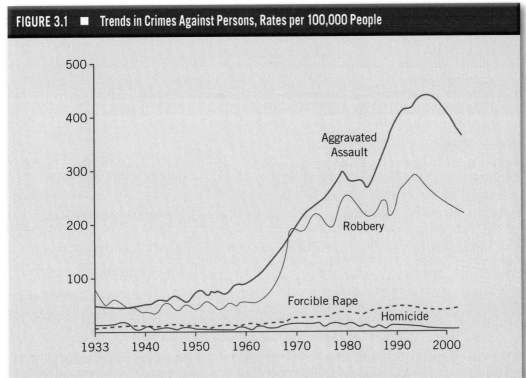

FIGURE 3.1 ■ **Trends in Crimes Against Persons, Rates per 100,000 People**

Source: President's Commission on Law Enforcement and the Administration of Justice. 1967, *The Challenge of Crime in a Free Society*, Washington, DC: Government Printing Office, p. 23, updated with yearly Uniform Crime Reports data.

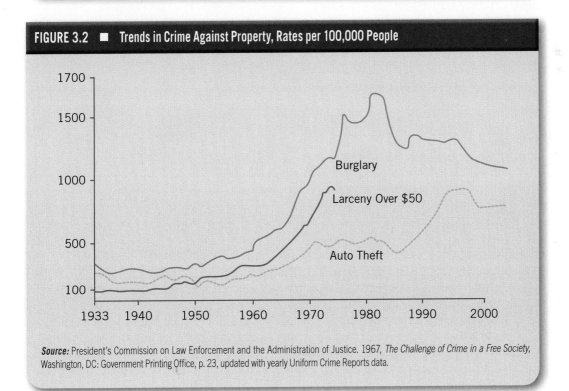

FIGURE 3.2 ■ **Trends in Crime Against Property, Rates per 100,000 People**

Source: President's Commission on Law Enforcement and the Administration of Justice. 1967, *The Challenge of Crime in a Free Society*, Washington, DC: Government Printing Office, p. 23, updated with yearly Uniform Crime Reports data.

Crime File 3.1

American Crime Problems From a Global Perspective

Issues

Transnational crime (i.e., crime that violates the laws of several international sovereignties or impacts another sovereignty) has grown incrementally over the past two decades, at a rate roughly corresponding to both the increase in international trade import-export figures and the developments in transportation and communications. Several events demonstrate the stark reality of transnational crimes: the destruction by a terrorist bomb of Pan American Flight 103 over Lockerbie, Scotland, in 1988; the 1993 terrorist bombing of the World Trade Center; the more recent conspiracy in New York City to destroy all Hudson River crossings and both FBI and United Nations headquarters; and the Bank of Credit and Commerce (BCCI) scam, with its estimated cost to U.S. taxpayers of between $200 billion and $1.4 trillion by the year 2021.

In each of these cases, U.S. law enforcement authorities responded vigorously, but with limited overall success. Our system has been developed to deal with criminality at the city/county level and, in some cases, at the national level. With respect to global crime, however, we lack readiness—in terms of education, research sponsorship, interagency cooperation (between the Departments of Justice and State), and a full commitment to a centralized and coordinated international effort.

Crime is not a strictly local or even national problem; although its impact is felt at the local level, much crime is internationally conditioned and coordinated. For instance, the connection between street crime and the importation and dissemination of drugs is well established. Similarly, an increase in fraud is commensurate with growth in the operational reach of commercial transactions. Profits from the international drug trade, "laundered" overseas and reinvested in American real estate, commercial, or entertainment enterprises, significantly affect U.S. citizens, who must pick up the burden for uncollected taxes on these transactions.

In addition, the impact of ethnic gang criminality on our "local" crime scene is readily apparent, for example, the wholesale trade in cocaine, controlled by illegal immigrants from Colombia; the importation of Chinese slave labor into the U.S. and exploitation of Chinese American businesses by Chinese gangs (triad-based); trade in arms and drugs by Jamaican gangs; burglaries by Albanian gangs; and involvement in the fuel distribution market and the international trade of weapons and nuclear materials by Russian gangs. These new ethnic gangs maintain intra-ethnic contacts, as well as relations with their countries of origin, and local law enforcement professionals are powerless to stop or control them.

Source: American Society of Criminology Task Force Report to U.S. Attorney General Janet Reno. *The Criminologist* (Special Issue), *20*(6), November/December 1995. Task force members were Gerhard O. W. Mueller (Chair), Paul Friday, Robert McCormack, Graeme Newman, and Richard H. Ward.

Web Research Project

Examine the issue of international crime. What is an additional, emergent crime problem not mentioned in Crime File 3.1? What strategies are needed for coping with this?

In 1968, the President's Commission on Law Enforcement and the Administration of Justice (1967) addressed the issue of American crime problem from a global perspective:

> There has always been too much crime. Virtually every generation since the founding of the Nation and before has felt itself threatened by the specter of rising crime and violence. A hundred years ago contemporary accounts of San Francisco told of extensive areas where "no decent man was in safety to walk the street after dark, while at all hours, both night and day, his property was jeopardized by incendiarism and burglary." Teenage gangs gave rise to the word "hoodlum," while in one central New York City area, near Broadway, the police entered "only in pairs, and never unarmed." A noted chronicler of the period declared that "municipal law is a failure . . . [and] we must soon fall back on the law of self-preservation." And in 1910

one author declared that "crime, especially its more violent forms, and among the young is increasing steadily and is threatening to bankrupt the Nation." (p. 101)

However violent crime may be in large cities today, both urban and rural areas of Sweden, Holland, and England were more violent during the Middle Ages (E. A. Johnson & Monkkonen, 1996). In *Hooligans,* Pearson (1982) remarks on the historical myth of a crime-free past in England and attributes it to the abundance as well as sophistication of modern statistics, nostalgia for the past, and cultural amnesia. The relationship of crime to the early history of many countries can be illustrated by Australia, a country that was settled as a penal colony for England. Gangs of "bushrangers" (horse rustlers) achieved notoriety, particularly the group led by Ned Kelly, whose reputation reached mythic proportions. This Robin Hood–like figure received support in opposing authority from small farmers who were nicknamed "cockatoos" or "cockys" because, like the bird, they scratched out a living from the ground. The cocky spirit was one of independence and defiance of authority as exemplified by Ned Kelly himself, who was obstinate to the end and was hanged at age 25. This spirit is expressed in Australia's most beloved song, "Waltzing Matilda," about a vagabond who steals a sheep and commits suicide rather than be caught (Levathes, 1985, p. 261):

> Up jumped the swagman
>
> And sprang into the billabong
>
> "You'll never catch me alive," said he.
>
> And his ghost may be heard
>
> As you pass by that billabong
>
> "You'll come a-waltzing Matilda with me."

Photo 3.2

Ned Kelly, an Australian outlaw, was the subject of that nation's beloved ballad "Waltzing Matilda." Here he wears his armored suit beaten out of plowshares (circa 1870).

Because systematic victim data have been available only since the early seventies, a relative comparison with UCR data before 1973 is not possible. Figures 3.3a and 3.3b and Table 3.4 present information from the National Crime Victimization Survey (NCVS).

Comparison of these trend lines with those from the UCR for the same period suggests similarities as well as differences. While most of the crimes in both measures are predominantly against property rather than against persons, the NCVS did not demonstrate the same steep rise in offenses in the seventies that the UCR reported.

In considering these figures, it is important to realize that, if we were to consider the full range of economic crimes such as the impact of corporate price fixing, then in fact every household has been touched by crime. Casting a wider net, were we to consider self-report data, particularly of minor offenses, we would conclude that criminality is pervasive. Despite problems in instruments used and in samples drawn, self-report studies provide much-needed evidence of the extensiveness of hidden criminality and law violation. One may control misunderstanding or overgeneralizations in referring to "criminals" by using operational definitions such as "those arrested" or "those identified by victims" or "those admitting to certain offenses."

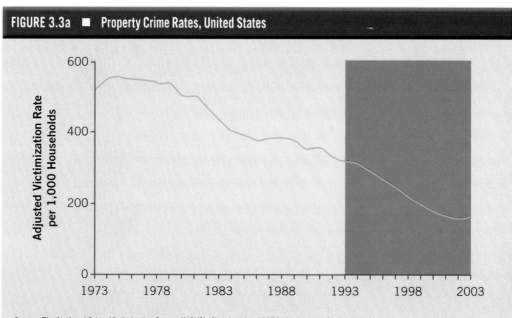

FIGURE 3.3a ■ **Property Crime Rates, United States**

Source: The National Crime Victimization Survey (NCVS). Ongoing since 1972, this survey of households interviews about 75,000 persons age 12 and older in 42,000 households twice each year about their victimizations from crime.

Note: Property crimes include burglary, theft, and motor vehicle theft. The darker area indicates data gathered after the NCVS redesign. The data before 1993 are adjusted to make them comparable with data collected since the redesign. The adjustment methods are described in Criminal Victimization 1973–1995. Estimates for 1993 and beyond are based on collection year while earlier estimates are based on data year. For additional information about the methods used, see Bureau of Justice Statistics, 2003.

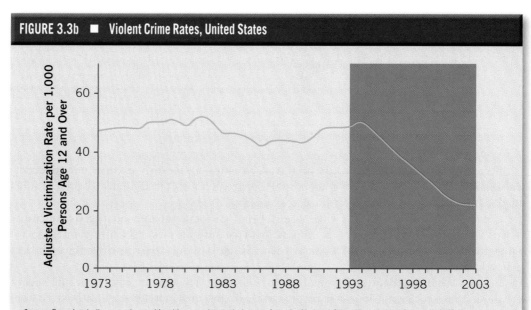

FIGURE 3.3b ■ **Violent Crime Rates, United States**

Source: Rape (excluding sexual assault), robbery, and assault data are from the National Crime Victimization Survey (NCVS). Ongoing since 1972, this survey of households interviews about 75,000 persons age 12 and older in 42,000 households twice each year about their victimizations from crime.

Note: The violent crimes included are rape, robbery, aggravated and simple assault, and homicide. The NCVS redesign was implemented in 1993; the area with the lighter shading is before the redesign and the darker area after the redesign. The data before 1993 are adjusted to make them comparable with data collected since the redesign. The adjustment methods are described in Criminal Victimization 1973–1995. Estimates for 1993 and beyond are based on collection year while earlier estimates are based on data year. For additional information about the methods used, see Bureau of Justice Statistics, 2003.

TABLE 3.4 ■ Criminal Victimization in the United States, 2007

	Number in Millions	Rate per 1,000*
All Crimes	22.9	
Violent Crime	5.2	20.7
Simple Assault	3.5	13.9
Aggravated Assault	.8	3.4
Robbery	.6	2.4
Rape/Sexual Assault	.2	1.0
Property Crime	17.5	148.5
Property Thefts	13.3	111.4
Household Burglary	3.2	26.9
Motor Vehicle Theft	.9	8.2

Source: From "Criminal Victimization 2007: National Crime Victimization Survey," *Bureau of Justice Statistics Bulletin,* 2008, Washington, DC: Bureau of Justice Statistics.

Age and Crime

Most of those arrested are young. Crime File 3.2 presents data on ages of those arrested for particular crimes. The peak arrest age for property crime is 16, while age 18 is the highest for violent crime. Overall, crime commission declines with age.

Journal Article Link 3.2
Examine literature regarding young people and crime.

Particularly glaring is the involvement of younger groups in serious property crimes. It is important to note that, while most persons arrested and convicted as adult criminals were first arrested as juveniles, most juvenile delinquents do not become adult criminals. Youthful offenders in urban areas are probably overrepresented in arrest statistics. Such areas have more efficient, formalized policing, while youth generally have less power than their elders to shield themselves from arrest. Juveniles also commit the types of crimes on which municipal police departments tend to concentrate. Excluding common youth offenses, such as curfew and runaway violations, and assuming juvenile offenders are often handled and recorded differently depending on police jurisdictions, the median age for arrested robbers, burglars, thieves, auto thieves, arsonists, and vandals is under 20 in all categories. Estimates of the average age of embezzlers, price-fixers, bribers, and the like considerably alter this age profile, however, since these crimes are committed by older criminals. The "graying of America," with a large proportion of the population becoming elderly, has led to forecasts of an increase in the number of older criminals (Wilbanks & Kim, 1984).

Criminologists had predicted a possible demographic time bomb as the number of people in the maximum crime-committing ages of 15–24 expanded at the end of the last century. From 1996 to 2000, there were 500,000 more males in this age group than there were a decade earlier—about a 20 percent increase. The homicide rate among 15- to 19-year-olds increased 154 percent between 1985 and 1991. This increase began with the advent of crack cocaine in the mid-1980s, along with a proliferation of guns to protect crack markets (Blumstein, 1995). The rate decreased beginning in 1992, in part due to the ending of this "crack epidemic."

Crime File 3.2

What Is the Relationship Between Age and Crime?

Participation in crime declines with age.

Arrest data show that the intensity of criminal behavior slackens after the teens, and it continues to decline with age. Arrests, however, are only a general indicator of criminal activity. The greater likelihood of arrests for young people may result partly from their lack of experience in offending and also from their involvement in the types of crimes for which apprehension is more likely (for example, purse snatching vs. fraud). Moreover, because youths often commit crime in groups, the resolution of a single crime may lead to several arrests.

The decline in crime participation with age may also result from the incapacitation of many offenders. When repeat offenders are apprehended, they serve increasingly longer sentences, thus incapacitating them for long periods as they grow older. Moreover, a RAND Corporation study of habitual offenders shows that the success of habitual offenders in avoiding apprehension declines as their criminal careers progress. Even though offense rates declined over time, the probabilities of arrest, conviction, and incarceration per offense all tended to increase. Recidivism data also show that the rates of return to prison tend to be lower for older than for younger prisoners. Older prisoners who do return do so after a longer period of freedom than younger prisoners.

Different age groups are arrested and incarcerated for different types of crime.

- Juveniles under age 18 have a higher likelihood of being arrested for robbery and UCR index property crimes than do members of any other age group.
- Persons between ages 18 and 34 are the most likely to be arrested for violent crimes.
- Among jail and prison inmates, property crimes, particularly burglary and public order crimes, are more common among younger inmates.
- Violent crimes were more prevalent among older inmates admitted to prison in 1982 but showed little variation among jail inmates of different ages.
- Drug crimes were more prevalent among inmates ages 25 to 44 in both prisons and jails.

Percentage of Arrestees Under 18 Years of Age, by Type of Crime

Most Serious Offense Charged	Juvenile Arrests as a Percent of Total Arrests
Total	16%
Violent Crime Index	15
Murder and non-negligent manslaughter	9
Forcible rape	16
Robbery	24
Aggravated assault	14
Property Crime Index	29
Burglary	29
Larceny-theft	28
Motor vehicle theft	29
Arson	51
Other assaults	19
Forgery and counterfeiting	4
Fraud	3
Embezzlement	7
Stolen property (buying, receiving, possessing)	19
Vandalism	39
Weapons (carrying, possessing, etc.)	23

Most Serious Offense Charged	Juvenile Arrests as a Percent of Total Arrests
Prostitution and commercialized vice	2
Sex offenses (except forcible rape and prostitution)	20
Drug abuse violations	12
Gambling	16
Offenses against family and children	5
Driving under the influence	1
Liquor laws	22
Drunkenness	3
Disorderly conduct	30
Vagrancy	8
All other offenses (except traffic)	10

Source: Juvenile Offenders and Victims: 2005 National Report, by H. Snyder and M. Sickmund, 2006, Pittsburgh, PA: National Center for Juvenile Justice, p. 26.

Web Research Project

Enter the phrase "age and crime" in a search engine and locate the "age-crime" debate. What is it and what is its importance?

Visit the Office of Juvenile Justice and Delinquency Prevention Web site (http://ojjdp.ncjrs.org) and note any trends or changes in crime by age.

Age-Crime Debate

An intramural academic war of sorts has broken out in criminology. It could be described as the **age-crime debate.** On one side of the debate are Gottfredson and Hirschi (1986, 1987), who view as a constant the "maturing out of" crime or desistance from crime as individuals age. They indicate the following:

> Further, this distribution is characteristic of the age-crime relation regardless of sex, race, country, time or offense. Indeed, the persistence of this relation across time and culture is phenomenal. As long as records have been kept, in all societies in which such records are available, it appears that crime is an activity highly concentrated among the young. (1986, p. 219)

They question the emphasis on career criminal research, incapacitation, and the recent "fetish" for longitudinal research that justifies a search for groups of offenders (career criminals) whose criminality does not decline with age (Blumstein, Cohen, & Farrington, 1988a, 1988b; L. E. Cohen & Land, 1987; Farrington, 1986). Blumstein and Cohen (1987), in a longitudinal study of those arrested for more serious crimes in the District of Columbia and Detroit in 1973, found that those who remained active in their twenties did not age out in their thirties, but only after age 45. Farrington (1986) suggests that offenses of different types peak at different times and that this represents "crime switching rather than replacement of one group of offenders by another" (p. 189). Steffensmeier (1989a) finds variation by age-specific type over time with the offenders becoming younger and younger, and also finds that some crimes such as embezzlement or fraud are less likely to decline with age.

The outcome of this age-crime controversy is claimed by the disputants to have important consequences for career criminal research (Tittle, 1988). Why do most criminals "mature out of" crime? Farrington (1986) suggests factors such as the influence of wives or girlfriends, the decline of gang or peer group support, and increased penalties, as well as increased legitimate opportunities as individuals reach their twenties. The decline in crime involvement is not explained by the physiology of aging since substantial decline in fitness does not occur until the late fifties or older. Social changes are more important than physiological changes in explaining this decrease. Steffensmeier and Allan (1990) identify a number of social changes that encourage conformity, including more legitimate access to material goods and excitement, changes in age-graded norms and anticipatory socialization, changes in lifestyle and peer groups, stronger social bonds, higher legal and social costs, and fewer illegitimate opportunities. Discussion of developmental theories in Chapter 6 will address this issue in depth.

Perhaps one of the best kept secrets in criminology is the existence of the "bible of juvenile justice research"—*Juvenile Offenders and Victims,* an annual report by Howard Snyder and Melissa Sickmund (2005), produced by the National Center for Juvenile Justice in Pittsburgh. This document, with tables and figures, reports the latest trends in juvenile justice.

Handbook Article Link 3.1
Read an article on gender and crime.

Journal Article Link 3.3
Examine literature regarding female offenders.

Gender Differences in Criminality

Table 3.5 presents some recent statistics on "women offenders" in the United States (Greenfield & Snell, 1999).

Of all demographic variables, gender is the best predictor of criminality; most persons arrested are males. In the United States, in the nineties, males represented about 83 percent of those arrested and, with the exception of primarily female offenses such as prostitution (in which "Johns," or customers, are seldom arrested), this difference holds for all criminal offenses. The male crime rate exceeds that of females universally, in all nations, in all communities, among all age groups, and in all periods of history for which statistics are available. Whereas in some more traditional countries the crime–gender arrest ratio may range from 200-to-1 to 1,000-to-1, in modernized societies the gap in gender variation in crime has been closing. Why such variation? Gender per se is not as important a variable as a particular culture's *conception* of gender. The female crime rate appears to be closer to the male level in countries in which females enjoy more equality and freedom and thus an increased opportunity to commit crime.

TABLE 3.5 ◼   **Characteristics of Female Offenders**

	Violent Offenders	All Arrestees	Convicted Felony Defendants	Correctional Populations
Female Offenders	2,135,000	3,171,000	160,500	951,900
Number as a Percentage of Each Category	14%	22%	16%	16%

- Based on the self-reports of victims of violence, women account for about 14 percent of violent offenders—an annual average of about 2.1 million violent offenders.
- Male offending equals about 1 violent offender for every 9 males age 10 or older, a per capita rate 6 times that of women.
- Three out of four violent female offenders committed simple assault.
- An estimated 28 percent of violent female offenders are juveniles.
- Three out of four victims had a prior relationship with the female offender.
- An estimated 4 in 10 females committing violence were perceived by the victim as being under the influence of alcohol or drugs at the time of the crime.

Source: L. A. Greenfield and T. L. Snell, *Women Offenders,* December 1999, Bureau of Justice Statistics Special Report, NCJ175688.

This universality of disproportionate male criminality can best be explained by the differential treatment of males and females. Traditionally, males are socialized to be dominant, active, and aggressive. In fact, chivalry and the law often require that the male take responsibility for what occurs. In many traditional societies, it is the husband who is punished for his wife's transgressions. Similarly, customary gender role socialization of females emphasizes passivity and subordination.

In explaining some of the social, psychological, and physical reasons for these differences, Steffensmeier (quoted in Blaum, 1991, p. 1) notes that "there is no female equivalent to the 'romanticized' rogue male." For some males, crime is considered macho and enhances status, while female crime is usually stigmatizing. Despite social change, social expectations of women still center on nurturing, beauty, virtue, and stereotypes of femininity that are incompatible with qualities valued in the criminal underworld. Steffensmeier (quoted in Blaum, 1991) also sees female criminals as more likely to be drug-dependent or to come from deprived family backgrounds. They are more likely to engage in crime because of intimate or romantic relationships and are often introduced to crime by a significant other. Male physical strength and agility also favor males' greater participation in certain crimes such as robbery and burglary. The threat of potential sexual victimization also limits females' mobility and access to criminal haunts.

Females tend to concentrate on less rewarding types of crime, such as shoplifting and employee theft, rather than on more lucrative, organized activities, such as burglary rings, drug cartels, and fencing networks. Even prostitution is usually controlled by males. Racketeering and corporate fraud are almost overwhelmingly male-dominated (Blaum, 1991). Girls exposed to a variety of familial risk factors are found to be far less likely to deviate than boys raised in similar circumstances (Dornfeld & Kruttschnitt, 1991).

The traditional handmaiden of sexism has been paternalism, a sort of sexual noblesse oblige in which males felt that they were responsible for protecting the dependent female. This policy is reflected in the law and its administration, since females generally receive much lighter sentences for the same offense, are viewed more favorably by judges and juries, and seldom receive the death penalty.

Recent literature on the subject of gender and crime note the **androcentric** (male-centered) **bias** in many delinquency and crime theories (Chesney-Lind, 1989). Burnett (1986) also notes that women have been left out of criminological scholarship and that a new era began with publications such as F. Adler's *Sisters in Crime* (1975), R. J. Simon's *Women and Crime* (1975), and F. Adler and Simon's *The Criminology of Deviant Women* (1979).

Using UCR data, R. J. Simon (1990) notes increases in female crime particularly in property offenses, especially white collar offenses involving small-to-medium amounts of money. While writers such as F. Adler (1975) were arguing that a gender convergence or closing of the gap between male and female crime rates was taking place, others such as Steffensmeier (1978) and Steffensmeier and Allan (1988) found no such change in the crime–gender ratio.

The National Institute of Justice has noted a rise in the number of female offenders at the turn of the twenty-first century (National Criminal Justice Reference Service [NCJRS], 2003). The NCVS for 2001 showed the percentage of estimated female perpetrators as increasing from 14 to 19 percent for violent crime. From 1990 to 1998,

Journal Article Link 3.4
Examine literature regarding women as criminals.

Photo 3.3

Maricopa County prison inmates participate in what is believed to be the nation's first female chain gang.

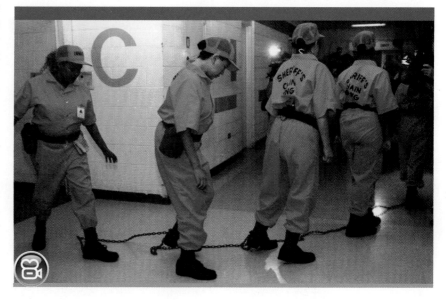

the number of women in prison grew 48 percent compared to 27 percent for men. Between 1988 and 1997, female delinquency cases increased 83 percent.

A major literature is developing regarding gender and crime. One example is a "power-control theory" of delinquency and gender (J. Hagan, Gillis, & Simpson, 1985, 1987), which proposes that male and female children react differently to parental power sharing. The authors hypothesized that balanced family structure (shared power by spouses) lessens the differences in delinquency between genders and that unbalanced family structures increase those differences. However, both Singer and Levine (1988) and Morash and Chesney-Lind (1991) found little support for this theory. For further review of this literature, the reader is referred to C. R. Mann (1984), I. L. Moyer (1990), Nagel and Hagan (1983), and Rosenbaum (1989b).

Social Class and Crime

Social class is not a category included in the UCR, yet the vast majority of those arrested or labeled as criminal are from lower social classes. Criminality for traditional crimes is higher among lower-class individuals, totally apart from bias in statistics or the administration of justice. Part of the excess rate is likely to be due to their lack of power and sophistication in shielding themselves from formal litigation proceedings. Traditional explanations of crime and social class view the relationship as an inverse one; that is, as social class becomes higher, the volume of crime commission decreases proportionately. Figure 3.4 attempts to depict this relationship schematically. Reckless (1967) proposes a bimodal theory of the distribution of crime commission in which the criminality curve has two modes (most frequently appearing cases) among the lower class and the upper class, though crimes of the latter are seldom reflected in national crime statistics.

The relationship between social class and criminality remains a subject of debate. The early self-report surveys (Nye, Short, & Olson, 1958; Short & Nye, 1958) found no relationship other than that lower-class offenders were more likely to be officially processed. Tittle, Villemez, and Smith (1978), in a literature review of major self-report studies, found no relationship between class and criminality. More recent research and reviews of self-report surveys suggest that much of this lack of difference by class may have been due to the measuring instruments, which tended to concentrate on rather trivial offenses. Lower-class youth were found to commit more serious crimes more often; and their offense profile was found to more closely follow that presented by official statistics (Elliott & Ageton, 1980; Hardt & Hardt, 1977; Hindelang, Hirschi, & Weis, 1979). Examining calls to police, Warner and Pierce (1991) found that poverty of area consistently increased the rate of assault, robbery, and burglary, although in examining

Journal Article Link 3.5
Examine literature regarding qualitative research on women in prison.

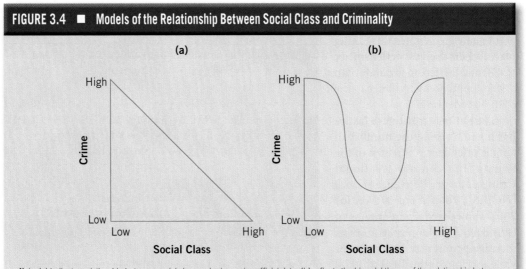

FIGURE 3.4 ■ **Models of the Relationship Between Social Class and Criminality**

Note: (a) indicates relationship between social class and crime using official data. (b) reflects the bimodal theory of the relationship between crime and social class.

delinquency, Larzelere and Patterson (1990) found that parental monitoring and discipline were more predictive than social class. It is important to caution, however, that official statistics undercount the typical crimes of higher socioeconomic groups, so even though the lower class has higher official crime rates, this does not indicate that individuals in this class are necessarily more criminal.

Race and Crime

Just as an androcentric bias was identified earlier, a Eurocentric bias may also exist in criminology, in part because African American criminologists in particular, but other ethnic minorities as well, have not played a significant role in the field (Young & Sulton, 1991). Eurocentric bias refers to the fact that the field of criminology is dominated by views reflecting those of European (white) descent and that such a bias may tend not to fully appreciate the interactions among racism, inequality, and the experiences of African Americans and other minorities in the criminal justice system.

Race is a relatively arbitrary, socially defined status. For example, in an early study, Herskovits (1930, p. 177) estimated that of the total number of black persons classified as "Negro" at that time in the United States, 15 percent were more "white" than black, 25 percent were equally "white" and black, and 22 percent were unmixed black. Thus, the concept of "race" is more a socially defined category than a taxonomically simple biological classification. As Sutherland and Cressey (1974) point out, "There is no avoiding the fact that at least 80 percent of the offenders contributing to the 'black' crime rate are part 'white'" (p. 132).

Given the increasing ethnic diversity of the United States, a racial and ethnic classification system that pigeonholes entire segments of the population obfuscates true understanding. Scientists have proven there is no biological basis for such traditional racial classifications. Students need to be aware of the inherent problems in this currently accepted system. The validity of this shorthand method of identifying different groups in an increasingly diverse society needs to be questioned by more criminologists (Mellow, 1996, p. 7).

The foregoing facts become important in light of cryptoracist theories that were rediscovered and treated by some as respectable in the 1970s. These were regenerated, long- discredited hereditary theories of racial inferiority to explain why African Americans, despite social changes in the 1960s, had failed to succeed (Skolnick & Currie, 1988, p. 12). Such theories obviously ignore the black experience in a nation that until relatively recently practiced institutionalized racism against African Americans.

To paraphrase one writer (Dr. Charles King), we have a society that crippled a people—then blamed them for limping (L. A. Ross & McMurray, 1996, p. 3). The long legacy of slavery, followed by "Jim Crow" laws (legalized or *de jure* segregation and discrimination) and then succeeded by *de facto* (in practice) discrimination, placed a generational burden on black Americans that far exceeded the milder forms endured temporarily by other ethnic groups. Much of the discrepancy between black and white crime rates can perhaps be explained by the fact that African Americans until relatively recently have been locked disproportionately into the lower class through a pseudo–caste system.

At the turn of this century, roughly 27 percent of those arrested in the United States were black, while blacks made up only about 12 percent of the population. In the early nineties, roughly one in four young black males in the United States was behind bars, on parole, or on probation. This was more than the number of black men enrolled in college. While 23 percent of black men in their twenties were under supervision, only 10 percent of Latinos and about 6 percent of whites were being similarly sanctioned. In Washington, D.C., estimates have been made that 70 percent of all black men have been arrested and served time in jail before the age of 35 (J. Marshall, 1992). Nationally, the disparity of rates between blacks and nonblacks was much greater for offenses of violence than for property offenses. This difference in arrest rates is generally taken to indicate equally disproportionate rates of crime commission. Most studies indicate that these differences are not a result of police discrimination.

In a book titled *The Myth of a Racist Criminal Justice System,* Wilbanks (1987) claims that although the criminal justice system was racist in the past, there is little racism or systematic discrimination in the criminal justice system today. One of his primary themes is that any disproportion in black arrest

Handbook Article Link 3.2
Read an article on race/ethnicity and crime.

Journal Article Link 3.6
Examine literature regarding race and crime.

Journal Article Link 3.7
Examine literature on African American violence.

and incarceration rates reflects actual higher offense rates among blacks. In critiquing this, C. M. Mann (1989, 1993) argues that racism in criminal justice is institutionalized in the same way that it is in other institutions in the United States such as education, politics, religion, and the economic structure. If our society is racist, do we not expect the criminal justice system to reflect this? Mann claims that Wilbanks ignores the informal aspects of the criminal justice system, or what Georges-Abeyie (1989) calls "petit apartheid realities," namely, stop-and-question and stop-and-frisk police practices that cause the police to be viewed by blacks as rude, insulting, and sometimes brutal. Claims of a nonracist system would have to be justified or supplemented with qualitative, observational research and actual accounts of minority experience.

Crime File 3.3, "Racial Profiling," describes a practice by some police departments of discriminatory practices in stopping and searching a disproportionate number of blacks and minorities, particularly in traffic stops.

Crime File 3.3

Racial Profiling

Crime profiling refers to attempts to construct typical characteristics of types of criminals (R. M. Holmes, 1989). It has been particularly useful in programs such as the FBI's Behavioral Science Unit and its tracking down of serial killers. However, it has been highly controversial when applied to traffic stops of those fitting the general profile. Phrases such as DWB (driving while black) or BWB (breathing while black) have been coined by black citizens who feel that they have been unfairly singled out for police attention simply for fitting the suspicious profile of being black.

Handbook Article Link 3.3
Read an article on racial profiling.

Charges of racial bias in the criminal justice system are certainly given support in cases such as the New York City Police Department's incidents involving Abner Louima and Amadou Diallo. Louima was sodomized and brutalized while in police detention, while Diallo was killed by many volleys from police revolvers when reaching for his identification. David Cole in his book *No Equal Justice: Race and Class in the American Criminal Justice System* (1999) indicates that, between 1995 and 1997, a total of 70 percent of those stopped on Interstate 95 in New Jersey and Maryland were blacks and Hispanics, even though they constituted only 17.5 percent of speeders. Similar findings were noted by Cole for Illinois.

Audio Link 3.2
Listen to a discussion of a cost-benefit analysis of racial profiling.

While blacks make up only 12 percent of the population of the United States, they represent over one half of the nation's prison population. In fact, one out of every three black men in his 20s is either in prison, in jail, on probation, or on parole. Much of this disparity is due to the "war on drugs." While the U.S. Public Health Service estimates that blacks represent 14 percent of U.S. illegal drug users, they are 35 percent of those arrested, 55 percent of those convicted, and 74 percent of those sentenced to prison for drug possession—a rate that is six times their representation in the population.

Racial profiling has what sociologists call a "self-fulfilling prophecy" quality about it. According to the developer of the concept, W. I. Thomas (Thomas & Swaine, 1928), "If men define situations as real, they are real in their consequences." Blacks are perceived by nonblacks as more criminal. They are arrested and incarcerated more than other groups. Studies of incarcerated populations show that a disproportion of prisoners are black; therefore, when authorities are profiling or looking for criminals, they concentrate on blacks. David Cole (1999) concludes by stating,

> Finally, and fundamentally, we need to think beyond policing. It sometimes appears that the only public resources that the majority is eager to supply to the inner cities are more (and more aggressive) police officers. But if similar levels of crime were occurring in white neighborhoods, and large numbers of white children were under criminal-justice supervision, isn't it likely that we would be hearing calls for different kinds of social investments, such as better schools, more job training, better after-care programs, and drug treatment? To restore legitimacy, the majority needs to show that it is willing to invest in something other than the strong arm of the law.

Web Research Project
What are some later developments in the issue of crime (racial) profiling? Have there been any new public policies for dealing with this issue?

Spohn and Cederblom's (1991) study found support for Kalven and Zeisel's (1966) "liberation hypothesis," which holds that racial discrimination in sentencing is significant primarily in less serious cases. While race was found to play no role in judicial decision making in Pennsylvania (Steffensmeier & Kramer, 1990), black murders of whites were found more likely to result in the death penalty in Kentucky (Keil & Vito, 1989). In the 1990s, federal drug laws featured more severe penalties for possession or sale of crack cocaine (favored by black dealers) than for powder cocaine (favored by whites). As a result, a disproportionate number of those given longer prison sentences were black.

Crime has in the past been primarily intraracial in nature; that is, in most cases, whites victimize whites and blacks victimize blacks. According to UCR arrest data, blacks represent 62 percent of robbers and exhibit particularly disproportionate rates for murder, rape, and assault (Bureau of Justice Statistics, 1988, p. 47). All of these crimes are relatively unsophisticated and command a great deal of police attention.

Statistics on crime by race are subject to countervailing pressures that may on the one hand overestimate and on the other underestimate the actual black crime rate. Blacks are more likely than whites to be arrested, indicted, convicted, and imprisoned. If convicted, they are less likely to receive probation, parole, or pardon. These factors may tend to exaggerate the black crime rate. In the past especially, many crimes by blacks against other blacks were ignored by the criminal justice system. A certain proportion of the rising crime rate beginning in the sixties reflected a greater willingness on the part of the police to respond to ghetto crime, which had previously been overlooked (see Gabbidon & Greene, 2005; Walker, Spohn, & DeLone, 1995).

Despite these offsetting trends, the crime rate of blacks is disturbingly disproportionate to that of the general population. Wolfgang's (1958) analysis of homicide in Philadelphia found the crime rate for nonwhite, 20- to 24-year-old males to be about 25 times the Caucasian rate for the same age group (see also Wolfgang, 1987). The few early self-report surveys suggested no significant differences by race with respect to admitted offenses (Gould, 1969; Hirschi, 1969; Voss, 1963). However, more recent research (Elliott & Ageton, 1980) again points to the tendency of many early measurement instruments to concentrate on trivial offenses. For more serious offenses, such as assault, robbery, and the like, black youths were significantly more persistent offenders, their rates in self-report surveys being similar to those in official studies.

A number of reviews of studies argue that there is no discrimination in the administration of justice (J. Hagan, 1987; S. P. Klein, Turner, & Petersilia, 1988; M. Myers & Talarico, 1987; Petersilia, 1983); however, the relationship is a subtle one as identified by other writers (Georges-Abeyie, 1984; Sampson, 1985). D. F. Hawkins (1986a, 1986b, 1987) found that racial differences between rates of arrest and imprisonment vary with the type of offense. The level of arrest failed to account for overincarceration of blacks for drug offenses, forgery, and driving under the influence, and an unexpected underincarceration for rape and robbery. Hawkins concludes that we must avoid the simplistic assumptions that blacks will be treated more severely than whites for all types of crime—the system of criminal justice is oppressive, but not without contradictions. A report by the National Council on Crime and Delinquency titled *And Justice for Some* (2000) found that black and Hispanic youth were treated more severely than white teenagers at each step of the juvenile justice system. Minorities were more likely than their white counterparts to be arrested, referred to juvenile court, detained prior to trial, formally processed by juvenile courts, found guilty in juvenile court, waived to adult criminal court, placed in juvenile prisons, and admitted to adult state prisons. The same report found that blacks charged with drug offenses are 48 times more likely than whites to be sentenced to juvenile probation. On a final note, it should be pointed out that the African American crime rate for insider trading, price fixing, defense procurement rip-offs, and other white collar crimes is very low. Crime File 3.4 discusses the plight of America's oldest minority, Native Americans.

Minority Groups and Crime

Race per se is not as crucial an explanatory variable in traditional crime commission as social class. Until recently, a large percentage of blacks were concentrated in lower-socioeconomic-class ghettos

Crime File 3.4

Native Americans and Crime

The indigenous peoples in the United States are members of about 550 federally recognized tribes including Cherokee (16.4 percent), Navajo (11.7 percent), Chippewa (5.5 percent), Sioux (5.5 percent), Choctaw (4.4 percent), Pueblo (2.8 percent), Apache (2.7 percent), and all others (51 percent; Greenfield & Smith, 1999, p. 1). The rate of victimization for American Indians was greater than that of all other races for rape and sexual assault, robbery, and aggravated and simple assault.

In 1999, the Bureau of Justice Statistics (BJS) issued a special report titled *American Indians and Crime* (Greenfield & Smith, 1999). The data were based on over 5 years of National Crime Victimization Survey data and reported that the rate of violent victimization among the nation's 2.3 million Native Americans is well above that of other American racial and ethnic groups and twice as high as the national average. Astoundingly, the rate of violent crime experienced by Native American women was nearly 50 percent higher than that reported by black males. (This report was updated by Perry in 2004.) American Indians were more likely to be victims of assault and rape/sexual assault committed by a stranger or acquaintance rather than an intimate partner or family member.

The BJS study indicated that the rate of Native American victimization was 124 violent crimes per 1,000 Native Americans, which was more than twice the rate of the nation as a whole (50 per 1,000). The average for whites was 49, for blacks 61, and for Asians 29 per 1,000. Native Americans, unlike other racial/ethnic groups, are more likely to be victims of interracial violence (that is, between races). Sixty percent of those committing crimes against Native Americans were whites; most of these offenses were attributed to racism and alcohol. The rate of murders committed by Native Americans (as of 1996) was 4 per 100,000, well below the then national average of 7.9 and the white rate of 4.9. Native Americans are, however, twice as likely as blacks and three times more likely than whites to be victims of rape or aggravated assault.

Native Americans are one of the smallest minority groups, yet they represent one of the largest percentages of prison inmates. There are 555 recognized Native American tribes in the United States and, despite the well-publicized success of gambling facilities on some reservations, a third of the country's two million Native Americans live below the poverty line. This figure is higher than for all other minority groups (Blackman & Simmons, 1995). In 800 or more treaties over the years, native tribes were guaranteed a reasonable level of education, health, and resources. However, such promises by the federal government were not honored and, instead, these people—who experienced near cultural and physical genocide—have also experienced high levels of unemployment, social disorganization, alienation, alcoholism, and crime.

Native American youth are the largest category of youth incarcerated under federal jurisdiction, about 65 percent (75) of the 124 confined in 1994 (Greenfield & Smith, 1999, p. 30). Similar to other indigenous peoples in other colonized countries, Native Americans were marginalized from the dominant society (Nielson, Fulton, & Tsosie, 2000, p. 5). Nielson (1998) states

> Native Americans have endured a century and more of government policies that ranged from genocidal to assimilative to supportive of limited sovereignty. The assimilative policies in general were, and still are (to the extent that they can still be found in American Indian law), important contributors to the development of social and economic conditions conductive [*sic*] to the development of Native American gangs. (p. 143)

Web Research Project
Visit the Web site http://www.bjs.gov and find the 2004 report by Steven Perry on *American Indians and Crime*. What are some additional facts of interest regarding Native Americans and crime?

that have traditionally exhibited high rates of breakdown. African Americans are disproportionately represented in the very largest cities. Early research by the "Chicago school" of sociology, most notably that of C. R. Shaw and H. D. McKay (1942) and their utilization of E. W. Burgess's "concentric zone theory" (1925), serves as an illustration of this relationship, which will be discussed in more detail in Chapter 6.

In examining certain areas for delinquency, C. R. Shaw and McKay (1942) report similar rates of delinquency in the same area of transition (zone II) despite changeover in racial and nationality groups. The concentric zone theory assumes that the highest crime rates are to be found in zone II. Despite this assumption, Nettler (1982, Vol. 2, p. 58) points out that Dutch, German, and Scandinavian settlers in the United States have had low crime rates in general, particularly for violent crimes. In addition, the low rates for Jews and Asians challenge the assumption that racial visibility, prejudice, and discrimination are sufficient explanations of criminality. It should be pointed out, however, that many of these groups were not lower-class immigrants, but instead had migrated during a period in which their craft, mercantile, and other skills were economically in demand (Flowers, 1988).

Journal Article Link 3.8
Examine literature regarding race-crime and ethnicity-crime.

The excessive violent crime rate for blacks in the United States stands in contrast to that of Latinos who are poorer, less educated, and have more menial jobs, but who also have lower rates of violent crime. Silberman (1978), in analyzing New York City crime rates, found the rate of black violent crime to be three times higher than the Latino rate—twice as high for homicide.

Since other minority groups which at one time were discriminated against were able to overcome difficulties and "rise from the ashes," so to speak, many ask the question, "Why haven't blacks been able to achieve the same success?" In 1968, in the aftermath of the worst series of urban riots in modern U.S. history, the Kerner Commission, the National Advisory Commission on Civil Disorders, addressed this issue by suggesting four reasons for differences between the immigrant and black experiences:

1. *The Maturing Economy:* When the European immigrants arrived, they gained an economic foothold by providing the unskilled labor needed by industry. Unlike the immigrant, the Negro migrant found little opportunity in the city. The economy, by then matured, had little use for the unskilled labor he had to offer.

2. *The Disability of Race:* The structure of discrimination has stringently narrowed opportunities for the Negro and restricted his prospects. European immigrants suffered from discrimination, but never so pervasively.

3. *Entry Into the Political System:* The immigrants usually settled in rapidly growing cities with powerful and expanding political machines, which traded economic advantages for political support. Ward-level grievance machinery, as well as personal representation, enabled the immigrant to make his voice heard and his power felt.

   By the time the Negroes arrived, these political machines were no longer so powerful or so well equipped to provide jobs or other favors, and in many cases were unwilling to share their influence with Negroes. (Kerner, 1968, p. 15)

4. *Cultural Factors:* Coming from societies with a low standard of living and at a time when job aspirations were low, the immigrants sensed little deprivation in being forced to take the less desirable and poorer-paying jobs. Their large and cohesive families contributed to the total income. Their vision of the future—one that led to life outside the ghetto—provided the incentive necessary to endure the present.

   Although Negroes worked as hard as the immigrants, they were unable to support their families. The entrepreneurial opportunities had vanished. As a result of slavery and long periods of unemployment, the Negro family structure had become matriarchal; the males played a secondary and marginal family role—one which offered little compensation for their hard and unrewarding labor. Above all, segregation denied Negroes access to good jobs and the opportunity to leave the ghetto. For them, the future seemed to lead only to a dead end. (Kerner, 1968, p. 15)

William Julius Wilson in *The Truly Disadvantaged* (1987) points out how the deindustrialization (loss of blue collar factory jobs) of the inner cities combined with racism, segregation, and poverty to condemn the black, inner-city poor to chronic unemployment and hopelessness. In 1989, an astonishing 35 percent of black males between the ages of 16 and 35 had been arrested at some point, and much of this was because of an aggressive nationwide crackdown on drugs that affected blacks much more than whites. Even though blacks represent about 12 percent of users, their visibility and greater involvement in trafficking in targeted urban areas has had a devastating effect. As noted earlier in this chapter, particularly problematic was the fact that much heavier sentences were levied against those using crack (a form of rock cocaine generally preferred by blacks) than against those abusing the same amount of powder cocaine (preferred by whites).

Decreasing job prospects in such poor, inner-city areas has created a vicious cycle in which

- Discrimination holds blacks back in the job market.
- The loss of blue collar jobs, erosion of real wages for low-skilled work, and weakened unions make unskilled work less available and attractive.
- Drug dealing becomes a fairly lucrative alternative employment option.
- Once caught, dismal job prospects for ex-cons make them less desirable marriage partners.
- This creates more female-headed households, a major cause of poverty (J. Marshall, 1992).

Does minority group status itself produce higher crime rates? In general, the answer to this question is no. Much depends upon the particular minority group and its specific values and cultural traditions. In the United States, for instance, the crime rate among Japanese Americans and Asians in general is lower than that of the general population. Many newer immigrant groups in the United States such as Cambodians, Koreans, and Vietnamese have low crime rates in part because of close extended family ties, a strong work ethic, and merchant skills (Launer & Palenski, 1988; Light & Bonacich, 1988). On the other hand, the crime rates for Algerians in France or Finns in Sweden or Latinos in the United States are higher than those of the general population.

Most immigrants to the United States have come from close-knit peasant societies, and the crime rate for this first-generational group is usually lower, with the exception of crimes peculiar to the area from which they emigrated. For instance, for the first-generation Italian immigrant, crime rates were lower, with the exception of murders and assaults. The areas of southern Italy and Sicily from which they came were experiencing at that time a wave of vendettas and violence, and this pattern was carried over into the new world. Similarly, Irish immigrants experienced higher rates for alcohol-related offenses; in the nineteenth century, Ireland reputedly had the highest alcoholism rate in the Western world.

It is not the parental group of immigrants that exhibits excess criminality; for many groups, it is the second generation that shows a marked upsurge in crime. Living in a strange new land and often being the victims of discrimination, the first generation tends to cling to old values. Moreover, this group may fear deportation. Wishing to be Americanized, the second generation often rejects many of these ways and attempts to assimilate the general values of U.S. culture. Unfortunately, in the area in which these individuals live (zone II, for instance, according to concentric zone theory), they also assimilate the criminal values of a high-crime area. Not being placed in such environments or possessing a higher parental social class may explain the relative success of Vietnamese immigrants in the United States.

Nettler (1982, Vol. 2, pp. 48–62) in a four-volume work, *Criminal Careers,* does an excellent job of summarizing much of the international research on ethnic migrants and crime. Care must be exercised in examining these data, since many of the studies refer to *Gastarbeiter* (guest workers), who are not immigrants as such, but rather temporary workers in the host community. Studies in Switzerland found the crime rate higher for foreigners than natives, particularly for violent crime. Ferracuti (1968) claimed that the crime rate of foreign workers increased as their numbers increased, although many of their crimes went unreported. Nettler (pp. 48–49) cites similar findings that suggest higher rates among Hungarians and Yugoslavs in Sweden; for Turks, Italians, Africans, and Mediterraneans in West Germany and Belgium; for Algerians in France; and for Irish, Asians, and West Indians in England. However, one chief problem in many of these studies is their failure to control for age and sex differentials, since many immigrant groups consist of a heavier population of young, single males, a group with a higher crime commission potential.

Regional Variation in Crime

Not only do crime rates, however difficult to measure, vary between nations, they also vary by region within a country (see Brantingham & Brantingham, 1984). Table 3.6 shows that in 2007 the rates for murder, robbery, aggravated assault, larceny, and burglary were highest in the U.S. South, which also had the highest overall crime rate. The Northeast had the lowest overall crime rate. The rates for aggravated rape were highest in the Midwest, while vehicle theft was highest in the West.

Urban/Rural Differences in Crime

Internationally, urban recorded crime rates are generally higher than rural crime rates, and, with few exceptions, this difference appears to have been the case since cities first appeared. Although crime rates tend to increase with the size of the community, there are some important exceptions. UCR statistics in general show a *positive relationship* in which, as size of community increases, the crime rate also increases (see Table 3.7).

Journal Article Link 3.9
Examine literature regarding crime in midsized rustbelt cities.

TABLE 3.6 ■ **Crime by Region, 2007 (rates per 100,000 population)**

	USA	Northeast	Midwest	South	West
Violent Crime	473.5	391.9	419.1	547.5	473.5
Property Crime	3,534.4	2,268.6	3,271.2	3,780.8	3,534.4
Murder	5.7	4.5	5.0	6.8	5.6
Rape	30.9	20.6	35.4	32.7	31.8
Robbery	149.4	151.2	132.1	157.1	152.5
Aggravated Assault	287.5	215.6	246.6	351	283.6
Burglary	729.4	429.9	692.1	903.8	727
Larceny-Theft	2,206.8	1,616.4	2,246.5	2,498.9	2,175.3
Motor Vehicle Theft	398.4	222.3	332.6	378.1	632.1

*Arson not included.

Source: *Crime in the United States, 2006,* FBI Uniform Crime Reports, 2007, Washington, DC: Government Printing Office.

TABLE 3.7 ■ **Crime Rates by Size of Community**

UCR Index Crime Rates per 100,000 Population	Violent Crime	Property Crime
Metropolitan statistical areas (MSAs)—urbanized areas that include at least one city with 50,000 or more inhabitants, or a Census Bureau–defined urbanized area of at least 50,000 inhabitants and a total population of at least 100,000	404.2	2,724
Non-MSA cities—these do not qualify as MSA central cities and are not otherwise included in an MSA.	197.1	1,724.5
Rural areas	236.1	1,753.7

Source: *Crime in the United States, 2006,* FBI Uniform Crime Reports, 2007, Washington, DC: Government Printing Office.

This same relationship with size of community also holds in the NCVS. Even if one assumes less reporting and recording of rural crime, the difference between rural and urban rates persists. In the 2004 data, however, the rates for property crime were actually higher for smaller than for larger cities, yet rural and suburban crime rates have actually been increasing faster than those of central cities since the sixties. Urbanism and its "way of life" are no longer confined to cities. The advent of modern communications and transportation has effectively erased many of the distinctions between rural and urban lifestyles, creating a truly urban society. The relatively high rates of violent crime in rural areas may be explained by certain criminalistic traditions in some areas that are much closer to frontier values and by a possible "subculture of violence" (to be explored in Chapter 4), which may explain high rural rates in the South as well as southern Appalachia.

Ferdinand (1991), in what he calls the "theft/violence ratio," notes that historical studies show a relationship between social structure and levels of property or violent crime. Theft was more prevalent in older, established cities with a preindustrial history, whereas violent crime was more common in rural, agricultural regions and in rapidly and newly industrializing cities.

Institutions and Crime

Sociologists define social **institutions** as relatively stable social patterns that serve a broad range of crucial functions in society; examples are the economy, family, church, state, and education. In contrast, associations are special-purpose organizations that serve a narrow range of interests; examples are corporations, unions, and professional societies.

The Family and Crime

The **family** is the primary or most important agent of socialization, particularly during childhood. The family has exclusive contact with the child during the period of greatest dependency and plasticity. Despite considerable popular literature on the subject, there is little, if any, scientific evidence on the subject of child rearing. Advocates of permissive or restrictive socialization to the contrary, the key appears to be firm but consistent discipline that is reinforced as well as understood by the child. The most important variables correlated with delinquency are probably poor home discipline, neglect, and indifference (J. Rosenbaum, 1989a).

Many U.S. studies of delinquency include under that label a significant number of activities, such as truancy, incorrigibility, and the like, that would not be criminal had they been committed by an adult. In a review of family factors associated with delinquency, Sutherland and Cressey (1974, pp. 203–218) as well as Hirschi (1983, pp. 53–68) point to moderate-to-high correlations between delinquency and immorality or criminality or alcoholism of parents, absence of one or both parents, a lack of parental control, unhappy home life, subcultural differences in the home, and economic pressures. The general process of family influence relates to the fact that the parental social class determines the residence, school, and associates of its offspring. Parental transmission of criminogenic attitudes or failure to train the child may influence delinquency. Similarly, a poor home environment may force the youth into the streets seeking peer primary group support.

Statistics on broken homes, ordinal positions (birth order) of siblings, and number of siblings, and their influence on crime and delinquency appear inconclusive (L. Rosen & Neilson, 1978; Sutherland & Cressey, 1974, pp. 216–217). It would appear that the quality of family interaction is more important than the family structure per se.

More sophisticated family studies of delinquents appear in early research by Sheldon Glueck and Eleanor Glueck (1950), who examined 500 delinquents and 500 nondelinquents and found that roughly 50 percent of the delinquents were from broken homes compared with about 29 percent of nondelinquents. Delinquents were more likely to have families characterized by physical illness, mental retardation, mental disturbance, alcoholism, and parental criminality. Such parents employed poor child-rearing practices, being either overly strict or overly permissive and exercising discipline inconsistently. Thus, defective family relations were perceived as a key causal variable in delinquency (F. E. Hagan & Sussman, 1988). Cathy Spatz Widom (1992) found that childhood abuse increased the odds of delinquency and future adult criminality by about 40 percent. Neglect alone, not just physical abuse, was significantly related to later violent behavior.

In the longest longitudinal study of delinquents—the "Cambridge-Somerville study"—begun in 1937, William McCord and Joan McCord (1958) found delinquents to be products of poor or weak parental discipline as well as a quarrelsome home environment. Family structure, that is, whether the home was broken or intact, was less salient than the nature of family interaction. All of the boys from quarrelsome environments had been convicted of crime (J. Q. Wilson & Herrnstein, 1985, p. 232). D. J. West and Farrington's (1977) longitudinal study of London working-class boys found the following associated with delinquency: low IQ, poor child-rearing practices, criminality of father, large family size, and low family income. Similar findings with respect to defective parental supervision and socialization have been suggested by Baumrind (1978), Hirschi (1969), Patterson (1982), Van Voorhis, Cullen, Mathers, and Garner (1988), and K. N. Wright and Wright (1995).

Loeber and Stouthamer-Loeber (1986), in an exhaustive analysis of the literature, summarize the relationship between family and delinquency as exhibiting (a) the most powerful predictors—lack of parental supervision, parental rejection, and lack of parent–child involvement; (b) medium predictors—background variables such as parents' marital relations and parental criminality; and (c) weaker predictors—lack of parental discipline, parental health, and parental absence. Research by other scholars confirms these findings (Farrington, Ohlin, & Wilson, 1986; R. E. Johnson, 1986; Laub & Sampson, 1988; Patterson & Dishion, 1985; K. N. Wright & Wright, 1995).

The proportion of children in American households living with both parents has declined since 1970. In that year, 90 percent of white children lived in intact households; in 1997, 74 percent did so. For black children, the figures were 64 percent (1970) and 35 percent (1997) (Thornberry, Lizotte, Krohn, Farnworth, & Jang, 1991). An estimated 40 percent of white children and 75 percent of African American children will experience parental separation or divorce by age 16, many of them multiple family transitions. Longitudinal studies of youth in Rochester, Denver, and Pittsburgh found substantial changes in family transition: Rochester—64.5 percent, Denver—49 percent, and Pittsburgh—30 percent. They found a consistent relationship between higher numbers of family transitions and higher delinquency (Thornberry, Smith, Rivera, Huizinga, & Stouthamer-Loeber, 1999).

Conservative writers who have more recently dominated the literature, such as Hirschi (1983) and J. Q. Wilson and Herrnstein (1985), seem to view material disadvantage and quality of family life as mutually exclusive explanations. Currie (1985) calls this belief that what goes on in the family is somehow separate from outside social forces that affect the family the **fallacy of autonomy** (p. 185). Those who commit this fallacy fail to view the family in a larger social context; have an obsessive concern with control rather than supportive social policies; and lend the impression of intractability of family problems, unresponsive to enlightened social policy.

Two influential works that challenge our criminological conception of family and crime are Daniel Moynihan's *Family and Nation* (1986) and Elliott Currie's *Confronting Crime* (1985). Moynihan reiterates his theme of the disintegration and siege of the American family, an issue that he claims should have concerned us in the sixties. He indicates that the individual has been the center of public policy in the United States rather than families: "This was a pattern almost uniquely American. Most of the industrial democracies of the world had adopted a wide range of social programs designed specifically to support the stability and viability of the family" (p. 5). While growth in federal entitlement programs since the sixties led to a major achievement—the virtual elimination of poverty among the elderly—by the 1980s the United States had achieved another unique distinction: It had become the first society in history in which people are more likely to be poor if they are young than if they are old.

The age bias of poverty is pronounced, affecting 24 percent of school-age children. The principal correlate has been a change in family structure—the rise of female-headed households and the **feminization of poverty.** In 1984, nearly half of the poor in the United States lived in female-headed

Photo 3.4

A mother and her children spend the night in a shelter after being evicted from their home in Texas. Twenty-four percent of school-age children live in poverty, which has been correlated with the rise of female-headed households and the feminization of poverty.

households (Moynihan, 1986, p. 96). The percentage of such households nearly doubled from 10 percent of all families in 1960. A flattening of the tax system and federal reduction in income maintenance programs constituted a federal family policy in reverse. While three fourths of median family income in 1948 was exempt from federal tax (a powerful national family policy), by 1983 less than one third of such income was exempt, although the new tax law promises to remove many low-income groups from the tax rolls. Of particular concern in fueling single parenthood are rising illegitimacy rates in the United States. Studies by the Alan Guttmacher Institute point out that 39 percent of 15-year-old mothers indicate that the fathers of their babies are 20 years old or older (J. Shapiro & Wright, 1995, p. 51). Under statutory rape laws, such sexual activity constitutes "unlawful sex with a minor."

While broken homes per se are an uncertain predictor of delinquency and crime, the stresses and lack of support systems that result in changed family functioning for the more impoverished and growing numbers of single mothers and children are of concern. Currie (1985) states,

> The real issue is whether we regard the evidence on the persistence of family problems and the continuity of troubling behavior from childhood to adult life as indicative of predispositions that are largely unrelated to their social context and that we are virtually powerless to alter. (p. 219)

Education and Crime

Journal Article Link 3.10
Examine literature regarding crime in urban ethnographic research.

The relationship between **education** (formal schools/schooling) and crime and delinquency is at least twofold. First, for adolescents in modern societies, schools—particularly high schools—represent a major factor in their self-esteem at a very important stage in their lives. Second, there is an inverse (negative) relationship between the amount of formal schooling individuals possess and arrest rates for traditional crimes.

The fact that traditional crime commission decreases with the amount of formal education simply reflects the fact that legitimate opportunities increase with formal education, as do occupational and corporate criminal opportunities, which are less likely to be criminally stigmatizing. It is not formal education per se that causes or prevents crime; rather, educational status reflects one's social class, location of residence, and exposure to criminal or delinquent opportunity.

Research on crime and delinquency has come to focus most heavily on family and education as critical variables (J. D. Hawkins & Lishner, 1987; J. Q. Wilson & Lowry, 1987). Moynihan (1986, p. 92) cites a study commissioned by the National Association of Elementary School Principals (1980) titled *The Most Significant Minority: One-Parent Children in the Schools,* which found that one-parent kids were twice as likely to drop out and showed significantly lower achievement in school.

Photo 3.5

Programs such as Head Start have been found to improve elementary school performance, and contribute to higher rates of graduation from high school, employment, and less crime and delinquency.

Fagan and Wexler (1987) argue that social influences outside the family, such as school, peers, and community, are very strong, and that the role of the family alone should not be overstated. Fagan, Piper, and Moore (1986) note that violent delinquents in inner cities differ from nondelinquents in their attachment to school, their peers, and weak maternal authority, among other variables. They indicate,

> Complex social, economic and political factors are contributing to the creation of a vast new class of poor persons who are younger, more poorly educated and more likely to give birth sooner. One of the predictable consequences of this phenomenon is the continuing isolation of inner-city communities and a hardening of the processes observed among these samples. . . . These findings suggest that delinquency policy should be linked with economic development policy. The infusion of material and social resources into inner-city neighborhoods may strengthen social institutions including schools and families and alter the familiar correlates of serious delinquency by providing for the natural controls which characterize lower-crime neighborhoods. (Fagan et al., 1986, p. 463)

Strong school bonding decreases the likelihood of delinquency. Denno (1985) indicates that a major predictor of delinquency is misconduct in school. An example of a highly successful program is Head Start, which targets preschool enrichment. Schweinhart and Weikart (1980), in evaluating such a program for disadvantaged black children, found better later elementary school performance, higher rates of graduation from high school, increased employment, and less crime and delinquency. F. Adler (1983), in a previously cited study of low-crime nations, analyzed 47 variables and found the only factor common to all low-crime countries was strong social controls outside the formal system of justice. This well illustrates the fact that crime and justice matters are not to be treated in isolation from general societal conditions.

Religion and Crime

Lee Ellis (1996), in a review of criminology texts, found that only 2 of 18 even mentioned **religion** as a variable in crime causation, despite the fact that a significant one third of the public believes it plays a significant role. Of major religious groups in the United States, Jews have the lowest official crime rate, followed by Protestants; Catholics have the highest rates. Why? There is a hidden variable: social class. Catholics in the United States include a significant proportion of low-income minorities, particularly in the Southwest. Of Protestant denominations, Presbyterians and Episcopalians have lower

Video Link 3.2
View a video of sexual abuse in the church.

crime rates, while Baptists have the highest. Again, social class rather than denominational affiliation is the explanation. L. Ellis (1985) reviewed over 60 studies in the research literature on this religion–crime connection. The majority of the studies confirm, as might be expected, that attendance at religious services reduced crime commission. This is a stronger relationship than that between religious denomination and crime.

War and Crime

War has an impact upon crime. Although it is an example of institutionalized violence, elements of war itself may be considered violations of international law. Social conflict theorists such as Simmel (1955) and Coser (1956) tell us that conflict with an outside group tends to increase the internal solidarity within groups; that is, as conflict with outside enemies increases, conflict within groups decreases.

During major wars, the domestic crime rate as a whole tends to decline. This probably reflects increased social solidarity, group cohesion against an outside enemy, and high employment. Juvenile delinquency tends to increase during such periods because of displacement of families and increased mobility. As noted earlier, female crime rates increase because of increased opportunity. A major form of crime that tends to increase during wartime is "white collar crime" such as black-marketeering, profiteering, wartime trade violations, violations of wage-price freezes, and the like (Sutherland & Cressey, 1974, pp. 240–241).

Archer and Gartner (1984, pp. 79–81), using their "Comparative Crime Data File," found that nations participating in World Wars I and II were more likely to experience postwar increases in homicide than control nations (those who had not participated). The differences are similar, but less pronounced, after smaller wars. They found data supporting a "legitimation of violence model" (p. 92) in which wars tend to legitimate the general use of violence in domestic society.

Economy and Crime

In summarizing the diverse literature on the relationship between trends in the **economy** and crime, Sutherland and Cressey (1974) draw the following conclusions:

- Serious crimes have a slight and inconsistent tendency to rise in periods of economic depression and to fall in periods of prosperity.
- The general crime rate does not increase significantly in periods of economic depression.
- Property crimes involving violence tend to increase in periods of depression, but property crimes involving no violence, such as larceny, show only a slight and inconsistent tendency to increase in depression periods.
- Juvenile delinquency tends to increase in periods of prosperity and to decrease during periods of depression (pp. 225–226).

Video Link 3.3
View a video of stimulus spending and crime rates.

Using data from the United States, Canada, England, Scotland, and Wales, Brenner (1978) examined historical data for all major crimes since 1900 and their relationship to employment/unemployment, per capita income, inflation, and other economic indices. He found in all five political areas that the rate of unemployment showed strong and significant relationships to increases in all major categories of crime (p. 562). There is a significant difference in these statistics before and after World War II. There has been a speeded up or a quicker reaction to unemployment since World War II, particularly an increase in violent crimes. The United States especially demonstrated inverse (negative) correlations between employment and incarceration rates (see also Cantor & Land, 1985).

Currie (Skolnick & Currie, 1988) argues that conservative criminologists tend to underemphasize the impact of economic forces on crime. He points out that, while little crime increase took place during the Depression, crime rose during the more prosperous sixties. Currie feels there is a strong, although subtle, relationship and that those who underemphasize the economy ignore three factors:

First, subgroups with high crime rates—such as young black men—do have high unemployment rates, even when overall unemployment is low. Second, unemployment has a different impact when it portends a lifetime of diminished opportunity. And finally, unemployment statistics do not reflect the quality of available work. (p. 471)

John Hagan (1993) points out that, while unemployment increases crime, the reverse also occurs; that is, individuals with criminal records become less employable.

Mass Media and Crime

A subject of continual heated debate is the role of the **mass media** in encouraging crime, particularly crimes of violence. Do comic books, violent video games, music, newspapers, magazines, movies, or television cause an increase of crime? This protracted debate is periodically fueled by crimes, especially brutal ones, which appear to have some link with the media coverage or fictionalization of criminal events.

Charles Manson was obsessed with the Beatles, while Mark Chapman, the assassin of John Lennon, carried a copy of *The Catcher in the Rye*. In the aftermath of the recent Virginia Tech massacre, yet another culprit was added to the witch hunt or moral panic regarding the media and crime (Ferguson, 2007). It was alleged that video games such as *Counter-Strike* or *Streetfighter II* may have inspired the shooter, Seung-Hui Cho. Two meta-analyses of video games by Sherry and another by Ferguson found no relationship between video games and aggressive behavior (cited in Ferguson, 2007). In contrast, more than two decades of research on video games includes the American Psychological Association's Resolution on Violence in Video Games and Interactive Media, passed in 2005, which concluded that exposure to violent media increases aggression in children and youth (Carll, 2007). For a subgroup of disturbed youth, such media predisposes them to violence by exposing them to violent video games that rehearse recipes for action. Such interactive games may be more potent than other media.

Two rival hypotheses exist with respect to media and violence: the **catharsis hypothesis** and the **precipitation hypothesis.** The former claims that exposure to media violence enables a vicarious letting-off-of-steam and thus has a calming effect. This notion comes to us from the Greek tragedies, in which it was assumed that audiences, as a result of identification with the terrible travail and violent experiences of the characters, would feel an emotional purging of their own frustration, anger, or desire. The precipitation hypothesis assumes that exposure to media coverage of violence, fact or fiction, will produce greater propensities to aggression and violence.

In a report titled *Television and Behavior,* the U.S. Department of Health and Human Services (1982) concluded on the basis of a review of the research literature that there is an association between the viewing of television violence and aggression. One finding that seems continually to present itself concerns the image of society that television creates. In an American Broadcasting Company (ABC, 1983c) poll of viewers, 51 percent of respondents thought television news (a) gives too much coverage to crime and violence and that this distorts the public view of what is really going on in the streets; (b) leads to the perception that crime is more rampant and a person more likely to be victimized than is in fact the case; and (c) brainwashes us into fear, suspicion, and feelings of vulnerability. Glaser (1978, p. 236) indicates that television networks exert pressure against federal sponsorship of research on the impact of television on violence and for the suppression of reports on such research for fear of public boycotts. The National Institute of Mental Health, in its review of the literature, concluded that violence on television was one factor in children's aggressiveness, although not necessarily in their violent behavior (ABC, 1983c).

One key to explaining media precipitation of violence is that portrayals of violence appear to have different effects on different viewers. Individuals vary in their vulnerability to suggestion (Belson, 1978), with physically aggressive boys both watching televised violence and practicing it (Bandura, 1973; Lefkowitz, Eron, Walder, & Huesmann, 1977; E. D. McCarthy, Langner, Gersten, Eisenberg, & Orzeck, 1975). Nettler (1982), in a review of such studies, concluded that "evidence points to the

Handbook Article Link 3.4
Read an article on mass media and crime.

Journal Article Link 3.11
Examine literature regarding kids, crime, and local television news.

possibility that persons in 'poor psychological and social health' are more vulnerable to lethal suggestions" (p. 265; see also Comstock, 1975; Liebert & Baron, 1972).

In the United States, the Surgeon General's Scientific Advisory Committee on Television and Social Behavior (1972) sponsored 23 independent research projects to examine the impact of television on violence. They claim,

> By the time the average American child graduates from high school, he had seen on television some 18,000 murders and countless highly detailed incidents of robbery, arson, bombing, forgery, and torture. One hundred forty-six research articles based on 50 studies involving 10,000 children had all shown that viewing violence increased aggressive behavior in the young. A review of the literature on the subject did not reveal a single study which showed that violence did not have such an effect. (Haskell & Yablonsky, 1983, p. 199)

Journal Article Link 3.12
Examine literature regarding children and crime on television.

In support of the precipitation hypothesis, Glaser (1978) states, "The $30 billion spent annually in the United States on advertising—about $5 billion of it on television—suggests that many people have faith in the impact of mass communication on conduct" (p. 235).

A contrary view is presented by a 14-nation international study of television violence, which found that Japan has the most violent programming and is relatively free of violent crime ("Japanese TV," 1992). This low crime rate appears unaffected by graphically violent comic books, tabloids, videos, and a thriving market in pornography.

The limits of free speech are tested by music groups such as 2 Live Crew, who sold almost 2 million records featuring a song about graphic, offensive degradation of women. Such degradation and abuse of women are presented under the guise of entertainment. Armstrong (1991) analyzed popular themes in both country music and rap music, finding that both deal heavily with traditional crimes of violence involving the focal concerns (see W. Miller, 1958) of trouble, toughness, smartness, excitement, fate, and autonomy. "Gangsta" rap has its counterpart in "outlaw" country music. Sociologist Elijah Anderson (1990) calls much gangsta rap an example of "oppositional culture" of the streets, a subculture with its own gangsta values, attitudes, and codes of street behavior. Rejecting values of civility and decency, it romanticizes violence and glorifies having a violent reputation. While clearly rap is a legitimate urban music form—and much of it is benign—many releases feature violence; obscene lyrics; and graphic videos celebrating guns, gangs, drugs, and the denigration of women. Nathan McCall, author of *Makes Me Wanna Holler: A Young Black Man in America* (1994), points out that never before have young people been bombarded nonstop by such violent language and imagery. In response to criticism, MTV (Music Television) and BET (Black Entertainment Television) began refusing to play videos featuring guns and gratuitous violence. Considerable debate continues regarding censorship versus freedom of expression with respect to rap music (Hamm & Ferrell, 1994; Krzycki, 1994). In 1992, Time Warner finally pulled its "Body Count" album by Ice T, which included the song "Cop Killer," a track some felt encouraged ghetto youth to kill police officers. Artists such as 2 Live Crew or Eminem are not characteristic of the genre, however, and many rap songs feature nonviolent themes.

Copycat crimes. The issue of differential impacts of the media on different subgroups can be illustrated by means of the notion of "copycat crimes." The term *copycat* is a slang expression for imitation; thus, **copycat crimes** are fads in crime and are often stimulated by media coverage or portrayals.

In the early days of television, there was tremendous concern about children imitating Superman by jumping from the tops of buildings. In the 1990s, when a 5-year-old girl in Norway was stoned and kicked by playmates and left to freeze to death in the snow, the Scandinavian network TV-3 dropped the live-action series "Mighty Morphin Power Rangers" from its broadcasts in Norway, Sweden, and Denmark (Mellgren, 1994). Other examples include the film *Money Train,* which portrays a man dousing a subway token booth with a flammable liquid and burning the attendant. This gruesome act was imitated in three similar incidents within 1 week in New York. In 1993, an episode of *Beavis and Butthead* was blamed by a mother for her 5-year-old starting a fire that killed his little sister.

Audio Link 3.3
Listen to a discussion of school shootings and copycat crimes.

The episode featured a character setting fire to others' hair by using a match to ignite spray from an aerosol can. The 1976 film *Taxi Driver,* in which a character played by Robert DeNiro attempts to kill the president, inspired Reagan's attempted assassin, John Hinckley, Jr. The film *The Program* inspired two teenagers to copy a scene in which drunken football players play "chicken" by lying in the middle of a highway. The boys were killed. The film *Natural Born Killers* was banned in Great Britain and Ireland because of copycat murders in the United States and France (Murr & Rogers, 1995, p. 47). The film *Child's Play* and its sequels featured a demonic, animated character called Chucky, and may have inspired two boys from Liverpool to lure a 2-year-old boy from a Liverpool shopping mall and subsequently to murder him (Kolbert, 1994, p. A13). The horrible dragging death of a black man, James Byrd, Jr., behind a pickup truck in Jasper, Texas, in 1998, perpetrated by three white men and widely publicized in the news media, was followed by two similar crimes in Illinois and Louisiana. Similar patterns with copycat school shootings and workplace violence have also been noted.

Photo 3.6

2 Live Crew appeared at the MTV Awards in Europe. From left to right: Fresh Kid Ice, Brother Marquis, Mr. Mixx, Luke Skyywalker.

Much like the tobacco industry, the purveyors of such violent themes claim no direct causal relationship between their product and harmful outcomes, yet the scientific studies are overwhelming in predicting harmful effects, even though in any particular case not all individuals are adversely affected.

On July 26, 2000, the American Medical Association, the American Academy of Pediatrics, the American Psychological Association, and the American Academy of Child and Adolescent Psychiatry issued a joint declaration stating the following:

> Viewing violence may lead to real life violence.... Children exposed to violent programming at a young age have a higher tendency for violent and aggressive behavior later in life than children who are not so exposed. (http://www.aap.org/advocacy/releases/jstmtevc.htm)

Noting that before he or she reaches age 18, the average child will witness more than 200,000 acts of violence on television including at least 16,000 murders, they indicated that the link between media violence and real violence has been quite clearly demonstrated by research, primarily in its role of leading to emotional desensitization.

Summary

In examining descriptions and statistical accounts of crime and criminals, it is important to examine the database or other sources of findings and conclusions. Official statistics, victim surveys, self-reports, and other sources all provide different pictures. Similarly, the various types of criminal activity being addressed, whether traditional or elite, will provide different findings. Although they differ slightly in definitions of crime, concentrating on incidents in the UCR and on victims in the NCVS, the two measures are viewed as converging, with each providing certain information that the other lacks.

Estimates of the prevalence of crime depend on the measure used, with the extent of crime increasing as we move from UCR to NCVS to self-reports to estimates of corporate crime and other forms of criminality. The bulk of UCR Part I or index crimes consists of property crimes; the most frequent arrests among Part II crimes are for service functions. Trends in crime as measured by the UCR demonstrate a major crime wave in the United States since the mid-sixties. Comparison of trends using the NCVS since the early seventies shows only a small increase, if not a stable pattern in criminal victimization. Despite the lack of good, representative self-report surveys of the general population, existing studies certainly suggest that crime is even more pervasive than is reported in the UCR and NCVS.

Official statistics on crime indicate that most of those arrested are young (15 to 19 years of age). This is particularly the case with serious property crimes. *Age* profiles obviously would be altered upward were we to have accurate estimates for corporate and "upperworld" violators. Of all demographic variables, gender is the best universal predictor of criminality; with the exception of prostitution, the male crime rate exceeds the female rate for all crimes, although the gap has been closing, particularly in developed societies. This difference in criminality by gender can best be explained by cultural and socialization differences rather than by innate genetic ones. While some self-report studies demonstrate less of a gap, others indicate patterns like those in official data.

Official statistics show an inverse relationship between social class and criminality (as measured by arrests); that is, as social class rises, criminality decreases. This relationship remains a subject of debate. Most recent self-report surveys indicate that patterns of lower-class admissions match those of official statistics. Such findings belie the possibility of high upper-class criminality, since the most typical upper-class offenses are not tapped by such sources.

In examining race and crime, the reader was informed of the precariousness of the scientific concept of race and that a goodly proportion of the black population in the United States might better be described as being of mixed race. UCR arrest statistics indicate a black crime rate, particularly for violent crimes, that greatly exceeds their proportion of the U.S. population. Despite countervailing biases in these statistics, it would appear that they are accurate descriptions of excess commission of these offenses. Most crime is intraracial in nature; that is, most whites victimize whites and most blacks victimize blacks. While self-report surveys show mixed results, more recent studies confirm higher rates among blacks for serious offenses.

Minority group status itself does not result in higher crime rates. Early research demonstrated high rates in the area of transition (zone II) despite turnover in the minority in residence. Some groups— Dutch and Japanese, for example—never appear to have had higher crime rates. The first generation of immigrants generally has lower crime rates than the native population. It is their offspring, becoming Americanized into the lower class, who experience higher rates. Blacks have had higher rates than other minorities in part because of a maturing economy, the disability of race, late entry into the political system, and cultural factors. Crime patterns of European immigrants are similar to general U.S. patterns, although many "immigrants" are really temporary guest workers.

Official U.S. crime rates vary by region, with the South highest for murder, rape, and burglary; the West highest for assault, vehicle theft, and larceny; and the Northeast highest for robbery. International variations are difficult to determine because of inadequacies in crime statistics. Urban crime rates are generally higher than suburban and rural ones, particularly for property crimes. Recently, suburban and rural rates have been increasing more rapidly than urban rates.

While much has been written regarding the impact of the family on crime, with variables such as poor home discipline, neglect, indifference, parental criminality, and others identified as correlates, the key appears to be the quality of the family interaction rather than its structure as such. The impact of education and crime is highly intercorrelated with social class.

Major external conflicts (*wars*) appear to decrease internal conflicts (crime), with the exceptions of female crime, juvenile delinquency, and certain white collar crimes. Studies of economic trends and crime show inconsistent results; however, since the end of World War II there has been a quicker crime increase, particularly in violent crimes, corresponding with dips in the economy.

The nature of the impact of media upon crime is unresolved. While some propose a *catharsis hypothesis* (media violence as a vicarious, tension-relieving function), others support a *precipitation hypothesis* (media violence as an encouragement of the acting out of fictional themes). Television portrayals of violence appear to create increased feelings of potential vulnerability on the part of the public. Surveys of the literature by the Department of Health and Human Services, the Surgeon General's Committee, and the National Institute of Mental Health all concluded that violence on television tends to increase aggressiveness in children. Media violence appears to have particular, albeit unpredictable, impacts on certain subpopulations of viewers. This point was illustrated by the occurrence of *copycat crimes,* imitation crimes based on media portrayals.

Criminology on the Web

Log on to the Web-based student study site at http://www.sagepub.com/haganintrocrim7e/ for author-created podcasts, e-flashcards, quizzes, and more.

KEY CONCEPTS

Age-Crime Debate

Androcentric Bias

Catharsis Hypothesis

Copycat Crimes

Crime Trends

Economy

Education

Fallacy of Autonomy

Family

Feminization of Poverty

Gender

Institutions

International

Mass Media

Minority Status

Precipitation Hypothesis

Race

Region

Religion

Social Class

Urban/Rural

Variations in Crime

War

REVIEW QUESTIONS

1. How does crime vary internationally? Where does the United States stand with respect to crime compared with other countries of the world?

2. What is the androcentric bias in criminology and what trends are taking place in the field to counteract it?

3. What is the issue of racial profiling? What impact has this had on rates of arrest and incarceration, and what is being done to remedy the problem?

4. How are Native Americans affected by crime, both as perpetrators and as victims? What explanations are there for this?

5. What is William Julius Wilson's major point in *The Truly Disadvantaged?* What are some steps that have been proposed to remedy this problem?

6. What is the role of the media in crime? What are some proposals for controlling media precipitation of crime?

7. What is the importance of crime typologies? How would you begin to answer the following question: "What causes crime?"

8. What role does age play in explaining crime rates?

9. What is transnational crime? What are some policy recommendations for dealing with this phenomenon?

10. What are some problems/limitations in using and interpreting international measures of crime?

WEB SOURCES

American Academy of Child and Adolescent Psychiatry
http://www.aap.org/advocacy/releases/jstmtevc.htm

American Indians and Crime
http://www.ojp.usdoj.gov/bjs/pub/pdf/aic02.pdf

Atlantic Monthly's Criminology Collection
http://www.theatlantic.com/atlantic/election/connection/crime/crime.htm

Florida State University's Criminal Justice Site
http://www.criminology.fsu.edu/p/cjl-main.php

International Crime and Surveys
http://www.uncjin.org/statistics.html

Megalinks in Criminal Justice
http://www.apsu.edu/oconnort/

National Criminal Justice Reference Service
http://www.ncjrs.gov/

National Institute of Justice
http://www.ojp.usdoj.gov.nij/

Office of Juvenile Justice and Delinquency Prevention
http://ojjdp.ncjrs.org

Sourcebook of Criminal Justice Statistics
http://www.albany.edu/sourcebook

TruTV's Crime Library
http://www.trutv.com/library/crime/

Uniform Crime Reports
http://www.fbi.gov/ucr/ucr.htm

WEB EXERCISES

Using this chapter's recommended Web sites, explore variations in crime.

1. What types of information are provided on the Atlantic Monthly site on crime?

2. Examine Tom O'Connor's Megalinks in Criminal Justice and find some sources on international variations in crime.

3. What types of information are available on *The Sourcebook of Criminal Justice Statistics*?

4. Perform an online search of "international crime," "international crime statistics," and "transnational crime." Can you draw any preliminary broad conclusions from this search?

SELECTED READINGS

Joanne Belknap. 1996. *The Invisible Woman: Gender, Crime and Justice*. Belmont, CA: Wadsworth.
> This book provides an excellent review and assessment of critical issues related to the issue of gender in criminology and criminal justice.

David Cole. 1999. *No Equal Justice: Race and Class in the American Criminal Justice System*. New York: New Press.
> Cole makes a strong case that inequality exists in the American criminal justice system primarily due to uneven enforcement as part of the war on drugs.

Erika Fairchild and Harry Dammer. 2000. *Comparative Criminal Justice Systems*. Belmont, CA: Wadsworth.
> This update of a classic provides an outstanding review of world criminal justice systems and issues related to international criminology.

Coramae Richey Mann. 1993. *Unequal Justice: A Question of Color*. Bloomington: Indiana University Press.
> As the title suggests, Mann makes strong arguments for the existence of racism within the criminal justice system in the U.S.

Alida Merlo and Peter Benekos, editors. 1999. *What's Wrong With the Criminal Justice System: Ideology, Politics and the Media*. Cincinnati: Anderson Publishing Company.
> This book reviews what is "wrong" with the criminal justice system and what can be done to make it "less wrong." It reviews three themes: the impact of ideology, the role of the media, and the politicization of crime and criminal justice.

Kathryn K. Russell. 1999. *The Color of Crime*. New York: New York University Press.

Russell examines the role of race in the operations of the American criminal justice system.

Howard Snyder and Melissa Sickmund. 1999. *Juvenile Offenders and Victims, 1999 National Report.* Washington, DC: Office of Juvenile Justice and Delinquency Prevention.

This annual report on the state of the U.S. juvenile justice system might better be titled "Everything you always wanted to know about the American juvenile justice system, but were afraid to ask." This "bible" of juvenile justice features excellent, multicolored graphics and is also available in a CD-ROM version. Contact the Office of Juvenile Justice.

Richard Terrill. 1999. *World Criminal Justice Systems: A Survey* (4th ed.). Cincinnati: Anderson.

Terrill provides an incisive review of criminal justice systems in countries throughout the world including updates on China and Russia.

William Julius Wilson. 1987. *The Truly Disadvantaged: The Inner City, the Underclass and Public Policy.* Chicago: University of Chicago Press.

In what has become a modern-day sociological classic, Wilson describes the "deindustrialization" and abandonment of areas of our inner cities and the accompanying social disorganization and crime.

Theory
Crime File 4.1 The Nacirema Undergraduate as Criminal: A Criminological "Why do it?"

Major Theoretical Approaches

Demonological Theory

Classical Theory

Neoclassical Theory
Crime File 4.2 "Designing Out" Gang Homicides and Street Assaults: Situational Crime Prevention
Crime File 4.3 Justifications for Punishment

Ecological Theory

Forerunners of Modern Criminological Thought

Economic Theory

The Theory–Policy Connection

Summary

Key Concepts

Review Questions

Web Sources

Web Exercises

Selected Readings

chapter 4

Cesare Beccaria

Early and Classical Criminological Theories

My object all sublime I shall achieve in time—To let the punishment fit the crime.

—*Gilbert and Sullivan,* The Mikado

The increasing popularity of the idea that much if not most crime and delinquency reflect innate and intractable predispositions has more to do, I think, with the larger social and economic trends in America in the last quarter of the twentieth century than it does with the meager and contradictory empirical evidence invoked to support it.

—*Elliott Currie,* Confronting Crime *(1985)*

In Chapter 1 theory was discussed as referring to a plausible explanation of reality, a reasonable and informed guess as to why things are as they appear. Theorizing represents a leap of faith, an *élan vital* (vital force) with which to shed light on the darkness of reality. The term *theory* is derived from the Greek *theoros,* to observe and reflect upon the meaning of an event. Representatives from the city-states of ancient Greece were sent to observe celebrations in honor of the gods and were asked to attempt to separate themselves from their personal views and try to conceive of what the gods wished. Without incisive theories, a field or discipline becomes a hopeless catalog of random and seemingly unrelated facts. However, theories are not laws or facts, though this is sometimes forgotten by those who become convinced of the correctness of a particular theory that they come to espouse. Thus, as powerful and persuasive as they may be, Freudian and Marxist theories, for example, are just that: theories; general or systemic models of how human personalities or societies function.

According to Turner (1974):

> Theorizing can be viewed as the means by which the intellectual activity known as "science" realizes three principal goals: (1) to classify and organize events in the world so that they can be placed into perspective, (2) to explain the causes of past events and predict when, where, and how future events will occur, and (3) to offer an intuitively pleasing sense of "understanding" why and how events should occur. (p. 2)

Theory

Theory in criminology refers to efforts to explain or understand crime causation. It is often viewed as an attempt to justify and excuse crime and/or as being wholly inadequate in guiding practical, existing social policy. On the contrary, explaining why or how things happen should not be confused with justifying or defending them. If we ignore this obvious fact, we risk killing the messenger bearing bad (though possibly important) news. The uninitiated find review and critique of these theories a futile exercise in self-flagellation in which criminologists parade their dirty laundry in bitter debates between warring camps of theorists. High hopes are raised for the discovery of a "key" to explaining all crime and criminality, though no such breakthroughs have occurred in the parent social sciences—sociology, psychology, political science, and economics—themselves, and many of the same competing schools do battle in these fields. Those who are uncomfortable with such a theoretical morass might best be advised to study chemistry or biology or auto mechanics— fields in which the theoretical and empirical turf is tidier; the subject matter of the social sciences is infinitely more complex and not likely to yield to a general and universally accepted theory in the near future.

In the meantime, what of the demands of applied theorists and practitioners for explanations with which to guide immediate policy? Some have abandoned pure theory as fruitless in providing guidelines for existing policy needs, yet they then propose therapies, treatments, and policies that surprisingly are based on one or the other of the pure theories they have rejected. In reality, criminology as an interdisciplinary field requires both pure and applied theory. The search for basic, underlying cause is important in itself for the mature development of the discipline, while obviously applied theories need not and cannot wait until ultimate laws are discovered before attempting to advocate existing policy programs. Many fields of learning utilize workable applied theory without having resolved the issue of ultimate causal theory.

Crime File 4.1 discusses the sociological classic *The Body Ritual of the Nacirema.* Read it and then, when reading our theory chapters, ask not only, "Do these theories explain the behavior of criminals?" but also, "Do they explain the behavior of Nacirema undergraduates?" Many Nacirema reject having their picture taken; but if you would like to see one, locate a "Rorrim" and look into it.

Crime File 4.1

The Nacirema Undergraduate as Criminal: A Criminological "Why do it?"

It was anthropologist Horace Miner who first brought the strange customs of the Nacirema to public attention nearly half a century ago. He prefaced his description of this tribe by explaining that anthropologists are trained in avoiding ethnocentric bias so that they are able to present objectively and without shock such extreme, exotic customs as exhibited by the **Nacirema.** In his classic *The Body Ritual of the Nacirema*, Miner (1956) describes a group whose land, **Asu,** lies between the Canadian Cree, the Carib of the Antilles, and the Yaqui Indians of Mexico. Miner indicates (1956):

> Little is known of the origin, although tradition states that they came from the east. According to Nacirema mythology, their nation was originated by a cultural hero Notgnihsaw, who is otherwise known for two great feats of strength—the throwing of a piece of wampum across the river Pa-To-Mac and the chopping down of a cherry tree in which the spirit of truth resided. (p. 503)

If you have not figured it out by now, Nacirema is American spelled backward, and Notgnihsaw is Washington. The entire piece is a clever pun that pokes fun at American culture as it might be viewed by someone who is wholly unfamiliar with the real reasons for certain American customs and beliefs. After a brief presentation of Miner's Nacirema, Hagan and Benekos (2000) tease some updates examining Nacirema undergraduates and invite you to apply the theories you are about to explore to analyze not criminals, but Nacirema undergraduates who are deviant or not performing as expected.

Nacirema Customs and Rituals

The Nacirema have an obsession with the human body and its care and adornment, without which they believe that their friends and lovers would desert them. Despite this they believe that their bodies are naturally ugly and condemned to disease and decay. Only by using powerful rituals and visiting specialized witch doctors do the Nacirema believe they can escape this corporeal destiny. Daily, secretive rituals in a room called the **moorthab** involve what Miner assumed was "worshipping" before a shrine box within which were stored various magical charms and solutions that the natives felt were necessary for living. Such potions were obtained from special shamans in return for very rich gifts. When ill, the Nacirema visit the witch doctor, who would often put magical wands in their mouth and inject them with supernatural needles.

Since Miner's analysis a major trend has been the large number of Nacirema youth who now attend college rather than work after their initial rite of passage. In the past only the children of the top leaders attended college, while now the majority of youth do so. In colleges large gifts are given to older members of the tribe known as Talkers or **Forp.** The Nacirema undergraduates work very hard at their assigned task, transcribing and later reciting or recounting shared tribal knowledge. New Nacirema inventions such as little boxes with brains called a **spotpal** can store and record information as well as access the **tenretni,** a magical stream of words, pictures and ideas that has accelerated even further changes in the land of Asu.

Theories of Nacirema Undergraduate Deviance

The ability of criminological theory to explain criminal/deviant behavior awaits theoretical developments in sociology, psychology, and the other social sciences that adequately explain normal behavior. Let us pretend for a moment that, instead of attempting to explain criminal behavior in society, we could apply these same theories to aberrant behavior by undergraduates in Nacirema colleges and universities. While space does not permit detailed coverage, a cursory visit should suffice.

What explains the poorly performing Nacirema undergraduate whose parents could purchase a new **rac** every year for the price they pay for their child to listen to a Forp? Despite the high prices, some students attend few classes, sleep, daydream, perform at terrible levels, and even eventually are asked to leave. Early theories might suggest that such students were possessed by demons, while later ones pursued more secular explanations. Some Nacirema undergraduates cheat on examinations or steal ideas from others for papers that they claim are their own work. Classical theories of criminology would explain that such deviants are rational and that the penalties for such behavior (flunking or expulsion) do not exceed the rewards (passing or graduating without doing the work).

Other theories would suggest that the students' social background or lower social class adversely impacts on them and explains their poor performance, while positivistic theories seek explanation in biological and psychological shortcomings. Are poor or deviant students

(Continued)

(Continued)

atavistic "throwbacks to the ape," born deviants whose biological inheritance dooms their chances of conformity and success? Perhaps mental deficiency, feeblemindedness, physical inferiority, mesomorphic body type, or brain disorders explain their zombie-like behavior and inadequate preparation, interest, and performance compared to their more accomplished peers. Many do a noncannibalistic imitation of the classic film *Night of the Living Dead.*

Most likely, the problem is not biologic at all, but psychological. Do they possess born, uneducable personalities; are they unconsciously repressing their sexual instincts; or are they victims of extroversion, inadequate behavioral conditioning (namely anti-intellectualism), or low IQs? A long-neglected explanation may be found in the work of the then-26-year-old genius and future president of the University of Chicago, Robert Maynard Hutchins, who in *Zuckercandl!* (1968) discusses the theories of the all but forgotten philosopher Alexander Zuckercandl, who, instead of trying to explain how people should behave, developed explanations for how people (in this case students) do behave. Zuckercandl's genius was to reverse Freudian theory. The latter enjoined us to "become conscious of our unconscious," while Zuckercandl indicates "we must become unconscious of our conscious," a motto to which many undergraduates might relate.

Modern sociological theories of crime view society and social groups as far more likely explanations of human (student) misconduct than those that have been mentioned. Edwin Sutherland's "theory of differential association" explains that individuals learn to become "poor students" due to an excess of contacts that advocate not taking seriously academic work and, as a result, students are predisposed to regard intellectual activity in a negative manner (Sutherland & Cressey, 1960). If students have a frequency and duration of such negative views (e.g., anti-intellectual background) and do not prefer or find meaningful academic pursuits, then such persons will tend to perform poorly. Nacirema society preaches that all members of the tribe should be able to become successful (own many gadgets) if they perform well as undergraduates, thus obtaining high-ranking tribal positions such as shamans, talkers, and listeners.

Robert Merton (1957, 1968), in his theory of anomie and modes of personality adaptation, explains that such high tribal positions and rewards are preached as available to all Nacirema if they pursue the legitimate means of being successful in college. Anomie (normlessness) ensues when a gap exists between the goal of success (high position and rewards) and adequate means for their achievement (undergraduate academic success). While most students demonstrate "conformist" personalities in that they accept this goal and adequate means exist for them to be successful in college, others find that they lack the adequate means in order to be successful. These students can become "innovators" and lie, cheat, and steal in order to graduate. Others are "retreatists." They reject the goal of success and spend their time rejecting the means (going to classes and getting good grades) as well. Alcohol and drugs may replace books as a primary orientation. "Ritualistic" students show up for class, but forget the purpose of doing so—getting good grades. They daydream and sleep in class and assume somehow they will learn what they consider irrelevant academic information through a process of osmosis. Finally, the "rebel" may use legitimate or illegitimate means to challenge the values and goals presented by the college.

Student as Renosirp

Another dimension of this analogy is that some students also display behaviors which incorporate norms that are characteristic of offenders in captivity. Rather than identifying with Forps, some Nacirema students cultivate anti-intellectual attitudes and values that enforce a code of silence and impose sanctions on those who violate what is described as the "**Renosirp** rules." Similar to inmate codes such as "don't talk to the screws," "don't trust the guards," and "Don't snitch" (Sykes, 1958), Nacirema undergraduates affect disinterest and ascholarly views. The student code may include the following: "Don't be smart, don't ask questions, don't answer questions, do your own time, and be cool." For example, one report found that 90 percent of college students say they would not turn in someone for cheating; that is, "don't snitch" (Kleiner & Lord, 1999, p. 57). As prisoners do their time by developing a prisoner subculture, students also develop a subculture which fosters their identity and legitimizes their roles as nonacademics and provides rationalizations for their deviant behaviors. Kleiner and Lord report that "plagiarism, copying, and similar deceits devalue learning" but that "the pressure to succeed . . . can drive students to consider extreme measures" (1999, p. 57). This suggests the "innovation" that Merton (1957, 1968) typologized as an adaptive mode to blocked means.

Students also use their dress to symbolize some of their codes. For example, part of the student uniform at the turn of the century was a well-worn piece of head gear called a **pac llab,** sometimes pulled down low over the forehead to conceal the eyes, thus avoiding eye contact with Forps and the dreaded possibility of being asked a question. When a pac is worn in reverse, it is done so to pay homage to Yogi Berra, a philosopher students have grown to admire and one who represents their worship of sports heroes.

By disdaining books, global events, and politics, students instead form identification with heroes of sporting events. These athletic stars represent models for success, fame, and wealth that surpass the financial achievement and professional status based on traditional adaptations that emphasize education, intellectual growth, and deferred gratification. By worshipping the athlete as a godlike figure, students seek vicarious success by adorning themselves with the colors and emblems of their favored sports gangs. This identification with these gladiator gangs represents another salient custom of the Nacirema undergraduate.

As prisoners do their own time, reject the rejecters, and try to avoid the guards, students also demonstrate behaviors that avoid socialization to college norms and values, that is, "collegization." The students as criminals distance themselves from Forps by sitting in the back of the classroom, by avoiding note-taking, by not participating in discussion, and generally by adhering to the Renosirp norms: "Don't get involved, don't ask questions, don't show emotion" (i.e., interest). Instead of engaging in intellectual growth and scholarship, these students have precipitated a co-optation of college values with an importation of working-class street values. Historically, college has reflected a middle-class orientation, and students assimilated appropriate norms and values. With the increasing number of high school graduates who would not have gone to college in the past now entering such institutions, a critical mass of students has facilitated a subculture that deflects socialization and assimilation of middle-class values and establishes norms to support alternative student behaviors (Miller, 1958).

One might ask, "Why don't all students behave in this manner?" Here again, criminological theory would say they would if they dared; but their attachments to parental units and Forps, belief in the academic model, commitment to accomplishing normative goals, and involvement in classes and intellectual exercises "bonds" (Hirschi, 1969) them to conformity. In contrast, those students who enter college as a default often manifest ritualistic behaviors (Merton, 1957, 1968) and lack commitment, involvement, beliefs, or attachments to ensure their bond to the scholarly community. In fact, as noted above, as a critical mass of nonscholar undergraduates is attained, the culture of the college becomes transformed. As with penal institutions, the question is often asked, "Who rules the joint?"

Criminological theory offers one approach to explaining this curious paradox in which students resist/reject pro-intellectual elements of college while drifting toward deviant norms that demonstrate the student as criminal. Theories of subculture, strain, social control, and differential association provide useful concepts in understanding how "normal" Nacirema students become "criminals" in academe.

Sources: Frank Hagan and Peter Benekos (2000, March). "The Nacirema Undergraduate as Criminal: A Theoretical Analogy." Paper presented at the Academy of Criminal Justice Sciences Meetings, New Orleans, Louisiana; Travis Hirschi (1969). *Causes of Delinquency.* Berkeley, CA: University of California Press; Robert M. Hutchins (1968). *Zuckercandl!* New York: Grove Press; Carolyn Kleiner and Mary Lord (1999, November). "The Cheating Game." *U.S. News & World Report,* 55–57, 61–66; Robert K. Merton (1957, 1968). *Social Theory* and *Social Structure* (Rev. ed.). New York: The Free Press; Walter Miller (1958, May). Lower Class Culture as a Generating Milieu of Gang Delinquency. *Journal of Social Issues,* 14, 5–19; Horace Miner (1956). Body Ritual Among the Nacirema. *The American Anthropologist,* 58, 503–507; Edwin H. Sutherland Donald C. Cressey (1960). *Criminology.* Philadelphia: Lippincott; Gresham M. Sykes (1958). *The Society of Captives: A Study of a Maximum Security Prison.* Princeton, NJ: Princeton University Press.

Web Research Project
What are some recent cases of actual campus crime among Nacirema college students? Search "campus crime."

Major Theoretical Approaches

This chapter explores many early theories that represent the historical legacy of the field. While many of the early theories have been discredited, their examination is warranted not only from the standpoint of gaining a sense of continuity of the discipline, but also because many expressions of these theories are resurrected in new forms in modern thinking. Sometimes these theories will be accepted or rejected on the basis of ideology rather than on the basis of empirical evidence (Blankenship & Brown, 1993, p. 171).

Table 4.1 presents an outline of the major theoretical approaches in criminology. The last type, sociological theory, is subdivided and described in more detail in Chapters 6 and 7. This division of criminological theories into types or schools of thought is primarily for purposes of convenient presentation since, in fact, some theorists demonstrate evolution in their views and may in fact exhibit theoretical conceptions that meld different types or schools of thought. The primary theoretical approaches in criminology (Table 4.1) are: the demonological, classical (neoclassical), ecological (geographic), economic, positivistic (biological and psychological), and sociological (the many subtypes of which will be discussed in Chapters 6 and 7). Discussion will begin with the demonological or supernatural approach to explaining crime causation, which is based in a superstitious and tradition-oriented past in which wrongdoers were perceived as controlled by otherworldly forces.

TABLE 4.1 ■ Major Theoretical Approaches in Criminology*

Theoretical School	Major Themes/Concepts	Major Theorists
Demonological	Criminal as evil, sinner, supernatural pawn	Traditional authority
Classical (Neoclassical)	Criminal as rational, hedonistic, free actor Incapacitation, punishment, deterrence	Beccaria, Bentham Wilson, G. Becker
Ecological (Geographic)	Group characteristics, physical and social ecological impacts on criminality Geographical and climatic impacts on criminality	Quetelet and Guerry Lieber and Sherin
Economic	Capitalism, social class inequality, and economic conditions cause crime	Marx, Bonger
Positivistic Biological	Physical stigmata, atavism, biological inheritance cause criminality Mental deficiency Feeblemindedness Physical inferiority Somatotypes—mesomorphs Brain disorders, twin studies, XYY syndrome, physiological disorders	Lombroso, Ferri, Garofalo Goring Goddard Hooton Sheldon Moniz, Christiansen, Jacobs
Psychological	Unconscious repression of sexual instincts Criminal personality, extroversion, inadequate behavioral conditioning, IQ	Freud, Eysenck, Skinner Hirschi, Hindelang
Sociological	Anomie, subcultural learning, elite dominance cause crime	Durkheim, Sutherland, Quinney

*See Tables 6.1 and 7.1 for greater detail regarding sociological theory in criminology.

Video Link 4.1
View a video of the issue of torture.

Photo 4.1
Under Iran's fundamentalist regime symbolized by the Ayatollah Khomeini, criminals or opponents of the regime were subjected to torture, death, or other forms of the "wrath of Allah."

Demonological Theory

Demonological or supernatural explanations of criminality dominated thinking from early history well into the eighteenth century; modern remnants still survive (see Huff, 1990). In a system of knowledge in which theological explanations of reality were predominant, the criminal was viewed as a sinner who was possessed by demons or damned by otherworldly forces. Humankind was viewed as at the mercy of the supernatural: fates, ghosts, furies, and/or spirits. Felonies (mortal sins) were viewed as manifestations of basically evil human nature reflecting either allegiance to the "prince of darkness" or an expression of divine wrath. The Salem witch trials in Puritan New England and the Spanish Inquisition serve as examples of the torture, hanging, burning at the stake, and other grim executions awaiting heretics, witches, and criminals. Such a world view perceived the violator's actions as deterministically controlled by forces beyond the individual's mastery. In Genesis (22:1–12), Abraham was ordered by God to sacrifice his son Isaac, although he was later released from this injunction. Appeasement of God or the gods, a world beyond human cognition and interpretable only by the clergy, the shaman, and other emissaries to the supernatural, was supported by a traditional worldview that looked to the "wisdom" of the past rather than attempt a rational interpretation of the present for guidance (see Fox, 1976, pp. 7–12).

In the Middle Ages in Europe, feudal lords established various means of determining guilt and innocence. God could indicate who is guilty or not by giving victory to the innocent in a trial by battle. Later, trial by ordeal was instituted in which the accused was exposed to dangerous tests and, if he or she survived, he or she had been protected by God. Women accused of witchcraft were tied up and thrown in the water. If they survived, they were innocent. Running the gauntlet and walking on fire were similar tests.

Application of the theological approach to crime control is not confined to the past but can be illustrated in the modern era by the ecclesiarchy (state–church fusion) in Iran under the Ayatollah Khomeini, in which criminals or opponents of the state were summarily subject to torture, death, or the "wrath of Allah."

Photo 4.2
Heretics were tortured and burned at the stake during the Spanish Inquisition.

The primary challenges to theological approaches to explaining reality were philosophical arguments that sought worldly, rational, secular explanations for human fate. The reasons for crime and criminality were to be found not in the supernatural, but the natural world. Table 4.2 provides a chronology of major developments in criminological theory.

Classical Theory

Classical theory in criminology refers to an approach that emphasizes free will and rationality on the part of the criminal actor.

Prior to the formulation and acceptance of classical theory, the administration of criminal justice in Europe was cruel, uncertain, and unpredictable. In England alone in the early nineteenth century, there were over 100 crimes punishable by the death penalty (Heath, 1963, p. 98). Penal policy was designed to control the "dangerous classes," the mass of propertyless peasants, workers, and unemployed. Emerging liberal philosophies espoused by such writers as Locke, Hobbes, and Rousseau advocated the "natural rights of man" and reason as a guide to regulating human conduct. This Enlightenment of the seventeenth and eighteenth centuries questioned the power of the clergy and aristocracy and gave birth to the American and French revolutions.

Cesare Beccaria

Italian Cesare Beccaria (1738–1794), actually Cesare Bonesana, the Marquis of Beccaria, was, along with British philosopher Jeremy Bentham (1748–1832), the principal advocate of the classical school of criminological theory. Beccaria's (1963) essay *On Crimes and Punishments,* originally published in 1764, had a profound impact on continental European as well as on Anglo-American jurisprudence. Beccaria was only 26 years of age when he wrote his treatise. His essential point is expressed in the concluding paragraph of this work:

> From what has thus far been demonstrated, one may deduce a general theorem of considerable utility, though hardly conformable with custom, the usual legislator of nations; it is this: In order for punishment not to be, in every instance, an act of violence of one or of many against a private citizen, it must be essentially public, prompt, necessary, the least possible in the given circumstances, proportionate to the crimes, dictated by the laws. (p. 99)

Beccaria was appalled by the arbitrary nature of the European judicial and penal systems of his time, which were unpredictably harsh, exacted confessions by means of torture, and were completely

Handbook Article Link 4.1
Read an article on classical perspectives.

Video Link 4.2
View a video of a mock discussion with Cesare Beccaria.

TABLE 4.2 ■ Chronology of Selected Important Developments in Criminological Theory

1750 B.C.	Hammurabi's Code	1950	Glueck and Glueck, *Unravelling Juvenile Delinquency*
1766 A.D.	Beccaria, *On Crimes and Punishments*	1951	Lemert, *Social Pathology*
1776	American Revolution	1951	Cohen, *Delinquent Boys*
1787	French Revolution	1958	Vold, *Theoretical Criminology*
1788	Bentham, *Moral Calculus*	1958	Miller, "Lower Class Culture and Delinquency"
1833	Guerry, *An Essay on Moral Statistics*	1964	Eysenck, *Crime and Personality*
1835	Quetelet, *Treatise on Man and the Development of His Faculties*	1967	Reckless, "Containment Theory"
1848	Marx, *Communist Manifesto*	1967	Clinard and Quinney, *Criminal Behavior Systems*
1859	Darwin, *Origin of Species*	1969	Hirschi, *Causes of Delinquency*
1863	Lombroso, *Criminal Man*	1970	Quinney, *The Problem of Crime*
1897	Durkheim, *Suicide*	1971	Chambliss and Seidman, *Law, Order and Power*
1910	Dugdale, *The Jukes*	1973	Bandura, *Aggression*
1913	Goring, *The English Convict*	1973	Taylor, Walton and Young, *The New Criminology*
1916	Bonger, *Criminality and Economic Conditions*	1975	Adler, *Sisters in Crime*
1920	Freud, *General Introduction to Psychoanalysis*	1979	Cohen and Felson, *Routine Activities*
1925	Park, Burgess, and McKenzie, *The City*	1980	Clinard and Yeager, *Corporate Crime*
1937	Tannenbaum, "Dramatization of Evil"	1984	Lea and Young, *Left Realism*
1937	Sutherland, *The Professional Thief*	1988	Daly and Chesney-Lind, *Feminist Theory*
1938	Merton, *Social Theory and Social Structure*	1991	Quinney and Pepinsky, *Peacemaking*
1939	Hooton, *The American Criminal*	1992	Clarke, *Situational Crime Prevention*
1939	Sutherland, "Differential Association"	1993	Messner and Rosenfeld, *Crime and the American Dream*
1940	Sheldon, "Somatotypes"		

subject to the whims of authorities. Since potential criminals had no way of anticipating the nature of the criminal law and its accompanying penalty if violated, punishment had little deterrent value. Beccaria was primarily interested in reforming the cruel, unnecessary, and unpredictable nature of punishment, feeling that it made little sense to punish lawbreakers with unjust laws (Vold, Bernard, & Snipes, 2002, p. 16). Beccaria was responsible for the abolition of torture as a legitimate means of exacting confessions. "Let the punishment fit the crime" is a succinct summation of Beccaria's argument.

Beccaria's work was radical for its time and, fearing repercussions, he published it anonymously. Even though he defended himself in the introduction to the book, claiming that he was not a revolutionary or an unbeliever, the book was condemned by the Catholic Church in 1777 for its rationalistic ideas (Vold et al., 2002). It was placed on the church's *Index of Forbidden Books* for over 200 years. Beccaria proposed the following principles for the proper operation of the criminal justice system:

- Laws should be made by legislatures and they should be specific.

- The role of judges is only to determine guilt and to follow strictly to the letter of the law in determining punishment. Judges should not interpret the laws.
- The seriousness of crime should be determined by the harm it inflicts on society and be based on the pleasure/pain principle.
- Punishment should be based on the seriousness of the crime and its ability to deter.
- Punishment should not exceed that which is necessary for deterrence.
- Excessive severity in punishment often increases crime that is then committed in order to avoid punishment.
- Punishment should be sure, swift, and certain.
- Capital punishment should be abolished as should the use of torture in order to gain confessions.
- Laws should be structured so as to prevent crime in the first place. It is better to prevent crimes than to punish them.
- All should be treated equally before the law (Beccaria, 1963).

Jeremy Bentham

Beccaria's British contemporary, Jeremy Bentham (1748–1832), borrowed from Beccaria the notion that laws should provide "the greatest happiness shared by the greatest number" (Beccaria, 1963, p. 8). Bentham graduated from Oxford University at the age of 12. In his will, the eccentric Bentham gave all of his original book manuscripts to the University College of London on one condition: The administrators had to embalm his body and put it on display for all to see. Such a wooden and glass box with a sitting, fully clothed Bentham is on display to this day. He saw the purpose of punishment to be deterrence rather than vengeance and, similar to Beccaria, was more interested in the certainty of punishment than in its severity.

Bentham has been called an advocate of "utilitarian hedonism" or "felicific calculus" or "penal pharmacy." Utilitarianism is a practical philosophical view that claims "we should always act so as to produce the greatest possible ratio of good to evil for all concerned" (Barry, 1983, p. 106). One of Bentham's best known contributions to criminology was his invention of the "panopticon" (from the Greek, meaning "all seeing"; Bentham, 1823). The panopticon, or "inspection house," was envisioned as a circular prison with a glass roof, featuring a central grand tower from which inspectors could observe all cells located around the perimeter. While prisons incorporating this design were built in both England and the United States, the plans were later found impractical and were modified (Hagan, 1996).

The classical theorists viewed individuals as acting as a result of "free will" and as being motivated by **hedonism.** The latter refers to a "pleasure principle," the assumption that the main purpose of life is to maximize pleasure while minimizing pain. Individuals are viewed as entirely rational in this decision-making process in which they will attempt to increase pleasure, even illicit desires, until the anticipated pain to be derived from a particular activity appears to outweigh the expected enjoyment. In a work titled *Seductions of Crime: Moral and Sensual Attractions in Doing Evil,* Jack Katz's (1988) research based on interviews with career criminals supports Beccaria's notion of the pleasure or thrill of evil outweighing the fear of punishment. Image, danger, glamour, and the excitement of crime overshadow any desire for a successful life in straight society. In assessing Katz's theory, McCarthy (1995)

Photo 4.3

Jeremy Bentham (1748–1832) borrowed from Beccaria the notion that laws should provide "the greatest happiness shared by the greatest number."

noted that such thrill-related property crime is influenced by age, gender, and strain associated with inadequate economic opportunities.

Critique of Classical Theory

The classical school and the writing of Beccaria in particular were to lay the cornerstone of modern Western criminal law as it became formulated from 1770 to 1812. The characteristics of modern Western criminal law—politicality, uniformity, specificity, and described penal sanctions—are in essence called for in Beccaria's essay. The *French Declaration of the Rights of Man* (Jacoby, 2004), which was passed by the revolutionary National Assembly of France in 1789, included the statement: "The law ought to impose no other penalties but such as are absolutely and evidently necessary; and no one ought to be punished, but in virtue of a law promulgated before the offense, and legally applied" (p. 215). The Eighth Amendment to the U.S. Constitution, prohibiting "cruel and unusual punishment," was also a Beccarian legacy.

Some recent analysis suggests that the importance of Beccaria's works may have been exaggerated and that he was actually less important than other social reformers of the eighteenth century such as Voltaire and Bentham (Newman & Marongiu, 1990). Beirne (1991) claims that Beccaria's famous treatise *Dei Delitti e Delle Pene* (*On Crimes and Punishments*) was the application to crime policy, not of rationality and humanism, but of the Scottish-inspired "science of man," which emphasized utilitarianism and determinism. He felt that Beccaria was less of an advocate of free will than has been supposed and that his writings exhibited much determinism (Beirne, p. 812).

The revolutionary and liberating impact of the ascendancy of classical theory in reforming Western jurisprudence is now taken for granted, but without the fundamental changes classical theory introduced, the remaining criticisms and subsequent modifications would not have been possible. However, classical theory contained the seeds of its own demise. While Justitia, the blind goddess of justice carefully weighing the evidence irrespective of the violator, is an appealing symbol, classical theory by its very insistence on equality of punishment proposes inequality. Should minors or the insane be treated in the same manner as others? Should repeat offenders be accorded the same sanctions as first offenders for an equivalent act? Thomas and Hepburn (1983) state:

> Contemporary criminologists tend to assign little importance to [classical theory's] concepts and ideas. Perhaps the two major reasons are that it focuses our attention on criminal law rather than criminal behavior and that it is based on a speculative set of philosophical premises rather than a sound theory that could be verified or refuted by the collection of systematic empirical evidence. (p. 137)

Application of the pure classical theory would rob judges of discretionary power, and seems to rest on a simplistic assumption of the ability to exactly measure individual conceptions of pain and pleasure. Recent revivals in the United States of determinate sentencing and mandatory punishments for specific offenses are remnants of classical theory. Although theoretically appealing because of the essential cookbook application of graduated punishment reflecting the seriousness of crime, implementation becomes problematic for reasons already described: the quantification of such acts and their perpetrators defies such a simplistic scheme (Hagan & Tontodonato, 2004).

Neoclassical Theory

Journal Article Link 4.1
Examine literature regarding psychological jurisprudence.

The **neoclassical** school basically admitted environmental, psychological, and other mitigating circumstances as modifying conditions to classic doctrine. The beginnings of this approach can be found in the later writings of Cesare Lombroso (1835–1909) and in those of his students, Ferri and Garofalo, to be discussed shortly. Beginning in the late 1960s, particularly in the writings of economist Gary Becker (1968), James Q. Wilson (1983a, 1983b), and Ernest Van den Haag (1966), a resurgence in neoclassical doctrine can be noted. Becker advocated a "cost/benefit" analysis of crime, reminiscent of hedonistic doctrine. Becker argues that individuals freely choose crime based on their estimate of their likelihood of being caught. Disappointed with criminology's overconcern with the search for basic causes of crime, Wilson (1975) proposed a policy analysis approach, applied research that is less concerned with finding "causes" and more concerned with "what works." These writers have sparked an interest in the abandonment of treatment and rehabilitation and in a return to the classical punishment model. Often ignored by devotees of such theories are the very limited categories of crime such theorists, in fact, address. Wilson (1975), for instance, quite clearly indicates that this call for incapacitation of offenders (criminals in jail can no longer victimize) is applicable to what we have described as conventional property offenders or common burglars and thieves. Although a more practical, policy-oriented approach is needed, what is disturbing in such theories is the relatively conservative ignorance of criminogenic, social structural conditions, as well as an often cavalier disregard for theoretical approaches to crime causation. While the neoclassicists argue that less theory and more action is needed, they at times ignore the fact that the basic theoretical underpinnings of their own theories are rooted in assumptions of eighteenth-century hedonism, utilitarianism, and free will. On balance, however, they make a key point: that one need not have a basic explanation of cause or wait for one in order to meet pressing policy needs that cannot wait for final explanation.

In another neoclassical theory, Cornish and Clarke's (1986) rational choice theory proposes that offenders weigh the opportunities, costs, and benefits of particular crimes. The argument by rational choice theorists is not that individuals are purely rational in their decision making, but rather that they do consider the costs and benefits. A number of factors may constrain choice, such as social factors, individual traits, and attitudes toward crime. Rational choice theorists also argue for a "crime-specific approach" to crime; that is, the circumstances involved in the typical burglary may differ from robbery or domestic assault. Offender characteristics are seen as combining with offense types in shaping offender choices. Rational choice theorists admit that much behavior is only partly rational, but that most offenders know quite well what they are doing. The criminal justice system must make crime less rewarding by increasing the certainty and severity of punishment. Crime is viewed as a matter of situational choice, a combination of costs, benefits, and opportunities associated with a particular crime. Increasing prevention or decreasing the opportunity to commit crime is viewed as an important means of deterring crime. "Situational crime control" could include target hardening (securing of entries, doors and locks), access control, entry screening, surveillance, better lighting, property identification, and other means of reducing criminal opportunity.

Research support has been mixed for rational choice theory. While consideration is given for the cost and benefit of crime, many criminals do not carefully plan their crimes. Changing such opportunity structures (e.g., creating defensible space and target hardening) may discourage potential offenders. Analyses of offenders' motivations, however, have shown that many act impulsively and fail to fully consider negative possibilities (Piliavin, Gartner, Thornton, & Matsueda, 1986; Tunnell, 1991). Crime File 4.2 presents an application of rational choice theory to controlling gang violence in Los Angeles.

Audio Link 4.1
Listen to a discussion of the three strikes law.

Other expressions of neoclassical theory can be found in the deterrence literature. Themes such as "just deserts," "three strikes and you're out," and mandatory sentencing policies all reflect the assumption that the criminal is a rational actor and will be deterred by more severe and certain punishment. The "just deserts" concept assumes that individuals must pay for their wrongdoing and that they deserve or "have it (the punishment) coming." Reflected in the biblical *lex talionis* (law of the talons), an eye for an eye and a tooth for a tooth, proper retribution is to be exacted for the wrongdoing. Deterrence policy assumes rationality on the part of the actor, wherein "specific deterrence" serves to discourage a particular individual from repeating a crime and "general deterrence" targets others. Legislation such as "three strikes and you're out," in which third-time offenders receive very severe punishment, has been found not to work since juries are often reluctant to convict a third-time offender and judges oppose such limitations on their discretion.

Video Link 4.3
View a video of mothers of three strikes law victims.

Crime File 4.3 discusses the various justifications for punishment. The deterrence argument best represents the classical and neoclassical explanations. These arguments are also applicable to the death penalty debate.

Crime File 4.2

"Designing Out" Gang Homicides and Street Assaults: Situational Crime Prevention

One of the leading theories of criminal opportunity is "situational crime prevention." Developed by criminologist Ronald V. Clarke, the theory is based on the assumption that crime can be reduced by pinpointing and blocking the forces that facilitate would-be offenders' criminal acts. Would-be offenders, the theory proposes, make rational choices in planning their criminal acts. For example, gangs may choose a particular street on which to commit a crime because they rationally determine that the way the street is situated provides them with ready access and exit, thereby creating an opportunity to more easily elude arrest.

Applying the model to gangs, the LAPD (Los Angeles Police Department) assumed that they did in fact make a "rational choice" about whether to engage in a particular act of criminal violence and whether to do so in a particular neighborhood setting. Evidence to support the theory has come from studies of residential burglary, shoplifting, and other crimes, but OCDS (discussed below) was an initial attempt to apply situational crime prevention to gang violence.

Issues and Findings

Discussed in This Brief. The use of a deceptively simple tactic, traffic barriers, to block automobile access to streets as a way of reducing gang violence. The tactic was used in a crime-plagued area of Los Angeles that had experienced the city's highest level of drive-by shootings, gang homicides, and street assaults. The National Institute of Justice (NIJ)-sponsored evaluation of Operation Cul de Sac (OCDS), as the program was called, examined whether the tactic could reduce gang crime.

Key Issues. OCDS was based on the theory of situational crime prevention, which postulates that crime occurs partly as the result of opportunity and can be reduced by first identifying and then blocking these opportunities rather than attempting to eliminate "root causes." The LAPD noted that in the OCDS target area, gang crime clustered on the periphery of neighborhoods linked to major roadways; police set up traffic barriers as a way to block the opportunities for crime the roadways created. The evaluation sought to determine whether these street closures could help to "design out" gang crime.

Key Findings. In its 2 years of operation, 1990 and 1991, OCDS appeared to reduce violent crime.

- The number of homicides and street assaults fell significantly in both years and rose after the program ended.
- Property crime decreased substantially during the first year of the program, but it also decreased in the comparison area where there was no OCDS, indicating that some factors other than the traffic barriers were responsible for the reduction in the OCDS site.
- In the second year of the program, property crime rose, suggesting the street closures affected only violent crime.
- Crime was not displaced to other areas. Violent crime fell, not only in the OCDS area, but also in contiguous areas. This may be because the areas of potential displacement are the turf of rival gangs. As such they would be off-limits to gangs that might want to enter new territory when the traffic barriers reduced their opportunities to commit crime on their own turf.
- Traffic barriers can be used as part of an approach to maximize neighborhood residents' defensible space by increasing their span of control. Zones configured with the barriers heighten the visibility of suspect activities. They can be particularly effective when combined with "natural guardians"—people who serve as informal sources of surveillance and social control.
- Although these findings indicate traffic barriers may work to reduce violent crime, it should be kept in mind that the experiment was conducted at only one site. Replications of OCDS and further evaluations are needed to fully test the effectiveness of the tactic.

Target Audience. Police chiefs, sheriffs, urban designers and planners, crime prevention organizers.

Source: James Lasley (1998). "Designing Out" Gang Homicides and Street Assaults. *National Institute of Justice Research in Brief,* November.

Web Research Project

Do a search using a search engine such as Google and locate recent developments on "designing out crime."

Crime File 4.3

Justifications for Punishment

The punishment of criminals has at least four justifications: retribution, deterrence (including incapacitation), rehabilitation, and protection and upholding the solidarity of society (Sutherland & Cressey, 1974, pp. 325–330). Retribution is the societal counterpart of individual revenge. When criminal laws were formulated, the state assumed responsibility for punishing offenders and forbade victimized parties from taking the law into their own hands. Criminals must pay their debt to society, not to the harmed party. Beginning as early as *lex talionis,* "an eye for an eye and a tooth for a tooth," criminals have been viewed as having to suffer in some way for justice to be served. Retribution is a moral motive for punishment, not simply a utilitarian one. Nazi hunters who are still searching for war criminals decades after World War II, when asked, "What good does it do?" reply, "It does justice." So public sentiment and outrage are the guideposts for enforcement, rather than any direct effect on future crime commission.

Deterrence refers to the belief that perceived punishment will serve as a warning and inhibit individuals (*specific deterrence*) and groups (*general deterrence*) from involvement in criminal activity. Based on the classical school of criminology and the writings of Cesare Beccaria (discussed earlier), the deterrence model assumes that if the pain (clear, swift, and certain punishment) outweighs any pleasure to be derived from the criminal act, then crime will be prevented. *Incapacitation,* the prevention of crime by keeping criminals behind bars for longer periods, is an additional example of special deterrence. In a revival of classical criminology, large and impressive bodies of literature have begun to accumulate on the issue of specific deterrence. Although inconclusive at this point, the research suggests the potentially positive impact of selective incapacitation of career criminals on lowering crime rates (Clarke, 1974; Greenberg, 1975).

Rehabilitation, which has been the watchword in the United States in the post–World War II period, assumes that the purpose of punishing criminals is to reform or resocialize them to conventional, law-abiding values. Even name changes indicate this philosophical shift: The field of penology is now called corrections, and prisons are correctional facilities. Nevertheless, there appears to be more talk about rehabilitation than programs to facilitate it. Martinson (1974), in "What Works?—Questions and Answers About Prison Reform," examined a large number of correctional programs and their claims of success in rehabilitation as well as their recidivism (repeating of crime) rates; he felt that there was little evidence that any significant programs in corrections had an important impact on reducing recidivism. Only later ("Martinson Attacks His Own Earlier Work," 1978; Martinson, 1979) did he retract this devastating critique by admitting that he may have suffered from "methodological fanaticism," in which substance was overlooked in the name of method and that some of the programs did have positive outcomes. With estimates of recidivism and reincarceration rates as high as 65 percent (Greenberg, 1975, p. 551), there seemed to be a decline in liberal optimism about the success of the rehabilitation model (Bayer, 1981). However, in defense of rehabilitation, some feel that it has never been given a decent chance. Badillo and Haynes (1972) indicate that in the early 1970s only about 5 percent of correctional budgets was used for rehabilitation programs and that rehabilitation has often been more a matter of talk than action (see Cullen & Gilbert, 1982). Glaser (1994) identifies a variety of programs that utilize penalties, fines, community services, restitution, and intermediate punishments that do indeed work.

Protection and the upholding of social solidarity as a goal of punishment reflect Durkheim's (1950) point made in Chapter 1—that a society reaffirms its values in reacting to and punishing wrongdoers. In this justification, the purpose of punishment is not to obtain revenge or deter or change the criminal; rather, it is an attempt to protect society from criminals and in so doing to reinforce group solidarity.

The Death Penalty Debate

Support for the death penalty in the United States appears to follow the crime rate; that is, higher crime rates bring greater support for the death penalty. During the low-crime period of the early 1960s, support dropped to a low of 42 percent in 1966. It steadily increased thereafter, a Gallup poll of 1988 showing 79 percent of Americans in favor. Retribution and vengeance appeared the major reasons for favoring the death penalty, since many were aware that such executions were not a deterrent (Dionne, 1990).

The subject of capital punishment has fueled heated debate among criminologists and the public at large.

Arguments in favor of the death penalty include:

- It is sanctioned in the Bible as well as by historical tradition as a culturally approved manner of dealing with heinous offenders.
- It is an effective deterrent in preventing cold-blooded murder.

(Continued)

Audio Link 4.2
Listen to a discussion of solitary confinement.

(Continued)

- It is more economical than the permanent, lifelong warehousing of the most dangerous criminals.
- It is the ultimate specific deterrence.
- In terms of retribution, those who kill innocent persons in cold blood deserve similar punishment; just deserts.
- Enactment of the death penalty where appropriate discourages private revenge and vigilantism.

Arguments against capital punishment include:

- The death penalty is irreversible; thus the rare execution of an innocent party cannot be undone.
- No matter the reason for execution, the state becomes a murderer, a cold-blooded killer.
- The death penalty is savage and has no place in civilized society. The society becomes as inhumane as the condemned.
- It has not been proven that the death penalty is a deterrent to crime.
- The adjudication of cases involving the death penalty is far more costly and raises difficulties in finding juries willing to find defendants guilty.
- Capital punishment has always been discriminatory. In the United States, the majority of those executed have been black (Bowers, 1974), and in 1982 blacks still represented over half of the death row population.
- All countries whose values resemble those of the United States have eliminated the death penalty.
- New scientific developments using DNA evidence have cleared a number of death row inmates.

In 1999 Governor George Ryan of Illinois put all executions on hold in Illinois. "Until I can be sure that everyone sentenced to death in Illinois is truly guilty—until I can be sure with moral certainty that no innocent man or woman is facing a lethal injection—no one will meet that fate" (Irvine, 2000, p. 5A). In Los Angeles, a police scandal revealed that 99 people were framed and convicted. Relying on DNA testing, at least 63 men convicted of murder or rape have been freed in the U.S. False confessions, bungled forensics, self-serving informants, and eyewitnesses identifying the wrong person have also taken place.

The research on capital punishment has indicated that the death penalty is no more of a deterrent than life imprisonment (Bowers & Pierce, 1975; Reckless, 1967; Schuessler, 1952; Sellin, 1959). Ultimately the capital punishment debate is a moral one that is likely to continue to be heated (Schmalleger, 1990; Van den Haag & Conrad, 1983).

Research Project
Search the concept of "punishment" and report on practices and their justification in various countries. Hint: Search the term "Sharia."

Ecological Theory

While some would point to Cesare Beccaria and his writing as the beginning point of criminology, his primary interest was not so much in the analysis of crime and criminals as in the reform of criminal law and punishment. Others point to the writings of Cesare Lombroso, to be discussed shortly, and view the century between the works of the two Cesares as a criminological Dark Age. On the contrary, the writings and research of A. M. Guerry (1802–1866) of France and Adolphe Quetelet (1796–1874) of Belgium qualify them as the "fathers of modern criminology" (Gibbons, 1982, p. 17; Vold et al., 2002, p. 22). Thomas and Hepburn (1983) best reflect this writer's view:

> It is hard to understand why so many criminologists persist in their apparent conviction that scientific criminology was not to be found until Lombroso. . . . Nevertheless, the wealth of scientific analyses published by those we can classify as members of the statistical [ecological] school are commonly ignored while the often absurd and poorly executed work of Lombroso is considered to be the first true criminological analysis. (p. 138)

Another explanation for the popularity and widespread acceptance of the Lombrosians and the relative obscurity of the early ecological theorists might be the fact that the latter were not translated into English until much later (Thomas & Hepburn, 1983, p. 152).

The **ecological** school of criminological theory is also referred to as the statistical, geographic, or cartographic school. *Ecology* is that branch of biology that deals with the interrelationships between organisms and their environment. Human ecology deals with the interrelationship between human organisms and the physical environment. This school was called *statistical* because it was the first to attempt to apply official data and statistics to the problem of explaining criminality. The labels *geographical* and *cartographic* have been assigned because writers in this group tended to rely on maps and aerial data in their investigations.

Andre M. Guerry and Adolphe Quetelet

Sometime after 1825, Andre M. Guerry (1802–1866) published what many regard as the first book in "scientific criminology," *An Essay on Moral Statistics* (Guerry, 1833; Vold, 1979a, p. 167). Guerry was more cartographic in his approach, relying exclusively on shaded areas of maps in order to describe and analyze variations in French official crime statistics. Since he employed these sections of maps and used them as his principal units of analysis, he is often viewed as the founder of the ecological or cartographic school of criminology (Thomas & Hepburn, 1983, p. 139). Comparing poverty with crime, Guerry found that the wealthier areas of France had higher property crime. Urban, industrial, northern regions had more property crime than rural, southern regions (Courtright & Mutchnick, 1999). He concluded that the higher rates were due to greater opportunity. Thus, burglary and theft occurred where more goods were available. Violent and personal crimes were higher in rural areas and southern regions. These rates were consistent annually.

Guerry was also credited with being a pioneer in comparative crime statistics in comparing English and French rates. Schafer (1969, p. 76) indicated that Guerry was the first to use "moral statistics" in that he applied cartographic methods to the state of morals in terms of crime (Courtright & Mutchnick, 1999, p. 3). Another adherent of this school was Henry Mayhew (1862), who in his *London Labour and the London Poor* made extensive use of official statistics and aerial maps.

Lambert Adolphe Jacques Quetelet (1796–1874) was the first to take advantage of the criminal statistics that were beginning to become available in the 1820s (Radzinowicz & King, 1977, p. 64; Beirne, 1987). He was the first scientific criminologist, employing an approach to his subject matter that was very similar to that of modern criminologists, and is the "father" of modern sociological and psychological statistics (Mannheim, 1965, pp. 96–98; Schafer, 1969, pp. 118–20; Thomas & Hepburn, 1983, pp. 140, 145). Challenging the classical school's view that individuals exercise free will in deciding their actions, Quetelet insisted on the impact of group factors and characteristics. In his *Treatise on Man and the Development of His Faculties* ([1835] 1969), which was translated into English in 1842, Quetelet noted that there was a "remarkable consistency" with which crimes appeared annually and varied with respect to age, sex, economic conditions, and other sociological variables. This consistency in group behavior, in crime rates, and the like, speaks against crime being solely a matter of individual choice. He argues (Quetelet, 1969):

> We can count in advance how many individuals will soil their hands with the blood of their fellows, how many will be swindlers, how many prisoners, almost as we can number in advance the births and deaths that will take place. . . . Society carries within itself, in some sense, the seeds of all the crimes which are going to be committed, together with the facilities necessary for their development. (pp. 299–308)

He described this constancy of crime as the annual "budget" of crime which must be paid by society with remarkable consistency. In a sense, the stage and script are provided by society and only the faces playing the individual characters change.

In his *Research on the Propensity for Crime at Different Ages* (1984), Quetelet viewed age as the greatest predictor of crime, with crime peaking at age 25. Courtright and Mutchnick (1999, p. 4) point out that, in examining poverty, relative economic inequality was the critical variable. Crime increases when an individual "passes in an abrupt way from a state of ease to misery and to insufficiency in satisfying all the needs which he has created" (Quetelet, 1984, p. 67). Schafer (1969, p. 76) even claims that, due to

Photo 4.4

Lambert Adolphe Jacques Quetelet (1796–1874) was the first to take advantage of the criminal statistics that became available in the 1820s.

his extensive use of crime statistics and statistical predictions, Quetelet was recognized by some as the "father of statistics."

Some of Quetelet's findings included the propensity for crime among younger adults and males, and the tendency of crimes against persons to increase in summer and property crimes to predominate in winter. In what is called his "thermic law" of crime, he claimed that crimes against persons increase in equatorial climates while property crimes are most prevalent in colder climates (Fox, 1976, p. 64). Social conditions such as heterogeneity of population tended to be associated with increased crime, as did poverty although the latter not in the manner usually supposed. Noting that some of the poorest provinces of France also had very low crime rates, Quetelet anticipated the concept of "relative deprivation" by suggesting, not absolute poverty, but a gap between status and expectation as a variable in crime causation (Quetelet, 1969, pp. 82–96).

Critique of Ecological Theory

The work of Guerry and Quetelet was done nearly half a century prior to the writings of Lombroso, to be discussed shortly, who is often viewed ("the Lombrosian myth") as "the father of criminology" (Lindesmith & Levin, 1937). Lombroso's principal work *L'Uomo Delinquente* (*The Criminal Man*), first published in 1876, emphasized the notion of "born criminality." Rather than representing progress in criminological investigation, the dominance of the early positivists such as Lombroso may have set the field on a half-century (plus) journey guided by arcane and ultimately useless concepts. The superordination of the early positivists may have represented an ideological coup d'etat in which medical concepts and psychologism (a reduction of analysis solely to the individual level) temporarily retarded the early mainstream sociological efforts of the ecologists. Pointing the finger at the individual, rather than social conditions, as had Guerry and Quetelet, was intellectually acceptable to the wealthy, who preferred to view criminality as an individual failing of the dangerous classes rather than as a societal shortcoming (Lindesmith & Levin, 1937; Radzinowicz, 1966; Vold et al., 2002).

On this point, Radzinowicz (1966) states:

This way of looking at crime [the ecological school's approach] as the product of society was hardly likely to be welcome, however, at a time when a major concern was to hold down the "dangerous classes" . . . who had so miserable a share in the accumulating wealth of the industrial revolution that they might at any time break out in revolt in France. . . .

It served the interests and relieved the conscience of those at the top to look upon the dangerous classes as an independent category, detached from the prevailing social conditions . . . a race apart, morally depraved and vicious. . . . (pp. 38–39)

The social statisticians with their emphasis on social facts, statistics, the use of official data, and external social factors were perhaps ahead of their time. Shortcomings in their analysis, such as lack of full awareness of the inadequacies of official statistics and appropriate use of statistics themselves, are excusable given their pioneering efforts and the state of knowledge of the time. The ecological school represented a critical transition from the philosophical and purely theoretical approach of Beccaria to the more scientific criminological approaches of the twentieth century.

Other Geographical Theories

The ancient origin of human interest in **astrology** and the assumed role of astrological bodies on human behavior represent just one of many attempts to predict human emotion and activity on the basis of outside physical forces: the moon, the weather, climate, and the like. The word "lunatic," from the Latin word *luna,* or moon, indicates the belief that human minds can be affected by phases of the moon. This is illustrated by legends and myths such as those about *les lupins* (werewolves) in French folklore. These creatures supposedly appeared on moonlit nights (Cohen, 1979, p. 87) and were dramatically presented in fiction in the introduction to the popular 1943 Universal Pictures film *The Wolf Man:*

> Even a man who is pure in heart
>
> And says his prayers by night
>
> Can become a wolf when the wolfbane blooms
>
> And the moon is full and bright.

Cohen (1979, pp. 84–89) cites studies of mental hospital records that claim more admissions of mental patients during new and full moons, as well as a study by a suicide prevention center and one by a coroner's office, both indicating more successful attempts at suicides around the full moon period. The most frequently cited study of this type is Lieber and Sherin's (1972) research on lunar cycles and homicides. They note that synodic cycles (phases of the moon) influence physical variables such as gravitation and atmospheric pressure, which, in turn, influence human behavior. For instance, tidal periodicity is greatest during the new and full moon because of stronger gravitational influences. Assuming such forces may also affect human behavior, Lieber and Sherin analyzed homicide statistics for Dade County (Miami), Florida, and Cuyahoga County (Cleveland), Ohio, and found a statistically significant difference at full and new moon periods for the Dade County figures and a high, but not statistically significant, relationship for Cuyahoga County. Indicating that a lunar influence may exist, they explain that the differences could be due to the fact that Florida is closer to the equator and would be more influenced by the gravitational pull of the moon. Other analyses of these same correlations, however, fail to support their hypothesis (Nettler, 1982, Vol. 1, p. 31; Pokorny, 1964; Pokorny & Jachimczyk, 1974). Most such studies do not show a relationship and, although more replications are needed, criminological interest in this line of investigation has waned. Criminologists interested in geographical and ecological impacts on criminality have focused their attention instead on the social as well as artificial environment. "The Chicago school" of sociology and its contribution to U.S. criminology will be detailed in the next chapter as an illustration of such an approach.

In examining a related line of inquiry, Fox (1976, p. 64) tells us that Quetelet's "thermic law" of crime was actually borrowed from Montesquieu, who claimed that criminality increases as one nears the equator, while drunkenness increases in proximity to the poles. Examination of official statistics both internationally and within the United States, France, Great Britain, and Canada seems to generally support this hypothesis (see Brantingham & Brantingham, 1984, pp. 251–296). While statistical analysis of official crime reports such as the UCR indicates that rapes and other violent crimes are more prevalent in warmer months and that property crimes such as shoplifting are heaviest in December (holiday season), these increases are more likely due to cultural rather than climatic effects (Cheatwood, 1988; LeBeau, 1988). Brantingham and Brantingham (1984), in analyzing spatial patterns in crime, indicate:

> Different crime patterns are associated with different demographic, economic, and social profiles. Homicide and assault are associated with high proportions of minority population, with poverty and low income, with low-status jobs and low education, and with income inequality. Robbery is highest in large, dense cities that rely on public transit and have high levels of pedestrian traffic. Burglary and theft rates are highest in cities with growing populations, with growing suburbs, and with low density. (p. 296)

In a related manner one can find that combinations of hot weather and foul air (polluted with airborne toxins such as ozone) may provoke violence, particularly family disputes.

The findings which link high levels of both pollution and crime do fit into the growing knowledge that, in the long run, many chemicals can cause nerve damage and behavioral changes. For example, scientists have known for years that mercury causes brain damage: The nineteenth-century "mad hatters" stammered, twitched and trembled from inhaling mercury vapors in London hat factories. Today, many factories use masks and protective hoods to shield workers from the worst effects of chemicals (Londer, 1987, p. 6).

While increased social interaction during warmer months in part explains increased violent crime, the most pronounced effect appears during heat waves, which might suggest that heat itself promotes aggression (Gladwell, 1990). Block and Block (1988) found that the rate of violent crime in the Uniform Crime Reports showed a stronger seasonal variation than that in the National Crime Victimization Survey. This may be because such incidents are less private and more likely to come to police attention in warmer months. Cohn (1990) found that assaults, burglary, collective violence, and rape increased along with a temperature up to about 85 degrees Fahrenheit. The relationship with homicide was uncertain, while there was no relationship between temperature and robbery, larceny, or motor vehicle theft.

While recognizing that climate itself is not a major factor but rather a precipitating or mitigating circumstance in deviant behavior, Lab and Hirschel (1988a, 1988b) emphasize studying the impact of the actual weather rather than seasonal or monthly data. After examining precipitation, humidity, temperature, and barometric pressure, they indicate that "not a single text or journal article substantiates the lack of a relationship between weather and crime" (1988a, p. 282). They conclude that the potential for weather conditions playing a part in criminal activity is related to the perspective of "routine activities" (LeBeau & Langworthy, 1986), in which criminal behavior is viewed as part of normal, everyday behavior. These social-ecological impacts on criminal behavior will be discussed as well as critiqued in greater scope in the discussion of the "Chicago school" in Chapter 6.

Forerunners of Modern Criminological Thought

The three thinkers who would have a critical impact upon the shaping of social ideas, as well as criminological inquiry, in the twentieth century did not even specifically address the issue of crime. Their ideas, however, would influence criminological theorists in a profound manner. The first figure was Karl Marx (1818–1883), whose *Communist Manifesto* (1848) and *Das Kapital* (*Capital;* 1867), the former coauthored with Friedrich Engels, emphasized the economic basis of societal conflict and would give birth to the economic school of criminology. The second was Charles Darwin (1809–1882), whose *On the Origin of Species* (1859) and *Descent of Man* (1871) contained theories of evolution, natural selection, and survival of the fittest that would heavily inspire the biological positivists, to be discussed shortly. The third was Sigmund Freud (1856–1939), whose many volumes dealing with unconscious sexual motivation would influence not only psychiatry but also the psychological positivists. These themes of economics, biology, and sex underlie a large number of the criminological theories to be discussed.

Economic Theory

Karl Marx

Audio Link 4.3
Listen to a discussion of economics and crime rates.

Karl Marx (1818–1883), the inspirational figure behind most **economic** criminological theories, was an economic determinist. He insisted that the economic substructure determines the nature of all other institutions and social relationships in society. In his view, the emergence of capitalism produces economic inequality in which the **proletariat** (workers) is exploited by the **bourgeoisie** (owners or capitalist class). This exploitation creates poverty and also is at the root of other social problems.

Since Marx did not specifically address the issue of crime, Marxist criminologists draw on his economic and philosophic writings and apply them to the crime issue.

Marx viewed the history of all existing societies as one of class struggle. Influenced by the writings of the German philosopher Hegel, Marx described this conflict as a dialectical process in which theses (existing ideas or institutions) spawn their opposites, or antitheses, until a final synthesis (new idea or social order) emerges. Thus for Marx, capitalism (thesis) breeds its own destruction by giving birth to a proletarian revolution (antithesis) and finally a new world order of socialism (synthesis). Since Marx applied Hegel's theory to the material world, this is often described as Marx's theory of dialectical materialism. For Marx the resolution of social problems such as crime would be achieved through the creation of a socialist society characterized by communal ownership of the means of production and an equal distribution of the fruits of these labors.

Willem Bonger

The foremost early Marxist criminologist was the Dutch philosopher Willem Bonger (1876–1940), whose most noted work was *Criminality and Economic Conditions* (1969), which first appeared in 1910. Bonger viewed the criminal law as primarily protecting the interests of the propertied class. In contrast to precapitalistic societies, which he claimed were characterized by consensus and altruism, capitalistic societies emphasized egoism (selfishness). Capitalism was viewed as precipitating crime by creating unequal access to the necessities of life as well as by viewing success in economic competition as a sign of status (Turk, 1969b). Bonger's work provides a very detailed review of a large number of works of the time that examined the impact of economic conditions on crime, a persistent theme since early times. In referring to the early Marxist orientation, Schafer (1969) indicates:

> Napolean Colajanni, Enrico Ferri and Willem Bonger and a number of others in the last 150 years represented the same "new" trends that our radicals seem to claim as their invention. The classical authors presented these proposals in a scholarly fashion quite often superior to that of our modern radicals; in fact almost nothing is said today in this line that was not already written in criminology a century ago. (p. 76)

Greenberg (1981, p. 11) points out that a large number of early Marxist thinkers did not seriously consider the crime issue, viewing it with typical "Marxian contempt for the lumpen proletariat the beggars, pimps and criminals" in capitalist society. Many writers with a distinctive Marxian and/or economic view of criminality are cited in the Bonger work (1969), although as Greenberg correctly indicates, Bonger is often mistakenly viewed as the only early Marxist criminologist.

Some of the basic claims made by Bonger regarding criminality included (Turk, 1969b, pp. 7–12):

- Notions of what constitutes crime vary among societies and reflect existing notions of morality.
- Criminal law serves the interest of the ruling class in capitalist systems and is enforced by force rather than by consensus.
- Hedonism (pleasure seeking) is natural among people, but capitalism encourages egoism (selfish individualism) to an extreme and to the disadvantage of the society and the poor.

Photo 4.5

For Karl Marx (1818–1883), the resolution of social problems such as crime would be achieved through the creation of a socialist society characterized by communal ownership of the means of production and an equal distribution of the fruits of these labors.

- All groups are prone to crime in capitalist society, but seldom are the crimes of the wealthy punished.
- Poverty resulting from capitalism encourages crime. The unequal distribution of rewards and encouragement of egoistic material accumulation encourage crime.
- Most crimes (other than those due to mental problems) would be eliminated in a socialist system in which the goods and wealth of a society would be equally distributed.

The writings of the early Marxist criminologists were more historical, analytic-inductive, and descriptive than empirical. The early Marxist theorists had the luxury of making theoretical predictions without empirical referents at the time. Marx and Bonger predicted the hypothetical benefits of a socialist state, comparing these with the evils of early capitalism, which were a grim reality. Shortcomings of socialism could not be observed. Modern radical and Marxist criminologists no longer have this luxury, as we will show later in discussing the conflict and radical schools of criminology.

The Theory–Policy Connection

Crime policies, various programs and activities aimed at controlling crime, do not occur in a vacuum, but are guided by contemporary theory. Explanations of the cause of criminal behavior (theory) sets the context of efforts to deal with crime and criminal behavior (policy). Often, the reaction to discussions of theory is: so what? Of what use are these theories? Even those who oppose theories as impractical are themselves espousing theories in their day-to-day operations. Table 4.3 outlines this theory and policy connection.

Demonological theory provided unpredictable, cruel, and inconsistent criminal justice policy. The writings of Beccaria, Bentham, and the classical school brought about a revolution in criminal justice policy and became the basis of judicial policy in the Western world. Beccaria's insistence that the criminal is rational and operates out of free will and is responsible in deciding his or her fate is a basic assumption of Western criminal law. Law and punishment are viewed as most effective if they are sure, swift and certain, graduated according to the seriousness of the offense, and the least necessary in the given circumstance. Neoclassical theory refined and further articulated theses assumptions and was reflected in mandatory sentencing policies and attempts to develop more strict

TABLE 4.3 ■ **The Theory and Policy Connection**

Theory	Key Notions of Crime Causation	Policies
Demonological	Supernatural forces (demons) determine criminality.	Torture, exorcism, brutality.
Classical/Neoclassical	Criminals are rational actors and responsible for their actions (free will).	The keystone of our criminal justice system, which assumes individual responsibility for one's actions. Let the punishment fit the crime.
Ecological	Group characteristics, social and physical environment affect crime.	Improvement of physical and social environment will decrease crime.
Economic	Capitalism and inequality cause crime.	Reduction of inequality and poverty will reduce crime.

deterrence policy. Neoclassical theory supported policies such as target hardening, creating defensible spaces, and deterrence. Programs such as "just deserts," "three strikes and you're out," and mandatory sentences all assumed a rational actor as criminal. Ecological theories posed a challenge to classical theories and viewed the latter as simplistic and solely concerned with the individual. It asked for theories and policies that would take into account higher or lower crime rates depending upon differing social and physical environments. Similarly, early economic theories targeted differential crime rates due to poverty and inequality. Both ecological and economic theory proposed policies to fight crime by improving social and physical environments that may be criminogenic.

Summary

A *theory* is a plausible explanation of a given reality. The major theoretical approaches in criminology are the demonological, classical, ecological, economic, positivistic (biological and psychological), and sociological approaches.

The earliest theories of crime causation were demonological in nature, seeking supernatural explanations for criminality. The criminal was viewed as possessed, sinful, or evil. The classical school of criminological theory, which developed in the eighteenth century, was reflected in the writings of Beccaria, Bentham, and later Garofalo and Ferri. Seeking rational explanations, classical theorists viewed the criminal as exercising free will, as motivated by hedonism (pleasure seeking), and as carefully weighing potential pleasure versus pain to be derived from an activity. Attacking the cruel and unpredictable penal methods of the time, classical theory inspired the reform of Western criminal law. Neoclassical theory admits extenuating circumstances (insanity, age, and the like) to the equal treatment for equivalent-offense notions of the classical school.

Ecological theory (sometimes called statistical, geographic, or cartographic) is concerned with the impact of groups and social and environmental influences on criminality. The earliest writers, Guerry and Quetelet, could be regarded as the "fathers of modern criminology" in that they employed statistics and scientific analysis in the investigation of their theories. Quetelet's "thermic law" hypothesized that violent crimes predominate in warmer climates, while property crimes increase in colder zones. Since he and Guerry extensively employed maps in their analyses, they are sometimes called the *cartographic school.* The work of this school was interrupted and for a time forgotten as a result of the popularity of the Darwin-inspired biological positivism of Lombroso. Other geographic theories relating to moon cycles, climate, weather, and the like have attracted considerable interest, but research verification has been inconclusive.

Three major thinkers who have inspired much criminological theory have been Marx (*economics*), Darwin (*evolution*), and Freud (*unconscious sexual motivation*). Early economic theories, based on Marx's writings, view crime as determined by the economic system in which capitalism creates inequalities that produce crime. Socialism is viewed as the solution to the crime problem. Bonger, a Marxist criminologist, suggests that egoism (selfishness), developed as a result of capitalism, causes criminality.

The four justifications for punishment are retribution, deterrence (including incapacitation), rehabilitation, and protection and maintenance of social solidarity. Arguments in favor or against the death penalty were briefly presented.

Criminology on the Web

Log on to the Web-based student study site at http://www.sagepub.com/haganintrocrim7e/ for author-created podcasts, e-flashcards, quizzes, and more.

KEY CONCEPTS

Astrology

Bourgeoisie

Classical Theory

Demonological Theory

Ecological Theory

Economic Theory

Hedonism

Neoclassical Theory

Proletariat

Theory

Thermic Law of Crime

REVIEW QUESTIONS

1. What are some of the basic concepts of the classical school of criminological theory? What is your opinion of the pros and cons of this theory?

2. What are some primary ideas of the Marxist approach to criminology?

3. What are some basic findings of Quetelet and Guerry and their ecological approach to criminology?

4. Who were the Nacirema? Do you think that their behavior is bizarre?

5. In examining arguments for or against the death penalty, how does classical and neoclassical theory play a role in this debate?

6. What are some of the major points raised by economic theory in explaining crime?

7. What was the "rational choice theory" of crime causation? What implications does this theory have for explaining crime?

8. Are there any parallels between demonological theories of crime and current society?

9. What is neoclassical theory and of what relevance is it to modern concepts of corrections?

10. What is the current status of geographic theories in criminology?

WEB SOURCES

Classical School Criminology
http://www.crimetheory.com/Theories/Classical.html

CrimeTheory
http://www.crimetheory.com

Criminal Justice Megalinks
http://www.apsu.edu/oconnort/

Criminology Theory Links on the Web
http://www.criminology.fsu.edu/crimetheory

WEB EXERCISES

Using this chapter's recommended Web sites, surf the listed theory sites.

1. Does Cecil Greek's site or the Criminal Justice Megalinks site provide the most information on theory?

2. What kinds of detail are available on the "Classical School Criminology" site?

3. How would you rank the offerings on the following sites: Crime Theory, Criminological Theory, Criminology Theory Links on the Web, Causal Theories of Juvenile Delinquency, and CrimeTheory Links?

4. Perform an online search on "neoclassical criminological theory." Hint: you may wish to use terms or concepts from that section in the book.

Ronald L. Akers and Christine S. Sellers. (2004). *Criminological Theories: Introduction and Evaluation* (4th ed.). Los Angeles: Roxbury.

> Akers and Sellers provide a succinct overview of major criminological theories in a clear, concise manner. This book is very readable for undergraduates.

Robert M. Bohm. (1997). *A Primer on Crime and Delinquency.* Boston: Allyn & Bacon.

> This brief (140-page) overview of major theories serves the purpose of giving a succinct picture of criminological theories. An excellent introduction for undergraduates.

Francis Cullen and Robert Agnew, editors (2003). *Criminological Theory: Past to Present* (2nd ed.). Los Angeles: Roxbury.

> This selection of 38 featured articles is one of the most informative anthologies available. The authors' clear, succinct introductions to the readings are superb and simplify some extremely complex theories.

Don C. Gibbons. (1994). *Talking About Crime and Criminals: Problems and Issues in Theory Development.* Englewood Cliffs, NJ: Prentice Hall.

> The strength of this work is its historical development of major theoretical views in criminology. This provides a context from which to view each theory.

J. Robert Lilly, Francis T. Cullen, and Richard A. Ball. (1995). *Criminological Theory: Context and Consequences* (2nd ed.). Thousand Oaks, CA: Sage.

> The authors provide a useful overview of major criminological theories both of a classical and contemporary variety.

Randy R. Martin, Robert Mutchnick, and Tim Austin. (1990). *Criminological Thought: Pioneers Past and Present.* New York: Macmillan.

> The authors provide detailed chapters on 15 pioneers in criminology who they feel have shaped the discipline into what it is today. These theorists are Beccaria, Lombroso, Durkheim, Freud, Park, Sheldon, Sutherland, Reckless, Merton, Albert Cohen, Ohlin, Sykes, Goffman, Becker, and Quinney.

George Vold, Thomas Bernard, and Jeffrey Snipes. (2002). *Theoretical Criminology* (5th ed.). New York: Oxford University Press.

> This classic work by George Vold has been updated for two editions by Thomas Bernard, and now Jeffrey Snipes has been added. It does a sound, readable job of reviewing criminological theory.

Frank P. Williams, III and Marilyn D. McShane. (1999). *Criminological Theory* (3rd ed.). Upper Saddle River, NJ: Prentice Hall.

> This succinct review of theory provides an excellent overview for undergraduates of the critical contributions made by each criminological theory.

Thomas Winfree, Jr., and Howard Abadinsky. (1996). *Understanding Crime: Theory and Practice.* Chicago: Nelson-Hall.

> Winfree and Abadinsky detail major classical and contemporary theories in American criminology.

Positivist Theory
  Precursors of Positivism
Biological Theories
  Cesare Lombroso
  Charles Goring
  The Jukes and Kallikaks
  Earnest Hooton
  Body Types
  A Critique of Early Biological Theories
More Recent Biological Theories
  Brain Disorders
  Twin Studies
  Adoption Studies
  Problems With Twin/Adoption Studies
  XYY Syndrome
  Other Biological Factors
  A Critique of Neobiological Positivism
Psychological Theories
  Freudian Theory
  Psychometry
  Hans Eysenck
  Crime File 5.1 Crime Profiling
  B. F. Skinner
  Albert Bandura
  Samuel Yochelson and Stanton Samenow
  Intelligence and Crime
  Crime File 5.2 The Insanity Defense
The Theory–Policy Connection
Summary
Key Concepts
Review Questions
Web Sources
Web Exercises
Selected Readings

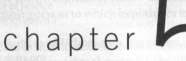

chapter 5

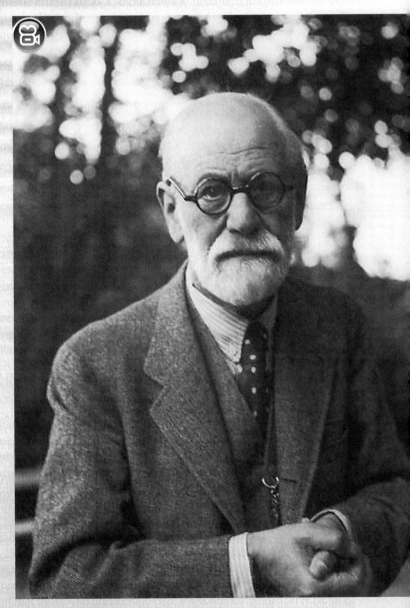

Sigmund Freud

Biological and Psychological Theories

This was not merely an idea, but a revelation. At the sight of that skull, I seemed to see all of a sudden, lighted up as a vast plain under a flaming sky, the problem of the nature of the criminal an atavistic being who reproduces in his person the ferocious instincts of primitive humanity. . . .

—*Cesare Lombroso (1911, p. xiv, in Wolfgang, 1960, p. 184)*

Positivist Theory

Positivism is a philosophical approach proposed by French sociologist Auguste Comte (1798–1857) and stated in the title of his work *A System of Positive Polity* (1877), originally published in 1851. Comte proposed the use of empirical (quantitative) or scientific investigation for the improvement of society. Taylor, Walton, and Young (1973, p. 22) indicate that the basic premises of **positivism** are: measurement (*quantification*), objectivity (*neutrality*), and causality (*determinism*). In applying Comte's approach, criminological positivists emphasize a consensus worldview, a focus on the criminal actor rather than the criminal act, a deterministic model (usually biological or psychological in nature), a strong faith in the scientific expert, and a belief in rehabilitation of "sick" offenders rather than punishment of "rational" actors. There are three elements to the positivistic approach, which stresses a scientific rather than a philosophical orientation:

1. Application of the scientific method
2. Discovery and diagnosis of pathology (sickness)
3. Treatment (therapy or corrections)

Through the systematic application of the scientific method, the positivists seek to uncover the basic cause of crime and then to prescribe appropriate treatments in order to cure the individual deviant. This section will examine precursors to positivism as well as the major types of positivism: early biological positivism, recent biological positivism, and psychological positivism. Table 5.1 outlines the major positivist theories that are presented in this chapter.

Precursors of Positivism

Prior to and competing with emergent positivism were various popular pseudosciences, some of which had existed since ancient times. **Astrology** had been used to predict human behavior by studying the alignment of the stars. The Copernican revolution and the realization that the earth did not occupy a fixed place in the universe discredited the basic premises of astrology, leaving it as a "hokum device" for fortune tellers and their superstitious clients.

Similar ideas whose time was rapidly passing were phrenology, physiognomy, and palmistry. **Phrenology** attempted to determine intelligence and personality on the basis of the size and shape of the skull and posited that certain areas of the brain corresponded to various psychological and intellectual characteristics. Writers such as Franz Gall (1758–1828) measured bumps on the head in order to identify brain development. Since sections of the brain do not completely govern specific personality characteristics and could hardly be analyzed by measuring configurations of skulls, phrenology was rapidly replaced by the more scientific methods of emergent positivism.

Physiognomy involved measuring facial and other body characteristics as indicative of human personality, while **palmistry** was concerned with "palm reading," analyzing a person's character and future by examining the lines on the palm. The fact that palmistry is not dead is illustrated by a brochure this writer received advertising a book by Paul Gabriel Tesla titled *Crime and Mental Disease in the Hand: A Proven Guide for the Identification and Pre-Identification of Criminality, Psychosis and*

TABLE 5.1 ■ Major Biological and Psychological Theories

Positivist Theories	Major Themes/Concepts	Major Theorists
Biological	Physical stigmata, atavism, biological inheritance cause criminality	Lombroso, Ferri, Garofalo
	Mental deficiency	Goring
	Feeblemindedness	Goddard
	Physical inferiority	Hooton
	Somatotypes—mesomorphs	Sheldon
	Brain disorders, twin studies, XYY syndrome, physiological disorders	Moniz, Christiansen, Jacobs
Psychological	Unconscious repression of sexual instincts	Freud, Eysenck, Skinner
	Criminal personality, extroversion, inadequate behavioral conditioning, IQ	Hirschi, Hindelang

Mental Defectiveness (1990). These theories have been discredited simply because their proponents were unable to provide any proof of accuracy in their forecasts and were rapidly overtaken by developments in modern biology and the social sciences.

Biological Theories

As previously indicated, the works of Charles Darwin, beginning in the mid-nineteenth century, had a profound impact on theory in the social sciences as well as on criminology. Concepts such as evolution, natural selection, survival of the fittest, and human genetic connections to a savage past captured the imaginations of theorists in the young social sciences, including criminology. Lombroso's concept of the appearance of criminals would be adopted by Hollywood casting directors as illustrated in Photo 5.1.

Handbook Article Link 5.1
Read an article about biological theory.

Cesare Lombroso

Cesare Lombroso (1835–1909) is sometimes called "the father of criminology." In this writer's and others' opinions (Lindesmith & Levin, 1937; Mannheim, 1965, pp. 96–98; Thomas & Hepburn, 1983), this is inaccurate, but he was certainly the most influential figure in **biological positivism.** Although he is best known for his early work, which gives an overly simplistic picture compared with his later, more sophisticated writing, Lombroso's ideas are important because of the large number of adherents and subsequent research he inspired. His most important work was *L'uomo Delinquente* (*The Criminal Man;* Lombroso, 1911), first published in 1876. Lombroso was highly influenced by Darwin's theory of evolution, and this led him to the development of his theory of **atavism** that criminals were "throwbacks" to an earlier and more primitive evolutionary period. Such born criminals could be identified by certain physical stigmata, outward appearances, particularly facial, which tended to distinguish them from noncriminals. He claimed to have made his discovery almost by serendipity during an autopsy of a criminal that he was performing in the course of his duties as a prison physician (Lombroso, 1911):

> This was not merely an idea, but a revelation. At the sight of that skull, I seemed to see all of a sudden, lighted up as a vast plain under a flaming sky, the problem of the nature of the criminal an atavistic being who reproduces in his person the ferocious instincts of primitive humanity and the inferior animals. Thus were explained anatomically the enormous jaws, high cheek-bones, prominent superciliary arches, solitary lines in the palms, extreme size of the orbits, handle-shaped or sessile ears found in criminals, savages, and apes, insensibility to pain, extremely acute sight, tattooing, excessive idleness, love of orgies, and the irresistible craving for evil for its own sake, the desire not only to extinguish life in the victim, but to mutilate the corpse, tear its flesh, and drink its blood. (p. xiv, in Wolfgang, 1960, p. 184)

Photo 5.1

Todd Hockney in the 1995 film *The Usual Suspects.*

Some examples of **physical stigmata** provided by Lombroso sound similar to characteristics Hollywood directors would search for in casting the villain on the silver screen: heavy jaw and

Photo 5.2

Cesare Lombroso (1835–1909) has been viewed by some as the "father of criminology." The author of this text feels that this is incorrect and that Quetelet and Guerry deserve this honor.

cheekbones, eye defects, large or small ears, strange nose shape, protruding lips, sloped forehead, and the like. Early biopositivism saw criminals as a separate group of people who were defective and biologically inferior. Classical prescriptions such as sure and certain punishment were seen as inapplicable with this group who were natural-born criminals (Akers & Sellers, 2004). Even though his analysis was sexist and flawed, Lombroso was also one of the first criminologists to address women and crime. In 1895, in his *The Female Offender,* he described women as inferior to men and even more likely to be atavistic. They were, however, more passive and less prone to crime. He saw women as less developed than men and criminal women as more masculine than other women.

While Lombroso's early work was well received at the time, it is not seriously regarded today. What remains, however, is his emphasis on observation, data collection, and the need to obtain positive facts to support theory. When his theories of atavism came under attack from mounting evidence to the contrary, Lombroso modified his theories, although maintaining that atavism existed in about a third of all criminals. His other categories were the insane criminal, the epileptic criminal, and the occasional criminal, hardly an exhaustive, comprehensive list of criminal types.

Lombroso's notions of biological determinism of criminality were very compatible with the ideological climate of the late nineteenth century, in which the philosophy of social Darwinism provided intellectual backing to the harsh realities of emergent industrial capitalism. **Social Darwinism** claimed that there is a "survival of the fittest" among human beings and social institutions. The success or failure of individuals competing in society was not to be interfered with, since success or failure was all part of a natural system of societal evolution. The Lombrosian model, which minimized the importance of social conditions such as inequality and ignored the extensive literature of the ecological school, blamed criminality on the individual rather than the society. This triumph of Lombrosian theory represented a "seizure of power of the medical profession who viewed criminology as a branch of medicine" (Bottomley, 1979, p. 44; Lindesmith & Levin, 1937, p. 669). Criminal behavior was viewed as the actions of "defective" individuals who were unable to adjust to an otherwise healthy society, the unfit in the struggle for survival. This social Darwinism applied Darwin's principles to human society.

The other two important figures in the Italian or continental school of positivists were Lombroso's students Enrico Ferri (1856–1929) and Raffaelo Garofalo (1852–1934). Ferri's *Criminal Sociology* (1917) was first published in 1878, and Garofalo's *Criminology* (1914) was originally published in 1884. Lombroso, Ferri, and Garofalo have been called "the holy three of criminology" by Stephen Schafer (1969, p. 123).

Enrico Ferri, Lombroso's son-in-law, proposed four types of criminals: insane, born, occasional, and criminal by passion. He proposed a multiple-factor approach to crime causation, admitting both individual and environmental factors. Often ignored in considering the diversity of Ferri's views was the fact that long before twentieth-century criminologists began to consider the shortcomings of official statistics, Ferri (1917, p. 77, cited in Vold, Bernard, & Snipes, 2002, p. 173) proposed his "law of criminal saturation." Like Parkinson, Ferri suggested that crime expands to fit the amount of control machinery assigned to it. Traveling widely as a visiting professor throughout Europe and South America and spreading the evangelism of the new positivism, he later became a supporter of Mussolini. Vold and Bernard (1979, p. 42) indicate: "The end of Ferri's career, ascent to Fascism, highlights one of the problems of positivistic theory, namely, the ease with which it fits into totalitarian patterns of government." Raffaelo Garofalo strongly advocated social Darwinism and the physical elimination of the "unfit" and their offspring, and also became a supporter of Mussolini's fascist regime.

Early positivism contributed theoretically to a scientific approach to criminology and inspired others to study the subject, but otherwise little remains of it in current criminological theory. Lombroso and his colleagues used poor sampling techniques, their findings were statistically insignificant, and they ignored the fact that physical stigmata were most likely environmental defects (due to poverty and malnutrition). Moreover, modern genetics negates their atavism theory. Their findings have been refuted by later investigators, particularly by their earliest and most vehement critic, Charles Goring.

Charles Goring

Charles Goring (1870–1919) in 1913 published *The English Convict,* the results of a study begun in 1902 of 3,000 English convicts and comparison groups of college students, hospital patients, and soldiers. He compared these "criminals" and "noncriminals" with respect to physical characteristics, personal histories, and mental qualities. The only differences he was able to discover were that the criminals in general were shorter and weighed less and, most important, were "mentally defective." While refuting Lombroso's theory of distinctive physical characteristics, he launched yet another search for hereditary mental deficiency as the cause of crime.

While Goring refuted Lombroso's notion of physical differences, his own methodology was critically flawed. Eschewing the then available Simon-Binet tests of mental ability, he used his own impressions in order to operationalize the mental ability of his subjects (Reid, 1982, p. 96). The nail in the coffin of Goring's theory was the advent of wide-scale mental testing of U.S. military conscriptees during World War I. Using Goring's definitions of **feeblemindedness,** nearly one third of the draftees would have been so classified; the standards for such tests were modified as a result. Other studies comparing mental age found no difference in performance between prisoners and the draft army, and one even found that the former performed better. As a result, the notion of feeblemindedness as a cause of criminal behavior was interred in the graveyard of outmoded criminological concepts (Vold et al., 2002, pp. 85–87). However, later we will examine modern psychometric approaches to crime and current efforts to identify and measure the "criminal personality" that represent more sophisticated revivals of this line of inquiry.

Photo 5.3

The theories of Charles Darwin (1809–1882) heavily influenced biological positivism in criminology.

The Jukes and Kallikaks

Other attempts to stress heredity as a source of criminality appeared in two case studies of generations of criminals who were claimed to be examples of degeneracy and depravity. Published only a year after Lombroso's *The Criminal Man,* Robert Dugdale's (1841–1883) *The Jukes* (1877) was a case study of generations of an American family. Tracing more than 1,000 descendants of Ada Jukes (a pseudonym), he found 280 paupers, 60 thieves, 7 murderers, 140 criminals, 40 venereal disease victims, and 50 prostitutes, as well as various other deviants—proof positive, he claimed, of inherited criminality.

A similar case study was conducted by Henry Goddard in his *Kallikak Family* (1912), which dealt with the offspring of one Martin Kallikak, a militiaman during the American Revolutionary War. Kallikak fathered a child out of wedlock to a "feebleminded barwench," a large number of whose descendants were feebleminded or deviant. The offspring of his marriage to a "respectable" woman were, on the other hand, all of the highest moral and mental standards. Goddard took these findings as proof positive of the real cause of crime feeblemindedness, or low mentality. He also was the first to use the term *moron.* The misreading of Darwin's evolutionary theory led to many perverse policies in Nazi Germany as well as in the United States. The IQ test movement, led by psychologists such as Goddard, fueled fears of high birth rates of Mediterranean immigrants to the U.S. and that the feebleminded would erode the

U.S. genetic stock. Eugenics, compulsory sterilization, and restrictive immigration policies attempted to control this perceived onslaught of feeblemindedness (Gamble & Eisert, 2004, p. 45).

Smith (1985) took a close look at Goddard's Kallikak research and reported that the photographs included in Goddard's book of the "bad" Kallikaks were retouched to make the Kallikaks appear more evil, that the methodology was unscientific, and that the historical data were simply not true (Haas, 1985). Rafter (1988) in *White Trash: The Eugenic Family Studies, 1887–1919,* as well as Gould in *The Mismeasure of Man* (1981), document Goddard's deceptions. Gould went to New Jersey and found some of the "bad" Kallikaks, who turned out not to be so bad after all. The popularity of this type of research can be explained by the fact that "it makes society's so-called superiors feel better about themselves" (Haas, 1985, p. 74). Fancher (1985) indicates, "The science of intelligence and its measurement has from the start been dominated by men who have been eager to show that the disenfranchised of society are at the bottom rung of the ladder because they are inherently inferior" (Haas, 1985, p. 74). A more detailed critique of biological positivism will follow shortly.

Earnest Hooton

Goddard attracted a major critic in the form of a neo-Lombrosian, Earnest Hooton (1887–1954), a Harvard anthropologist who in *Crime and the Man* (1939) claimed, on the basis of a very detailed and extensive study of physical differences between criminals and noncriminals, that he had discovered the cause of criminality: physical inferiority. His 12-year study of 14,000 prisoners and 3,200 college students, firemen, and others led him to conclude (1939):

> Criminals are organically inferior. Crime is the resultant of the impact of environment upon low grade human organisms. It follows that the elimination of crime can be effected only by the extirpation [eradication] of the physically, mentally, and morally unfit, or by their complete segregation in a socially aseptic environment. (p. 309)

Some physically distinguishing characteristics of Hooton's damned included: tattooing; thin beard and body hair, but thick head hair; straight hair; red-brown hair; blue-gray and mixed eye color; thin eyebrows; low and sloping foreheads; thin lips; pointed and small ears; and long, thin necks with sloping shoulders. These findings and their interpretations could be regarded with a tolerant, mild curiosity if they had appeared in Lombroso's 1876 work, but these were released in 1939 by a professor from one of America's finest universities. Positivism's compatibility with totalitarianism was again illustrated. In the same year that Hooton's work appeared, Hitler had already built experimental gas chambers in mental hospitals and in a 2-year period "extirpated" (murdered) 50,000 non-Jewish Germans, a grim prophecy of what was in store for millions of Jews, Eastern Europeans, and groups the Nazis considered to be *Untermenschen* ("subhuman").

Since many criticisms of biological positivism apply in general to all such theories, a detailed critique will be presented at the conclusion of this section.

Body Types

Imagine that, as part of your college freshman orientation, you were asked to report to the college infirmary and have nude photographs taken as part of a research project. From the 1940s through the 1960s, all freshmen at many Ivy League and other prestigious universities complied with just such a request as part of a study on body shape and intelligence. The Smithsonian Institution as late as 1995 was in possession of such photographs, which most likely included Secretary of State Hillary Rodham Clinton, ABC's Diane Sawyer, former New York Governor George Pataki, and former President George W. Bush.

Advocates of attempts to discover distinctive body types and relate them to crime include Ernst Kretschmer (1926), William Sheldon (1940), and Sheldon Glueck and Eleanor Glueck (1950). In the best known of these efforts, Sheldon (1940) proposed three **somatotypes,** body builds that relate to personality characteristics (temperaments). Endomorphs have soft, round, plump physiques and tend

Journal Article Link 5.1
Examine literature regarding biological correlates of criminality.

to be relaxed, easygoing, and extroverted; mesomorphs are hard and muscularly built, and are aggressive, assertive, extroverted, and action-seekers; and ectomorphs are thin and fragile of form as well as introverted, sensitive, and subject to worrying. Comparing judgmental samples of "problem" youths with college males, Sheldon claimed that the problem youths tended to be mesomorphic.

Similar studies by the Gluecks (1956) found delinquents to be more mesomorphic than nondelinquents and suggested that this body type may be more suited to the delinquent role, while endomorphs were too slow and ectomorphs too frail to occupy it. In similar research Cortés (1972) found 57 percent of his delinquent sample mesomorphic, while only 19 percent of nondelinquents had such body builds. McCandless, Persons, and Roberts (1972) were unable to find any relationship between body type and self-reported delinquency.

Describing much of biocriminology as a "frightening slice of historical criminology," McCaghy (1976b) points out:

> Today scarcely a year goes by without some revelation concerning the possible connection between a biological characteristic and human behavior. In fairness to the scientists of today they are generally far less sweeping in their claims than were the researchers of a few decades ago. But the probing of every nook and cranny of the human system goes on. One can only sense the public's anticipation that someday a pill or a swipe of a scalpel will put an end to thievery, homosexuality, and all sorts of behavior. (p. 11)

G. K. Chesterton (1935) felt that such early positivism gave draconian powers to control agents who could use such theories as weapons against the poor (Jenkins, 1982). Jenkins quotes Chesterton and offers a summary of his position (1935, pp. 171–177):

> Eugenics, psychology and criminology were pseudo-sciences, and were class ideologies which obviously ignored the interests of the poor. "Crime is not a disease," he wrote. "It is criminology that is a disease." The methodological foundations of the new penology were especially shoddy. Criminal anthropology meant "that very poor men, and especially poor men more or less in the hands of police, can safely have their ears pulled, their skulls measured, their teeth counted, tested or pulled out, so as to establish by scientific methods a sort of composite photograph of all criminals, which was really a composite photograph of all very poor men." He imagines what might happen if a criminologist attempted to apply the same methods to a corrupt American millionaire. The "galleries of criminal types" therefore lacked something vital; "the defect being the entire absence of any types of anti-social activity [among those] who had ever [earned] more than £ 200 a year." This class bias made criminology and sociology no more valid than astrology or alchemy.

The fact that the search goes on can be illustrated by means of a brief treatment of more recent examples of biological positivism. But first a critique of early biological positivism from Lombroso to the Gluecks is in order.

A Critique of Early Biological Theories

Here are some common problems that nearly all of the early biological theories share:

- They suggest that one can genetically inherit a trait or propensity (to violate criminal laws) that is socially defined and culturally relative.
- Any biological differences that are found are likely to explain only a minor proportion of criminal behavior compared with social and cultural factors.
- The biological theories seem to share a conservative consensus worldview, an unquestioned acceptance of official definitions of criminality, and the social class bias that crime is primarily to be found among "the dangerous class."
- Most of these studies reflect Reid's "dualistic fallacy" notion, which assumes the mutual exclusivity of criminals (defined as prisoners) and noncriminals (defined as nonprisoners).

- Most of their analyses are plagued by weak operationalization of key concepts such as "feeblemindedness," "inferiority," and "crime."
- Not all biological differences are inherited; many may be due to prenatal environment, injury, and inadequate diet (Vold et al., 2002, p. 99).
- Modern genetics has simply bypassed many of these simplistic theories. Most modern biologists speak against notions of the inheritance of acquired characteristics, emphasizing instead selective adaptation and mutation (p. 100).
- Many of these studies are based on small and/or inappropriate samples.
- As a result of the dominance of this approach, criminological theory was very likely led down the wrong path. The popularity of many of these theories related to their conservative and individualistic emphasis and to their compatibility with authoritarian and simplistic solutions to the crime problem.

Biology applied to human behavior has a disturbing legacy of misuse. "Social Darwinism" borrowed portions of evolutionary theory and twisted them into justifying class differences (Goode, 2000, p. D9). Social Darwinists argued that the struggle for power and wealth was, to use the words of sociologist Herbert Spencer, "a survival of the fittest." Associations of such theories with Nazi genocide and racist propaganda also led to disrepute.

On balance, however, it should be pointed out that the early biological theorists made some important contributions:

- The commitment of the early positivists to testing their theories by means of experiments, the collection of empirical data, and the employment of statistics are continuing features of modern criminology.
- As the following discussion of modern biological (biosocial) theories illustrates, one cannot rule out the biomedical approach simply because of this school of thought's association in the past with simplistic theories or the political abuse of such theories by fascist regimes. In *Taboos in Criminology,* Sagarin (1980) views the exploration of many of these subject areas as unfortunately representing topics that have recently been declared "untouchable," fruitless," or "mined-out."
- The early positivist approach did influence Western criminal codes and led to modifications in the classical model. Special treatment of juvenile offenders, indeterminate sentences for career criminals, extension of the insanity ruling, probation, corrections, and rehabilitation were all positivist contributions (Radzinowicz and King, 1977, p. 62).

Pierre Van den Berghe in an article titled "Bringing Beasts Back In" (1974) argues that modern criminology, in rejecting early positivism, had swung the other way and was ignoring the biological basis of human behavior. A brief examination of some of the research of modern biological theorists may provide some support for his point. Modern biological theories replace simplistic biological determinism with biological approaches that take into account the interplay of biological and socio-environmental factors (Shah & Roth, 1974). Whether criminality can be explained by human nature (genetics, inherited characteristics) or nurture (environment, learning, socialization) is a continuing debate among criminological thinkers.

More Recent Biological Theories

Audio Link 5.1

Listen to a discussion of the neurochemistry of violent behavior.

Shah and Roth (1974), in their review of criminology's **nature versus nurture controversy** (whether criminality is explained by genetics or environment), detail a variety of research including biochemical effects, brain disorders, endocrine and hormonal problems, nerve disorders, and other factors that can hardly be ignored in explaining at least a restricted number of individual cases of criminality (Marsh & Katz, 1985).

The newest biological theories (some advocates prefer the name *biosocial* theories) focus on a broad range of biological factors including genetic and environmental. Factors such as head injuries, diets, exposure to toxins such as lead, and birth complications are viewed as affecting the nervous system. "No one argues that there is a gene leading directly to crime" (Cullen & Agnew, 2003, p. 3). Most of these theories recognize that interaction exists between biological factors and the environment and suggest that some biological factors partially account for some crime in some types of offenders.

Brain Disorders

While the early phrenologists were convinced of their ability to map areas of the brain that controlled aspects of personality, modern attempts to probe the brain were begun by the Portuguese physician Antonio Moniz, who beginning in 1935 performed prefrontal lobotomies (destruction of portions of the frontal lobes of the brain) as a last resort for nonresponsive mental patients. McCaghy (1976b) reports:

> His subjects were twenty mental patients who had been unaffected by other treatments; according to Moniz fifteen showed some degree of improvement as a result of the operation. One lobotomized patient was later to pump five bullets into Dr. Moniz, but the operation and variations of it were widely hailed as the answer to many behavioral problems. (p. 28)

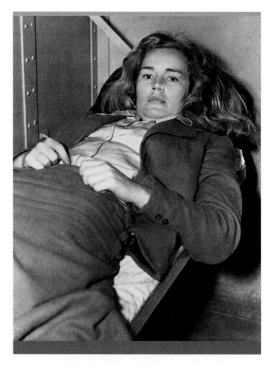

Photo 5.4

Actress Frances Farmer was forced into psychiatric treatment in the 1940s and allegedly treated with psychosurgery.

An American version used ice picks inserted through eye sockets to destroy brain tissue.

Psychosurgery, surgical alterations of brain tissue in order to alter personality or behavior, became quite popular. Roughly 50,000 such operations were performed in the United States alone from the mid-thirties to the mid-fifties (McCaghy, 1976b). Lobotomized patients were indeed more controllable with respect to behavior, but were often described as resembling hollow shells of human beings, zombies, or human vegetables, devoid of a full range of normal emotions. The impact of lobotomies was illustrated by the fate of the hero in Ken Kesey's novel *One Flew Over the Cuckoo's Nest.*

Consider an illustrative case of the misapplication of such a drastic procedure. In the forties actress Frances Farmer was forced into psychiatric treatment, allegedly for alcoholism and other related problems. Her real "problem" was radicalism, which was "treated" with psychosurgery (Jenkins, 1984, p. 180). It should be pointed out that Farmer, her family, and the hospital all denied that she had received a lobotomy.

Vernon Mark and Frank Ervin (1970) in *Violence and the Brain* proposed the use of psychosurgical procedures in order to control brain malfunctions, particularly those that may trigger aggressive behavior. Although they have been known to produce some positive results (Brown, Wienckowski, & Bivins, 1973), biomedical and surgical approaches to criminality represent a last resort, a "quick fix" that, although applicable in rare, special cases, has little to offer as a general solution to the crime problem.

By the 1950s drug therapy began to replace lobotomies. A more sophisticated form of psychosurgery called cingulotomy was used on a selective basis with consenting patients after other treatments were exhausted. The procedure involves passing an electrode needle through two small holes in the skull and searing a tiny lesion in the cingulum, a bundle of nerve fibers linking the emotional centers of the brain with the thought centers in the cortex (Beck & Cowley, 1990).

Twin Studies

In one of the more celebrated cases of researcher fraud, Sir Cyril Burt, a British psychologist, has conducted studies of twins that claimed to prove the inherited nature of intelligence. After his death, researchers discovered that he had faked his data (Wade, 1976) although others have charged that his critics had been guilty of character assassination (Fletcher, 1991). Studies of twins and adoptees are ingenious ways of attempting to address the "nature vs. nurture" debate, that is, whether criminality is inherited or learned. Such studies are ex post facto in nature. They begin with criminals who have a twin and then attempt to find the other twin in order to discover whether he or she is also criminal (Christiansen, 1968; Dalgard & Kringlen, 1975; Lange, 1931; Rosanoff, Handy, & Plesset, 1934). Such studies often compare monozygotic (MZ) with dizygotic (DZ) twins. Monozygotic (identical) twins are produced by a single egg and therefore exhibit the same hereditary environment, while dizygotic (fraternal) twins are produced by separate eggs and reveal less biological similarity.

Although findings have been mixed, Dalgard and Kringlen's (1975) study of all twins born in Norway between 1900 and 1935 concluded that the significance of hereditary factors in registered crime is nonexistent. They examined 33,000 twins in order to turn up 139 pairs in which one or the other committed crime as measured by a national crime registry. Their study and others found greater concordance (*similar patterns with respect to criminality*) among monozygotic than among dizygotic pairs. A review of studies conducted from 1929 to 1961 by Mednick and Volavka (1980) found roughly 60 percent concordance among MZ twins and about 30 percent among DZ pairs. Christiansen's (1968) study of 3,586 male twins found 52 percent MZ concordance and 22 percent among DZs.

Adoption Studies

Related in design and execution to twin studies are adoption studies. The assumption underlying such studies is that, if the behavior of children more closely matches that of their biological parents than that of their adoptive parents, this finding would support the argument for a biological base of human behavior. Schulsinger (1972), for instance, found criminality in adopted boys to be higher when biological fathers had criminal records. Hutchings, Mednick, and Sarnoff (1977) studied 1,145 male adoptees born in Copenhagen between 1927 and 1941; they found 185 adoptees with criminal records and determined that the criminality of the biological father was a major predictor of the child's behavior. Crowe (1974), however, discovered no differences between adoptees and a control group, except that the former demonstrated a higher proportion of psychopathic personalities. However, he admits problems with small samples as well as the fact that other environmental influences may have been responsible for the higher psychopathy among adoptees.

Problems With Twin/Adoption Studies

There are a number of methodological problems associated with twin and adoption studies, despite painstaking research and admirable scholarship on the part of those who have conducted them:

- Most studies involve a small number of cases, since they attempt to combine two phenomena: twins/adoptees and crime.
- Some studies are subject to unsystematic and uncontrolled samples (Dalgard and Kringlen, 1975, p. 230).
- Often the operationalization of DZ and MZ relies on official records rather than on blood-serum group samplings. The latter is a far more accurate means of distinguishing between identical and fraternal twin patterns.
- Official records are the major source of data on the dependent variable (crime commission).
- A shift of only a few cases (which may have been misdiagnosed) can erase the DZ-MZ differences.
- Higher concordance among MZs (identical twins) may still be due to more similar environmental treatment, as identical twins are likely to be treated the same.

XYY Syndrome

In the late fifties in England, speculation began regarding males who possessed an XYY chromosome pattern, an extra male chromosome. Of the 46 chromosomes most humans possess, males receive an

X chromosome from their mother and a Y chromosome from their father, while females receive two X chromosomes, one from each parent. Beginning with papers by Jacobs, Brunton, Melville, Brittain, and McClemont (1965), in which a large number of the 197 Scottish inmates they studied were found to be "double Ys," the hypothesis was proposed of a "double male" or "supermale syndrome." This theory held that the possession of an extra Y chromosome caused males to be unusually tall, to suffer severe acne as adolescents, and to be predisposed to aggressive and violent behavior. During the late 1960s, defense attorneys for brutal murderers in France and in Australia and for Richard Speck, the murderer of eight Chicago nurses, employed as part of their defense the claim that their clients were XYYs. Only in the Australian case was the accused acquitted (Sarbin & Miller, 1970).

While early research suggested that a larger proportion of XYYs could be found in prisons than among the general public (less than one double Y per 1,000 live male births), further research has found no difference (Shah & Roth, 1974, p. 137; Witkin et al., 1976). Because of the relative rarity of the syndrome, studies demonstrating a large number of cases are difficult. A Danish study by Witkin et al. (1976) did not support the aggression hypothesis and even found that incarcerated XYYs showed less aggression while in prison than did the other inmates. Earlier reviews of the research by Fox (1971) and Sarbin and Miller (1970) essentially agreed with these findings and found XYYs when institutionalized to have less serious offense records than others. While more research is required in this area, the negative findings have considerably lessened interest in the **XYY syndrome** as a cause of criminality.

Gamble and Eisert, in addressing the controversy surrounding the application of biological and psychological approaches to crime and delinquency, note that it is obvious that all human behavior has a biological base; without sensory surfaces, the environment could not affect us. "How could it be that psychology, the science of behavior and mental processes—how people learn, behave and think—is not relevant to understanding delinquency?" (Gamble & Eisert, 2004, p. 43). Sex, age, and race are three of the strongest predictors of crime, and they are all biological constructs. There is little comparison between early biopositivism, such as the work of Lombroso, and modern biological researchers who examine very complicated connections between hormones, brain chemistry, and other physiological processes that affect personality, learning, and delinquency (Pollock, 1999).

Other Biological Factors

Further theoretical and empirical work in the tradition of biological criminology continues to raise interesting hypotheses and some explanations of individual cases of criminality (see Fishbein, 1990; Walters & White, 1989a; Walsh, 2002).

In the 1970s ideas proposed by Edward Wilson (1975) in his book *Sociobiology* attracted adherents. Basically, the sociobiological perspective insists on the genetic base of human behavioral differences. Individuals are born with different potentialities; their reactions to the social environment are modified by biochemistry and the cellular reactions of the brain. Each individual's unique genetic code and nervous system react differently to the same environmental stimuli (Jeffrey, 1978, p. 162). One study suggested that birth trauma–induced brain injury among high-risk children may contribute to violent criminality (Kandel & Mednick, 1991). White et al. (1991) found that children with attention deficit hyperactivity disorder (ADHD) are at high risk for delinquency. A variety of other biological factors have been explored, primarily by means of the limited case-study approach, and require more study before definitive conclusions can be drawn.

Variables such as diet, environmental pollution, endocrine imbalance, and allergies have been claimed to have criminogenic influence. In the "Twinkie defense" we see claims that sugar consumption (too little or too much) is a causal agent in crime. Hypoglycemia (low blood sugar) also has been claimed to be linked to impaired brain function and violent crime.

Explorations of endocrine imbalance have found an obvious connection with sexual functioning, but no clear relationship with crime. Theories of relationship between male hormones (testosterone levels) and criminality have been found inconclusive, although injection of the female hormone estrogen has been found to decrease male sexual potency (Mednick & Volavka, 1980). Dalton's (1961)

Audio Link 5.2
Listen to a discussion of genetics and rape.

study "Menstruation and Crime" found that nearly half of the crimes of her sample of female inmates had occurred during menstruation or premenstruation. The PMS (premenstrual syndrome) defense emerged in the early 1980s when two women had their murder charges reduced to manslaughter on the basis that severe PMS reduced their mental capacity (Rittenhouse, 1991).

Cerebral and neuro-allergies to food substances have also been suggested as potential crimino-genic factors (Schauss, 1980; Wunderlich, 1978). In Schauss's study comparing nutritional differences between delinquents and nondelinquents, the surprising major difference was that delinquents drank more milk. Similar investigations of the effects of environmental pollution on aberrant behavior indicate that substances such as lead, mercury, and other poisonous substances can adversely affect human behavior and life itself.

Neurological studies have suggested that criminals are more likely to exhibit abnormal electroen-cephalogram (EEG) patterns (a measurement of brain waves), although studies of association of such patterns with criminality have presented mixed findings (Moyer, 1976). Denno (1985) reports:

> Considerable evidence indicates that many biological and developmental disorders associated with delinquency (for example, learning and reading disabilities) may be attributable, in part, to minor central nervous system (CNS) dysfunction which is linked, most predominantly, to complications occurring before and after birth. (p. 713)

She also points out that biological factors are more predictive of female behavior and environmental factors more predictive of male behavior. While female behavior is more subject to cultural and social constraint, male behavior is more susceptible to environmental forces (Denno, 1990). Researchers who would deny any biological influence on human behavior ignore the obvious—gender effects. Males traditionally have higher crime rates than females universally except for a few crimes such as prostitution.

It is important that criminologists keep an open mind on this matter and not view such studies as a taboo area (Sagarin, 1980) nor mistake more modern biological studies for their more primitive Lombrosian ancestors. Ellis (1982), in his review of the genetics and criminal behavior literature of the 1970s, offers prudent conclusions:

> Sensing the weight of this accumulating evidence, especially throughout the past decade, along with several other types of less direct evidence not treated in this article (e.g., the discovery of a growing number of neurological and neurochemical correlates of criminal and psychopathic behavior), many scientists have concluded since the start of the 1970s that some significant genetic factors are probably, or at least very possibly, causally involved in criminal behavior variability. . . . However, it seems important to quickly insert and underscore the point that none of these scientists in any way excluded the possibility of environmental factors also being involved. (pp. 57–58)

In fact, nearly all of them specifically entertained hypotheses about which one or more of those environmental factors might be within the very same reports in which they acknowledge possible or probable genetic influences.

In the last 20 years, new developments in genetics and molecular science have further argued that biology has some impact on the way people behave. Evolutionary psychology might involve laboratory experiments, cross-cultural studies, and other approaches. Such researchers approach the mind as an ancient engineering project, developing and testing out hypotheses about what "design problems" needed solving and what universal mental structures might have been designed, by the pressures to survive and reproduce, to solve them (Goode, 2000, p. D9).

Ellis and Walsh (1997) review five evolutionary (or gene-based) theories of criminality. These all believe that genetic factors predispose people to various criminal behaviors and natural selection has operated on humans to favor certain tendencies toward criminal and antisocial behavior.

- Evolutionary theory of rape argues that a substantial proportion of males in most populations would employ rape, especially when the prospects of being punished are low.

- Evolutionary theory of spousal and dating assault is associated with mainstay copulatory access.
- Another evolutionary theory attempts to explain child abuse and neglect.
- A **"cheater theory"** that claims that some males have been naturally selected to make lower parental investment than women by seeking numerous partners. Cheater theory views more deviant males as developing an alternative reproduction strategy to support their reproduction. "Cads" use force or deception in order to impregnate females and reproduce a greater number of offspring. Their approach to women reflects a "born to take advantage" view of sexual involvement.
- The r/K continuum theory of crime, which indicates that criminals and psychopaths are at one end of a continuum of reproduction. The K end (qualitative end) has persons who proliferate their genes by major investment in a small number of offspring, while the r end (quantitative end) proliferate offspring and neglect them (Lilly, Cullen, & Ball, 2007, p. 288). The latter propagates criminals and psychopaths (Ellis, 1987).

Ellis and Walsh (1997) indicate that the theories are too new to have been fully tested. They have, however, received much play in the media. They note that "decades of careful empirical testing will be required to assess the merit of many of these hypotheses" (p. 229). Ellis and Walsh (2000) summarize "perinatal, health, morphologic and internal biological correlates of crime that they maintain are well established or established variables correlated with crime commission" (pp. 302–303). These are low birth rate and shorter gestation, minor physical anomalies, maternal smoking during pregnancy, accidental injuries, hypoglycemia, premenstrual syndrome, body type, physical attractiveness (lack of), and skin conductivity (less perspiration under threatening conditions).

Ellis and Walsh (2000) also summarize well-established hormonal, neurotransmitter and miscellaneous biochemical, and neurological correlates of crime. These include serotonin levels (lower neurotransmitter activity), cholesterol (lower for criminals), monoamine oxidase (MAO; an enzyme that is lower among criminals), abnormal brain patterns, slow brain response to external stimuli, and greater pain tolerance.

A literature review examining the link between learning disabilities and criminality commissioned by the Law Enforcement Assistance Administration concluded that no such connection had been proven (Murray, 1976). Some studies have claimed that brain dysfunction (damage) is associated with violence, suicide, and the likelihood of being processed by criminal justice authorities (Monroe, 1978).

Biosocial criminologists argue that it is time to abandon the "naïve" nature versus nurture debate and view human behavior as an interaction between biological and environmental factors. The argument is not that these biological factors influence human behavior. They do.

The issue is whether they influence not just antisocial behavior, but criminal behavior and also to what extent they influence what types of crime. Biosocial criminology has three broad areas: behavioral genetics, evolutionary psychology, and neuroscience (Walsh & Hemmens, 2008, p. 274.)

Audio Link 5.3
Listen to a discussion of genes and behavior.

Behavioral Genetics

This area examines the degree to which genetics and environment influence various human characteristics. "There are no genes for criminal behavior, but there are genes that lead to particular traits such as low empathy, low IQ, and impulsiveness that can increase the likelihood of criminal behavior when combined with the right environments" (Walsh & Hemmens, 2008, p. 274). Behavior geneticists use twin and adoption studies to examine the influence of genes and environment and although the advocates of such studies are very confident of the role of genetics in criminal behavior, our previous analysis concluded that their findings were inconclusive. Individual genes have indirect effects on behavior by way of their effects on traits. Individual genes account for only a small effect on criminal behavior but contribute to traits linked to criminality, not directly to criminality itself. A literature review by Ellis and Walsh (2000) found that 93 percent of the 72 studies they reviewed found that genes affected criminal behavior. Other studies found that the influence of genes on antisocial behavior was

Video Link 5.1
View a video of genetics and violence.

modest. While the majority of delinquents had little genetic vulnerability, a small minority were particularly vulnerable, with both environmental and genetic risk factors. Chronic offenders with early onset of delinquent/criminal behavior were most likely to exhibit genetic effects (Moffit & Walsh, 2003). Genetic theories are viewed as possibly underpinning many more environmentally oriented, mainstream criminological theories.

Evolutionary Psychology

This approach attempts to use human evolution to explain human behavior and crime. Various traits that underlie criminal behavior are adaptations developed in human evolution. The specific criminal acts are of course not the adaptation. There are no criminal genes for burglary or car theft. Evolutionary psychologists use the term *cheating* to refer to criminal, deviant, or antisocial behavior. Cheating involves defaulting on rules of cooperation in society (Walsh & Hemmens, 2008, p. 278). Parenting and mating are important parts of evolution. Rowe (2002) tells us that the most useful traits for parenting are altruism, empathy, nurturance, and intelligence. Mating behavior, including the maximization of partners, is more typical of males, whereas parenting effort is more typical of females. Ellis and Walsh (2000), in a review of 51 studies relating number of sexual partners to criminal behavior, found 50 of them to be positive. That is, a high number of sexual partners was related to higher rates of criminal behavior. Furthermore, cultures that emphasized mating efforts exhibit low-level parental care, hypermasculinity, and transient bonding, which are viewed as antisocial traits in Western cultures.

Gender is the single most important trait in distinguishing violence in individuals. This is an international phenomenon. Bjorklund and Pellegrini (2002) argue that human females may have evolved superior inhibitory capabilities in the aggression domain. This may have arisen in the gender differences in the parental investment in offspring. While males achieved reproductive success by impregnating as many females as possible, females have far greater responsibilities in rearing offspring. This is a greater investment in reproduction. Women may have evolved greater inhibitory capacity that would affect aggressive behavior.

The Neurosciences

While our more primitive brain networks are inherited at birth, higher development depends on environmental developments after birth. Positive and negative experiences impact on brain development As Fishbein (2002, p. 111) explains: "There are many aspects of biochemistry that influence human behavior, and each one operates through effects of the central nervous system (CNS), which comprises the brain and spinal cord and the peripheral autonomic nervous system (ANS)." Neurotransmitters act as chemical messengers conveying information in the form of electrically charged signals. They regulate emotion, mood, hunger, thirst, and a number of behavioral and psychological processes. The neurotransmitter dopamine system is involved with aggression or violent behavior. Its overproduction has been known to produce antisocial behavior and violence. Low serotonin levels also affect impulsive-aggressive behavior. "Put simply, a deficit in serotonin activity jeopardizes the ability to inhibit urges, increasing the likelihood that underlying hostility or negative mood will lead to aggression or another inappropriate behavior" (Fishbein, 2002, p. 112). A number of other hormones and enzymes have been targeted as influencing behavior such as the neurotransmitter norepinephrine, the enzyme monoamine oxidase, testosterone and other male hormones called androgens, and stress hormones such as cortisol, prolactin, and ACTH. In addition, damage to the prefrontal cortex of the brain may result in antisocial behavior.

Differences in the physiological activity of the nervous system have been found to be associated with violence and drug abuse. Stimulation—seeking, impulsiveness, aggressiveness, attention deficit hyperactivity disorder (ADHD), and lack of empathy are often associated with serotonin and dopamine system abnormalities. Persons prone to violence, psychopathy, or drug abuse have low levels of central nervous system and autonomic nervous system activity (Fishbein, 2002, p. 114). Imaging studies suggest that people exhibiting inappropriate behavior may have prefrontal brain function deficits.

Such insufficient control over emotions may encourage deviant behavior. The human prefrontal cortex is involved in planning, the inhibition of behavior and the maintenance of attention. It examines the long-term consequences of behavior. Impulsive actions may occur when the prefrontal cortex is not sufficiently active. ADHD may be due to deficits in the prefrontal cortex. Serotonin, dopamine, and norepinephrine affect the brain operations that are relevant to antisocial behavior.

Childhood Mental Disorders. There is little relationship between mental disorders and crime and delinquency; however, there are three disorders for which a clear link between mental disorder and delinquency exists. These are: ADHD, oppositional defiant disorder (ODD), and conduct disorder (Gamble & Eisert, 2004, p. 57). Antisocial personality disorder could also be added to this list. ADHD involves developmentally inappropriate inattention, hyperactivity, and impulsiveness. ODD involves defiant, disobedient, and hostile behavior toward authorities, while conduct disorders involves the violation of age-appropriate societal rules. The comorbidity (having both or all three) among these disorders is high. Comorbidity of youth with ADHD and conduct disorder presents a strong prognosis for antisocial behavior.

A Critique of Neobiological Theories

While recent biological positivistic research is more sophisticated and less grandiose than the early theories discussed previously, most examples given are of limited case studies. Illustrative cases can be found that show connections with criminality, but often just as many cases can be brought forth of individuals with the same "claimed causative agent" whose behavior is normal. Much of this research is limited by small samples, is prone to the dualistic fallacy, is overreliant on incarcerated subjects, and often employs poor sampling procedures. While biological factors undoubtedly have an impact on particular individuals and their commission of certain crimes, biological explanations tend to be limited and appear to offer less exposition than social and cultural factors. Goldkamp (1987) notes that once-taboo biological explanations have enjoyed a resurgence of respectability and that "the long frustrated ghost of Lombroso must be smiling at last" (p. 129).

Neobiological research that continues attempts to draw such literature into the criminological mainstream, such as Wilson and Herrnstein's *Crime and Human Nature* (1985), has not received the same laudatory reviews in criminology and criminal justice as it has in the popular press (Austin, 1986; Gibbs, 1985).

In 1992 a National Institute of Mental Health–sponsored conference on genetics and crime scheduled to take place at the University of Maryland was cancelled when protestors threatened to disrupt the proceedings. The conference later took place at a rural conference center on Maryland's eastern shore, despite being disrupted by protesters who accused the conference of being racist and of condoning genetic research on criminal behavior (Wheeler, 1995). This concern regarding such theories is that they attempt to provide scientific justification for conservative ideological policies current at a given time. Such findings of a biological cause of violence support a political climate in which the blame for crime, poverty, and other social ills is placed solely on the individual and not on national social policies. Nelkin (1995) indicates:

> The idea of a criminal gene also implies a hope of controlling crime, not through the uncertain route of social reform, but through biological manipulation. . . . But, at its core, the debate over crime has more to do with social than biological causes; we must deal with the real sources of crime: social conditions that are so strongly associated with violence. Biological predisposition is not necessary to explain why a child who suffers racism and violence, without much hope of escape, might become indifferent to human life. (p. A17)

The ghost of Lombroso and the extremes of fascist genocide still haunt biocriminology, painting it with a tarred brush. Fear still exists that repressive policies, may emerge from the use of these policies, and vigilance is warranted. Fishbein (2001) indicates that in order for the biological perspective to be fully evaluated, four things must be accomplished: first, an estimation of the extent of biological disorders in the antisocial population. Additionally, an identification of causal mechanisms must take place, an assessment of the biological and environmental factors undertaken and determination as to whether therapy works. In speaking to this, Lilly et al. (2007, pp. 299–304) point out that although evidence exists that both biological and social disadvantage are high among offenders, differences with the general population are unclear. Do the physiological characteristics cause the antisocial behavior, or are they merely associated with it? Correlation does not prove causation. Perhaps the relationship is in reverse, in that the antisocial behavior produces the physiological characteristic. Taking a probability approach rather than a cause-effect approach may make more sense. That is, biological factors may examine vulnerabilities that may amplify the antisocial effects of certain environments. Biosocial theories have the most immediate promise in dealing with substance abuse (Lilly et al., 2007, p. 300).

Biocriminology, while presenting the aura of science, has often been criminologically naïve, often forgetting that criminal behavior varies from one society to the next. Following labeling theory, the study of biosocial factors could just as easily focus on the labelers or law constructors as the violators. Biosocial approaches may have more success in explaining alcoholism than white collar crime. Focusing on specific categories of behavior may prove more successful than attempting to describe all antisocial behavior. The fear still remains that success in areas most explainable by genetic and biological theories will support policies that ignore the role of social conditions and crime. Success of the biological model may move the field closer to a medical model than a criminal justice model. Identification of criminals as biologically different than the rest of us has the repressive potential to suggest that people with bad biology are causing all the trouble (Lilly et al., 2007, p. 304.)

It should be said that an appreciation of these biosocial theories is necessary if we are to understand all factors having an impact on crime. Explanation of crime requires a consideration of biological, psychological, and sociological factors. It is clear that biological and psychological factors play a role in some crime, especially in understanding the behavior of chronic offenders.

Psychological Theories

Video Link 5.2
View a video to explore the criminal mind.

Various psychological, psychiatric, and psychoanalytic theories of criminality have in common the search for criminal pathology in the human personality. Although the approaches overlap, psychology is the study of the individual human mind, personality, and behavior, while psychiatry is a branch of medicine that deals with the diagnosis and treatment of mental disorders. Psychoanalysis, originally based on the writings of Sigmund Freud, is an applied branch of psychological theory that employs techniques such as free mental association and dream therapy in order to diagnose and treat mental problems; the therapist assists the patient in probing the unconscious in search of sources of mental pathology. Most such theories tend conservatively to take for granted the existing social order and to scrutinize the human psyche for explanations of individual deviation. Much of this approach can be illustrated with the continual quest for "the criminal personality," measurable traits that enable the distinguishing of criminals from noncriminals. Many adherents of this approach also concentrate more on applied therapy and rehabilitation of identified criminals and less on pure theoretical explanations of crime causation (see Bartol & Bartol, 1986; Hollin, 1989).

Freudian Theory

Video Link 5.3
View a video on mentally ill men and women in jails and prisons.

While Charles Darwin was the intellectual forefather of many biological positivistic theories, other early psychological and psychoanalytic approaches were based on the writings of Sigmund Freud (1856–1939). While Freud did not address his writings specifically to the crime issue, his theories of personality as well as psychopathology have been applied to explanations of criminal behavior. He emphasized the instinctual and unconscious bases of human behavior.

Freud viewed the human personality as being made up of three parts: id, ego, and superego. The id is the instinctive, natural, or animalistic self. It is totally selfish and seeks to maximize pleasure. Expressions of this pleasure principle (or libido) are the life or love instinct (eros) as well as the death instinct (thanatos). The superego is the socialized component of the personality, the part developed in order to function and gain acceptance in human society. Repressing the pleasure-seeking instinct, the superego is in constant conflict with the id. The ego is the mediator or "referee" in this contest (Freud, 1930).

Psychoanalytic adherents of Freudian theory view much criminality as unconsciously motivated and often due to repression (hiding or sublimation into the unconscious) of personality conflicts and unresolved problems experienced in early childhood. Hostility to male authority symbols (the Oedipus complex) originates when the male child's id, desiring sexual relations with the mother, is blocked by the father. Overly harsh toilet training, premature weaning, or other unpleasant sex-related episodes contain the seeds of unconscious motivation for later adult criminality. Some hold that the inability to control instincts due to inadequate ego and superego development causes criminality (Friedlander, 1947). Crime represents a substitute response (displacement reaction); that is, when original goals are blocked, they are sublimated (displaced) and replaced by substitute goals. Crime may be committed because of the unconscious desire to be caught and punished (an expression of the thanatos complex, or death wish).

Relying extensively on case studies, Freudians document examples of the operation of the Oedipus or Electra complex, the death wish, inferiority complex, frustration-aggression, birth trauma, castration fears, and penis envy, in which crime is a substitute for forbidden acts (Vold et al., 2002). While it has had a profound impact on Western thought, Freudian theory, dealing as it does with abstract notions of the human psyche, has not lent itself well to empirical analysis. Most of his hypotheses have been neither verified nor refuted. Cullen and Agnew (2003) indicate that there are two general types of psychological theory in criminology: those that focus on traits and those that focus on learning theory. The former searches for individuals with certain traits that cause them to be more prone to crime, while the latter examines crime as a learning process.

Psychometry

Psychometry is the field that seeks to measure psychological and mental differences between criminals and noncriminals. This search for the distinctive criminal mind or personality could also be described as taking the form of a criminological "wild goose chase." While it originates in the work of Goddard, described earlier, modern and more sophisticated tests have been employed in the attempt to discern basic mental and psychological differences. In an early literature review, Schuessler and Cressey's (1953) examination over a 25-year period of such studies was unable to find conclusive evidence of specific personality characteristics related to criminality. Later reviews by Waldo and Dinitz (1967) of the literature from 1950 to 1965 confirmed Schuessler and Cressey's conclusion, as does a later survey by Tennebaum (1977).

In this tradition of mental testing, or searching for traits in criminals, the Gluecks (1950) conducted a survey of 500 delinquent boys and 500 nondelinquent boys, and found the former to be more assertive, defiant, destructive, hostile, and ambivalent toward authority. Even these differences were small and may have been an illustration of post hoc error, where differences observed after the fact (the official labeling of delinquency) are assumed to be the cause of behavior. Similarly, some research suggests that incarcerated criminals suffer greater emotional disorders than the general population, a likely reaction to confinement. Other studies attempting to link "psychopathology" and crime have also been inconclusive. Inheriting this tradition of identifying psychological traits and behavioral characteristics has been crime profiling, which is described in Crime File 5.1.

Hans Eysenck

Hans Eysenck (1977) in *Crime and Personality* merges a number of streams in social scientific thought in proposing a theory of criminality. Borrowing from psychologist B. F. Skinner (1971), as well as from the

Crime File 5.1

Crime Profiling

Crime profiling refers to attempts to construct typical characteristics of certain offenders. Also called psychological profiling or crime investigation analysis, it has been produced on various levels in the social sciences for years. Mystery writers from Agatha Christie to Arthur Conan Doyle had their detectives surmise behavior of typical offenders on the basis of past cases. Profilers of figures such as Hitler, the Boston Strangler, and the Mad Bomber also systematically applied, with various levels of success, profiles. Many became fascinated with profiling as a result of Hollywood depictions, particularly of the work of the FBI Behavioral Science unit. Retired members of this unit wrote books and novels depicting their successes. Robert Ressler's *Whoever Fights Monsters* and John Douglas's *Mindhunter, Journey Into Darkness,* and *Obsession* all piqued the interest of the public in the subject.

Profiling involves three goals according to Holmes and Holmes: a social and psychological assessment of the offender, a psychological evaluation of the suspected offender's belongings, and suggestions for the most efficient way for police to go about interviewing the subject once he or she is apprehended.

Crime profiling has been found to be particularly useful in investigating arsons, bombings, kidnappings, murders, child molestations, and serial murders and rapes. While success in many high-profile cases has added to the mystique, hype, and public fascination with crime profiling and while it is a very useful investigative tool, care must be taken in assuming greater precision and predictability than the technique is capable of at this time.

Source: Robert K. Ressler and Tom Schachtman (1992). *Whoever Fights Monsters.* New York: St. Martin's Press; Jack Douglas and Mark Olshaker (1996). *Mindhunter.* New York: Simon & Schuster; Jack Douglas and Mark Olshaker (1998). *Obsession.* New York: Simon & Schuster; Ronald Holmes and Stephen Holmes (1996). *Profiling Violent Crimes* (2nd edition). Thousand Oaks, CA: Sage.

Web Research Project
Visit the companion Web site and click on and read the article by Kocsis titled "Criminal Psychological Profiling: Validities and Abilities." Does the author see any problems with criminal profiling?

Journal Article Link 5.2
Examine literature regarding psychological profiling.

Audio Link 5.4
Listen to a discussion from an FBI profiler.

---

classical school of criminology, he views human conscience and guilt as merely conditioned reflexes, simple reactions to the apprehension of pleasure and pain. Eysenck claims the extroverted (outgoing) personality is more delinquent or criminal than the introverted (inhibited) personality. His disciple, Gordon Trasler (1962), feels that conditioned anxiety reaction (fear of punishment) inhibits individuals from crime. Extroverts, however, are less responsive to this conditioning. Viewing the labeling of deviant acts as nonproblematic, Eysenck feels that society is too permissive in its child-rearing practices and is unwilling to rationally apply the knowledge of modern psychology in the area of **behavioral modification,** which attempts to encourage positive behavior through the application of pleasure and pain (Taylor et al., 1973, p. 49). Kraska (1989) explains:

> Eysenck's theory could best be termed as a "biologically rooted conditioning theory" (Eysenck, 1980). He maintains that individuals refrain from law breaking to the extent that they are adequately socially conditioned and acquire an internalized conscience. This conditioning takes place in early childhood when one learns moral habits and develops a conscience governing his or her conduct. Thus, the undersocialization of the conscience is the key to antisocial and criminal behavior. (p. 2)

Hindelang's (1971) self-report survey of 234 high school boys supported Eysenck's theory that extroverts are more delinquent, particularly among the most normal or middle-neurotic group.

B. F. Skinner

Perhaps the most influential proponent of the branch of applied theory called behavioral psychology is B. F. Skinner (1953), who in his *Science and Human Behavior* views behavior as primarily a response to consistent conditioning or learning reinforced through expected rewards and punishments. Through behavioral modification (sometimes in laboratory settings called operant conditioning), which is widely used in juvenile corrections, unacceptable behavior can theoretically be engineered toward acceptable behavior. While an apparently effective therapeutic strategy, Skinner's approach is a pure behaviorist approach: That is, it says "behavior causes behavior"; it is less concerned with addressing the issue of the underlying origin of crime, criminal law, or conditions in the social order that act as prior conditions to the transmission of behavior. As an applied theory or therapeutic strategy, it is attractive, despite its shortcomings as pure theory.

Albert Bandura

Albert Bandura's (1973) social learning theory looks at the thought processes of the person and external sanctions. By observing others, individuals "learn" how to engage in aggression. This might include exposure to such models, aversive treatment by others, or positively anticipating participating in such actions. The reinforcement or punishment of such actions is important (Cullen & Agnew, 2003).

Samuel Yochelson and Stanton Samenow

More recent advocates of the existence of a distinctive *criminal personality* are psychiatrist Samuel Yochelson and clinical psychologist Stanton Samenow (1976), who, on the basis of their 14-year therapeutic work with 240 hard-core criminal and delinquent subjects at St. Elizabeth's Hospital for the criminally insane in Washington, D.C., claim to have challenged prevailing sociological and economic theories of crime causation. In a revival of early biological and psychological positivism, they argue that socioenvironmental constraints on individual criminality are irrelevant, that there is a "criminal personality," and that such individuals freely choose to become criminal (1976, p. 199). Feeling that their criminal patients were conning them by using current theories in the social sciences to rationalize their criminality, they claim that criminals were victimizers of society rather than its victims (Vold & Bernard, 1986, p. 253).

Proposing a therapeutic treatment technique rather than a theory of crime causation, they make some of the following points:

- The criminal personality is imprinted at birth and is relatively unaffected by the family.
- Criminal personalities seek the excitement of crime.
- They are exploitative and selfish in interpersonal relationships.
- They are amoral, untrustworthy, intolerant of others, and manipulative, lack empathy, and are in a pervasive state of anger.
- They lack trust and refuse to be dependent.
- In all, Yochelson and Samenow claim to have discovered 52 criminal thinking patterns (Yochelson & Samenow, 1976).

Yochelson and Samenow propose a treatment program similar to programs like Alcoholics Anonymous (AA), Synanon, the Delancey Street Foundation (Fox, 1985, p. 255), and the program proposed in Glasser's (1965) *Reality Therapy,* in which criminals must confront their antisocial thoughts. AA calls this a rejection of "stinkin' thinkin'," in which the subjects abandon past excuses and rationalizations. Criminals are expected to totally reject their former criminal personalities and assume personal responsibility for their wrongdoing.

Despite its promise as a behavioral therapy, Yochelson and Samenow's theory has a naïve "old-wine-in-new-bottles" flavor about it; it seems a revival of the biological "grunts and bumps" theories of

Audio Link 5.5
Listen to a discussion of predicting violence.

the past. They cite little convincing empirical evidence of success for their treatment and fail to refute evidence regarding environmental and social influences. Among their methodological problems, their operationalization of basic terms is unclear. In response to Yochelson and Samenow's claim that they have refuted criminological theory of environmental influences on crime, Vold et al. (2002) retort:

> It does not appear, however, that the study demonstrates that point. It is certainly possible that providing a criminal with insight into the root causes of his behavior does not change that behavior. That is very different than saying crime does not have root causes. (p. 159)

Rather than restricting its focus to specific types of offenders, the search for "the criminal personality" of which Yochelson and Samenow's theory is the most recent example is too globally ambitious in trying to explain all types of criminals. A related example of the attempt to identify personality traits associated with criminality appears in the writings of Glen Walters and his "lifestyle theory" (Walters & White, 1989b). Criminal lifestyles represent a choice by individuals reflecting three elements: conditions, choice, and cognition. These choices reflect limits placed by biological and environmental conditions. Criminal behavior is a general pattern of life expressed in characteristics such as impulsiveness, irresponsibility, and a continuing willingness to violate societal rules. Conditions such as impulsiveness and low IQ interact with environmental situations such as attachment to significant others. Cognitive styles as a result of conditions and choice include many thinking errors, disregard for one's victims, a sense of entitlement, present-orientation, and discontinuity in thinking patterns. Such patterns produce behavioral styles that conflict with the law such as rule breaking, intruding into the lives of others, self-indulgence, and irresponsibility (Walsh & Hemmens, 2008, pp. 242–243.) Walter's theory very much contains the same problems as Yochelson and Samenow detailed above.

Intelligence and Crime

The first intelligence tests were developed in France at the beginning of the twentieth century by Alfred Binet. He assembled a series of tasks involving basic reasoning skills and assigned an age level to each in terms of its difficulty. The age associated with the most difficult task became the "mental age," and general intellectual ability was calculated by subtracting the chronological age. If the mental age was behind the chronological age, the child was identified for special education programs. German psychologist W. Stein revised the method by dividing mental age by chronological age and multiplying the score by 100, thus creating the intelligence quotient (IQ). The test gained wide popularity particularly with Henry Goddard (Curran & Renzetti, 2001, p. 69). Hirschi and Hindelang (1977) charge that, because of the discrediting of much of the early work on intelligence and crime by Goddard and Goring and others, the field of criminology has ignored the strong evidence of a link between IQ and crime. On the basis of an extensive literature review, they argue that the textbooks have been wrong on this subject and that

- IQ is more important in predicting official delinquency among white boys than is social class;
- IQ is a better predictor of delinquency than is the father's social class, especially among black boys;
- all other things being equal, the lower the IQ, the higher the recidivism; and
- there is a roughly nine-point deficit in the IQs of delinquents compared to nondelinquents.

Unable to find contrary conclusions in current research, Hirschi and Hindelang (1977) conclude that IQ is at least as good a predictor of delinquency as race and social class. Wolfgang, Figlio, and Sellin's (1978) study found arrested juveniles in their Philadelphia cohort to have lower IQs, but also found that race was a more important predictor, while the contribution of IQ to criminality, independent of race and class, was also indicated in a literature review of such studies by Herrnstein (1983). Research by Gordon (1987) similarly claims to have demonstrated that black–white differences in juvenile delinquency rates were best predicted by IQ rather than by socioeconomic variables. IQ, similar to race, is a slippery concept. In the 1960s, psychologist Arthur Jensen claimed that African

Americans were inferior in intellect due to inherited genes. Prompted by this finding, psychologist James Flynn reviewed decades of IQ data and discovered his "Flynn effect," that IQ tests had increased .3 point every year over the last 30 years and as much as 25 points in some countries. One third of the gap between black and white Americans had been erased in 30 years (Holloway, 1999).

Richard Herrnstein and Charles Murray in *The Bell Curve* (1994) once again resurrected this theme of IQ. They see blacks as performing more poorly on IQ tests than whites, thus demonstrating less intelligence. They view this poor performance as predominantly genetically caused and contend that it is relatively unaffected by social programs or public policy. This school of psychometry has a long tradition of discovering that Jews are not very smart, that Mediterranean people are genetically inferior to Nordic ones, and that the average mental age of white military enlistees in World War I was 13 (Holt, 1994, p. A15). Vold and Bernard (1986, p. 82) cite the rather seamy history of the IQ controversy, indicating that blaming low IQ for delinquency has a long tradition. In the 1820s the high delinquency of the Irish was attributed to their inferior racial stock (Finestone, 1976), and at the turn of the century early IQ tests were utilized to show the inferiority of Southern European and Eastern European immigrants. The IQs of Italian American children, with a median of 84, were 16 points below the national norm (Pinter, 1923), about the same as those of black children today.

Modern advocates of a relationship between IQ and delinquency and crime do not, as did earlier writers, insist that intelligence potential is entirely inherited, viewing it as an acquired as well as inherited entity. Given criticisms of cultural bias in intelligence testing, they insist that, although no test is culture-free, one obtains similar results from a variety of measures. IQ remains a critical variable in explaining traditional crime and delinquency and may even shed light on white collar crimes, whose perpetrators are likely to have higher IQs. Little research, however, has taken place with respect to the latter.

One of the better explanations for the IQ–delinquency relationship examines school failure due to lower IQ, which breeds frustration and humiliation. Alienation from school and its prosocial influences increases delinquency. Lynam et al. found that 75 percent of the relationship between IQ and delinquency was mediated by school failure (Lynam, Moffitt, & Stouthamer-Loeber, 1993). Gamble European and Eisert (2004) argue that schools should be made as rewarding for lower IQ students as it is for higher IQ ones. This contrasts sharply with the current emphasis on school performance and testing, which will probably alienate low IQ students even further from school settings.

A full and detailed account of modern psychological and psychiatric approaches to crime exceeds the intentions of this volume. Much of this literature, which has often been given short shrift by criminological theorists, is important, but it also has been of an applied theoretical or clinical nature, proposing treatment rather than postulating causes (see Kutchins, 1988). Schafer (1976, chapters 8 and 9) provides excellent coverage of such work by Erik Erikson (1950) on identity crisis, family therapy, reality therapy (Glasser, 1965), gestalt therapy (Perls, 1970), and other important therapeutic approaches. Similarly, Jacks and Cox (1984) provide an excellent anthology of psychocriminology. On this point Fox (1976) states:

> Crime is so complex that a single theory or small constellation of theories is difficult to operationalize and evaluate through controlled research. There is disagreement between the research and the clinical viewpoints, most graphically demonstrated by the demand for solid research by (pure research) sociologists and experimental psychologists, on the one hand, and the more pragmatic clinical viewpoint (applied research) held by psychiatrists, clinical psychologists, and social workers on the other. . . . These disparate viewpoints will probably never become congruent. It is apparent that both are needed. (p. 416)

Some more recent promising psychological research on crime examines the role of ADHD and delinquency. Other research points to the presence of nonsecure parental attachment in childhood and psychological risk for negative development outcomes such as aggression and delinquency. Childhood oppositional defiant disorder and conduct disorders have also been found to contain predictive power. Kenneth Dodge has examined the role of "hostile attributional bias" (attributing hostile intent to others) as related to aggression (Dodge, 1991).

Crime File 5.2 discusses the insanity defense.

Crime File 5.2

The Insanity Defense

Handbook Article Link 5.2
Read an article on mental illness and crime.

Video Link 5.4
View a video on the crime of insanity.

In order for an individual to be held guilty or responsible for violating the criminal law, he or she must exercise *mens rea,* or proper criminal intent. Exceptions to this rule are cases of negligence or strict liability, such as the felony murder doctrine. Anglo-American common law is based on the classical theory of criminology that assumes that individuals are rational actors and thus will respond in kind to threats of punishment. Individuals are to be held responsible for their conduct, but what if the individual is insane?

Basic Decisions. The **M'Naghten rule** is named after an 1843 English decision regarding Daniel M'Naghten, an individual who, suffering from severe delusions of persecution (paranoia), attempted to shoot Sir Robert Peel (the famous founder of the London police, who were nicknamed "Bobbies" in his honor). M'Naghten missed and killed Peel's secretary. This decision held that individuals who are insane, unable to distinguish between right and wrong, cannot be held responsible for their actions. The M'Naghten rule became the basis of psychiatric justice in Anglo-American criminal law, such individuals being held "not guilty due to insanity." It is sometimes referred to as the "right-wrong test" or the **NGRI defense** (not guilty by reason of insanity). Individuals who are successfully defended under this rule are usually institutionalized for long-term psychiatric treatment. In 1897 the federal courts as well as many states added a modification called the "irresistible impulse" test to the "right-wrong" test. Accused persons could not be found guilty if they had a mental disease that prevented them from controlling their conduct (Morris, 1987, p. 1).

The Durham decision was a 1954 decision in the U.S. Court of Appeals, District of Columbia, regarding Monte Durham, a burglar who had been found guilty in a lower court when psychiatrists indicated that there was no clear evidence that Durham could not tell right from wrong, even though he had a long history of psychotic and deranged behavior. The appeals court overturned the previous decision and ruled that individuals are not guilty due to insanity if their acts are the *product* of mental disease or defect. This rule was valid in the federal system from 1954 to 1972.

The Brawner Test. In 1972 the federal system rejected Durham and adopted a standard suggested by the Model Penal Code. This is used by half of the states, although it is subject to increasing restriction in the wake of the Hinckley trial. John Hinckley, the attempted assassin of former President Reagan, had been tried under what is called the Brawner test (*United States v. Brawner*), which holds:

> A person is not responsible for criminal conduct if at the time of such conduct as a result of mental disease or defect he lacks *substantial capacity* [emphasis mine] either to appreciate the wrongfulness of his conduct or to conform his conduct to the requirements of the law. (Morris, 1987, p. 2)

Although the Brawner test dominated federal and state practice until the Hinckley trial, after the adverse reaction to this trial, a new standard resembling the original M'Naghten rule was adopted, which placed the burden of proof of insanity on the defendant. Table 1 summarizes these insanity defense standards.

TABLE 5.2 ■ Insanity Defense Standards

Test	Legal Standard Because of Mental Illness	Final Burden of Proof	Who Bears Burden of Proof
M'Naghten	Didn't know what he or she was doing or didn't know it was wrong	Varies balance of probability	Defense
Irresistible Impulse	Could not control his or her conduct	Beyond reasonable doubt	Prosecutor
Durham	The criminal act was caused by his or her mental illness	Beyond reasonable doubt	Prosecutor
Brawner Test	Lacks substantial capacity to appreciate the wrongfulness of his or her conduct or to control it	Beyond reasonable doubt	Prosecutor
Present Federal Law	Lacks capacity to appreciate the wrongfulness of his or her conduct	Clear and convincing evidence	Defense

Source: Norval Morris (1987), "Insanity Defense," *Crime File,* National Institute of Justice, p. 3.

Guilty but Mentally Ill. By the early eighties a number of states (Morris, 1987, p. 2) had abolished the NGRI defense and substituted the "guilty, but mentally ill" rule in which convicted individuals undergo civil commitment until cured and then serve out their time in jail. Other states such as Oregon have established psychiatric review boards both to determine release and to monitor follow-up counseling. Over half of the states have joined the federal government in tightening insanity defense standards.

The Twinkie Defense. "You are what you eat!" "Be careful not to eat too many Twinkies, or you might become a killer!" In San Francisco a riot resulted when under a "diminished responsibility" argument a jury found Dan White not guilty of first degree murder in the 1978 killing of that city's mayor, William Mosconi, and Harvey Milk (a gay city supervisor). Despite the fact that White gave a full confession, psychiatrists convinced the jury that his overindulgence in junk food (Twinkies and Cokes) diminished his ability to premeditate (NBC, 1983).

Growing steroid (synthetic growth hormone) abuse by athletes and bodybuilders in order to "bulk up" has led to the increasing documentation of adverse side effects including "bodybuilder's psychosis," which involves bizarre and violent behavior. It has led some to speculate on yet another tool for defense attorneys—"the dumbbell defense" (Monmaney & Robins, 1988, p. 75).

In 1988 a New York jury ruled that Reuben Pratts, a Vietnam veteran, was not guilty of a murder he confessed to because he suffered from chronic posttraumatic stress disorder (PTSD) resulting from traumatic experiences in Vietnam. Pratt had experienced flashbacks to his wartime experiences (French, 1989; Palmer, 1990). In 1990 lawyers first began using the "Prozac defense" to argue that their clients were not legally responsible for their crimes. Prozac, an antidepressant drug, was blamed for, among other things, a mass murder-suicide in Louisville, Kentucky. The first trial resulted in reduced charges because the jury believed that the

Photo 5.5

Dan White was found not guilty of first degree murder in the killings of San Francisco Mayor William Mosconi and City Supervisor Harvey Milk after psychiatrists convinced the jury that his overindulgence in junk food diminished his ability to premeditate.

accused was suffering from side effects of Prozac. So far, the success in other cases has been mixed ("Eli Lilly," 1990; Marcus, 1991). Similar defenses arguing that drugs such as Halcion, a widely used sleep remedy, were responsible for violence and murder have been unsuccessful (Cowley, 1992).

A related defense strategy has been called the "abuse excuse" (Dershowitz, 1994). In the celebrated case of Lorena Bobbit, a physically abused wife who cut off her husband's penis, the defendant was found not guilty by reason of insanity because of her history of abuse. "'Roid rage" (due to steroids), "black rage" (due to racism), "fetal trimethadione syndrome" (due to a mother's use of a drug during pregnancy), "adopted child syndrome," and "abused child syndrome" (used in the Menendez trial) have all become part of the "I am a victim of _____" excuse (Slade, 1994, p. B12). In a claimed "cultural insanity" defense, a black bank robber standing trial claimed that he was the victim of long-term exposure to white racism that drove him insane. He claimed he was a victim of "posttraumatic stress disorder" due to unwarranted exposure, victimization, and repetitive confrontation with white racism. The first use of this defense was by two black men on trial for the beating of white truck driver Reginald Denny during the 1992 Los Angeles riots. The men claimed that pent-up rage due to racism caused them to attack Denny (Forsthoffer, 1999).

Despite the bizarre nature of these and similar cases and the media attention they arouse, the reader should be aware that such cases are rarities and attract interest for that very reason. Morris (1987, p. 1) points out that another reason for the rare use of the insanity defense is that a person found not guilty by reason of insanity may be held in a mental hospital longer than

(Continued)

(Continued)

he or she would be imprisoned if convicted. The insanity defense is raised in only 1 percent of all felony cases, and in only about a quarter of these is it successful. For additional information on the insanity defense, the reader is referred to Hermann, 1983; Moran, 1985; Morris, 1987; Simon and Aaronson, 1988; and Toch and Adams, 1991.

The Psychopath

Psychopath, sociopath, and **antisocial personality** are all terms referring to the same phenomenon—the inadequately socialized personality. While at one time such persons were viewed as having innate psychological defects, the concepts of psychopath, sociopath, and antisocial personality imply that such personalities are learned through socialization. Harrington (1972) describes the roots of psychopathy in the following manner:

> Persons diagnosed as psychopathic begin as rejected, cruelly or indifferently treated children, or may possibly have suffered early brain damage, detected or not. They strike back at the world with aggressive, unrestrained, attention-drawing behavior. (Why one person emerges from a disordered childhood inhibited and neurotic and another, the psychopath, with the opposite tendencies remains unclear.) Since conscience is instilled by early love, faith in the adults close by, and desire to hold their affection by being good, the child unrewarded with love grows up experiencing no conscience. Uncared for, he doesn't care, can't really love, feels no anxiety to speak of (having experienced little or no love to lose), does not worry about whether he's good or bad, and literally has no idea of guilt. (p. 15)

A psychopath never really develops the full range of human emotions. Some general characteristics linked with the phenomenon include the lack of inhibition, guilt, fear, conscience, and superego. Such individuals' lack of empathy is illustrated by mass killers such as Charles Manson, who was described as viewing other people as furniture or objects in the world around him.

Hervey Cleckley (1976), in *The Mask of Sanity,* identifies the following traits as characteristic of psychopathy: unreliability, insincerity, superficial charm, inability to learn from mistakes, impersonal sexual behavior, and an incapacity to love. The background of Charles Manson is instructive. Manson was born to a 16-year-old prostitute who did not know the identity of the father. She was sent to prison when he was 4, and he spent the next 4 years with relatives who gave little love or affection. His mother finally returned and took up her old ways. At the age of either 9 or 12, he was sent to a reform school; by age 32 he had spent nearly his entire life in correctional institutions in which he had been exposed to a considerable amount of violence. He had become totally institutionalized to prison. When he was finally eligible for release on parole, he pleaded with officials to permit him to remain in prison (Scheflin & Opton, 1978, pp. 28–29). Manson attracted a small, devoted following, but imagine a sociopath operating in a larger arena.

The following description of Adolf Hitler by Albert Speer, his armaments minister, illustrates both the shallowness and the inexplicable charm of the psychopath:

> Hitler could fascinate, he wallowed in his own charisma, but he could not respond to friendship. Instinctively, he repelled it. The normal sympathies that normal males and females enjoy were just not in him. At the core, in the place where the heart should be, Hitler was a hollow man. He was empty . . . the man's drive—his iron will, his demonism—fascinated even while it repelled . . . I was enthralled. (Harrington, 1972, p. 32)

In October 1999, Luis Garavito confessed to killing 140 children over a 5-year period in Colombia. Donald Black in *Bad Boys, Bad Men: Confronting Antisocial Personality Disorder* (1999) describes Columbine killers Eric Harris and Dylan Klebold as "antisocial personalities," cold and calculating with no regard for the consequences. Others who have been described as fitting the mold are Ted Bundy, who murdered 50 women, and Andrew Cunanan, who killed five including designer Gianni Versace (Barovick, 1999).

The actual definition and diagnosis of psychopathy are elusive; there is considerable disagreement and confusion within the psychiatric profession itself regarding the concept. Many critics view it as a "wastebasket concept," a catch-all, a diagnosis of convenience or of last resort. If some inexplicably horrible crime defies our sensibilities, the person who commits it is labeled a psychopath.

Web Research Project
Search under the term "insanity defense" and find a recent case in which this defense has been introduced.

It is important that, in explaining the forest (environmental and sociological influences, to be discussed in the next chapter), we do not ignore the individual trees (biological and psychological differences among individuals). Monahan and Splane (1980) indicate: "What the field of criminology needs, it appears to us, are sociologists who use psychological intervening variables without embarrassment and psychologists who are aware of the social roots of the individual processes they study" (p. 42). Psychological theories are very important in explaining microcriminology. Why do some individuals respond to the same environment differently than do others? Nicole Rafter in *The Criminal Brain: Understanding Biological Theories of Crime* (2008) reviews the new biocriminology and argues that the new writers view biology and environment as interdependent (Beaver, 2009; Walsh, 2009; Walsh & Beaver, 2008). Mainstream criminologists are still not eager to embrace the new biocriminology (Monaghan, 2009).

Tom Gamble and Amy Eisert (2004, pp. 64–65) note that biopsychological explanations of criminal behavior are, unfortunately, often viewed as politically regressive compared to social explanations. If such approaches do lead to regressive policies, then they are being used inappropriately. Biopsychological approaches can have positive policy consequences. Because a disorder has biological roots does not mean that it is more difficult to alter the behavior than those having social roots. Nearsightedness and mental disorders are two examples. That many mental disorders are not the fault of the individual or his or her family reduces the stigma attached to the illness. The fact that some addictions and sexual preferences may be biologically based rather than "lifestyle" choices may prove the way to greater acceptance or toleration. Biopsychological approaches will never replace social etiology. Conclude Gamble and Eisert (2004):

> Looking into the next century, the question that remains is not whether biological or psychological or sociological approaches will be most successful in understanding crime and delinquency, but rather how successful will we be in describing how biological, psychological and sociological factors interact to produce delinquent behavior. (pp. 64–65)

The Theory–Policy Connection

The most obvious policy outcome of biological and psychological theories in criminology was the shift in focus from "Why is there crime in society?" to "Why is this individual a criminal?" The criminal as patient and crime as a disease became predominant and rehabilitation the emphasis. Table 5.3 outlines some elements of this theory–policy connection.

The policy implications of these theories, which were heavily influenced by the writings of Charles Darwin, were conservative (Lilly et al., p. 28). Positivism, with its emphasis on scientific investigation, also helped sponsor a reform orientation. Rehabilitation and reformation replaced punishment as the primary motive of corrections. In fact, the very terminology changed from penology to corrections. Its emphasis on therapy and treatment and scientific evaluation were positive developments. **Psychological positivism** placed a greater emphasis on counseling and improving the lot of potential criminals. The paradigm was shifted from punishing the criminal to rehabilitating him or her. The creation of a separate juvenile court system recognized the redemptive possibilities of criminal justice policies. An increase in discretion and individualized treatments reflected this rehabilitation orientation. Biopsychological explanations of criminal behavior are often viewed as politically and socially repressive. If the use of such findings leads social policy away from improving environmental conditions, then the approaches are being used for repressive ends (Gamble & Eisert, 2004, p. 63).

> Although the notion of biological determinism might diminish notions of free will and criminal responsibility, it also could be used to suggest that criminals are incurably wicked and beyond the powers of therapeutic interventions to save. Were this mode of thinking to prevail, it could have disquieting implications, given that the correctional system is disproportionately populated by the nation's poor and minority citizens.

TABLE 5.3 ■ **The Theory–Policy Connection**

Theory	Basic Assumptions About Crime Causation	Policies
Early Positivism	Social Darwinism Born criminals Feeblemindedness Psychosurgery Eugenics	Emphasis on treatment and rehabilitation Psychosurgery Sterilization
Recent Biological	Crime as a medical problem Genetic predisposition Biological determinism Physical inferiority	Emphasis on treatment and rehabilitation Restricted immigration
Recent Psychological	Crime as a medical problem Psychological determinism Criminal personality IQ	Emphasis on treatment and rehabilitation Juvenile court Indeterminate sentencing Probation Counseling

Seeing the disadvantaged as biologically deficient risks acquitting "the rest of us" of any responsibility for social inequality and the role it plays in fostering criminogenic conditions (Lilly et al., p. 304). Biopsychological explanations can also have positive consequences. Many mental disorders with strong biologically based causation can be effectively treated. Biopsychological explanations will never replace social etiology. The question becomes, how do psychological, biological, and sociological factors interact to produce crime and delinquency?

Summary

Positivistic theory was based on three elements: (1) use of the scientific method in order to (2) diagnose individual pathology and thus enable (3) the prescribing of treatment. The criminal is viewed as sick. Precursors to positivism included astrology, phrenology, physiognomy, and palmistry, none of which criminologists take very seriously today.

Biological positivism proposed the notion of "the born criminal." Lombroso viewed criminals as atavistic beings (savage "throwbacks" to earlier human ancestors); he proposed the identification of physical stigmata as a means of identifying such persons. *Social Darwinism* is a philosophy that posits a "survival of the fittest in society" among human groups and their institutions. Ferri and Garofalo extended and modified the biopositivist tradition, which was well accepted by conservative and totalitarian political structures, since the blame for crime rested on the individual and not society. The text presents a critique of early positivism. Goring's research, which was highly critical of Lombrosian theory, proposing instead "inherited mental deficiency" (*feeblemindedness)* as the explanation of crime, has been discredited by more sophisticated mental tests. Other biopositivist theories include: case studies of the Jukes and the Kallikaks, Hooton's notion of "physical and mental inferiority" of criminals, and Sheldon's "somatotypes" (body types). The text also presents a more detailed critique of early biological positivism.

More recent biological positivism is more sophisticated in addressing the "nature vs. nurture" argument; it generally views criminality as produced by a combination of genetics and environment. Such research includes variables such as brain disorders, biochemical effects, endocrine and hormonal abnormalities, and nerve disorders. Twin and adoption studies have produced mixed findings, but suggest that monozygotic (identical) twins are more similar (concordant) in their criminal behavior than are dizygotic (fraternal) twins. A critique of such studies includes the point that many twins experience similar environmental influences. The XYY syndrome (supermale phenomenon) has been largely discredited. Other areas of inquiry in this tradition include sociobiology (which insists on the genetic base of human behavior). A further critique is presented. Three broad areas of biosocial criminological theory are presented: behavioral genetics, evolutionary psychology, and neuroscience. Cheater theory holds that males have a greater interest in mating and little interest in child rearing and that they use illegitimate means to maximize their offspring.

Psychological positivism reflects psychological, psychiatric, and psychoanalytic theory; much of the early work was based upon Freud's writings. The latter includes a tripartite personality system consisting of the id (instinctual self), ego (mediator), and superego (socialized self). According to Freudian theory as applied to criminology, the basis of deviance can be found in repressed sexual motivations deeply hidden in the individual's subconscious.

The discussion of psychiatry and the law details the M'Naghten rule (NGRI; not guilty by reason of insanity, or the "right-wrong test") and the Durham decision (innocent because of mental defect), as well as the "irresistible impulse test" (mental disease at the time of the act) and the "substantial capacity test" (lack of capacity to appreciate criminality of the act). Many states, in reaction to cases such as Hinckley, are passing guilty but mentally ill laws and abolishing NGRI. The "Twinkie defense" (poor nutrition obviates guilt) was briefly discussed as yet another bizarre defense. The concept of psychopath-sociopath-antisocial personality is used as a catch-all constructed to describe individuals who exhibit a variety of characteristics including the lack of empathy, guilt, fear, conscience, and superego. Actual diagnosis of psychopathy has been unreliable and of questionable validity.

Psychometry (mental testing) attempts to discover personality characteristics of criminals. Various literature reviews of these efforts find the evidence inconclusive. Research by the Gluecks suggests that differences do exist between delinquents and nondelinquents, as does Eysenck's research, the latter attributing crime to extroverted personalities who lack adequate societal conditioning (training).

Skinner's theory of behavioral modification (modeling behavior by means of rewards and punishments) has had a major impact on clinical programs in corrections. The continual search for a distinctive "criminal personality" is illustrated in the work of Yochelson and Samenow, who on the basis of their clinical studies identify specific traits and propose a therapeutic technique similar to that of Alcoholics Anonymous in which criminals are challenged to take personal responsibility for their actions and to reject rationalizations. Although perhaps clinically useful, as a theory of crime causation the theory leaves much to be desired.

Hirschi and Hindelang revive the IQ controversy by insisting that, on the basis of their literature review, criminologists have been too unappreciative of the role of IQ in crime and delinquency.

The text provides a critique of psychological positivism, along with a rejoinder that criminologists, while concerned with sociological forces in crime causation, cannot afford to ignore individual factors.

◉ **Criminology on the Web**

Log on to the Web-based student study site at http://www.sagepub.com/haganintrocrim7e/ for author-created podcasts, e-flashcards, quizzes, and more.

KEY CONCEPTS

Atavism

Behavioral Modification

Biological Positivism

Cheater Theory

Durham Decision

Feeblemindedness

IQ and Crime

"Law of Criminal Saturation"

Lobotomy

M'Naghten Rule

Monozygotic Concordance

"Nature-Nurture" Controversy

NGRI Defense

Palmistry

Phrenology

Physical Stigmata

Physiognomy

Positivism

Psychological Positivism

Psychometry

Psychopath/Sociopath/Antisocial
  Personality

Social Darwinism

Somatotypes

XYY Syndrome

REVIEW QUESTIONS

1. You have been exposed to a brief historical account of early theories in criminology. What do you consider to be the beginnings of criminology and who is/are the parent(s) of criminology: Beccaria, Quetelet and Guerry, or Lombroso?

2. Some feel that modern sociobiology is a far cry from the primitive early biological positivism. Do you agree or disagree? Defend your answer.

3. What theories do you feel were influenced by the forerunners of modern criminological thought?

4. What are the basic elements of the positivist approach? How does this differ from classical theory?

5. Positivism, particularly early biological positivism, has been described as a "frightening slice of American criminology." G. K. Chesterton felt so strongly that he described early criminology as a pseudoscience. Explain why there has been such a strong reaction to biological positivism by such critics. What do you think about these criticisms?

6. What is the "nature vs. nurture" controversy and how are twin and adoption studies designed to address this issue? What has been resolved by research?

7. What is the claimed relationship between intelligence and crime according to Hirschi and Hindelang? What are some criticisms of this claimed relationship?

8. What are some characteristics of a psychopath?

9. How does the M'Naghten decision differ from the Durham decision?

10. How has the "guilty but mentally ill" standard affected the insanity defense?

WEB SOURCES

Australian Institute of Criminology
http://www.aic.gov.au

Centre for Crime and Justice Studies (UK)
http://www.crimeandjustice.org.uk/

Crime Times
http://www.crimetimes.org

Crime Theory
http://www.crimetheory.com

Criminal Justice Links
http://www.criminology.fsu.edu/cj.html

Criminal Justice Megalinks
http://www.apsu.edu/oconnort/

Criminal Psychology
http://www.geocities.com/CapitolHill/Lobby/6027

Diagnostic and Statistical Manual
http://www.apa.org/science/lib.html

Western Society of Criminology
http://www.sonoma.edu/cja/wsc/wscmain.html

WEB EXERCISES

Using this chapter's recommended Web sites, surf the listed sites.

1. Call up the Western Society of Criminology site that features online journal articles and see if you can find one that fits this chapter. If not, what kinds of theory-related articles did you find?

2. What is and what does the Diagnostic and Statistical Manual site feature?

3. What does the Criminal Psychology site have to say about "antisocial personality"?

4. What types of information are available on the Centre for Crime and Justice Studies (UK) and Australian Institute of Criminology sites?

5. Using your Web browser, search the terms "criminal personality" and "biology and crime."

SELECTED READINGS

Curt H. Bartol and Anne M. Bartol. (1986). *Criminal Behavior: A Psychosocial Approach* (2nd ed.). Englewood Cliffs, NJ: Prentice Hall.

This is one of only a few criminology theory books devoted exclusively to psychological approaches.

Lee Ellis and Anthony Walsh. (2000). *Criminology: A Global Perspective.* Boston: Allyn & Bacon.

This is one of the only introductory texts to feature a biological positivistic perspective.

Dian H. Fishbein, editor. (2000). *The Science, Treatment, and Prevention of Antisocial Behaviors.* Kingston, NJ: Civic Research Institute.

This volume features recent articles on biological and psychological theories of crime.

Thomas J. Gamble and Amy C. Eisert. (2004). Delinquency Theory: Emerging Explanations from the Biopsychological Perspective, in *Controversies in Juvenile Justice and Delinquency* (pp. 43–68), edited by Peter J. Benekos and Alida Merlo. Cincinnati: LexisNexis.

This article contains an excellent presentation of the latest psychological theories of crime and delinquency.

Stephen J. Gould. (1981). *The Mismeasure of Man.* New York: Norton.

Gould launches a major criticism of eugenic studies such as those of the Jukes and Kallikaks and exposes their methodology and findings as fraudulent.

Frank E. Hagan and Pamela Tontodonato. (2004). Delinquency Theory: Classical and Sociological Explanations, in *Controversies in Juvenile Justice and Delinquency* (pp. 21–39), edited by Peter J. Benekos and Alida V. Merlo. Cincinnati: LexisNexis.

This article gives general coverage of classical and sociological theories of crime causation featuring early as well as contemporary selections.

Joseph Jacoby, editor. (2004). *Classics of Criminology* (3rd ed.). Chicago: Waveland Press.

This is an excellent compendium of 57 classic and contemporary articles on criminology and criminological theory.

B. F. Skinner (1953). *Science and Human Behavior.* New York: Macmillan.

This is Skinner's classic work on behavioral conditioning.

George Vold, Thomas Bernard, and Jeffrey Snipes. (2002). *Theoretical Criminology* (5th ed.). New York: Oxford University Press.

This work provides broad coverage of the variety of criminological theories.

Anthony Walsh and Lee Ellis, editors. (2003). *Biosocial Criminology: Challenging Environmentalism's Supremacy.* New York: Nova Science Publishers.

The authors present the latest literature on biological criminology, which they feel has been given short treatment in criminological theory.

Major Sociological Theoretical Approaches in Criminology

Anomie Theories

   Émile Durkheim and Anomie

   Merton's Theory of Anomie

   Robert Agnew's General Strain Theory (GST)

   Subcultural Theories

   Cohen's Lower-Class Reaction Theory

   Cloward and Ohlin's Differential Opportunity Theory

Social Process Theories

   The Chicago School

   Shaw and McKay's Social Disorganization Theory

   Sutherland's Theory of Differential Association

   Crime File 6.1 Designing Out Crime

   Miller's Focal Concerns Theory

   Matza's Theory of Delinquency and Drift

Social Control Theories

   Reckless's Containment Theory

   Hirschi's Social Bond Theory

   Gottfredson and Hirschi's General Theory of Crime

   John Hagan's Power-Control Theory

Developmental and Life Course (DLC) Theories

   Farrington's Antisocial Potential (AP) Theory

   Sampson and Laub's Life Course Criminality

The Theory–Policy Connection

Summary

Key Concepts

Review Questions

Web Sources

Web Exercises

Selected Readings

chapter 6

Émile Durkheim

Sociological Mainstream Theories

Positive criminology accounts for too much delinquency. Taken at their terms, delinquency [crime] theories seem to predict far more delinquency than actually occurs. If delinquents were in fact radically different from the rest of conventional youth. . . . Then involvement in delinquency would be more permanent and less transient, more pervasive and less intermittent than is apparently the case. Theories of delinquency yield an embarrassment of riches, which seemingly go unmatched in the real world.

—*David Matza,* Delinquency and Drift *(1964)*

The early classical, biological, and psychological traditions in criminology theory were similar in their relatively conservative view of society (the consensus model) as well as in their search for the cause of crime in either lack of fear of deterrence, defective individual genetics, or the psyche. The individual criminal was the unit of analysis. The only departures from this deviant behavior approach to criminality were found in the writings of the economic theorists (Marx and Bonger) and the ecologists (Quetelet and Guerry). Economic and ecological theories constitute the groundwork for the preeminence of sociological approaches to criminological theory beginning in the 1930s in the United States. Societal conditions, groups, social disorganization, and conflict have become additional units of analysis. Crime is perceived as a status (definition) as well as behavior (pathology), and sociological criminology in general takes a more critical stance toward the society itself as generator of criminal conduct.

Major Sociological Theoretical Approaches in Criminology

Table 6.1 is a more detailed outline of the sociological theories that were briefly presented in Table 4.1, Major Theoretical Approaches in Criminology. These include mainstream sociological theories: anomie, social process, social control, and developmental and life course theories.

Discussion will begin with the mainstream tradition and the views of late-nineteenth-century sociologist Émile Durkheim and the "anomie theories" that he inspired. Other representatives of this approach are Robert Merton, Richard Cloward and Lloyd Ohlin, and Albert Cohen.

TABLE 6.1 ■ **Major Theoretical Approaches in Mainstream Criminology (Sociological)***

Theoretical School	Major Themes/Concepts	Major Theories
Sociological Mainstream	**Crime reflects consensus mode**	
Anomie Theory	Anomie (normlessness) lessens social control	Durkheim
	Anomie (gap between goals and means) creates deviance	Merton
	Differential social opportunity	Cloward and Ohlin
	Lower-class reaction to middle-class values	Cohen
Social Process	Social disorganization and social conditions	Shaw and McKay
	Routine activities	Cohen and Felson
	Crime is learned behavior, culturally/subculturally transmitted	Sutherland
	Local concerns of lower class	Miller
	Subterranean values, drift techniques of neutralization	Matza
Social Control	Containment theory	Reckless
	Social bonds weakened, reducing individual stakes in conformity	Hirschi
	Low self-control and self-interest	Gottfredson and Hirschi
Developmental/ Life Course	Antisocial potential	Farrington
	Longitudinal studies	Blumstein
	Life course criminality	Sampson and Laub

*See Table 7.1 for other theoretical approaches in criminology.

Anomie Theories

Émile Durkheim and Anomie

Handbook Article Link 6.1
Read an article on strain theories.

The writings of French sociologist Émile Durkheim (1858–1917) were in sharp contrast to the social Darwinist, individualist, and psychological and biological positivist theories dominant in the late nineteenth century. The works of Durkheim represented a return to the thinking and orientation of the statistical/ecological theories advocated by Quetelet and Guerry, an approach that had been preempted by the popularity of Lombroso and the early biological positivists.

In his works, which included *The Division of Labor in Society* (1964), originally published in 1893, and *Suicide* (1951), first released in 1897, Durkheim insisted on the primacy of groups and social organizations as explanatory factors of human misconduct. As we said in Chapter 1, he viewed crime as a normal phenomenon in society because group reactions to deviant actions assist human groups in defining their moral boundaries. In his doctoral dissertation, *The Rules of Sociological Method* (1950), which was completed in 1893, Durkheim defined the sociologist's role as that of systematic observers of "social facts," empirically observable group characteristics that affect human behavior. Durkheim's analysis of suicide clearly demonstrated his hypothesis of group influences on individual propensity to suicide. In *Suicide* (1951), he identified several types, which included altruistic ("selfless" suicide), egoistic (self-centered suicide), and anomic (suicide due to "anomie" or a state of normlessness in society). The latter concept is Durkheim's principal contribution to the field of criminology.

The term **anomie** appeared in the English language as early as 1591 and generally referred to a disregard for law (Fox, 1976, p. 115). Anomie, from the Greek *anomia* ("without norms"), as used by Durkheim involves a moral malaise; a lack of clear-cut norms with which to guide human conduct (normlessness). It may occur as a pervasive condition in society because of a failure of individuals to internalize the norms of society, an inability to adjust to changing norms, or even conflict within the norms themselves.

Social trends in modern urban-industrial societies result in changing norms, confusion, and lessened social control over the individual. Individualism increases and new lifestyles emerge, perhaps yielding even greater freedom but also increasing the possibility for deviant behavior. The close ties of the individual to the family, village, and tradition (what Durkheim calls "mechanical solidarity"), though confining to the individual, maintained social control. In modern societies (characterized by "organic solidarity") constraints on the individual weaken. In a theme that would influence many later criminological theories, Durkheim viewed anomie in modern societies as produced by individual aspirations and ambitions and the search for new pleasures and sensations that are beyond achievement even in times of prosperity (Durkheim, 1951, p. 256).

This notion of anomie would influence a number of criminological theories, constituting a theoretical school of thought within mainstream or conventional criminology that began with the work of Robert Merton in the late thirties and continued with Richard Cloward and Lloyd Ohlin and Albert Cohen in the post–World War II period. Chronologically preceding these later developments in the anomie tradition were the work of "the Chicago school" of sociology and another major approach, the social process school of thought. These theories were less concerned with the origin of crime in society and concentrated instead on the social process (learning, socialization, subcultural transmission) by which criminal values were transmitted to individuals by groups with which they were affiliated.

Merton's Theory of Anomie

As part of the jointly sponsored American Society of Criminology and Academy of Criminal Justice Sciences' Criminology and Criminal Justice Oral History Project, Robert Merton described how he developed some of his theories (American Society of Criminology, 2004). He was interested in examining "what is it about our society and cultural institutions, not just individual characteristics such as feeblemindedness, that causes deviance?" There is a dysfunction between the American dream (a cultural value) of success and social structure (means of achieving). Class and ethnic structures provided differential access. Merton wanted to look at deviance in addition to conformity in society and explain differential rates. Functionalism had concentrated on positive functions of things, and so he wanted to explore their dysfunctions.

Robert Merton's theory of anomie first appeared in 1938 in an article titled "Social Structure and Anomie." Modifying Durkheim's original concept, Merton (1957, pp. 131–194) viewed anomie as a condition that occurs when discrepancies exist between societal goals and the means available for their achievement. This discrepancy or strain between aspirations and achievement has resulted in Merton's conception being referred to as strain theory. According to this theory, U.S. society is firm

in judging people's social worth on the basis of their apparent material success and in preaching that success is available to all who work hard and take advantage of available opportunities. In reality the opportunities or means of achieving success ("the American dream") are not available to all. Merton (1938) states:

> It is only when a system of cultural values extols, virtually above all else, certain common symbols of success for the population at large while its social structure rigorously restricts or completely eliminates access to approved modes of acquiring these symbols for a considerable part of the same population, that antisocial behavior ensues on a considerable scale. (p. 78)

Thus, according to Merton's theory of anomie, antisocial behavior (crime) is produced by the very values of the society itself in encouraging high material aspirations as a sign of individual success without adequately providing approved means for all to reach these goals. This discrepancy between goals and means (strain) produces various **"modes of personality adaptation,"** different combinations of behavior in accepting or rejecting the means and goals. Given this high premium placed on individual success without concomitant provision of adequate means for its achievement, individuals may seek alternate (nonapproved) means of accomplishing this goal. American fiction, the Horatio Alger stories of "rags to riches," the media, and literature constantly pound home the theme of success. Social Darwinism (the theme that the capable or fit will succeed) and the "Protestant work ethic" (the attachment of religious value to work) have been persistent philosophies. These values are generally accepted by persons of all social classes.

One of the essential premises of this approach is that organization and disorganization in society are not mutually exclusive, but rather many of the cultural values that have desirable consequences ("manifest functions") often contain within them or produce undesirable consequences ("latent functions"; Merton, 1961).

Modes of Personality Adaptation

Merton describes five possible modes of personality adaptation that represent types of adjustments to societal means and goals: the conformist, the innovator, the ritualist, the retreatist, and the rebel. All except the conformist are deviant responses. The *conformist* accepts the goal of success in society and also the societally approved means of achieving this status, such as through hard work, education, deferred gratification, and the like. Acceptance of the goals does not indicate that all actually achieve such satisfactory ends, but that they have faith in the system.

The *innovator* accepts the goal of success, but rejects or seeks illegitimate alternatives to the means of achieving these aims. Criminal activities such as theft and organized crime could serve as examples, although societally encouraged activities such as inventing could also provide illustrations. An interesting example is the case of Fred Demara, Jr., well known through the book *The Great Imposter* (Crichton, 1959). A high school graduate, Demara was disappointed that people had to spend much of their lives preparing usually for only one occupation. Forging credentials and identities, he launched into careers as a college professor, Trappist monk, penitentiary warden, and surgeon in the Canadian Navy, to mention just a few.

Photo 6.2

J. K. Rowling, author of the *Harry Potter* series, has had a "rags to riches" experience. She went from living on welfare to becoming a multimillionaire within 5 years.

The *ritualist* is illustrated by the "mindless bureaucrat" who becomes so caught up in rules and means to an end that he or she tends to forget or fails to place proper significance on the goal. This individual will compulsively persist in going through the motions with little hope of successful achievement of goals.

The *retreatist* represents a rejection of both societally approved means and ends. This adaptation might be illustrated by the advice of Timothy Leary, the prophet of psychedelic drugs in the sixties, who preached, "tune in, turn on, and drop out." Chronic alcoholics and drug addicts may eventually reject societal standards of jobs and success and choose the goal of "getting high" by means of begging, borrowing, or stealing.

The *rebel* rejects both means and goals and seeks to substitute alternative ones that would represent new societal goals as well as new methods of achieving them, such as through revolutionary activities aimed at introducing change in the existing order outside normal, societally approved channels.

A Critique of Merton's Theory

Merton's theory, well received in sociology and in criminology, became the basis of a number of subcultural theories of delinquency, to be discussed shortly. Criticisms of the theory include:

- His assumption of uniform commitment to materialistic goals ignores the pluralistic and heterogeneous nature of U.S. cultural values.
- The theory appears to dwell on lower-class criminality, thus failing to consider law breaking among the elite. Taylor, Walton, and Young (1973, p. 107) express this point: "Anomie theory stands accused of predicting too little bourgeois criminality and too much proletarian criminality."
- The theory is primarily oriented toward explaining monetary or materialistically oriented crime and does not address violent criminal activity.
- If Merton is correct, why does the United States now have lower property crime rates than many other developed countries?

While many writers (Hirschi, 1969; Johnson, 1979; Kornhauser, 1978) have concluded that Merton's theory does not hold up empirically, more recent research by Farnworth and Lieber (1989) argues in favor of its durability. They indicate that strain (anomie) theory combines psychological and structural

Photo 6.3

Fred Demara, Jr., known as "the Great Imposter," masqueraded as a civil engineer, a doctor of applied psychology, a lawyer, and a sheriff's deputy, in addition to many other positions.

explanations for crime and thus avoids purely individualistic explanations, and that the research of the critics failed to examine the gap or strain between economic goals and educational means. Farnworth and Lieber found this a significant educational predictor of delinquency in their sample of juveniles, and concluded that the theory is "a viable and promising theory of delinquency and crime" (1989, p. 273).

Classic strain theory (as it is sometimes called given the strain or discrepancy between goals and means) has had additional conflicting support. Research did not find higher delinquency among those with the greatest gap between aspirations and expectations. Those with low aspirations and low expectations had the highest offense rates. Other studies, however, have shown support (Agnew, Cullen, Burton, Evans, & Dunaway, 1996; Cullen & Agnew, 2003, p. 119).

There have been a variety of efforts to revise strain theory. One revision involves using the concept of relative deprivation, one's felt sense of deprivation relative to others, such as a reference group. Another alteration is to view adolescents as pursuing a number of goals besides those involving money and status. These might include popularity with peers and romantic partners, good grades, athletic prowess, and even positive relationships with parents (Agnew et al., 1996; Cullen & Agnew, 2003).

Robert Agnew's General Strain Theory (GST)

A persistent writer in the strain tradition has been Robert Agnew (1992, 1995, 1997). He views strain as due to negative relationships in which individuals feel that they are being mistreated. These negative relationships may take a variety of forms: others preventing the achievement of goals such as monetary success; activities that threaten to remove valued relationships such as the loss or death of a significant other; and the threat of negatively valued stimuli such as insults or physical assault. For some, such activities increase the likelihood for anger and frustration, as well as the likelihood that crime becomes a means of resolving these emotions. Agnew and White (1992) claim that delinquency was higher among those experiencing negative life events, for example, parental divorce or financial problems. It was also higher for those with interactional problems with teachers, parents, and others. Why some react to the strain by committing crime and others do not needs to be specified.

Agnew (1992, 1994) has revised traditional anomie (strain) theory by going beyond Merton's presumed economic strain and identifying other sources of strain. His general strain theory (GST) views strain as a more general phenomenon than the discrepancy between aspirations and expectations. Strain can also take place when others take something of value from us or when one is confronted with negative circumstances. A psychological state of negative affect is critical and includes disappointment, frustration, and anger. Delinquency becomes a means to regain what one has lost or been prevented from obtaining (instrumental), or retaliatory (a means of striking back), or escapist (a means of getting away from anger and strain). Thus, Agnew identifies three major types of strain. In addition, other types of strain may:

- Prevent one from achieving positively valued goals.
- Remove or threaten to remove positively valued stimuli that one possesses.
- Present or threaten to present one with noxious or negatively valued stimuli.

In a test of GST, Paternoster and Mazerolle (1999) obtained mixed findings. Negative relationships with adults, dissatisfaction with friends, and school life and stress were related to delinquency. Such strain may, however, be managed by other strategies such as drug use, compensatory success in school, athletics, or after-school jobs. Strain does weaken conventional social bonds and strengthen unconventional bonds.

An extension of Merton's theory has been offered by Steven Messner and Richard Rosenfeld in their *Crime and the American Dream* (1994) and their institutional anomie theory. The hunger for wealth is viewed as insatiable, and all social institutions become subservient to the economic structure. Culturally induced pressure to accumulate material rewards combined with weak controls by noneconomic institutions produces an institutionalization of anomie (Chamlin & Cochran, 1995) and an institutionalization of the use of deviant means for success. The unimpeded pursuit of monetary success is the "American dream." Economic institutions predominate, subordinating all other institutions such as the family, church, or school, reducing their power particularly in the socialization of children.

Subcultural Theories

Merton's modification of Durkheim's notion of anomie began the "anomie tradition" in U.S. criminology, with further influential theoretical work by writers such as Richard Cloward and Lloyd Ohlin and Albert Cohen that directed itself toward **subcultural theories** of delinquency.

Merton's theory had a major impact on many of the more sociologically oriented theories of crime and delinquency. A major area of theoretical focus from the thirties through the sixties in U.S. criminology related to juvenile gangs, as studies of citations in recent criminology textbooks (Schichor, 1982) and frequently cited books and journal articles (Wolfgang, 1980) show.

Cohen's Lower-Class Reaction Theory

Photo 6.4

Many anomie theories of crime directed themselves to explaining gang behavior.

Albert Cohen was an undergraduate student of Robert Merton and later a graduate teaching assistant for Edwin Sutherland at Indiana University. In the Oral History Project tapes (American Society of Criminology, 2004), he explains that, despite having been a 1939 Phi Beta Kappa graduate of Harvard, he had been turned down for aid in graduate schools because he was Jewish. He even received a letter from a department chair of a state university saying it was not their policy to hire Jews. Fortunately, Sutherland had offered him a teaching assistantship at Indiana.

Albert Cohen's (1955) *Delinquent Boys* presents a theory about lower-class subcultural delinquency. According to his theory, delinquency is a **lower-class reaction** to middle-class values. Lower-class youths use delinquent subcultures as a means of reacting against a middle class–dominated value system in a society that unintentionally discriminates against them because of their lower-class lifestyles and values. Unable to live up to or accept middle-class values and

judgments, they seek self-esteem by rejecting these values. Cohen (1955, p. 25) carefully qualifies his remarks by indicating that this theory is not intended to describe all juvenile crime.

He views much lower-class delinquency as nonutilitarian, malicious, and negativistic. Much theft, for instance, is nonutilitarian, performed for status purposes within the gang rather than out of need. Maliciousness is expressed in a general disdain for middle-class values or objects and a negative reaction to such values. The delinquent gang substitutes its own values and sources of self-esteem for the middle-class values it rejects. Some examples of middle-class values include: ambition, individual responsibility, verbal skills, academic achievement, deferred gratification (postponement of rewards), middle-class manners, nonviolence, wholesome recreation, and the like. The gang subculture, as depicted in Photos 6.4 and 6.5, offers a means of protection and of striking back against values and behavioral expectations the lower-class youth is unable to fulfill.

A Critique of Cohen's Theory

Major criticisms of Cohen's theory relate to:

- His overconcentration on lower (working)-class delinquency.
- His assumption that lower-class boys are interested in middle-class values (Kitsuse & Dietrick, 1970).
- Cohen, like other subcultural theorists, fails to address ethnic, family, and other sources of stress as well as the recreational ("fun") aspects of gang membership (Bordua, 1962).
- By emphasizing the nonutilitarian nature of many delinquent activities, Cohen tends to underplay the rational, for-profit nature of some juvenile criminal activities.

Cohen's theory fits into the anomie tradition in that he views lower-class delinquency and gang membership as a result of strain or a reaction to unfulfilled aspirations. A related subcultural theory by Walter Miller disagrees with this strain hypothesis and argues instead in the social process tradition of Shaw, McKay, and Sutherland that lower-class delinquency represents a process of learning and expressing values of one's membership group. Miller's theory will be discussed in detail shortly.

Cloward and Ohlin's Differential Opportunity Theory

Audio Link 6.1
Listen to a discussion of Lyndon Johnson's war on poverty.

An extension of the works of both Merton and Sutherland (to be discussed) appeared in Richard Cloward and Lloyd Ohlin's (1960) *Delinquency and Opportunity. A Theory of Delinquent Gangs.* According to their theory of **differential opportunity,** working-class juveniles will choose one or another type of subcultural (gang) adjustment to their anomic situation depending on the availability of illegitimate opportunity structures in their neighborhood. Borrowing from Merton's theme, Cloward and Ohlin view the pressure for joining delinquent subcultures as originating from discrepancies between culturally induced aspirations among lower-class youths and available means of achieving them through legitimate channels. In addition to legitimate channels, Cloward and Ohlin stress the importance of available illegitimate opportunities, which may also be limited, depending on the neighborhood. Neighborhoods with highly organized rackets provide upward mobility in the illegal opportunity structure. Individuals occupy positions in both legitimate and illegitimate opportunity structures, both of which may be limited. Illegitimate opportunities are dependent on locally available criminal traditions.

Delinquent Subcultures

Cloward and Ohlin identified three types of illegitimate juvenile subcultures: criminal, conflict, and retreatist. The criminal subculture occurs in stable slum neighborhoods in which a hierarchy of available criminal opportunities exists. Such a means of adaptation substitutes theft, extortion, and property offenses as the means of achieving success. Disorganized slums (ones undergoing invasion-succession or turnover of ethnic groups) are characterized by a conflict subculture. Such groups, denied both legitimate and illegitimate sources of access to status, resort to violence, "defense of turf," "bopping," and/or "the rumble," as a means of gaining a "bad rep" or prestige. The retreatist subculture is viewed by Cloward and Ohlin as made up of "double failures." Unable to succeed either in the legitimate or illegitimate opportunity structures, such individuals reject both the legitimate means and ends and simply drop out; lacking criminal opportunity, they seek status through "kicks" and "highs" of drug abuse. These subcultures become the individual's reference group and primary source of self-esteem. According to this theory, delinquent gang members do not generally reject the societal goal of success, but, lacking proper means to achieve it, seek other opportunities.

A Critique of Differential Opportunity Theory

Cloward and Ohlin's theory, building as it had on other respected theories, was well received in the field of criminology. Criticisms of the theory have generally involved:

- This theory focused exclusively on delinquent gangs and youths from lower- and working-class backgrounds, ignoring, for instance, middle-class delinquent subcultures.
- It is doubtful that delinquent subcultures fall into only the three categories they identified. In fact, much shifting of membership and activities among members appears common (Bordua, 1961; Schrag, 1962).
- The orientation and specialization of delinquent gangs, even if the analysis is restricted to the United States, appear far more complex and varied than their theory accounts for.

Despite criticism, Cloward and Ohlin's ideas were very influential in the field and comprised a broader theory than that of Albert Cohen (1955). Where Cloward and Ohlin viewed delinquency as an anomic reaction to goals, means discrepancy, and the particular form of adaptation dependent on available illegitimate opportunities, Cohen perceived delinquency as a reaction of lower-class youth to unobtainable middle-class values.

The implications of Cloward and Ohlin's theory were not lost on policymakers. By improving legitimate opportunities, delinquency could be controlled. Then-Attorney General Robert Kennedy read *Delinquency and Opportunity* and was impressed and asked Lloyd Ohlin to assist in drafting legislation that resulted in the passage of the Juvenile Delinquency Prevention and Control Act of 1961. Such community action programs later became the basis of President Lyndon Johnson's "War on Poverty" (Vold, Bernard, & Snipes, 2002).

Social Process Theories

Social process theories emphasize criminality as a learned or culturally transmitted process and are presented as an outgrowth of the Chicago school of sociology in the works of Henry Shaw and David McKay, Edwin Sutherland, Walter Miller, and David Matza.

The Chicago School

In 1892 the first American academic program in sociology was begun at the University of Chicago, marking the inception of sociology's **Chicago school.** Names associated with this school would constitute a virtual Camelot of sociology: Park, Burgess, Wirth, Shaw, McKay, Thrasher, Zorbaugh, Anderson, Mead, Faris, Dunham, Thomas, Znaniecki, Cressey, and Sutherland, to mention just a few. Originally begun by sociologist Albion Small, the school would have a primary influence on

Audio Link 6.2
Listen to a discussion of the new gangs of New York.

the development of sociology as a distinctive American discipline in the twenties and thirties with Robert Park, Ernest Burgess, and Louis Wirth as the primary mentors. This group would develop a comprehensive theoretical system urban ecology that would generate a remarkable number of urban life studies (Stein, 1964).

Human Ecology

Like Durkheim, Park (1952) saw that freedom from group constraints often also entailed freedom from group supports. While Durkheim referred to this as anomie, Park used the notion of "individualization due to mobility." Ecology is a field that examines the interrelationship between human organisms and environment. Park's theory was based on **human ecology,** looking at humans and the environment and, more specifically, at urban ecology, viewing the city as a growing organism, heavily employing analogies from plant ecology. According to Park, the heterogeneous contact of racial and ethnic groups in the city often leads to competition for status and space, as well as conflict, accommodation, acculturation, assimilation, or amalgamation—terms all quite similar to concepts in botany (plant biology), such as segregation, invasion, succession, and dispersion. One of Park's key notions was that of **natural areas,** subcommunities that emerge to serve specific, specialized functions. They are called "natural" since they are unplanned and serve to order the functions and needs of diverse populations within the city. Natural areas provide institutions and organizations places to socialize its inhabitants and to provide for social control. Such natural areas include ports of embarkation, Burgess's "zone of transition," ghettos, bohemias, hobohemias, and the like. Burgess's (1925) concentric zone theory, which views cities as growing outward in concentric rings, served as the graphic model for the Chicago school's theory of human ecology. Figure 6.1 presents Burgess's concentric zone theory.

Wirth's (1938) theory of urbanism as a way of life viewed the transition from the rural to the urban way of life as producing social disorganization, marginality, anonymity, anomie, and alienation because of the heterogeneity, freedom, and loneliness of urban life. The Chicago school expressed an anti-urban bias in its analysis and nostalgia for the small Midwestern towns in which most of its theorists had originated.

Using Park's concept of natural areas as a building block, Chicago school students were enjoined to perform case studies of these areas in order to generate hypotheses as well as, it was hoped, generalizations. Park (1952) expressed this hope:

> The natural areas of the city, it appears from what has been said, may be made to serve an important methodological function. They constitute, taken together, what Hobson has described as "a frame of reference," a conceptual order within which statistical facts gain a new and more general significance. They not only tell us what the facts are in regard to conditions in any given region, but insofar as they characterize an area that is natural and typical, they establish a working hypothesis in regard to other areas of the same kind. (p. 198)

This empirical orientation, as opposed to armchair theorizing, was the chief contribution of the Chicago school. Among the students inspired to perform field research were Clifford Shaw and David McKay, and Edwin Sutherland.

Shaw and McKay's Social Disorganization Theory

Handbook Article Link 6.2
Read an article on social disorganization theory.

Ironically, although Clifford Shaw and Henry D. McKay are pointed to as members of the Chicago school, they never enjoyed faculty status at the University of Chicago, but performed their research while employed by the Illinois Institute for Juvenile Research in Chicago. Snodgrass (1972) indicates that neither Shaw nor McKay received his doctorate because of foreign language requirements, but they worked closely with many faculty and students from the university (Carey, 1975). The lasting contribution of Shaw and McKay's ecological studies in the thirties was their basic premise that crime

FIGURE 6.1 ■ Burgess's Concentric Zone Theory

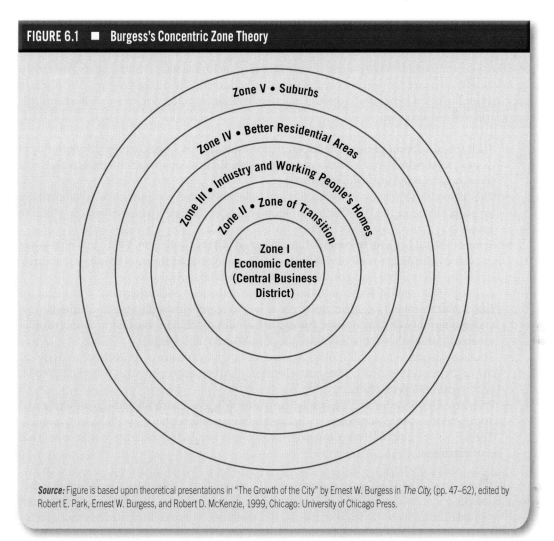

Source: Figure is based upon theoretical presentations in "The Growth of the City" by Ernest W. Burgess in *The City,* (pp. 47–62), edited by Robert E. Park, Ernest W. Burgess, and Robert D. McKenzie, 1999, Chicago: University of Chicago Press.

is due more to social disorganization in pathological environments than the deviant behavior of abnormal individuals (Gibbons, 1979).

In the tradition of the statistical school of criminological theory, Shaw and McKay made extensive use of maps and official statistics to plot the ecological distribution of forms of **social disorganization** such as juvenile delinquency (Shaw, 1929; Shaw & McKay, 1942). Using Burgess's concentric zone theory as a schema, as well as Park's notion of natural areas, they were able to document the ecological impact on human behavior. For instance, one transitional area (an area undergoing invasion/ succession) was shown to exhibit very high crime rates despite considerable change in its ethnic makeup. Such areas breed criminogenic influences that predispose occupants to crime and social disorganization. In other research, Shaw utilized ethnographic and autobiographical field methods in order to provide case studies of criminals and delinquents (Shaw, 1930; Shaw, McKay, & MacDonald, 1938). Imposing concentric circles on mapped areas of Chicago on which rates of social disorganization had been plotted, Shaw (1930) was able to demonstrate the highest rates of truancy, crime, delinquency, and recidivism in Zone II (area of transition), while such rates declined as one moved farther out from the rings. Criminal attitudes and social pathology were viewed as culturally transmitted within the social environment.

A Critique of Social Disorganization Theory

The human ecologists' insistence on ecological and social conditions having criminogenic impacts on otherwise normal individuals would inspire later criminologists such as Sutherland. Their stress on field studies and an empirical orientation would provide credibility to the fledgling disciplines of sociology and criminology and win them greater academic acceptance. A number of shortcomings, however, have been identified:

- Their theories at times border on ecological determinism: that an area or physical environment causes social pathology. Concentration on the geophysical environment tends to make the social structure and institutions secondary.
- The attempt to borrow an organic analogy and adapt biological concepts, such as competition, invasion, succession, and the like, to criminology saddled the field with unnecessarily primitive concepts.
- Some of the studies tend to commit the **ecological fallacy** (Robinson, 1950), in which group rates are used in order to describe individual behavior. Aggregate statistics do not yield accurate estimates if the intended unit of analysis is the behavior of individuals.
- Although Shaw and McKay studied other cities, the theories and conceptions of the Chicago school (such as the concentric growth of cities) were perhaps applicable to Chicago, a city undergoing fantastic urbanization during the twenties and thirties, but may not apply to other urban communities, particularly since the post–World War II period.
- These theories assume stable ecological areas, which in fact do not exist. Such areas disappeared in the post–World War II decentralization of urban areas (Bursik, 1988, pp. 523–524; Schuerman & Kobrin, 1986).
- Problems in operationalizing (measuring) key concepts, such as delinquency rate and disorganization, arise when there is a heavy reliance on official statistics (Pfohl, 1985, p. 167).
- There is an overemphasis on consensus in community and a lack of appreciation of political conflict (Bursik, 1988, p. 524).

In defense of Shaw and McKay, Brantingham and Brantingham (1984, p. 312) point out that they were not as guilty of falling into the ecological fallacy trap as many of their followers, since they supplemented many of their statistical studies with case studies. This was illustrated by ethnographic works such as Shaw, McKay, and MacDonald's (1938) *Brothers in Crime* and Shaw's (1930) *The Jack Roller*. Focusing on group or social process, the urban ecologists Shaw and McKay in particular were influential in shifting criminological analysis from an overconcentration on the individual deviant and instead toward the criminogenic influences of social environments.

Routine Activities Approach

A resurgence and rediscovery of interest in the ecological and social disorganization theories of crime have been rekindled by formulations such as Cohen and Felson's (1979) and Felson's (1983) "routine activities approach" to crime causation. This approach says, "The volume of criminal offenses will be related to the nature of normal everyday patterns of interaction. . . . There is a symbiotic relationship between legal and illegal activities" (Messner & Tardiff, 1985, pp. 241–242). In summarizing the routine activities approach, Felson (1987) indicates:

> (1) It specifies three earthy elements of crime: a likely offender, a suitable target, and the absence of a capable guardian against crime. (2) It considers how everyday life assembles these three elements in space and time. (3) It shows that a proliferation of lightweight durable goods and a dispersion of activities away from family and household could account quite well for the crime wave in the U.S. in the 1960s and 1970s without fancier explanations. Indeed modern society invites high crime rates by offering a multitude of illegal opportunities. (p. 911)

Journal Article Link 6.1
Examine literature regarding women's stalking victimization and routine activities theory.

Journal Article Link 6.2
Examine literature regarding defensible space and routine activities of place.

Koenig (1991) explains that one of the origins of this theory was the Hindelang et al. (1977) lifestyle exposure theory, which proposes that the probability of crime varies by time, place, and social setting. An individual's lifestyle places him or her in social settings with higher or lower probabilities of crime.

Other works that illustrate the reaffirmation of social disorganization theory are by Simcha-Fagan and Schwartz (1986), who include social disorganization and subcultural approaches in explaining urban delinquency. Similarly, Byrne and Sampson (1986) indicate that the social-ecological model is based on the premise that community has independent impacts on crime that are not able to be separated from the individual level. White (1990) found that neighborhood burglary rates varied by highway accessibility; ease of entry increases both familiarity with escape routes and vulnerability. In "Deviant Places," Stark (1987) argues that "kinds of places" explanations are needed in criminology in addition to the "kinds of people" explanations. He codifies 30 propositions from over a century of ecological explanations of both the Chicago school and the moral statisticians of the nineteenth century.

Stark identifies five aspects of high-deviance areas: density, poverty, mixed use, transience, and dilapidation. These elements create criminogenic conditions for crime. His propositions include that density is associated with interaction between least and most deviant populations, higher moral cynicism, overcrowding, outdoor gatherings, lower levels of supervision of children, poorer school achievement, lower stakes in conformity, and increased deviant behavior. Crowding will increase family conflict, decrease the ability to shield wrongdoing, and thus increase moral cynicism. While Stark's hypotheses are too numerous to cover here, his systematic extraction of propositions from over a century of social disorganization/ecological research represents a reaffirmation and resurgence of such literature (see also Taylor & Harrell, 1996). Crime File 6.1 reports on environmental crime prevention and "Designing Out Crime."

Sutherland's Theory of Differential Association

Perhaps the most influential general theory of criminality was that proposed initially in 1934 by Edwin Sutherland (1883–1950) in his theory of **differential association.** Simply stated, the theory indicates that individuals become predisposed toward criminality because of an excess of contacts that advocate criminal behavior. Due to these contacts a person will tend to learn and accept values and attitudes that look more favorably on criminality.

Sutherland's theory was strongly influenced by Charles Horton Cooley's (1902) theory of personality, the **"looking-glass self."** Cooley viewed the human personality as a "social self," one that is learned in the process of socialization and interaction with others. The personality as a social product is the sum total of an individual's internalization of the impressions he or she receives of the evaluation of others—"mirror of alters." "Significant others," people who are most important to the individual, are particularly important in this socialization process. Thus, in Cooley's perception, the human personality is a social self, a product of social learning and interaction with others. Sutherland was also influenced by Shaw and McKay's (1942) notion of social disorganization and cultural transmission of crime, as well as by French sociologist Gabriel Tarde's (1912 [1890]) concept of imitation as the transmitter of criminal values. Similarly, in Sutherland's explanation of criminality, crime is a learned social phenomenon, transmitted in the same manner that more conventional behavior and attitudes are passed on.

In explaining how he developed the theory, Sutherland indicated that he was not even aware that he had done so until, in 1935, Henry McKay referred to "Sutherland's theory": "I asked him what my theory was. He referred me to pages 51–52 of my book" (Sutherland, 1956b, p. 14). The first edition of Sutherland's text was published in 1924; while the 1934 edition to which McKay referred contained the nexus of a theory, it was in the 1939 edition that Sutherland outlined its major propositions. These were slightly modified in the 1947 edition and have remained essentially the same in subsequent editions, which have been coauthored or (since Sutherland's death in 1950) authored by Donald Cressey.

Handbook Article Link 6.3
Read an article
on peers and crime.

Crime File 6.1

Designing Out Crime

Issues

Our failure to bring crime under control through a wide range of modifications to the criminal justice system has blinded us to the successful efforts continuously being made by a host of private and public entities—municipalities, schools, hospitals, parks, malls, bus companies, banks, department stores, taverns, offices, factories, parking lots—to bring a wide range of troublesome and costly crimes under control. In most cases these successes are achieved by identifying ways to reduce opportunities for highly specific kinds of crime, the approach advocated by environmental crime prevention.

The essential tenets of environmental crime prevention, of which Crime Prevention Through Environmental Design (CPTED) and Situational Crime Prevention are the best known examples, are to:

- Increase the difficulty of committing crime (e.g., credit card photos).
- Increase the perceived risks (e.g., burglar alarms).
- Reduce the rewards associated with criminal acts (e.g., PIN for car radios).
- Reduce the rationalizations that facilitate crime (e.g., simplify tax forms).

While the federal government gave some support to CPTED in the 1970s, interest in environmental crime prevention has languished in our country. One reason for this loss of support was the concern that blocking opportunities for crime would result in its displacement to some other target, time, or place (i.e., the net amount of crime would remain the same, although its manifestations would be different). This belief was bolstered by criminological theories that generally failed to recognize important situational determinants of crime, such as the availability of tempting goods to steal and the absence of adequate guardianship of vulnerable property and persons.

In recent years, however, new criminological theories have emphasized the role of opportunities in crime causation. These theories, which include routine activity theory and rational choice theory, argue that as the number of opportunities for crime increases, more crimes will be committed; conversely, as opportunities are reduced, so crime will decline. Whether or not displacement takes place depends on the ease with which offenders can obtain the same criminal rewards without greatly increased effort or risks. Somebody who has developed the habit of shoplifting from the supermarket will not inevitably turn to some other form of crime, involving greater risk of detection and more severe penalties, if the store takes effective preventive action. In fact, particular crimes serve special purposes for the offender. A thwarted rapist will not turn to mugging or drug dealing.

Policy Recommendations

- **Federal Crime Prevention Department:** A crime prevention department should be established in the Department of Justice along the lines of similar units now functioning in a number of European countries. This unit would have a research and dissemination role and would also initiate action to "design out crime" that more naturally falls to central government than to state or local agencies. For example, the department could ensure the security of the phone system, of credit cards, or of ATM cards through federal influence on manufacturers and service providers at an industry level. Important preventive initiatives that currently need federal government sponsorship include development of effective personal alarms for repeat victims of domestic violence and the use of PIN numbers for VCRs and other electronic devices that are targets for burglary.
- **Crime Prevention Extension Service:** A Crime Prevention Extension Service, linked to local universities, along the lines of the successful agricultural model, should be developed within the Department of Justice. Its mandate should be to deliver expert crime prevention advice to small businesses and local communities. Such a service would complement rather than compete with the work of the police, especially as community policing ideas take hold.

Sources: American Society of Criminology Task Force Report to Attorney General Janet Reno, *The Criminologist* (Special Issue), *20*, 6, November/December 1995. Task Force members on "Designing Out Crime" included: Ronald V. Clarke, chair; Patricia Brantingham, Paul Brantingham, John Eck, and Marcus Felson.

Web Research Project

Search the name "Oscar Newman." What is his main thesis regarding the impact of defensible space on crime?

The nine propositions of the differential association theory are these (Sutherland, 1947, pp. 6–7):

- Criminal behavior is learned.
- Criminal behavior is learned in interaction with other persons in a process of communication.
- The principal part of the learning of criminal behavior occurs within intimate personal groups.
- When criminal behavior is learned, the learning includes: (a) techniques of committing the crime, which are sometimes very simple, and (b) the specific direction of motives, drives, rationalizations, and attitudes.
- The specific direction of motives and drives is learned from definitions of the legal codes as favorable or unfavorable.
- A person becomes delinquent because of an excess of definitions favorable to violation of law over definitions unfavorable to violation of law.
- Differential association may vary in frequency, duration, priority, and intensity.
- The process of learning criminal behavior by association with criminal and anticriminal patterns involves all of the mechanisms that are involved in any other learning.
- While criminal behavior is an explanation of general needs and values, it is not explained by those general needs and values since noncriminal behavior is an expression of the same needs and values.

Differential association theory is not directed at the issue of the origin of crime in society, but concentrates instead on the transmission of criminal attitudes and behavior. It is a behavioristic theory—"previous behavior causes subsequent behavior"—and contains elements of a "soft social determinism," that is, exposure to groups does not cause but predisposes individuals to criminal activity or causes them to view it more favorably. Why, then, do not all with similar exposure become similarly criminal? Sutherland's notion of variations in contacts provides for individual reaction to social groups and exposures.

Contacts in Differential Association

Contacts in differential association vary according to frequency, duration, priority, and intensity. *Frequency* deals with the number of contacts; *duration* with the length of time over which an individual is exposed to such contacts. The sheer length and volume of association with criminogenic influences affect different people in different ways. Humans are not robots responding in a predictable manner to a given number of influences. *Priority* refers to the preference individuals express toward the values and attitudes to which they are exposed, while *intensity* entails the degree of meaning the human actor attaches to such exposure. While Sutherland (1947) admits an inability to reach a quantitative or exact measurement of these modalities, a very general example should illustrate their operation: What explains the good child in the bad environment? Despite a great frequency and long duration of exposure to criminal attitudes, such individuals fail to prefer such values and attach greater meaning to noncriminal attitudes that, although less frequently available, may be found in "significant others," perhaps role models such as teachers, coaches, peers, and the like.

A Critique of Differential Association

Because it is a general theory of criminality and is relatively compatible with many other criminological explanations of crime, differential association theory enjoyed widespread acceptance in the field. It was not, however, without critics. Donald Cressey, Sutherland's coauthor, explains that since Sutherland's principal propositions are presented in only two pages in his textbook, the theory is often misinterpreted by some critics, most notably Vold (1958, p. 194). Among these claimed errors of interpretation, Cressey (Sutherland & Cressey, 1974, pp. 78–80) mentions the following:

- The theory is concerned only with contacts or associations with criminal or delinquent behavior patterns. (It actually refers to both criminal and noncriminal behavior, as demonstrated by the use of terms such as "differential" and "excess" of contacts.)

Handbook Article Link 6.4
Read an article on social learning theory.

- The theory says persons become criminals because of an excess of associations with criminals. (It actually says that criminal attitudes can be learned from the unintentional transmission of such values by noncriminals.)
- Using the 1939 version of the theory, critics believe the theory refers to "systematic criminals." (This has been modified since the 1947 version to refer to all criminal behavior.)
- The theory fails to explain why persons have the associations they have. (It does not pretend to do so.)

Cressey (1960) also addresses other criticisms that he feels are misinterpretations; however, a number of shortcomings have been identified:

- While Sutherland traces the roots of criminality to culture conflict and social disorganization, a comprehensive theory of criminality should provide more explanation of the origin of crime.
- Since it is a general theory, it is difficult to either empirically prove or disprove it by means of research, and reformulations are necessary in order to permit testing (see Burgess & Akers, 1966; DeFleur & Quinney, 1966).
- The theory fails to account for all forms of criminality.
- The theory fails to acknowledge the importance of non-face-to-face contacts such as media influences (Radzinowicz & King, 1977, p. 82).

Despite these and other criticisms, differential association remains important as a useful general theory of criminality even though it may fail to specify the process for each individual case of criminality. The theory of differential association remains one of the most cited theories in modern criminology and will probably remain so until a more acceptable general theory of criminality appears. It has also received support in recent research (Matsueda, 1988; Orcutt, 1987), although Warr and Stafford (1991) indicate that attitude is not as important as actual peer behavior and group pressures to conform.

A variation of differential association can be found in Ronald Akers and Robert Burgess's (Burgess & Akers, 1966) differential reinforcement (social learning) theory.

> "Differential reinforcement" refers to the balance of anticipated or actual rewards or punishments that follow or are consequences of behavior. Whether individuals will refrain from or commit a crime at any given time . . . depends on the past, present, and anticipated future rewards and punishments for their action. (Akers, 1994. p. 98)

Akers's theory combines Sutherland's concept with behavioral conditioning and even classical concepts of rewards and punishments and has found considerable empirical support (Burgess & Akers, 1966, pp. 102–104).

Miller's Focal Concerns Theory

Walter Miller's (1958) ideas appeared in an article titled "Lower Class Culture as a Generating Milieu of Gang Delinquency." Miller limits the applicability of his theory to "members of adolescent street corner groups in lower class communities" (1958, p. 5). Unlike Cohen, who viewed such delinquency as a lower-class reaction to middle-class values, Miller views such activity as a reflection of the **focal concerns** of dominant themes in lower-class culture. These are "areas or issues which command widespread and persistent attention and a high degree of emotional involvement" (p. 7). Faced with a chasm between aspirations and the likelihood of their achievement, lower-class youth seek status and prestige within one-sex peer units (gangs) in which they exaggerate focal concerns already in existence in lower-class culture. Thus gang delinquency, rather than representing an anomic reaction to unobtainable middle-class goals, represents, in the tradition of social process theory, a pattern of subcultural transmission or learning of values prevalent in the local environment.

Photo 6.5

Lower-class youth seek status and prestige within gangs.

The focal concerns of lower-class culture emphasize trouble, toughness, smartness, excitement, fate, and autonomy. Getting into *trouble* often confers prestige and a means of obtaining attention. The "class clown" and the "bad dude" become attention-getting roles. *Toughness,* "machismo," having physical prowess, or being able to handle oneself are highly prized characteristics among lower-class males. The "hard guy" is preferable to the "chump," "wimp," or "sissy." *Smartness* "involves the capacity to outsmart, outfox, outwit, dupe, 'take,' or 'con' another" (p. 7). This is illustrated by the "streetwise" game of "playin' the dozens" (Berdie, 1947), a highly ritualistic game of razzing, "ranking," or "cappin' on someone's Mom" practiced by lower-class black males in particular. Extremely foul insults are traded by two antagonists, the themes usually relating to sexual matters and female relatives of one's opponent. Such insults are rhythmically presented one-liners whose object is to leave the opponent speechless or "humbled out." "Playin' the dozens" is also known as *signification.* A young man engaging in such activities, creating poetry of the streets, would be regarded in conventional society as having a "bad mouth."

The theme of *excitement* emphasizes the quest for skill, danger, risk, change, or activity. Rather than a subject of control and planning, the future is perceived as a matter of *fate,* luck, or good fortune. Gambling's popularity in lower-class culture makes it the "poor person's stock exchange." *Autonomy* (independence) looms as a dominant concern in lower-class culture, particularly among males, even though it is less likely to be achieved given their narrow occupational and life options. "Being one's own man," that is, being free from authority, "the man," and external constraint, is a strong value.

A Critique of Miller's Theory

- Like the other subcultural theories, Miller's theory also ignores middle- and upper-class delinquent/criminal activity.
- By focusing exclusively on the lower class, Miller and others in this tradition are perhaps most responsible for the criticism that mainstream sociology ignores deviance of the powerful.
- Miller's theory rests heavily on the assumption of the existence of a distinctive lower-class culture that holds values and attitudes distinct from, if not at odds with, dominant middle-class values. The pluralistic nature of U.S. society makes it quite uncertain that such a distinctive value system, solely based on class, indeed exists.

Miller's theory views criminogenic influences as learned or transmitted as part of subcultural values. Similarly, the writings of David Matza present delinquency as part of a general social process of learned cultural values rather than as an anomic reaction to unobtainable goals.

Matza's Theory of Delinquency and Drift

The theories of David Matza are presented in his book *Delinquency and Drift* (1964) and in a coauthored article with Gresham Sykes (Sykes & Matza, 1957), titled "Techniques of Neutralization." Matza's theories, including that of **delinquency and drift,** are an example of **soft determinism,** which holds that, although human behavior is determined to some extent by outside forces, there still exists an element of free will or individual responsibility (Matza, 1964, pp. 5–7). Humans are neither entirely constrained nor entirely free, nor is the individual entirely committed to delinquent or nondelinquent behavior. Matza (p. 28) explains the drift theory of delinquency:

> The delinquent exists in a limbo between convention and crime responding in turn to the demands of each, flirting now with one, now the other, but postponing commitment, evading decision. Thus he drifts between criminal and conventional action.

Subterranean Values

Rather than being wholly committed to delinquency, most delinquents are dabbling in it and are acting out **subterranean values** of society (pp. 63–64) that exist alongside more conventional values in a pluralistic society such as the United States. Conventional society attempts to control the expression of these values and reserve it for the proper time and place; in a sense, it is the practice of "morality with a wink." The delinquent, rather than being committed to goals that are alien to society, exaggerates society's subterranean values and acts them out in caricature. Sykes and Matza explain (1957):

> The delinquent may not stand as an alien in the body of society but may represent instead a disturbing reflection or caricature. His vocabulary is different, to be sure, but kicks, big time spending and rep have immediate counterparts in the value system of the law abiding. The delinquent has picked up and emphasized one part of the subterranean values that coexist with other, publicly proclaimed values possessing a more respectable air. (p. 717)

Thus, while conventional mores disapprove of subterranean values, they often represent "hidden" patterns or themes in the culture. Illicit sexual behavior, slick business practices, a dislike of work, substance abuse, and media violence as a popular form of entertainment are examples. Delinquents simply have poor training and timing in the expression of subterranean values. The pervasiveness of subterranean values might be illustrated by the attempt of conventional members of "straight" society to appear "hip," "with it," and "streetwise."

Techniques of Neutralization

Sykes and Matza's (1957) term **"techniques of neutralization"** refers to rationalizations or excuses that juveniles use to neutralize responsibility for deviant actions. In drift situations, offenders can lessen their responsibility by exaggerating normal legal defenses (for example, self-defense or insanity) or by pointing to the subterranean values prevalent in society. They identify five techniques of neutralization:

1. Denial of responsibility, such as appeals based on one's home life, lack of affection, and social class.

2. Denial of harm to anyone, such as defining stealing as "borrowing" or drug abuse as harming no one but the offender.

3. Denial of harm to the victim, in which the assault is justified since the person harmed was also a criminal.

4. Condemning the condemners, reversing the labeling process by claiming that authorities are more corrupt than the offender, and are hypocritical as well.

5. Appeal to higher authority, which claims that the offense was necessary in order to defend one's neighborhood or gang.

As an illustration of the techniques of neutralization, the song "Gee, Officer Krupke" from the musical *West Side Story* has members of the Jets arguing that they are victims of "a social disease." Sykes and Matza (1957) explain:

> The delinquent both has his cake and eats it too, for he remains committed to the dominant normative system and yet so qualifies its imperatives that violations are "acceptable" if not "right." Thus the delinquent represents not a radical opposition to law abiding society but something more like an apologetic failure, often more sinned against than sinning in his own eyes. We call these justifications of deviant behavior techniques of neutralization; and we believe these techniques make up a crucial component of Sutherland's "definitions favorable to the violation of the law." It is by learning these techniques that the juveniles become delinquent, rather than by learning moral imperatives, values or attitudes standing in direct contradiction to those of the dominant society. (p. 668)

A Critique of Matza's Theory

Matza provides a transition between Sutherland's social process theories and the social control theories to be discussed next. By combining deterministic models with the notion of free will, he avoids the overly deterministic nature of many earlier theories and explains why the majority of individuals who find themselves in criminogenic settings do not commit crime. His concept of neutralization enables him to escape the problem inherent in previous subcultural theories of delinquency, which rested on the premise that delinquent values were at variance with conventional values. Some possible shortcomings of Matza's views include the following:

- While some research has shown offenders to be prone to rationalizing their behavior (Ball, 1980; Regoli & Poole, 1978), Hindelang (1970) found different value systems among delinquents. Obviously more research is needed.
- In order for his theory to be correct, empirical evidence must demonstrate that Matza's neutralization takes place during the period of drift preceding the act, a concept that may be difficult to operationalize.

Hamlin (1988) argues that the notion of rational choice in neutralization theory has been misplaced and that such rationalizations are utilized after the fact only when behavior is called into question (see Minor, 1981, 1984, for additional analysis).

The transitional nature of Matza's theories with social control approaches can be found in his notion of drift, in which individuals become temporarily detached from social control mechanisms. This release from group bonds is the basic unit of analysis in social control theories.

Social Control Theories

The final grouping of mainstream sociocriminological theories to be discussed is referred to as *social control theories* and is represented by the work of Walter Reckless and Travis Hirschi.

Video Link 6.2
View a video on music, criminal behavior, and crime.

Social control theories address the issue of how society maintains or elicits social control and the manner in which it obtains conformity or fails to obtain it in the form of deviance. As Gibbons (1979, p. 113) points out, this once major area of sociological investigation is still viable. While one aspect of the concept dealt with penology or corrections, another aspect, the subject of this discussion, was concerned with socialization and learning processes, the internalization of societal norms (inner controls), and external influences (outer controls; Clark & Gibbs, 1965). Although a number of writers have contributed to social control theories, this presentation will concentrate primarily on the formulations of Walter Reckless and his associates (1956, 1961; Reckless & Dinitz, 1967) and Travis Hirschi (1969).

Reckless's Containment Theory

One of the earliest and best known examples of social control theory was Walter Reckless's (1961) containment theory. Like his contemporary Sutherland, Reckless was a product of the Chicago school of sociology and one of the mainstream pioneers in U.S. criminology (Gibbons, 1979). Reckless wrote an early textbook called *The Crime Problem* in 1940, and in a much later edition began to state his theories. **Containment theory** basically holds that individuals have various social controls (containments) that assist them in resisting pressures that draw them toward criminality. This theory attempts to account for social forces that may predispose individuals to crime as well as for individual characteristics that may insulate them from or further propel them toward criminality. Various social pressures, treated in previously discussed deterministic theories, exert pushes and pulls on the individual; these pressures interact with containments (protective barriers), both internal and external to the individual, and these containments add the element of free will in resisting criminality. Thus the presence or absence of social pressures interacts with the presence or absence of containments to produce or not produce individual criminality.

The basic elements of Reckless's containment theory (Reckless & Dinitz, 1967; Reckless, Dinitz, & Kay, 1957) can be summarized:

Layers of Social Pressures

- External pressures push an individual toward criminality. Variables impinging on an individual include poor living conditions, adverse economic conditions, minority group membership, and the lack of legitimate opportunities.
- External pulls draw individuals away from social norms and are exerted from without by bad companions, deviant subcultures, and media influences.
- Internal pressures push an individual toward criminality; they include personality contingencies such as inner tensions, feelings of inferiority or inadequacy, mental conflict, organic defects, and the like.

Containments

- Inner containments refer to the internalization of conventional behavioral values and the development of personality characteristics that enable one to resist pressures. Strong self-concept, identity, and strong resistance to frustration serve as examples.
- Outer containments are represented by effective family and near support systems that assist in reinforcing conventionality and insulating the individual from the assault of outside pressures.

Reckless and his colleagues (Reckless, Dinitz, & Kay, 1957) felt that the theory was helpful in explaining both delinquency and nondelinquency, as indicated by the title of one article, "The 'Good

Boy' in a High Delinquency Area." Individuals may become predisposed toward criminality because of strong external pressures and pulls and weak inner and outer containments, while others with these same pressures may resist because of a strong family or through a strong sense of self. Weak containments plus strong external pressures provide the conditions for individual criminality. The attractiveness of containment theory is its general ability to subsume variables discussed in other more specific theories as well as its attempt to link the deterministic and free will models and to intersect socioeconomic factors, as well as biological and psychological factors, with individual biography.

A Critique of Containment Theory

Reckless and associates (Reckless, Dinitz, & Kay, 1957; Reckless, Dinitz, & Murray, 1957; Scarpitti, Murray, Dinitz, & Reckless, 1960) have attempted to verify his theory. In one study, they had teachers in a high-delinquency area nominate "good boys"; they found strong self-images as well as more conventional behavior among this group 4 years later. But critics call for more research, indicating that poor operationalization and weak methodology have plagued these studies (Schrag, 1971; Schwartz & Tangri, 1965). As a very general sensitizing theory that attempts to account for both criminogenic forces and individual responses, the containment theory is a useful descriptive model, but actual empirical specification of the process is problematic.

Hirschi's Social Bond Theory

Travis Hirschi (1969) in *Causes of Delinquency* presented his **social bond theory,** which basically states that delinquency takes place when a person's bonds to society are weakened or broken, thus reducing personal stakes in conformity. Individuals maintain conformity for fear that violations will rupture their relationships (cause them to "lose face") with family, friends, neighbors, jobs, school, and the like. In essence, individuals conform not for fear of prescribed punishments in the criminal law, but rather from concern with violating their groups' mores and the personal image of them held by those groups. These bonds to society consist of four components: attachment, commitment, involvement, and belief.

Journal Article Link 6.3
Examine literature regarding effects of social bonds on probationers.

Attachment refers to a bond to others (such as family and peers) and important institutions (such as churches and schools). Weak attachment to parents and family may impair personality development, while poor relationships with the school are viewed as particularly instrumental in delinquency. *Commitment* involves the degree to which an individual maintains a vested interest in the social and economic system. If an individual has much to lose in terms of status, job, and community standing, he or she is less likely to violate the law. Adults, for instance, have many more such commitments than do juveniles. *Involvement* entails engagement in legitimate social and recreational activities that either leaves too little time to get into trouble or binds one's status to yet other important groups whose esteem one wishes to maintain. Finally, *belief* in the conventional norms and value system and the law acts as a bond to society. Like Reckless's containment theory and Matza's delinquency and drift, Hirschi's social bond theory combines elements of determinism and free will; individual choice still enters the equation.

A Critique of Social Bond Theory

Social bond theory has been relatively well received because as a general theory it subsumes and is supported by many more specific findings with respect to relationships between crime/delinquency and particular variables. School performance, family relationships, peer group attachments, and community involvement as predictors of norm violation have been stock items in criminological research. Research by Hirschi (1969), a partial replication by Hindelang (1973), and review of studies by Bernard (1987) provide some strong support for control theory. Strong parental attachments, commitment to conventional values, and involvement in conventional activities and with conventional peers were found to be predictive of nondelinquent activity. While Agnew (1985) found that social control variables explained only 1 to 2 percent of future delinquency and that cross-sectional studies exaggerated the importance of Hirschi's theory, Rosenbaum (1987) found that the theory explained some types of delinquency better than others. The theory accounted for more female than male crime and for more drug use than violence or property offenses. Variations of social control theory have been offered by Briar and Piliavin (1965), who theorize that individuals evaluate the risk of being caught and punished once bonds are weakened, and Glaser (1978), who combines elements of differential association, control, and classical theory. While Hirschi's social control appears to be quite useful in explaining the general process of commitment/ noncommitment to delinquency, more research is certainly needed in order to specify and modify it. Hirschi's theory is not concerned with societal origins of crime but with individual deviation from given societal norms.

Gottfredson and Hirschi's General Theory of Crime

As a successor to his social bond theory, Hirschi joined with Michael Gottfredson in proposing another theory. Combining elements of classical, positivistic, and social control theories, Gottfredson and Hirschi (1990; Hirschi & Gottfredson, 1990) claim to have developed a "general theory of crime." This general theory is that "low self-control" in the pursuit of "self-interest" causes crime. Deficiencies in parenting distinguish those who express this trait, who express themselves in greater deviance and criminality, from those who do not. Those with high self-control would be less likely to become involved in such activity. Surprisingly, Hirschi and Gottfredson also claim that this same "self-control" theory explains white collar crime (Hirschi & Gottfredson, 1987) and that the causes of white collar crime are not distinct from the causes of other crimes (see Cullen et al., 1991; Daly, 1989).

Glaser (1990) notes that Gottfredson and Hirschi's general theory of crime is "usefully complemented, and not contradicted, by differential association, deviant subculture, and social learning theories. These theories explain why socially disorganized neighborhoods provide the greatest opportunities, social support, and learned rationalizations for persons to express low self control in street felonies" (p. 2).

While Gottfredson and Hirschi's general theory of crime is a very ambitious effort, it is regarded as severely flawed in relation to what later in the next chapter is described as the "global fallacy," the tendency to make a useful specific theory of crime explain all crime. Is this theory intended to explain corporate price fixing, insider trading on Wall Street, or international terrorism? Hirschi and Gottfredson also rely upon the Uniform Crime Reports (UCR) for their measurement of white collar crime. This is a "baffling" (Reed & Yeager, 1991) error since, as any student of criminology is aware, the UCR measures only the white collar crimes of fraud, forgery, and embezzlement, and even these tend to be less serious cases (Steffensmeier, 1989b). The UCR is a worthless measure of white collar crime. Reed and Yeager (1996) further point out that Hirschi and Gottfredson test their theory by focusing on white collar crimes that most resemble conventional crimes. When Reed and Yeager examined the theory using organizational (corporate) offenders, they found it inadequate.

Finally, as discussed in Chapter 3 in the section on the family, there is a tendency in this theory to commit what Currie (1985, p. 185) calls the "fallacy of autonomy," to assume that what happens in the family (poor parenting creating low self-control) is somehow separate from other social policies, inequality, racism, unemployment, and social neglect.

John Hagan's Power-Control Theory

John Hagan (1989) in his **power-control theory** of crime, attempts to rectify a major shortcoming in delinquency theory: its almost total ignoring of female offenders. Viewing much delinquency as

Journal Article Link 6.4
Examine literature regarding a critique of general theory.

risk-taking or fun, children who are exposed to strong parental controls will avoid risk, which lessens delinquency. According to John Hagan, power relationships between father and mother influence the control exercised over sons and daughters.

In traditional patriarchal households, boys are exposed to fewer controls than girls and are, therefore, greater risk takers and more delinquent than girls. In more equalitarian family structures, both sexes are subject to similar social controls and have more similar delinquency levels. Cullen and Agnew (2003) indicate that the empirical validity of Hagan's thesis is still in doubt. The theory does not appear to address single-parent families or more serious, violent crime.

Developmental and Life Course (DLC) Theories

Developmental and life course theories address three ideas:

1. The development of offending and antisocial behavior.

2. Risk factors of committing delinquency/crime at different ages.

3. Effects of life events on life course development (Farrington, 2003).

Video Link 6.3
View a video of the teenage brain.

David Farrington, one of the leading advocates of these theories, carefully delimits their purpose when he indicates that the theories are intended to explain "crimes of theft, burglary, robbery, violence, vandalism, minor fraud and drug use" as exhibited in official records and self-reports. Thus, DLC theories are intended to apply to offending by lower-class, urban males in Western society (Farrington, 2003). **Developmental theories** in criminology began in the 1980s with the work of Alfred Blumstein and associates (Blumstein , Cohen, Roth, & Visher, 1986) with longitudinal studies of criminal careers. Large-scale longitudinal studies during the 1990s supplied the raw material for developmental theories. These included studies in Denver, Pittsburgh, Rochester, New Zealand, and Montreal (Huizinga, Wylie Weiher, Espiritu, & Esbensen, 2003).

Farrington (2003, pp. 223–224) identifies 10 assumptions about offending that DLC theories must explain:

1. Offending prevalence peaks between 15 and 19 years of age.

2. Onset offending peaks between ages 8 and 14 and desistance occurs between ages 20 and 29.

3. Early onset portends long criminal duration and the commission of many offenses.

4. There is continuity in offending from childhood to adolescence to adulthood. High offenders in one period tend to be high offenders in the next, even though most eventually desist from crime.

5. Chronic offenders have an early onset, high offense frequency and long criminal careers.

6. Offenders are versatile rather than specialized, with violent offenders indistinguishable from other frequent offenders.

7. Offenders are versatile at crimes as well as antisocial behavior such as bullying, truancy, and heavy drinking.

8. Crimes in the teenage years tend to take place in groups, while offenses after age 20 are committed alone.

Audio Link 6.3
Listen to a discussion of juvenile crime.

9. Prior to age 20, revenge, excitement or anger may motivate offenders; after this age, utilitarian motives predominate.

10. The onset of different types of offenses occurs at different ages. Shoplifting takes place sooner than burglary, which occurs before robbery. Diversification in crime increases to age 20, after which specialization increases (Piquero, Oster, Mazerolle, Brame, & Dean, 1999). Gang membership as depicted in our photograph has its onset in the teens and desistance in the early twenties.

Journal Article Link 6.5
Examine literature regarding patterns of crime across the life course.

Desistance (quitting criminal activity) after age 20 is predicted by life events such as marriage, employment, military service, and better residential environments. The task of DLC theories is to specify risk factors and protective factors for persistence or desistance after age 20. Farrington (2003) denotes a variety of DLC theories that are too detailed to cover other than in a cursory manner in this text. This includes Farrington's integrated cognitive antisocial potential (ICAP) theory, which features his key variable of "antisocial potential" (AP).

Farrington's Antisocial Potential (AP) Theory

Farrington's ICAP on **antisocial potential** (AP) theory posits that relatively few people have high AP or potential to commit antisocial acts. Long-term AP involves impulsiveness, strain, and life events, while short-term AP depends on situational and motivating factors. Desires for material goods, peer status, excitement, and sexual experience combined with antisocial means of satisfying these needs that are denied legitimately result in high AP. Attachments, the socialization process, and other factors associated with the individual and his or her social environment affect AP.

Other DLC theories include Catalano and Hawkins's (1996) social development model (SDM), which explores the balance between antisocial and prosocial bonding. The prosocial pathway rewards prosocial behavior, while the antisocial pathway leads to antisocial bonding. Offending in teenage years is affected primarily by bonding to antisocial peers, while life events such as marriage and moving out of the city leads to desistance. Terrie Moffit (1999) distinguishes between "life-course persisters" (LCP) and "adolescence-limited offenders" (AL). LCP is predicted by neuropsychological problems such as hyperactivity, impulsivity, low self-control, and childhood temperament (Farrington, 2003, p. 241). LCPs fail to learn prosocial behavior, while AL offending is only temporary.

Marc LeBlanc (1996) proposed an integrated control theory which argues that bonding and personality influence modeling and constraints that influence offending. Social disorganization, rational choice, self-control, and opportunities all influence crime commission. Terence Thornberry and Marvin Krohn's (2001) "interactional theory" sees offending (onset, duration, and desistance) as affected by other life course trajectories such as attachment to parents and commitment to school and work. Social class, race, and neighborhood influence behavioral trajectories. Causal processes (poverty, ineffective parents) interact with negative temperament and neuropsychological defects. Rolf Loeber and associates (1993), on the basis of their Pittsburgh longitudinal study, suggest different pathways to crime and delinquency. These include an "authority conflict pathway" that features stubborn behavior that leads to disobedience and defiance and a "covert pathway" characterized by lying and property destruction and street property crime. The "overt pathway" involves aggressive acts. Perhaps the best known of recent DLC theories is found in the writings of Robert Sampson and John Laub.

Sampson and Laub's Life Course Criminality

Robert Sampson and John Laub, in *Crime and Deviance Over the Life Course* (2003), look at social bonds as they affect adult offending and examine continuities and change in criminality over time. What accounts for persistence or desistance in adult criminal behavior? Sampson and Laub find the answer in social interaction with adult institutions of social control, particularly jobs and marital relations that serve as inhibitors of crime. The "life course" is defined as "pathways through the age differential life span" during which events take place that influence life stages, transitions, and turning points (Sampson & Laub, 2003). Trajectories and transitions are key components in life course theory. The timing and ordering of significant life events affect criminality. A "trajectory" is a long-term pathway such as work–life, marriage, parenthood, self-esteem, and criminal behavior. "Transitions" are specific events that take place in these trajectories such as first job or first marriage. The same event followed by adaptations may lead to different trajectories (Sampson & Laub, 2003).

Sampson and Laub contend that childhood antisocial behavior is associated with a variety of later adult misconduct, such as offenses in the military, educational failure, employment instability, and marital discord. Furthermore, they posit that "social bonds to adult institutions of informal social control (e.g., family, education, neighborhood, work) influence criminal behavior over the life course despite an individual's delinquent and anti-social background" (Sampson & Laub, 2003). The importance of social controls varies across the life course. In childhood and adolescence, family, school, and peer groups are important; in early adulthood, higher education, training, work, and marriage take precedence; while in later adulthood, work, marriage, parenthood, and community become important.

In developing their theory, Sampson and Laub did a secondary analysis (reanalysis) of longitudinal data gathered by Sheldon Glueck and Eleanor Glueck (1950), which had begun in the 1930s. The Gluecks had followed matched cohorts of 500 delinquent and nondelinquent boys. Involving detailed follow-up with parents, teachers, and officials, they were interviewed at age 14, age 25, and age 32 and measured on a wide range of variables including biological, sociological, and psychological. In examining the Glueck data, Laub and Sampson found, as Hirschi described it, that when social bonds are weakened, delinquency increased. In addition, adult bonds such as marriage and jobs also explained criminality beyond earlier delinquency. Strong social relationships were also found to build "social capital" (what Hirschi called "stakes in conformity") that inhibit deviance.

In 2003, Sampson and Laub (2003) published the longest longitudinal study in criminological history. The follow-up on the Glueck's data tracked their cohort literally from age 7 to age 70. Examining whether they could identify a distinct offender group whose crime persisted with increasing age and the effect of individual, childhood, and family background on offending trajectories, they came up short, finding that crime declined with age eventually for all offenders. Desistance worked for even the most active offenders and life course persisters. Even childhood background predictors were ineffective in predicting long-term offending trajectories. All offenders were "life course desisters" in that all desisted, but at different times in the life course.

DLC theory has been very popular and influential both in the field of criminology as well as in juvenile justice policymaking. It has been endorsed by the Office of Juvenile Justice and Delinquency Prevention as a component of its comprehensive plan for delinquency prevention. It has also been adopted by states such as Washington and Pennsylvania.

The Theory–Policy Connection

Mainstream sociological theories are primarily concerned with how criminal values are transmitted. For Merton, crime originates in the American dream itself, a strain between generally accepted criteria for success and the lack of adequate means for many to achieve this success. This permeates American society and its institutions. Criminality is learned and culturally transmitted. It is due to social disorganization in pathological environments. Routing activities theory became quite popular as an explanation in the field of private security. By reducing and guarding targets better, crime is, we hope, reduced. Table 6.2 outlines some of these theory–policy connections.

TABLE 6.2 ■ **The Theory–Policy Connection**

Theory	Basic Assumptions About Crime Causation	Policies
Anomie	Anomie lessens social control and creates deviance	Policies to create greater opportunities and to improve neighborhoods
Social Process	Social disorganization, routine activities, subculturally transmitted, subterranean values	Eradication of slums, "War on Poverty"
Social Control	Containment theory, weakened social bonds, and poor self control	Programs to strengthen the family, Head Start Job Corps, Comprehensive Employment and Training Act
Developmental/ Life Course	Crime over the life course: onset, persistence, desistance	Expanded opportunities, strengthen Institutions

If socially disorganized slums cause crime, then policy strategies to eradicate such blight and to improve community in such settings became predominant. LBJ's War on Poverty was heavily based on strain theory. Increasing opportunity was emphasized. Cloward and Ohlin, whose writings heavily influenced the Kennedy administration, developed a Mobilization for Youth Program that emphasized improving opportunities for disadvantaged youth. The programs were all optimistic ones and assumed that crime could be stamped out if only more opportunities could be created. Other programs of this era were Head Start, Job Corps, the Comprehensive Employment and Training Act, and affirmative action. Programs to strengthen the family derived from social control theories. The emphasis was on strengthening institutions rather than punishment and deterrence.

Summary

Theory is necessary for capturing the essence of criminology. The major sociological theoretical approaches in criminology are mainstream theories (anomie, social process, and social control approaches) and critical theories (labeling, conflict, and radical [Marxist] theories).

Émile Durkheim is the father of the "anomie tradition," which also includes Merton's notion of "anomie and personality adaptations," Cloward and Ohlin's "differential social organization," and Cohen's theory that delinquency is a "lower-class reaction" to middle-class values. While Durkheim viewed anomie as a state of normlessness, a moral malaise experienced by individuals when they lack clear-cut guidelines, later theorists such as Merton adapted the theory to refer to a situation that results from a gap between societal goals and the means provided to achieve them. This, according to Merton, results in "modes of personality adaptation": conformity, innovation, retreatism, ritualism, or rebellion. Cloward and Ohlin argue that the juvenile-subculture gang response to anomie depends on the "differential social organization" (legal and illegal opportunity structures) in the neighborhood. Depending on the type, one of three juvenile delinquent subcultures may emerge: the criminal, conflict, or retreatist. Cohen's theory of delinquency presents it as "lower-class reactions to unobtainable or rejected middle-class values" such as ambition, verbal skills, nonviolence, and the like. He views much delinquency as nonutilitarian, malicious, and negativistic.

The *social process tradition* concentrates on learning, socialization, and subcultural transmission of criminal values. Originating in the work of the "Chicago school" of sociology in the twenties and thirties, and in particular with the works/ideas of Burgess ("concentric zone model"), Park ("natural areas"), and Wirth ("urbanism as a way of life"), human ecology was seen, at least initially, as an organizing perspective. This approach examines the interrelationship between humans and their physical/social environment. Included among better known Chicago school criminologists are Clifford Shaw, David McKay, and Edwin Sutherland.

Making extensive use of maps and official statistics, Shaw and McKay viewed delinquency as reflecting the "social disorganization" of areas in which individuals lived, so delinquency was less a matter of individual abnormality and more a matter of "cultural transmission" or social learning. Concern that Shaw and McKay committed the "ecological fallacy" (attributed group characteristics to individuals) may be alleviated by the fact that they performed a number of case studies of criminals. Cohen and Felson's (1979) "routine activities approach" views crime as related to everyday, normal activities such as the proliferation of consumer goods and the lack of guardians. Sutherland's "differential association theory," the most popular theory in U.S. criminology, states that individuals become predisposed toward criminality because of an excess of contacts that advocate criminal behavior, contacts that vary according to frequency, priority, intensity, and duration. Differential association aims at describing the process by which crime is transmitted but does not address itself to origins of crime. Miller's theory of delinquency views it as reflecting "the focal concerns of the lower class," such as an emphasis on trouble, toughness, smartness, excitement, fate, and autonomy.

David Matza's "delinquency and drift" theory claims that individuals are often in a limbo or uncommitted status between delinquent and nondelinquent behavior. He and Gresham Sykes view delinquents as acting out "subterranean values" (underground values that exist along with more conventionally approved values) and utilizing "techniques of neutralization" (rationalizations) in order to justify their behavior.

Social control theories argue that individuals deviate when removed or weakened. Reckless's "containment theory" views containments (Walter Reckless) or social bonds (Travis Hirschi) as individuals resisting or giving in to various pressures based on social controls (self-concept or close support systems). Hirschi's "social bond theory" states that delinquency arises when bonds to society are reduced and the individual has fewer stakes in conformity. These bonds consist of *attachment, commitment, involvement,* and *belief.*

Developmental and life course criminality theories are an attempt to track the onset, persistence, and desistance of criminal behavior. They represent an effort to track crime commission longitudinally.

Criminology on the Web

Log on to the Web-based student study site at http://www.sagepub.com/haganintrocrim7e/ for author-created podcasts, e-flashcards, quizzes, and more.

KEY CONCEPTS

Anomie	Differential Association Theory	Power-Control Theory
Antisocial potential	Differential Opportunity	Social Bond Theory
Chicago School	Ecological Fallacy	Social Control Theory
Cohen's Lower-Class Reaction Theory	Human Ecology	Social Disorganization Theory
Containment Theory	Looking-Glass Self	Soft Determinism
Delinquency and Drift	Miller's Focal Concerns	Subcultural Theories
Desistance	Modes of Personality Adaptation	Subterranean Values
Developmental/Life Course Theory	Natural Areas	Techniques of Neutralization

REVIEW QUESTIONS

1. How does Merton's concept of anomie differ from that of Durkheim? What is your assessment of the usefulness of Merton's anomie/strain theory in explaining crime in the United States?

2. What contribution did the Chicago school of sociology make to the study of criminology?

3. What are Sutherland's differential association theory's assumptions regarding crime causation?

4. What is Miller's notion of delinquency reflecting the focal concerns of the lower class? How does this differ from Albert Cohen's notion of delinquency being a "lower-class reaction to middle-class society"?

5. David Matza had three important concepts: delinquency and drift, subterranean values, and techniques of neutralization. Discuss each of these and explain how they explain delinquency/crime.

6. Discuss Reckless's containment theory. What are some containments that enable individuals to overcome the various layers of social pressures?

7. What is the major premise of social bond theory? How do these bonds vary for each individual? What have been some criticisms of this theory?

8. What is your assessment of Gottfredson and Hirschi's general theory of crime?

9. How do mainstream sociological theories differ from the earlier classical, economic, ecological, and positivistic theories?

10. What is routine activities theory? Give an example of the practical application of this theory.

WEB SOURCES

Crime Connections on the Web
http://www.justiceblind.com/links.html

Criminal Justice Education Page
http://www.cjed.com/

Criminal Justice Megalinks
http://www.apsu.edu/oconnort/

Crime Theory.com
http://crimetheory.com

WEB EXERCISES

Using this chapter's recommended Web sources, surf the listed sites.

1. How much detail is provided on the "Durkheim and Suicide" site?

2. What does the Criminal Justice Education page feature?

3. Choose a mainstream criminological theory and search some of these sites for information.

4. Perform an online search on "Robert K. Merton" and "Edwin H. Sutherland." Who had the most listings and, if you had a choice, on which theories do you think you could find the most useful information for a term paper?

Ronald Akers and Christine Sellers (2004). *Criminological Theories*. Los Angeles: Roxbury.

This book provides excellent coverage of major classical and contemporary theories.

Francis Cullen and Robert Agnew, editors (2003). *Criminological Theory: Past to Present* (2nd ed.). Los Angeles: Roxbury.

This anthology has a selection of 38 classic and contemporary articles and features excellent introductory essays to each section.

Daniel Curran and Claire Renzetti (2001). *Theories of Crime* (2nd ed.). Boston: Allyn & Bacon.

This book provides an excellent discussion of criminological theories.

Frank E. Hagan (1987). The Global Fallacy and Theoretical Range in Criminological Theory. *Journal of Criminal Justice, 2,* 19–31.

This is a classic article by the author discussing the need to limit the scope of most theories.

Robert K. Merton (1961). *Social Theory and Social Structure* (Rev. ed.). New York: Free Press.

This work contains many of the seminal ideas of a giant in American criminology, Robert Merton.

Steven F. Messner and Richard Rosenfeld (1994). *Crime and the American Dream*. Belmont, CA: Wadsworth.

In a well-received update of Merton's strain theory, Messner and Rosenfeld discuss an institutionalization of deviant means to economic success as pervasive in American society. This theory can be used to explain white collar crime.

George B. Vold, Thomas J. Bernard, and Jeffrey B. Snipes (2002). *Theoretical Criminology* (5th ed.). New York: Oxford University Press.

This remains the standard reference on past and current developments in criminological theory.

Frank Williams III and Marilyn McShane (1998). *Criminological Theory* (3rd ed.). Englewood Cliffs, NJ: Prentice Hall.

This is an excellent general reference work that covers the entire range of criminological theory.

Mainstream Versus Critical Criminology

Critical Criminology

Labeling Theory

   Lemert's "Secondary Deviance"

   A Critique of Labeling Theory

   John Braithwaite's Shaming Theory

Conflict Criminology

   Austin Turk

   William Chambliss and Richard Quinney—
   Conflict Theory

   W. E. B. Du Bois

   Jeffrey Reiman

Feminist Criminology

New Critical Criminology

   Left Realism

   Peacemaking

   *Crime File 7.1 Incorporating Restorative and
   Community Justice Into American
   Sentencing and Corrections*

   Postmodernism

Radical "Marxist" Criminology

   Richard Quinney—Radical Criminology

   William Chambliss

Conflict Versus Marxist Criminology

Critiques of Conflict and Radical Criminology

   Klockars's Critique

Integrated Theories of Crime

   Delbert Elliott's Integrative Theory

   Terence Thornberry's Interactional Theory

Criminal Typologies

   A Critique of Typologies

   A Defense of Typologies

   Criminal Behavior Systems

   *Crime File 7.2 Some Sociological Typologies
   of Criminal Behavior*

Theoretical Range and Criminological
Explanation

   The Global Fallacy

The Theory–Policy Connection

Summary

Key Concepts

Review Questions

Web Sources

Web Exercises

Selected Readings

chapter 7

Sociological Critical Theories and Integrated Theories

The whole political process of law making, law breaking, and law enforcement becomes a direct reflection of deep-seated and fundamental conflicts between interest groups and their more general struggles for the control of the police power of the state.

—*George Vold,* Theoretical Criminology *(1958, pp. 208–209)*

Mainstream Versus Critical Criminology

The general characteristics of mainstream criminology, although subject to variation in individual anomie, social process, or social control theories, include the following (Gibbons, 1979, pp. 77–79; Gibbons & Garabedian, 1974):

- An emphasis on criminal behavior rather than on the criminalization of behavior. Emphasis had been on the criminal rather than on the social control machinery.
- A consensus worldview in which the existing society and its operations are perceived as relatively viable or unquestioned.

- A critical, sometimes cynical stance with respect to societal institutions, combined with a liberal optimism on reform measures.
- A mild pessimism regarding the perfectibility of the criminal justice system, but willingness to work within the established social order.
- Advocating the rehabilitation of offenders and their adjustment to the status quo.
- A positivistic orientation that stresses objectivity and empirical analysis.

Much contemporary criminological theory fails to address the full range of criminal behavior and confines its theorizing to the measurable, official crime and delinquency and lower level white collar crimes. It is unclear whether John Gotti, Charles Keating, Oliver North, or Osama bin Laden lack self-control, have no bonds to society, or have IQ deficits; and it is unclear whether executives at General Electric, North American Rockwell, General Motors, or Ford fit traditional "lambda" profiles (rates of offending), have IQ or genetic deficits, or should be the subjects of "three strikes and you're out" provisions.

Critical criminology consists of a variety of perspectives that challenge basic assumptions of mainstream criminology. It is espoused by a group of U.S. thinkers who emerged in the sixties and seventies and who have been variously labeled as representatives of "conflict," "radical," "new," "critical," or "Marxist" criminology. Inciardi (1980, p. 7) explains:

> The perspective is new and radical in that it departs somewhat from the mainstream or traditional criminological emphases on the nature and etiology of criminal behavior; it is conflict oriented and critical in that it focuses more fully on value and cultural differences, social conflicts, racism, and sexism as sources of crime and deviance in contemporary society; and it is Marxist in that a number of its representatives argue that law and, by extension, crime and the structure of individual and group interactions which support legal codes flow from the manner in which the relations of economic production are organized.

Critical Criminology

Critical criminology consists of five major types of theoretical approach: the labeling (societal reaction) perspective, conflict theory, and the feminist, new critical, and radical (Marxist) viewpoints. While each of these approaches will be detailed shortly, here are some common characteristics of critical criminology:

- Crime is a label attached to behavior, usually that of the less powerful in society.
- More powerful groups in society control this labeling process in order to protect their vested interests.
- The conflict model rather than the consensus model explains the criminalization process.
- Crime is often a rational response to inequitable conditions in capitalistic societies.

More extreme statements advocate a critical philosophy and practical revolutionary action (**praxis**) as opposed to value-free scientific inquiry.

While views of critical criminologists diverge, making it difficult to identify unitary themes, critical criminologists perceive themselves as making a radical break with a consensus, ameliorative, and essentially conservative worldview. Critical criminologists view their mainstream counterparts as handmaidens or social technicians for the status quo and see themselves as champions of the underdog and sometimes as prophets of a new social order. They feel that mainstream theories seemed to ignore economic, racial, and sexual inequality. Milovanovic (1996) sees more recent critical criminology as an outgrowth of radical and feminist criminology of the 1970s and 1980s. Table 7.1 outlines the major approaches to critical theory. It also includes an outline of integrated theories.

TABLE 7.1 ■ Critical and Integrated Criminological Theories

Theoretical School	Major Themes/Concepts	Major Theorists
Critical		
Labeling	Crime reflects the conflict model	
	Societal reaction	
	Dramatization of evil	Tannenbaum
	Secondary deviance	Lemert
	Crime as label, status	
Conflict	Imperatively coordinated associations	Dahrendorf
	Pluralistic model	Vold
	More powerful groups define criminal law	Turk
Feminist	Feminist criminology	Daly
	Androcentric bias	Chesney-Lind
	Patriarchy	Messerschmidt
New Critical	Left realism	Young
	Realistic social policy	DeKeseredy
	Peacemaking	Milovanovic
	Postmodernism	Quinney, Pepinsky, Henry, Pfohl
Radical	Capitalism causes crime	Quinney, Chambliss
	Attempts to link theories	
Integrated	Integrated theory of juvenile delinquency	Elliott
	Interactional theory of delinquency	Thornberry

Note: See Tables 4.1, 5.1, and 6.1 for other theoretical approaches in criminology.

The discussion of critical criminology will begin first with the labeling (societal reaction) theory, followed by conflict, feminist, new critical, and radical criminological theory.

Labeling Theory

If men define situations as real, they are real in their consequences.

—*W. I. Thomas,* The Child in America *(1928)*

Although there were earlier precedents, labeling theory (sometimes called "the societal reaction perspective") became a major criminological approach in the sixties, primarily in the United States. Labeling theorists base their point of view on symbolic interactionism, a school of thought that emphasizes the subjective and interactional nature of human experiences. Derived from the writings of George Herbert Mead and Charles Horton Cooley and expressed later in the work of Herbert Blumer,

Handbook Article Link 7.1
Read an article on labeling and symbolic interaction theories.

George Homans, and Harold Garfinckel, with variations called exchange theory, ethnomethodology, and role theory, the emphasis in symbolic interactionism is on analysis of subjective meanings of social interaction as perceived from the standpoint of the actor. Individuals perceive the meaning of their activity through the reaction of others.

Labeling theory says that individuals are deviant mainly because they have been labeled as deviant by social control agencies and others. The notion of deviance is not inherent in the act itself, but rather in the reaction and label attached to the actor; that is, crime is a label and not an act. Frank Tannenbaum called the process of attaching a label to deviants "the dramatization of evil" (1938). He viewed this criminalization process as

> a process of tagging, defining, identifying, segregating, describing, emphasizing, making conscious and self-conscious; it becomes a way of stimulating, suggesting, emphasizing, and evoking the very traits complained of. (pp. 19–20)

Along with Edwin Lemert (1951), Howard Becker (1963, 1964), Edwin Schur (1969, 1971), and others, Tannenbaum and the labeling theorists attempted to shift criminological inquiry from the deviant act to the machinery of social control and societal reaction. In a sense, this reverses the usual process of analysis; rather than assuming that criminal behavior causes societal reaction, it posits that societal reaction causes criminal behavior.

Schrag (1971, pp. 89–91) summarizes some of the basic assumptions of labeling theory:

- No act is intrinsically criminal.
- Criminal definitions are enforced in the interests of the powerful.
- A person does not become a criminal by violation of the law, but only by the designation of criminality by authorities.
- Due to the fact that everyone both conforms and deviates, people should not be dichotomized into criminal and noncriminal categories.
- The act of "getting caught" begins the labeling process.
- "Getting caught" and decision making in the criminal justice system are a function of the offender as opposed to offense characteristics.
- Age, socioeconomic class, and race are the major offender characteristics that establish patterns of differential criminal justice decision making.
- Labeling is a process that eventually produces identification with a deviant image and subculture and a resulting "rejection of the rejecters."

Lemert's "Secondary Deviance"

Two important concepts in labeling theory are Edwin Lemert's (1967, p. 17) notions of primary deviance and secondary deviance. **Primary deviance** refers to the initial deviant act itself, while **secondary deviance** is concerned with the psychological reorganization the individual experiences as a result of being caught and labeled as a deviant. Once this stigma or discrediting mark or status is attached, the individual may find it very difficult to escape the label and may come to identify with this new deviant role.

Deviant behavior, then, is viewed as having been created in society by control agencies representing the interest of dominant groups (Piven, 1981, p. 490). For Lemert the usual approach to analyzing deviance is reversed. He states (Lemert, 1967):

> This is a large turn away from the older sociology which tended to rest heavily upon the idea that deviance leads to social control. I have come to believe that the reverse idea, i.e., social control leads to deviance, is equally tenable and the potentially richer premise for studying deviance in modern society. (p. v)

Cullen and Agnew (2003, p. 271) give the example of sexual assault on women. In the past, designation of rape was reserved for victimizations by strangers in which physical injuries have occurred—what Estrich (1987) said the criminal justice system at the time considered to be "real rape." The women's rights groups challenged this and broadened the label to include "date rape," or sexual assaults committed in intimate relationships. A new reality was constructed, and rape was broadened to include a wider range of victimizations.

Sociologist Howard Becker (1963) has coined the term *moral entrepreneurs* to describe agents or officials who are concerned with creating and labeling new categories of deviance in order to expand the social control function of their organization. In Becker's view, deviance, rather than being inherent in the quality of the act, is so designated only by societal reaction and the subsequent labeling or stigmatization process.

A Critique of Labeling Theory

Some of the criticisms of the labeling perspective include the following:

- Labeling theory is overly deterministic and denies individual responsibility. Akers (1967) very dramatically states:

  Those of this school come dangerously close to saying that the actual behavior is unimportant. . . . One sometimes gets the impression from reading this literature that people go about minding their own business, and then "wham" bad society comes along and slaps them with a stigmatized label. Forced into the role of deviant the individual has little choice, but to be deviant. This is an exaggeration of course, but such an image can be gained easily from an overemphasis on the impact of labeling. (p. 46)

- Violators of societal rules are not passive robots of societal reaction.

Photo 7.1

Carjacking is regarded as *mala in se.*

- Some acts are universally regarded as intrinsically "wrong" (Wellford, 1975, p. 334). While labeling theorists have concentrated on public order crimes where the model may be more appropriate, they tend to generalize to all forms of deviance. Murder, forcible rape, aggravated assault, and robbery are more universally regarded as *mala in se*. Schur (1971) observes, "borderline forms of deviance seem to be especially good candidates for labeling analysis and those deviations on which widespread consensus exists (homicide, incest, and so on) less promising candidates" (p. 14).
- The societal reaction approach pays inadequate attention to the causes of the initial deviant act, almost as if to say that the social control agencies cause crime.
- While labeling theorists citing self-report surveys argue that nearly everyone commits crime, their argument seems to suggest that labels are attached capriciously, almost randomly. In fact, offenders involved in serious crimes are more likely to be labeled.
- Wellford (1975), on the basis of a review of Schrag's assumptions and the existing empirical evidence in criminology and the social sciences, concluded, "The assumptions underlying the theory are at significant variance with the data as we now understand it, or are not crucial to the labeling perspective" (p. 343).

Robert Bohm (1997, pp. 116–118) offers some additional criticisms of labeling theory:

- In siding with the "underdog," labeling theorists tend to romanticize the offender as someone reacting to an unjust society. Most of these offenders are victimizing others from the same social group and are committing real harm and suffering.
- Labeling is not a theory, but a "sensitizing concept."
- It does not explain primary deviance. The label does not create the initial act. In some instances people develop criminal self-images without ever being labeled. Bohm (1997, p. 117) indicates: "Furthermore, if the delinquent label is so stigmatizing, why do most delinquents not engage in adult criminality. . . . Why do most criminals stop their illegal activities when they reach middle age?"
- Labeling theory ignores individual differences among criminals, for example, low-risk versus high-risk criminals.
- The theory has a simplistic view of the criminalization process and the differential power of label makers.
- Does the process of labeling create more crime than it prevents? The answer is unknown.

By focusing primarily on the social control machinery, labeling theory has obvious inadequacies as a general theory of criminality; but this focus on societal reaction corrected an overly conservative, positivistic approach to criminological theory a tradition to be even further challenged by conflict criminology. While critics are correct that studies of the enforcement laws and administration of justice for traditional crimes (murder, rape, and the like) do not indicate bias (Wellford, 1975), this does not therefore repudiate the labeling point of view since it still does not speak to the conflict perspective that acts committed by the poor are more likely to be labeled criminal than acts committed by the wealthy. Labeling theory appears to have some validity with respect to areas of deviant behavior, such as mental illness, and in highlighting the lack of stigmatization in many areas of organizational and occupational crime, but it has clearly been repudiated when attempts have been made to apply it to traditional and universally condemned crimes such as murder.

John Braithwaite's Shaming Theory

John Braithwaite's (1989a) **shaming theory** argues that stigmatizing shaming of offenders makes matters worse and increases crime. Such a process makes the offender an irredeemable outlaw, irreconcilable with the community. In a sense the person is made into a permanent persona non grata and has little choice but to associate with similarly stigmatized persons. Braithwaite calls for "reintegrative shaming," efforts to reintegrate the offender back into the community of respectables. He claims that this is practiced in Japan and is one of the reasons for that country's low crime rate. Significantly, Braithwaite (1989b) applies his theory to organizational offenders, an admirable effort given criminological theory's obsession with juvenile delinquency. Acceptance back into the conventional society reinforces conventional social bonds and reduces recidivism.

Conflict Criminology

Conflict theory in sociology has a long tradition, beginning with Georg Simmel (1955) in his *Conflict and the Web of Group Affiliations,* originally published in English in 1908. Criminological expressions of this tradition can be traced to Marx and Bonger, discussed in Chapter 4, and later to Ralf Dahrendorf (1959) and George Vold (1958). In Chapter 1 we made a distinction between the consensus model, which views criminal law as originating in agreement of the majority, and the conflict model, which points to a conflict of interest among groups in which the dominant group controls the legal machinery of the state. The initial edition of Vold's *Theoretical Criminology* was the first to be extensively based on the conflict approach. Thorsten Sellin's (1938) notion of "culture conflict" as an explanation of crime is also part of this tradition. Sellin viewed criminal law as originating in cultural or normative conflict in which more powerful groups in society are able to make laws that reflect their norms and values.

Ralf Dahrendorf (1959) reformulated Marxian theory in *Class and Class Conflict in Industrial Society,* proposing a more pluralistic conflict theory that depicts numerous groups competing for power, influence, and dominance. While Dahrendorf did not specifically speak to the crime issue, his theoretical work influenced much of the conflict tradition in criminology.

George Vold's (1958) *Theoretical Criminology,* subsequent editions of which were posthumously updated by Thomas Bernard (Vold, 1979; Vold & Bernard, 1986; Vold, Bernard, & Snipes, 2002), builds on the work of Dahrendorf. Vold proposed that society is made up of a variety of continually competing interest groups and that conflict is one of its essential elements (Vold, 1979, p. 204), with more powerful groups able to have the state formulate laws in their interests. In Vold's view, many criminal acts represent challenges by subordinate groups to the existing dominant group's control, although he seems to restrict this explanation to issues related to political-ideological conflicts such as political reform movements, union conflicts, civil rights disputes, and the like. Crime, then, can be explained as a product of intergroup conflict that expresses the political struggle of these groups. While Vold's theory does not adequately explain irrational, personal, violent acts, his emphasis on the conflict basis of criminal law had a profound impact on later theories.

As indicated previously, a number of tags are used to refer to the "new" or emergent conflict criminology in the seventies. It is at times difficult to distinguish between "conflict" criminology (which, as expressed by Vold and Dahrendorf, proposes a pluralistic model with a variety of competing groups) and "radical" criminology (which generally espouses an orthodox, neo-Marxian, ideological view). Austin Turk has been one of the more persistent advocates of conflict criminology. Many figures to be discussed, particularly William Chambliss and Richard Quinney, demonstrate theoretical evolution from early conflict-orientation to later, more Marxian conceptions. The pluralistic conflict approach assumes that different class, racial, ethnic, and subculturally distinct interest groups vie for political dominance and the assistance of the legal machinery of the state in order to protect their interests (Hills, 1971). Unlike the situation with the Marxian model, no one group dominates completely.

Austin Turk

Austin Turk (1969a, 1972, 1980) has been a prolific writer in the conflict perspective. His basic position can be summarized in the following propositions:

- Individuals are different in their understandings and commitments.
- Divergence leads to conflict.
- Each conflicting party tries to promote his or her own views.
- This leads to a conscious struggle over the distribution of resources.
- People with similar beliefs tend to join forces and develop similar understandings and commitments.
- Continuing conflicts tend to become routine and develop into stratification systems.
- Such systems exhibit economic exploitation, sustained by political domination in all forms.
- The relative power of conflicting parties determines their hierarchical position as well as changes in the distribution of power.
- Human understandings and commitments are dialectical, characterized by continual conflict (Turk, 1980, pp. 82, 83).

Turk's theory, while abstract, alerts us to the political nature of criminal law as well as to the pluralistic conflict basis of such norms.

William Chambliss and Richard Quinney—Conflict Theory

Other statements of conflict theory are in the early works of William Chambliss (with Robert Seidman, 1971) and Richard Quinney (1970); their later writings would evolve into more radical perspectives. Chambliss and Seidman viewed criminal law as representing the interests of the most powerful forces in society and deviance as a political rather than moral question (Chambliss & Seidman, 1971, p. 4). Richard Quinney (1970, pp. 15–23) in *The Social Reality of Crime* presented six propositions describing the relationship between crime and the social order:

- Crime is a definition of human conduct created by authorized agents in a politically organized society.
- Criminal definitions describe behaviors that conflict with the interests of segments of society that have power to shape public policy.
- Criminal definitions are applied by segments of society that have power to shape the enforcement and administration of criminal law.
- Behavior patterns are structured in segmentally organized society in relation to criminal definitions, and within this context persons engage in actions that have relative probabilities of being defined as criminal.
- Conceptions of crime are constructed and diffused in the segments of society by various means of communication.
- The social reality of crime is constructed by the formulation and application of criminal definitions, the development of behavior patterns related to criminal definitions, and the construction of criminal conceptions.

Critics of Quinney's formulations argue that his propositions oversimplify reality and that many represent statements rather than necessarily empirically supported propositions (Manning, 1975). A more detailed critique will be provided at the conclusion of this chapter.

W. E. B. Du Bois

Chambliss, in a classic study in conflict criminology, examined the vagrancy laws in fourteenth-century England that made it illegal to give alms to anyone who was able, but unemployed. Due to the plague, there was a vast need for labor. W. E. B. Du Bois, pictured in Photo 7.2, did a similar analysis and, as pointed out by Gabiddon (1999), he represents a neglected conflict criminologist.

While most of his activities were associated with civil rights activity, Du Bois's academic publications included his book *The Philadelphia Negro* (Du Bois, 1899 [1973]). As Gabiddon indicates, his most important work concerning conflict theory was "The Spawn of Slavery: The Convict-Lease System in the South" (Du Bois, 1901). According to Russell (1992), Du Bois may also be considered the "founder of black criminology." Similar to Chambliss's study of vagrancy laws, Du Bois discussed the enactment of the Black Codes and convict lease system by the Southern oligarchy as a means to compensate for lost labor and profits as a result of Emancipation. The courts meted out two forms of justice: different sentences for whites and blacks. As Gabiddon (1999) indicates: "African-Americans were criminalized to secure the necessary labor for aristocracy" (p. 4). The fact that his work may have represented one of the earliest scientific works on crime but was ignored by early American criminology illustrates the "Eurocentric bias." The latter refers to the dominance of criminological discourse by writers of European descent and the ignoring of works by those of African descent (Green, 1979; Ross & Edwards, 1998; Young & Sulton, 1991).

Photo 7.2

W. E. B. Du Bois (1868–1963) may be considered the "founder of black criminology."

Jeffrey Reiman

In *The Rich Get Richer and the Poor Get Prison,* Jeffrey Reiman (1998) argues a conflict perspective that includes the following propositions:

- Acts that are not treated as crimes pose at least as great a danger to the public as those that have been criminalized.
- Acts that are criminalized are generally those of the poor.
- The system often fails to treat as criminal the dangerous acts of the wealthy and powerful.
- The failure of the criminal justice system in fighting street crime conveys an important ideological message, that the greatest danger to the average citizen is from below him or her on the economic ladder.
- Crime in the suites should be prosecuted in the same manner as crime in the streets, and all acts should be prosecuted in proportion to the actual harm they produce.

Reiman (1984) concludes: "Every step toward economic and social justice is a step that moves us from a system of *criminal* justice to a system of criminal *justice*" (p. 162).

Feminist Criminology

Feminist criminology comes in a variety of forms, but shares in common the general theme that "malestream" (male-mainstream; McDermott, 1992; Renzetti, 1993) approaches to criminology express an androcentric bias and exclude women from their analysis. Emphasizing various perspectives including Marxist, interactionist, and critical theory, feminist writers view dominant empirical positivism

Journal Article Link 7.1
Examine literature regarding race, abuse, and female criminal violence.

Photo 7.3

Feminist criminology created new crimes and criminals to study. Gender is now being incorporated into theory in a big way.

Journal Article Link 7.2
Examine literature regarding masculinity and violence.

as failing to include gender as a central force, blind to its ideological bias, and ignoring females. Their view is that much nonfeminist research is sexist due to cultural beliefs and to a preponderance of perspectives that assume traditional gender roles. This bias expressed itself in the past, particularly on topics such as rape and domestic violence.

A huge literature now exists on feminist theory in criminology (see Belknap, 1996; Bowker, 1998; Chesney-Lind & Shelden, 1998; Daly & Tonry, 1997; Dobash, Dobash, & Noaks, 1995; Messerschmidt, 1997; Miller, 1998; Muraskin, 2000; Rafter & Maher, 1995; Simpson & Ellis, 1995). Three general areas of crime have received the most attention in feminist theories: the victimization of women, gender differences in crime, and gendered justice (the differential treatment of females in the justice system; Bohm, 1997, p. 133). Crime is examined as it is related to gender-based inequality. There are actually a variety of approaches under the rubric of feminist criminology.

A leader in feminist criminology, Meda Chesney-Lind, explains that the earliest feminist criminology dealt with victimization issues, for example, sexual assault (American Society of Criminology, 2004). Feminist criminology "created" new crimes and criminals to study. Later added as a topic were women as workers in the system. This was an outgrowth of the civil rights movement and increasing numbers of women in policing. Women in corrections and the courts were other concerns, as were women offenders. The field had ignored gender in early theories and currently is incorporating it into theory in a big way.

A basic distinction can be made between "liberal feminists" and "critical" or "radical feminists." The liberal feminists are represented by pioneering works in the 1970s such as Rita Simon's *Women and Crime* (1975) and Freda Adler's *Sisters in Crime* (1975). Simon predicted an increase in female crime as opportunities increased. Adler also foresaw an increase in female criminality. She assumed that, as women assumed more assertive positions in society, they would participate in more previously "masculine" activities including crime. While the relationship appears logical, little support for this thesis was found. This liberation thesis was additionally not supported in that the greatest increase in female offenders was among those not achieving greater occupational equality. Radical feminists argued that the liberal feminists understated the role of patriarchy (male dominance) and its ability to continue to control and victimize women (Cullen & Agnew, 2003, p. 343).

Radical feminism is the dominant approach today in feminist criminology. Its major theme is patriarchy (male power and domination in society). "Patriarchy defines women as subjects, with men having the right of control. Sexism defines the value of women in terms of the family (unpaid housework as natural) and gives men control over reproduction" (Williams & McShane, 1994, p. 236). Male violence against women, especially domestic violence and rape, was traditionally ignored and helped bolster the patriarchal system (Danner, 1989). To overcome the androcentric bias in criminological theory it was deemed necessary to develop gender-specific theories.

In what is called the "generalizability problem," the question is raised as to whether theories of men's behavior apply to women (Daly & Chesney-Lind, 1988). Reviews of the criminological literature suggest that the answer is in the affirmative. This does not mean that gender-specific theories are not needed (Cullen & Agnew, 1999, p. 344). A "gendering" of traditional crime theories with larger structural (patriarchal) conditions holds promise.

James Messerschmidt in *Masculinities and Crime* (1993) claims that, even though feminists brought gender to the center of criminological theory, their vision of men is stereotypical. He views crime for some males as a way of "doing gender," exerting their manliness when other means are unavailable. This exertion of masculinity varies by age, class, race, and the like. Critics ask, "How does this relate to female crime?" While certainly a long overdue development, feminist criminology has been criticized for overfocusing on gender as its central theme. What about differences between white and black women? More research is needed, but feminist criminology will remain an active subject of inquiry and a permanent fixture in the field of criminology.

Encouraging women to examine crime through their own experiences with sexism, feminist research is at times in opposition to the scientific method (Simpson, 1989). While liberal feminism emphasizes affirmative action, it is viewed as not challenging "white, male, capitalist privilege" (Daly & Chesney-Lind, 1988). Socialist feminism sees capitalism and patriarchy as creating inequality and crime (Messerschmidt, 1986). Radical feminism views male aggression and control of female sexuality as the basis of patriarchy and the subordination of women. Rape, for example, is defined as a crime of male power and the use of violence to control and dominate women.

New Critical Criminology

New critical criminology includes emerging perspectives such as left realism (DeKeseredy, 1988), peacemaking, and postmodernism (Schwartz & Friedrichs, 1994). Such perspectives view the causes of crime as due to class, ethnic, and patriarchal (male-dominant) relations endemic in society (DeKeseredy & Maclean, 1993, p. 362).

Left Realism

Left realism questions the conservative approaches to crime control that emphasize prisons, more police, and longer sentences and argues instead for greater public access to, and involvement with, the police (DeKeseredy, 1988; Kinsey, Lea, & Young, 1986). Unlike Marxists, the left realists accept the reality of street crime and do not view it as a sort of revolutionary activity of the oppressed. With its primary expression in Britain (Jock Young) and Canada (Walter DeKeseredy), left realism attempts to translate radical ideas into realistic social policy (Williams & McShane, 1994, p. 166). Realists recognize that crime is a real problem that exists in socialist as well as capitalist societies, but insist on social justice as an important policy objective.

The term "realism" comes from the attempt to translate radical ideas into realistic social policy. Williams and McShane (1994) indicate that work by Tony Platt (1985) in the U.S. also reflected the attempt to make the perspective of practical use to policymakers without losing a critical perspective. Realists also believe that crime control is something to be taken seriously since it affects all social classes including the poor and working class. Proposals for police initiatives have included democratic forms of control over the police and community participation in the formulation of crime prevention schemes. Finally, left realists have one major goal: to emphasize "social justice as a way of achieving a fair and orderly society" (Matthews & Young, 1986, p. 6; Williams & McShane, 1994, p. 167).

Some of the proposals of Lea and Young (1984) contain some of the basic elements of left realism:

- Demarginalize offenders and instead of prisons, emphasize community service and restitution.
- Preemptive deterrence (before the fact) through citizen groups.
- Minimal use of prisons.
- Transform the "police force" into a "police service."
- Criminologists should be realistic about crimes (Beirne & Messerschmidt, 2000, p. 231).

Video Link 7.1
View a video of community policing.

Journal Article Link 7.3
Examine literature regarding a critique of left realism.

Photo 7.4

One of the basic elements of left realism is to demarginalize offenders and emphasize community service and restitution.

Peacemaking

Journal Article Link 7.4
Examine literature regarding peacemaking criminology.

A traditional role of policing has always been "peacekeeping." Advocates of what we will call "peacemaking" theory go beyond this and propose that crime can be eliminated once we establish peace and justice. **Peacemaking theory** or the "peacemaking movement" has its origins in the writings of Richard Quinney (1980, 1988) and Harold Pepinsky (Pepinsky & Quinney, 1991) and combines criminology with a transcendental or religious approach (Martin, Mutchnick, & Austin, 1990, p. 399). In *Providence* (1980), Quinney's thinking moved to a theological level. For example, he states: "Our historical struggle is thus for the creation of a social and moral order that prepares us for the ultimate of divine grace the kingdom of God fulfilled. Peace and justice through the Kingdom of God" (Quinney, 1980, p. 114).

In *Peacemaking* (Quinney, 1988, p. 67) he states: "Crime can be ended only with the ending of suffering (only when there is peace) through the love and compassion found in awareness." Peace can end suffering, which can end crime. Peacemaking is also a nonviolent approach to criminal justice. It assumes that violence cannot be overcome with more violence. Pepinsky (Pepinsky & Quinney, 1991, p. ix) calls for an " 'expressive criminology' of compassion, forgiveness and love . . . a continuing movement for a world of peace and social justice."

Quinney's (1991) progression to spiritualism as an approach to crime views social justice as the solution to the crime problem. Individuals must transcend their selves and understand that there is suffering in the world and that crime is suffering. Crime will cease when social justice is accomplished and suffering ended. Programs such as "restorative justice" fit with the peacemaking theme very well. Such an approach seeks to mediate conflict, assist victims, and reintegrate offenders into the community (Cullen & Agnew, 2003, p. 273). Crime File 7.1 describes some restorative justice programs.

It is as difficult to critique "peacemaking theory" as it is to criticize the religious beliefs of others. In a way, the field of criminological theory has come full circle from theological to philosophical to scientific and now back to a less demonological, humanistic theological approach. The message is the same message of prophets of old love one another and do God's work. But it is a theme that underlies much sociological thinking regarding crime and deviance; a more just social order (justice) is

Crime File 7.1

Incorporating Restorative and Community Justice Into American Sentencing and Corrections

Video Link 7.2
View a video of restorative justice in the community.

Programs based on restorative and community justice principles have proliferated in the United States over the past decade in concert with tough-on-crime initiatives like three-strikes, truth-in-sentencing, and mandatory minimum laws. Restorative justice and community justice represent new ways of thinking about crime. The theories underlying restorative justice suggest that government should surrender its monopoly over responses to crime to those most directly affected—the victim, the offender, and the community. Community justice redefines the roles and goals of criminal justice agencies to include a broader mission—to prevent crime, address local social problems and conflicts, and involve neighborhood residents in planning and decision making. Both restorative and community justice are based on the premise that communities will be strengthened if local citizens participate in responding to crime, and both envision responses tailored to the preferences and needs of victims, communities, and offenders.

In contrast to this bottom-up approach, recent changes in sentencing law are premised on retributive ideas about punishing wrongdoers and on the desirability of controlling risk, increasing public safety, and reducing sentencing disparities. Restorative and community justice goals of achieving appropriate, individualized dispositions often conflict with the retributive goal of imposing certain, consistent, proportionate sentences.

What Is Restorative Justice?

Restorative justice has evolved from a little-known concept into a term used widely but in divergent ways. There is no doubt about its appeal, although the varied uses of the term cause some confusion. The umbrella term "restorative justice" has been applied to initiatives identified as restorative by some but not by others. Examples are sex-offender notification laws, victim impact statements, and murder victim survivors' "right" to be present at executions. Most advocates of restorative justice agree that it involves five basic principles:

Journal Article Link 7.5
Examine literature regarding restorative justice policy.

- Crime consists of more than violation of the criminal law and defiance of government authority.
- Crime involves disruptions in a three-dimensional relationship of victim, community, and offender.
- Because the crime harms the victim and the community, the primary goals should be to repair the harm and heal the victim and the community.
- The victim, the community, and the offender should all participate in determining the response to crime; government should surrender its monopoly over that process.
- Case disposition should be based primarily on the victim's and the community's needs—not solely on the offender's needs or culpability, the dangers he presents, or his criminal history.

The original goal of restorative justice was to restore harmony between victims and offenders. For victims, this meant restitution for tangible losses and emotional losses. For offenders, it meant taking responsibility, confronting shame, and regaining dignity.

This notion has evolved, with the major recent conceptual development the incorporation of a role for the community. Many people still associate restorative justice primarily with victim–offender mediation or, more broadly (but mistakenly), with any victim-oriented services. The more recent conceptualization—that offenses occur within a three-dimensional relationship—may change the movement.

All three parties should be able to participate in rebuilding the relationship and in deciding on responses to the crime. The distinctive characteristic is direct, face-to-face dialogue among victim, offender, and increasingly, the community.

What Is Community Justice?

The concept of community justice is less clear. It can be portrayed as a set of new organizational strategies that change the focus of criminal justice from a narrow, case processing orientation; operations are moved to neighborhood locations that offer flexible working hours and services, neighborhoods are assigned their own officers and are provided with more information than is standard practice, and residents may identify crime problems and define priorities for neighborhood revitalization. Most experience with community justice is in the context of community policing, but prosecutors, judges, and correctional officers are increasingly rethinking their roles and goals.

Restorative Justice Practices

Although something akin to restorative justice has long been observed in premodern and indigenous societies, restorative justice principles, in the form of victim–offender reconciliation programs, appeared in Western industrialized countries only in the 1970s. The first program was established in 1974 in Kitchener, Ontario. By the 1990s, such programs had spread to all Western countries—at least 700 in Europe and 300 in the United States.

(Continued)

(Continued)

Victim–Offender Mediation. Victim–offender mediation is the most widespread and evaluated type of restorative program. Offenders and victims meet with volunteer mediators to discuss the effects of the crime on their lives, express their concerns and feelings, and work out a restitution agreement. The agreement is often seen as secondary to emotional healing and growth. Victims consistently report that the most important element of mediation is being able to talk with the offender and express their feelings, and offenders also emphasize the importance of face-to-face communication. Advocates believe that developing an offender's empathy for the victim has preventive effects.

In many countries, victim-offender mediation is widely used. In Austria, for example, it became an official part of the juvenile justice system as early as 1989. Public prosecutors refer juveniles to mediation, probation officers coordinate cases, and social workers serve as mediators. If an agreement is reached and completed, the case is dismissed.[1] In the United States, most programs are operated by private, nonprofit organizations; handle largely juvenile cases; function as diversion programs established and operated (or at least initiated) by corrections departments, police, or prosecutors; and are used as a condition of either probation or dropping charges. Most studies of mediation programs report high rates of success.[2]

Advocates are beginning to challenge the assumption that mediation is not suitable for violent or sexual crimes. Increasingly, in the United States and Canada, for example, victims and offenders meet in prisons. These meetings are not oriented to a tangible goal such as a restitution agreement, nor does the offender obtain benefits like early release or parole consideration. Usually the meetings are held because the victim wants to meet the offender and learn more about what happened to reach beyond fear and anger and facilitate healing. The results of a Canadian survey indicated that 89 percent of victims of serious, violent crimes wanted to meet the offender.[3]

Serious violent crimes are usually mediated on a case-by-case basis, but the need for permanent programs is growing. Such programs are offered, for example, by the Correctional Service of Canada in British Columbia and the Yukon Territory and by the Texas Department of Criminal Justice.

Family Group Conferencing. Family group conferencing is based on the same rationales as victim–offender mediation, with two main differences. Conferencing involves a broader range of people (family, friends, coworkers, and teachers), and family members and other supporters tend to take collective responsibility for the offender and for carrying out his or her agreement. The other difference is that conferencing often relies on police, probation, or social service agencies for organization and facilitation.

Family group conferences originated in New Zealand, where they became part of the juvenile justice system in 1989. There, the new juvenile justice model, which incorporates Maori traditions of involving the family and the community in addressing wrongdoing, has four dispositional options:

- An immediate warning by the police
- "Youth Aid Section" dispositions in which a special police unit may require, for example, an apology to the victim or community service
- Family group conferencing
- Traditional youth court sentencing

About 60 percent of juvenile offenders receive a warning or go the Youth Aid Section, 30 percent go to conferencing, and 10 percent go to youth court.[4]

By the mid-1990s, family group conferencing had been adopted in every state and territory of Australia. In South Australia, it is used statewide as a component of the juvenile justice system and resembles the New Zealand approach. In Wagga Wagga, New South Wales, conferences (originally part of a police diversion program) were organized and facilitated by police officers who were often in uniform.[5] Responsibility was transferred to juvenile justice agencies in 1998, and trained community members now facilitate conferences. In Canberra, the Federal Police set up a program called the Reintegrative Shaming Experiment, which involved more than 100 trained police officers.

There is evidence that conferencing can be successful. A recent evaluation of the Bethlehem, Pennsylvania, Police Family Group Conferencing program revealed that typical police officers were able to conduct conferences in conformity with restorative justice and due process principles if adequately trained and supervised, and that very high percentages of offenders, victims, and other participants were pleased with the process.[6] Evaluation of Canberra's Reintegrative Shaming Experiment showed similar results.[7]

Sentencing Circles. Sentencing circles originated in traditional Native Canadian and Native American peacemaking. They involve the victim and the offender, their supporters, and key community members, and they are open to everyone in the community. They attempt to address the underlying causes of crime, seek responses, and agree on offenders' responsibilities. The process is based on peacemaking, negotiation, and consensus, and each circle member must agree on the outcomes.

Sentencing circles are so named because participants sit in a circle, and a "talking piece" (a feather, for example) is passed from person to person. When participants take the talking piece, they explain their feelings about the crime and express support for the victim and the offender. Separate circles often are held for the offender and the victim before they join in a shared circle.

In Minnesota, sentencing circles are used not only in Native American communities but also in rural white, suburban, and inner-city black communities. Community Justice Committees, established by citizen volunteers, handle organizational and administrative tasks and provide "keepers" who lead the discussions. Judges refer cases, and the committees make the final decision on acceptance. The agreements reached are presented to the judge as sentencing recommendations. In some cases, the judge, prosecutor, and defense attorney participate in the circle, and then the agreement becomes the final sentence.

Reparative Probation and Other Citizen Boards. Reparative probation in Vermont involves a probation sentence ordered by a judge, followed by a meeting between the offender and volunteer citizen members of a Reparative Citizen Board. Together they draw up a contract, based on restorative principles, which the offender agrees to carry out. Fulfilling the contract is the only condition of probation.

Vermont's program is different from most other restorative justice initiatives in the United States. Designed by the state's Department of Corrections, it operates statewide, handles adult cases, and involves a sizable number of citizen volunteers. Compared with family group conferencing or sentencing circles, the Reparative Citizen Boards work faster, require less preparation, and can process more cases; however, they involve fewer community members. For example, offenders' and victims' families and supporters usually are not present.

Citizen boards also may be established to adjudicate minor crimes. For example, a Merchant Accountability board in Deschutes County, Oregon, consists of local business owners who adjudicate thefts of property valued at $50 or less and some more serious cases involving property valued at between $51 and $750. Under an agreement with the district attorney, the police refer all minor shoplifting cases directly to the program. If offenders decide to participate, they are typically ordered by the board to pay fines, make restitution, or both.

Manitoba's Restorative Resolutions Project offers an alternative to custodial sentences for offenders who otherwise are likely to face a minimum prison sentence of 6 months. Offenders and project staff develop sentencing plans, and victims are encouraged to participate. The plans are presented to judges as nonbinding recommendations. Most plans require restitution, community service, and counseling or therapy. A recent evaluation revealed that offenders who participate have significantly fewer supervision violations and slightly fewer new convictions than those in comparison groups.[8]

The Future of Restorative and Community Justice

How deeply restorative and community justice ideas will penetrate the traditional justice system remains to be seen. So far, restorative justice approaches are used much more for juveniles than for adults, and for minor offenses rather than for serious crime. Experience with community justice has consistently shown that generating citizen involvement and building relationships with the community is a challenge. Both movements have spread rapidly, however, and both are increasingly reaching out to encompass adult offenders, more serious crime, and disadvantaged urban communities where, arguably, the need is greatest.

Notes

1. Lösching-Gspandl, Marianne, and Michael Kilchling (1997). Victim/Offender Mediation and Victim Compensation in Austria and Germany—Stock-Taking and Perspectives for Future Research. *European Journal of Crime, Criminal Law and Criminal Justice, 5,* 58–78.

2. Umbreit, Mark (1994). Victim Meets the Offender. *The Impact of Restorative Justice and Mediation.* Monsey, NY: Criminal Justice Press. It should be noted that evaluations of restorative justice conducted in the United States are usually not based on experimental and control groups, do not often measure recidivism rates, and seldom use sophisticated research designs.

3. Gustafson, Dave (1997). Facilitating Communication Between Victims and Offenders in Cases of Serious and Violent Crime. *The International Community Corrections Association Journal on Community Corrections, 8,* 44–49.

4. Maxwell, Gabrielle, and Allison Morris (1993). *Family, Victims, and Culture: Youth Justice in New Zealand.* Wellington, New Zealand: Social Policy Agency and Institute of Criminology, Victoria University of Wellington.

5. Wundersitz, Joy, and Sue Hetzel (1996). Family Conferencing for Young Offenders: The South Australian Experience, in *Family Group Conferences: Perspectives on Policy and Practice, ed.* John Hudson et al. Monsey, NY: Criminal Justice Press, 1996.

6. McCold, Paul, and Benjamin Wachtel (1998). *Restorative Policing Experiment: The Bethlehem Pennsylvania Police Family Group Conferencing Project.* Pipersville, PA: Community Service Foundation. This evaluation was sponsored by the National Institute of Justice.

7. Sherman, Lawrence, Heather Strang, Geoffrey Barnes, John Braithwaite, Nova Ipken, and Min-Mee (1998). *The Experiments in Restorative Policing: A Progress Report to the National Police Research Unit in the Canberra Reintegrative Shaming Experiments (RISE).* Canberra: Australian Federal Police and Australian National University.

8. Bonta, James, Jennifer Rooney, and Suzanne Wallace-Capretta (1998). *Restorative Justice: An Evaluation of the Restorative Resolutions Project.* Ottawa: Solicitor General of Canada.

Source: Leena Kurki (1999). "Incorporating Restorative and Community Justice Into American Sentencing and Corrections (Sentencing and Corrections Issues for the 21st Century)." National Institute of Justice Research in Brief, Papers from the Executive Sessions on Sentencing and Corrections No. 3, September.

Web Research Project

Use the companion Web site and click on the article by Lemley titled "Designing Restorative Justice Policy." What does the article add to your understanding of the notion of restorative justice?

necessary before one can achieve law and order. In this writer's opinion, peacemaking in the final analysis is an admirable social movement, a utopian *Weltanschauung* (worldview) more than it is an attempt to explain specific types of crime.

Postmodernism

Postmodernism is a movement that attacks modernity. Bohm (1997, pp. 134–135) tells us that postmodernism began in the late 1960s as a rejection of "modern" or Enlightenment scientific rationality as the predominant philosophy for gaining knowledge and achieving progress. Unconscious, free-floating signs and images, and the rejection of knowledge and languages' ability to create hierarchy and domination, were viewed as critical. Postmodernists argue for a plurality of interpretations of the law and an abandonment of standard theories of crime causation (Bohm, 1997). The latter assumes that people can control objects, nature, and reality whereas the former assumes that objects now have more and more control over us (Schwartz & Friedrichs, 1994, p. 223). Having originated in the field of literary and linguistic analysis, postmodernists examine how knowledge is constituted, the significance of language and signs, and how metaphors and concepts capture reality and set the context and conditions in which crime occurs. Media and technology create a "hyperreality" in which simulations and reality become confused. Modernity has become a force not for liberation, but subjugation, oppression, and repression.

Criminologists who have been identified with postmodernism are Dragan Milovanovic (1992), Stuart Henry (Henry & Milovanovic, 1993), and Stephen Pfohl (1993). Stuart Henry and Dragan Milovanovic (1996) describe three elements of postmodernism in criminology:

- Crime is the ability to impose one's will on others.
- Some persons construct harms to others in the expression of power and control in which others are objectified as "separate, dehumanized entities" (p. 175).
- Law definers must be provided with "liberating life narratives" (p. 224).

While postmodernism has generated controversy and empirical research, its literature has been described as "gratuitously obscure, incoherent, and undisciplined" (Schwartz & Friedrichs, 1994, p. 228; see also Michalowski, 1993).

Radical "Marxist" Criminology

Richard Quinney—Radical Criminology

Perhaps the foremost spokesperson for radical criminology is the same Richard Quinney who was at one time a more moderate conflict theorist and is now a peacemaker. For Quinney then, an orthodox Marxist crime was the result of capitalism, and the crime problem could be resolved only by the establishment of a socialist state (Quinney, 1974a, 1974b, 1974c, 1977). In his critical theory of crime control in the United States, he provides the following propositions:

- U.S. society is based on an advanced capitalist economy.
- The state is organized to serve the interests of the dominant economic class, the capitalist ruling class.
- Criminal law is an instrument of the state and the ruling class to maintain and perpetuate the existing social and economic order.

- Crime control in capitalist society is accomplished through a variety of institutions and agencies established and administered by a governmental elite, representing ruling class interests, for the purpose of establishing domestic order.
- The contradictions of advanced capitalism, the disjunction between existence and essence, require that the subordinate classes remain oppressed by whatever means necessary, especially through the coercion and violence of the legal system.
- Only with the collapse of capitalist society and the creation of a new society based on socialist principles will there be a solution to the crime problem.

For Quinney and other Marxist criminologists, crime is a necessary outcome of inequality in capitalistic societies. Criminal law originates in conflict of interest in which the most powerful ruling class (capitalists or bourgeoisie) makes the laws and controls the criminal justice machinery. Marxist criminologists often reject the positivistic tradition of analyzing crime causation through objective and empirical analysis. Instead, they advocate an ideological commitment to Marxist philosophy wherein their task is to provide descriptive and analytical examples to serve as evidence for a precon-firmed social reality that capitalism causes crime.

William Chambliss

Radical criminologists argue that, by concentrating on the crimes of the poor rather than on racism, imperialism, and inequality, criminologists become conservative handmaidens of state repression (Platt, 1974). Advanced industrial capitalism creates "surplus people" (Spitzer, 1975), an underclass that is unneeded in the system of production. Among William Chambliss's (1975b) later views regarding capitalism and crime are these:

- As capitalist societies industrialize and the gap between the bourgeoisie and the proletariat widens, penal law will expand in an effort to coerce the proletariat into submission.
- Crime diverts the lower classes' attention from the exploitation they experience and directs it toward other members of their own class rather than toward the capitalist class or the economic system.
- Crime is a reality that exists only as it is created by those in the society whose interests are served by its presence.
- Crime is a reaction to the life conditions of a person's social class.
- Socialist societies should have much lower rates of crime because the less intense class struggle should reduce the forces leading to the functions of crime.

Similar perspectives have been enunciated by many others, including Gordon (1973), Krisberg (1975), and Taylor, Walton, and Young (1973, 1975). In their *Critical Criminology,* Taylor, Walton, and Young (1975, p. 49) called for the use of Marxism as the method of analysis in a "materialistic criminology" whose purpose is to expose the basis of social control in capitalistic societies. The tenets of Marxist theory, rather than representing subjects for empirical analysis, now become foregone conclusions, ideological dictates requiring illustration rather than proof.

Radical or Marxist criminologists view praxis (*practical critical action*) as more important than the objective analysis of their theoretical formulations. "They view 'intellectualism' as a negative quality due to the 'academic repression' and 'elitism' associated with intellectuals. Praxis is then the most important factor in the struggle to replace capitalism with socialism" (Pelfrey, 1980, p. 96).

Cullen and Agnew (1999) do a nice job of reviewing other writers who, although they are not Marxists, support the general theme of the harmful impacts of economic structure on youth opportunities. David Greenberg (1993), for example, indicates that an economic system that is unable to provide full-time jobs for teenagers consigns them to schools and prolonged adolescence. Peer group activities

Photo 7.5

Inner-city minority neighborhoods push youth toward criminal subcultural adaptations such as drug markets.

requiring consumption increases adolescent theft. "Masculine status anxiety" strikes those who are unable to gain employment and assume traditional male roles. Structural conditions of the economy may block the American dream (Cullen & Agnew, 2003). Colvin and Pauly (1983) argue that parents' class position influences how they discipline their children. Those in "dead-end jobs" are more coercive in socialization, often alienating their children and reducing parental bonds.

John Hagan in *Crime and Disrepute* (1994) views the capital disinvestment in inner-city minority neighborhoods as creating what William Julius Wilson (1987) calls "the truly disadvantaged." Such neighborhoods rob their youth of opportunities for legitimate advancement and push them toward criminal subcultural adaptations such as drug markets. Inequality and racism in postindustrial capitalism presents no solution to the crime problem unless there is large-scale investment in such inner-city neighborhoods (Cullen & Agnew, 2003).

Conflict Versus Marxist Criminology

While the two are often confused, conflict criminology posits a pluralistic conflict model (a diversity of conflicting parties), places less emphasis on capitalism alone as the source of crime, favors objective research, does not reject the legal order, and advocates reform rather than revolution (Bohm, 1982; Friedrichs, 1980b, p. 39). Marxist or radical criminology, on the other hand, advocates a singularistic conflict model (capitalistic class control), names capitalism and inequality as the sources of crime, holds Marxist theory as a fact to be illustrated rather than a subject for empirical investigation, rejects the legitimacy of the existing legal order, and advocates revolutionary overthrow of the system.

Critiques of Conflict and Radical Criminology

While conflict criminology has done much to reverse overconcentration on criminal actors and unquestioned acceptance of the consensus model of criminal law and to point to the criminal justice system as a possible transgressor, it has been criticized for ignoring the consensual basis of much criminal law and for assuming rather than demonstrating discrimination in traditional law enforcement.

Radical (Marxist) criminology has attracted a barrage of critics. In Geis and Meier's (1979) survey of leading criminologists, nearly 40 percent of the respondents indicated the emergence of Marxist ideology in criminology as a "less healthy development" in the field. Comments such as "ideology whether in theory or method is pretentiously seen as 'new paradigms,' 'theories,' 'methods'"; "the substitution of ideology for science"; "nonscientific voices"; "Marxist rhetoric and ideological narrowness" (Geis & Meier, 1979, pp. 180–181) were offered. Toby (1980) states that much of "the New Criminology is the Old Baloney," that this tradition, "far from being new, is the explicit assertion of a relativism and a sentimentality that is as old as sympathy for members of the oldest profession." Sparks (1980) criticizes radical criminologists for the lack of attention to solid research that would critically test their theoretical assumptions.

Klockars's Critique

The definitive, though most controversial, critique of Marxist criminology appears in Carl Klockars's (1979) "The Contemporary Crisis of Marxist Criminology," which in turn has stirred considerable commentary (Akers, 1980; Friedrichs, 1980a; Mankoff, 1980). Klockars's critique can be paraphrased in the following way:

- Marxist criminology resembles an untrustworthy social movement, since it ignores Russian gulags (Solzhenitsyn, 1975), Cuban domestic repression, and other abuses within socialist states. By giving a social movement a higher priority than academic inquiry, Marxist criminologists abandon science for ideology and are untrustworthy as objective scholars.
- Marxist criminology as a social movement operates on predictable, orthodox lines. After class, the legal order and capitalism are blamed for everything; these themes are reiterated ad nauseam.
- In their subjective zeal for advocating social revolution, Marxist criminologists find evil in everything associated with the American state, legal, and economic system, ignoring good laws. In their mystical transcendence of reality, they destroy their academic credibility. They dramatize and stress issues (e.g., that politicians are corrupt or businesses dishonest, as if these were startling revelations), insulting the intelligence of the general public.
- All of the problems of justice are collapsed into the economic interest of classes.
- American Marxist criminologists criticize society from "a moral ground set so high and so far removed from any extant social reality that it loses all perspective" (Klockars, 1979, p. 484).
- They elevate Marx from a social philosopher to the status of prophet or saint. By describing the ideal of Marxism, they avoid responsibility for the present depredations of existing Marxist states.
- Marxist criminology resembles a new religion in which its "true believers" are unwilling to test, evaluate, or objectively examine their theories or beliefs.

While the Klockars critique pulls no punches, it is difficult to apply these points to all writers within the Marxist tradition, although his criticisms appear on target on the whole. As Akers (1980) states:

Compared to a socialist ideal system, the real American system looks unjust, repressive, and controlled by a tiny capitalist elite. Compared to the Soviet Union, China, Vietnam, North Korea, East Germany or Cambodia, to name some socialist alternatives, or to Iran, South Korea or Chile, to name some nonsocialist alternatives, American society looks pretty good. (p. 138)

The collapse of Soviet communism may have in part reduced interest in Marxism.

Integrated Theories of Crime

A primary criticism of most theories that have been discussed is their tendency to associate crime with a single cause, for example, some biological or psychological defect. Critics of these approaches merely had to demonstrate the presence of these conditions in equal proportions among noncriminals in order to refute these assumptions. This single-factor deficiency has led some writers (Glueck & Glueck, 1950; Healy, 1915) to propose a multifactor approach in which crime is assumed to be produced by multiple factors (biological, psychological, and sociological) with different combinations of variables coming into play, depending on the type of crime being examined. This approach

Journal Article Link 7.6
Examine literature regarding theoretical integration.

is appealing in that multiple factors are indeed involved in any causal explanation of criminality; however, the identification of factors associated with a process does not constitute a causal theory. In that sense, the multifactor approach is atheoretical (without theoretical content).

Albert Cohen (1951) has provided a succinct critique of the multifactor approach, which may be paraphrased:

- Advocates of this approach confuse causal theories that employ a single variable with those that propose a single theory. Simply listing correlations of factors associated with crime does not represent a theory, while a single theory may utilize multiple factors.
- Due to the emergence of easily available, sophisticated statistical programs that enable the calculation of multiple correlations, researchers forget that correlation does not equal causation. Since variables account for a certain proportion of variance in crime, this does not mean that they substantively cause that amount of crime.
- This approach falls into the "evil causes evil" fallacy: evil outcomes require evil causes, which represents a conservative, consensus view of crime as an evil intrusion into an otherwise healthy society.

Integrated theories attempt to combine various theoretical traditions into one theory. Such integrated theories are more than the identification of factors involved in crime, but attempt to theoretically link these factors into theories that explain crime. Messner, Krohn, and Liska in *Theoretical Integration in the Study of Deviance and Crime* (1989) were very influential in identifying this genre of theories. Cullen and Agnew (2003, p. 207) tell us that the most common strategy for formulating integrated theories is to temporally order theories "end to end." That is, a theorist might link theories by showing how a high level of strain might lead individuals to join subcultures, which then leads to crime.

In order to illustrate the usefulness of theories, Figure 7.1 presents an intensive aftercare model used in research sponsored by the Office of Juvenile Justice and Delinquency Prevention for guiding serious, chronic juvenile offenders. This particular model links strain (anomie) theory, social learning theory, and social control theory.

They also point out that many theories, although not specifically identified as integrated theories, possess that quality in attempting to link different theories (Cullen & Agnew, p. 208). Shaw and McKay of the Chicago school attempted to bring together elements of strain, learning, and social control theory. Cohen's lower-class reaction theory tried to tie together strain and differential association, as did Cloward and Ohlin. While most of the integrated theories have been at the micro level attempting to explain individual deviance, some have also been at the macro level looking at the impact of cultural and social structural forces. While a number of theories have been identified as being examples of integrated theories (see Akers, 1994), two examples that are on nearly every list are Delbert Elliott, Suzanne Ageton, and Rachelle Cantor's (1979) integrative theory of juvenile delinquency and Terence Thornberry's (1987) interactional theory of delinquency.

Delbert Elliott's Integrative Theory

Delbert Elliott (1985) combines strain (anomie), social control, and learning theory. Delinquency (as measured by self-reports in the National Youth Survey) is due to:

1. Strain due to the gap between aspirations and achievements as well as other sources of strain such as the family and school (strain theory).

2. Attachment and commitment to family and school (social control or bonding theory).

3. Exposure to, preference for, and identification with deviant peers (learning theory).

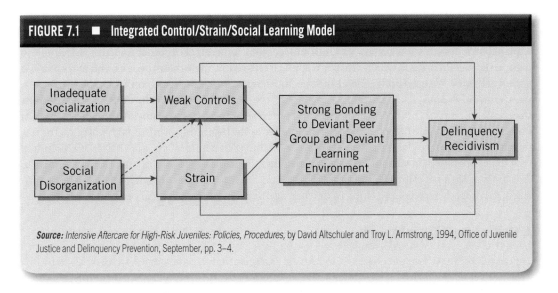

FIGURE 7.1 ■ Integrated Control/Strain/Social Learning Model

Source: Intensive Aftercare for High-Risk Juveniles: Policies, Procedures, by David Altschuler and Troy L. Armstrong, 1994, Office of Juvenile Justice and Delinquency Prevention, September, pp. 3–4.

Elliott has found support for his theory using the National Youth Survey. In this he found that bonding and strain variables had little effect themselves on delinquency. Bonding to delinquent peers had the major effect. Thus, social learning appeared most significant (Akers, 1994, p. 189). Social learning rather than social bond theory was most predictive. "The social bonding proposition that strong attachment to others prevents delinquent behavior, even when that attachment is to unconventional peers, is not supported" (Akers, 1994, p. 190). Thus, in an attempt to integrate social control with social learning theory, only the latter survives. More research, of course, is still needed.

Terence Thornberry's Interactional Theory

Thornberry's (1987) interactional theory attempts to combine social structure, social control, and social learning theories. Social structure, such as race, social class, and community, affects social control and learning. The weakening of bonds to society predisposes adolescents to delinquency. Now the learning factor occurs, in which identification and association with delinquent peers happens. These relationships are reciprocal and not unilinear (one way), that is, family attachment may affect school commitment, but the opposite is also true (Cullen & Agnew, 2003). Thornberry also adds a life course dimension to his theory indicating that the importance of different variables changes over the life course. Akers (1994, p. 192) tells us that Thornberry, Lizotte, Krohn, Farnworth, and Jang (1991) did not find support for their hypothesis about reciprocal relationships between parental attachment and school commitment.

Criminal Typologies

One limitation of many discussions of crime and of theories of crime causation is the global manner in which the concept of crime is employed. To expect criminologists to address the question, "What causes crime?" is comparable to asking medical pathologists to answer the query "What causes sickness?" Asking "What *type* of sickness?" or "What *type* of crime?" is the next logical step in approaching these questions. While the only thing most sicknesses have in common is that they have produced an unhealthy biological state, the only thing most crimes have in common is that they are, in a given place, at a given time, a violation of criminal law. Thus cancer, polio, and the common cold probably have about as much in common as shoplifting, embezzlement, and murder.

While it is important that the field of criminology continue the theoretical work of explaining crime and criminal behavior as a whole, it is also important and perhaps more expeditious in the short run to explain particular criminal behaviors. Until an acceptable general theory is developed, it is desirable to delimit the specific areas to which a theory is applicable, to coordinate these theories, and to try to build a general theory. We need both general and specific theories and must avoid confusing the two.

Criminal typologies are attempts to classify types of crimes and criminals. These attempts may represent one of the oldest theoretical and practical approaches to crime. Although the work of Lombroso (discussed in Chapter 5) is often pointed to as the beginning of criminal typologies, the tradition of attempting to classify lawbreakers precedes him (Schafer, 1969, pp. 140–182; 1976). Criminal typologies are based on various criteria. Crime File 7.2 outlines a few of the better known efforts to develop typologies of criminals or criminal behavior. These typologies are not intended to be memorized; they merely serve as an exhibit of the many different attempts to classify criminal behavior.

A Critique of Typologies

Typologies can have two purposes: (1) to be used as a scientific classificatory system, or (2) to be utilized as an educational tool. The former effort is exemplified by taxonomical classifications in biology where life forms are sorted into categories such as phylum, species, and the like on the basis of physical characteristics. Related to this tradition are prison classification systems (Fox, 1976) that attempt to line up criminal offense records with treatment regimens. This effort has obviously been limited by inadequacies of offense records themselves for the purposes of classifying individuals. Many critics of the typological approach expect typologies to meet rigorous taxonomical refinement. Their critiques of typologies include:

- Specific offenses vary according to time and place.
- Some offenders exhibit great diversity, participating in more than one behavior system, or may in fact change their offense profiles.
- No typology can contain purely homogeneous types.
- The number of career criminals specializing in one type of offense is smaller than has been suggested by the typologies developed thus far (p. 354).
- Some typologies attempt to make types of crimes and criminals more distinct from each other than they really are, thus oversimplifying reality (Conklin, 1972, p. 16).
- No single typology is useful to group all offenders (Thomas & Hepburn, 1983, p. 262).
- Typologies overemphasize unique aspects and minimize similarities among types (Thomas & Hepburn, 1983, p. 262).

A Defense of Typologies

The real value of criminal typologies is their educational benefit in providing a useful, illustrative scheme, a practical device that, although subject to abstraction and overgeneralization, enables us to simplify and make sense of complex realities. Any ideal types are prone to oversimplification, but without them the categorical equivocations in discussing reality become overwhelming. The first purpose of typologies as classificatory systems requires empirical verification using actual quantitative research, while the second purpose recognizes that concepts or typologies as ideal types have qualitative, heuristic value. They sensitize or alert us to and are useful in explaining critical features of reality even though as ideal or constructed types they obviously oversimplify that same reality.

Criminal Behavior Systems

As an organizing scheme, this text will make use of a variation of a typology of **criminal behavior systems** originally developed by McKinney (1966) and elaborated by Clinard and Quinney (1986) in

Crime File 7.2

Some Sociological Typologies of Criminal Behavior

Gibbons's "Criminal Role Careers"

1. Professional thieves
2. Professional "heavy" criminals
3. Semiprofessional property offenders
4. Naïve check forgers
5. Automobile thieves, "joyriders"
6. Property offenders, "one-time losers"
7. Embezzlers
8. White collar criminals
9. Professional "fringe violators"
10. Personal offenders, "one-time losers"
11. Psychopathic assaultists
12. Statutory rapists
13. Aggressive rapists
14. Violent sex offenders
15. Nonviolent sex offenders
16. Incest offenders
17. Male homosexuals
18. Opiate addicts
19. Skid Row alcoholics
20. Amateur shoplifters

Schafer's "Life Trend" Typology of Criminals

1. Occasional criminals
2. Professional criminals

3. Abnormal criminals
4. Habitual criminals
5. Convictional criminals

Lombroso's Types of Criminals

1. Born criminals
2. Criminaloids
3. Occasional criminals
4. Criminals by passion

Abrahamsen's Types of Criminals

1. Acute criminals
   a. Situational
   b. Associational
   c. Accidental

2. Chronic offenders
   a. Neurotic
   b. Psychopathic
   c. Psychotic

Glaser's Types of Crime

1. Predatory crime
2. Illegal performance offenses
3. Illegal selling offenses
4. Illegal consumption offenses
5. Disloyalty offenses
6. Illegal status offenses

Sources: Don C. Gibbons (1982), *Society, Crime and Criminal Behavior*, 4th ed., Englewood Cliffs, NJ: Prentice Hall, p. 225; Stephen Schafer (1976), *Introduction to Criminology*, Reston, VA: Reston, pp. 107–108; Gina Lombroso-Ferrero (1972), *Criminal Man According to the Classification of Cesare Lombroso*, Montclair, NJ: Patterson Smith, p. 100; David Abrahamsen (1960), *The Psychology of Crime*, New York: Columbia University Press, p. 14; Daniel Glaser (1978), *Crime in Our Changing Society*, New York: Holt, Rinehart and Winston, p. 15.

Web Research Project
Using the companion Web site, click on and read the article by Dowden and Andrews titled "What Works for Female Offenders: A Meta-Analytic Review." Discuss a few things that work.

their now-classic work, *Criminal Behavior Systems: A Typology.* This typology is based on constructed types "that serve as a means by which concrete occurrences can be compared and understood within a system of characteristics that underlie the types." Clinard and Quinney (1986, p. 15) identify nine types of criminal behavior:

1. Violent personal crime

2. Occasional property crime

3. Occupational crime

4. Corporate crime [added to the typology later]

5. Political crime

6. Public-order crime

7. Conventional crime

8. Organized crime

9. Professional crime

These types are based on four characteristics:

1. The criminal career of the offender

2. Group support of the criminal behavior

3. Correspondence between criminal and legitimate behavior

4. Societal reaction and legal processing of offenders

Clinard and Quinney admit that there are undoubtedly other ways of delineating crime into types, taking into account these four characteristics; however, the typology serves useful purposes that permit the ordering of presentation of research on various forms of crime. Rather than using legal categories for the organization of materials, the purpose is to derive as few categories of crime, based on behavior similarities, as possible, in order to simplify analysis. Chapters 8 through 13 will concentrate on crime and criminal activity, making use of a variety of elements of this typology.

Theoretical Range and Criminological Explanation

This presentation of theories in criminology can only introduce major themes and schools of thought, leaving more formalized and sophisticated exposition to upper level theory texts (see Vold et al., 2002); detailed explication of the general theories' applications to types of criminal behavior would require far more space than is possible in this volume. However, to summarize the interrelationship between descriptions, criminal behavior typologies, and general theory, some final points may prove fruitful.

Theoretical range, or scope in this writer's view, refers to the unit(s) of analysis and levels of explanation that may be sought in a particular theory. In their *New Criminology,* Taylor et al. (1973, pp. 270–278) provide an example of theoretical range when they describe the formal requirements or scope of a general theory in criminology. Such a model must describe:

1. The wider origins of the deviant act.

2. Immediate origins of the deviant act.

3. The actual act.

4. Immediate origins of the societal reaction.

5. Wider origins of the deviant reaction.

6. The outcome of the societal reaction on the deviant's further action.

7. The nature of the deviant process as a whole.

Allen, Friday, and Roebuck (1981) address this issue of theoretical range:

> What has been, and is, remiss in theoretical criminology in the opinion of many, is the spurious attempt to explicate all crime on the basis of one unitary, universal theory. Any theory that attempts to explain all crime, it is charged, cannot escape being a general theory of all human behavior, because criminal behavior encompasses a wide and divergent body of conduct. A general theory of crime would have to explain too much and therefore would explain too little. The essential questions are: what kinds of criminals and what kinds of circumstances, result in the commission of what kinds of crime? In short, the development of criminal typologies, in this view offers the most plausible approach to the etiology of crime. (p. 39)

The Global Fallacy

Williams and McShane (1988) point out:

> The sheer variety of behavior defined as criminal also presents a problem. When we use the term "crime," the reference is often to a wide range of illegal behavior. . . . Thus, theories of crime and criminal behavior must encompass a wide range of human activity. This is the reason that some criminologists advocate the limiting of theories to a very specific behavior. (p. 134)

A long-recognized limitation of many discussions of crime as well as theories of crime causation, particularly early ones, relates to the global (or broad) manner in which the concept of crime is employed. The only thing most crimes hold in common is the fact that they are at a given point in time defined or viewed as violations of criminal law. The **global fallacy** refers to the tendency to attempt to generalize relatively specific explanations to all types of crimes (Hagan, 1987c). Many individual theories are not invalid in themselves, but are either too globally ambitious or are interpreted as such. A perfectly appropriate theory for explaining burglary may not apply at all to inside trading, nor should it be expected to do so.

Ferdinand (1987, p. 855) calls this the domain of theory or the area of crime or delinquency that a theory intends to explain. An excellent illustration of the global fallacy is the neoclassical "general theory of crime" proposed by Gottfredson and Hirschi (1990; Hirschi & Gottfredson, 1987, 1989). They maintain that all crime is due to an individual's lack of self-control and that white collar crime (defined solely as the Uniform Crime Reports' inadequate measures of only embezzlement, fraud, and forgery) does not require any different explanation than street crime. This is a "baffling" (Reed & Yeager, 1991) disregard of elementary research findings on white collar crime, which will be discussed in Chapter 10. Do corporations, state terrorists, the Mafia, or Iran-Contra conspirators all lack self-control?

The range of theories may be at the general (macroscopic) level, addressing a broad issue such as "How does crime originate in society?" or at the specific (microscopic) level, addressing a question such as "What causes specific individuals to commit specific crimes?" Merton (1968, p. 45) advocates development of "theories of the middle range," proposing explanations aimed at describing specific activity between macroscopic and microscopic levels. Gibbons (1992, p. 8) also argues that "progress lies in the direction of theories focused on specific forms of lawbreaking." All of the major theoretical views in criminology in these last two chapters were seen as subject to certain shortcomings; in many instances, the criticisms were as much based on what the theories failed to cover as they were on what they did address.

Figure 7.2 presents a summary schema that compares the major theoretical views in criminology in terms of whether they address the following range of activities:

1. Origin of crime in society.
2. Immediate factors of transmission of criminal values.
3. Individual criminality.
4. Prevention of individual criminality.
5. Prevention of crime in society.

In addition, Figure 7.2 indicates types of criminal behavior addressed by each theory discussed in this book.

While the author's analysis of the presence or absence of features of each theory may be debated, and you can judge for yourself as we examine each type of criminal behavior, none of the general theories apply to all of the types of criminal behavior. Many specific theories discussed, such as Wolfgang and Ferracutti's subculture of violence and Cressey's theory of embezzlers, represent "theories of the middle range," more of which are needed to build more crime-specific explanations. Until more all-encompassing, all-purpose theories concerned with all types of crimes are developed, the middle range, crime-specific theories appear to be taking a fruitful direction. In the discussion of typologies it was suggested that the first response to "What causes crime?" is "What type of crime?" Perhaps a criminological Einstein or Galileo will yet arrive to provide an acceptable general theory. Until then more Sutherlands, Mertons, and Hirschis will hopefully provide needed "middle range" theoretical explanations.

The purpose of the four theory chapters in this text is a modest one, and that is to acquaint the beginning student with criminological theory. More detailed exposition would require too much space here and would be appropriate in a course on theory. The reader is referred to some excellent works on this subject by Akers (1994); Curran and Renzetti (2001); Lilly, Cullen, and Ball (1995); Vold et al. (2002); Williams and McShane (1994); and Cullen and Agnew (1999).

The Theory–Policy Connection

Critical theories share in common their critical view of society as responsible for causing, creating, and defining crime. Its focus is on the society rather than the individual and asks "Why do some societies or groups have higher crime rates than other groups?" Labeling (societal reaction) theory assumes that crime is a label and reflects societal reaction to crime. The impact of Prohibition or the later "Wars on Drugs" indicate the criminalization or decriminalization of activities. Conflict theory posits that the enforcement of laws are influenced by the values and interests of the most powerful groups and asks the question: "Whose behavior is singled out for attention (criminalization)?" The weak penalties and nonenforcement of white collar crimes illustrate policies that reflect this perspective. Table 7.2 outlines some of the policy implications of critical theories as well as integrated theories.

FIGURE 7.2 ■ Range of Major Theoretical Views in Criminology

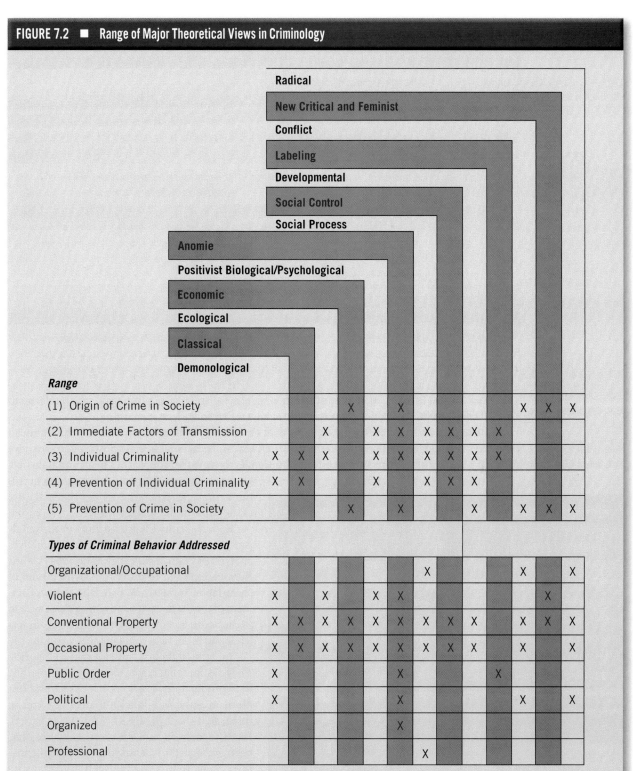

Source: *Theoretical Range in Criminological Theory*, paper presented by Frank E. Hagan at the Academy of Criminal Science Meetings, Las Vegas, Nevada, April 1995.

TABLE 7.2   ■   The Theory–Policy Connection

Theory	Basic Assumptions About Crime	Policies
Critical	Crime reflects the conflict model	
Labeling	Crime is a label and reflects societal reaction to crime	Reversal of normal assumption that crime causes societal reaction, but rather societal reaction creates crime. New labels lead to criminalization or decriminalization (e.g., Prohibition, War on Drugs)
Conflict	Enforcement of laws are influenced by values and interests of the most powerful groups	Weak penalties and nonenforcement of white collar crimes. Whose behavior is targeted?
Feminist	Androcentric bias	Incorporation of greater female roles in criminal justice
New Critical	Left realism Peacemaking Postmodernism	Attempts to elicit citizen participation in programs (e.g., community policing)
Radical	Marxism Capitalism causes crime	Laws reflecting class inequality
Integrated	Crime is explained by a combination of various theories	BARJ (Balanced and Restorative Justice) programs involve restoring the offender to good standing in society.

Audio Link 7.1

Listen to a discussion of community policing.

Photo 7.6

A well-known white collar crime case is that of Martha Stewart, who was convicted of insider trading. She served 5 months in prison and an additional 5 months of house confinement.

Other critical theories such as feminist theory, left realism, peacemaking and postmodernism have all been reflected in policy proposals and changes. Feminist theory, charging androcentric bias in criminal justice, has successfully changed a lot of thinking in criminal justice and this has been reflected in greater female presence in the field. Left realism, peacemaking and postmodernism have fostered a variety of reforms that question traditional conservative approaches to criminal justice. Community policing and community corrections reflect peacemaking policies that make use of the community to fight crime.

Radical crime theory assumes that capitalism and class inequality cause crime. Chambliss's analysis of English vagrancy laws (1964) claims they were passed in order to provide a pool of cheap labor. Similarly, W. E. B. DuBois saw enactment of the Black Codes and convict lease system in the South as a means of compensating for lost labor and profits as a result of Emancipation.

Integrated theories are not critical theories. They assume that crime causation is explained by a combination of various theories. One example of an ambitious policy initiative based on an integrated theory is BARJ (pronounce barge). BARJ (Balanced and Restorative Justice) has been embraced by Pennsylvania's juvenile justice system. In this program, community justice programs utilize citizen involvement and community building in addition to the traditional victim–offender relationship as a means of restoring justice in the juvenile justice system.

Summary

Mainstream criminology (anomie, social process, and social control theories) has been viewed as emphasizing the study of the criminal rather than of social control agencies, thus stressing positivism, a consensus worldview, and liberal reformism. In response to this, in the sixties and seventies in the United States, critical criminology emerged, which consists of the labeling, conflict, and radical perspectives. Critical criminology stresses the conflict model, inequality, the process of assigning criminal labels, and, in some cases, ideology.

Labeling theory (societal reaction approach) is derived from symbolic interactionism (a stress on subjective meanings of social interaction). Labeling theory assumes that individuals are criminal because they have been labeled as such by social control agencies; that is, societal reaction causes criminality. Schrag's summary of the basic assumptions of this school of thought was presented along with Lemert's concept of "secondary deviance;" the latter refers to continued deviance once an individual has been caught and labeled.

Conflict theory advocates a "pluralistic conflict model" of criminal law. It is represented in the writings of Dahrendorf and Vold and, in conflict criminology, in the works of Austin Turk and the early work of Richard Quinney and William Chambliss, as well as that of Jeffrey Reiman. According to conflict criminology, a variety of groups compete for control of the law-making and enforcement machinery in order to protect their vested interests. New critical theories include left realism, peacemaking, and postmodernism. Feminist theory represents a large and growing theoretical perspective in criminology.

Radical "Marxist" criminology, as presented in the later writings of Quinney and Chambliss, views crime as a result of capitalism, with the criminal law representing the interests of the capitalist class. The Marxist prescription for solving the crime problem is the collapse of capitalism and the creation of a socialist state. Major critiques of radical criminology, as well as of each of the other theoretical approaches, were presented in this chapter.

Criminal typologies (attempts to classify criminals or criminal behavior) have two purposes: (1) a scientific classification system, and (2) a heuristic (practical) scheme. While many criticisms have been levied against such typologies as pure scientific classes, the heuristic benefit of using criminal typologies as organizing schemes for presentation or discussion purposes remains. After a brief review of other typologies, Clinard and Quinney's typology of criminal behavior systems was presented. This examines nine criminal behavior systems: violent personal, occasional property, occupational, corporate, political, public-order, conventional, organized, and professional crime from the standpoints of criminal career, group support, correspondence with legitimate behavior, societal reaction, and legal processing.

Theoretical range refers to the units of analysis and level of explanation that may be sought in a particular theory. This range may focus on the macroscopic level, for example, general theories of the origin of crime or on the microscopic explanations of individual criminality. Merton's concept of "theories of the middle range" argues for explanations aimed at describing specific activity between the macro- and microscopic.

The "range of major theoretical views" attempts to summarize the theoretical range (origins, immediate factors, individual criminality, individual prevention, and societal prevention) of each theory and its ability to address different types of criminal behavior. This illustrates the view that the question "What causes crime?" must first be met with the question "What type of crime?" The answer to the first question awaits the development of an acceptable general theory of criminology.

Criminology on the Web

Log on to the Web-based student study site at http://www.sagepub.com/haganintrocrim7e/ for author-created podcasts, e-flashcards, quizzes, and more.

KEY CONCEPTS

Criminal Behavior Systems
Criminal Typologies
Critical Criminology
Feminist Criminology
Global Fallacy

Labeling Theory
Left Realism
Peacemaking Theory
Postmodernism
Praxis

Primary Deviance
Radical "Marxist" Criminology
Secondary Deviance
Shaming Theory
Theoretical Range

REVIEW QUESTIONS

1. How do critical criminological theories differ from mainstream criminological theories? What do you see as the strengths and weaknesses of each?

2. How does labeling theory reverse the usual approach to crime and criminality? What are some pros and cons of this "societal reaction" approach to crime?

3. Compare conflict criminology with radical criminology. What is the major difference between the two?

4. Discuss feminist criminology. What is the difference between liberal and more radical feminist criminology?

5. How do the concepts of malestream, androcentric bias, and patriarchy influence traditional criminology according to feminist criminologists?

6. What is the basic assumption of left realism?

7. Discuss the notions of peacemaking and restorative justice. How do these approaches differ from "just deserts," "retributive justice," and "three strikes and you're out" approaches?

8. What is postmodernism and postmodernist criminology?

9. What is the goal of integrated theories in criminology? Give an example.

10. What is the "global fallacy" in criminological theory?

WEB SOURCES

Conflict Theory
http://www.criminology.fsu.edu/crimtheory/conflict.htm

Crime Theory
http://www.crimetheory.com

Critical Criminology Division Home Page
http://www.critcrim.org

Gender and Criminology
http://www/kelt.awebconcepts.com.au/ecrgend1.htm

Left Realist Criminology
http://www.malcolmread.co.uk/JockYoung/leftreal.htm

Peacemaking and Crime
http://www.peacemakingandcrime.blogspot.com

Reintegrative Shaming Experiments
http://www.aic.gov.au/criminal_justice_system/rjustice/rise.aspx

Restorative Justice
http://www.restorativejustice.org

WEB EXERCISES

Using this chapter's recommended Web sites, examine critical criminological theory and integrated theory.

1. What items are featured on the Critical Criminology Division home page?

2. What have been some results of "Restorative Shaming Experiments"? Does the Restorative Justice site add any additional information?

3. Which of the remaining critical sites seems the most useful: Conflict Theory, Gender and Criminology, Left Realism Theory, or Peacemaking and Crime?

4. Perform an online search on the general topic of postmodern criminology. Were you able to come up with anything? If so, in your own words, what do you think postmodern criminology entails?

SELECTED READINGS

Joanne Belknap. (1996). *The Invisible Woman: Gender, Crime, and Justice.* Belmont, CA: Wadsworth.

This work presents an excellent example of feminist criminology.

John Braithwaite. (1989). *Crime, Shame, and Reintegration.* Cambridge, UK: Cambridge University Press.

Braithwaite outlines his shaming and reintegration of offenders conceptualizations.

Meda Chesney-Lind and Randall G. Shelden. (1998). *Girls, Delinquency, and Juvenile Justice* (2nd ed.). Belmont, CA: Wadsworth.

The authors present a more radical feminist perspective in examining gender and delinquency.

Stuart Henry and Dragan Milovanovic. (1996). *Constitutive Criminology: Beyond Postmodernism.* London: Sage.

This work provides an example of postmodernist thinking in criminology.

James Inciardi, editor. (1980). *Radical Criminology: The Coming Crisis.* Beverly Hills, CA: Sage.

Author Inciardi puts together a very readable anthology that represents some of the earlier thinking on radical criminology.

Harold Pepinsky and Richard Quinney, editors. (1991). *Criminology as Peacemaking.* Bloomington: University of Indiana Press.

The emergent "peacemaking" perspective is presented in a series of articles.

Jeffrey Reiman. (1998). *The Rich Get Richer and the Poor Get Prison* (5th ed.). Boston: Allyn & Bacon.

Reiman's highly cited work examines inequalities in the criminal justice system.

Martin D. Schwartz and David O. Friedrichs. (1994, May). Postmodern Thought and Criminological Discontent: New Metaphors for Understanding Violence. *Criminology, 32,* 221–246.

One of the few understandable attempts to appraise the postmodernist criminological perspective.

Ian Taylor et al. (1973). *The New Criminology: For a Social Theory of Deviance.* New York: Harper and Row.

British writers present one of the earliest views of radical (critical) criminology.

History of Violence in the United States

   Murder and Mayhem

   Types of Multiple Murders: Multicide

   Crime File 8.1 International Violent Crime

   Crime File 8.2 The Virginia Tech Massacre

   Victim Precipitation

   Crime File 8.3 The D.C. Snipers, the BTK Killer, and the Red Lake Massacre

   Typology of Violent Offenders

Legal Aspects

Homicide and Assault Statistics

Patterns and Trends in Violent Crime

   Workplace Violence

   Crime File 8.4 Workplace Violence: Issues in Response

   School Violence

   Guns

   Crime File 8.5 Deadly Lessons: The Secret Service Study of School Shooters

Sexual Assault

   Acquaintance Rape

   Crime File 8.6 The Problem of Acquaintance Rape of College Students

   Amir Versus Brownmiller

   Rape as a Violent Act

Robbery

   Crime File 8.7 Name That Bank Robber

   Conklin's Typology of Robbers

Domestic Violence

   Child Abuse

   Spouse Abuse

   Elder Abuse

   Kidnapping

Criminal Careers of Violent Offenders

   Culture of Violence

   Subculture of Violence

   Career Criminals/Violent Offenders

Societal Reaction

Theory and Crime

Summary

Key Concepts

Review Questions

Web Sources

Web Exercises

Selected Readings

chapter 8

Violent Crimes

Violence is as American as cherry pie.

—*H. Rap Brown, black militant of the sixties*

Violence by human beings against other human beings has scarred history from earliest times. In addition to hundreds of smaller conflicts, the twentieth century witnessed two major world wars with casualties in the millions and devastation, such as that at Hiroshima and Nagasaki, that is unparalleled in human history. Mass genocide of populations by the Nazis, human purges in which millions disappeared as in Stalin's Russia, and continued torture of political opponents in many countries throughout the world made it a frightening century indeed. The first decade of the twenty-first century finds that such activities have not abated. We have already witnessed terrorism on a massive scale; genocide in Darfur; and threats of chemical, radiological, and nuclear violence, none of which suggests that the human capacity for violence is lessening.

While writers such as Konrad Lorenz (1966) and Robert Ardrey (1963) argue that humans have a "killer instinct," a natural predisposition toward violence and aggression, most social scientists reject this view, arguing instead that individuals learn violence, like nonviolence, through socialization. Anthropological studies have discovered wide variations in the degree of violence prevalent in human cultures, with a few cultures in which violence is unknown. Japan's transition from a violent, warlike society before and during World War II to a pacifist society in the postwar period suggests that violence is not an inevitability. Just as violence can be learned and assumed to be a natural part of a culture, it probably also can be unlearned.

History of Violence in the United States

In their report to the National Commission on the Causes and Prevention of Violence, titled *Violence in America: Historical and Comparative Perspectives*, Hugh Davis and Tedd Gurr (1969) indicate that we ignore history when we view our present levels of violence as unusual. They claim that violence in the United States is rooted in *six historical events* that are deeply imbedded in our national character (pp. 770–774):

1. Revolutionary doctrine expounded in the Declaration of Independence
2. A prolonged frontier experience, which tended to legitimize violence and vigilante justice

3. A competitive hierarchy of immigrants that has been highly conducive to violence

4. A pervasive fear of governmental power, which "has reinforced a tendency to define freedom negatively as freedom from" (p. 772)

5. The Industrial Revolution and the great internal migration from countryside to city, which has produced widespread social dislocation

6. Unmatched prosperity combined with unequal distribution and unequal opportunity, which has produced a "revolution of rising expectations" in which improved economic rewards can coincide with relative deprivation, generating frustration and violence

Glaringly absent from this list is the bitter legacy of slavery and subsequent racially motivated violence against blacks. The burning cross of the KKK (Ku Klux Klan) symbolized the bombings, lynchings, murders, shootings, arsons, mutilations, and other violent tactics used against African Americans as well as others. In the 1950s and 1960s, the bombing of churches and murders of civil rights workers, often in collusion with local police officers, aroused the United States to oppose racism (Revell, 1988, p. 10; W. C. Wade, 1987). While item two above mentions frontier violence, elaboration of this theme would note that the almost eternal war by white settlers against Native American tribes was of genocidal proportions, a holocaust that wiped out entire tribes as part of stealing their land.

Photo 8.1

The burning cross of the KKK symbolizes the violent tactics used against African Americans and others.

Historian Richard Brown (1969) sums up much of this:

> Violence has formed a seamless web with some of the noblest and most constructive chapters of American history: the birth of the nation (Revolutionary violence), the occupation of the land (Indian wars), the stabilization of frontier society (vigilante violence), the elevation of the farmer and the laborer (agrarian and labor violence), and the preservation of law and order (police violence). The patriot, the humanitarian, the nationalist, the pioneer, the landholder, the farmer, and the laborer (and the capitalist) have used violence as a means to a higher end. (pp. 69–70)

Violence may indeed reflect a society's values. For example, Americans value "life, liberty, and the pursuit of happiness," while their less violent next-door neighbors, the Canadians, reflect a less revolutionary view of society and applaud "peace, order, and good government." The United States has inherited a violent cultural tradition, but as a relatively young country, its tradition may not be that much different from the early histories of older civilizations of Europe and Asia. The United States is not alone in being plagued by violent crime.

Murder and Mayhem

In 1966, former Eagle Scout leader and engineering honor student Charles Whitman murdered his wife and mother and then, with a small arsenal of weapons and ammunition, climbed to the top of a tower at the University of Texas. Taking aim with deadly accuracy, he randomly killed 16 persons and wounded another 30 before being killed himself by police.

Photo 8.2

Charles Manson was convicted and sentenced to death in 1971, but was spared execution when California abolished the death penalty. He has been denied parole 11 times and is currently an inmate at Corcoran State Prison.

On August 8, 1969, devotees of a cult mesmerized and run by Charles Manson brutally murdered pregnant actress Sharon Tate and four other guests at her home and 2 days later murdered two members of the La Bianca family in an apparent attempt to foment a race war. Particularly frightening in the incident was Manson's Rasputin-like ability to inspire undying devotion in his followers, most of them young female drifters.

After the dismembered remains of 11 victims were found in his apartment in July 1991, Jeffrey Dahmer admitted killing 17 boys and young men, primarily in the Milwaukee area, over a 13-year period. After luring victims to his apartment to take nude pictures, watch videos, and have sex, he drugged, killed, and dismembered them. He often took pictures of his victims and boiled some of their skulls in order to preserve them.

Each of these cases is an example of multiple murders. While cases such as these attract much public attention, they are relatively rare and make up only a very small proportion of the incidents of violent crime. Media, fictional, and popular accounts of violent crime tend to focus on the dramatic tales of murder and mayhem that make our blood curdle as much as the latest Stephen King novel. The post–World War II period has had no shortage of material for such chronicles.

Video Link 8.1
View a video of
Jeffrey Dahmer.

Types of Multiple Murders: Multicide

R. M. Holmes and DeBurger (1988, p. 19) estimate that between 3,500 and 5,000 persons may be slain per year in the United States by multiple murderers, and that even though such killings are not new, they appear to have increased since the sixties. Much of the gap in our academic knowledge of multiple murder is being addressed by more recent scholarship (Egger, 1984; J. A. Fox & Levin, 1985; Hickey, 2006b; Jenkins, 1988; Leyton, 1986). Criminologists agree that at least three different types of multiple murders (multicide) exist: serial murder, mass murder, and spree murder (Bureau of Justice Statistics [BJS], 1988; see Crime File 8.1).

Crime File 8.1

International Violent Crime

The United States, while having higher rates than other developed countries with respect to murder and rape, does not stand alone with respect to violent crime (Rounds, 2000). Some recent examples serve as illustrations:

- Schools in France closed their doors to protest rising levels of school violence. Teachers and students joined in a strike to protest conditions.
- Eight London youths in a racially motivated gang rape brutally raped and attacked an Austrian tourist before throwing her into a canal. She survived the attack.

Homicides in Cities (per 100,000 population, average, 1995–1997)

City	Rate	City	Rate
Vienna	1.8	Copenhagen	4.6
Ottawa	1.9	Rotterdam	5.0
London	2.2	Amsterdam	7.9
Paris	3.3	New York	16.8
Berlin	3.8	Washington, D.C.	64.1
Stockholm	4.1		

- In 1996 Rio de Janeiro recorded 53.3 murders per 100,000 compared with 20 for Los Angeles. A garrison state existence is becoming more common as the wealthy retreat behind their own private fortresses.
- Avid soccer fan and fan of the *Nightmare on Elm Street* movies, 19-year-old Italian Pietro Marso, along with friends, murdered his parents in hopes of receiving his inheritance. His explained motive was to obtain a brilliant life, with expensive cars and good quality clothes (Cowell, 1994). This lust for material success was dubbed "the Verona syndrome."
- At least 103 people were beheaded in Saudi Arabia in 1999, one more executed than in the United States, which uses mostly lethal injections.
- Overall, five people are kidnapped per day in Colombia, mostly children and teenagers. This is the highest rate in the world (Faiola, 1999).
- In India parents of two lovers killed them for breaking a marriage taboo. This is called an "honor killing" (to preserve the honor of the clan). They had broken the taboo of marrying another from the same village (Bearak, 1999).

Sources: Bearak, B. (1999, April 9). A tale of two lovers, and a taboo recklessly flouted. *New York Times*, p. A7. British Home Office, 1998. *Legal Infrastructure of the Netherlands in International Perspective: Crime Control* (p. 7), by Frans van Dijk and Jaap de Waard, 2000, Amsterdam: Ministry of Justice, the Netherlands; Faiola, A. (1999, May 3). Bogota's street of the damned loses daughter to kidnappers. *Erie Morning News*, p. 11A.

Web Research Project

Search the term "murder" and locate some interesting international examples. How do these examples compare with our discussion of U.S. homicides?

Serial Murder

Serial murder is the killing of several victims in three or more separate incidents over weeks, months, or even years. Herman Mudgett, Juan Corona, Wayne Henley, John Wayne Gacy, Ted Bundy, and David "Son of Sam" Berkowitz are just a few of the "Jack the Rippers" who have shocked us in modern times. In April 1989, the 13th victim of cult slayings was discovered in Matamoros, Mexico, the work of cult "godfather" Adolpho de Jesus Costanzo and cult "witch" Sara Aldrete, who allegedly ritualistically sacrificed victims in order to "provide a 'magical shield' for members of a drug-smuggling ring" ("13th Victim," 1989, p. A1). In a case reminiscent of the classic movie *Arsenic and Old Lace,* directed by Frank Capra, in 1988 Dorothea Puente, a boardinghouse landlady in Sacramento, California, was charged with poisoning at least eight of her elderly boarders and collecting their social security checks. In Philadelphia in 1987, police arrested Gary Heidnik and an accomplice, charging them with running a "Little Shop of Horrors" (T. Johnson, 1987, p. 29). Heidnik, who had a history of psychiatric problems, attracted women to his house and imprisoned, tortured, sexually abused, murdered, and cannibalized them. Police have accounted for at least six victims. In 2007, Buffalo, New York's infamous "bike path rapist," Altemio Sanchez, was arrested, convicted, and sentenced to 75 years to life in prison. He was so named because many of the numerous murders and rapes that he committed happened on bike paths (Staba, 2007, p. A26).

Between 1982 and 1984, the bodies of 48 women, mostly prostitutes, were found along the banks of the Green River near Seattle. In November 2003, the "Green River Killer" was finally caught. Gary Ridgeway pleaded guilty to strangling 48 young women and was sentenced to life without parole. Similarly, it is estimated that from 1955 to 1966 the "Zodiac killer" in San Francisco killed over 50 people in the Bay Area. No one was ever prosecuted in these cases. The Green River and Zodiac killers terrify the public, which seems transfixed by Hollywood movies such as *Silence of the Lambs* and characters such as Hannibal Lecter.

A variety of typologies (taxonomies) of serial killers have been proposed. One of the most accepted typologies of serial killers is that of R. M. Holmes and DeBurger (1988) who identify the following types:

1. Visionaries: Believed to be suffering from some sort of psychosis, they kill in response to voices or visions.

2. Mission-oriented: Their mission in life is to rid society of "undesirables," for example, prostitutes.

3. Hedonists: These are "thrill seekers" who murder for creature comforts, profit, or pleasures in life, as well as lust murderers.

4. Power/control: These killers enjoy power and control over helpless victims and enjoy watching them suffer and beg for mercy (pp. 55–60).

The FBI Behavioral Science Unit, which has conducted extensive investigations and crime profiling of such murders, indicates that the victims often represent someone in the killer's adolescence who inflicted some perceived pain on them (Douglas & Olshaker, 1995, 1997).

Organized serial killers usually plan their offenses, transport their victims, and keep "trophies" (victim belongings); they are normal in appearance and socially competent. *Disorganized serial killers* are usually socially and sexually incompetent. Both types often return to the gravesite or dumpsite, often to satisfy their sexual fantasy through masturbation.

Many serial killers displayed what is called "the terrible triad" as children: bedwetting, firestarting, and cruelty to animals. Most were products of dysfunctional families.

Mass Murder

Mass murder is the killing of four or more victims at one location on a single occasion. Thomas Hamilton (who in 1996 slaughtered 16 kindergartners and their teacher in Scotland), Richard Speck

Journal Article Link 8.1
Examine literature regarding serial murder.

Audio Link 8.1
Listen to a discussion of past and present serial killers.

Crime File 8.2

The Virginia Tech Massacre

The Virginia Tech massacre occurred on April 16, 2007, with 32 people killed (5 faculty and 27 students) it represents the worst mass murder in American history. In two separate attacks approximately 2 hours apart, Seung-Hui Cho, a South Korean with U.S. permanent-resident status who had lived in the country since the age of 8, used two firearms, a .22-caliber Walther P22 semiautomatic handgun and a 9-mm semiautomatic Glock 19 handgun, both of which Cho was able to purchase despite his documented mental problems, which did not appear on the instant background check.

Cho shot and killed two students at 7:15 a.m. in a dormitory. He then mailed a package of writings and a video recording to NBC News. Contained in the package was an 1,800-word manifesto along with photos and videos. In the manifesto Cho compared himself to Jesus Christ and expressed his hatred for "rich kids" and "deceitful charlatans." About 2 hours later he entered a classroom building, chained the three main entrance doors shut, and proceeded to murder students and faculty. He would unleash a deadly volley of 170 rounds, killing an additional 30 people and wounding many more. The hollow-point bullets that he used inflicted greater injury to the victims. Finally, when police arrived, Cho killed himself.

Cho fit the classic profile of a school shooter. He was a "loner," and he felt that he was picked on and made fun of. He was bullied due to speech difficulties. He also had a history of mental problems. Several of his professors found his writing for class assignments to be very disturbing and had encouraged him to obtain counseling. Earlier in 2005, he had been declared mentally ill by a Virginia special justice and ordered to seek treatment.

Virginia Governor Tim Kaine created an eight-member panel to review the massacre, Cho's mental history, and the school's delay in alerting students to the danger after the first incident. However, no academic expert on multiple murders was included on this panel.

The incident reignited the gun control debate, with National Rifle Association lobbyists charging that Virginia Tech's gun-free zone prevented the victims or anyone else from defending themselves. Other countries were aghast at such violence, but indicated that they foresaw little chance of any changes in U.S. gun laws or gun culture.

Source: http://www.nytimes.com/2007/04/16/us/16cnd=shooting.html.

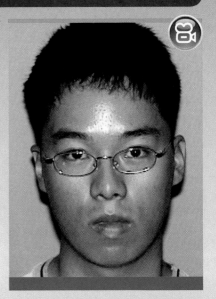

Photo 8.3

Seung-Hui Cho was responsible for the worst mass murder in American history, killing 32 people and wounding many others before committing suicide.

Web Research Project

Why do you think that the United States is plagued with more school shootings than all other developed countries combined? What policies do you think would help lessen the number of school shootings?

(who murdered eight Chicago nurses), James Huberty, Charles Whitman, and George Hennard are examples of mass murderers. In 1984, Huberty killed 21 and wounded a dozen others at a McDonald's restaurant in San Ysidro, California. In 1991, Hennard drove his pickup truck through the windows of Luby's Cafeteria in Killeen, Texas, and at point-blank range shot and killed 22 and wounded 23, making the Killeen massacre the worst in U.S. history (although 87 died in an arson fire at the Happy Land nightclub in New York City in 1990).

In 1993, a Brooklyn man, Colin Ferguson, opened fire in a crowded commuter train on Long Island, killing 6 people and wounding 17. Crime File 8.2 describes the Virginia Tech massacre of 2007, the worst mass murder in U.S. history at that time.

Spree Murder

The **spree murderer** kills at two or more locations with almost no time break between murders (BJS, 1988, p. 4; Crockett, 1991).

One of the most celebrated cases of "spree murder" was that of Andrew Cunanan, the murderer of Italian fashion designer Gianni Versace in Miami Beach, Florida, in 1997. Cunanan, who was HIV-positive, left San Diego on April 24, 1997, and 5 days later murdered his ex-lover David Madson and friend Jeffrey Trail. The largest unsuccessful manhunt in U.S. history ended on July 15 with Cunanan's suicide. After the first two killings, he also killed Lee Miglin, a wealthy Chicago developer; William Reese, a cemetery worker in New Jersey; and Versace. In 1999, Mark Barton, a day trader (Internet stock speculator) in Atlanta, distraught over heavy stock losses, killed 9 people and wounded 13 others before killing himself. At the time Barton said, "It's a bad trading day and it's about to get worse." Barton had also killed his wife and two small children. Crime File 8.3 discusses the D.C. Snipers Case, the BTK Killer, and the Red Lake Massacre.

African Americans and Serial Murder

Wayne Williams, age 23 and himself black, terrorized the African American community of Atlanta, murdering an estimated 28 young blacks over a 2-year period ending in 1981. Hating poor young blacks, whom he regarded as racially inferior, Williams lured them into his company with promises of fame in the entertainment business and then murdered them when they agreed to perform homosexual acts (see Detlinger, 1983).

As a black serial murderer, Williams appears to be an anomaly since most media portrayals feature white killers. In a thorough analysis of this issue, Jenkins (1992a) concludes that blacks are proportionately as likely as whites to be serial murderers. He indicates that

> for centuries, the lives of African-Americans have often been blighted by stereotypes, usually negative, and frequently associating them with crime and violence (Lynch and Patterson, eds., 1992; Rose and McClain, 1990). This paper has considered an area where stereotypes imply a diametrically opposite image, and Blacks appear disproportionately free of involvement in the most serious of violent crimes. However, this image is false; and this apparently favorable stereotype is both as inaccurate and as pernicious as any of the more familiar racial slurs. Significantly, the very failure to draw attention to Black serial killers might in itself arise from a form of bias within the media and law enforcement. (pp. 16–17)

African Americans make up a sizeable proportion of serial killers, and this has practical consequences for the fate of those blacks and other minorities who are most likely to fall victim to this type of predator. Underestimating minority involvement in serial homicide can thus lead to neglecting the protection of minority individuals and communities who stand in greatest peril of victimization.

As previously indicated, while bizarre murders and mass murders attract media and public attention, they represent the rare and dramatic rather than the typical violent crime.

Victim Precipitation

Victimology is the study of victims of crime, a group that in the past has been neglected by the criminal justice system. In examining violent crime, Lombroso (1911) was one of the first to note that passionate criminals often acted under the provocation of victims. In many violent crimes such as assault and voluntary manslaughter, a flip of the coin separates the victim from the offender, with both parties being active participants. In many violent crimes, victims contribute to their own harm (Von Hentig, 1948). Benjamin Mendelson (1963), one of the pioneers of victimology, developed a typology of victims in terms of their degree of guilt in the perpetration of crime.

Video Link 8.2
View a video on Jack the Ripper.

Video Link 8.3
View an *America's Most Wanted* video on empowering victims of crime.

Crime File 8.3

The D.C. Snipers, the BTK Killer, and the Red Lake Massacre

The D.C. Snipers

On October 25, 2002, the nightmare ended for the Washington, D.C., area when an alert truck driver spotted a car with two sleeping occupants at a highway rest stop in Maryland. The car fit the description of one possibly used by those involved in the sniping attacks of 19 and deaths of 13 randomly chosen victims mostly in the Washington, D.C., metropolitan area. Arrested were John Muhammed, age 41, and Lee Malvo, age 17. The latter, who was nicknamed "sniper," was the trigger man in the incidents. Since the shootings occurred in a post-9/11 environment, authorities had feared a terrorist connection. The motive, however, turned out to be an attempt by the perpetrators to extort money from authorities in return for a cessation of the shootings. The sequence of murders represented a hybrid between serial and spree murder. Begun on February 16, 2002, in Tacoma, Washington, and ending with the capture of the perpetrators on October 25, 2002, the relatively short time period between murders had the quality of spree murders.

Photo 8.4a
D.C. sniper John Muhammed, 41.

Photo 8.4b
D.C. sniper Lee Malvo, 17.

The case terrified the Washington, D.C., area and mesmerized the nation because of the bizarre qualities of the incidents. A tarot card with the encryption "I am God" was left behind at the scene of one of the crimes. As it turned out, unreliable witnesses had the police concentrate on a white van and white driver as the suspect. Many were shocked when the actual perpetrators turned out to be African Americans even though proportionately blacks are as likely as whites to be serial murderers.

In a trial for one of the murderers, Lee Malvo was convicted of murdering FBI analyst Linda Franklin and given a life sentence. John Muhammed was given the death penalty for having mentored the young Malvo. Muhammed was executed in 2009. (A Supreme Court ruling in 2005 ruled out the death penalty for juveniles.) It was alleged at the trial that Malvo was trained as a sniper by the older Muhammed. The killings were an effort to extort $10 million from the government.

Photo 8.4c
BTK killer Dennis Rader, 59.

The BTK Killer

The self-described BTK killer (bind, torture, and kill) terrorized the Wichita, Kansas, area for over 30 years. Arrested on February 25, 2005, was Dennis Rader, a 59-year-old city worker in suburban Park City, Kansas. He was charged with and admitted to 10 counts of first degree murder. Eight of these took place in 1974 and two later. The killer had resurfaced after nearly 25 years of silence and inactivity. He taunted the police and media with letters and pieces of evidence from murder victims such as a photocopy of one victim's driver's license and pictures of the victims.

Rader was married with two grown children. He had a degree in Criminal Justice from Wichita State University, and was a former Cub Scout leader and president of his church congregation. Analysts speculate that he had a great need for attention and may have had a felt need to tell his story. Perhaps this need for attention plus arrogance and feelings of superiority did him in. On August 18, 2005, Rader was sentenced to 10 consecutive life sentences.

The Red Lake Massacre

On March 21, 2005, at a school on the Red Lake Chippewa Reservation, Minnesota, the worst school shooting since Columbine took place. Jeff Weise, age 16, first killed his grandfather and his grandfather's girlfriend, stole his grandfather's police cruiser, guns, and bulletproof vest, and drove to the school. There he gunned down and killed a security guard, a teacher, and five students, and wounded seven others. After a brief shoot-out with police, he committed suicide.

Police investigated Weise's fascination with Hitler and a neo-Nazi Web site where he posted messages and logged on as "Todesengel," German for "angel of death." He dressed as a goth and was isolated from other students. He was described as a loner who was often made fun of. Previously his father had committed suicide and his mother was in long-term treatment for a serious automobile accident. Eerie parallels were drawn between Weise and Eric Harris and Dylan Klebold, the Columbine killers. All had been the subject of school discipline, were goths, and were admirers of Hitler and the Nazis. They had also asked their victims if they believed in God and mocked them.

Photo 8.4d

Members of the community showed their support for those killed in the shooting.

Sources: Laura Parker and Larry Copeland (October 9, 2002). Md. Sniper's Profile Puzzles Investigators. *USA Today*, pp. 1A-2A; Courtney Cloyd, Glen Sharp, and Andrew Murr (December 13, 2004). Stalking a Serial Killer. *Newsweek*, pp. 31–32; Associated Press (February 27, 2005). Police say they have arrested the BTK serial killer. *Erie Times-News*, p. 4A; Patrick Driscoll (March 23, 2005). Nazi Web Link Eyed in Killings. *USA Today*, p. 1A.; Tom Kenworthy and Patrick O' Driscoll (March 23, 2005). Red Lake Community Shaken by "Darkest Day." *USA Today*, p. 4A.

Web Research Project

Do an online search for more recent examples of mass murder. How are these examples similar to or different from the three mentioned here?

- The completely innocent victim, such as a child or an unconscious person
- The victim with minor guilt, such as a woman who provokes a miscarriage
- The victim as guilty as the offender, such as in cases of suicide and euthanasia
- The victim as more guilty than the offender, such as those who provoke someone to commit a crime
- The victim as most guilty, such as the aggressive victim who was killed in self-defense
- The simulating or imaginary victim, such as paranoids, hysterics, or senile persons

Wolfgang (1958, p. 252), in his study of criminal homicide in Philadelphia, views victim precipitation as present in incidents in which the victim initiated the altercation by being the first to use or threaten violence. Victim precipitation is common in murder and assault; it may also be common in other crimes.

Typology of Violent Offenders

John Conrad (cited in Spencer, 1966; Vetter & Silverman, 1978, p. 65) has proposed a very useful typology of violent offenders:

- Culturally violent offenders are individuals who live in subcultures (cultures within a culture) in which violence is an acceptable problem-solving mechanism. The "subculture of violence" thesis, to be explored shortly, is used as a means of explaining the greater prevalence of violent crime among low-income minorities from slum environments of large central cities.
- Criminally violent offenders use violence as a means of accomplishing a criminal act, such as robbery. Mental illness or brain damage characterizes pathologically violent offenders. (Discussions later in this chapter focusing on psychiatry and the law and on psychopathy will further elaborate on the mentally disturbed violent criminal.)
- Finally, situationally violent offenders commit acts of violence on rare occasions, often under provocation, such as in domestic disputes that get out of hand. These incidents are frequently described as "crimes of passion," in which the individual temporarily loses control and often expresses regret for the actions later. While the media focuses on the criminally and pathologically violent offender, the culturally and situationally violent offenders are the most common.

Legal Aspects

Table 8.1 provides a brief outline of the key legal features of violent crimes. Although violent crimes were at one time treated under tort law as a matter of private wrong to be settled by the parties involved, today the state has assumed authority and jurisdiction in cases of harmful, violent personal behavior. Private revenge is forbidden and is replaced with retribution under the law. The violent act is viewed as a threat to the well-being of the society, and the culprit must pay his or her debt to that society.

Homicide and Assault Statistics

Journal Article Link 8.2
Examine literature regarding Durkheim's theory of homicide.

Nearly all murders arise from some form of aggravated assault and, although the latter generally is not taken as seriously as murder, in reality, there is a thin line separating the two. Both offenses entail the use of violence as a means of resolving some grievance; in the case of murder, the victim dies. In our previous discussions of the shortcomings of crime statistics, it was pointed out that official police statistics such as the UCR underestimate the actual rate of crime commission. While this is true, the accuracy of these statistics varies according to the type of crime, and homicide statistics are one of the most accurate (see Reidel & Zahn, 1985).

Generally, homicide is regarded as the most serious crime. A body is present; there may be witnesses; and, as a result, such a crime is very likely to be reported to the police. In addition, homicide is the type of crime that the public, the media, and the police place a high priority on solving. Because of all of these factors, homicide has the highest "clearance by arrest" proportion of all UCR offenses. Clearance means that, as far as the police are concerned, the person responsible for the crime has been accounted for through arrest or incarceration.

Assault statistics are less accurate, and figures on rape have been notoriously poor until relatively recently. In fact, police and criminal justice professionals have applauded the recent rise in the rape rate, not because more rapes are occurring but because there is a greater willingness on the part of victims to report the crime to the police. The least accurate violent crime statistics relate to intrafamily violence such as spouse abuse, child abuse, and incest. Such offenses, which will be described in detail shortly, have in the past been regarded as family secrets.

TABLE 8.1 ■ Legal Aspects and Definitions of Violent Crime

Assault: Threatening to do bodily harm to a person or placing him or her in fear of such harm. Assault is attempted, but uncompleted, battery.

Battery (Aggravated Assault): An offensive, unconsented to, unprivileged, and unjustified offensive bodily contact. Battery includes *mens rea*, meaning that the contact was intentional or resulted from wanton misconduct, and indicates that bodily harm takes place.

Forcible Rape: Forcible and unlawful sexual relations with a person against her or his will. Rape is defined in common law as carnal knowledge of a female forcibly and against her will.

Statutory Rape: Sexual relations with a victim under the age of consent.

Murder: Killing that is calculated, in cold blood, or with malice aforethought.

First Degree Murder includes the following elements:

1. Intent to effect death with malice aforethought
2. Deliberate act
3. Premeditated act

Second Degree Murder includes:

1. Intent to effect death with malice aforethought
2. Without deliberation or premeditation. In essence, in most states second degree murder is any murder that is not defined as first degree.

   Felony Murder Doctrine: If in the act of committing a felony the death of one of the victims is brought about, this is murder even though it does not necessarily result from intent, deliberation, or premeditation.

   Manslaughter: Homicide without malice aforethought.

   Voluntary: Intentional (nonnegligent) killing without malice aforethought: often described as homicide "in hot blood" and often results from provocation.

   Involuntary: Unintentional (negligent) killing without malice aforethought, for example, vehicular homicides.

For additional details on legal definitions and the nature of the criminal law, see Katkin (1982) *The Nature of Criminal Law* or any text on criminal law.

Sources: Adapted from Bureau of Justice Statistics (1983), *Report to the Nation on Crime and Justice: The Data* (pp. 2–3), Washington, DC: Government Printing Office; Federal Bureau of Investigation (1992), *Crime in the United States*, 1991 (p. 340), Washington, DC: Government Printing Office; and Henry C. Black (1991) *Black's Law Dictionary*, St. Paul, MN: West.

Web Research Project
Review the concept of criminal law. What are some current issues in this field?

Figure 8.1 shows the overall trend with respect to homicide statistics in the United States, while Table 8.2 provides comparative international figures on the rate of homicide in selected countries.

Although the willful homicide rate declined from its peak in 1933, this dip may be misleading. Faster ambulances, better communications, transportation, and emergency room service meant better treatment for seriously injured persons, so that many who previously would have been homicide statistics were surviving. By the seventies, however, the sheer volume of violence had surpassed these extraordinary means of patching up the victims. It is also important to note that, although prior to the thirties the United States had no national crime statistics, fragmentary information suggests that at the turn of the twentieth century we had violent crime rates equal to present levels. In 1916, Memphis had a homicide rate that was seven times greater than its rate in 1969, and Boston, Chicago, and New

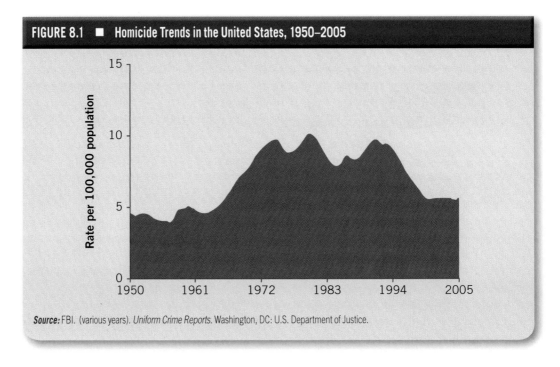

FIGURE 8.1 ■ Homicide Trends in the United States, 1950–2005

Source: FBI. (various years). *Uniform Crime Reports.* Washington, DC: U.S. Department of Justice.

TABLE 8.2 ■ Interpol Homicide Statistics (per 100,000)

Country	Homicide Rate (year)
United States[a]	5.5 (2004)
Australia	3.6 (2000)
Bahamas	27.1 (2001)
Brazil	23 (2001)
Canada	4.1 (2001)
Denmark	4.1 (2003)
France	3.6 (2000)
Ireland	1.6 (2002)
Jamaica	43.7 (2001)
Japan	1.1 (2002)
Russia	22.4 (2002)

Source: Interpol. *International Crime Statistics, 2003.* Lyons, France: Interpol Secretariat, 2004.

a. U.S. statistics were not included in the Interpol report and are taken from the Uniform Crime Reports for 2005.

York during and after World War I had higher rates than they did in 1933, when the first national statistics were published (National Commission on the Causes and Prevention of Violence, 1969, p. 20).

Table 8.2 presents Interpol homicide statistics for selected countries. These statistics do not include those for many developing nations, which have notoriously inaccurate crime statistics. Countries with the highest homicide rates in 2004 were Jamaica (43.7), the Bahamas (27.1), Brazil (23), and Russia (22.4). The U.S. rate of 5.5 was higher than that of other developed countries such as 4.1 for Denmark and Canada and 3.6 for France. The lowest rates were in Japan (1.1) and Ireland (1.6). Inexplicably, beginning in 2004, Interpol has no longer made these statistics available to the public. The UCR for 2006 showed a national aggravated assault rate of 299.3 in the United States, with the highest rate in the South (351) followed by the West (283.6), the Midwest (246.6), and the Northeast (215.6). For homicide in 2006, the national rate was 6.1 with the highest rate once again in the South (6.8), followed by the same order: West (5.6), Midwest (5.0), and Northeast (4.5).

Patterns and Trends in Violent Crime

Despite the association of crime with urbanization, crime was basically a rural rather than an urban problem during ancient times and the Middle Ages. Walled cities were built to provide protection from marauding highwaymen (V. Fox, 1976, p. 41). In the United States, it was not until the 1960s that urban rates for homicide exceeded those of rural areas (Glaser, 1978, p. 210), and in Canada the rural homicide rate still exceeded the urban rate in the 1970s (Schloss & Giesbrecht, 1972, p. 22).

Journal Article Link 8.3
Examine literature regarding the global impact of gangs.

Racial disparity in arrest rates is highest for crimes of violence. The black arrest rate for homicide is about eight times the white rate in the United States. Wolfgang's classic study *Patterns in Criminal Homicide* (1958, pp. 33, 66) found that the overall murder rate for white males in Philadelphia was 1.8 per 100,000 and, for white males age 20–24, it was 8.2 per 100,000. These same rates for black males were 24.6 and 92.5, respectively. In the Northeast and Midwest, the highest rates are for recent black migrants from the South to large cities. Internationally, lower-class slum backgrounds are significantly associated with high rates of violence.

Nettler (1982, Vol. 2, pp. 32–39) found that a number of studies indicate that countries with greater inequalities in income distribution have higher murder rates. International homicide statistics are questionable, however, since totalitarian regimes do not report statistics for government murder of citizens. Studies by Messner (1980) and Atkinson (1975) found moderate relationships between inequality in income distribution and homicide rate ($r = .40$ to $.59$).

Suppose you have been tipped off that you are likely to be the victim of a violent crime. Whom would you avoid? When? Where? Surprisingly, you are most likely to be stabbed, shot, beaten, or abused in your own home or in the home of one of your friends, relatives, or acquaintances. Saturday nights are lethal, as is the month of December, when friends and relatives get together and drink; summer months are even more deadly. Alcohol is a contributing factor in the majority of homicides, assaults, and rapes; it serves as a disinhibitor, causing individuals to have less rational control over their emotions as well as less awareness of the consequences of their actions (see Collins, 1981; Fagan, 1990). In the United States, more people are killed by drunk drivers every year than by all other types of killers.

Homicide rates began to decline in the early nineties; however, the rate of homicide by juveniles particularly black, urban juveniles exploded. From 1990 to 1993, the murder rate for Americans age 25 and older dropped 10 percent while the rate for 18- to 24-year-olds increased by 14 percent and the rate for juveniles overall increased 26 percent. Blumstein (1994) explains that juvenile involvement in crack cocaine distribution in the mid-eighties led to an arms race on the streets in which fistfights became shootings. Further exacerbating the situation were high levels of poverty, single-parent households, educational failures, and economic hopelessness. By 1993, the crack cocaine epidemic had peaked and crime rates began to decline.

After a sharp increase in the late 1980s, the homicide rate fell steeply beginning in 1992, falling to rates similar to the 1960s. Beginning in the 1990s, UCR data displayed a disturbing trend: growth in the number of homicides in which the relationship between victim and offender is "unknown." The popular press presented this as murder among strangers, but that is inaccurate. Most of the "unknown" murder victims were involved in drugs or gangs. In 1993, for instance, 47 percent of victims were killed by family members, friends, neighbors, or acquaintances, and 39 percent involved "unknown" relationships. Gun-related homicides were on the rise in the early nineties, particularly among juveniles (Glick, 1994). Female murder rates remained constant or declined. Women perpetrate a small proportion of murders and, when they do murder, they generally do so in self-defense and in marital/intimate relationships. "The availability of shelters and other supportive services may be providing some women avenues to escape self-defensive homicide as well as lethal victimization by their partners" (Benekos, 1995, p. 234).

The American cultural tradition of violence, combined with certain subcultures in which resorting to aggression is legitimized, presents a strong predisposition to violence in the United States. These are the raw materials of typical interpersonal homicides. The spark that sets off this kindling could be any number of interpersonal conflict situations, such as arguments over money, love triangles, threats to masculinity, and the like. While such disputes do not ordinarily lead to violence, the addition of two other fuels to the fire can spell danger: alcohol and guns. A surprising number of homicides are what some cynical observers call "public service killings" in which criminals kill other criminals, theoretically doing the public a favor in eliminating them. In Baltimore in 2007, fully 91 percent of those slain had an arrest record, much of it related to drug trafficking (K. Johnson, 2007). Similar patterns have been found in Milwaukee (77 percent), Newark (85 percent), and Philadelphia (75 percent) (p. 3A).

Workplace Violence

As of the early 2000s, **workplace violence** is the number two factor in on-the-job deaths and number one among women. Women were more than three times as likely to be murdered on the job than men. Since about 1980, the violent victimization rates of women and men have converged. From the mid-nineties onward, the rate of female violent victimization was about two thirds of the rate for men. It is worth noting that 9 of 10 female murder victims were killed by males (Craven, 1996; see the National Institute for Occupational Safety and Health Web site for online resources on violence against women, http://www.odc.gov/niosh/injury/traumaviolence.html).

The National Crime Victimization Surveys (NCVS) for 1992–1999 indicate that on average about 1.7 million violent victimizations occur annually while people are at work or on duty. Simple assault was the most common (about 1.5 million victimizations), followed by 396,000 aggravated assaults, 51,000 rapes and sexual assaults, 84,000 robberies, and 1,000 homicides (Warchol, 1998). Of the occupations examined, law enforcement had the highest rate of workplace victimization, followed by corrections officers, taxi drivers, private security, and workers in bars. The majority of workplace violent crime was committed by a stranger. Intimate perpetrators were identified as less than 1 percent, while 35 percent of the offenders were acquaintances. The workplace crime rate declined 44 percent from 1993 through 1999 (Duhart, 2001). Reasons for this decline are only speculative, but relate to the growing concern and attention by employers to the seriousness and prevalence of such incidents. Crime File 8.4 examines the FBI's report on workplace violence.

Although men make up 80 percent of the victims, murder on the job is the leading cause of workplace death for women. Despite public perceptions, worker-to-worker violence (such as the highly publicized U.S. Postal Service slayings) is relatively rare—4 percent of all workplace homicides ("Postal Violence Said to Be a Myth," 2000). Despite the use of the term "going postal," postal employees were at no greater threat of assault than other employees. Particularly vulnerable to assault (but not homicide) are health care workers and government workers. A survey by the Society for Human Resource Management of 500 human resource professionals in 1993 found 75 percent of violent incidents in the workplace were fistfights, 17 percent were shootings, 7.5 percent were stabbings, 6 percent were rapes or other sexual assaults, and less than 1 percent were explosions (Swisher, 1994, p. 11S). Protective measures to discourage victimization have included bulletproof glass in taxicabs and at hospital admission desks, escort services for evening workers, better illumination at night, more careful screening of employees, and limits on the amount of cash on hand.

Crime File 8.4

Workplace Violence: Issues in Response

What Is Workplace Violence?

On August 20, 1986, a part-time letter carrier named Patrick H. Sherrill, facing possible dismissal after a troubled work history, walked into the Edmond, Oklahoma, post office where he worked and shot 14 people to death before killing himself.

Though the most deadly, the Edmond tragedy was not the first episode of its kind in this period. In just the previous 3 years, four postal employees were killed by present or former coworkers in Johnston, South Carolina; Anniston, Alabama; and Atlanta, Georgia. The shock of the Edmond killings raised public awareness to the type of incident now most commonly associated with the phrase "workplace violence"—murder or other violent acts by a disturbed, aggrieved employee or ex-employee against coworkers or supervisors. An early appearance of the phrase itself in Nexis, a database of articles in many U.S. newspapers, was in August 1989, in a *Los Angeles Times* account of yet another post office shooting.* As a result of this seemingly new trend, mass murders in the workplace by unstable employees have become media-intensive events. In fact, the apparent rise in such cases may have been an impression created by this increased media attention. Still, the frequency of episodes following the Edmond post office killings was startling.

However, contrary to popular opinion, sensational multiple homicides represent a very small number of workplace violence incidents. The majority of incidents that employees/managers have to deal with on a daily basis are lesser cases of assaults, domestic violence, stalking, threats, harassment (to include sexual harassment), and physical or emotional abuse that make no headlines. Prevention programs that do not consider harassment in all forms and threats are unlikely to be effective. While agreeing on that broader definition of the problem, specialists have also come to the consensus that workplace violence falls into four broad categories. They are:

TYPE 1: Violent acts by criminals, who have no other connection with the workplace, but enter to commit robbery or another crime.

TYPE 2: Violence directed at employees by customers, clients, patients, students, inmates, or any others for whom an organization provides services.

TYPE 3: Violence against coworkers, supervisors, or managers by present or former employees.

TYPE 4: Violence committed in the workplace by someone who doesn't work there, but has a personal relationship with an employee—an abusive spouse or domestic partner.

Employers' important roles in violence prevention can include:

- Adopting a workplace violence policy and prevention program and communicating the policy and program to employees.
- Providing regular training in preventive measures for all new/current employees, supervisors, and managers.
- Supporting, not punishing, victims of workplace or domestic violence.
- Adopting and practicing fair and consistent disciplinary procedures.
- Fostering a climate of trust and respect among workers and between employees and management.
- When necessary, seeking advice and assistance from outside resources, including threat assessment psychologists and other professional social service agencies and law enforcement.
- Employees have a right to expect a work environment that promotes safety from violence, threats, and harassment. They can actively contribute to preventive practices by doing the following:
  o Accept and adhere to employer's preventive policies and practices.
  o Become aware of and report violent or threatening behavior by coworkers or other warning signs.
  o Follow procedures established by the workplace violence prevention program, including those for reporting incidents.

Source: Excerpts from Critical Incident Response Group (2004). *Workplace Violence: Issues in Response.* Quantico, VA: FBI Academy.

*Despite a number of highly publicized post office incidents, a Postal Service Commission reported in 2000 that postal employees are actually less likely to be homicide victims than other workers. The phrase "going postal" is a "myth."

Web Research Project

Search for some recent articles on incidents of workplace violence. Do these incidents illustrate some of the points made in the Crime File?

Video Link 8.4
View footage from the Columbine shootings.

School Violence

The United States is not unique in having school violence, but it leads the world in this type of violence. U.S. school shootings have included the following (R. Lawrence, 2006):

- April 16, 2007: Seung-Hui Cho kills 32 faculty and students at Virginia Tech University, the worst mass murder in U.S. history.
- March 2005: A 16-year-old kills his grandfather and his grandfather's girlfriend, a security guard, a teacher, and five classmates in Red Lake, Minnesota.
- February 2000: A 6-year-old boy shoots and kills a 6-year-old classmate at Buell Elementary School in Mount Morris Township, Michigan.
- April 20, 1999: Two students at Columbine High School in Littleton, Colorado, kill 12 fellow students.
- May 21, 1998: Two students kill or injure more than 20 by opening fire at a high school in Springfield, Oregon.
- April 24, 1998: A teacher is killed at an eighth-grade dance in Edinboro, Pennsylvania.
- March 24, 1998: Four students and a teacher are killed and 10 wounded in Jonesboro, Arkansas, when two boys, ages 11 and 13, open fire from nearby woods.

Journal Article Link 8.4
Examine literature regarding violence against girls in school.

It is critical that a balanced picture of school violence be drawn. Less than 1 percent of school-age homicides occur in or around schools. Despite the highly visible, horrific cases such as Columbine, school-related violence actually decreased during the 1990s. Richard Lawrence (2006, p. 1) speculates that the relative infrequency of school violence is the very reason it is frequently covered in the media. He also adds that, while such media coverage has distorted the actual risk of violent death at school, it has had a positive effect on enhancing safety and preventing bullying and other threatening behavior (p. 5). The Bureau of Justice Statistics (2004) reports that school crime decreased by 50 percent between 1992 and 2002. Crime File 8.5 reports on the findings of the U.S. Secret Service study of school shooters.

Some elements of a solution to school shootings that have been suggested include the following:

- Better control over the access to guns. This is opposed by one of the most powerful lobbies in the United States—the National Rifle Association (NRA).
- Identify and help troubled youths before they attack.
- Involvement of parents is critical.
- More and earlier intervention and prevention programs.
- Antibullying programs.

Guns

In the 1990s, the Brady Bill finally became law despite opposition by the NRA. This law required a waiting period plus a record check before purchase of a firearm. This was later altered due to NRA pressure to instant background checks. Much emotion surrounds the **gun control** debate, with opponents of control arguing that regulation would hurt only the law-abiding, who would be unable to protect themselves from the criminal. The law-abiding, however, are also of concern. The very weapon purchased to protect the family against outside intruders all too often causes the death of a loved one. Morris and Hawkins (1970) describe the issue quite succinctly. A major precipitating condition of murder in the United States is the possession of a gun. They state, "Easy access to weapons of this kind may not merely facilitate violence, but may also stimulate, inspire, and provoke it" (p. 72).

Why such continuing opposition to handgun control, despite clear public support for such measures? The NRA is probably the single most powerful lobby in Washington, representing a $2 billion per year business with a generous campaign donation policy and a strategic direct mail effort. Yet

Crime File 8.5

Deadly Lessons: The Secret Service Study of School Shooters

In 2002, the U.S. Secret Service released its report on a study of 37 school shootings involving 41 attackers. They reviewed investigative, school, court, and mental health records and interviewed 10 attackers.

What We Know

Attackers Talk About Their Plans. Prior to most incidents, the attacker told someone about his idea or plan. In more than three fourths of the cases examined in the Safe School Initiative, the attackers told a friend, schoolmate, or sibling about his idea of a possible attack before the action. In nearly all cases, the person who was told about the impending incident was a peer, and rarely did anyone bring the information to an adult's attention.

Attackers Make Plans. Incidents at targeted schools are rarely impulsive. In almost all of the incidents, the attacker developed the idea about harming the target before the attack. In many cases, the attacker formulated the idea of the attack at least 2 weeks in advance and planned out the incident. Quick efforts to inquire and intervene are extremely important because the time span may be short between the attacker's decision to attack and the actual incident.

There Is No Stereotype or Profile. There is no accurate or useful profile of the "school shooter." The personality and social characteristics of the shooters varied considerably. They came from a variety of ethnic and racial backgrounds and varied in age from 11 to 21 years. Few had been diagnosed with any mental disorder prior to the incident, and less than one third had histories of drug or alcohol abuse. Thus, profiling is not effective for identifying students who pose a risk for targeted violence at the school.

A fact-based approach may be more productive in preventing school violence than a trait-based approach. This study indicates that an inquiry based on a student's behaviors and communications will be more productive than attempts to determine risk by attending to students' characteristics or traits. The aim should be to determine if the student appears to be planning or preparing for an attack. If so, how far along are the plans, and where or when would intervention be possible?

Attackers Had Easy Access to Guns. Most attackers had used guns previously and had access to gun(s) used in the attack from their own home or that of a relative.

School Staff Are Often First Responders. Most shooting incidents were not resolved by law enforcement intervention. More than half of the attacks ended before law enforcement responded to the scene—despite law enforcement's often prompt response. In these cases faculty or fellow students stopped the attacker, or the attacker stopped shooting on his own or committed suicide.

Attackers Are Encouraged by Others. In many cases, other students were involved in some capacity. The attacker acted alone in at least two thirds of the cases. However, in almost half of the cases, friends or fellow students influenced or encouraged the attacker to act. Advanced knowledge among students about the planned incidents contradicts the assumption that shooters are "loners" or that "they just snap."

Bullying Can Be a Factor. In a number of cases, bullying played a key role in the decision to attack. A number of attackers had experienced bullying and harassment that were longstanding and severe. In those cases, the experience of bullying appeared to play a major role in motivating the attack of the school. Bullying was not a factor in every case, and clearly not every child who is bullied in school will pose a risk. However, in a number of cases, attackers described experiences of being bullied in terms that approach torment.

Warning Signs Are Common. Most attackers are engaged in some behavior prior to the incident that caused concern or indicated a need for help. In more than half of the cases, the attacker's behavior caught the attention of more than one person. A significant problem in preventing targeted violence is determining how best to respond to students who are already known to be in trouble. This study indicates the importance of giving attention to students who are having difficulty in coping with major losses or perceived failures, particularly when feelings of desperation and hopelessness are involved.

Source: Excerpts from: "Preventing School Shootings: A Summary of a U.S. Secret Service Safe School Initiative Report." *NIJ Journal, 248,* 2002: 11–15, NCJ 190633.

Web Research Project
Using a Web browser, search the term "school shooters" and locate some recent cases. Do these shooters fit or not fit the Secret Service profile?

more Americans have been killed with guns by their fellow citizens in the twentieth century than have been killed in all of the wars this nation has ever fought.

A popular defense of the pro-gun lobby in the United States is "Guns don't kill people; people kill people." In reality, people with guns (particularly handguns) do kill people, and it is no coincidence that the United States has both the highest homicide rate by far of any developed nation and the largest armed civilian population in the world. This widespread ownership of firearms combined with a culture and subculture of violence foments lethal combinations.

In February 1998, Great Britain instituted a total ban on private ownership of handguns in response to public outcry over the 1996 massacre of 16 schoolchildren and their teacher in Dunblane, Scotland. At the same time, there remains an estimated 50–70 million firearms in the United States in private hands. On Mother's Day, May 14, 2000, a Million Mom March on Washington took place, calling for stronger background checks, licensing of handgun owners, comprehensive gun registration, safety child locks, limits on handgun purchases, and reasonable cooling-off periods before newly purchased firearms could be taken into possession.

In light of success in suing tobacco companies for causing a public health menace, public bodies, particularly cities, have filed suit against gun manufacturers. In a typical action, the city of Chicago filed a $433 million lawsuit against 38 gun retailers, distributors, and manufacturers, alleging they had created a public nuisance by knowingly saturating the city with illegal firearms. The city of New Orleans charged that the manufacturers had created "unreasonably dangerous" products (Witkin, 1988). In February 1999, a Brooklyn jury awarded Steven Fox $500,000 for injuries suffered in a shooting. The jury found 15 companies guilty of "negligent distribution" of guns and ordered three to pay Fox for his injuries.

While heated arguments continue between opponents and proponents of gun control, more research is needed on the potential impacts of various policy options (Zimring & Hawkins, 1992). As an example, a large survey of convicted felons by J. D. Wright and Rossi (1986) suggested the following:

- Rather than reducing crime in violent urban neighborhoods through gun control, the violence endemic to such impoverished areas must be reduced, thus reducing the need for carrying weapons.
- The theft of firearms must be reduced.
- The informal market for guns must be interdicted.
- Mandatory sentences for crimes with guns are ineffective and do not serve as a deterrent.
- The control of "Saturday night specials" (cheap handguns) would simply encourage criminals to switch to more lethal weapons (Bonn, 1987).

It is quite difficult to target public policy objectives when there is no consensus regarding the essential nature of the problem of firearms and high violent crime rates in the United States. A promising intermediate strategy has tested positive in the Kansas City Gun Experiment. Police directed patrols at gun crime "hot spots" and were able to reduce gun crimes by seizing illegally carried weapons (Sherman, Shaw, & Rogan, 1995). The project was replicated in Indianapolis with much the same results. Sheley and Wright (1995), in interviews with juvenile inmates and students in inner-city high schools, found that the primary motivation for gun possession by these groups was fear, not criminal activity, gang membership, or drug trafficking. By 2009, the U.S. had become the Colombia of gun trafficking, the shopping center of choice for terrorists and drug smugglers (McKinley, Jr., 2009, pp. A1, A14.) "Sending straw buyers into American stores, cartels have stocked up on semi-automatic AK-47 and AR-15 rifles converting some to machine guns, investigations in both countries [the U.S. and Mexico] say. They have also bought .50 caliber rifles capable of stopping a car and Belgian pistols able to fire rounds that will penetrate body armor" (p. A14). For more information on gun control, see J. B. Jacobs (2002), Nisbet (2001), and Spitzer (2003).

Sexual Assault

On the night of July 10, 1991, at St. Kizito's coed boarding school in Meru, Kenya, 71 teenage girls were raped by their classmates and 19 others died when the boys attacked the girls for refusing to join them in a strike against the school's headmaster. A report on the front page of the *Kenya Times* called the rape of the St. Kizito coeds a "common occurrence" sanctioned by the principal and his staff. The paper quoted the deputy principal as saying, "The boys never meant any harm against the girls. They just wanted to rape" (Heise, 1991).

Handbook Article Link 8.1
Read an article
on sexual offenses.

This attitude is not uncommon. In some countries, rape is viewed as a man's right or as a crime against the honor of the woman's family or husband, not an offense against the woman. The word itself is derived from the Latin for "theft." A large percentage of rapes are perpetrated against children. In fact, more than half of rape victims are under 18, and the younger the victim the more likely the attacker is to be a relative or acquaintance and the less likely the rape is to be reported. In some countries, women are often blamed and punished for the rape (Heise, 1991). Susan Estrich, a rape victim herself, maintains in her book, *Real Rape: How the Legal System Victimizes Women Who Say No* (1987), that little has changed in the way most rape cases are handled by the courts and that judges still use their personal views to decide the victim's claim (Berger, 1988). This all depends on whether the rape is viewed as "real rape" or "simple rape." (These terms are put in quotes to indicate that such distinctions are made by ill-informed persons.)

Misperceptions of rape falling into two categories, real and simple, held back attempts to deal with rape as a problem. *Real rape* is aggravated rape involving violence, weapons, and attackers, and is recognized as rape by the courts. *Simple rape* is everything else, including date rape, and is dismissed as not "real rape." Victims of simple rape are viewed with suspicion, as not really victims, particularly if the victim did not physically resist. In order for it to even be considered rape in this mistaken view, a woman would have to demonstrate that she really fought back (show some physical damage to her), or else it would be assumed that she cooperated and it really was not a rape (Berger, 1988, p. 65).

The women's movement has been instrumental in altering public and official views of rape and rape victims. Susan Brownmiller (1975) in *Against Our Will: Men, Women and Rape* claims that the criminalization of rape took place with the emergence of a money economy in which violation of virginity posed potential economic loss for the family since the future bride was now tainted goods.

Statistics regarding the extent of rape have been notoriously poor. Women have been reluctant in the past to report rapes for a variety of reasons, including:

- The stigma attached to rape, which alleges that the victim either invited the attack or cooperated in it.
- Sexist treatment of many rape victims, who are in effect mentally raped a second time by the criminal justice system (the police, defense attorneys, and judges).
- Legal procedures that have permitted defense attorneys to probe the victim's sexual past in potentially humiliating ways.
- The burden of proof, which has been shifted to the victim so she must show that the attack was against her will and that she resisted the assault.

Table 8.3 presents recent rape rates in selected countries using Interpol statistics. Caution should be exercised in interpreting these statistics given their historically questionable nature as well as cultural reluctance to record such data. Given these limitations, the highest rape rates were for the Bahamas (55.9), Jamaica (35), the United States (32.1), and France (17.5). The U.S. rate was higher than that of other developed countries. The lowest rates were in Saudi Arabia (.3), Japan (1.9), Russia (5.6), and Brazil (8.4).

Only recently have a significant proportion of rape victims been willing to report rapes and undertake prosecution of their attackers. The growth of rape crisis centers, featuring counseling and

TABLE 8.3 ■ **International Rape Statistics, Interpol Data (per 100,000 population)**

Country	Rape Rate (year)
United States[a]	32.2 (2004)
Bahamas	55.9 (2000)
Brazil	8.4 (2001)
Denmark	8.7 (2003)
France	17.5 (2003)
Ireland	13.7 (2003)
Israel	9.4 (1998)
Jamaica	35 (2001)
Japan	1.9 (2002)
Russia	5.6 (2002)
Saudi Arabia	.3 (2002)

Source: Interpol, *International Crime Statistics, 2003*, Lyons, France: Interpol Secretariat, 2004.

a. U.S. statistics were not included in the Interpol report and are taken from the Uniform Crime Reports for 2005.

Web Research Project
Locate an article on rape in another country. How was rape viewed there and were there any proposed solutions?

support services for victims, has been instrumental in this greater willingness to prosecute. Other factors that account for an increase in the tendency to report rape include:

- More female police officers.
- Better training of police in sensitive handling of rape cases.
- Changes in rape laws in many states, which, for instance, prevent defense attorneys from probing into the victim's prior sexual behavior.

Rape was the fastest growing of all UCR index crimes in the seventies, increasing 85 percent from 1970 to 1979. However, the NCVS victim data available for part of this same period show virtually no change in the rape rate. The increase in the seventies rape rate was primarily due not to an increase in the number of rapes, but rather to a growth in the willingness to report such crimes.

Figure 8.2 presents violent crime rates for the United States since 1973 using NCVS data. Data prior to 1992 were adjusted to make them more similar to data collected after that date using redesigned methodology.

The Department of Health and Human Services ("Philadelphia Crime Statistics Questioned," 1998) reported that half of all rape victims were assaulted before their 17th birthday and that an estimated 17.7 million women in the United States—nearly 18 percent—have been victims of rape or attempted rape.

For 2006, the Midwest had the highest rape rate in the United States (35.4), followed by the South (34.1), the West (31.8), and the Northeast (20.6). Convicted rapists generally present the same demographic profile as other violent and conventional property criminals; that is, for the most part they are 15–24 years old, unmarried, from lower-class and minority backgrounds (particularly black), and

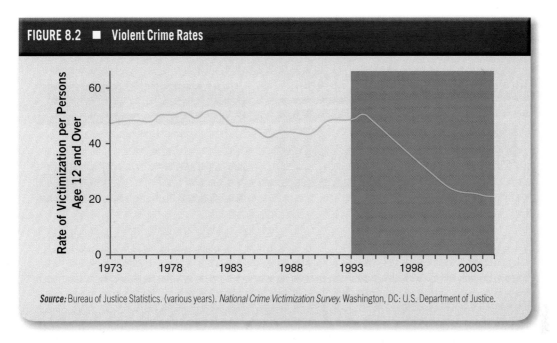

FIGURE 8.2 ■ Violent Crime Rates

Source: Bureau of Justice Statistics. (various years). *National Crime Victimization Survey.* Washington, DC: U.S. Department of Justice.

they choose victims of the same race. Amir (1971), in his Philadelphia study of forcible rape, found that most rapists had backgrounds in property offenses rather than in violent criminal activity. He examined 646 cases in 1958 and 1960 and found that rape rates were 12 times higher among blacks than among whites. Amir's research was based solely on official police reports, however, and for that reason has been subject to dispute.

Acquaintance Rape

Many rapes take place within established dating relationships, and the vast majority are never reported. Victims often fear the publicity, notoriety, and family reaction that pressing charges would entail.

While official data suggest that about half of all rapes involve strangers, this is partially offset by the victims' greater willingness to report stranger-precipitated incidents. Victimization surveys suggest that over 80 percent of rapes and attempted rapes are by strangers (McDermott, 1979). Amir (1971, p. 143) found that 58 percent of single-offender and 90 percent of group rapes were planned rather than spontaneous events and that rape by more than one attacker took place in over 40 percent of the cases. But studies of attempted rapes reported in anonymous surveys of high school and college women indicate that many of these incidents are unlikely to be reported even to other household members, let alone to victim-survey interviewers. Such potential "date rapes" do not meet the stereotype of the attacker as a stranger on a dark street. Contrary to Amir's research, which concluded that most rapes were by strangers, these findings suggest that attacks by intimates and acquaintances make up the majority of rapes (Christensen & Gregg, 1970; Kirkpatrick & Kanin, 1957). Crime File 8.6 reports on campus acquaintance rape.

Gang rape is usually perceived as something that occurs across the tracks among the underclass, but Ehrhart and Sandler's *Campus Gang Rape* (1985) and Sanday's *Fraternity Gang Rape* (1990) tell us of the close-to-home nature of such activity. Belknap (1990) notes,

There exists a strong misogynistic, sexist, racist rape culture at most universities which appears to be particularly virulent among fraternities. . . . The male entitlement acquired through fraternity or male athletic team membership helps these young men believe that they are special and superior not only to women, but also to other men. (pp. 285–286)

Crime File 8.6

The Problem of Acquaintance Rape of College Students

Rape is the most violent crime on American college campuses today. Most rapes are unreported, perhaps giving the impression that current efforts in fighting campus rapes are adequate. College women are more at risk for rape and other forms of sexual assault than for noncollege women of the same age. It is estimated that almost 25 percent of college women have been victims. Rape rates vary to some extent by school, type of school, and region, suggesting that certain places within schools are more rape-prone than others. Some features of the college environment—frequent unsupervised parties, easy access to alcohol, single students living on their own, and the availability of private rooms—may contribute to high rape rates of female college students. There are no data on the number of college students raped after unknowingly ingesting Rohypnol, GHB, or similar drugs.

Ninety percent of college women who are victims of rape or attempted rape know their assailant. The attacker is usually a classmate, friend, boyfriend, or other acquaintance (in that order). Most acquaintance rapes do not occur on dates, rather they occur when two people are in the same place (e.g., at a party or studying together in a room). Congress enacted the Student Right-to–Know and Campus Security Act of 1999 [U.S.C/ 1092 (f) (1)] covering all colleges and universities receiving federal funds and a 1992 amendment to the act requires campuses to spell out rape victim's rights and to annually publish information on prevention programs. A 1998 amendment added important reporting obligations and renamed the act the Jeanne Cleary Disclosure of Campus Security Policy and Campus Crime Act. Jeanne Cleary's rape and murder on a college campus brought to light some of the inadequacies in the college reporting of crime.

College students are the most vulnerable to rape during the first weeks of the freshman and sophomore years. In fact, the first few days of the freshman year are the riskiest, limiting the value of any rape prevention programs that begin after that. Research has shown that rapes of college women tend to occur after 6 p.m., and the majority occur after midnight.

Source: Rana Sampson. Acquaintance Rape of College Students. Problem-Oriented Guide for Police, #17, August 2003. http://www.cops.usdoj.gov.

The practice of campus gang rape is called "playing train" or "pulling train" (one man follows another) by perpetrators such as fraternity members. A common pattern is described by Ehrhart and Sandler (1985):

> A vulnerable young woman, one who is seeking acceptance or who is high on drugs or alcohol, is taken to a room. She may or may not agree to have sex with one man. She then passes out, or is too weak or scared to protest and a train of men have sex with her. Sometimes the young woman's drinks are spiked without her knowledge and when she is approached by several men in a locked room, she reacts with confusion and panic. Whether [she is] too weak to protest, frightened, or unconscious, as has been the case in quite a number of instances, anywhere from two to eleven or more men have sex with her. In some party invitations [to men] the possibility of such an occurrence is mentioned with playful allusions to "gang bang" or "pulling train." (pp. 1–2)

As if being raped is not degrading enough, consider the additional and final victimization of also being infected with AIDS. Unfortunately, most schools do not formally punish either date rapists or gang rapists, choosing to keep such activity quiet. Parents of a Lehigh University freshman who was raped and murdered in 1986 forced the passage of the "Cleary Law," which requires the publication of campus crime rates in order to inform parents and students of potential danger. Adopted in 1992, the Crime Awareness and Campus Security Act requires colleges and universities to disclose their crime statistics or risk losing federal aid.

Amir Versus Brownmiller

Chappell and Fogarty (1978) in the literature review *Forcible Rape* indicate two early major landmark works in the field: Menachem Amir's (1971) *Patterns in Forcible Rape* (1971) and Susan Brownmiller's (1975) *Against Our Will: Men, Women and Rape* (1975). Amir's work, at the time the most widely cited source in rape literature, espoused a now controversial theory of **victim precipitation,** in which the victim's behavior is seen as contributing to the incident. This theory was viewed as perpetuating many of the common myths regarding rape. Brownmiller's work put forth a theory that rape is "nothing more or less than a conscious process of intimidation by which all men keep all women in a state of fear" (p. 15). Chappell and Fogarty indicate that much of the pre-1972 literature such as J. M. MacDonald's (1971) *Rapists and Their Victims* suffered from male chauvinist biases prevalent at that time.

Rape as a Violent Act

Rape is often perceived primarily as a sexually motivated act, but most authorities on rape identify it as primarily a violent act in which sexual relations are merely a means of expressing violence, aggression, and domination. While our discussion will give consideration to arguments as to whether rapists are sexually or violently motivated or both, the classification of rape as a crime of violence looks not at the motivation of the offender but at the perception of the act by victims. Similar to the argument that robbery is really a property crime, sex (rape) or money (robbery) may be the motivation, but the tool employed and perceived by victims is violence or threats of violence and intimidation. For this reason, the author views rape as a crime of violence. On the basis of their study of over 500 convicted rapists, Groth and Birnbaum in *Men Who Rape* (1979) identify three types of rape:

- The anger rape, in which sexual attack becomes a means of expressing rage or anger, involves far more physical assault on the victim than is necessary. Groth and Birnbaum (1979) claim that 40 percent of their subjects were anger rapists.
- The power rape, in which the assailant primarily wishes to express his domination over the victim, is viewed as an expression of power rather than a means of sexual gratification. Thus, the rapist generally uses only the amount of force necessary to exert his superordinant position. The majority, about 55 percent, of Groth and Birnbaum's offenders were of this type.
- In the sadistic rape, the perpetrator combines the sexuality and aggression aims in psychotic desires to torment, torture, or otherwise abuse his victim. About 5 percent were of this type.

Glaser (1978, p. 364) proposes four categories of rapists: naive graspers, meaning stretchers, sex looters, and group conformers. *Naive graspers* are usually sexually inexperienced youths with an unrealistic conception of female erotic arousal. Awkward in relating to the opposite sex, they hold high expectations that their crude advances will be met with affection by their victims. They possess a strong desire for affection but little respect for their victim's autonomy in resisting such advances. *Meaning stretchers* are the most typical rapists, the date rapists. They stretch the meaning of, or misinterpret, a woman's or date's expressions of friendliness and affection as indicating that the female desires coitus even when she says no. *Sex looters* have little desire for affection and/or little respect for the victim's autonomy and callously use women as sex objects. This type figures in the stranger-precipitated rape that is most likely to be reported to the police. *Group conformers* participate in group rapes or gang bangs, often following the leader, a sex looter, out of a need for conformity and a perverted notion of demonstrating masculinity.

In an examination and criminal profiling of 41 convicted "serial rapists" (defined in the project as those who had committed 10 or more rapes), Hazelwood, Burgess, and associates (A. W. Burgess, Hazelwood, Rokous, Hartman, & Burgess, 1987; Hazelwood & Burgess, 1987; Hazelwood & Warren, 1989) found that 76 percent had been sexually abused as children. The majority of the serial rapes had not been reported to authorities. Such typologies may not accurately reflect the motivations of typical rapists at the time of the offense. They are based on ex post facto (after the fact) case studies of incarcerated offenders and interviews with offenders, which makes them prone to **post hoc error,** the assumption that since one variable is observed before another, it must be the cause of that outcome. They are also unnecessarily steeped in psychiatric assumptions regarding offender motivations. In addition, incarcerated rapists are more likely to be of the "stranger" variety and perhaps either more violence prone or more willing to use violence than the nonstranger rapists.

Opinions on victim resistance appear to be mixed. While resistance (particularly screaming rather than physical defense) increases the likelihood of escape, it also increases the possibility of injury.

The criminalization of forced marital intercourse (Sigler & Haygood, 1988), or "marital rape," underscores the view of rape as a violent rather than sexual crime. In *State v. Rideout* (1969), an Oregon court challenged the marital immunity defense, that is, the husband's right to force involuntary intercourse. This is now recognized as rape in many states.

Finally, an examination of prison inmates is telling. One in five violent offenders serving time in state prison report having been victimized a child. Two thirds of prisoners convicted of rape or sexual assault had committed their crime against a child. Inmates who victimized children were less likely to have a prior criminal record. Three in 10 child victimizers had multiple victims, and in the vast majority of cases they knew the victim prior to the incident. In fact, one third committed their crime against their own child (Greenfield, 1996).

Sexual Predators

Video Link 8.5
View a video of the Jaycee Dugard kidnapping.

Another category of sexual offenders is the violent sexual predators who prey upon children. In February 2004, the nation was shocked to see security camera videotape footage of the abduction of 11-year-old Carla Bruscia. She was subsequently raped and murdered. In 1994, Megan Kanka, a 7-year-old New Jersey girl, was kidnapped, raped, and murdered by a convicted sex offender who lived across the street from her. Unbeknownst to the family, the man had a long record of sexual attacks on children. In response to this incident, New Jersey, and later other states and the U.S. Congress, passed "Megan's Law," requiring states to inform local communities when known high-risk sex offenders are being released into the community.

In some instances, sexual predators have still been able to slip through the net. Some move their residence without notifying authorities. Mapsexoffenders.com, launched in 2005, has mapped publicly available state sex-offender registries onto Google Maps. With 34 U.S. states mapped by August of 2005, people could type in their address and a map of their neighborhood pops up, along with red flags that can be clicked on in order to view the predator's name, address, photo, and list of offenses. The U.S. Justice Department also inaugurated the Sex Offender Public Registry (http://www.nsopr .gov), which links individuals directly with 24 state databases (Campo-Flores, 2005).

A related offense, stalking, is deliberately and without justification following or surveilling (or both) another person. It also includes threatening another person with immediate or future bodily harm, sexual assault, or confinement.

Stalking

Stalkers may include spurned lovers, ex-spouses, admirers, obsessive fans, and the like. California passed the first antistalking law in 1990, and at least 20 other states have followed suit with laws prohibiting the willful, malicious, and repeated following, threatening, or harassing of another person.

The National Violence Against Women Survey funded by the National Institute of Justice and Centers for Disease Control (Tjaden, 1997) defines stalking as "a course of conduct directed at a specific person that involves repeated physical or visual proximity, nonconsensual communication, or verbal, written, or implied threats" sufficient to evoke fear in a reasonable person (p. 1). The survey discovered that stalking was a much bigger problem than previously believed, with about 1.4 million victims annually. Controlling behavior and physical, psychological, and sexual abuse by a woman's former intimate partner was the major mode of conduct. Nearly half of the incidents were reported to the police, and about 25 percent resulted in a restraining order. The former partner's motivation was to keep the victim in a relationship.

Robbery

Robbery involves theft through violence or the threat of violence. P. J. Cook (1983) describes robbery as the "quintessential urban crime" (pp. 1–2): The six largest cities in the United States experience one third of the nation's robberies. Robbery is more concentrated in larger cities than any other crime. The 2006 U.S. robbery rate was 160.4 offenses (per 100,000). Robbery was highest in the South (157.1), followed by the West (152.5), the Northeast (151.2), and the Midwest (132.1.). The average loss for 2006 robberies was $1,268, with the total loss for all robberies estimated at $567 million. The highest losses were for bank robberies ($4,330 each, on average), and the lowest were for convenience store robberies (average of $761). Loss to residence per robbery was $1,469.

For a number of years, Los Angeles has led all cities in the United States in bank robberies, accounting for one fifth of the national total (Florida and New York are runners-up). The city has a lot of new branch banks that are not built like many of the older fortress-like banks back East, and there are numerous convenient freeways for quick escape. Despite relatively low takes (the average heist in 2003 was $4,767); poor odds for success (85 percent of bank robbers are caught); stiff sentences (armed robbery carries a 25-year maximum); and easy detection because of witnesses, pictures, and police attention, many robbers are still not deterred. Bank robbers are not always the craftiest. One opened a bank account in his real name and then robbed the same bank. Another was mugged outside the bank he had just robbed and called the police to complain. Crime File 8.7 is an FBI article reporting on the fact that often-goofy names given to robbers attract attention and perhaps greater rates of apprehension.

Table 8.4 reports international robbery statistics for selected countries. Based on Interpol statistics, countries with the highest robbery rates were the Bahamas (457), France (210), Russia (148.9), and the United States (142.2). Those with the lowest rates were Saudi Arabia (.1), Japan (5.5), Denmark (59.1), and Ireland (74.7).

The typical robbery is rather minor. Most robberies are of individuals rather than of commercial establishments and for amounts considerably less than one might assume. A disproportionate number of robbers are young black males, while a large proportion of victims are white males over 21. The unusual interracial nature of robbery, unlike other violent crimes, can be explained by the fact that robbers are primarily interested in money and adult white males are perceived as good targets. Haran's (1982) examination of armed robberies in New York City found that, in the sixties, 60 percent of the robbers were older white males, but that, in the seventies, 61 percent were black males, 58 percent of them under age 26. However, Glaser (1978, p. 254) points out that official statistics overlook both juvenile offenders and victims.

Perhaps the biggest cause of fear in victims of robbery is the risk of personal harm in street robberies. Roughly one third of victims are harmed to some degree, with 2 percent requiring inpatient hospital care. Feeney and Weir (1975) indicate that while resisting robbery greatly increases the possibility of suffering injury, screaming and yelling may cause the robber to quit and do not tend to increase the likelihood of harm. Many armed robbers are primarily interested in intimidating the victim and in fact may employ unloaded or even fake weapons, whereas unarmed robbers are far more likely to attack their victims (Conklin, 1972, pp. 113–116).

In September 1992, a Maryland woman was dragged to her death by thieves stealing her car. They forced the woman from behind the steering wheel and threw her 2-year-old child, car seat and all, from the moving vehicle. In a new variation of robbery called carjacking, armed car thieves approach occupied vehicles and commandeer them. *Carjacking* (robbery auto theft) is defined as completed or attempted robbery of a motor vehicle by a stranger to the victim. It differs from other

Crime File 8.7

Name That Bank Robber

Catchy Monikers Help Nab Culprits

We call her the "Ponytail Bandit." She's a pretty young woman whose shoulder-length blonde ponytail sprouts from the back of her baseball cap in images captured by banks she allegedly robbed in three states in May. In each case, she approaches bank tellers, demands money, and then waits, arms crossed, slouching slightly, as tellers comply.

That she doesn't seem to fit the stereotype of a bank robber—brazen, masked, gun-slinging—has drawn a lot of interest in her story, whatever it may be. But the bottom line is the woman captured on grainy bank surveillance tapes in Texas, California, and Washington is a bank robber committing a crime.

With her distinguishing moniker, the "Ponytail Bandit" joins the ranks of hundreds of other bank robbers sought or captured by the FBI and local law enforcement agencies. We play a direct or indirect role in all bank robbery investigations and compile detailed statistics on bank and armored car heists, which last year amounted to more than $72 million in stolen loot.

Bank "Bandits"

- Ponytail Bandit
- Goofy Hat Bandit
- Irreconcilable Differences Bandit
- Grocery Store Bandit
- Paint-by-Numbers Bandit
- Pepper Spray Bandit
- Impersonator Bandit

Here's a look at a few creative ways we're working with local authorities and banks to catch robbers or prevent heists before they happen:

- In Baltimore, Special Agent Jeff Cisar worked with public and private entities last year to create a website specifically to spotlight suspected bank robbers in the Baltimore, Washington, D.C., and Northern Virginia regions. The site, bankbandits. org, is seeded with a catalog of surveillance images from area bank robberies and offers a way to provide tips online.
- In Los Angeles, Special Agent Stephen May took a page from Agent Cisar's playbook and, through partnerships with other agencies, created labankrobbers.org, which covers seven counties in Southern California. The site, like Baltimore's, features dozens of still surveillance images, as well as a category of "serial bandits" distinguished by colorful monikers assigned by Agent May—"Bad Rug Bandit," "Banana Bandit," "Paparazzi Bandit."
- In Seattle, Special Agent Larry Carr is the man behind some of the most unique names assigned to bank robbers—"Attila the Bun," "Groundhog Day Bandit," "Hollywood Bandit." The catchy names serve dual purposes—a hook for media attention to help solve robberies and a tool to help investigators track serial suspects. Meanwhile, to help prevent bank robberies in the first place, Agent Carr developed a tactic designed to put potential robbers on their heels—shower them with immediate personal service. Agent Carr's "Operation Safecatch" tutors bank employees on how to recognize potential trouble and then take action—which in many cases simply means breaking the ice with a "customer" before they get their nerve up to act. Carr credits the tactic, which throws potential robbers off their game plans, for a recent drop in bank robberies in the Seattle area.

Agent Cisar in Baltimore said the push to catch bank robbers by posting all of their images on the Net has been effective—both in leading to captures and aiding local police investigators and private industry in an era of tightening resources.

"We want everyone looking at these sites," he said. "You might recognize someone."

Source: Quoted directly from "Headline Archives: Name That Robber," Federal Bureau of Investigation, June 25, 2007, http://www.fbi.gov/page2/june07/bandits062507.htm.

TABLE 8.4 ■ **International Robbery Statistics, Interpol Data (per 100,000 population)**

Country	Robbery Rate (year)
United States[a]	142.2 (2003)
Australia	121.7 (2000)
Bahamas	457.2 (1999)
Canada	88.2 (2001)
Denmark	59.1 (2003)
France	210.0 (2000)
Ireland	74.7 (2002)
Jamaica	80.9 (2001)
Japan	5.5 (2002)
Mongolia	26.5 (2001)
Russia	148.9 (2002)
Saudi Arabia	.1 (2002)

Source: Interpol. *International Crime Statistics, 2003.* Lyons, France: Interpol Secretariat, 2004.

a. U.S. statistics were not included in the Interpol report and are taken from the Uniform Crime Reports.

Web Research Project

Search for the term "robbery" and report on some recent cases.

The Interpol definition of robbery is "robbery and violent theft." The UN definition is "the taking away of property from a person, overcoming resistance by force or threat of force."

The United States defines robbery as "the taking or attempting to take anything of value from the care, custody, or control of a person or persons by force or threat of force or violence and/or putting the victim in fear."

vehicle theft because the victim is present and the thief uses or threatens force. Between 1993 and 2002, the NCVS estimates an average of 38,000 attempted or completed carjackings in the United States. Half of these were successful (Klaus, 2004). Carjacking has been a federal crime since 1992. Some experts speculate that carjacking has increased in response to new antitheft devices, particularly on expensive automobiles.

Conklin's Typology of Robbers

Conklin (1972) developed a **typology of robbers** based on interviews with 67 convicted robbers in Massachusetts prisons, as well as 90 victims. He classified robbers as professional robbers, opportunist robbers, addict robbers, and alcoholic robbers. *Professional robbers* have a long-term commitment to crime, their major source of livelihood. They are very rational about crime and plan their operations carefully. *Opportunist robbers* are the most common type of bandit. Having little commitment to, or specialization in, robbery, they are all-purpose property offenders. Their engagement in robbery is infrequent and relatively unplanned. Often young and from lower-class, minority backgrounds, such offenders often operate in groups. Gabor and Normandeau (1989, pp. 273–282), in studying armed robbers in Montreal, found that most did not fit the stereotype of the meticulous professional. Most were under 22 years of age, wore no disguise, and usually stole less than $100. Nearly half of the robbers either did not plan ahead at all or planned their robbery far less than an hour in advance. The robbers indicated that they viewed armed robbery as the fastest and most direct means of getting money compared with burglary and fraud. Younger robbers in particular claimed that they enjoyed the thrill, excitement, status, and feeling of power associated with the crime. In 2004, a Belfast, Ireland, bank robbery netted 22 million pounds (about $42 million at the time) and rivaled some of the biggest robberies in history. In the post–World War II chaos, $7.4 billion disappeared from the German Reichsbank. One billion was robbed from Iraq's Central Bank in 2003 and $168 million was taken from a British bank in Beirut by guerrillas in 1976 ("Big Belfast Heist," 2004).

Addict robbers are addicted to substances such as heroin or other drugs and commit robbery to support their expensive habits. Most drug abusers are interested in safe and quick criminal gain, and are less likely to be involved in robbery than in burglary and sneak thievery. Such offenders are less likely to use weapons and therefore more likely to use physical force as a means of intimidation. Alcoholic robbers have little commitment to robbery; they engage in unplanned robberies on occasion in order to support their habit. Many claim to have been intoxicated at the time of their offense.

Perhaps the most feared type of robber (though not a separate type) is the *mugger,* a "strong-arm" robber who generally does not use a weapon. Many muggers are semiprofessional; they do some planning and specialization, though not to the extent of the professional robber. *Mugging,* an American slang term for robbery, may refer to everything from purse-snatching to brutalization or murder of the victims. Working in groups, muggers may start by carefully surveying the scene and the mark (victim). The actual techniques employed vary from the "yoke," a method of grabbing the victim from behind around the neck, to the use of knives or guns to scare the victim. Young black offenders are more likely to commit purse snatches and street robberies, netting small amounts, while adult white offenders are more likely to participate in commercial robberies.

Richard Wright and Scott Decker in *Armed Robbers in Action: Stickups and Street Culture* (1997) conducted field research on 86 uncaught robbers in St. Louis. They indicate the following:

> The offenders in our study typically compel the cooperation of intended victims through the creation of a convincing illusion of impending death. They create this illusion by catching would-be victims off guard and then using tough talk, a fierce demeanor and the display of a deadly weapon to scare them into unquestioning compliance. (p. 128)

The armed robbers often chose to commit their crimes while under perceived pressure to continue illicit action, for example, gambling, drug use and drinking binges, or self-indulgent street culture. Most preyed on other local criminals, particularly drug dealers. Pressed for cash, most chose the first rather than the best available target. Not all offenders targeted such criminal victims; many chose law-abiding citizens. Carefully observing dress and demeanor of potential targets, they chose victims who were most likely to turn over their possessions without protest. This might explain why mostly black robbers chose white victims, whom they regard as less likely to offer resistance compared to black victims. This might also explain why robbery is the only interracial crime that often involves a black perpetrator and a white victim.

High robbery rates are not an inevitable product of urbanization in advanced capitalistic societies, as can be illustrated by Japan. In the late seventies, New York City had 11,000 robberies per million residents, while Tokyo had about 40 robberies per million population (Nettler, 1982, Vol. 2, p. 34).

Domestic Violence

In our previous discussion, we indicated that one stands the greatest chance of being kicked, stabbed, shot, or otherwise brutalized within the refuge of one's own home. Child abuse may include excessive physical assault, neglect, and/or sexual molestation. **Spouse abuse** usually involves physical assaults by husbands against their wives, although the reverse is not unheard of.

Child Abuse

While corporal punishment is an approved-of disciplinary practice in many societies, *child abuse* is defined as excessive mistreatment, either physical or emotional, of children beyond any reasonable explanation (Kempe & Kempe, 1978). In 1996, another sad case of child abuse and death caught the public eye, the death of 6-year-old Elisa Ezquierdo. Of particular concern was the failure of New York City's child protection system. Rescuers who pulled little Elisa from her bed found deep red blotches, welts, and cigarette burns over her entire body; bruises near her kidney, face, and temples; and ghastly wounds around her genitals. Despite repeated expressions of concern in reports to child care workers by other family members, Elisa had been put in the care of her deranged mother. The mother believed the child was possessed by the devil.

In a nationwide survey of 2,143 families, Gelles (1978) discovered that during the previous year, over 20 percent of children in the United States had been assaulted by their parents by having objects thrown at them or by being kicked, bitten, or hit with fists. A 1995 Gallup poll of 1,000 U.S. parents estimated that 5 percent of parents punish their children by punching, kicking, or throwing the child down or hitting with a hard object on some part of the body other than the bottom. The poll also found that 1.3 million children were sexually abused. These figures are much higher than those cited by the National Center on Child Abuse and Neglect, which estimated over 200,000 victims of physical abuse and 130,000 victims of sexual abuse. The federal statistics are based on reported cases (Lewin, 1995, p. A17).

The child batterer strikes the defenseless. Historically, he or she is exercising a traditional prerogative of parents. Infanticide was a parental privilege in many ancient societies, and childhood was simply not regarded as a particularly important stage in life. In the United States, it was not until 1866 that state protection of abused children began, using SPCA (Society for the Prevention of Cruelty to Animals) authority for the removal of a child from an abusing household.

The extent of homicide and brutal assault and torture vented upon child victims was illustrated by a study by Raffali (1970). Following up on 302 battered children reported by some New York City hospitals, he discovered that 1 year later, 35 had died and 55 had suffered permanent brain damage from their injuries. The fact that most instances are never reported to police or come to the attention of authorities would suggest a large "dark figure" of child abuse. Gelles and Straus (1979) put this statistic as high as 1.9 million per year who are physically abused. Abusive activities seem to take place more often in families that have a foster parent or stepparent present (Zalba, 1971), with boys more frequently the targets than girls until age 12, after which girls are more subject to attack.

A virtual statistical epidemic in reports of child abuse has occurred in the United States since the sixties, primarily because of increased efforts at detection and reporting. Emergency room personnel, for instance, receive special training in spotting the "battered child syndrome," which includes a variety of symptoms such as lethargy; fear of parents; subdural hematoma (blood and swelling next to bones or skull); multiple broken bones demonstrating various stages of healing (thus multiple incidents); and suspicious bites, bruises, and the like that cannot be reasonably explained by parents (Fontana, 1973, pp. 28–29). Between 1976 and 1985, reports of child abuse grew from 669,000 reports per year to 1.9 million. It is important to note that more than half of the latter figure were unfounded; that is, the abuse was determined not to have taken place (Whitman, 1987, p. 39). In a reversal of previous rulings, the U.S. Supreme Court in *Coy v. Iowa* ruled that children in sexual abuse cases must confront their alleged abusers "face-to-face" (Lauter, 1988, p. 6A). Such a ruling may discourage prosecutions of real batterers.

Studies of spouse and child batterers suggest a frightening although not inevitable link in which former child abuse victims grow up to become child or spouse abusers themselves. Not all abused children become scarred irreparably or turn into future abusers, but not surprisingly, many do (see Koski, 1988). Researchers have found that the abused and neglected, particularly males, exhibit a higher frequency of arrest for adult violent offenses (J. Miller, 1989; Widom, 1989). Widom (1992) found that childhood abuse or neglect increased the odds of future delinquency or adult criminality by about 40 percent.

Journal Article Link 8.5
Examine literature regarding family violence and dating violence.

In addition to a history of abuse, some other characteristics of child abusers include family isolation from helping resources in periods of crisis, disappointment with the child, and some crisis that precipitates maltreatment (Kempe & Kempe, 1978, p. 24).

According to Bakan (1975, p. 100), hostility toward children is generally associated with two age-maturity distortions. First, the adult may ascribe to him- or herself the role of a younger person. Second, he or she may ascribe to the child a maturity beyond the child's years. Studies of child abusers and their personality characteristics, however, are often plagued by post hoc error. Gelles (1977) indicates that psychological conditions that are identified as being present after an abuse incident tend to be viewed as the cause of the incident. Abusers are often described as being depressed and paranoid, but these conditions could be results of the incident rather than its cause.

Child battering may constitute assault or even homicide, but most cases in the United States are handled in family or juvenile courts as categories of child neglect or abuse. Such courts have an orientation toward rehabilitation rather than toward imposing penalties or imprisonment. Often children are removed from the home for their own protection and temporarily placed in foster homes until their parents are adjudged fit, but a major objective in the past has been to maintain the family unit. This could account for the one third of abused children who later suffered death or brain damage in Raffali's (1970, p. 301) study.

Austria, Croatia, Cyprus, Denmark, Finland, Italy, Latvia, Norway, and Sweden bar parents from spanking their children (Straus, 1999). Germany and Great Britain in 2000 were also considering placing limits on corporal punishment. Murray Straus in his book *Beating the Devil Out of Them: Corporal Punishment in American Families* (1994) indicates that, while working as a sanction in the short run, corporal punishment increases the probability of violence and future crime.

Spouse Abuse

Photo 8.5

The O. J. Simpson trial helped to call public attention to spouse battering.

Despite the fact that ex–football star O. J. Simpson won his criminal case and was found not guilty of the murders of his ex-wife Nicole Brown Simpson and Ron Goldman, tape recordings of a 9-1-1 call for assistance by Nicole, as well as pictures and police testimony, documented Simpson's history of battering Nicole. This history was also a factor in the civil action brought against him by the Brown and Goldman families. That "wrongful death" suit was successful, and Simpson was ordered to pay several million dollars in damages. This called public attention to the widespread nature of an all-too-common form of violence: spouse battering.

Straus, Gelles, and Steinmetz (1980) claim a "sexual symmetry" in spouse abuse: that husbands and wives are equally likely to batter each other. The conventional wisdom at the time assumed the male to be the major aggressor. In a 1975 survey and again in a 1985 national survey, Straus and Gelles (1986) found men and women about equally likely to be the assailant. In a more extreme view, Thibault (1992) even claims that anti-male sexist stereotypes tend to ignore female-initiated violence, giving women a license to batter their children and male partners in the home. Even if this is the case, violence by men tends to cause more serious injury to victims than does

violence by women. Studies such as "The Marriage License as a Hitting License" (Straus et al., 1980) and "The Family as a Cradle of Violence" (Steinmetz & Straus, 1978) illustrate the intimate nature of intrafamily violence, particularly with respect to spouse abuse.

While family conflict studies show equal rates of domestic assault by men and women, victimization and clinical studies show much higher rates of assault by men (M. P. Johnson, 1995). Other studies have shown that much of women's violence really consists of "self-defense" or "fighting back" (DeKeseredy, Saunders, Schwartz, & Alvi, 1997). The NCVS estimates that in 1998 about 1 million "intimate partner crimes" were committed by current or former spouses, boyfriends, or girlfriends. About 85 percent of such crimes were committed against women. The rate of intimate partner crimes has been declining at a rate of 4 percent per year for males and 1 percent per year for females. In 1998, women represented almost three out of four victims of the 1,830 murders by intimate partners in the United States (Rennison & Welchans, 2000). While men are killed by women in self-defense, the greatest danger to women is when they decide to leave an abusive relationship. The men are the pursuers, not defenders.

Some of the reasons for the decline in intimate partner homicide may be better police and prosecution procedures, greater female participation in the workforce, better services for victims, and better legal advocacy including "protection from abuse" orders (Dugan, Nagin, & Rosenfeld, 2000). Later age at marriage could be added to this list.

In their self-report survey of 2,143 husbands and wives, Straus et al. (1980, pp. 274–275) indicate that one out of every six couples admitted having done one of the following in the past year: threw something at spouse, pushed, grabbed, shoved, slapped, kicked, bit, hit with fist or other object, beat up, or threatened with or used a knife or gun against the spouse. In line with our earlier discussion of victim precipitation of many assaults, they found the most common situation was one in which both spouses used violence, although husbands employed the most dangerous and injurious forms of violence and were greater repeaters. Often family members such as women and older persons, who are generally thought of as victims, become assailants and spontaneously strike back, sometimes with lethal consequences (Kratcoski, 1988; Kuhl, 1985).

Traditionally in Western society, "a man's home is his castle," and wife beating has been the prerogative of the "master of the house" as has abuse of children. In codifying common law in the eighteenth century, English jurist William Blackstone determined "a rule of thumb" for wife abuse. While Somers (1994) claims it is a fable, under this rule a husband had the right to physically discipline an errant wife as long as the stick used was no thicker than his thumb (Straus & Gelles, 1986, p. 465). Many traditional societies approved of husbands murdering their wives for serious transgressions such as adultery, although the "double standard" did not permit the reverse.

Although the list is subject to post hoc error, G. Newman (1979) identifies the following as characteristic of wife abusers:

- Alcohol abuse
- Hostility
- Dependence on their wives
- Excessive brooding over trivial events
- Belief in societal approval of battering
- Economic problems
- A sudden burst of anger
- Present military service
- Having been a battered child (pp. 145–146)

A study by the Police Foundation (1977) found that, in the 2 years preceding a domestic assault or homicide, the police had been at the address of the incident five times or more in half of the cases. In the **Minneapolis Domestic Violence Experiment** (Sherman & Berk, 1984), a randomized field experiment demonstrated that arrested domestic offenders were about half as likely to commit repeat violence as nonarrested offenders. Replications of this experiment indicated that the effect varied by

Video Link 8.6
View a transcription and listen to Nicole Simpson's 911 call.

Journal Article Link 8.6
Examine literature regarding criminal justice interventions for domestic violence.

marital status and employment status as well as race of the perpetrator. The experiment certainly illustrated that there are things police can do to intervene in and prevent family violence (Berk & Newton, 1985; Binder & Meeker, 1988).

Elder Abuse

Handbook Article Link 8.2
Read an article
on elder abuse.

With longer life expectancies and, consequently, larger populations of elderly in modern societies, a growing problem of abuse of the elderly has presented itself. Koenig (1991) notes that our knowledge of elder abuse is probably where our knowledge of child and spouse abuse was two or three decades ago. Elder abuse may involve physical, sexual, or emotional abuse; neglect or desertion; and taking and misusing an elderly person's money or property. In 2008, an estimated 2.1 million older Americans were victims of physical, psychological, or other forms of abuse or neglect. These were reported cases. It is estimated that there may be five times as many unreported cases ("Elder Abuse," 2008).

Kidnapping

The most famous U.S. case of kidnapping—the holding of individuals hostage for ransom purposes—is the notorious 1932 Lindbergh case when the 20-month-old son of Charles Lindbergh was kidnapped for ransom and murdered. This led to the passage of the "Lindbergh Law" making kidnapping a federal offense. Kidnapping is relatively rare in the United States, while world figures by region show the following for 1995 through 1998: South America 6,755 kidnapped, Asia/Far East 617, Europe 271, Africa 211, Middle East 118, and North America 80. Only 1 in 10 kidnappings worldwide ends in the death of the person abducted (Auerbach, 1998; Whitelaw, 1999). The Philippines, Colombia, Pakistan, Brazil, and Mexico top the international list for kidnap-for-ransom (Kohut, 1997). Foreign businesspeople are favorite targets; ransoms vary from $50,000 to $100,000 in the Philippines to much larger sums (sometimes $1 million or more) in Latin America. While rare, child abduction by strangers is the stuff of real nightmares.

Criminal Careers of Violent Offenders

Conservative writers in the early 1990s predicted a massive wave of violent super-predators. W. J. Bennett, DiIulio, and Walters in *Body Count: Moral Poverty and How to Win America's War on Crime* (1996) predicted that the United States was about to experience unprecedented violence, blaming this on moral poverty rather than joblessness, racism, and inequality. Not only did their forecast never come true, but crime declined the most in the very areas that they predicted would be worst. Most violent offenders, such as murderers, assaulters, and forcible rapists, do not have criminal careers or extensive backgrounds in and commitment to violent crime as a major component of their lives. Most do not view themselves as criminals nor associate with other criminals. The major exceptions to this are people incarcerated for robbery, whose crimes, except for their violence, resemble those of conventional property criminals.

As seen in Chapter 3, various cultural and subcultural values and attitudes regarding violence have an impact on the relative frequency of violent crimes and their prevalence in various countries, regions within countries, urban/rural areas, social classes, races and ethnic groups, ages, and sexes.

Culture of Violence

Entire cultures can have a predisposition toward the use of violence to resolve grievances. Due to a lack of centralized law enforcement in the last century, areas such as Sardinia and Sicily were characterized by vendettas, which required personal revenge for wrongs against oneself or one's kin.

In some cases, whole families killed each other off, responding in kind to the need to avenge past harm to relatives. In addition, in the 1950s the nation of Colombia experienced what has been called *violencia Colombiana,* in which during a 10-year period 200,000 persons were killed in a nation of only 10 million, an astounding rate of 1 out of every 50 Colombians. In the 1980s, Colombia had a homicide rate estimated at an unbelievable eight times the U.S. rate (Rosenberg, 1991). Moreover, Wolfgang and Ferracuti (1967, p. 280) claim that in the 1960s in Mexico City, the risk of death from homicide was greater than the risk of death from bombing during the London blitz in World War II. With the collapse of communism in Albania, vendettas have returned to that country, most of them involving family feuds over land reform. Local blood feuds going back as far as the fifteenth-century Ottoman Empire have been rekindled. In the past, such violence was officially approved and avengers received light sentences (Post & Field, 1992).

Subculture of Violence

Marvin Wolfgang and Franco Ferracuti (1967), in their now-classic *The Subculture of Violence,* refer to the "culture within a culture" that exists among some ethnic and lower-class groups and demonstrates favorable attitudes toward the use of violence as a means of resolving interpersonal grievances. In such subcultures, violence is viewed as a necessary means of upholding one's masculinity. "Quick resort to physical combat as a measure of daring, courage, or defense of status appears to be a cultural expectation, especially for lower socioeconomic class males of both races" (p. 189). The southern United States has traditionally had higher rates of homicide than other regions of the country. This has led some to view the region as imbued with a subculture of violence. Not coincidentally, the South also has the highest rates of firearm ownership. A rival explanation for the higher murder rates in the South may relate to the fact that poorer emergency medical services exist there than in other regions of the country (Doerner, 1988; Doerner & Speir, 1986).

A basic tenet of the "subculture of violence" thesis is that in such subcultures, violence is not viewed as undesirable conduct, and little guilt or disapproval is experienced when aggression is used. Erlanger (1974), in an empirical assessment of this concept, indicates the surprising lack of examples in the ethnographic literature, citing Liebow's (1967) *Tally's Corner* and Whyte's (1955) *Street Corner Society* as examples. Other such literature, which he does not cite, certainly does lend credibility to the theory, for example, J. Allen's (1977) *Assault With a Deadly Weapon* and C. Brown's (1964) *Manchild in the Promised Land.* In a reanalysis of data originally gathered for the President's Commission on Violence, Erlanger concludes that on attitudinal measures of approval/disapproval of violence, lower-class and minority groups were no different from the general society, and that the social and economic deprivation experienced by these groups is primarily a result of social structural factors—for example, poverty and racism—rather than the product of group pathology. In essence, while there is no greater attitudinal approval of violence, the lack of sophistication with respect to other means of resolving grievances results in higher rates of violent behavior. Other researchers have indicated that blacks and Latinos had lower tolerance of violence than the general population. However, as Shoemaker and Williams (1987) note, "Demographic and residential variables explained more of the variance in violence tolerance and experiences with violence than did ethnic background" (p. 464).

Luckenbill (1991) and Best and Luckenbill (1982) propose a model of murders as "character contests" in which the parties involved attempt to save face and demonstrate character at each other's expense. The stages of this character contest involve the following (Savitz, Kumar, & Zahn, 1991):

- A personal offense (statement or gesture) in which a person feels that he or she has lost face (self-image)
- An assessment that interprets the action as offensive
- Retaliation or demonstration of strength of character ("face") by showing anger or contempt

- A working agreement that violence is an appropriate means of settling the matter
- A battle in which the offender has or obtains a weapon and attacks the victim
- Termination, when the target falls and the contest is over (pp. 20–21)

In examining a sample of murders, Savitz et al. (1991) concludes that over half of the killings fit Luckenbill's model, although many cases lacked sufficient detail to establish whether they fit the model.

Luckenbill and Doyle (1989), as an addendum to this model, proposes the hypothesis that "disputatiousness" (likelihood of being offended or seeking reparation through protest) increases if a person is attacked by an equal in a public place.

Machismo, the code of conduct requiring that males defend their sense of honor, is particularly virulent in Latin American cultures. In Brazil, for instance, some courts until recently refused to convict husbands of killing unfaithful wives, although the reverse did not apply. The view is that a man should not be punished for defending his honor. Bourgois (1988) describes a "culture of terror" in the underground drug economies of U.S. central cities in which regular displays of violence are necessary for success in the street-level, drug-dealing world. What outsiders view as senseless violence may be viewed as public relations, "a curriculum vita that proves their capacity for effective violence and terror."

Career Criminals/Violent Offenders

Petersilia, Greenwood, and Lavin (1977), basing their research on interviews with 49 incarcerated robbers, found that such individuals committed roughly 214 offenses apiece, although these crimes were nonspecialized and were just as likely to involve conventional, nonviolent property crime as they were to involve violence. They divided career criminals into two types: the *intensives* and the *intermittents.*

Intensives have continuing criminal involvement from an early age and commit on the average 51 crimes per year. *Intermittents* are irregular in their offense patterns, committing an average of 5 crimes per year, generally with lower takes from their victims. In a previously discussed longitudinal study of Philadelphia delinquents by Wolfgang, Figlio, and Sellin (1978) it was indicated that roughly 6 percent of the 1945 male birth cohort were "chronic offenders," accounting for 52 percent of all the crimes committed by this group. J. M. Chaiken and Chaiken (1982) in *Varieties of Criminal Behavior* used self-reports and official records in a survey of 2,200 inmates in California, Michigan, and Texas, in which they identified "violent predators" who commit a highly disproportionate amount of crime, consisting of a combination of robbery, assault, and drug dealing. These criminals began taking drugs as juveniles; committed violent crimes before age 16; were addicted to multiple drugs; and perpetrated an exceptionally high level of robberies, property crimes, and assaults in order to support their addictions. Most were unmarried, had few other family obligations, and were irregularly employed. Their distinctive characteristic was multiple drug use, for example, heroin with barbiturates or alcohol or amphetamines with alcohol. The California inmates who had been addicted admitted to, on the average, 34 robberies, 68 burglaries, and 72 thefts per year, while the same figures for those not using drugs were 2, 3, and 8 per year.

Societal Reaction

As our discussion has suggested, most violent crime is intimate; a large proportion of violent offenders are not career criminals and their crimes reflect situational or subcultural reactions to interpersonal disputes. Studies indicate that a high proportion of crimes are committed by a small portion of the criminal population, the chronic or career offenders. Therefore, social policies to identify and specially process these career criminals hold much promise. The creation of special career criminal bureaus by police departments and district attorney offices, using computerized information on up-to-date offense records that are shared with the courts, can assist in preventing such career felons from slipping through the cracks in the system.

As previously cited, a Police Foundation (1977) survey in Kansas City found that in the 2 years preceding an assault or homicide, police had answered calls about domestic disturbances in 85 percent of the cases and at least five times in half of the cases. Subsequent replication studies specified that the use of arrest depended on offender characteristics such as race, unemployment, and marital status. Thus, early use of crisis intervention teams could help reduce the high rates of domestic violence. Social programs related to substance abuse and family crisis intervention address two key components of the violence equation.

A key explanatory variable in elucidating the very high interpersonal violence rates in the United States compared to other developed countries is the widespread availability and ownership of handguns. Although the majority of the population favors stricter legislation and control, the public has remained relatively passive in this regard. Until strong, active public pressure is felt, Americans will continue to murder one another at a rate that bewilders most of the civilized world.

Violent crime, particularly by strangers, has had a profound impact on urban life in the United States. In many respects, urban wastelands, including some downtown areas in the evening hours, are grim reminders of an erosion of the urban vitality that is the hallmark of civilized societies. Until society can control violent crime, our culture will fail to realize its full potential.

Theory and Crime

The application of various theoretical explanations to areas of violent crime offers many opportunities for the criminologist to explain criminality in individuals as well as in society. See if you can use your own creativity and imagination in applying some of the theories that we have learned to specific crime and criminality. If we examine theories related to the "subculture of violence" or "culture of violence," we gain explanation in understanding why some cultures have very high levels of violence while others are very nonviolent. Similarly, some subcultures such as in the Southern U.S. or lower-class African American subcultures have higher violent crime rates than other groups. These values and attitudes are learned as part of group membership and are reflected in different behavior than that for other groups.

Many of the biological and psychological theories of crime have the greatest explanatory power when applied to individual cases and particularly to multiple murderers. Charles Manson, for example, was an "antisocial personality." Not knowing who his father was, he was the son of a teenage prostitute who shuffled him between various family members during his youth. He had spent most of his life in reformatories or jails and as a young adult emerges as a "sociopath," lacking the warmth, love and emotional equipment of other adults. Other multiple murderers exhibit similar biological and/or psychological deficits. A lot of Freudian theory seems to apply to the abnormal psychosexual childhood socialization of these individuals.

Much of the high violent crime rate in the United States consists of domestic violence, which is often dictated by alcohol, drugs and poor conflict resolution skills. The majority of violent offenders are not career criminals, but reflect situational or subcultural reactions to interpersonal disputes. Social programs related to substance abuse and family conflict resolution skills hold much promise in reducing this type of violence.

Summary

Violence is an ignominious blot on the history of civilization, particularly in the twentieth century. Some writers claim that violence is instinctual in humans, but most social scientists view it as a culturally learned phenomenon. The National Commission on the Causes and Prevention of Violence identified six factors that may explain the high level of violence in the United States: the Declaration of Independence, the frontier experience, immigrant competition, fear of government power, movement from rural to urban/industrial centers, and relative deprivation amidst affluence. Violence has also been intimately tied to major historical changes throughout U.S. history, although other young countries have had similar experiences.

Social and technological changes have had impact on various fads and fashions in crime, many of which, such as train robbery, cattle rustling, and grave robbing, have now largely disappeared. Brief accounts of murder and mayhem of the past indicate that horrible, violent criminals of the present such as Whitman, Williams, and Manson are not mere modern aberrations. Multiple murders may take the form of serial murder, mass murder, or spree murder.

Victim precipitation is quite common in many violent crimes. Mendelson's types of victims include the completely innocent victim, one with minor guilt, one who is as guilty as the offender, one who is more guilty than the offender, the victim as most guilty, and the imaginary victim. Conrad's typology of violent offenders includes culturally violent offenders, criminally violent offenders, pathologically violent offenders, and situationally violent offenders. Haskell and Yablonsky's types include violence by the state as well as by syndicates, both of which will be discussed in later chapters.

The close relationship between homicide and assault was described; in the case of the former, the victim dies. Homicides have the highest clearance rate by police because of their serious nature, the presence of witnesses, and the high priority they are given. Rape statistics have been notoriously underreported, but have improved in accuracy as a result of better support for victims.

The overall trend in homicide in the United States has been a decline after a peak in 1933, a rise in the sixties, and new highs from the seventies to the present. The dip in the forties and fifties may actually have been due to better medical treatment procedures, which masked a rise in potentially lethal violent assaults. Fragmentary historical evidence suggests even higher rates prior to 1933. The United States possesses by far the highest homicide rate among economically developed countries, although this rate is lower than those of many developing countries. Domestic homicide in the United States in the twentieth century exceeded the combined fatalities of every war the country has ever fought.

Patterns of violent crime indicate perpetration and victimization associated with large cities, males, youths, the lower class, and ghetto blacks. Unlike robbery, most violent crime occurs between intimates. A large proportion of violent crimes are committed by repeaters. Some relationship between social inequality and homicide rates is suggested. Surprisingly, one's own home is the most likely setting for one's murder, and intimates/acquaintances are the most likely perpetrators. Alcohol consumption has a high association with violent crimes such as homicide, assault, and rape and particularly with vehicular homicide. Factors associated with the typical homicide include a backdrop of cultural/subcultural traditions of violence, personal dispute, alcohol, and guns, the last being the most telling.

Workplace violence is now the second leading cause of occupational fatalities. The United States has the largest armed civilian population in the world, and a relationship between firearm possession and homicide rate is strongly suggested.

Estrich distinguishes between "real rape" (aggravated, involving violence) and "simple rape" (all other types). The latter is still not recognized in the courts.

The reluctance to report rape has been due to stigma, sexist treatment by the criminal justice system, prosecutorial invasion of privacy, and shifting the burden of proof onto the victim. Increased reporting has been spurred by victim centers, female officers, better trained police, and changes in the law. While UCR data show a precipitous rise in rape in the seventies, victim data show no such increase. Rapists are generally young, lower-class, unmarried, and disproportionately black. Other analysis suggests that victims underreport rapes by offenders whom they know, in which case rape may more closely resemble most other violent crimes. Factors involved in most rapes are violent values, machismo, sexist views of women as legitimate victims, conducive dating-game circumstances, and alcohol. While Amir views rape as a sexual and sometimes victim-precipitated act, Brownmiller views it as entirely a method of male intimidation and violence against women. While both authors take extreme positions, there is little argument that rape is a violent crime, regardless of offender motivation. Many typologies of rapists suffer from after-the-fact analysis and post hoc error. More research is needed.

Robbery rates show great recent increases according to the UCR but stability according to victim surveys. Robbery is more likely to be interracial and to involve strangers than other violent crimes, although official statistics overlook large numbers of juvenile offenders and victims.

The majority of robberies do not involve direct physical harm, although "strong-arm" robbery (mugging) and victim resistance (other than screaming) increases its likelihood. Conklin's typology of robbers includes professional, opportunist (the most common), addicted, and alcoholic robbers.

Recent research has suggested that arrest of domestic assaulters can deter repeat offenses, at least for certain types of assaulters.

The largest "dark figure" of violent crime has been spouse and child abuse; there has been a virtual statistical epidemic of such crimes since the sixties. Although post hoc error also operates in this area, it appears that many of those abused as children are likely to become future abusers.

Most violent offenders do not make a career of such violations and have little commitment to crime. Cultures/subcultures of violence may serve to reinforce predispositions to use violence in resolving grievances. Wolfgang and Ferracuti's subculture of violence thesis is used to explain the disproportion of such crimes among certain lower-class minorities and the high rates in the U.S. South. Violent predators who are persistent offenders are labeled career criminals; they are responsible for a disproportionate number of crimes of violence. Programs to identify, isolate, and expedite incarceration of such offenders are viewed as a promising strategy to decrease the rate of violent crime. Recent research suggests a variety of types and patterns of *career criminality*. Better programs in family crisis intervention, alcohol treatment, and career offender rehabilitation are viewed as trends in societal reaction.

Criminology on the Web

Log on to the Web-based student study site at http://www.sagepub.com/haganintrocrim7e/ for author-created podcasts, e-flashcards, quizzes, and more.

KEY CONCEPTS

Assault

Battery

Child Abuse

Culture of Violence

Factors in Rape

Forcible Rape

Gun Control

Manslaughter

Mass Murder

Minneapolis Domestic Violence Experiment

Patterns/Trends in Violent Crime

Post Hoc Error

Rape as Violent Act

Serial Murder

Spouse Abuse

Spree Murder

Subculture of Violence

Types of Career Criminals

Types of Robbers

Types of Violent Offenders

Victim Precipitation

Workplace Violence

REVIEW QUESTIONS

1. One of the explanations for the United States having higher violent crime rates than other developed countries is that we have had an extensive history of violence, some would say a culture of violence. Discuss this theory.

2. Discuss the three types of multiple murder and give examples of each.

3. Discuss the nature and characteristics of workplace violence.

4. Discuss the pros and cons of gun control. What side of this issue do you tend to favor? Defend your choice.

5. What explains the greater willingness of women to report rape in the United States? What explains the growing recognition of "acquaintance rape"?

6. What is the claimed "sexual symmetry" of interpersonal violence? Do you agree with this notion? Explain.

7. Discuss the "subculture of violence" thesis. Give some examples of countries, regions, and places where this is found.

8. What were some of the findings of the Secret Service Study of School Shooters?

9. Discuss some of the findings of Wright and Decker with respect to armed robbers.

10. What patterns do you see in our international comparisons of murder, robbery, and rape? What do you think explains these differences?

WEB SOURCES

Armed Robbery Training Manual
http://www.armedrobbery.com

Child Abuse Prevention Network
http://www.child-abuse.com

Domestic Violence Handbook
http://www.domesticviolence.org

Homicide Research Working Group
http://homicideworkinggroup.cos.ucf.edu/

Internet Crime Archives of Mass Murderers
http://www.mayhem.net/Crime/archives.html

Pavnet: Violence Prevention Initiatives
http://www.einet.net/review/24168-582613/Pavnet_Online.htm

School Violence
http://www.ncdjjdp.org/cpsv/

truTV's Crime Library
http://www.trutv.com/library/crime/serial_killers/predators/

Workplace Violence 101
http://www.fbi.gov/publications/violence.pdf

WEB EXERCISES

Using this chapter's recommended Web sites, explore the topic of violent crime.

1. What types of programs are being promoted by Pavnet: Violence Prevention Initiatives?

2. What suggestions are made for reducing "School Violence" and "Workplace Violence" by those sites?

3. What is the Homicide Research Working Group?

4. What is featured on the Armed Robbery Training Manual?

5. Using your Web browser, search the terms multiple murder and sexual predators.

Ann Auerbach. 1998. *Ransom: The Untold Story of International Kidnapping.* New York: Henry Holt and Company.

> This is one of the few books to explore the neglected topic of international kidnapping. While there is little kidnapping in the U.S., it is quite common in some other countries.

Donald Black. 1999. *Bad Boys, Bad Men: Confronting Anti-Social Personality.* New York: Oxford University Press.

> The psychopath, sociopath, antisocial personality remains a little understood complex. Black explores this concept and provides interesting examples.

Hugh Davis and Ted Gurr. 1969. *Violence in America: Historical and Comparative Perspectives.* New York: The New American Library.

> This work brings together many of the materials provided to the *Violence Commission* (1968) regarding the history of violence in America.

Jack Douglas and Mark Olshaker. 1997. *Journey Into Darkness.* New York: Scribner's.

> Douglas, a retired member of the FBI's Behavioral Science Unit, gives accounts of crime profiling and investigation of serial murderers.

James Alan Fox and Jack Levin. 1985. *Mass Murder: America's Growing Menace.* New York: Plenum Press.

> The world of serial murder is explored by two criminologists.

Nicholas Groth and H. Jean Birnbaum. 1979. *Men Who Rape: The Psychology of the Offender.* New York: Plenum Press.

> This highly cited classic features their typology of rapists as well as excellent examples.

Peggy R. Sanday. 1990. *Fraternity Gang Rape.* New York: New York University Press.

> Sanday explores the all-too-neglected misogynist world of college fraternities and the hidden rapes that occur on campus.

Murray A. Straus. 1994. *Beating the Devil Out of Them: Corporal Punishment in American Families.* San Francisco: Lexington.

> One of America's best known family sociologists reports on a survey of spanking and concludes that it has negative impacts for future criminality and other antisocial conduct.

Marvin Wolfgang and Franco Ferracuti. 1967. *The Subculture of Violence.* London: Tavistock.

> This is the classic work on the concept of the subculture of violence.

Richard T. Wright and Scott H. Decker. 1997. *Armed Robberies in Action: Stickups and Street Culture.* Boston: Northeastern University Press.

> Wright and Decker report on their field study of uncaught robbers in the field.

Occasional Property Crimes
  Shoplifting
  Vandalism
  Crime File 9.1 Graffiti
  Motor Vehicle Theft
  Check Forgery
Conventional Property Crimes
  Burglary
  Fencing Operations
  Stings
  Larceny-Theft
Arson: A Special-Category Offense
Criminal Careers of Occasional and Conventional
    Property Criminals
Societal Reaction
Professional Crime
  The Concept of Professional Crime
Characteristics of Professional Crime
  Argot
  A Model of Professional Crime
  Edelhertz's Typology
Scams
Big Cons
  Maurer's *The Big Con*
  Identity Theft
  Ponzi Schemes
  Crime File 9.2 Identity Theft
  Pyramid Schemes
  Crime File 9.3 The Bernie Madoff Affair: "One Big Lie"
  Religious Cons
  The PTL Scandal
  Crime File 9.4 Emerging Patterns of Professional Crime
  Crime File 9.5 Nigerian Letter Scams
Boosters
Cannons
  Crime File 9.6 Shoplifting
Professional Burglars
  The Box Man
The Professional Fence
Paper Hangers
  Crime File 9.7 Intellectual Property Theft
  *Crime File 9.8 Busting the Biggest Band of Cable Pirates
    in U.S. History*
Professional Robbers
Professional Arsonists
Professional Auto Theft Rings
Professional Killers
  Crime File 9.9 Car Cloning: A New Twist on an Old Crime
Criminal Careers of Professionals
Societal Reaction
Theory and Crime
Summary
Key Concepts
Review Questions
Web Sources
Web Exercises
Selected Readings

chapter 9

Property Crime

Occasional, Conventional, and Professional

The more things available for theft and the lower the probability of penalty, the more larceny.

—*Gwynn Nettler (1982)*

There's a sucker born every minute.

—*P. T. Barnum*

Because that's where the money is.

—*Willie Sutton (when asked why he robbed banks)*

Offenses against property were among the first to be punished under formal legal systems. The basic offense—theft—was referred to under English common law as larceny, defined simply as the taking of the property of another without the owner's consent. Although the specific legal definition varies by country and state, the various forms of larceny include embezzlement, the receipt of stolen goods, shoplifting, employee theft, burglary (breaking and entering with intent to steal), robbery (stealing by means of force or threat of force), forgery (the fraudulent use of commercial instruments), auto theft, vandalism, and arson (the willful burning of a dwelling or property).

Larceny can be committed by a variety of criminal types, ranging from the most amateur to the most highly organized or professional criminal. Table 9.1 depicts the range of criminals involved in property offenses (though actual placement on the continuum is obviously more problematic than is suggested by a simple schema).

At the far end of the continuum is the **career criminal**. The notion of *career criminality* is explained by Clinard and Quinney (1973):

> The characteristics of a fully developed criminal career include identification with crime and a conception of the self as a criminal. There is group support for criminal activity in the form of extensive association with other criminals and with criminal norms and activities. Criminality progresses to the use of more complex techniques and frequent offenses, and ultimately crime may become a sole means of livelihood. Those who have careers in crime generally engage in some type of theft of property or money. (p. 57)

TABLE 9.1 ■ Range of Career Criminal Involvement in Property Crime

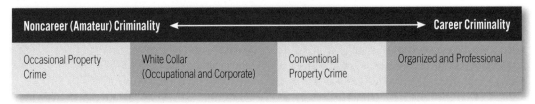

Noncareer (Amateur) Criminality ←		→ Career Criminality
Occasional Property Crime	White Collar (Occupational and Corporate)	Conventional Property Crime	Organized and Professional

White collar property offenses will be covered in Chapter 10, and organized (syndicate) criminality will be detailed in Chapter 12. This chapter will concern itself with comparing and discussing three different types of criminal behavior systems: occasional property criminals, conventional property offenders, and professional criminals.

Occasional offenders are the opposite of career criminals; conventional criminals are usually unsuccessful aspirants to careers in crime (Clinard & Quinney, 1973, pp. 57, 132). **Occasional property criminals** steal or damage property on an infrequent basis. They account for most, but not all, auto theft, shoplifting, check forgery, and vandalism. These offenses are committed relatively irregularly and rather crudely, with little skill or planning. In contrast, **conventional criminals** tend to commit crimes of theft/larceny and burglary on a more regular basis and, although they are at the bottom rung of the ladder or continuum of career criminality, they exhibit elements of career criminality.

In Table 9.1 occasional property criminals represent the noncareer end of the continuum; organized and professional criminals generally exhibit characteristics of career criminality; conventional offenders exhibit the rudiments of such characteristics; and white collar and occupational offenders, because of their commitment to the conventional world, fall closer to the noncareer criminality pole of the continuum. UCR index statistics show that property crimes outnumber violent crimes 9 to 1.

Occasional Property Crimes

Most occasional property criminals lack a past official history of criminality. They exhibit little progressive knowledge of criminal techniques or of crime in general. In contrast to career criminals or even conventional criminals, crime is not their sole or major means of livelihood, and they do not view themselves as criminal. Not identifying with criminal behavior, they have little of the vocabulary or "street sense" of the conventional criminal.

Under the category of occasional property offenses, discussion will center on shoplifting, vandalism, motor vehicle theft, and check forgery. (Professional crimes of these types will be discussed later.) Surprisingly, there have not been many studies that focus on occasional and ordinary property offenders (Hepburn, 1984; Shover, 1983).

Handbook Article Link 9.1
Read an article on theft and shoplifting.

Shoplifting

The polite term for **shoplifting** used by the retail trade industry is "inventory shrinkage"—quite literally, goods have disappeared or shrunk from the total of accountable inventory. The slang term "five-finger discount" is a less polite term for this same process. While shoplifting is perhaps as ancient as trade, the post–World War II emergence of a consumer society and of large retail chains has created both a greater desire and a greater opportunity for retail theft. Inventory shortage costs in the United States account for about 2 percent of retail sales; the actual proportion may be considerably higher, depending on location, product, and clientele. Hollinger and Davis (2002) estimate the average retail firm loses 1.7 percent of gross revenues to "inventory shrinkage." Shrinkage refers to the losses that are due to shoplifting, employee theft, vendor fraud, and administrative error. About one third of these losses are due to shoplifting. The average shoplifter is caught with $200 worth of merchandise, and the U.S. retail industry loses an estimated $10 billion per year to shoplifting. Many shoplifters are not motivated by need, as illustrated in the case of actress Winona Ryder, shown in Photo 9.1.

The classic study on shoplifting is Mary Owen Cameron's (1964) *The Booster and the Snitch: Department Store Shoplifting*, which was based on store records and arrest data in the late 1940s; later research by L. E. Cohen and Stark (1974) supports her findings (see also Klemke, 1992). Cameron distinguishes between "boosters" (or **"heels"**)—professional shoplifters— and "snitches," amateur shoplifters.

Boosters (to be discussed later in this chapter) are like other professional criminals in carefully planning and skillfully executing their thefts and in concentrating on expensive items that can be quickly converted to cash by prearrangement with a "fence" (dealer in stolen goods). On a continuum of shoplifters, between the booster and the snitch are "shadow" professionals (Hellman, 1970; Stirling, 1974, p. 120), individuals who in an avocational manner supplement their legitimate incomes by stealing. The majority of shoplifters are **snitches,** amateurs or individuals who do not view themselves as criminals. According to Cameron (1964), most are females and the vast majority have no official history of previous recorded criminal involvement. One example is the highly publicized case of Oscar nominee Winona Ryder, who was arrested at Saks Fifth Avenue in Beverly Hills for shoplifting. She stole $5,560.40 worth of designer merchandise. Acquitted of burglary, she was convicted of grand theft and vandalism. Authorities believe that most shoplifters have the money to pay for their stolen items. One illustration of that is the case of Claude Allen. Convicted in 2006, Allen was at the time an assistant to George W. Bush on domestic policy. He perpetrated a fraudulent return scheme at Washington, D.C.–area stores. Taking purchased merchandise to his car, he would then return to the store with his receipt and select duplicate items of what he had just bought and request a refund for them. His take in 2005 was estimated at $5,000. His salary that year at the White House was $161,000 (Rohrlick, 2007).

Sensormatic Electronics Corporation, the largest maker of electronic antitheft devices, conducted a review of 166,000 theft reports from 101 retailers that showed that sneakers, earrings, and compact discs were among the most common items stolen (Huang, 1999). The rate of male shoplifters was only slightly higher than the female rate. While there were more adult shoplifters than juveniles, the latter were more frequently caught. Shoplifting increases during the Christmas shopping season, while shoplifting arrest rates increase in March (spring break). An annual survey conducted by Richard

Photo 9.1

Actress Winona Ryder reacts to her sentencing for shoplifting at a Saks Fifth Avenue store on Rodeo Drive. On December 6, 2002, she was sentenced to 3 years' probation and 480 hours of community service, fined $11,300, and ordered to undertake drug and psychological counseling.

Audio Link 9.1
Listen to a high school student's discussion of shoplifting.

Hollinger of the University of Florida on behalf of ADT Security Services estimated the losses for inventory shrinkage in 2006 to be $40.5 billion and that 47 percent of this ($19 billion) was due to shoplifting ("Shoplifting Costs U.S. Retailers," 2007).

Most snitches steal small, inexpensive items for their own personal use. In most instances, they have on their person sufficient funds to cover the stolen items. Such snitches come from all walks of life. Nettler (1982, Vol. 3) indicates, for instance, that "theories of poverty and low education and shoplifting would surprise store owners in university towns who experience three times the amount of theft as stores in other neighborhoods" (p. 106). Most snitches simply do not anticipate being caught. In the past when snitches were apprehended, most stores avoided lawsuits or possible adverse publicity by releasing the offenders after brief admonishment. When apprehended, snitches usually attempt to rationalize or excuse their behavior. For the middle-class offender with a psychology bent, "kleptomania"—a compulsion to steal—becomes a handy rationalization.

Adventure, excitement, need, greed, or simply available opportunity or inadequate security may prove to be more likely reasons. Cameron (1964) claims that most snitches, when caught and faced with an unacceptable criminal self-image, cease shoplifting.

Sensormatic Electronics assumes that shoplifting tends to increase during recessions—"when the going gets tough, some of the tough go shoplifting" (A. Newman, 1990). Sensormatic produces a variety of equipment, including security tags attached to articles that will set off alarms if the article is taken from the store without the tag having been removed by a salesperson. Dabney, Hollinger, and Dugan in "Who Actually Steals? A Study of Covertly Observed Shoplifters" (2004) used closed-circuit television to observe shoppers and recorded their demographic and behavioral characteristics. A significant number (8.5 percent) were determined to be shoplifters. They found that juveniles and young adults were not more likely to shoplift. Middle-aged shoppers (35–54) were the most common shoplifters. More research is needed in order to confirm this finding. Recently, a number of states have passed antishoplifting statutes that enable retailers to stick shoplifters with some of the cost of security. "Civil demand" letters are sent to the accused shoplifters or their parents asking for payment of a $100 to $200 penalty in addition to the returned merchandise in exchange for the retailer not suing for civil damages (Schellhardt, 1990).

Most retail thefts involve employees pilfering goods. Sixty percent of inventory shrinkage is thought by retail experts to be due to employee theft, 30 percent to outside shoplifters, and 10 percent to paper errors (Rupe, 1980). Far more research on shoplifting is required to gain a definitive picture of its varieties. A mail survey of 850 employees at 50 different grocery companies by the Food Marketing Institute in 1994 found 44 percent admitted to some form of theft, although only 1 percent said they stole money from their employers (Boccella, 1994). Of those admitting theft, 32 percent stole and ate food, 20 percent stole merchandise, 3 percent shortchanged customers, 2 percent gave refunds for unpurchased items, and 1 percent stole money. Those who were about to quit their jobs stole seven times as much as others. Males ages 16–30 on the night shift had the highest rates, while women over 30 were the most honest. A rather interesting case is one of a shoplifting ring that involved a head football coach, two teachers, and two students at Green Run High School, Virginia Beach, Virginia. In 1999, a 17-year-old honor student who was employed at a Big K Mart gave away merchandise or undercharged her coconspirators who brought goods to her checkout line (Brush, 1999).

Vandalism

Vandalism involves the willful destruction of property without the consent of the owner or agent of the owner. The term is derived from the Vandals, a barbaric Teutonic tribe that sacked Rome in the fifth century, senselessly destroying many priceless works of art. Clinard and Quinney (1973) explain:

Vandalism or the willful destruction of property is widespread in American society. It constitutes one of the largest categories of juvenile delinquency but occurs at all ages. It is associated with affluence for it virtually never occurs in less developed countries (except as a part of rioting) where the destruction of goods in limited supply is inconceivable. Vandalism in the United States is widespread against schools, parks, libraries, public transportation facilities, telephone and electric company facilities, traffic department equipment and housing. In one year, the public school system of Washington, D.C., reported a loss of 28,500 window panes, replaced at a cost of $118,000. (p. 59)

A. L. Wade (1967) identifies three basic **types of vandalism:** wanton, predatory, and vindictive.

Wanton vandalism consists of destructive acts that have no purpose and produce no monetary gain. These are the most common acts of vandalism, "senseless" destruction practiced by juveniles "just for the hell of it" or for fun. *Predatory vandalism* comprises destructive acts for gain, such as "trashing" or destroying vending machines in order to steal their contents. *Vindictive vandalism* is undertaken as an expression of hatred, such as of a particular racial or ethnic group; examples are swastikas painted on synagogues, Ku Klux Klan attacks on black churches, or antibusing groups' assaults on school buses.

Most acts of wanton vandalism are committed by juveniles, who regard these acts as an extension of play activity, "goofing off," or "raising hell." In some U.S. communities, the evening before Halloween is called "devil's night," when juveniles play tricks that may not be restricted to throwing eggs at houses or soaping windows. A. L. Wade (1967) describes the typical pattern of wanton vandalism as consisting of

- Hanging around, waiting for something to happen;
- An initial exploratory gesture of vandalism by one member;
- Mutual conversion of others to participate;
- An escalation of destructive behavior from minor to major property damage; and
- After-the-fact feelings of guilt and remorse combined with pleasure at having done something "naughty."

Such vandalism is rationalized by the offenders as not really criminal, since they did not plan or intend it, and realized no monetary gain. Often urban public facilities—for example, some big-city subway systems—are "graced" with the unrequested graffiti of freelance "artists" or anyone who can afford a can of spray paint. New York City had great success in eliminating such graffiti through a program begun in 1984 that involved more patrolling of trains, targeting known offenders, developing logo intelligence files, special protection for clean trains in "lay-up" yards, and undercover operations in train storage areas (Kelling, 1988a). Crime File 9.1 reports on patterns of graffiti.

In 1994 at Millersville State University (in Pennsylvania), students gathered for the annual "Naked Coed Relays" in which naked students raced around a stadium track, then went on a vandalism rampage when they discovered the stadium locked ("Naked Vandalism," 1994). In July 1998, more than 1,500 people attending a State College, Pennsylvania, arts festival (many of them Penn State students and alumni) went on a rampage after the bars were closed. They tore down street lights, smashed storefronts, set bonfires, and even stripped naked and burned their underwear (Hoover, 2002; D. Kinney, 1998). The former president of Ohio State University claimed that Buckeye fans looked for any excuse to riot. Describing a culture of rioting, she said she witnessed people doing "disgusting things, unbelievable things. . . . They think it's fun to flip cars, to really have absolute drunken orgies" ("Former Ohio State President," 2007). A bit more sinister, more organized example of wanton vandalism was "Chaos Days 95" in Hanover, Germany ("Germany's Punks," 1995), in which thousands of self-described punks converged on the city, drinking, fighting, and generally trashing the place. In some cities, "tagging" is practiced: gangs tag or mark their territory with their

Crime File 9.1

Graffiti

For many people, graffiti's presence suggests the government's failure to protect citizens and control lawbreakers. There are huge public costs associated with graffiti: An estimated $12 billion a year is spent cleaning up graffiti in the United States. Graffiti contributes to lost revenue associated with reduced ridership on transit systems, reduced retail sales, and declines in property value. In addition, graffiti generates the perception of blight and heightens fear of gang activity.

Graffiti is not an isolated problem. It is often related to other crime and disorder problems, including

- Public disorder, such as littering, public urination, and loitering;
- Shoplifting of materials needed for graffiti, such as paint and markers;
- Gangs and gang violence, as gang graffiti conveys threats and identifies turf boundaries; and
- Property destruction, such as broken windows or slashed bus or train seats.

Types of Graffiti

There are different types of graffiti. The major types include

- Gang graffiti, often used by gangs to mark turf or convey threats of violence, and sometimes copycat graffiti, which mimics gang graffiti;
- Tagger graffiti (see "tagging," defined below), ranging from high-volume simple hits to complex street art;
- Conventional graffiti, often isolated or spontaneous acts of "youthful experience," but sometimes malicious or vindictive; and
- Ideological graffiti, such as political or hate graffiti, which conveys political messages or racial, religious, or ethnic slurs (Weisel, 2004).

Video Link 9.1
View a video of differing opinions on graffiti.

colors, nicknames, club names, and symbols. This type of vandalism, in most instances, is an extension of wanton vandalism. Such senseless vandalism is not restricted to juveniles. Drunken fans of a team winning a Super Bowl, World Series, or NCAA championship sometimes "trash" downtown areas as part of the celebration. Recent football (soccer) "hooliganism" in England has cost not only property, but human lives. The United Kingdom's National Criminal Intelligence Service carried the following advertisement on its Web site in 1999:

Are You a Fan With Intelligence?

Do You Know Anyone Planning Trouble?

Do You Know Where They Will Be Meeting Before or After the Game?

Do You Know How They Will Get There?

Do You Know Who Their Mates Are?

Anyone with information, please call our 24-hour confidential Football Hooligan Hotline on Freephone 0800–515495

One variation of predatory vandalism and theft is "bibliotheft" and destruction, in which students intentionally steal and destroy library reference materials. Rather than take notes or make copies of materials, students tear out the needed information, thus raising the cost of library materials and denying others the opportunity to use such references.

Motor Vehicle Theft

P. Harris and Clarke (1991) note a serious lack of scholarly attention devoted to auto theft compared to that devoted to other crimes. In 2006, the highest vehicle theft rate in the United States was in the West (632.1), followed by the South (378.1), the Midwest (332.6), and the Northeast (222.3). Using Interpol statistics for selected countries, the U.S. rate of 433 was lower than Australia (726), Canada (547), and Denmark (440). The lowest rates were for developing countries that have low ownership rates, although Japan, a developed country, had a low rate of 49. Table 9.2 lists Interpol auto theft rates for selected countries.

After falling during the nineties, auto theft rates have crept up in the first decade of this century ("Hot Spots for Stolen Cars," 2004). Confusion exists regarding the most targeted vehicles. It depends on whose list one uses. Comparing insurance claims with overall insured vehicles, the Insurance Institute for Highway Safety indicates that the Cadillac Escalade EXT had the highest theft rate followed by the Nissan Maxima, Cadillac Escalade, Dodge Stratus, and Dodge Intrepid ("Most Stolen Cars," 2004). The National Insurance Crime Bureau (NICB) provides a list that is most often referred to by the media. The 2007 bureau listing shows the 1995 Honda Civic as the most often targeted, followed by the 1991 Honda Accord, the 1989 Toyota Camry, 1997 Ford F-150, and the 1994 Chevrolet C/K 1500 (P. Stern, 2008). McCaghy, Giordano, and Henson (1977) have proposed a typology of auto theft that includes the following:

- Joyriding: The vehicle is temporarily "borrowed," usually by juveniles, not for theft purposes, but for temporary adventure and enjoyment.
- Short-term transportation: The vehicle is stolen as a temporary means of transportation and then abandoned.
- Long-term transportation: The car is stolen for the purpose of providing a relatively permanent means of transportation for the thief.
- Profit-motivated: Highly organized auto theft rings permanently alter the vehicle's identity; "chop shops" cannibalize the auto for parts; and "auto strippers" abandon the hulk after removing the valuable parts.

Because of state laws requiring auto insurance, as well as insurance regulations requiring police reports for reimbursement, auto theft is the most regularly reported of larcenies (about 90 percent are reported to police). *Joyriding* is occasional property crime committed almost exclusively by juveniles on an unplanned, unskilled, and sporadic basis. A car is stolen, either by "hot wiring" (jumping the ignition) or by finding keys left in the ignition. The car is then temporarily used for cruising and abandoned when it runs out of gasoline. The intent is not to strip the vehicle of parts nor to permanently possess it. Most offenders view their activity as a prank and rationalize that, since they had not intended to actually steal the car and were simply borrowing it, their behavior was not really criminal. In contrast to the occasional property criminal, profit-motivated offenders are for the most part either conventional criminals or professionals. Such profit-oriented auto thieves may range from sporadic amateur thieves (the hubcap crooks) to full-time professionals in auto theft rings.

Check Forgery

As defined in the UCR, *forgery* is "making, altering, uttering, or possessing, with intent to defraud, anything false which is made to appear true." *Fraud* involves the conversion or obtaining of money or property under false pretenses. Both fraud and forgery may vary from simple actions, to be discussed in this section, to elaborate professional "con" games, to be examined later in this chapter.

The classic study on check forgery was that of Edwin Lemert (1953), "An Isolation and Closure Theory of Naive Check Forgery," in which he makes the distinction between "naive check forgers" and "systematic check forgers."

TABLE 9.2 ■ International Auto Theft Rates

Since the United Nations groups all of its crimes of theft into one category, it has no statistics for auto theft. Interpol defines auto theft as "theft of motor cars." The U.S. definition includes theft of any motor vehicle that is self-propelled and runs on a surface rather than on rails. This includes motorcycles, motor scooters, and the like; however, the overwhelming number of U.S. motor vehicle thefts are thefts of autos or trucks. Probably the most important factor in the rate of motor vehicle theft is the number of motor vehicles per capita in the country. Developed nations in which automobile ownership is widespread generally had the highest rates of auto theft.

International Auto Theft Rates, Interpol Statistics (per 100,000 population)	
Country	**Auto Thefts**
United States[a]	433.4 (2003)
Australia	726.2 (2000)
Brazil	88.5 (2001)
Canada	547.6 (2001)
Denmark	440.5 (2003)
France	388.5 (2002)
Ireland	3.8 (2002)
Japan	49.2 (2002)
Mongolia	.4 (2001)
Russia	33 (2002)
Saudi Arabia	89.8 (2002)

Source: *International Crime Statistics, 2003.* Lyons, France: Interpol Secretariat, 2004.

a. U.S. statistics were taken from the *Uniform Crime Reports for 2003* (Federal Bureau of Investigation, 2004).

Web Research Project

Search the concept of "car theft" and examine how smartphones such as the iPhone or the Palm Pre may pose a new threat to break-ins to cars.

The majority of check forgers—those passing bad checks—are **naive check forgers**. Faced with a financial crisis, such as an alcoholic binge, gambling debts, or creditors demanding immediate payment, they resolve this crisis by writing checks for which there are no covering funds. "Closure" is what Lemert calls this use of bad checks to solve personal problems, since it is a last resort for solving a financial crisis. In his study of naive check forgers, Lemert found that such offenders did not identify themselves as criminals nor did they associate with criminals. While most amateur forgers were from middle-class backgrounds, many were also unemployed, divorced, or alcohol abusers, conditions that tended to isolate them and bring about closure.

In contrast to the amateur, the **systematic check forger** or "paper hanger" is a professional, making a good portion of his or her living passing bad checks. Most check artists work alone and associate very little with other criminals.

While there are different types of vandals, auto thieves, check forgers, and shoplifters, the majority of these are described as occasional property offenders because, in contrast to conventional property offenders (to be discussed next), most commit their crimes sporadically, infrequently, and crudely. They also lack identity with criminal lifestyles.

Conventional Property Crimes

Figure 9.1 illustrates the decline in property crime in the United States using the NCVS since 1973. This includes burglary, theft, and motor vehicle theft. (In the FBI classification, burglary and motor vehicle thefts are listed separately from other miscellaneous larceny or theft.)

U.S. property crimes have continued to plummet since 1974, and the United States now has lower burglary and motor vehicle theft than countries such as England, Denmark, and Sweden. Why? Some point to a growing "couch potato" factor wherein more people stay home due to VCRs and cable TV. Two-car garages, which have increased in the last 20 years, protect cars and bicycles. A more affluent society has reduced demand for many stolen items, since most people already have them. The use of credit cards and less cash carrying has caused pickpocketing and purse snatching to drop 53 percent from 1991 to 1997. Private security measures have added additional protections.

Conventional property criminals commit crimes of theft/larceny and burglary on a fairly persistent basis. Their activities constitute rudimentary forms of career criminality. Most such offenders identify with criminal behavior and associate with other criminals. They are often described as "semiprofessionals" or "minor leaguers" in the world of crime. They begin their careers in crime as juvenile delinquents and, even though most juvenile delinquents do not graduate to adult criminality, conventional property criminals do. Most conventional offenders exhibit a diversified offense record including theft, larceny, robbery, burglary, and the like. Lacking the skill and organization of more successful career criminals, they are more likely to be arrested and imprisoned. The majority will retire from conventional crime in their mid-twenties.

Journal Article Link 9.1
Examine literature regarding property crime and juveniles.

FIGURE 9.1 ■ Property Crime Rates, 1973–2006

Note: Property crimes include burglary, theft, and motor vehicle theft.

Journal Article Link 9.2
Examine literature regarding burglary.

Journal Article Link 9.3
Examine literature regarding property crime in Portugal.

Burglary

Burglary, the unlawful entry of a structure in order to commit a felony or theft, may include actual forcible entry, unlawful entry where no force is used, or attempted forcible entry. In 2006, the U.S. burglary rate of 748.7 was the lowest in over two decades, about a third lower than in 1989. The highest rate was in the South (903.8), with the West (727), Midwest (692.1), and Northeast (429.9) following. That same year, burglary offenses cost victims an estimated $4 billion with average loss per burglary set at $1,834. About two thirds of burglaries were residential in nature and 63 percent took place during the day. Fifty-seven percent of commercial burglaries occurred at night. The average losses incurred due to residential and nonresidential burglaries were about the same (FBI, 2004). As a general rule, burglars are nonviolent and choose this particular brand of thievery as a relatively safe, nonconformist means of obtaining their booty. Table 9.3 presents international burglary rates for selected countries. The highest burglary rates were found in Australia (2280.8), Canada (980.9), and the United States (740.5), while the lowest were for the Bahamas (173.9), Ireland (265.5), and Russia (425.6).

Types of Burglars

Marilyn Walsh (1977), in *The Fence,* provides an interesting typology of burglars, a continuum from most organized to least organized.

Walsh's types of burglars are professionals, known burglars, young burglars, juvenile burglars, and junkies. The professional, "skilled," or "master" burglar exhibits the characteristics of professional criminal behavior. Such offenders are highly skilled, undertake extensive planning, and concentrate on "big jobs," since burglary is often their sole livelihood. Known burglars are far less sophisticated, professional, or successful, even though burglary may represent a major source of their livelihood. Their operations are generally much cruder, and they rely less upon organization than do more professional burglars. Being older and more experienced than other amateur burglars, the known burglars are so called because they are known to the police, which suggests that they are less successful than professionals. They are an excellent illustration that "practice does not always make perfect" (see Rengert & Wasilchick, 1985; Shover, 1973).

Young burglars are usually in their late teens or early twenties, have less planning or organization in their operations than professionals, and are well on their way to becoming professional or known burglars. *Juvenile burglars* are under 16 years of age and prey on local neighborhood targets chosen by chance or occasion; such juveniles often operate under the supervision of older fences and burglars. Finally, *junkies* (drug addicts) are simply opportunist burglars and are the least skilled of such thieves.

Other analyses of burglars have basically supported Walsh's distinctions (Pope, 1980; Repetto, 1974; Scarr, 1973). Although not constituting a distinct typology as such, Repetto's case study of 97 burglars provides some interesting profiles. Juvenile offenders were generally unskilled, concentrated on local easy targets of small gain, and viewed crime more as a game than as a commitment to a way of life. The 18- to 25-year-old offenders, despite previous convictions, continued to burglarize because they found low-risk targets. Many in this group were drug users. Their targets were more likely to be outside their neighborhoods and produced higher gains; they made more extensive use of fences. Older offenders (over 25) had extensive incarceration histories, continued at burglary because of its low risk, exercised better planning, and had fewer but higher quality targets. Such individuals were more highly committed to criminal careers. Drug users were likely to perform more burglaries than nonusers, but were more likely to work near their own neighborhoods and to be more reckless or unplanned in their operation. In contrast, the non–drug user performed fewer but better planned burglaries. In a statistical analysis of burglaries in California, Pope found that those with no criminal records concentrated on nonresidential targets, while those with records preferred residential sites. He concludes that "unlike violent crimes in which there is an interactive pattern [between type of burglar and type of burglary], burglary and other property crimes as well, may reflect more opportunity than choice" (p. 50; see also R. Wright & Decker, 1996). R. Wright and Decker (1996) interviewed 105 active burglars using a "snowball sample" (snowball sampling relies on referrals from initial subjects

TABLE 9.3 ■ International Burglary Rates

For most crimes, U.S. rates are much higher than European rates. Burglary is the only crime explained in the Bureau of Justice Statistics' comparison for which U.S. rates are less than double those of European countries. A number of European countries had higher rates than the United States. Since the United Nations groups all theft into one category, it has no statistics for burglary. The Interpol definition of burglary is "breaking and entering." The U.S. definition is "the unlawful entry of a structure to commit a felony or theft. The use of force to gain entry is not required to classify an offense as a burglary."

International Burglary Rates, Interpol Data (per 100,000 population)	
Country	Burglaries
United States[a]	740.5 (2003)
Australia	2280.8 (2000)
Bahamas	173.9 (1999)
Canada	908.9 (2001)
Denmark	1876 (2003)
France	683.6 (2002)
Ireland	265.5 (2002)
Mongolia	464.7 (2001)
Russia	425.6 (2002)

Source: Interpol. *International Burglary Rates, 2003* (per 100,000; Interpol, 2004).

a. U.S. statistics were taken from the *Uniform Crime Reports for 2003.* (Federal Bureau of Investigation, 2004).

Web Research Project
Read http://www.stephenjdubner.com/journalism/silverthief.html about the exploits of Blane Nordahl, the "cat burglar." How typical of burglars is Nordahl?

to generate additional subjects) and an ex-offender with high status among St. Louis criminals. They found that two thirds of their sample averaged 10 or fewer burglaries in a year, while 7 percent averaged 50 or more per year. Most viewed themselves as "hustlers" and committed a variety of other offenses. Most offenders followed a script in which they worked with others, searching the master bedroom first and the living room last. Most do not remain long, fearing apprehension. They dispose of their bounty quickly through fences or acquaintances for a mere fraction of its value.

The biggest news on burglary in the United States is its decline from 3.8 million reported cases in 1981 to 1.9 million in 2006—over a 30 percent drop. In New York City during this period, burglary decreased more than 50 percent. While other crimes also declined during this period, the decrease in burglaries was the most persistent. Explanations have included changes in tax write-off laws for burglary losses, better security awareness, proliferation of guns (which favors robbery), replacement of heroin with crack (the latter requiring more and quicker money), and better police investigations.

Fencing Operations

Of great importance in the crime of burglary as well as other property crime is the burglar's connections with **fences,** dealers in stolen property. "Professional or master fences," full-time specialists in stolen property, are essential to the operation of professional burglars and are detailed later in this chapter. Amateur burglars and less professional burglars are more likely to deal with a "neighborhood fence" or an "outlet fence" (Blakey & Goldsmith, 1976, pp. 1530–1535). The Neighborhood Watch

Program, featuring community involvement and the engraving of valuables to make them more traceable and more difficult to fence, is one strategy that has met with some success (Gilham, 1992).

Related to fencing are pawnshops, which purchase or give loans for items placed as collateral. These are often used for fencing hot goods. Recent database software designed for pawnshops, which permits monitoring of inventory and downloading periodic reports for police, may be closing one avenue of unloading hot goods (Krane, 2000).

Stings

Because conventional burglars shop around for fences, "sting operations" (police antifencing programs) have been relatively successful. In these efforts, the police pose as dealers in stolen goods. These operations, first introduced at the federal level in 1974, have resulted in a high conviction rate and produced a subsequent decline in property crimes in the areas in which they have operated.

An interesting early sting operation was PFF, Inc., which ran for 5 months in Washington, D.C., in 1975 ("The Sting," 1976). PFF, Inc., stood for Police-FBI Fencing, Incognito, and was headquartered in an abandoned warehouse. Undercover agents hung Playboy centerfolds at the entrance; behind them, a camera photographed entering customers and videotaped each transaction. Their customers assumed the proprietors were Mafiosi, and one subject filled out an application for a "hit man" job in which he supplied information on a hitherto unsolved murder. Running out of "buy money"—funds with which to purchase the stolen property—PFF announced a formal party to which they invited their good customers. The customers checked their guns at the door and on entering were arrested and escorted out the back door to jail.

"Operation Road Spill" was an FBI sting operation of auto thieves conducted in South Kearney, New Jersey, until it was closed down in September 1994. Posing as an auto leasing agency whose shady dealers were willing to buy "hot" cars, agents purchased 120 cars, valued at $4 million, for $140,000 before arresting 30 men and seeking another 15. Half of the thieves were members of a loosely knit Brooklyn car-theft ring known as the Flatbush Pulley Gang for the device they used to pull out ignition locks. The gang had about 100 members (McFadden, 1994).

Larceny-Theft

Most conventional property offenders tend to hustle or to be generalists in theft. Some may concentrate on burglary, but they are also opportunists, taking advantage of a given occasion to commit any variety of larceny-theft.

Larceny-theft makes up the largest category of the traditional UCR index. It was for this reason in 2004 that the FBI decided to eliminate reporting the traditional crime index. The feeling was that the index was inflated by the inclusion of larceny-theft. Larceny-theft is the most underreported of former index crimes with only about a third reported to the police. With an estimated 6 million offenses, larceny-theft represents 59 percent of the 2006 UCR crime index total and 60 percent of all property crimes in 2006.

The highest larceny-theft rate in the United States was in the South (2,499), followed by the West (2,247), the Midwest (2,175), and the Northeast (1,616). Estimates of the value of losses from larceny-theft are conservative since many cases of small value are not reported to the police. The categories of larceny-theft and estimated average losses for 2006 were theft from buildings, $1,170; theft from motor vehicles, $734; theft of motor vehicle parts or accessories, $522; purse snatching, $440; pocket-picking, $443; bicycle theft, $2,263; and theft from coin-operated machines, $317 (FBI Uniform Crime Reports Statistics, 2007).

The pervasive nature of larceny-theft can be illustrated by airport baggage thieves. Some airports in Africa and Latin America are known for their high rates of theft, with airport security personnel and even government officials working with the thieves.

Larceny-theft as a category in the UCR has been correctly described as a "garbage can" (McCaghy, 1980, p. 164), a "wastebasket concept," a catch-all for miscellaneous property crime. It covers a large variety of offenses and lumps together relatively minor offenses with major professional crime.

Arson: A Special-Category Offense

It is an irresistible although admittedly bad joke to say that arson is a hot topic or a burning issue, but only since the late seventies have U.S. law enforcement officials devoted attention to this matter to a degree proportional to its seriousness. Defined by the UCR as "any willful or attempted malicious burning of a structure, vehicle, aircraft, or property of another," **arson** was added to the crime index (Part I crimes) in 1979 by congressional statute. Because arson differs from other crimes, statistics are acquired from fire services and the insurance industry as well as from law enforcement agencies. Being that fires of suspicious or unknown origin are not included in the statistics and only those determined through investigation are included, the actual number of arsons is probably higher than indicated by the UCR. For 2006, a total of 64,043 arson offenses were reported in the United States (FBI Uniform Crime Reports Statistics, 2007).

One fire department official called arson "the cheapest crime in the world to commit. All you need is a box of matches" ("Arson for Hate and Profit," 1977, p. 22). While the actual annual cost of arson in the United States can only be estimated, such guesses have ranged from $1 billion to $15 billion. The full cost of arson increases when we include the following (National Criminal Justice Reference Service, 1979):

- Death and injury of innocent citizens and firefighters
- Increased insurance premiums
- Increased taxes to support fire, police, and court services
- Inferior education facilities during reconstruction of burned-out schools
- Erosion of tax base as property values fall
- Loss of jobs at burned-out factories and businesses
- Lost revenue to damaged stores and shops

In the early twenty-first century, about 17 percent of arsons were cleared by arrest. With roughly 51 percent of those arrested under age 18, arson has a higher percentage of juvenile involvement than any other index crime (FBI Uniform Crime Reports Statistics, 2007). Arson is viewed as a special-category offense because of the varying motives of its perpetrators. Based on these motives of arsonists described by Boudreau et al. (Boudreau, Kwan, Faragher, & Denault, 1977) and Inciardi (1970), McCaghy (1980) proposes a typology of arson.

Profit-motivated arson is illustrated by insurance fraud, in which structures are purposely torched in order to collect on their insured value. Insurance companies themselves have in the past encouraged such practices by insuring questionable properties for large amounts as well as by not performing sufficient investigation before honoring claims. Arson often serves as an index of urban decay, as owners unload their deteriorating properties on the companies that insure them. Profit-motivated arson is most

Photo 9.2

Arson has been given serious attention from U.S. law enforcement since the late 1970s.

often committed by "white collar criminals," occupational and corporate offenders, often in conjunction with professional "torches." This crime may also be committed, as in New York City, by welfare recipients seeking city assistance in gaining better accommodations, by drug addicts, or even by conventional criminals dubbed "mango hunters" by New York City police. This latter category of offenders burns structures in order to expose and facilitate the stealing of fire-resistant plumbing and fixtures ("Arson," 1977, p. 22).

Revenge arson may take place out of spite or jealousy, as a means of getting even. The burn-for-hate category includes fired employees who seek revenge against employers or the jealous suitor who burns down a nightclub in which his girlfriend was socializing with someone else. This category could also serve as an example of vindictive *vandalism arson,* in which arson is an expression of hatred toward a particular group or individual. Vandalism accounts for the most arsons in deteriorating urban areas. Any structure may be torched in an extension of play activity, but abandoned properties are particular targets. *Crime concealment arson* is a way to dispose of murder victims or physical evidence or to draw attention away from a crime being committed elsewhere. *Sabotage arson* covers fires set during labor or racial strife, prison riots, or other civil disturbances.

Excitement arson is often carried out by pyromaniacs, individuals who have a morbid fascination with setting and observing fires. Freudians would assign a sexual basis to pyromania, in which such individuals experience erotic satisfaction by means of arson.

Revenge arsonists resemble violent personal criminals in acting out strong emotions, but arson for crime concealment may involve conventional property offenders or corporate or organized offenders. Arson as sabotage resembles political criminality, and vandalism-inspired arson is associated with the occasional property offender. Finally, pyromania is the realm of the psychotic or psychopathic offender. Some of the California fires of 2007 were set by arsonists, and it is suspected in some instances they were serial arsonists.

Criminal Careers of Occasional and Conventional Property Criminals

The distinction between occasional and conventional offenders does not lie in the legal categories of the offenses that they commit but in the way they commit crimes and the degree of their identification with the criminal world (see Table 9.4).

Most snitches, amateur shoplifters, wanton vandals, joyriders, and naive check forgers do not view their activities as criminal, have short or no official criminal records, do not commit crime as a means of livelihood, and are not "streetwise" or "crimewise" in the language of the criminal life. On the other hand, most conventional property offenders exhibit minor levels of career criminality. Many have early histories of truancy, vandalism, street fighting, delinquent gang membership, and contact with the law (Clinard & Quinney, 1973, p. 149).

The criminal careers of most conventional criminals peak in their late teens and rapidly decline after they reach their mid-twenties. Reaching ages where the full weight of criminal penalties falls on them as adult offenders and not being skilled enough to make a living at crime, most simply curtail their level of criminal involvement. Middle age and maturity, rather than any specific rehabilitation regimen, eventually reform the conventional criminal. Werthman (1967) indicates,

> After a few years of this existence [street hustling], these boys are really at the end of their "delinquent" careers. Some get jobs, some go to jail, some get killed, and some simply fade into an older underground of pool rooms and petty thefts. Most cannot avoid ending up with conventional jobs, however, largely because the "illegitimate opportunities" available simply are not that good. (p. 170)

TABLE 9.4 ■ Occasional Versus Conventional Criminals

Clinard and Quinney's (1986) Occasional Versus Conventional Criminals	
Occasional	**Conventional**
Do not view their activities as criminal	Identify with crime
Short or no criminal record	Early history of delinquency/crime
Crime is avocational	Crime is a vocation
Not streetwise	Streetwise
Short criminal careers	Longer criminal careers
Often operate alone	Usually operate in groups
Committed to legitimate society	Only partially committed to legitimate society

Occasional property offenders do not require criminal associations in order to commit their crimes, as they are fairly easy to commit, requiring little training or skill. In contrast, conventional property offenders often operate in groups or gangs in which they learn many of the techniques of crime from their peers. While most occasional offenders maintain their commitment to conventional society and reject criminal identities, conventional offenders are only partially committed to legitimate society.

Societal Reaction

The societal reaction to occasional property offenders is relatively mild. Since most offenders lack a previous criminal record, charges are usually dismissed, or the individual is given immediate probation or a suspended sentence. In contrast, societal reaction toward conventional property offenders is relatively strong and, until recently, even stronger than that against professional, organized, and corporate criminals (Cullen, Link, & Polanzi, 1982). Conventional offenders are of a different social class than those who make and enforce the law. This may in part explain the stronger legal processing of such offenders when compared with that of corporate offenders.

Occasional property offenders strongly identify with middle-class societal values and reject criminal identification. Because of this, most offenders are deterred from future activity once they are confronted with legal action or arrest. Relative leniency with offenders is often justified, since few have previous criminal records of any substance and most are unlikely to progress into a career of crime. Since many such offenders come from the same social class as those who make and enforce the laws, they fail to fit popular public stereotypes of criminals. Diversion of such offenders into restitution programs or accelerated rehabilitation dispositions also relieves the caseload burden of the courts.

Most conventional criminal offenders tend to identify with criminal behavior and are less likely to be deterred by the threat of arrest or the stigmatization of the label of criminal. For some, criminal processing enhances their "bad" reputation on the street. The attachment of the criminal label and record may also begin a criminalization process that isolates the individual from more conventional associates and reinforces a criminal identity. Because of their relative lack of skill and organization, most conventional criminals are eventually arrested and a large proportion are imprisoned. About half of all prison inmates in the United States are conventional criminals. It is to this population that the term "revolving door of justice" has been applied because some end up doing the equivalent of "life" on the installment plan; conventional rehabilitation plans are relatively unsuccessful with this group of offenders (Martinson, 1974). Most of these offenses are outgrowths of deprived lower-class environments and subcultures, and legal processing appears to have only minimal effects.

Tunnell (1991), in a study of property offenders, found most were motivated by the need for quick cash for drugs, alcohol, and living expenses. They also expressed a sense of accomplishment and of winning a game. While they feared being caught and jailed, most overcame this, and basic deterrence policy had little effect.

The predatory street crime pattern is typical of juveniles, who commit such offenses for years, then cease such activity because of the threat of jail, the availability of conventional job opportunities, the development of simple maturity, or some combination of these factors. One researcher, Mercer Sullivan (1989), claims that solutions calling for employment programs or "getting tougher" with such offenders are too simplistic and that the campaign for selective incapacitation particularly misses the mark: Most offenders are not identified as serious offenders until they are at least 20 years old—"over the hill in terms of street crime." A strategy to treat such individuals as hard-core, career criminals earlier and to incapacitate them through imprisonment, assuming, as demonstrated in a Wolfgang and associates cohort study (Wolfgang, Figlio, & Thornberry, 1978), that 5 to 8 percent account for over 50 percent of juvenile crime, ignores the fact that most teen crime careers are short; thus, such policies risk incarcerating juveniles at the very time when most would be ending their criminal careers.

Comparing white, African American, and Latino juveniles, Mercer Sullivan (1989) notes that they begin in conflict-oriented gangs that prey only on each other. Growing older, white teenagers are able to obtain part-time jobs through their parents' contacts. These jobs occupy their time and supply money, and street robberies would not be tolerated by the neighbors. African Americans and Latinos lack such employment contacts, and many become the self-employed of the streets, experimenting with muggings and burglaries, sometimes at the expense of their neighbors. Local muggers, since they tend to operate close to home, are not tolerated, and most are eventually arrested.

While the crimes of occasional property offenders point out the pervasiveness of violations among those in otherwise respectable society, offenses of conventional property offenders and the relative lack of success of correctional efforts with such offenders continue to disappoint crime control policy planners. Conventional property offenders fail to respond to the very policies, such as threat of jail and stigmatization, that appear to work very well in discouraging occasional property offenders. Conservative approaches that aim to reform individuals without concomitant efforts at social reform are likely to continue to fail.

Professional Crime

In 1996, a rebellion broke out in formerly communist Albania. Citizens blew up bridges, attacked police stations, and looted military arsenals. Armed with military weapons including tanks and patrol boats, they took over areas of the country. The reason for the revolt was anger on behalf of thousands of Albanians who had been swindled in a huge Ponzi scheme (illegal pyramid scheme). Many of those involved in organizing the scam were high government officials. Reports claim that nearly every Albanian was victimized by the scheme (Nelan, 1997; D. R. Simon & Hagan, 1999, p. 88). The Albanian government collapsed as a result of a professional crime, the Ponzi scheme, to be discussed in this chapter.

On the rainy night of October 29, 1964, Roger Clark dropped off Jack Murphy and Allan Kuhn at the American Museum of Natural History in Manhattan. Jack "Murph the Surf" Murphy (see Photo 9.3) and his accomplice scaled an 8-foot fence, climbed a ladder from a courtyard to a fourth-floor ledge, opened a window, lowered themselves into the museum's Morgan Hall of Minerals and Gems, and proceeded to steal 24 precious stones. These included the Midnight Star Sapphire; the DeLong Star Ruby; and the celebrated sapphire Star of India. The latter was a most extraordinary gem. Weighing more than 563 carats, it was the largest such stone in the world and was estimated to be worth more than $1 million in 1986 (Preston, 1986, pp. 210–211).

The "Star of India" burglary bore an uncanny resemblance to the plot of the then-current film *Topkapi,* which was about a jewel theft at the Topkapi Palace Museum in Istanbul. Murphy and his accomplices claimed to have been inspired by the film ("Museum Jewel Robbery," 1964, p. 23). Prior to the Star of India burglary, they had cased the Guggenheim Museum and the Metropolitan Museum of Art before settling on the American Museum of Natural History as their target. Ten days of reconnaissance at the museum included a dry-run nighttime burglary. But despite their elaborate plans and successful heist, the trio were shortly arrested and convicted. Of the three thieves, Murphy attracted the most press attention because of his glamorous Miami beach-boy lifestyle. This same "charmer" would later carry out other jewel thefts, including the strong-arm robbery of a $25,000 diamond ring from actress Eva Gabor, and would eventually be convicted of the double murder of two women, for which he served prison time until 1984. Murphy was even the object of a made-for-television movie starring Robert Conrad in 1975. "Murph the Surf" and his accomplices represent the more glamorized, romanticized view of professional criminals. But what, in fact, are professional criminals?

The Concept of Professional Crime

In sociology, the concept of **profession** refers to occupations that require esoteric, useful *knowledge* acquired after lengthy training, and a claimed *service* orientation and code of ethics that permit occupations to attempt to obtain *autonomy* or independence of operation and various concomitants such as high prestige and remuneration (F. E. Hagan, 1975). This knowledge-service-autonomy dimension of the professions is inapplicable to criminals and criminal activity. Mack (1972) prefers the term "able criminal," while J. F. Klein (1974) suggests the concept "grifter" (one who acquires money illicitly) as a more appropriate construct. Some field studies of criminal subjects indicate that they do not use the category of "professional criminal" (Letkemann, 1973; Prus & Sharper, 1977; Steffensmeier, 1986). However, labeling certain types of criminals as "professionals" is widespread in the literature, which justifies treating professional crime as a separate category, though we will consider other constructs. Cressey (1969, p. 45) warns that many skilled criminals are simply slightly better than other crooks at lying, cheating, and stealing and that we should be careful about calling them "professionals." Studies of career criminals by Petersilia, Greenwood, and Lavin (1977) suggest the term "intensives" for those who commit more sophisticated crimes and face lesser chances of arrest or conviction. Therefore, while the **professional criminal** is not truly a professional in the sociological sense, the term is appropriate in reference to those who earn a considerable portion of their livelihood in criminal pursuits.

Photo 9.3

Murph the Surf plays guitar in the chapel of Raiford Prison, where he found religion.

Characteristics of Professional Crime

The benchmark in the United States for analysis of professional criminal behavior was Edwin Sutherland's (1937) *The Professional Thief,* a work based on interviews and a detailed case study of a professional thief/confidence man with the pseudonym "Chic Conwell." In his original formulation, Sutherland saw the professional thief as characterized by crime as a sole means of livelihood, careful planning, reliance on technical skills and methods, and a migratory lifestyle. Professional thieves were found to have a shared sense of belonging, rules, codes of behavior, and a specialized language (pp. 3–4).

Professional crime is a sociological rather than a legal construct. What distinguishes professional crime from other crimes is not the legal definition of the behavior, but the way the crimes are performed. Clinard and Quinney (1973) identify the following features of professional crime:

1. Crime is the criminal's sole livelihood and is engaged in for economic gain.
2. The criminal career is highly developed.
3. Considerable skill is involved.
4. High status in the criminal world is bestowed on professional criminals.
5. Professional criminals are more successful than others at avoiding detection and imprisonment (p. 246).

Argot

Acts of professional criminals differ from those of less professional criminals only in the distinctive manner in which they are committed. Professional crime primarily involves the relatively safe and consistent stealing of large sums of money on a systematic, rational, planned, skillful, and nonviolent basis. Professional criminals attempt to avoid "heat," or the daring and bravado characteristic of many amateur criminals, which tend to attract public attention and often subsequent police action. Distinctive **argot** or specialized language is also characteristic of the world of professional crime. Arthur Judge's *The Elizabethan Underworld* (1930), Henry Mayhew's *London's Underworld* (1862), and McMullan's *The Canting Crew: London's Criminal Underworld 1550–1700* (1984) provide some of the argot of seventeenth-century Elizabethan professional criminals (see also L. Taylor, 1984). Sutherland's (1937) informant, "Chic Conwell," provides us with the criminal argot of the United States in the early twentieth century.

Early Twentieth-Century U.S. Argot

Cannons: professional pickpockets

Heels: sneak thieves who operate in stores and offices

Boosters: professional shoplifters

Pennyweighting: stealing from jewelry stores by substituting fake lookalikes

Hotel prowling: stealing from hotel rooms

The con: confidence game or swindle

Hanging paper: passing bad checks, money orders, and other commercial paper

The shake: extorting money from others who are criminally involved

Maurer (1964) quotes a professional pickpocket who was asked to explain in court what he had done:

> Well, Judge, your honor I was out gandering around for a soft mark and made a tip that was going to cop a short. I eased myself into the tip and just topped a leather in Mr. Bates' left prat when I blowed I was getting a jacket from these two honest bulls. So I kick the okus back in his kick and I'm clean. Just then this flatfoot nails me, so here I am on a bum rap. All I crave is justice, I hope she ain't blind. (p. 55)

A Model of Professional Crime

Table 9.5 depicts a continuum **model of professional crime.** The purpose of this model is to underline the fact that there are no hard-and-fast divisions between professional and amateur crime; the professionalism of criminal activity is a matter of degree rather than of kind. Thus, the greater the extent to which an individual's criminal activity involves key factors—crime as a sole livelihood, an extensive criminal career, skill, high status in the criminal world, the successful avoidance of detection or immunity from prosecution, and a criminal subculture and organization—the greater the likelihood that such activity can be labeled "professional crime." The concept, however, is an *ideal type,* a heuristically useful overgeneralization that is unlikely to exist in a pure form.

Professional crime is similar to legitimate occupations such as entertainment and professional sports in that it is a "skyrocket profession." For those who succeed, it can be a glamorous life of fast living; however, it shares another characteristic with these fields, and that is that "many are called, but few are chosen." Many semiprofessional criminals who might occupy a middle ground on the continuum are unable to survive with crime as their sole livelihood—they are not good enough at it. Most eventually leave the world of crime because they simply cannot make a living; they are less skillful, enjoy less status in the criminal world, and are less successful at avoiding detection than other criminals. Although they belong to criminal subcultures to a degree, these subcultures lack the network of talent and successful contacts that the more professional criminal enjoys. Finally, less professional offenders generally commit crimes that require or involve less planning, and they have less success in avoiding prosecution and incarceration.

It was previously indicated that career criminality includes characteristics such as identification with crime, extensive association with other criminals, progression and specialization in crime,

TABLE 9.5 ■ A Model of Professional Crime

Amateur Crime		Professional Crime
Occasional	(Source of Income)	Full-time
Short Duration	(Extent of Career)	Lifelong
Little	(Level of Skill)	Extensive
Low	(Status)	High
Unsuccessful	(Avoidance of Detection)	Successful
None	(Association With Criminal Subculture)	Extensive
None	(Level of Planning)	Extensive
No	(Employment of "the Fix")	Yes

and crime as a sole means of livelihood. Professional and organized criminals best fit this model, though, as noted above, differences between these types are often a matter of degree rather than kind. Similarly, criminal pursuits such as videotape/motion picture pirating, drug sales, prostitution, and pimping are pursued by criminals of various degrees of sophistication, with career criminals generally more persistent and successful in their activities. Professionals tend to "freelance" or be less tied to relatively permanent criminal organizations than their organized-crime counterparts. In the last analysis, however, criminologists have not arrived at a consensus in categorizing these activities.

Edelhertz's Typology

Edelhertz (1970) has developed a typology of white collar crime. Table 9.6 presents one of Edelhertz's categories, which is more professional than occupational or corporate in nature. The main distinction between professional crime and occupational/corporate crime is that in the former, the sole purpose of the business is to perform criminal activity, while in the latter, crime is incidental to a legitimate business or professional service. Some offenses listed in Table 9.6 are not simply professional crimes but, following the model, become so to the degree that they involve the characteristics of professional criminal activity.

TABLE 9.6 ■ **Examples From Edelhertz's (1970) Categories of White Collar (Professional) Crime**

White Collar Crime as a Business, or as the Central Activity:

1. Medical or health frauds
2. Advance fee swindles
3. Phony contests
4. Bankruptcy fraud, including schemes devised as salvage operations after insolvency of otherwise legitimate business
5. Securities fraud and commodities fraud
6. Chain referral schemes
7. Home improvement schemes
8. Debt consolidation schemes
9. Mortgage milking
10. Merchandise swindles:
    a. Guns and coins
    b. General merchandise
    c. Buying or pyramid clubs
11. Land fraud
12. Directory advertising schemes
13. Charity and religious frauds
14. Personal improvement schemes:
    a. Diploma mills
    b. Correspondence schools
    c. Modeling schools
15. Fraudulent application for, use, or sale of credit cards, airline tickets, etc.
16. Insurance fraud:
    a. Phony accident rings
    b. Looting of companies by purchase of overvalued assets, phony management contracts, self-dealing with agents, intercompany transfers, etc.
    c. Fraud by agents writing policies to obtain advance commissions
    d. Issuance of annuities or paid-up life insurance, with no consideration, so that they can be used as collateral for loans
    e. Sales by misrepresentation to military personnel or those otherwise uninsurable
17. Vanity book and song publishing schemes
18. Ponzi schemes
19. False security fraud, i.e., Billy Sol Estes or De Angelis types of schemes
20. Purchase of banks, or control thereof, with deliberate intention to loot them
21. Fraudulent establishing and operation of banks or savings and loan associations
22. Fraud against the government:
    a. Organized income tax refund swindles, sometimes operated by income tax "counselors"
    b. Aid frauds, i.e., totally worthless goods shipped
    c. FHA (Federal Housing Authority) frauds:
        (1) Obtaining guarantees of mortgages on multiple family housing far in excess of value of property, with foreseeable inevitable foreclosure
        (2) Home improvement funds
23. Executive placement and employment agency frauds
24. Coupon redemption frauds
25. Money-order swindles

Scams

Scam is a criminal slang term used to refer to various criminal techniques, "hustles," or operations. Many criminal operations that may be described as examples of professional crime or sometimes even as white collar crime (since they are committed by stealth, nonviolently, and by persons of apparent respectability) are in fact semiprofessional in nature. They involve little skill, and they prey on gullible victims. **Confidence (con) games** meet this definition. These may be called "confidence" because they rely on winning the confidence of the victim in order to steal from him or her. Another possible origin of the term *con* is the word "cony," meaning dupe or victim. Sometimes called "flim flam" or "bunko" or "short cons," such scams come in an infinite variety, although some of the more common ones are disturbingly familiar and successful. It is surprising to many that most short con artists at the present time are women, as are most of their victims.

The **pigeon drop** is one of the best-known simple confidence swindles. Here is a description of an actual pigeon drop that has been repeated hundreds of times with remarkable success:

> A 61-year-old widow (the "mark" or victim) gave her $3,596 in life savings to two con artists in a downtown five-and-dime with the assumption that she would receive this back and $6,000 besides. A woman (one of the con artists—the "catch woman") approached the widow while she was shopping in a department store, began talking to her and a few seconds later a third woman (accomplice) appeared. The third woman said she had found a wallet with $20,000 in cash and a note saying that the money should be delivered to Castro in Cuba. She showed them some loose twenties (the rest of the roll of money actually consists of a "Michigan bankroll" of phony money), and the three women discussed the situation for a while. Then the woman with the wallet convinced them that she would phone "Attorney Burger," a fictitious man, who she said worked in City Hall.
>
> After a fake phone conversation with him, she reported that he asked all three to put up some money to show good faith ("good faith deposit") until the money could be divided three ways, with $2,000 for the lawyer. The woman who had first spoken to the victim said she could come up with $4,000 from a recent insurance claim, and she supposedly went to get it. Then the other woman took the victim to a bank for her life savings, which she turned over to them in the ladies' room of a five-and-dime. She was told to report to "Attorney Burger at City Hall" at 4:30 PM. She arrived at the appointed time and came to the grim realization that she had been duped. "I've read about that one in the papers, but I never thought it could happen to me," she told police. ("Con Game Nets $3,600 From Widow," 1972, p. 11A)

Many victims are so humiliated that they do not even report their victimization to the police. Short cons such as the pigeon drop usually prey not upon the affluent, but on middle-aged, retired, and widowed working-class types, particularly females.

The **badger scam** also preys on the naivete of victims.

> An alert teller at a savings and loan association alerted police when an obviously distressed 81-year-old man withdrew his life savings of $10,622—in cash. He had been visited on a number of occasions by a 19-year-old girl who had indicated that she represented a Bible Institute. During the last visit, a man feigning the role of an outraged father burst into the apartment and accused the victim of having illicit relations with his daughter. His paternal rage could, however, be forgotten for the right price. (Langway & Smith, 1975, p. 67)

In the **bank examiner's scam,** swindlers pretend to be government investigators who are seeking the cooperation of the victim in order to catch a dishonest bank teller. The mark (victim) is asked to withdraw money and turn it over to the investigators, who will mark it in order to apprehend the dishonest employee. Obviously, government agencies are not so hard up that they have to use the money of private citizens in order to conduct their undercover operations.

"Too-good-to-be-true" opportunities for easy money, and "get-rich-quick" schemes lure victims. The following examples are illustrative:

- Postal fraud is widespread. It includes skipping town with payments for orders, precollecting fees for fake franchises, and offering to arrange a "guaranteed" business loan or employment for an "advance fee"—then failing to deliver as promised. Beware of paying advance fees for estates that have been left to you by unknown people, chain letters, work-at-home schemes, and sales of far-off land.

- Hundreds of companies are sending out millions of postcards or e-mails guaranteeing rewards for "prize winners" if the recipients send money for shipping and postage. The prizes are either nonexistent or virtually worthless.

- Circus grifting (dishonest carnival games) is another example of a short con. Grifters will often work with a "shill," or plant, who pretends to be a winning customer. Such con games were quite common in traveling carnival shows that worked Midwestern county fairs. Hidden pedals for gambling wheels; switched rings for the ring toss; loaded dice for the craps game; fake "two-headed" ladies or men purported to be half-human, half-ape; and willing customers ever eager to part with their money embody P. T. Barnum's slogan, "There's a sucker born every minute."

- In this same vein of "short cons," readers should beware of snake oil salespersons or offers to increase breast size or sexual prowess or to sell diplomas (Stewart & Spille, 1988). In one home-repair scam that targeted the elderly, the swindler would short out electric wires, start fires near furnaces, or release bugs or mice in order to create problems in need of solution. In a similar scam, someone claiming to be waterproofing the customer's roof with a "clear, silicone-based liquid" would spray plain water on the roof ("Home Repair Scam," 1986).

- By 1992, the Federal Communications Commission finally began cracking down on the abuse of "900" telephone numbers, which charge the unwary caller exorbitant fees per minute for "phone sex," "advice," "astrology"—anything to keep "suckers" on the line. Needless to say, when telephone solicitors call claiming to be checking credit card numbers, one should never reveal one's account number.

- Toll call fraud, the unauthorized use of others' calling card numbers, is facilitated by con artists who stake out airports with binoculars and sell telephone numbers.

- The three-card Monte is the floating card sharp's equivalent of the shell game and is common in Times Square in Manhattan. Monte dealers deftly shuffle three bent cards atop a cardboard box while palming a bettor's selection (whose chance of winning is zero). They often have six to eight confederates (shills) in the crowd who fake winning (see McNamara, 1995).

Peter Maas in *King of the Gypsies* (1975) describes the elaborate gypsy con game of **boojo,** in which superstitious victims, usually visitors to fortune-telling parlors, are conned into turning over their life savings in cash in order to have a "curse" removed. The gypsy con artist removes the cash instead. In one boojo operation, a Warren, Missouri, farm couple was warned that a darkness or hex had been placed on them by their family and not to talk to them. A long string of evil omens and misfortunes were forecast, including fatal cancer and gangrene. Each could be removed by paying the gypsy woman money. The family lost over $150,000 in the scam before they finally wised up (O'Connor, 1987, p. 41).

A recurring scam for the past couple of decades has been the "Nigerian fee fraud scam" or 419 scam (named for the applicable section of Nigerian criminal law). The victim is contacted via mail or e-mail by someone claiming to be a Nigerian official with an offer to share millions of dollars if the victim would provide an account in which funds can be hidden, after which the victim is promised a share in the payoff. Once the victim's account number is obtained, the con artists drain the account. The victims pay processing fees, travel costs, and bribes in anticipation of rewards. The scam artists ask for more and more money to assist in the transfer of funds.

Many scams are on the edge of slick business practices. Used car dealers have been known to leave spouses alone to talk over the deal in a "bugged" office. Or after the agreement seems to be made and the papers are almost filled out, they use "the close and bump." Pretending to take the papers with the agreed-upon price to the boss in the back room, the salesperson then says that the boss wants $100 more.

Big Cons

Maurer's *The Big Con*

David Maurer in *Whiz Mob* (1964) and *The Big Con* (1940) describes the following steps in "the big con" (1964, pp. 15–16):

1. Putting up the mark (investigating and locating likely victims)
2. Playing the con (gaining the confidence of the victim)
3. Roping the mark (steering the victim to meet the inside man or woman)
4. Telling the tale (showing the victim how he or she can make big money dishonestly)
5. Giving the convincer (permitting the victim to make a profit)
6. Having the victim invest further
7. Sending the victim after more money
8. Playing the victim against the big store and fleecing him or her
9. Getting the victim out of the way
10. Cooling out the mark (having the victim realize that he or she cannot turn to the law)
11. Putting in the fix (bribing or influencing action by the law)

While not all big confidence games involve all the steps Maurer describes, each example to be discussed demonstrates variations of these steps. Wealthy marks, such as business executives, entertainment personalities, and—recently—wealthy professionals (e.g., doctors and dentists who are hunting for "tax shelters" or even greater affluence), are ideal targets for those on the lookout for "fingering the score."

Identity Theft

In 2005, an article titled "Grand Theft Identity" (S. Levy & Stone, 2005) noted that 40.5 million people in the United States had their credit card hacked; 800,000 had their information sold to con artists; and hundreds of thousands had personal data compromised through theft from legitimate vendors such as banks, schools, and businesses. Crime File 9.2 reports on some of the activities of identity theft.

Ponzi Schemes

Perhaps the most legendary swindler of all times was Charles "Get Rich Quick" Ponzi (shown in Photo 9.4), whose modus operandi has now inherited his name—Ponzi schemes. A **Ponzi scheme** pays off early investors with money obtained from later investors in a doomed enterprise. In 1919, Ponzi discovered that postal return coupons (international reply coupons, or IRCs) could be purchased overseas and redeemed in the United States at anywhere from 100 to 300 percent profit. He offered investors 40 percent profit in 90 days. He then paid off his first investors sooner and with larger dividends than promised. Once the word got around, investors were beating down his door. Many preferred to reinvest rather than to withdraw their money. As investors multiplied, he ran out of the product (coupons) and simply operated a **pyramid scheme** in which early customers were paid off with money obtained from later ones. Ponzi lived like a king and was rumored to have taken in over $15 million. When it was discovered

Crime File 9.2

Identity Theft

What Is Identity Theft?

Identity theft involves acquiring key pieces of someone's identifying information, such as name, address, date of birth, social security number, and mother's maiden name, in order to impersonate them. This information enables the identity thief to commit numerous forms of fraud that include, but are not limited to, taking over the victim's financial accounts; opening new bank accounts; purchasing automobiles; applying for loans, credit cards, and social security benefits; renting apartments; and establishing services with utility and phone companies.

The following information provides the actions recommended by the U.S. Postal Inspection Service and the financial industry to help reduce the likelihood of becoming a victim of identity theft. The last section provides names and phone numbers of the agencies referred to throughout this pamphlet.

Preventive Actions

- Promptly remove mail from your mailbox after delivery.
- Deposit outgoing mail in post office collection mailboxes or at your local post office. Do not leave in unsecured mail receptacles.
- Never give personal information over the telephone, such as your social security number, date of birth, mother's maiden name, credit card number, or bank PIN code, unless you initiated the phone call. Protect this information and release it only when absolutely necessary.
- Shred preapproved credit applications, credit-card receipts, bills, and other financial information you don't want before discarding them in the trash or recycling bin.
- Empty your wallet of extra credit cards and IDs, or better yet, cancel the ones you do not use and maintain a list of the ones you do.
- Order your credit report from the three credit bureaus once a year to check for fraudulent activity or other discrepancies.
- Never leave receipts at bank machines, bank counters, trash receptacles, or unattended gasoline pumps. Keep track of all your paperwork. When you no longer need it, destroy it.
- Memorize your social security number and all of your passwords. Do not record them on any cards or on anything in your wallet or purse.
- Sign all new credit cards upon receipt.
- Save all credit-card receipts and match them against your monthly bills.
- Be conscious of normal receipt of routine financial statements. Contact the sender if they are not received in the mail.
- Notify your credit-card companies and financial institutions in advance of any change of address or phone number.
- Never loan your credit cards to anyone else.
- Never put your credit card or any other financial account number on a postcard or on the outside of an envelope.
- If you applied for a new credit card and it hasn't arrived in a timely manner, call the bank or credit card company involved.
- Report all lost or stolen credit cards immediately.
- Closely monitor expiration dates on your credit cards. Contact the credit-card issuer if replacement cards are not received prior to the expiration dates.
- Beware of mail or telephone solicitations disguised as promotions offering instant prizes or awards designed solely to obtain your personal information or credit-card numbers.

Internet and Online Services

- Use caution when disclosing checking account numbers, credit card numbers, or other personal financial data at any Web site or online service location unless you receive a secured authentication key from your provider.
- When you subscribe to an online service, you may be asked to give credit card information. When you enter any interactive service site, beware of con artists who may ask you to "confirm" your enrollment service by disclosing passwords or the credit card account number used to subscribe. Don't give them out!

Audio Link 9.3
Listen to a discussion of identity theft.

Video Link 9.2
View a video of a speech from former trickster Frank Abagnale.

Who to Contact for Copies of Your Credit Report

- Equifax

  P.O. Box 105873
  Atlanta, GA 30348-5873
  Telephone: 1-800-997-2493

- Experian Information Solutions (formerly TRW)

  P.O. Box 949
  Allen, TX 75013-0949
  Telephone: 1-888-397-3742

- TransUnion

  P.O. Box 390
  Springfield, PA 19064-0390
  Telephone: 1-800-916-8800

Action Steps for Identity Theft Victims

- Contact all creditors, by phone and in writing, to inform them of the problem.
- Call your nearest U.S. Postal Inspection Service office and your local police.
- Contact the Federal Trade Commission to report the problem.
- Call each of the three credit bureaus' fraud units to report identity theft. Ask to have a "Fraud Alert/Victim Impact" statement placed in your credit file asking that creditors call you before opening any new accounts.
- Alert your banks to flag your accounts and contact you to confirm any unusual activity. Request a change of PIN and a new password.
- Keep a log of all your contacts and make copies of all documents. You may also wish to contact a privacy or consumer advocacy group regarding illegal activity.
- Contact the Social Security Administration's Fraud Hotline.
- Contact the state office of the Department of Motor Vehicles to see if another license was issued in your name. If so, request a new license number and fill out the DMV's complaint form to begin the fraud investigation process.

Report Identity Theft to:

- Equifax Credit Bureau, Fraud
  1-800-525-6285

- Experian Information Solutions (formerly TRW)
  1-888-397-3742

- TransUnion Credit Bureau, Fraud
  1-800-680-7289

- Federal Trade Commission
  1-877-FTC-HELP

- Local Police Department
- U.S. Postal Inspection Service (*local post office*; See federal government phone list)
- Social Security Administration, Fraud Hotline
  1-800-269-0271

Source: U.S. Postal Inspection Service. http://www.usps.gov/postalinspectors

Web Research Project
What steps are being taken to reduce "identity theft"?

Photo 9.4

Charles Ponzi invented the eponymous scheme in which early investors are paid off with money obtained from later investors in a doomed enterprise.

that Ponzi had a criminal record that included forgery, the house of cards fell. Investors demanded their money, but of course, he had spent it all. After serving various sentences, he eventually died in 1971 in a Brazilian charity ward, but the legacy of Ponzi's technique lives on. Ponzi schemes prey on greedy victims who want something for nothing (Nash, 1975, Vol. 2, pp. 337–341).

In a Ponzi scheme of immense proportions, the Foundation for New Era Philanthropy, which declared bankruptcy in 1995 and had been run by John G. Bennett, Jr., scammed about 300 churches, museums, universities, and philanthropists—including former Treasury Secretary William Simon and venture capitalist Laurance Rockefeller—out of as much as $500 million. Paying off early investors with money from later investors, Bennett convinced organizations such as Drexel University and the Philadelphia Orchestra to invest millions. The money would be held for a period of 6 months, at which time it would be matched by an anonymous donor, thus doubling the gift. No such anonymous donors existed (Stecklow, 1996).

The explosion of financial services, deregulation, and the bewildering number of new investments available to the public all contributed to the resurgence of Ponzi schemes in the 1990s. Many of the new Ponzi schemes relied on the "herd instinct" for new investors. During its initial stages, the scheme may zero in on members of a specific pro football team, a law office, or a military base and then rely on initial victims to enthusiastically recruit new customers.

In 2000, Ponzi schemer Patrick Bennett was sentenced to 30 years in prison for swindling thousands of small investors out of $700 million. He sold phony "lease-backed securities," investments in leases for photocopying fax machines and other office equipment. Many of these were sold to duplicate sets of investors, or the contracts did not even exist ("$700 Million Ponzi Schemer," 2000).

As of this writing, the FBI is still seeking Shalom Weiss, who vanished rather than face his sentence of 845 years in February of 2000. His Ponzi scheme is believed to have siphoned $450 million from an insurance company that went bankrupt. Weiss and accomplices bought the company with a check and then loaned themselves the money to cover the purchase price. Once in charge, they bought worthless stocks and mortgages in deals that lost the insurer millions. Most of this money ended up in Weiss bank accounts. Crime File 9.3 reports on the Bernie Madoff affair, the biggest Ponzi scheme in history.

Pyramid Schemes

A *pyramid scheme* resembles the familiar chain letter that asks you to send a dollar to the first name on the list, add your name to the bottom, duplicate four copies of the new list, and recruit four new members to continue the chain. Assuming that the chain is not broken, you could reap, for example, $256 in return for your original dollar investment—*if* the four people you recruit obtain four others each (16) and they secure four each (64), and they in turn find four others (256), who all mail a dollar to you (the name at the top of the list). The problem with such schemes is that they generally break down before reaching the bottom of the pyramid, and thus usually provide rewards only to the initial organizers.

In 1994, the Friends' Network, a massive pyramid game, was closed down in at least 13 states. The con was spread by word of mouth among friends and relatives. Players gave a $1,500 "unconditional gift" (which they believed was tax free) to the person who recruited them. Then they each recruited one more new player. The money was passed up the chain to the name at the top of the pyramid.

Crime File 9.3

The Bernie Madoff Affair: "One Big Lie"

"One big lie" was the way Bernie Madoff described the asset wing of his investment firm. His scheme was discovered in December 2008 and was described at the time as the biggest Ponzi scheme in history. The losses to his investors were estimated at $65 billion. Although Madoff claimed to have acted alone, investigators had a hard time accepting this explanation given the complex nature of the transactions. Thousands of investors lost millions, some their life savings, and in one case a person committed suicide because he had trusted Madoff. His hedge fund Ascot Partners paid out unbelievable, steady, double-digit returns year after year that begged investigation or discovery. One observer indicated: "It is virtually impossible to have returns like Madoff reported, and it should have been a major warning signal" (Lenzner, 2008). Madoff had even been Chairman of the Board of Directors of the National Association of Securities Dealers and very respected on Wall Street. He pled guilty to an 11-count indictment that included securities fraud and perjury.

Supposedly, in December 2008, Madoff informed his sons that he was about to give millions in bonuses earlier than scheduled and admitted that his investment firm was one big Ponzi scheme. His sons reported him to authorities and he was arrested. In a typical (though huge) Ponzi scheme, Madoff admits to having made no investments with his clients' money since 1991. Others place the date for him beginning his scam as the mid-1980s. Among the many victims of Madoff's operations were Steven Spielberg, First Manhattan Company, art collector Norman Braman, Elie Wiesel, Kevin Bacon, John Malkovich, and Zsa Zsa Gabor along with numerous charities and civic organizations.

On June 30, 2009, a federal judge in Manhattan sentenced Madoff (who was 71 at the time) to the maximum of 150 years. This was comparable to the types of sentences given in the past to terrorists, traitors, and the most violent criminals. The federal system does not have parole.

Sources: Lenzner, Robert. "Bernie Madoff's $50 Billion Ponzi Scheme." Forbes.com, December 12, 2008; "The Madoff Affair." *PBS Frontline.* May 12, 2009.; "Inside the Madoff Scandal." *The Wall Street Journal video,* March 13, 2009.

Video Link 9.3
View a video on Bernie Madoff.

In 1991, the Securities and Exchange Commission charged Melvin Ford and his International Loan Network based in Washington, D.C., with operating a pyramid scheme that defrauded more than 40,000 investors. Promising investors returns of 500–1,000 percent in as little as 180 days, the organizers charged clients $125 to $1,000 for initiation fees and promised them bonuses for anyone they recruited, setting up a pyramid. Later members could join additional income-opportunity programs. The International Loan Network had not been paying promised returns, and its only source of income was the continued recruitment of new investors (Salwen, 1991, p. B8).

Also in 1991, the Federal Trade Commission and the Securities and Exchange Commission began an investigation of Nu Skin International, which encourages pyramidal distributorships (over 100,000 in 1991), for possible pyramid scheme violations. If signing on new distributors becomes more lucrative and important than selling a product, a violation exists (Springer, 1991).

In 1999, William Koop ran a multimillion-dollar pyramid scheme that ripped off at least 80 investors in 28 states. One investor gave him $2.5 million and was promised an 800 percent return, or $20 million, in 45 days. Koop claimed that he had knowledge of super-secret bank trades and that the investments were guaranteed (Lefer, 1999). The only problem was that no such trades existed and the prime bank notes were fictitious financial instruments.

Religious Cons

Today, a new source of big money in professional crime appears to be burgeoning religious cults. Although most are probably sincere operations, a number appear to be interested in capturing the minds, bodies, and assets of their members. The son of L. Ron Hubbard (author of *Dianetics* [1963] and founder of Scientology, a pseudoreligious movement) claimed that the organization was simply a front or con for the private aggrandizement of Hubbard, who used most of the organization's money to buy drugs (American Broadcasting Company [ABC], 1983b). An IRS audit in the early 1970s proved that Hubbard had skimmed millions of dollars from the church, laundering the money through dummy corporations in Panama and then hiding it in Swiss banks. Hubbard employed his own private police force, the "Guardians Office," to attack and harass enemies and defectors from the organization. His son also claimed that the guardians on one occasion broke into an IRS office in an attempt to steal income tax records.

Televangelists such as W. V. Grant, Larry Lea, and Robert Tilton have hidden behind the constitutional right to religious freedom in their pocketing of immense amounts of tax-free funds. Grant raised $350,000 per month for an orphanage in Haiti but actually gave it between $2,000 and $4,000. Lea solicited funds for, and claimed to be building, a church at Auschwitz that was actually being built by another group. Tilton, the self-proclaimed "Apple of God's Eye," had run "preacher scams" since his days as a college prankster and was an associate of savings and loan scammer Herman Beebe (ABC, 1991).

In this era of Elmer Gantry evangelism, mystic cults, charlatans, channelers, spoon-benders, and faith healers, psychics are

> joined by people whose honesty is difficult to assess because they appeal to messages from supernatural sources to justify their asking for our money. . . . [One] evangelist . . . told Americans that God would "take" him to another world if he did not receive $4.5 million in donations by April, 1987. (Nettler, 1989, p. 73)

In 1986, magician James Randi exposed psychic spoon-bender Uri Geller as a fake and also debunked faith healer Peter Popoff as using the old trick of "calling out" to people in an audience and listing their names, occupations, ailments, and other surprising personal information. Randi showed an audience on the Johnny Carson show how Popoff simply used confederates beforehand to garner information and transmit it to him through a tiny earpiece (Randi, 1988). Randi (in Jaroff, 1988) reports,

> Popoff says that God speaks directly to him because he's an anointed minister. Three things amaze me about that. First of all, it turns out that God's frequency—I didn't know that he used a radio—is 39.170 MHz, and that God is a woman and sounds exactly like Popoff's wife, Elizabeth. (p. 72)

Affinity group fraud involves individuals claiming to be fellow members of the victim's religion, ethnic group, or professional group and claiming to want to help one of their fellow members. The victim is more likely to trust a person who is from the same group. In 1999, a trial of Tampa-based Greater Ministries International Church charged conspiracy, money laundering, and mail fraud in a massive Ponzi scheme that allegedly defrauded more than 17,000 investors of as much as $200 million. Many of the victims were fundamentalist Christians, including Mennonites. They were promised their money would double in installment payments made over 17 months or less. Investors were quoted Luke 6:38: "Give and it shall be given unto you." They were told their investments were "gifts" and payments to investors were "blessings" and not subject to taxes (Department of Corporations, 1999).

The PTL Scandal

In 1987, a major scandal broke, toppling the PTL (Praise the Lord) Ministry empire of televangelist Jim Bakker and his wife, Tammy Faye. The Bakkers were charged by the IRS with drawing $9.3 million in excess pay and by others (ABC, 1987) with perpetrating the biggest religious fraud in history, with as much as $100 million of the church's funds siphoned off for their personal use. Their posh lifestyle came complete with an air-conditioned doghouse. The last straw in the collapse of the Bakker operation was a sex scandal. Jim Bakker was removed from the ministry and defrocked (J. Carey, 1988). Crime File 9.4 describes some other examples of scams.

On a final note, Crime File 9.5 reports on the "Nigerian letter scam," a pervasive Internet crime.

Crime File 9.4

Emerging Patterns of Professional Crime

Boiler rooms, slammers, taps, mile busters, and dirt pile swindles are all part of the jargon of contemporary professional crime. Some of these activities include phony accident claims, art theft, boiler room frauds, the investigator scam, abusive tax shelters, oil and gas investment and lease lottery frauds, gold and silver investment frauds, business opportunity and franchise frauds, commodity and penny stock frauds, vacation time-sharing scams, precious metals financing programs, and "dirt pile" gold swindles (F. E. Hagan, 1991).

- Phony accident claims (insurance fraud) persist to this day. SEPTA (Southeastern Pennsylvania Transportation Authority) paid out $47 million in injury claims in 1990 (about 18 percent of fare collection). In one publicized bus crash, 11 "passengers" filed suit, even though the bus was unoccupied (Stieg, 1990b, p. 16A). In 1989, the estimated loss to insurance companies due to fraudulent claims was estimated at $17 billion ("Super Sleuths," 1991, pp. 3–4).

- Calling it "Showtime" or "Let's Go Make a Movie," "crashers" or "cappers" fake car accidents in California to the tune of an estimated $500 million a year in bogus insurance claims. Crashers use "hammers" (crash cars that will hit other cars). Some rings use Hollywood stunt drivers, as well as crooked doctors and "ambulance-chasing" lawyers. The doctors, lawyers, and cappers split the settlement three ways. In a recent and more deadly insurance scam, truckers on California freeways are set up for rear-end collisions by criminals seeking large insurance settlements (K. Hall, 1992).

- Art theft is now estimated by Interpol as the second largest international trafficking crime after drugs, and only 10 percent of such cases are ever solved (Plagens, Starr, & Robins, 1990, p. 50). Art theft internationally is estimated at $860 million to $2.6 billion (Del Piano, 1993). The International Foundation for Art Research began keeping track of art theft in 1976. By 1979, there were 1,300 recorded thefts, and by 1989 this number had risen to over 30,000 cases on file. Various groups, many of them made up of professional criminals, are involved. French police speculate that the Japanese Yakuza (organized crime groups) participated in recent French and U.S. thefts, as well as in networks organized by French and Italian antique dealers. They also claim that the international art, narcotics, and arms underworlds overlap (Dickey, 1989, pp. 65–66). Periodic explosions in art prices contribute to the increase in thefts as does the willingness of insurance companies to negotiate with thieves. Thieves have long looted archeological sites and sold their booty to a waiting international art market (Carley, 1991). Extortion and bribery are part of the world of art forgery. Even the sloppiest imitations find their way into auction houses (Carlisle, 1998). In one auction, the forged works were dated 4 years after the artist's death.

  - On March 18, 1990, the biggest art theft in history took place at the Isabella Stewart Gardner Museum in Boston. A team of robbers stole treasures valued in excess of $300 million. At 1:24 a.m., two men in uniform convinced a security guard to open the museum's side door and tricked him into summoning his partner. Both guards were handcuffed and their mouths were duct taped. Then the thieves roamed the museum stealing masterpieces. The crew used police-band radios, lookouts, transfer vehicles, and other signs of planning (R. Blumenthal, 1994). A provision to the 1994 crime bill makes art theft a felony offense.
  - Boiler room frauds are the basis of many schemes. *Boiler room* is slang for a rented office with banks of telephones operated by high-pressure operators or salespersons (called slammers in the trade) who solicit funds or tout products with outrageous promises. Such telemarketing fraud can catch the unwary by surprise (F. E. Hagan & Benekos, 1992, p. 6). These phone solicitors often use "sound-alike charities" or names for their organizations that sound very similar to known legitimate organizations. They use lists of suckers, or "taps" as they are called in the fundraising game, since they can be tapped again and again.
  - Selling bogus products and services via 900 phone numbers is a major new scam that began in the late 1980s. The primary aim is to keep the victim on the line for expensive charges for each minute of service. Phone sweepstakes, in which "winners" are encouraged to call a 900 number for details, and employment scams are current favorites. In the latter, customers are given information they could have obtained for free in the local newspaper. It turns out that, with few exceptions, 900 numbers are havens for rip-off artists.

Web Research Project
What are some new "scams" or new examples of scams that may not have been discussed in this chapter?

Crime File 9.5

Nigerian Letter Scams

"Nigerian letter scams" refer to international scams conducted by letter or, more often, e-mail requesting assistance in transferring money from Nigeria (or variations from other countries) to your country. For your trouble, the scammers offer the participant "advance fees" or "transfer taxes" to pave the way. Named "419 frauds" after the section of the Nigerian criminal code describing such fraud, the scam begins with the receipt of an unsolicited letter, fax, or e-mail containing an offer to assist in money laundering or a business proposal. Variations of the proposal are endless. The victim is next asked to provide an advance fee or to grant banking privileges in order to move the money. Additional fees are requested due to complications. Such fraud is one of the biggest businesses in Nigeria with much complicity from high government officials. This makes the recovery of funds lost in such schemes nearly impossible.

Visit the 419 Coalition Web site listed below for more information. Featured below is one variation of a Nigerian letter scam received by the author. The letter features errors as they appeared in the original.

Subject: complement of the season
Date: Tue. 6 Jan 2004
From: JESSICA SAVIMBI jessicasavimbi02@fsmail.net
To: undisclosed-recipients

Dear Friend,
I pray to God Almighty this message reaches you in wonderful spirit. How are you and your family?
I am Mrs. Jessica E. Savimbi, from Angola. I got your contact and profile W.W.W. and coupled with the information i gathered about you from the external trade department of my country [sic] chamber of commerce and industry in my country.
There is an information I would like to keep very confidential. There is sum amount of money my late Husband UNITA Rebel leader Jonas Malheiro.Savimbi deposited in a Bank in africa under a suspense account for safe keeping before he was killed by Angolan Military force. . . . The money in question is $4,500,000.00 U.S. DOLLARS.
What I would want you to do is to assists me to get the fund transfer into your account in your country for investment prior to your advise, and we shall open a small account in a Bank on your name, and transfer the money to your country, through the account. I will give you 20% of the money for your assistance. There is no risk in this transaction. I will use the remain balance of the money for an investment in your country for the future of my children. If you are interested, and can maintain the very confidential of this transaction, you e-mail me immediately through my email address for more clarification, and also note that I am a refugees in Nigeria, because of the killing of late husband,and the civil war going on in my country. Kindly reply by to my email address.
I can speak very little english, and my son also speaks english very well. Thank you very much.

Mrs. Jessica Savimbi

Source: Nigerian letter scam Web site: http://home.rica.net/alphae/419coal/

Web Research Project
Search for other articles and examples of Nigerian letter scams and 419 fraud.

Boosters

While con artists represent the "aristocracy of the professional criminal world," "boosters" (professional shoplifters) or "heels" represent the lowest class. Among professional criminals, boosting is viewed as requiring less skill or talent and thus enjoys less status. One confidence man said of a booster, "While he is undoubtedly a professional thief, I should have been ashamed to be seen on the street with him. . . . My reputation would have suffered in the eyes of my friends to be seen in the company of a booster" (Adams, 1976, p. 76). Mary Owen Cameron's classic, *The Booster and the Snitch* (1964), distinguishes between the booster (professional) and snitch (amateur) shoplifter. Boosters carefully plan their operations for big "scores" in order to minimize risks and to make sure they will be able to sell their booty to a fence (dealer in stolen property). Snitches, on the other hand, often commit their crimes on the spur of the moment, with little planning, and take enormous risks in order to "five-finger discount" relatively inexpensive items for their own personal use.

Boosters often rely on a variety of equipment and special paraphernalia such as "booster boxes" (boxes with slots or removable sections); special scissors or razor blades for removing labels; and special booster bags, coats, pants, and skirts with hidden compartments. Overweight shoplifters may employ the "crotch walk," in which goods are actually held between their legs and hidden by long coats or dresses.

Professional shoplifters usually work in groups, with each individual having an assigned role. The "stall" "throws a hump" or creates a commotion in order to attract the attention of the store personnel, while the "clout" steals the goods and possibly turns them over to a "cover," who may actually carry the booty out of the store.

In what sounds like Charles Dickens's *Oliver Twist,* a school for shoplifters, complete with a "how-to" manual, was discovered by police in New York City in 1987. About 75 boys ages 11 to 14 were believed to be involved. They were trained in avoiding security at suburban shopping malls and sent on expeditions with shopping lists in hand (A. Hamilton, 1987).

Since statistics on shoplifting are poor, it would be hazardous to estimate the proportion perpetrated by professionals. However, it is clear that professionals account for only a small segment. In 2007, two of the shopping industry's largest trade groups joined forces with the FBI's Gangs and Criminal Enterprises Division to create a database to track retail crime gangs (Mui, 2007). The retail industry estimates that it lost $37.5 billion to theft and fraud in 2005, a 20 percent increase from 2004. In a practice known as "e-fencing," gangs often sell their stolen property at 70 percent of their value on online auction sites. They travel the East Coast with U-Hauls full of stolen goods. Crime File 9.6 provides some further detail on the world of shoplifting.

Cannons

Professional **pickpockets** (sometimes called "**cannons**," "dips," or "picks") require exceptional dexterity and an awareness of the art of misdirection. Most pickpockets work with a "stall" who "puts up" (sets up) the mark. This is usually accomplished by tripping against, bumping, or otherwise distracting the subject, while the "tool" or "claw" or "mechanic" actually hooks or steals. Pickpockets usually work in groups of two, three, or four, with a specific role for each. One may select *marks,* another may locate the valuables or money on the person ("fanning") and maneuver him or her into position, and another pickpockets the item and passes it off to yet another (Inciardi, 1977, 1983, 1984).

According to Stirling (1974, p. 105), authorities on the subject claim that South Americans are the world's most skillful pickpockets. In the late eighties, some large airports in the United States were plagued by gangs of thieves who specialized in pickpocketing and stealing from luggage-laden travelers. Speculation was that the gangs were graduates of the infamous "School of the Seven Bells," a pickpocketing school in Colombia. In order to graduate, students must steal items from the instructor's coat, to which seven bells are attached. If a bell rings, the student flunks (Fry, 1986, p. 7D).

Since ancient times, pickpockets have worked crowds at parades, carnivals, sporting events, and the like. In medieval Europe, even during public executions of pickpockets, cannons worked

Crime File 9.6

Shoplifting

Goods Sold

Perhaps the principal factor determining a store's shoplifting rate is the type of goods sold. For obvious reasons, furniture stores have much lower shoplifting rates than, say, convenience or drug stores. The following table shows the most common items stolen in the United States. These include tobacco products (particularly cigarettes); health and beauty products (such as over-the-counter analgesics, decongestants, popular remedies, and birth control products); recorded music and videos; and apparel ranging from athletic shoes to children's clothing, with an emphasis on designer labels.

High-Risk Merchandise, by Store Type (based on items recovered from shoplifters)

Store Type	Merchandise
Auto parts	Small accessories, dash covers, compact disc players, driving gloves
Book	Cassette tapes, magazines
Consumer electronics/computers	CDs
Department	Clothing: shirts, particularly Hilfiger and Polo
Discount	Clothing, undergarments, CDs
Drug/pharmacy	Cigarettes, batteries, over-the-counter merchandise
General merchandise	Earrings
Grocery/supermarket	Over-the-counter remedies, health and beauty aids, cigarettes
Home center/hardware	Hand tools
Music	CDs
Shoes	Sneakers
Specialty	Bed sheets
Specialty apparel	Assorted clothes with designer labels
Sporting goods	Nike shoes
Theme park	Key chains, jewelry
Toy	Action figures
Video	Video games
Warehouse	Pens, movie videos

The acronym CRAVED captures the essential attributes of these "hot products": They are concealable, removable, available, valuable, enjoyable, and disposable. The last of these attributes, disposability, may be the most important in determining the volume of goods shoplifted. Those shoplifting for a living or to support a drug habit—who account for a disproportionate share of shoplifters—must be able to sell or barter what they steal. Unfortunately, little is known about the market for shoplifted goods.

Store Layout and Displays

Research provides little guidance, but common sense suggests certain store layout and display features contribute to shoplifting. Most of these relate to the staff's ability to supervise shoppers, and stores at greater risk include those with

- Many exits, particularly where they are accessible without passing through the checkout;
- Passageways, blind corners, and hidden alcoves;
- Restrooms or changing rooms;
- High displays that conceal shoppers (and shoplifters) from view;
- Crowded areas around displays of high-risk items; and
- Aisles that staff cannot easily survey from one end of the store.

Store areas at greater risk of shoplifting include, as mentioned, those with the most desirable goods. In addition, goods on the ground floor and near entrances are at greater risk of theft, because the shoplifter is in the store for less time and is thus at less risk of getting caught.

Source: "Shoplifting," by R. V. Clarke, September 8, 2003, *Problem-Oriented Policing Guide* No. 11, Washington, DC: U.S. Department of Justice.

the crowds. Derby Day, Super Bowl week, and world's fairs all attract a large influx of cannons. The next time you attend a large sporting event, see if, with the practiced eye of a bunko squad detective, you can spot the cannons in the crowd. Look for the people who are continually watching the crowd rather than the event. Unless they are security personnel, they may very likely be cannons attempting to "set a mark."

Related to, but less skillful than, pickpockets are "cutpurses," those who attempt to surreptitiously steal women's purses by cutting the purse straps. If such a theft involves rough stuff, such as shoving or physical force, the "cutpurse" has crossed the boundary from pickpocket to strong-arm robber or mugger. A related sneak thief is a "moll buzzer," who attempts to steal unattended purses in public places. The number of expert or class cannons has considerably declined from the thousands during the pre–World War II period to fewer than a thousand today (Inciardi, 1977, p. 21).

Professional Burglars

An example of professional burglary crews in operation was revealed when 22 were arrested in burglaries of Kennedy Airport cargo in 1994 in New York. Three major rings were stealing merchandise whose value was estimated in the tens of millions of dollars from air-cargo warehouses. Although they were not members of organized crime "families," the thieves paid tax or tribute to local organized crime groups (Firestone, 1994).

Shover (1973) interviewed 143 successful career burglars, some of whom were professionals. Such burglars typically work in groups, although there might be constant turnover in the members from job to job. Critical connections for a professional burglar's success are tipsters, fixers, and fences. *Tipsters* provide information on likely targets in return for a portion of the take, while *fixers* are attorneys and bondsmen who use bribery to fix or ward off prosecution. *Fences* or criminal receivers readily convert the burglar's booty into more portable cash. Many burglars rationalize their activity by claiming that most people are insured anyhow and that, when they read reports of their burglaries, the amount lost often is inflated by the victim in order to cheat the insurance company.

One of the more flamboyant professional burglars was a former paratrooper dubbed "Spiderman" (Derrick James), whose exploits resemble Hitchcock's 1950s classic, *To Catch a Thief,* starring Cary Grant. Without any equipment, he scaled luxury high-rises in South Florida and very selectively stole expensive jewelry. He seldom left the scene in disarray, and often the victims did not even know they had been burglarized until months later when they tried to find a particular item to wear. While 10 floors up was standard, he once broke in on the 30th floor (Pressley, 1998). He rated his own police task force, involving 25 police agencies. Upon his arrest in 1998, police contend he was responsible for over 100 burglaries worth over $8 million.

Picking a lock is just one of the many skills of an able burglar. The best lock pickers may practice daily and are the first to buy and master the latest "burglar-proof" locks when they become available. Some take locksmith correspondence courses advertised in magazines, and some are even licensed locksmiths. On entering premises, the skilled burglar will often stick a small object such as a matchstick into the lock so that it will jam if the occupants unexpectedly return and insert their key.

Plate (1975, p. 20) describes some burglaries of jewelry firms in Manhattan in the seventies that involved such feats as breaking through two concrete walls and opening two huge safes without leaving even a fingerprint. In one job, the front windows were sprayed with black paint, and in another the main cable serving a protection-service alarm system was cut, affecting thousands of Manhattan customers.

"Chic Conwell," the pseudonym for the professional thief in Sutherland's 1937 book, used the term "hotel prowler" to refer to burglars who specialize in stealing from hotel rooms. Such sneak thieves are particularly active in convention towns; they may pay off hotel employees, who act as accomplices. One hotel prowler told Plate (1975, p. 50) that, after first obtaining a master key from an accomplice, he would wait until 2:00 a.m., maintaining that at that hour few conventioneers were in their rooms sleeping and, if they were, they were so "bombed" they would not notice his presence. The only town he had problems in was Philadelphia, where he claimed the streets were so dead at night that conventioneers stayed in their rooms and drank.

The Box Man

At the top of the hierarchy of burglars are safecrackers or "box men." Chambliss's (1975a) edition of professional safecracker Harry King's autobiography, *Box Man* (reissued as *Harry King: A Professional Thief's Journal* [King & Chambliss, 1984]), reveals that King ranks professional safecrackers, although a dying breed, with the big con artists as high-status criminals within the professional criminal hierarchy.

Safecracking has engendered a constant escalation of technology, first to secure safes and second, in reaction, to develop better ways to open them. **Box men** are really professional burglars who specialize in breaking into safes. Between 1890 and 1940, professional burglary gangs flourished, hampered only by the newly developed burglar alarm. Telephones and automobiles were also beginning to narrow the apprehension gap (Rosberg, 1980, pp. 44, 52). As more and more sophisticated safes were developed, the methods employed to break into them improved. Since dynamite often damaged the safe contents, a core drill (a diamond-tipped construction device) provided more sophisticated means of entry, as did burning bars (oxygen lances that burn at temperatures up to 7,000 degrees Fahrenheit).

The Professional Fence

The dilemma of a thief who makes a "big score," but who lacks connections needed to dispose of the goods, was brought home to the author one evening in Cleveland while I was walking across a parking lot of a neighborhood shopping center. Two shady-looking characters blocked my path with their car, and the driver said, "Hey, sport, I have a bunch of cashmere sport coats in the back; and if we can find one that fits ya, I'll give you one helluva deal." Sure enough, glancing into the back seat I could see at least 20 boxes with a recognizable name in men's clothing on them. When I indicated a lack of interest, they shrugged, saying, "Suit yourself, sport," and drove off. Such amateurs without connections are not only in the business of stealing, but also in the even riskier business of soliciting unscreened customers in order to dispose of "hot" goods. More experienced and professional thieves would have quickly disposed of the property with a reliable fence or receiver of stolen property.

A **fence** is an individual who buys and sells stolen property. Legitimate operators of pawnshops, secondhand and antique shops, junkyards, and other general merchandisers may knowingly add stolen goods to their inventory, but a professional fence does this on a regular basis. J. Hall (1952) distinguishes between the "lay receiver" (customer), the "occasional receiver" (a rare buyer), and the "professional receiver" (a specialist in stolen property).

Professional burglars could not operate on a long-term basis without reliable relationships with fences willing to buy large quantities of stolen property on short notice. Klockars (1974) describes how "Vince Swaggi" (a pseudonym), the professional fence he studied, was able to sell a lot of factory seconds and other legitimate merchandise to customers who assumed the goods were "hot" (stolen) and thus a bargain. Similar findings caused Steffensmeier (1986) to entitle his case study *The Fence: In the Shadow of Two Worlds*. Fences and other professional criminals may also obtain a certain degree of immunity by acting as informants to the police. The importance of the fence to property criminals and sneak thieves is well illustrated by the relative success of police fencing sting operations that, after being in operation for only a short time, are able to arrest large numbers of thieves.

The classic professional fence was an Englishman by the name of Jonathan Wild, who operated in the early eighteenth century (Klockars, 1974). Wild placed advertisements in the newspaper and claimed that he was a "thief-taker," that he could recover stolen goods. He paid thieves higher-than-usual prices for their booty and then sold the goods to the victims at considerable profit. Wild was a "double dealer," building quite a reputation for turning in thieves as well as for fencing their goods. Finally, when some thieves accused him of being a fence, he was tried, found guilty, and hanged in 1725.

Blakey and Goldsmith (1976, pp. 1530–1535) identify four types of fences: the neighborhood fence, the outlet fence, the professional fence, and the master fence, in rising order of sophistication. The *neighborhood fence* is usually a small merchant who occasionally deals in stolen goods, while the *outlet fence* regularly sells "hot" merchandise along with legitimate stock. Using a legitimate company as a front, the *professional fence* is a major distributor of stolen articles. The *master fence* is involved at all levels, from organizing the theft to contacting customers in advance to distributing the goods; theft of art and museum masterpieces and their sale to wealthy private collectors may serve as one example that requires a master fence. Plate (1975, p. 65) tells us that most fences determine the price for items they buy and sell by using the Sears wholesale catalog or, if pressed, by calling the manufacturer directly, pretending to be an interested customer.

Paper Hangers

Paper hanging (passing bad checks and other documents) is a persistent form of professional crime. In the United States, cash is becoming the poor man's credit card. A larger proportion of transactions are conducted by means of checks and credit cards, which create a ripe situation for the forger. Lemert (1958) distinguishes between *naive check forgers* and professionals, or *systematic forgers;* the former are amateurs and only occasional offenders, while the latter make an illegal business of forging checks (see J. F. Klein & Montague, 1977). Lemert also found that forgers often operate independently and are less a part of the world of professional criminals than some other offenders. Sutherland (1937) claims that forgers and counterfeiters are considered marginal in professional crime, perhaps because such operators are often loners or *technicians* and thus do not share the professional criminal subculture (see Bloom, 1957).

Old-fashioned counterfeiting of money has not disappeared and may even be proliferating with recent technology—so much so that the U.S. government is considering changing the currency by, for instance, printing bills of different denominations in different colors. Many developed countries have based the design of their paper currency on that of U.S. currency. Color copiers, bleaching dollar bills and using the paper to print bills of higher denominations, and other ingenious methods have been used to make counterfeit money (Gladwell, 1990). In 1992, C-notes ($100 bills) began popping up around the globe, notes so authentic looking that they fooled currency-handling equipment at the Federal Reserve. Speculation has it that an unfriendly foreign government or a terrorist organization may have been the culprit (Wartzman, 1992).

Color copiers have accelerated the number of counterfeit bills in circulation. The amount has doubled every year and was estimated at $1.6 to $2 billion by the end of the 1990s. In order to fight counterfeiting, particularly *superbills* (near-perfect fake $100 bills), the U.S. Mint changed the $100 bill, the favorite of counterfeiters, first. These security features have since been incorporated into lower-denomination bills as well.

Crime File 9.7

Intellectual Property Theft

It's an age-old crime—stealing. But it's not about picking a pocket or holding up a bank. It's robbing people of their ideas, ingenuity, and creative expressions—what's called intellectual property, or IP. What does IP include? Things like music, movies, books, software, video games, even designer clothes and perfume. How important is IP to our economy? Consider this: IP is the single largest sector of the American economy, accounting for nearly 5% of the country's GDP, according to the International Intellectual Property Alliance. The U.S. leads the world in creating and exporting IP and IP-related products.

How big is the IP piracy problem? Big, and getting bigger all the time, thanks to electronic technologies like broadband, CD/DVD burners, MP3 recorders, and P2P digital file swapping networks on the Internet. And to add to the challenge, much of the theft takes place overseas, where laws are often lax and enforcement more difficult. All told, IP theft costs U.S. businesses upwards of $250 billion a year. And it robs the nation of hundreds of thousands of jobs and as much as a billion dollars a year in lost tax revenues. That's why the FBI, an international organization with agents stationed throughout the world, takes its responsibility to protect intellectual property very seriously. Case in point, literally. Recently, the FBI teamed with several Metro Atlanta Police Departments to shut down what has been called the largest counterfeit and pirated music CD operation in the southeast U.S. and the largest such ring ever broken up by law enforcement. Six defendants arrested in October, 2003, were charged with churning out 60,000 illegal CDs a week—well over 3 million in all. Who got burned? Some of the biggest recording artists around, including Britney Spears, Santana, Lenny Kravitz, Pink, and the Beastie Boys.

Headquartered in Atlanta, the operation allegedly included nine production plants, seven label production plants, a business office, and three storefronts in Macon and the Atlanta metro area. Once manufactured and packaged, the CDs were sold in bulk to wholesalers/dealers for $2 each and then sold to the public for about $5 at flea markets and other retail locations. The total cost of illegal CDs: about $50 million.

You can help. To report an IP violation or theft, contact the national Intellectual Property Rights Coordination Center, an interagency clearinghouse for combating IP crime run jointly by the FBI and Immigration and Customs Enforcement. You can file a complaint online at http://www.ice.gov/graphics/enforce/pr/iprform.htm.

Or call (202) 927–0810.

Source: Quoted directly from "Protecting America's Creative Genius," FBI Headline Archives, 2003. Available at http://www.fbi.gov/page2/dec03/ip122103.htm.

Web Research Project
Find some additional articles on counterfeit products.
Visit the FBI's Intellectual Property Crime Web site, http://www.CBP.gov/xp/cgov/newsroom/news_releases/archives.

Secret Service and CIA officials have speculated that a possible source of the "superdollar" may have been the government of Iran, which was sold the same intaglio printing presses used at the U.S. Mint before the fall of the Shah in 1979.

A growing area of professional fraud and counterfeiting relates to phony credit cards, records, tapes, and spare parts. The latter are produced in foreign factories and are reasonable facsimiles of the real thing. These pose some minor problems for those who purchase phony, inexpensive, "hot" watches from "Duke the Goniff" at the Greyhound station. They present major problems if they happen to be unsafe parts for airplanes, elevators, and manufacturing machinery.

Credit card counterfeiting was relatively unknown before 1990, and by 1998 was estimated to cost $1.5 billion per year (Wallace, 1998). Counterfeiters thwart each technological security advance in credit cards. Holograms were simply made by counterfeiters themselves, and embedded magnetic strips were defeated by "skimming" to read the information and then using laptop computers for

Crime File 9.8

Busting the Biggest Band of Cable Pirates in U.S. History

Consider: A New Jersey street vendor who sold hot dogs, relish, and "hot" cable box de-scramblers. A crook who "innocently" told our undercover agents, "The only way the FBI can get me is if they were standing right here, right now watching me do this." A corrupt lawyer who crashed his own car and even shot himself in the neck—just to get out of entering a plea in court. They're all part of "Operation Cable Trap"—a strange case from our files that smashed what turned out to be the nation's largest cable piracy ring ever.

At the center of it all was Joe Smith (an alias), a Florida crook with ties to the mob. For years, he sold illegal cable boxes that unscrambled signals and provided "free" cable service. His partners in crime? Corrupt cable company insiders who provided the boxes, other thieves who stole them, technicians who modified them, and crooks who sold them, to name just a few. We learned of the ring in 1991, when we got a tip about that duplicitous hot dog stand. To bust the ring, we created "Prime Electronics and Security Inc." in a 2,000 square foot warehouse in Kenilworth, New Jersey. We filled it to the brim with cable boxes and staffed it with undercover agents and a local police officer. We didn't have to wait long for our criminal customers to come calling.

To unearth elements of the scheme, we even agreed to help the crooks launder their illicit proceeds across the world. As a result, we soon turned up a massive money laundering operation based in the Cayman Islands, which led to another major investigation—"Operation Hot Money." Smith's pirating days came to an end in 1995, when he ordered up a thousand cable boxes from our "business." At the same time, agents in many other states executed 39 related search and seizure warrants on his criminal partners. Ultimately we seized $15 million in ill-gotten gains from the ring.

These cable pirates didn't have to walk the plank, but they did land in the brig. In the end, 34 members of the ring were convicted—including Smith, who pled guilty. The investigation also generated dozens of spin-off cases involving not just money laundering, but public corruption/bribery, insurance fraud, and bank fraud.

Source: Federal Bureau of Investigation (2005). Busting the Biggest Band of Cable Pirates in U.S. History. Retrieved March 27, 2007, from http://www.fbi.gov/page2/feb05/cabletrap020205.htm.

Web Research Project
Find some recent articles on "cable piracy." Are there any new wrinkles in this illegal activity or any new law enforcement strategies for dealing with it?

duplication. Chinese organized crime groups dominate the trade in California. According to Wallace, groups such as the Wo Hop To Triad, Wah Ching, and United Bamboo are active.

Law enforcement of the future is faced with increasing sophistication in this area. For example, how do businesses control checks that an hour or two after being cashed decompose and disappear? Advanced laser printers and color copiers make counterfeit documents such as checks, letterheads, and business cards more difficult to detect. In 1998, Canadian Mounties and the U.S. Secret Service busted a group with believed links to Russian and Asian organized crime groups that used scanners and inkjet printers to create over $2 million in Canadian $100 and $50 bills. They also used thermal, silkscreen printers to produce exact replicas of credit cards from six banks. Police were surprised that Canadian currency was being produced, since most false notes are usually American ("Mounties," 1998).

G. R. Gordon (1991) indicates that there is little or no international law dealing with international financial crimes. Many countries have bank secrecy laws. Most counterfeit credit cards are produced in Hong Kong, where it is not a crime to manufacture them. In other countries it is illegal to use, but not to possess, such cards. As Gordon notes, the need for multilateral agreements to fight such transnational crime is obvious. Malaysia, Thailand, and Hong Kong account for 44 percent of global credit

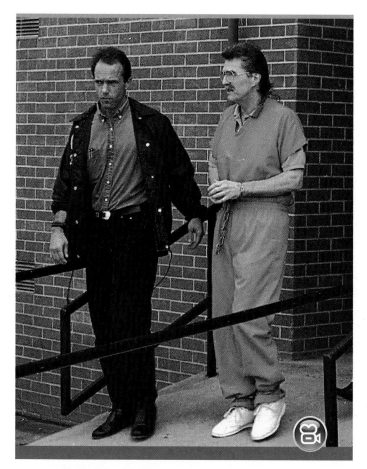

Photo 9.5

Patrick Mitchell is escorted by a U.S. Marshal. He claims he and his gang robbed over 100 banks across the United States and Canada since 1980.

card counterfeiting and are known as the "plastic triangle" (Duckworth, 1991). Crime File 9.7 reports on the burgeoning area of intellectual property theft.

Although they are very costly, enforcement officials are calling for the adoption of "smart cards," which have a small semiconductor memory circuit, but even these would only buy time until counterfeiters cracked the technology. In "Fraud Masters: Professional Credit Card Offenders and Crime," Jerome Jackson (1994) examines the working habits of a gang of credit card thieves. Part of their reason for choosing this form of offending was the knowledge that there was little risk of being caught, reported, or prosecuted.

A growing racket is that of "credit doctors" who sell clean credit references to people with bad credit (Reibstein & Drew, 1988). Using computers, credit thieves can steal your good credit and sell it, along with your social security number and credit references, to someone who has a similar name. Needless to say, your credit will not remain good for long.

"Video piracy," the massive production of fake videotapes, became a burgeoning industry during the late eighties. We are not talking here about people making tapes for their own personal use, but about well-organized syndicates that can often get "knock-off copies" to the market before the originals, causing the manufacturers to lose about $1 billion a year as a result. An estimated 15 percent of the movie videos on display in U.S. stores are illegal (Pauly, Friday, & Foote, 1987). Theft of cable services was estimated to cost U.S. cable companies $3 billion annually in the early 1990s. Crime File 9.8 gives an account of cable pirates.

Another example of the dangers of counterfeit products was the accusation in 1987 (J. Anderson & Van Atta, 1987) that bogus bolts made of cheaper alloys have either been found in, or are believed to be in, the nation's airplanes, buildings, bridges, nuclear power plants, and military hardware.

In 1873, archaeologist Heinrich Schliemann discovered the ancient remains of the fabled City of Troy. Many of the "treasures" found at the site were later determined to be hoaxes, bought from dealers or brought from other sites. Art is now a major item of investment, but "art forgery" is not clearly defined in the law (Haywood, 1987, p. 7). Interpol, the U.S. Treasury Department, and the New York Police Department's Fine Arts Squad estimate that some 10,000 stolen works of art are on the market. The identification of forgeries has found even major art museums deceived (Dutton, 1983; Savage, 1976).

Professional Robbers

Professional robbers differ from most other professional criminals in that they threaten, and are willing to use, force if necessary. Also, in contrast to others, they need little specific training or skill to be "stick-up artists." Professional "heavy" criminals (Gibbons, 1977, p. 2), such as robbers, tend to band together for particular jobs, but only on a short-term basis. This is in sharp contrast to the Jesse James–type gangs of the Wild West or the Depression-era Bonnie-and-Clyde-type groups.

In 1994, the clock finally ran out for Patrick Mitchell, head of a bank robbery ring known as the "Stopwatch Gang" or "the presidential robbers." The former name was given them because the gang tried to complete their jobs in 2 minutes; the latter name came from the masks of presidents they wore during robberies. This gang is estimated to have pulled more than 100 bank jobs since 1980 (Harrist, 1995). Also arrested that same year was Johnny Madison Williams, Jr., one of the most successful bank robbers in U.S. history. Responsible for 56 bank robberies, which he carefully documented in a hand-written log, Williams was known as "the Shootist" because he always fired shots in the air at the start of his robberies. The 8 years before his arrest constituted the longest unsolved string of bank robberies ever investigated by the FBI ("A Confession Ends Career of Robbery," 1994). The most successful type of robbery today is probably the hijacking of trucks (Glaser, 1978, p. 446). Writers such as Abadinsky (1983), Teresa (1973b), and Walsh (1977) have all found increasing cooperation between professional and organized criminals, particularly in the area of truck hijacking.

One major difference between amateur robbers and professional robbers is that the former tend to rob individuals while the latter tend to concentrate on commercial establishments. Letkemann (1973), based on his interviews with bank robbers, describes how many, in planning their jobs ("casing the joint" in gangster English), were aided by the fact that many branch banks were architectural clones of each other. Such similar layouts plus practiced impression management—a persona that robbers utilize to verbally intimidate bank personnel and customers—made most stick-ups routine. Robbers try to show they mean business in order to enhance cooperation. In short, no one gets hurt. Professional robbers, in contrast to most other criminals, do not require as much subcultural support from other professional criminals to acquire skill and technique or to plan and execute their operations.

Gangs of professional jewelry thieves called "the Colombians" stalk, set up, and rob members of America's jewelry industry. Many are from Colombia and five other Latin American nations, and they consist of 2,000 thieves organized in teams of 10 to 20 individuals. They particularly target salespeople on the road. The gangs are especially violent, and they survey and trail likely victims for days before attacking (Annin & Rhine, 1999).

Professional Arsonists

Most arson is committed by amateurs, individuals who do not make a career out of burning down structures. However, there are professional arsonists. In the late seventies, Morris Klein, a member of a ring of torch artists, boasted, "I can make concrete burn" (Karchmer, 1977). Klein's ring, which sold a complete package of arson services to businesses, was responsible for hundreds of fires in several states. For a percentage of the insurance settlement, he could mobilize a team of engineers, torches, and insurance experts. He was

> a fire broker who scouted around for troubled firms to sell. . . . If the business kept fumbling, Klein would approach the owner with an arson scam proposal. . . . He informed clients that their buildings were burning with the code message "The sky is red" (Karchmer, 1977).

Professional Auto Theft Rings

While most of us carefully guard and secure our valuables, one of our most expensive investments, our automobile, is often left unguarded on the streets or in a parking lot. While most automobiles are stolen by amateur juveniles for joyriding purposes, a significant number are stolen by auto theft rings and either chopped up for parts or refinished, complete with new papers and serial numbers, and sold

Journal Article Link 9.4
Examine literature regarding auto theft in New Jersey.

to a waiting market. While professionals organize such rings, the job with the greatest risk—actually stealing the cars—is done by young car thieves who may even be given shopping lists for specific makes or models (D. Savitz, 1959).

Some car thieves are professionals who possess standard burglary tools and master keys. Plate (1975, p. 28) indicates that Porsche master keys go for several thousand dollars. Master keys may be duplicated from those bought from showroom employees, for instance. Another tool is a "slam hammer" or "bam-bam" instrument, which is usually used to take dents out of cars. Thieves can use this tool to heist a car by inserting one end of the small hammer over the door lock of a car, which enables the entire door lock to be removed in seconds. Looking at the code number on the lock and using an auto code key book and key cutter, the thief can prepare the exact key for the auto in less than 2 minutes (pp. 29–30).

The thief's job ends when the car is left at the drop-off spot, usually a local shopping center. Runners or "gophers" transport the car to the shop. New plates can easily be obtained from states with lax inspection laws. Using a die tool, new numbers can be etched into the VIN number (vehicle identification number), which can be found die-cast on various parts of the car. Dishonest junkyard owners can also furnish registration cards to serve as false ownership credentials. Other auto theft rings operate "chop shops," where the stolen auto is immediately cannibalized for parts and sold to legitimate repair shops, which can then, because of low overhead, underbid competitors on repair work. Professional auto thieves vary greatly in their operations, sophistication, and organization.

In 1992, auto thieves in Dade County, Florida, claimed to be earning $15,000 per week stealing vehicles. Many of the vehicles not on their way to chop shops were driven onto container ships and sold for millions of dollars in Latin American countries such as the Dominican Republic, which is awash in luxury automobiles from the United States. Jail overcrowding often means that, when car thieves are caught, few get jail time ("Auto Thieves," 1992). A stolen vehicle will often net double its price overseas, and auto theft is frequently viewed as victimless since it is usually covered by insurance. "The low apprehension, prosecution, and conviction rate of auto thieves makes this crime a booming industry, with high profits and low risks" (Beekman & Daly, 1990, p. 16). When federal authorities were able to successfully bust these container ship operations, the thieves shifted gears. They exported legitimate cars, removed their VIN numbers when they arrived at the foreign port, and sent these back to the United States to be placed on "born again" stolen vehicles (p. 17).

International auto theft rings are estimated to operate an $8 billion industry increasingly associated with crime syndicates (Ragavan & Kaplan, 1999). Of the 1.4 million stolen vehicles in 1997, at least 200,000 ended up overseas. Vehicle theft is one of the lowest crimes on the law enforcement priority list. There is only a 14 percent arrest rate for auto theft, the lowest of any crime category. Mexico is the principal port of destination for stolen U.S. autos. In 1997, the cities with the highest auto theft rates were Miami, Jersey City, Fresno, Memphis, and New York. European experts document similar activity in Europe and even claim that it is now as big an international problem as drug smuggling (August, 1997). Russian and Chechen syndicates transport stolen vehicles to the former Soviet Union. In some cases, owners of the vehicles sell them to the thieves and then report them as stolen in order to collect the insurance. Crime File 9.9 reports on the latest form of car theft—car cloning.

Professional Killers

Professional assassins, "hit men," are popular subjects of fiction and, undoubtedly, a few do in fact exist, for instance, in the shady world of international espionage. Most organized crime executions appear to be assigned to members in addition to their ordinary tasks. Even members of Murder Inc. (the Brownsville Boys, an organized crime group of the 1920s to 1940s) did not spend all their time performing executions. The literature is either scant or unreliable in providing an accurate picture of professional killers.

Crime File 9.9

Car Cloning: A New Twist on an Old Crime

A gang of professional car thieves trolls the parking lot of an upscale retail mall until they find what they want: a shiny new SUV. Within seconds they've disabled the alarm, hot-wired the engine, and driven away.

Now, how to sell the car for big bucks?

They COULD do the traditional . . . and sell it for parts or as hot merchandise. But that would only bring in a few thousand dollars. Instead, they turn to the most lucrative scam on the block that will help fetch top dollar for their prize.

Here's how cloning works:

After leaving the mall, the thieves head for a neighboring state. They seek out a large car dealership and look for a car that's the exact make and model (and even the same color) of the stolen one.

Then, they jot down the vehicle identification number (or VIN) stamped on the top of the dashboard and drive off.

Later, they make an exact replica of the VIN tag, pull the old tag out of the car, and pop in the new one.

Voilà, a clone is born: two identical cars, one identification number.

Now, one final step—the thieves use a little forgery to get a real title or other ownership documents from the motor vehicle office in the neighboring state. Then, it's no problem to sell the vehicle to an unsuspecting victim for nearly full price. And since it's legally registered and not reported stolen, it's nearly untraceable.

"Right now, cloning is by far our biggest car theft challenge," says Supervisory Special Agent Ryan Toole, who leads our vehicle theft program at FBI Headquarters in Washington. "The good news is, it's preventable."

Here's how: by linking motor vehicle departments in every state. "If states could talk to one another electronically, you couldn't register a car in Maryland, for example, using a VIN from a car in Virginia," says Toole. "The system wouldn't allow it."

That's why we're on something of a crusade—working hand-in-hand with the American Association of Motor Vehicle Administrators—to get all 50 states and the District [of Columbia] to participate in the "National Motor Vehicle Title Information System" that would create such linkages. So far, only a handful of states are connected.

It's difficult to estimate how many cars are cloned, but we know it's a significant percentage of the 1.3 million cars stolen in the U.S. each year. Overall, the total price tag for auto theft is about $8 billion annually. One in three stolen cars never makes it back home.

In addition to tackling cloning, we continue to work with our local and state partners on auto-theft task forces that focus on dismantling larger rings, from the street level to the upper echelons of the criminal hierarchy.

"We're never going to stop the small-time thieves who just want a joyride," Toole says. "But we can make and are making a difference in taking down the big-time syndicates."

Source: Quoted directly from "Car Cloning," *FBI Headline Archives,* 2007. Available at http://www.fbi.gov/page2/march07/carcloning032907.htm.

Criminal Careers of Professionals

Reviews of the use of the term *professional criminal* in the field of criminology (Staats, 1977; Winslow, 1970) point out the heavy reliance on case studies and popular sources. Anthologies such as Bruce Jackson's *In the Life* (1972) and Duane Denfield's *Streetwise Criminology* (1974) are illustrative. These provide firsthand accounts, primarily by incarcerated criminals, of their lives in crime. While these are revealing, it is unclear how typical such accounts may be. Given this methodological limitation, much of the description of criminal careers of professional criminals is limited and certainly requires more investigation.

Indicating that professional criminals may not be specialists in any one area of crime, Plate (1975) identifies 10 characteristics of professional criminals:

1. They seek anonymity.
2. They are often on speaking terms with police as informants, as bribers, or simply as those who work in a related area.
3. They are not necessarily members of organized crime, although they cooperate in some cities.
4. They are usually not drug addicts.
5. They take arrest and imprisonment in stride, often putting money away for a rainy day.
6. They do not leave fingerprints.
7. When possible, they will run through a crime (practice it) beforehand.
8. They are well aware of the law and police clearance rates.
9. Most avoid gaudy display or conspicuous consumption.
10. Many are stable family members. (pp. 7–10)

While some professional criminals, such as hired killers or professional robbers, are into "heavy" crimes, most attempt to avoid rough stuff; to avoid "heat"; to operate through wit, guile, cunning, technical skill, and "grifting." Most professional criminals look with disdain on the tactics and sense-less violence of amateur criminals. Professionals plan and carefully choose their victims in order to maximize the score and minimize risks.

A criminal does not usually simply decide one day that he or she is going to be a professional crimi-nal. Recognition, skill, and contact with other hustlers and professionals are prerequisites; without this contact, the required knowledge and experience for a successful move into professional crime is less likely. Recruits into the world of professional crime may come from the ranks of hotel workers, waitresses, and cab drivers, as well as of pimps, fences, and promising conventional property crimi-nals. While in earlier times, professional training schools such as Fagin's in *Oliver Twist* did in fact exist, today the training appears to be much more informal, although the nation's prisons seem to be a major training ground for some.

A leading explanation of professional criminality is Sutherland's (Sutherland & Cressey, 1978, pp. 80–83) "differential association" theory. In explaining patterns of professional criminality, the the-ory points to criminal contacts (values and attitudes) as essential to the learning process. Some profes-sional criminals, particularly cannons, con artists, and professional burglars, participate in an informal apprenticeship of jobs, learning very specific skills and making the indispensable contacts with fixers and fences without which they would have great difficulty in operating. Letkemann (1973) sees this "crime as work" orientation among professional criminals as involving not only the learning of technical skills, but, just as important, social and organizational skills such as victim management. A vital com-ponent of professional criminality is the shared subculture that requires frequenting common haunts (bars, restaurants, and the like) in order to discover "what's going down" or "what's happening."

Compared with other categories of criminal activity, professional crime is rare and perhaps becom-ing rarer. Generally, most professional criminals come from better economic backgrounds than con-ventional or organized criminals. Many begin their careers at a later age. This varies, however, with the area of criminality. Safecrackers and bank robbers, for instance, appear to require early juvenile crime experience (Conklin, 1981, p. 265). Maurer (1964) describes the professional criminal as one who approaches crime in a businesslike manner, expecting to earn his or her living from it. Professional criminals know and are known to other professional criminals. Such criminals highly identify with criminal activity and are proud that they are good at their work. Many professional criminals rational-ize their activity, feeling that all people are crooked or involved in what Al Capone called "the legiti-mate rackets." As Mel Weinberg, the Abscam consultant, put it, "I'm a swindler . . . the only difference between me and the congressmen I met on this case is that the public pays them a salary for stealing" ("The Man Behind Abscam," 1980, p. 1A). Con artists justify their behavior on the basis of the dishon-est behavior of many of their victims, who may be trying to avoid taxes or buy stolen goods.

Informants from the ranks of professional crime as early as the time of Sutherland's (1937) "Chic Conwell" have indicated that, while cannons tend to restrict their criminal activities to their specialty of pickpocketing, most others "hustle" or engage in a variety of offenses, even though they may prefer their specialties. Boosters and paper hangers appear to be more similar to cannons in attempting to stick to their specialties.

Societal Reaction

Most experts on the subject see professional criminality as declining since its heyday during the Depression (Inciardi, 1975; J. F. Klein, 1974; Shover, 1973). This decline may simply represent an increase in semi-legal enterprises (Roebuck & Windham, 1983). Bank robbery for the most part has passed into the hands of amateurs. There appear to be fewer big-time con artists around than previously. Pickpockets, although still around, have been replaced by muggers. Female involvement in bunko operations has increased considerably in the post–World War II period. Inciardi (1975) suggests that the decline of professional crime began in the 1940s with the application of modern communication and scientific methods to the field of criminal investigation. Computerized information, fingerprints, regional cooperation in law enforcement, and greater professionalization of criminal justice "raised the ante" for a career in crime.

Other observers (Chambliss, 1975a; F. E. Hagan, 1991; Staats, 1977) see professional crime not as declining, but as shifting into other areas of operation. The "professional street crime" characteristic of an industrial society has given way to "white collar professional crime." While professional burglars and pickpockets have declined in numbers, there has been an increase in the number of sophisticated con artists. In a paper titled "The Ghost of Chic Conwell: Professional Crime and Fraud in the Twenty-First Century" (F. E. Hagan, 2000), this writer expressed the opinion that Chic Conwell (professional crime) is not dead but has moved into cyberspace.

As previously described, most professional criminals attempt to commit crimes that are difficult to track, and their operations are often characterized by use of specialists for each element of the job. Since most criminals, even the best, are eventually caught, the more sophisticated professionals will attempt to forestall action by victims or the criminal justice system by "putting in **the fix**"—gaining the cooperation of corrupt officials: crooked judges, court administrators, lawyers, or police officers. The latter, for instance, could convince victims of the futility of attempting to proceed with a case and offer immediate compensation. According to the President's Commission on Law Enforcement and the Administration of Justice (1967), two essential elements that explain the success of professional crime are "the fix" and "the fence" (p. 154). Yet another element should be added—a steady demand for stolen goods. Without a ready market, much professional theft would dry up.

The traditional public view of burglary and thievery is that the perpetrators of these crimes are unorganized and are reduced to crime to support a drug or gambling habit. However, 90 percent of the dollar value of objects taken is accounted for by crime rings, although those rings are responsible for only 10 percent of the incidents of burglary (Pennsylvania Crime Commission, 1980, p. 167).

While the very nature of postindustrial society and its reliance on banking, international trade, computers, and instantaneous communications have created fertile ground for fraud and professional crime, efforts in public and private investigation have begun to meet the challenge. Most noteworthy in this regard are beefed-up efforts by the Federal Bureau of Investigation, the National White Collar Crime Center, and the Association of Certified Fraud Examiners.

Most discussions of white collar crime by organizations such as the U.S. Chamber of Commerce, or federal law enforcement agencies such as the FBI, concentrate on areas we have discussed in this chapter as professional crime, particularly the work of confidence artists. This would suggest that the most serious white collar crimes are committed by "hustlers," "fast buck artists,"—actors who are clearly unacquainted with standard business practices. In reality, such operations, although a serious concern, are relatively minor compared with pervasive and economically more costly operations that are incidental sidelines of legitimate business enterprise, where many of the same tactics are employed.

Theory and Crime

Property crime in general involves the illegal pursuit of things of economic value. Separating professional property crime from occasional and conventional property crime is a matter of the skill involved rather than the act itself. Routine activities theory is a particularly useful model for explaining shoplifting. Using the three elements of the approach—likely criminals, likely targets, and guardianship—security managers can explain changes in rates of shoplifting. Vandalism and graffiti have a number of manifestations, but programs designed to reduce targets and increase guardianship have been quite successful. One example was the virtual disappearance of graffiti on the New York City subway system. Better security devices (guardianship) on motor vehicles precipitated declines in auto theft rates.

Much conventional property crime, such as burglary and larceny, peaks in a person's teens and rapidly declines after individuals reach their mid-twenties. Developmental/life course theories examine the trajectories over time of such criminality, looking at the onset, persistence, and desistance of criminal careers.

Illustrated in 2009 by the Bernie Madoff affair, professional criminal activity involves some of the more costly economic crimes. Anomie theory and institutional anomie theory argue that the emphasis on material success to an extreme without the provision of legitimate opportunities for their achievement contains the roots of deviance and crime. Professional criminal activity can also be examined by means of differential association theory wherein individuals learn and become socialized to criminal activity.

Summary

Offenses against property, among the earliest to be punished under formal legal systems, include a wide variety of violations usually labeled larceny (theft). These offenses can be committed by a variety of criminal types, two of which are discussed and contrasted in this chapter: occasional property criminals and conventional property criminals. *Career criminality* is characterized by identification with crime, criminal self-concept, group support, association with other criminals, progression in criminality, and crime as a sole means of livelihood. While occasional property offenders are the antithesis of career criminals, conventional property violators are on the bottom rung of the ladder of career criminality. Occasional property offenders commit their crimes relatively infrequently, irregularly, crudely, and without identifying themselves as criminals. Conventional criminals commit their offenses more regularly and tend to aspire to career criminality.

Occasional property criminal behavior includes most, but not all shoplifting, vandalism, motor vehicle theft, and check forgery. Cameron distinguishes between two types of shoplifters: *boosters* (professionals) and *snitches* (amateurs). The majority of snitches have no previous criminal history, do not identify with criminality, and are deterred from future activity when threatened with formal legal processing. *Vandalism,* the willful destruction of the property of others, has been identified by Andrew Wade as consisting of three types: wanton (senseless), predatory (criminal), and vindictive (hateful). Wanton vandalism by juveniles is the most common type and usually represents an extension of play activity. Motor vehicle theft also consists of a variety of types: joyriding, short-term transportation, long-term transportation, and profit. Joyriders, who borrow a car for temporary adventure, illustrate well the occasional property criminal.

Check forgers have been distinguished by Lemert as consisting of two types: naive check forgers and systematic check forgers. The former, who are occasional property criminals, write bad checks as a means of resolving a temporary crisis. The bad check writing is a result of closure, or limited possibilities for solving this problem.

Conventional property criminals are those who commit larceny-theft and burglary on a fairly persistent basis, constituting a rudimentary form of career criminality. Such offenders are less skilled and organized than their professional counterparts and represent about half of the prison inmates in the United States. Most will eventually reduce or cease their "careers" by their mid-twenties. *Burglary* involves the unlawful entry of a structure in order to commit a felony or theft. This may include actual forcible entry, unlawful entry without force, or attempted entry. As a rule, burglars attempt to avoid violence. Walsh identifies types of burglars. These include professionals, known burglars, young burglars, juvenile burglars, and junkies, in decreasing order of sophistication and organization. Other characteristics of burglars and burglary were described separately by Scarr, Repetto, and Pope. Indispensable to property criminals and particularly burglars is the *fence*, a dealer in stolen property. Successful police "stings," or antifencing operations, were described. Most conventional offenders are nonspecialists; they "hustle," or take advantage of various criminal opportunities.

Larceny-theft, which includes a broad category of property crimes, makes up over half of the index offense total and as a category constitutes a "wastebasket concept," a catch-all. Property offenses are more characteristic of youthful offenders, who tend to commit crimes in groups. Arson, which has since 1979 been included as a UCR index offense, involves any willful or attempted malicious burning of another's property. Arson is described as a special-category property offense because of the variety of motivations involved, including (according to McCaghy's typology) profit, revenge, vandalism, crime concealment, sabotage, and excitement.

Comparisons of criminal careers of occasional versus conventional property criminals demonstrate that only the latter exhibit any level of commitment to criminality, and even they are often youthful offenders whose property criminality peaks at age 16, halves by age 20, and continues to decline thereafter. Societal reaction to occasional offenders is relatively mild, but it is relatively strong against conventional property offenders. Anthropological field research, such as that by Shover, suggests that programs aimed at identifying and getting tough with "career criminals" must be careful to examine the interplay among employment, threatened incarceration, and aging of offenders lest they get tough at the very time that most will mature out of crime.

In sociology, *professionals* are those in occupations who possess useful knowledge and claim a service orientation for which they are granted autonomy. In this light, the term may be an inappropriate tag with which to designate skilled, able grifters or intensive career criminals. It is so widely used in the literature, however, that not to use the concept would be more confusing than to employ it. Sutherland's classic 1937 work on the subject, *The Professional Thief*, describes some characteristics of professional criminals as including crime as sole livelihood, planning, technical skills, codes of behavior, high status, and an ability to avoid detection. Professional crime is a sociological rather than a legal entity.

The *argot* (specialized jargon) of the professional criminal world uses Depression-era U.S. terms. Some examples include cannons (pickpockets), heels (sneak thieves), boosters (shoplifters), and the con (confidence games). A continuum model of professional crime presents crime as professional to the degree that it possesses the following characteristics: sole livelihood, extensive career, skill, high status, avoidance of detection, criminal subculture, planning, and "the fix." *The fix* refers to the ability to avoid prosecution by compromising the criminal justice process. The term *scam* refers to various criminal techniques or hustles. Professional crime differs from occupational/corporate crime in that, in the former, crime is the sole purpose of a business. Some examples of professional crime from Edelhertz's typology were presented, most of which are examples of fraud.

Some professional crime might be described as semiprofessional in that it involves less skill and planning. Sometimes called *bunko* or *flim flam* or *short con* operations, these scams include the pigeon drop, the badger scam, the bank examiner's scam, postal frauds, circus grifting, boojo (a gypsy con game), and various home-improvement frauds.

The *big con* involves far more skill, more elaborate planning, higher-status victims, and much larger rewards for the criminal. *Ponzi schemes* are frauds in which early investors in a nonexistent product are paid high dividends on the basis of money obtained from later investors. *Pyramid schemes* require investors to seek a chain of other investors in order to reap a promised high return. Examples of big con operations also included religious cons.

Various professional criminal trades were discussed in the chapter. These included boosters, cannons, professional burglars, box men, fences, paper hangers, robbers, arsonists, and auto thieves.

Descriptions of careers of professional criminals are methodologically limited by the need to rely on case studies and popular sources for many accounts. Most professional criminals seek anonymity, know the police and members of organized crime, are very deliberate in plying their trade, and avoid conspicuous consumption. They avoid rough stuff and "heat" and attempt to minimize risks. Requiring skill and contact with others, most seek subcultural support as suggested in Sutherland's "differential association" theory.

The professionalization of criminal justice has appeared to reduce many of the previous opportunities available in professional crime. The President's Commission on Law Enforcement and the Administration of Justice pointed to the importance of two essential elements that explain the success of professional crime: "the fence" and "the fix." The high cost of legal defense also may be responsible for a portion of the decline of such crime.

Criminology on the Web

Log on to the Web-based student study site at http://www.sagepub.com/haganintrocrim7e/ for author-created podcasts, e-flashcards, quizzes, and more.

KEY CONCEPTS

Argot	Conventional Criminality	Pigeon Drop
Arson	Fence	Ponzi Schemes
Badger Game	Fix, The	Profession
Bank Examiner's Scam	Heel	Professional Crime
Boojo	Larceny-Theft	Pyramid Schemes
Booster	Model of Professional Crime	Scams
Box Man	Motor Vehicle Theft	Shoplifting
Burglary	Naive Check Forgers	Snitch
Cannon	Occasional Property Criminality	Systematic Check Forgers
Career Criminality	Paper Hanging	Types of Arson
Characteristics of Professional Crime	Pennyweighting	Types of Vandalism
Confidence Games		

REVIEW QUESTIONS

1. Compare and contrast occasional property criminals with conventional property criminals.

2. What are some different types of shoplifting? What are some trends in shoplifting?

3. What are the three types of vandals and how do they differ in their motivations?

4. What are some trends and patterns in auto theft both nationally and internationally?

5. Discuss international patterns in burglary. Why do you think property crimes are declining in the U.S.?

6. What are some trends and new wrinkles in the criminal field of "paper hanging"?

7. How do the operations of professional robbers differ from those of amateurs?

8. What are "affinity frauds" and how do they operate?

9. Distinguish between boosters and snitches.

10. Discuss the operations of professional burglars. What are some of their critical connections and typical operations?

Association of Certified Fraud Examiners
http://www.cfenet.com

Coalition Against Insurance Fraud
http://www.insurancefraud.org

Department of Justice's Computer Crime and Intellectual Property Section
http://www.usdoj.gov/criminal/cybercrime/index.html

International Association of Auto Theft Investigators
http://www.iaati.org

James Randi Educational Foundation
http://www.randi.org/

National Check Fraud Center
http://www.ckfraud.org/

Ponzi Schemes
http://www.businesspundit.com/the-10-nastiest-ponzi-schemes-ever/

Pyramid Scheme Alert
http://www.pyramidschemealert.org/

U.S. Chamber of Commerce
http://www.uschamber.com/issues/accomplishments/ip.htm

Using this chapter's recommended Web sites, explore the area of property crime.

1. What did you learn from the James Randi educational foundation site?

2. What is the Department of Justice doing about computer crime and intellectual property theft?

3. How does the National Check Fraud Center work?

4. What types of services are offered by the Association of Certified Fraud Examiners?

5. Using your Web browser, search the areas of "private security" and "internet scams."

Mary Owen Cameron. 1964. *The Booster and the Snitch.* New York: The Free Press.
> This is the classic study of shoplifters using a secondary analysis of the records of retail theft of private department stores.

James Inciardi. 1970. The Adult Firestarter: A Typology. *Criminology, 8,* 145–155.
> An excellent typology and analysis of arsonists.

Darrell Steffensmeier. 1986. *The Fence: In the Shadow of Two Worlds.* Totowa, NJ: Rowman & Littlefield.
> Sociologist Darrell Steffensmeier's case study of the world of professional fence "Sam Goodman."

Kenneth Tunnell. 2000. *Living Off Crime.* Chicago: Burnham.
> Kenneth Tunnell's in-depth interviews of jailed property criminals opens the door to a little-explored world.

Richard Wright and Scott Decker. 1994. *Burglars on the Job.* Boston: Northeastern University Press.
> Wright and Decker interview burglars in the field to gain information on their careers, modus operandi, and motivations.

Harry King and William Chambliss. 1984. *Harry King: A Professional Thief's Journal.* New York: Wiley.

> This update of *Box Man* is a case study of a safecracker, Harry King.

David W. Maurer. 1964. *Whiz Mob.* New Haven, CT: College and University Press.
> Linguist Maurer examines the argot, roles, and lifestyles of professional con artists.

David Simon and Frank Hagan. 1999. *White Collar Deviance.* Boston: Allyn & Bacon.
> Chapter 5 looks at the world of professional crime, defined as professional white collar deviance.

Edwin H. Sutherland. 1937. *The Professional Thief.* Chicago: University of Chicago.
> Sutherland's account of the professional criminal world of "Chic Conwell" is the beginning point in the analysis of professional crime.

Richard T. Wright and Scott H. Decker. 1997. *Armed Robbers in Action: Stickups and Street Culture.* Boston: Northeastern University Press.
> This is one of the few studies of unincarcerated robbers.

White Collar Crime—The Classic Statement
  Related Concepts
The Measurement and Cost of Occupational and
Corporate Crime
The History of Corporate, Organizational, and
Occupational Crime
Legal Regulation
  Occupations and the Law
  Organizations and the Law
Occupational Crime
  Crimes by Employees
  Crimes by Employees Against Individuals (the
  Public)
  Crimes by Employees Against Employees
  Crimes by Employees Against Organizations
  Crimes by Individuals (or Members of Occupations)
Corporate Crime
  Crimes by Organizations/Corporations Against
  Individuals (the Public)
  Crime File 10.1 Financial Crimes: FBI Releases
  Annual Report to the Public
  Crime File 10.2 The Great Savings and Loan Scandal:
  The Biggest White Collar Crime in U.S. History
  Crime File 10.3 The Donora Fluoride Death Fog:
  A Secret History of America's Worst Air Pollution
  Disaster
  Crimes by Organizations Against Employees
  Crimes by Organizations (Corporations) Against
  Organizations
  Crime File 10.4 Pirates of the Internet: Criminal Warez
  Groups
Criminal Careers of Occupational and Organizational
Offenders
  Corporate Environment and Crime
  Corporate Concentration
  Rationalizations
Societal Reaction
  Why the Leniency in Punishment?
  Theory and Crime
Summary
Key Concepts
Review Questions
Web Sources
Web Exercises
Selected Readings

chapter 10

White Collar Crime

Occupational and Corporate

Seldom do members of a profession meet, even be it for trade or merriment, that it does not end up in some conspiracy against the public or some contrivance to raise prices.

—*Adam Smith (1776/1953, p. 137)*

We have no reason to assume that General Motors has an inferiority complex or Alcoa Aluminum Company a frustration-aggression complex or U.S. Steel an Oedipus complex or Armour Company a death wish, or that Dupont wants to return to the womb.

—*Edwin H. Sutherland (1956a)*

The best way to rob a bank is to own one.

—*William Crawford, California Savings and Loan Commissioner (quoted in Pizzo, Fricker, & Muolo, 1989, p. 318)*

White Collar Crime—The Classic Statement

Although previously discussed in the popular literature, the concept of **white collar crime** was first introduced in the social sciences by Edwin Sutherland in a 1939 presidential address to the American Sociological Association. Defining white collar crime as "a crime committed by a person of respectability and high social status in the course of his occupation" (Sutherland, 1940), his address was important in that it was the first major statement on white collar crime in academic criminology. Volk

Handbook Article Link 10.1
Read an article on
white-collar crime.

(1977) describes Sutherland's pioneering effort as "the sign of a Copernican revolution in Anglo-Saxon criminology" (p. 13), a radical reorientation in theoretical views of the nature of criminality. Mannheim (1965) felt that if there were a Nobel Prize in criminology, Sutherland would deserve one for his effort. It certainly represented, to use Kuhn's (1962) notion, "a paradigm revolution," a new model that served to radically reorient future theoretical and empirical work in the field.

Sutherland's (1949) investigation using records of regulatory agencies, courts, and commissions found that of the 70 largest industrial and mercantile corporations studied over a 40-year period, every one violated at least one law and had an adverse decision made against it for false advertising, patent abuse, wartime trade violations, price fixing, fraud, or intended manufacturing and sale of faulty goods. Many of these corporations were recidivists with an average of roughly eight adverse decisions issued for each. On the basis of his analysis, it becomes obvious that, although he used the general label *white collar crime,* Sutherland was in fact primarily interested in organizational or corporate crime.

Sutherland maintained that while "crime in the streets" attracts headlines and police attention, the extensive and far more costly "crime in the suites" proceeds relatively unnoticed. Despite the fact that white collar crimes cost several times more than other crimes put together, most cases are not treated under the criminal law. White collar crime differs from lower-class criminality only in the implementation of criminal law that segregates white collar criminals administratively from other criminals (Sutherland, 1949). Furthermore, white collar crime is a sociological rather than a legal entity. It is the status of the offender rather than the legal uniqueness of the crime that is important (Geis, 2007).

The hazard of identifying white collar crime simply by official definitions is demonstrated by Hirschi and Gottfredson (1987, 1989). They (erroneously) dispute the usefulness of the label "white collar crime" because the four UCR measures of white collar crime (fraud, embezzlement, forgery, and counterfeiting) that they measure show that most offenders are middle class and differ little from traditional offenders. Steffensmeier (1989b) correctly responds that UCR offense categories are not appropriate indicators of white collar crime.

Related Concepts

One of the earliest scholars to discuss types of behaviors that later would be described as "white collar crime" was Edward Ross (1907) in an article that appeared in the *Atlantic Monthly.* Borrowing a term used by Lombroso (Lombroso-Ferrero, 1972), Ross refers to **criminaloids** as "those who prospered by flagitious [shameful] practices which may not yet come under the ban of public opinion" (A. E. Ross, 1907, p. 46). Describing the criminaloid as "secure in his quilted armor of lawyer-spun sophistries" (p. 32), Ross views such offenders as morally insensible and concerned with success, but not with the proper means of achieving it. C. Wright Mills (1952) uses a similar notion, "the **higher immorality,**" to characterize this moral insensibility of the power elite. Mills felt this was a continuing, institutionalized component of modern U.S. society, involving corrupt, unethical, and illegal practices of the wealthy and powerful.

A variety of other terms have been proposed as substitutes, synonyms, variations, or related terms for white collar crime, including "avocational crime" (Geis, 1974a), "corporate crime" (Clinard & Quinney, 1986), "economic crime" (American Bar Association, 1976), "elite deviance" (D. R. Simon, 1999), "the criminal elite" (Coleman, 1994), "occupational crime" (Clinard & Quinney, 1986; G. Green, 1990), "organizations crime" (Schrager & Short, 1978), "professional crime" (Clinard & Quinney, 1986), and "upperworld crime" (Geis, 1974b). (See also Albanese, 1995; Blankenship, 1995; Friedrichs, 1995; Jamieson, 1995; Schlegel & Weisburd, 1994.)

This chapter will concentrate on two key types of criminal activity: occupational criminal behavior and corporate (organizational) criminal behavior. **Occupational crime** refers to personal violations that take place for self-benefit during the course of a legitimate occupation, while **corporate (organizational) criminal behavior** refers to crimes by business or officials, committed on behalf of the employing organizations. Although organizational crime refers to crime on behalf of the organization, it becomes corporate (business) crime when it is done for the benefit of a private business. Thus, much of what ordinarily would be branded as corporate crime in a free enterprise economy is

FIGURE 10.1 ■ Relationship Among Elite Deviance, White Collar Crime, and Economic Crime

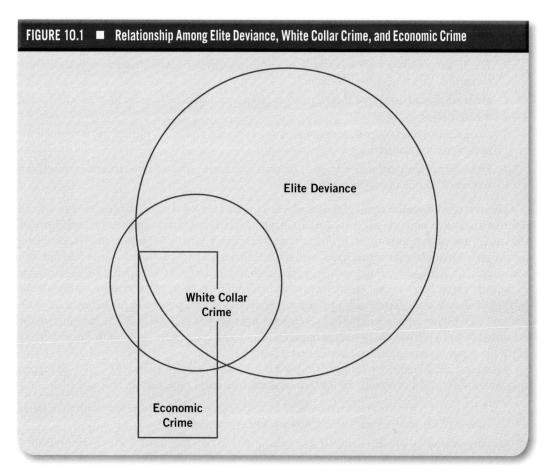

labeled organizational crime when committed by state bureaucrats in socialist systems. The organizational, economic crimes discussed in this chapter are also distinct from political crimes by government, which will be discussed in Chapter 11; the latter have more to do with efforts to maintain power, ideology, and social control than with economic advantage.

Figure 10.1 depicts the relationship among the many definitions of white collar crime and is an attempt by this author to address the debate among writers as to whether the best term for this subject of study is elite deviance (D. R. Simon, 1999), white collar crime (Coleman, 1994), or economic crime (various writers). This author views elite deviance as the broadest term, while white collar crime focuses on elite "crimes" but also includes nonelite activities, for example, employee theft and lower-level occupational crime. When observers ignore the status of the offender, economic crime can include minor fraud, embezzlement, and the like, even when it is not committed by individuals of high status. The issue is not which of these concepts is best, but rather how each taps a different dimension of "white collar crime."

The Measurement and Cost of Occupational and Corporate Crime

Even in societies that permit a measure of freedom of information, the collection of accurate data on most occupational and corporate crimes is difficult. Our primary sources of data (discussed in Chapter 1), such as official statistics (the UCR), victim surveys (the NCVS), and self-reports, generally do not include much information on corporate or upper-level occupational crimes.

Handbook Article Link 10.2
Read an article
on organizational crime
and illegalities.

Journal Article Link 10.1
Examine literature
regarding occupational
crime.

Problems faced by researchers who attempt to examine occupational crime include the following:

1. The higher professions are self-regulating, and very often codes of silence and protectionism rather than sanctions greet wrongdoers.
2. Many employers simply ask for resignations from errant workers in order to avoid scandal and recrimination.
3. Occupational crime statistics are not kept on a systematic basis by criminal justice agencies or by professional associations.
4. Probes of occupational wrongdoing by outsiders are usually greeted by secrecy or a professional version of "honor among thieves."

The cost of white collar crimes far exceeds the cost of traditional crimes as recorded in official police statistics and as previously discussed in Chapter 1. The Senate Subcommittee on Investigations (Senate Permanent Subcommittee, 1979) estimated that cost at roughly $36 billion in 1976. Estimates for the early 1980s place the figure at upward of $50 billion, a costly sum considering that FBI estimates for all UCR property crimes such as burglary, larceny, and robbery were in the $10 billion range in the early 1980s. Much higher estimates of costs incurred from white collar crime have been made by the Judiciary Subcommittee on Antitrust and Monopoly, which put the figure between $174 billion and $231 billion annually in the late seventies (Clinard & Yeager, 1980, p. 8). By the nineties, the estimated cost of $500 billion for bailing out savings and loan companies alone, with 5 to 40 percent of the losses due to fraud, justifies even higher estimates.

Among others, Clinard and Yeager (1978, pp. 255–272) and Geis and Meier (1977, pp. 3–4) suggest that there are a number of reasons for the lack of research on corporate crime in the past:

1. Many social scientists are inexperienced in studying corporate crime, which often requires some sophistication in areas of law, finance, and economics.
2. Corporate violations often involve administrative and civil sanctions to which criminologists have limited exposure.
3. Enforcement is often carried out by state and federal regulatory agencies rather than by the usual criminal justice agencies.
4. Funds for such studies have not been generally available in the past.
5. Corporate crime is complicated by the very complexity of corporations.
6. Research data are not readily available because of the imperviousness of the corporate board room.
7. Corporate crime raises special problems of analysis and research objectivity.

Despite these obstacles, rising public concern about corporate wrongdoing has encouraged increased research into corporate crime.

Simpson (2003, p. 4) points out:

In my opinion, the main difficulties confronting [white collar] crime scholars have to do with access, measurement, quality and type of available data, and the unit of analysis. To conduct research with executives or other corporate personnel, one must penetrate the corporate veneer and there is a skill in 'marketing' the research problem to guardians at the gate. Corporations are image sensitive and fear any kind of exposure that may cause reputational damage. Thus, just getting a foot in the door is often a major accomplishment. Getting executives to talk about crime is an even greater challenge. If one company allows research access, the next issue relates to sampling. With whom does one talk or survey? Usually access is limited to select personnel and not to the entire company. Thus, representativeness of respondents is an additional issue. Finally, it is important to recognize that getting into one company does not guarantee access anywhere else. The situation is a far cry from the relative ease of gathering self-reports from high school and university students.

The History of Corporate, Organizational, and Occupational Crime

Current publicity and concern with corporate, organizational, and occupational crime sometimes create the false impression that such activities did not exist in the past. Nothing could be further from the truth. In fact, history is replete with examples of past corporate wrongdoing; the current business climate probably sets higher moral expectations than ever before.

In the early history of capitalism and the Industrial Revolution, fortunes were made by unscrupulous "robber barons," who viewed the state and laws as negotiable nuisances. Cornelius Vanderbilt, the railroad magnate, when asked whether he was concerned with the legality of one of his operations, was said to have stated, "Law! What do I care about Law. Hain't I got the power?" (quoted in Browning & Gerassi, 1980, p. 201).

Journalistic "muckrakers," or specialists in exposing what Becker (1954) calls "sex, sin, and sewage" (p. 145), preceded criminologists in analyzing abuses in high places. Works such as Lincoln Steffens's *The Shame of the Cities* (1904) and Upton Sinclair's *The Jungle* (1906) dramatically focused on and aroused public interest in corruption and abuse in public and private organizations. John Kenneth Galbraith tells the story of John D. Rockefeller, the founder of the family fortune, and a lecture he was fond of giving to Sunday school classes: "The growth of a large business is merely the survival of the fittest. . . . The American Beauty rose can be produced in the splendor and fragrance which bring cheer to its beholder only by sacrificing the early buds which grow up around it" (quoted in Peter, 1977, p. 87). Browning and Gerassi in *The American Way of Crime* (1980) claim that the period between the Civil War and World War I was probably the most corrupt in U.S. history and describe this time as a "dictatorship of the rich." No one valued private property more than the industrial magnates who were stealing it (p. 210). Jay Gould, a captain of industry, gobbled up railroads through stock manipulation, rate wars, the falsifying of profit records, and the intimidation of competitors by means of hired thugs such as the Hell's Kitchen mob (pp. 133–136). G. Myers in *The History of Great American Fortunes* (1936, pp. 13, 17) reports an episode in which Russell Sage (a New York financier and politician) and his business associates masterminded a swindle against their creditors; after it succeeded, Sage conned his own partners out of their proceeds from the caper.

Political corruption, bribery, kickbacks, and influence peddling among political officeholders federal, state, and local have been rife since the very beginnings of the republic. The widespread acceptance of such corruption has given rise to a number of humorous comments, for example, the description of Mayor Curley of Boston as having been so crooked that when they buried him, they had to screw him into the ground. Another cynical remark claims that it was so cold the other day, the politicians had their hands in their own pockets. States such as Illinois, New Jersey, and Louisiana have had particular problems with public corruption in the first decade of the twenty-first century. Louisiana was described thus: "Half of Louisiana is under water and the other half is under indictment" ("Jindal", 2009). In Illinois, Governor Rod Blagojevich was arrested for demanding payoffs in return for appointment to President Obama's vacated senate seat and in New Jersey, more than 130 officials were found guilty of corruption over a 7-year period.

Audio Link 10.1
Listen to a discussion of lessons learned from white collar criminals.

Photo 10.1

John D. Rockefeller (1839–1937) founded the Standard Oil Company and eventually became America's first billionaire.

Photo 10.2

Bernard Ebbers is currently serving a 25-year prison term at Oakdale Federal Correctional Complex in Louisiana.

In the post–Civil War period in the United States, political machines were epitomized by "Boss" Tweed's Tammany Hall (New York City's Democratic Party), in which widespread vice and corruption were combined with political favoritism and voter fraud. More than one political election was won with stuffed ballot boxes or the "graveyard vote." S. Ross in *Fall From Grace: Sex, Scandal, and Corruption in American Politics From 1702 to the Present* (1988) documents the fact that political scandals have struck in nearly every decade since before the American Revolution. In 2007, a total of 11 New Jersey officials were charged with taking thousands of dollars in bribes in exchange for promising municipal business to undercover officers posing as insurance brokers (Chen, 2007). More than 100 public officials in New Jersey had been convicted of federal corruption charges between 2001 and 2007.

The first years of the twenty-first century were racked with corporate scandals. On July 14, 2005, Bernard Ebbers, CEO of WorldCom, was sentenced for what was described as the largest corporate fraud in U.S. history, an $11 billion accounting fraud (see Photo 10.2). Around the same time, stiff sentences were handed out to other corporate executives. Adelphi Communications founder John Rigas received 15 years in prison and his son Timothy 20 years for conspiracy, bank fraud, and securities fraud. Others included Tyco International CEO Dennis Kozlowski and Chief Financial Officer Mark H. Schwartz, who were convicted of grand larceny, conspiracy, securities fraud, and falsifying business records. They were found guilty of looting the company of over $600 million to pay for extravagant lifestyles. Former CEO Richard Scrushy of HealthSouth was acquitted on June 28, 2005, on 36 counts of conspiracy, false reporting, fraud, and money laundering. He pointed the finger at 15 HealthSouth executives, who all pleaded guilty. Former chief financial officer of Enron, Andrew Fastow, pleaded guilty to two counts of conspiracy and received a 10-year sentence, and former Enron Treasurer Bill Gilson, Jr., received a 5-year sentence for his role in the fraud. Enron founder Ken Lay was found guilty on fraud charges in May 2006, but he died before further charges could be prosecuted (Eichenwald, 2006).

Sometimes described as the poster girl for white collar crime, lifestyle guru Martha Stewart was convicted of insider trading. She served 5 months in prison and an additional 5 months of house confinement for conspiracy, obstruction of justice, and making false statements. Her broker, Peter Bacanovic, also served a 5-month sentence. Although Stewart was a client rather than a corporate executive, focus on her case may have been misleading. A 2004 study (AccountingWEB, 2004) by KPMG, a corporate fraud investigation firm, found that, of the 100 fraud cases they had been asked to investigate from 2002 to 2003, senior managers or directors committed over two thirds of the crimes. In 72 percent of the cases, the perpetrators were all males. Thirteen percent involved males and females, and only 13 percent of the cases involved females only. The finance department was the most likely targeted area with 40 percent of the cases. The *Wall Street Journal* (cited in Tolson, 2002) said "the scope and scale of corporate transgressions is greater than anything Americans have seen since the years before the Great Depression."

A widely cited typology of white collar crime is the one proposed by Edelhertz (1970). He identifies the following:

1. Crimes by persons operating on an individual ad hoc basis (e.g., income tax violations, credit card frauds, bankruptcy frauds)

2. Crimes committed in the course of the occupations of those operating inside business, government, or other establishments, in violation of their duty of loyalty and fidelity to employers or clients (e.g., embezzlement, employee larceny, payroll padding)

3. Crimes incidental to, and in furtherance of, business operations, but not central to the purpose of the business (e.g., antitrust violations, commercial bribery, food and drug violations)

4. White collar crime as a business, or as the central activity (pp. 19–20). (This is covered in this text under the label "professional crime"; it refers to activities such as medical and health frauds, advance fee swindles, and phony contests.)

Eliminating Edelhertz's item 4 as more appropriately an example of professional crime, Table 10.1 proposes an **occupational/organizational crime grid,** which classifies the crimes in terms of both perpetrators and victims. Goff and Reasons (1986) have proposed a similar model for organizational crime.

TABLE 10.1 ■ The Occupational/Organizational Crime Grid

	Crime Committed by (Perpetrator)		
Crime Committed Against (Victim)	Individual (Public, Consumer)	Employee	Organization (Corporation, State)
Individual (Public, Consumer)	(1) Merchant vs. Consumer Professional vs. Client	(2) Individual Corruption Payoffs	(3) Production of Unsafe Products Deceptive Advertising
Employee	(4)	(5) Sweetheart Contracts	(6) Occupational Health and Safety Violations Environmental Hazards on Job
Organization (Corporation, State)	(7) Insurance Fraud Tax Fraud	(8) Embezzlement Inside Trading	(9) Industrial Espionage Unfair Competition Patent Violations

Type	Description
(1)	Individual vs. Individual (Public)
(2)*	Employee vs. Individual (Public)
(3)	Organization vs. Individual (Public)
(4)*	Individual vs. Employer
(5)*	Employee vs. Employee
(6)	Organization vs. Employee
(7)*	Individual vs. Organization
(8)	Employee vs. Organization
(9)	Organization vs. Organization

*These crimes may not have direct corporate or occupational ramifications.

While many crimes in fact defy placement in mutually exclusive, homogenous categories, these types offer a useful scheme for organizing the presentation of occupational crime and organizational/corporate crime in this chapter.

Legal Regulation

Audio Link 10.2
Listen to a discussion of legislation and corporate wrongdoing.

Occupations and the Law

In Western societies, the legal regulation of occupations is often "self-regulation." Although laws and codes of ethics purportedly exist to protect the public from harmful occupational activity, much self-governance has been used instead to protect the interests of members of the occupation. The more developed professions attempt to convince legislatures that they possess highly sophisticated, useful, esoteric knowledge; that they are committed to serving societal needs through a formal code of ethics; and that they therefore should be granted autonomy, since they and only they are in a position to evaluate the quality of their service. In fact, the actual legal codes that control occupational practice tend to be formulated by the occupations themselves in order to dominate or monopolize a line of work. Playwright George Bernard Shaw (1941) in *The Doctor's Dilemma* has one of his characters state that "all professions are a conspiracy against the laity" (p. 9). More developed occupations (professions) virtually control the law-making machinery affecting their work. Professional organizations and their political action committees are quite effective in blocking legislation that may be detrimental to their interests.

An example of professional power is the AMA (American Medical Association), which Friedson (1970) describes as "professional dominance" and Harmer (1975) as "American Medical Avarice." The AMA as a lobbying organization appears more concerned with guarding profit, competition, and private enterprise in the business of medicine than supporting legislation that would improve the quality of medical care delivery. According to the 1968 Report of the National Advisory Commission on Health Manpower (cited in Skolnick & Currie, 1982), the health statistics of certain groups in the United States, particularly the poor, resemble those in a developing country.

Occupational crime can be controlled by professional associations themselves, by traditional criminal law, by civil law, and by administrative law. Actions by professional ethics boards can include suspensions, censure, temporary or permanent removal of license and membership, and the like. Traditional criminal prosecution also occurs, such as for larceny, burglary, and criminal fraud; civil actions by the government may include damage and license suspension suits. Administrative proceedings may call for taking away licenses, seizing illegal goods, and charging fines.

The FBI in its early history was involved primarily in investigating and enforcing white collar crimes, such as false purchases, security sales violations, bankruptcy fraud, and antitrust violations; only later did it become preoccupied with its gangbuster image (Lowenthal, 1950, p. 12). As late as 1977, however, the House Judiciary Committee charged that the FBI was soft on white collar crime and that its idea of white collar crime was small-scale fraud (D. R. Simon & Swart, 1984).

Organizations and the Law

A corporation is a legal entity that permits a business to make use of capital provided by stockholders. Although the federal government has had the power to charter corporations since the 1791 *McCulloch v. Maryland* decision, it rarely uses it; most chartering is done by the states. Corporations have been considered legal "persons" since a Supreme Court decision in 1886 (Clinard & Yeager, 1980, pp. 25–28).

In the United States, beginning in the nineteenth century, certain business activities were defined as illegal. These included restraint of trade, deceptive advertisements, bank fraud, sale of phony securities, faulty manufacturing of foods and drugs, environmental pollution, and the misuse of patents and trademarks (Clinard & Quinney, 1986, p. 207). In the late nineteenth century, concern grew about the development of monopolies, which threatened to control economies and stifle competition and thereby jeopardized the very philosophy of free-market enterprise.

The Sherman Antitrust Act (1890) was the first of many regulatory laws passed to control corporate behavior. This law forbids restraint of trade and the formation of monopolies; it currently makes price fixing a felony, with a maximum corporate fine of $1 million, and authorizes private treble (triple) damage suits by victims of price fixing. For the most part, the policing of corporate violations is done by federal regulatory agencies—for example, the Federal Trade Commission (FTC), which was set up in 1914 at the same time as the Clayton Antitrust Act and the Federal Trade Act. There are over 50 federal regulatory agencies with semi–policing functions with respect to corporate violations. Among these agencies are the Civil Aeronautics Board (CAB), the Environmental Protection Agency (EPA), the Federal Communications Commission (FCC), the Food and Drug Administration (FDA), the Federal Power Commission (FPC), the Interstate Commerce Commission (ICC), the National Labor Relations Board (NLRB), the Nuclear Regulatory Commission (NRC), the Occupational Safety and Health Administration (OSHA), and the Securities and Exchange Commission (SEC). Some areas regulated by these agencies and discussed in this chapter are air safety, air and water pollution, unfair advertising, safe drugs and healthy food, public utility services, interstate trucking and commerce, labor-management practices, nuclear power plants, health and safety in the workplace, and the sale and negotiation of bonds and securities.

Regulatory agencies have a number of sanctions they can use to force compliance with their orders: warnings, recalls, orders (unilateral orders, consent agreements, and decrees), injunctions, monetary penalties, and criminal penalties (Clinard & Yeager, 1980, p. 83). In addition to criminal proceedings, acts such as the Clayton Act (Section 4) permit "treble damage suits" by harmed parties. Guilty companies, with their batteries of lawyers and accountants, generally have more expertise, time, and staff to devote to defense than the Justice Department, under its Antitrust Division, has for prosecution. Indefinite delays and appeals are not uncommon.

If the government appears to have a solid case, corporations are permitted to plead **nolo contendere,** or "no contest," to charges. This is not an admission of guilt, and thus enables corporations to avoid the label of criminal. Consent decrees amount to a hand slap; that is, the corporation simply agrees to quit committing the particular violation with which it was charged.

A number of criticisms have been levied against federal regulatory agencies and their efforts against corporate crime:

1. Lacking sufficient investigative staff, the agencies often rely on the records of the very corporations they are regulating to reveal wrongdoing.

2. The criminal fines authorized by law are insignificant compared with the economic costs of corporate crime and become, in effect, a minor nuisance, a "crime tax," a "license to steal," but certainly not a strong deterrent.

3. Other criminal penalties such as imprisonment are rarely used and, when they are, tend to reflect a dual system of justice: Offenders are incarcerated in "country club" prisons or are treated in a far more lenient manner than traditional offenders.

4. The enforcement divisions of many regulatory agencies have been critically understaffed and cut back, as in the Reagan administration's EPA and other agencies, to inoperable levels.

5. The top echelons of agency commissions are often filled with leaders from the very corporations or industries to be regulated, creating potential conflicts of interest.

6. Relationships between regulators and regulated are often too compatible, with some agency employees more interested in representing the interests of the corporations they are supposed to be regulating than in guaranteeing the public well-being. The fact that many retiring agency employees are hired by the formerly regulated companies lends support to this argument.

In reviewing the state of regulation of illegal corporate activity, Clinard and Yeager (1980) state,

> One may well wonder why such small budgets and professional staffs are established to deal with business and corporate crime when billions of dollars are willingly spent on ordinary crime control, including 500,000 policemen, along with tens of thousands of government prosecutors and officials. (p. 96)

Gross (1980) in his book *Friendly Fascism* answers their question by letting us in on what he calls "dirty secrets":

> We are not letting the public in on our era's dirty little secret: that those who commit the crime which worries citizens most—violent street crime—are, for the most part, products of poverty, unemployment, broken homes, rotten education, drug addiction, alcoholism, and other social and economic ills about which the police can do little if anything. . . . But, all the dirty little secrets fade into insignificance in comparison with one dirty big secret: Law enforcement officials, judges as well as prosecutors and investigators, are soft on corporate crime. . . . The corporation's "mouthpieces" and "fixers" include lawyers, accountants, public relations experts and public officials who negotiate loopholes and special procedures in the laws, prevent most illegal activities from ever being disclosed and undermine or sidetrack "overzealous" law enforcers. In the few cases ever brought to court, they usually negotiate penalties amounting to "gentle taps on the wrist." (pp. 110, 113–115)

While every year the FBI publishes its Uniform Crime Reports to give an annual account of primarily street crime, no such annual report exists to measure the far more costly corporate crime. Robert Mokhiber (1999), editor of the Washington, D.C.–based *Corporate Crime Reporter,* ranked the "Top 100 Corporate Criminals of the 1990s." These were only the tip of the iceberg in that the majority of corporate wrongdoing is handled under civil law. This list includes only those who were caught and criminally fined. The 100 corporate criminals fell into 14 categories of crime (http://www.corporatepredators.org/top100.html): environmental (38), antitrust (20), fraud (13), campaign finance (7), food and drug (6), financial (4), false statements (3), illegal exports (3), illegal boycott (1), worker death (1), bribery (1), obstruction of justice (1), public corruption (1), and tax evasion (1). The top 10 corporate criminals of the 100 identified by Mokhiber were the following:

Journal Article Link 10.2
Examine literature regarding controlling transnational corporations.

1. **F. Hoffman-LaRoche Ltd.** The Swiss pharmaceutical company pleaded guilty and paid a record $500 million criminal fine for fixing prices on vitamins.

2. **Daiwa Bank Ltd.** The bank pleaded guilty to 16 federal felonies and paid a $340 million criminal fine. It pleaded guilty to two counts of conspiracy to defraud the United States and the Federal Reserve Bank, misprision (concealment) of felony, 10 counts of falsifying bank records, 2 counts of wire fraud, and 1 count of obstructing a bank examination.

3. **BASF Aktiengesellschaft** This German pharmaceutical company pleaded guilty and agreed to a $225 million criminal fine for fixing prices on vitamins.

4. **SGL Carbon Aktiengesellschaft** The world's largest producer of graphite and carbon products pleaded guilty to price fixing and paid a $135 million fine.

5. **Exxon Corporation** Exxon pleaded guilty to criminal charges related to the 1989 Exxon Valdez oil spill and paid a $125 million fine. This was the largest criminal recovery obtained in an environmental case.

6. **UCAR International Inc.** The largest producer of graphite electrodes in the United States pleaded guilty to fixing prices and paid a $110 million criminal fine.

7. **Archer Daniels Midland** Pleading guilty to charges of price fixing of lysine and citric acid markets, the company paid a $100 million fine.

8. (tie) **Banker's Trust** The bank was fined $60 million for making false reports of financial performance, having made false entries in books and records.

8. (tie) **Sears Bankruptcy Recovery Management Services** Sears pleaded guilty to bankruptcy fraud and agreed to pay a $60 million fine. The company had already paid over $180 million in restitution to 188,000 debtors and $40 million in civil fines to 50 state attorneys general. Sears had systematically misled those in bankruptcy into believing they had to pay certain debts.

10. **Haarman and Reimer Corporation** A subsidiary of the German Bayer AG, the corporation pleaded guilty and agreed to pay a $50 million fine for fixing prices on the citric acid world-wide market.

Occupational Crime

Crimes by Employees

Although there are cases of overlap, both "crimes by employees" and "crimes by individuals" can be examples of occupational crime committed in the course of a legitimate occupation for one's own benefit. While the types of activities to be discussed in this section are executed by employees (those who work for someone else), those to be examined in "crimes by individuals" will primarily be crimes by professionals.

Edelhertz's Typology

One attempt to delineate white collar crime is the widely cited typology and examples provided by Edelhertz (1970, pp. 73–75) (see Table 10.2).

While Edelhertz had two other types of white collar crime in his classification, many of those listed in his "crimes by persons operating on an individual . . . basis" are not necessarily occupational in nature, except that the victims often happen to be organizations (business or the state). Some examples that he gives include bankruptcy frauds and violations of Federal Reserve regulations by pledging stock for further purchases, flouting margin requirements. His category of "white-collar crime as business, or as the central activity" better fits the definition of professional crime as defined in Chapter 9. Edelhertz's category A fits our discussion of "occupational crime," while category B better fits our definition of "corporate crime."

Crimes by Employees Against Individuals (the Public)

Self-aggrandizing *crimes by employees against the public* (type 2 in Table 10.1) take the form of political corruption by public servants or office holders (public employees), or commercial corruption by employees in the private sector. These activities are distinguished from corporate or organizational criminal activities of the same type by the fact that in this case the employee personally benefits from the violation.

Public Corruption

"Cigar smoke, booze, and money delivered in brown paper bags"—this is how Hedrick Smith envisions the backroom world of politics in the PBS telecast *The Power Game* (Smith, 1989). The list of occupation-related crime on the part of political employees or office holders may include furnishing favors to private businesses such as illegal commissions on public contracts, fraudulent licenses, tax exemptions, and lower tax evaluations (Clinard & Quinney, 1973, p. 189). As an example, health

TABLE 10.2 ■ **Edelhertz's (1970) Typology of White Collar Crime**

Edelhertz's typology of white collar crime details a variety of offenses:

A. Crimes committed in the course of their occupations by those operating inside business, government, or other establishments in violation of their duty of loyalty and fidelity to employer or client.

   1. Commercial bribery and kickbacks (i.e., by and to buyers, insurance adjusters, contracting officers, quality inspectors, government inspectors and auditors)

   2. Bank violations by bank officers, employees, and directors

   3. Embezzlement or self-dealing by business or union officers and employees

   4. Securities fraud by insiders trading to their advantage by the use of special knowledge

   5. Employee petty larceny and expense account fraud

   6. Frauds by computer, causing unauthorized payments

   7. "Sweetheart contracts" entered into by union officers

   8. Embezzlement or self-dealing by attorneys, trustees, and fiduciaries

   9. Fraud against the government:
      a. Padding of payrolls
      b. Conflict of interest
      c. False travel, expense, or per diem claims

B. Crimes incidental to and in furtherance of business operations, but not the central purpose of the business

   1. Tax violations

   2. Antitrust violations

   3. Commercial bribery of another's employee, officer, or fiduciary (including union officers)

   4. Food and drug violations

   5. False weights and measures by retailers

   6. Violations of Truth-in-Lending Act by misrepresentation of credit terms and prices

   7. Submission or publication of false financial statements to obtain credit

   8. Use of fictitious or overvalued collateral

   9. Check kiting to obtain operating capital on short-term financing

   10. Securities Act violations (i.e., sale of nonregistered securities to obtain operating capital, false proxy statements, manipulation of market to support corporate credit or access to capital markets)

   11. Collusion between physicians and pharmacists to cause the writing of unnecessary prescriptions

   12. Dispensing by pharmacists in violation of law, excluding narcotics trafficking

   13. Immigration fraud in support of employment agency operations to provide domestics

   14. Housing code violations by landlords

   15. Deceptive advertising

   16. Fraud against the government:
      a. False claims
      b. False statements
         (1) Statements made to induce contracts
         (2) Aiding fraud
         (3) Housing fraud
         (4) Small Business Administration fraud, such as bootstrapping, self-dealing, cross-dealing, etc., or obtaining direct loans by use of false financial statements
      c. Moving contracts in urban renewal

   17. Labor violations (Davis Bacon Act)

   18. Commercial espionage

Web Research Project

Using a keyword search, examine "white collar crime." What are some types discussed in the articles?

inspectors in New York City ("City Inspectors," 1988) turned the Department of Health into the Department of Wealth and doubled or tripled their salaries by extorting payments from restaurants, threatening to cite them for health code violations if they did not pay up.

In 1999, eight federal food inspectors were arrested in a bribery and kickback scheme that permitted wholesalers to cheat their suppliers. The scheme involved the inspectors grading fruits and vegetables as low quality, gaining lower prices for the wholesalers who then turned around and sold the items as Grade A produce. Some of the inspectors earned over $100,000 a year in payoffs (Weiser, 1999).

Mark Twain (1899) once said, "There is no distinctly American criminal class except Congress" (p. 98). The use of public office for private gain defines political corruption. Twain was not quite accurate in his observation in that such behavior is widespread internationally. The Transparency International Corruption Perception Index (CPI), developed by Berlin-headquartered Transparency International Incorporated, rates countries on the basis of seven surveys of businesspeople, political analysts, and the general public. The Corruption Perception Index for the year 2008 (http://www.transparency.org/policy_research/surveys_indices/cpi/2008) ranged from a high of 10 (highly clean) to 0 (highly corrupt) (Hagan, 2005). Some selected country ranks and scores included:

Corruption Perception Index 2008		
Rank	Country	CPI Score
1	Denmark	9.3
1	New Zealand	9.3
1	Sweden	9.3
9	Canada	8.7
16	United Kingdom	7.7
18	USA	7.3
55	Italy	4.8
70	Colombia	3.8
180	Somalia	1.0

After a 1999 report commissioned by the European Parliament indicated that 2–10 percent of the value of business transactions involves bribery, 20 European commissioners resigned en masse after criticism of their failure to do anything about it (Partridge, 1999). Dolive (1999), in examining systematic corruption in Italy, Japan, and Russia, claims that the corrupt politicians in those countries were the initiators and perpetuators of systematic corruption. Rather than the system being dependent on particular individuals who are at times exposed or removed, successors continue the system of corruption, which is the driving force of both the economy and politics.

- In France, in the Elf scandal (named for state-owned oil company Elf Aquitane), slush funds and corruption were traced to former foreign minister Roland Dumas and, in 2000, to former interior minister Charles Pasqua. Also implicated was former German Chancellor Helmut Kohl's political party (Ignatius, 2000).
- In Russia, a group of gangster capitalists called the "oligarchs" looted the assets of the state during the 1990s.
- In the United States, officials at Citibank operated as the "private bank" for unsavory figures such as Raul Salinas, brother of the former president of Mexico, who is now in prison for

murder; Asif Zardari, former husband of former Pakistani Prime Minister Benazir Bhutto, who is in prison for kickbacks; and two daughters of former Indonesian President Suharto, who allegedly stole billions of dollars from that country.

- A Swiss investigation in 1999 uncovered evidence that Mabetex, a construction company, paid $10–$15 million to Russian officials, including then-President Boris Yeltsin, in order to obtain contracts (LaFraniere, 1999).

Joel Henderson and David Simon in their book *Crimes of the Criminal Justice System* (1994) document widespread and persistent corruption and wrongdoing throughout the criminal justice system.

Police Corruption—The Mollen Commission

Between 1992 and 1993, the Mollen Commission, named after a former New York City deputy mayor for public safety, conducted an investigation of corruption in that city's police department and focused attention on police wrongdoing the likes of which had not existed for 20 years (since the Knapp Commission of the 1970s revealed widespread corruption, particularly associated with narcotics enforcement). The investigation was ordered because five New York City officers had been arrested by Suffolk County (Long Island) police for selling cocaine; Mollen Commission hearings featured informants from within the ranks who revealed police extortion practices, theft and reselling of drugs, rolling of drunks, robbing of dead people, snorting cocaine while on duty, and indulging in brutality particularly in poor sections of the city. Often higher-ups in the department had blocked investigations ("NYC's Mollen Commission," 1993). A blue wall of silence and loyalty to peers can take precedence over concerns about graft and violation of oath of office (see also Kappeler, Sluder, & Alpert, 1994).

Certainly police wrongdoing was not limited to the nation's largest department:

- In New Orleans in 1993, over 50 officers were convicted on charges including murder, rape, assault, and drug trafficking. One was convicted of killing another police officer while robbing a convenience store.
- In Jersey City, New Jersey, police officers were accused of, among other things, participating in what police investigators called "Operation Boneyard": stolen and illegally parked cars were towed to the city car pound and converted to city property without the owners being notified.
- In 1987, nearly 100 Miami police officers (1 in 18 on the force) were believed to be involved in serious corruption primarily related to drug trafficking ("Miami Police Scandal," 1987).
- In Cleveland, an FBI sting operation resulted in the arrest of 23 police officers who had served as security guards for illegal gambling dens and warned them of impending raids ("FBI Gambling Sting," 1991).
- In Detroit, former Police Chief William Hart was convicted of helping embezzle $2.6 million from a special police fund to give his girlfriend lavish gifts ("Detroit's Former Chief Convicted," 1992).

In 1999, at Rampart Community Police Station in Los Angeles, 20 police officers were involved in systematic corruption. Their crimes included planting illegal drugs on innocent people, planting guns on suspects who had been shot by police, burglarizing the homes of petty criminals, and framing roughly 100 people. In a plea bargain, one officer testified that prisoners were routinely railroaded by fabricated evidence and police lies. As a result, a large number of previous convictions have been overturned. In 2000, a jury acquitted four New York officers who mistakenly gunned down innocent citizen Amadou Diallo with 41 rounds, killing him instantly. In 2000, New York City police officer Justin Volpe was sentenced to 30 years in prison for torturing Abner Louima in the bathroom of a Brooklyn station house by sticking a broom handle up his rectum, doing considerable physical harm.

As a result of cover-ups by police during the initial stages of these incidents, juries have become more skeptical of testimony by police officers and other witnesses.

While public preoccupation with police corruption is viewed defensively by police, for most people the police officer symbolizes the law and engenders higher public expectations of proper conduct (Barker & Carter, 1986). Coleman (1994) explains that "police officers simply have more opportunities to receive illegal payments than other public employees" (p. 45) because they are asked to enforce inadequate vice laws that try to control very profitable black markets.

Police corruption is mirrored in other agencies of government, in industry, in labor, and in the professions. In Pennsylvania, a large-scale police raid of Graterford Prison by the state police, correctional officers, and U.S. Customs officers closed down wide-scale drug trafficking in the prison. Thirteen guards were arrested because they were believed to have been instrumental in the drug overdose deaths of 11 inmates ("Drug Raid," 1995). In 1988, an undercover investigation in Philadelphia city jails (T. Jacoby, 1988) found over 30 guards involved in, among other offenses, smuggling drugs, money, and weapons into the prison; helping inmates escape; and taking bribes from reputed mobsters. The fact that not much has changed at Graterford was demonstrated by the arrest of four guards in 2007 who were charged with taking cash and drugs in exchange for smuggling drugs to inmates. A fifth guard was charged with helping a murderer with an attempted escape ("Feds: Philly Guards Gave Inmates Drugs," 2007).

Judgescam—Operation Greylord

In 1983, Federal Bureau of Investigation agents revealed that for 3 years they had posed as lawyers and criminals to run a "sting" operation on the Cook County, Illinois, criminal justice system. The sting was code-named "Operation Greylord" (referring to the powdered wigs historically worn by judges). This was the largest and most successful investigation into judicial misconduct in U.S. history, and by the fall of 1987 it had resulted in convictions of 61 persons, including 11 judges, as well as police officers, lawyers, and court officials, with additional trials and indictments ongoing (Bensinger, 1987).

In 1998, the *Pittsburgh Post Gazette* published a 10-part series that alleged that federal agents and prosecutors repeatedly broke the law in the pursuit of convictions (Moushey, 1998). Investigators claimed to have found examples of prosecutors lying, hiding evidence, distorting facts, engaging in cover-ups, paying for perjury, and setting up innocent people in order to obtain indictments, guilty pleas, and convictions. Some criminals walked free as a reward for conspiring with the government.

Photo 10.3

President Richard Nixon gives his resignation speech at the White House following the Watergate scandal.

Watergate

Perhaps no one event evokes images of official corruption, deceit, and subterfuge as much as Watergate. This story began with the discovery of an illegal break-in at the Democratic National Committee Headquarters located in the Watergate complex in Washington, D.C. The burglary was carried out by agents in the employ of then-President Richard Nixon.

Nixon certainly was not the first U.S. president to be involved in crooked practices (see Chambliss, 1988a). He was, however, the first to be driven from office in disgrace because of the extent of his activities and the first to be saved from certain criminal prosecution through a full pardon before-the-fact (issued by his successor, President Gerald Ford). At the time, President

Video Link 10.1
View a video about Watergate.

Nixon's attitude toward the probe appeared in one of the later-to-be-released "missing tapes": "I don't give a shit what happens. I want you to stonewall it. Let them plead the Fifth Amendment, cover up, or anything else if it'll save the plan" (cited in "The Case of the Doctored Transcripts," 1974).

Among the offenses of the Watergate team were burglary, illegal surveillance, attempted bribery of a judge (Ellsberg case), selling ambassadorships in return for illegal campaign donations, maintenance of an illegal "slush fund," destruction of evidence, use of "dirty tricks" in political campaigns planned by the FBI director and the president, requests by U.S. Attorney General John Mitchell (the nation's top law enforcement officer) for IRS audits on opponents, use of the CIA and FBI to attempt to halt the investigation, perjury, withholding information, altering evidence, and deliberate lying to the American public by the nation's top officeholder (D. R. Simon, 1996, p. 3).

Abscam

Abscam (Arab or Abdul Scam) was an FBI sting operation in which agents posing as rich oil sheiks bribed a number of members of the U.S. Congress. Whether we call it *baksheesh* (Middle East), *bustarelle* (Italy), *pot de vin* (France), *mordida* (Latin America), or just plain bribes (North America), kickbacks and corruption are apparently both widespread and international in scope. Individuals in their occupational roles may give or receive bribes for their own personal benefit (occupational crime) or for the benefit of the organization/corporation (organizational/corporate crime). Bribery, influence peddling, and corruption are acceptable patterns of international commerce and are not even illegal in many countries.

Particularly revealing in the Abscam operation was the relative ease with which foreign agents were able to bribe members of the U.S. Congress. Although many regard such federal sting operations as entrapment (causing a crime to happen that would not have occurred if the stimulus had not been put there by the government), others perceive such "aggressive tactics" as the only means of ferreting out "upper-world crime."

Private Corruption

Commercial bribery and kickbacks (in which the individual personally benefits) can take place in a variety of ways. Buyers for large retail chains may accept gifts or cash in return for placing orders. At the expense of the general public in the form of higher prices, insurance adjusters, contracting officers, and quality control inspectors may all be willing to accept bribes in return for overlooking their duties to employers.

Auto dealers can be both perpetrators and victims of *sharp practices*. In analyzing what they call **"coerced crime,"** Leonard and Weber (1970) describe how the four major domestic auto producers pressure their roughly 30,000 dealers (who are technically independent proprietors) into bilking their customers. These dealers commit "coerced crime" because in order to retain their franchises, they must meet minimum sales quotas, and in order to meet these, they must often employ "shady practices." The latter include forcing accessories on the customer, service gouging, high finance charges (at times even employing loan sharks), overcharging for parts, misuse of "book time" (preset and inflated charges for labor time on repairs), and odometer (mileage meter) tampering.

Crimes by Employees Against Employees

While a variety of crimes like theft may be committed by an employee against another employee for personal benefit (type 5 in Table 10.1), many such violations would not necessarily be occupationally related and, therefore, would not be appropriate examples for the occupational/organizational crime grid. But one type of violation that certainly fits is the *sweetheart contract* in labor–management negotiations, which involves labor officials and negotiators secretly making a deal with management to the disadvantage of the workers whom the labor officials represent. For example, the union president and representatives might make a deal with management to take a bribe of $50,000. They then might indicate to the workers that they have examined the company books and found that management can only afford a 20 cent per hour raise rather than the 50 cents originally promised. Depending on the size of the workforce, management could save millions of dollars.

Another example is workplace violence perpetrated by a fellow employee. Such perpetrators take out their frustrations usually associated with loss of job on their fellow workers and supervisors. While murder is the most highly publicized form of workplace violence, other forms include assaults, rapes, and suicides, as well as psychological and mental health episodes. Drug and alcohol abuse may create hazardous work conditions. Hostile, intimidating, and offensive work environments may also foster sexual harassment, sexual assault, and other psychological and emotional damage.

Crimes by Employees Against Organizations

Organizations are vulnerable to a variety of offenses that employees can commit against them (type 8 in Table 10.1). In this section, we will briefly focus on employee pilferage, computer crime, and embezzlement, but employee crimes obviously include many types of offenses discussed under Crimes by Employees Against Individuals (Public), corporate bribery, and the like.

Embezzlement

One form of stealing from one's employer is through **embezzlement,** which is theft from an employer by an individual who has reached a position of financial trust. The classic work on the subject is Donald Cressey's *Other People's Money* (1953), which contains interviews with 133 incarcerated embezzlers. He proposes the following explanation of why trust violators steal:

1. Individuals who have achieved a position of trust are faced with what they conceive of as a nonshareable financial problem.

2. They feel they can resolve this problem by violating their position of trust, that is, by "temporarily borrowing" from their employer.

3. This rationalization of "borrowing" eventually breaks down as embezzlers realize they have been discovered and cannot make repayment in time (p. 30).

Gambling, sexual affairs, and high living are often the factors behind the unshareable nature of the financial problem.

The typical embezzler does not fit the stereotype of the criminal. Most are middle-aged, middle-class men who have lived relatively respectable lives and lack a history of criminal or delinquent activity; however, in *Women Who Embezzle or Defraud,* Zeitz (1991) notes increased embezzlement by women as managerial and executive positions open up for females. One example is the case of Dorothy Hutson, a Merrill Lynch stockbroker who systematically cheated investors out of $1.4 million and used the money to finance Las Vegas and Lake Tahoe gambling junkets (Siconolfi & Johnson, 1991).

In one of the larger embezzlements, Phar-Mor, Inc., a discount drugstore chain, disclosed that two executives had allegedly embezzled more than half of the company's net worth. The company estimated its losses at $350 million ("Phar-Mor Discloses," 1992). In December of 1995, Michael Modus, former president of Phar-Mor, was sentenced to nearly 20 years in prison for fraud, tax evasion, and embezzlement. In 2000, Merrill Lynch discovered a $40 million embezzlement had been perpetrated by a former employee who stole from elite, private banking clients by using the name of a dead person to transfer the securities from Arab International Bank to Swiss bank accounts (Huang, 2000).

Cressey's analysis of embezzlers has been criticized by Schuessler (1954, p. 604), who claimed that it was limited to an ex post facto (after the fact) study of only caught embezzlers and that his descriptions may not be characteristic of most embezzlers. Nettler's (1974) study found embezzlers to be motivated by greed and temptation as well as by the opportunity to commit the crime. Unlike Cressey, Nettler did not find that a non-shareable problem was a necessary component of embezzlement.

Smigel and Ross in *Crimes Against Bureaucracy* (1970) indicate that individuals—particularly employees—feel less guilt the bigger the victim organization. Many individuals, who would consider themselves criminals were they to steal from other persons, rationalize their theft from large, impersonal organizations by saying that "they can afford the loss." According to Smigel and Ross, the very size, wealth, and impersonality of large bureaucracies, whether governmental or business, provide a rationalization for those who wish to steal from such organizations. The "Robin Hood myth" holds that theft from such organizations really hurts no one, since the victim is a large, wealthy organization. Combined with this is a certain public antipathy toward the large corporation or big government. Obviously, the Robin Hood rationalization breaks down when we consider the higher cost of goods consumers must pay because of "inventory shrinkage."

A 1997 study of the crimes of 1,324 employees by the Ethics Officer Association and the American Society of Chartered Life Underwriters and Chartered Financial Consultants found that 48 percent of U.S. workers admitted to unethical or illegal activities in the previous year. This included cheating on expense accounts, discriminating against coworkers, participating in kickbacks, forging signatures, trading sex for sales, and violation of environmental laws. Over half (57 percent) indicated that they felt more pressure to be unethical than 5 years ago, and 40 percent believed that it had gotten worse over the past year (Jones, 1997).

Cameron, in her classic work on retail theft, *The Booster and the Snitch* (1964), suggested that "inventory shrinkage" (loss of goods) in retail establishments was primarily caused by employee theft rather than shoplifting. Store security personnel concur, estimating that as much as 75 percent of such loss is due to employee theft. A familiar story relates to security personnel who suspected that an employee was "ripping off" the company because every day he left work with a wheelbarrow full of packages. Every day they carefully checked the packages to no avail. When finally discovered, the employee had stolen over a thousand wheelbarrows. Employees can be quite ingenious in illegally supplementing their wages at the expense of their employer.

Some common techniques in employee retail theft include the following:

1. Cashiers who ring up a lower price on single-item purchases and pocket the difference, or who ring up lower prices for "needy" friends going through the checkout.

2. Clerks who do not tag some sale merchandise, then sell it at the original price and pocket the difference.

3. Receiving clerks who duplicate keys to storage facilities and return to the store after hours to help themselves.

4. Truck drivers who make fictitious purchases of fuel and repairs, and split the gains with truck stop employees.

5. Employees who simply hide items in garbage pails, incinerators, or under trash heaps until they can be retrieved later (McCaghy, 1976b, p. 179).

Abuses of expense accounts, travel allowances, and company cars are additional means by which employers are robbed of organizational income.

Crimes by Individuals (or Members of Occupations)

Crime in the Professions

Medicine. Medical quackery and unnecessary operations may very well kill more people every year in the United States than crimes of violence. A House subcommittee estimated that the American public was the victim of 2.4 million unnecessary surgical procedures per year, which resulted in a loss of $4 billion and in 11,900 deaths (Coleman, 1994, p. 37). A Harvard study (Gerlin, 1999) estimates that 1 million American patients are injured yearly by hospital errors and 120,000 die as a result. This is equivalent to a jumbo jet crash every day and is three times the 43,000 people killed each year in U.S. automobile accidents. Americans may be becoming overdoctored, having twice the per capita number of surgeons, anesthesiologists, and operations as England and Wales, yet higher mortality rates. Jesilow, Pontell, and Geis (1985) estimate that U.S. physicians defraud federal and state medical assistance programs of up to 40 percent of all program monies.

Some violations that physicians may become involved in include practices such as fee splitting (in which doctors refer patients to other doctors for further treatment and split the fee with them). "Ping-ponging" doctors refer patients to other doctors in the same office, "steering" entails directing patients to particular pharmacies, and "gang visits" involve billing for unnecessary multiple services ("White-Collar Crime," 1981). Quinney's (1963) analysis, "Prescription Violations by Retail Pharmacists," reveals higher numbers of violations among pharmacists who see themselves as businesspeople rather than as professionals. If clients (whom the professional views with concern for their health and the provision of ethical service) are seen as customers (whose greater consumption equals greater profit), then more frequent occupational violations are likely to ensue.

With the end of the Cold War in the nineties, the FBI reassigned agents from counterespionage activity to the investigation of health care fraud, and this began to show dividends. In a 1992 under-cover operation, FBI agents arrested 82 pharmacists and physicians for cheating private insurance companies and Medicaid. Some of the schemes involved pharmacists filling prescriptions with generic drugs, billing for brand-name products, and charging payers (insurance or Medicaid) multiple times for the same prescription or for prescriptions that were never written or filled. In 1994, the Public Citizen's Health Research Group claimed that some 420,000 Caesarean baby deliveries are performed unnecessarily each year in the United States. It is currently the most common surgery performed in the country. In 1970, C-sections accounted for only 5.5 percent of births, but were nearly 25 percent by 1988. The most Caesareans are performed in for-profit hospitals (Neergaard, 1994). Another survey of 449 programs in adult and pediatric critical care found that 39 percent used the bodies of people who had just died to teach medical procedures, but only 10 percent required that the patient's family give consent (Kolata, 1994).

In 1993, National Medical Enterprises, an operator of psychiatric and acute care hospitals, agreed to pay $125 million to settle charges by three major insurers for filing fraudulent claims. The company also faced charges by 130 former psychiatric patients who claimed that the company held them against their will, misdiagnosed patients, physically abused them, and administered unnecessary medications and treatments in order to run up bills (Kerr, 1993). In 1998, Allstate Insurance Company sued 45 doctors, lawyers, chiropractors, and others for alleged involvement in systematically staging fake auto accidents and filing phony insurance claims (Abram, 1998). In 2007, as many as 30,000 Medicaid providers were charged with cheating the Internal Revenue Service in seven states and failing to pay more than $1 billion in federal taxes in 2006. This amounted to 5 percent of Medicaid providers in those states (R. Wolf, 2007). In 2008, two pharmaceutical giants announced that they would begin posting an online database of doctors receiving payment from them. Such payments typically include speaking fees to doctors for giving speeches about products to other doctors while the group dines at the drug company's expense and consulting fees. In the latter, experts advise the company on new product development. The concern is that such money could taint doctors' research plans or clinical judgment (Carey, 2008).

Finance. Wrongdoing has certainly not been limited to the health and medical professions. "The Great Savings and Loan Scandal," to be discussed, was the biggest financial public policy failure in U.S. history, with estimated costs of $500 billion. I*n The Greatest-Ever Bank Robbery: The Collapse of the Savings and Loan Industry,* Mayer (1990) indicates,

> What makes the S&L outrage so important a piece of American history is not the hundreds of billions of dollars, but the demonstration of how low our standards for professional performance have fallen in law, accounting, appraising, banking and politics—all of them. (p. 298)

The federal government has sued many of these professionals and their firms for collusion in S&L collapses. In *The Big Six: The Selling Out of America's Top Accounting Firms,* Mark Stevens (1991) asks, if CPA firms are truly independent of the clients they audit (who foot the bill), "how can accountants be truly independent of the cash register that pays their bills?" Berton (1991) notes, "Many legislators and the General Accounting Office, an arm of Congress, are rapidly losing confidence in accountants because their independence seems tarnished and they still duck the job given them by government of protecting the public against financial fraud" (p. A12).

Law. Illegal and unprofessional activities by lawyers may include "ambulance chasing," that is, soliciting and encouraging unnecessary lawsuits (such as fraudulent damage claims) in order to collect commissions (M. H. Freedman, 1976; Reichstein, 1965). Describing the practice of law as sometimes constituting a "con game" against clients, Blumberg (1967) mentions activities in which the lawyer collects fees for defending a client and then simply "plea bargains" to expedite the case, with little concern for the client's well-being. Other legal rackets include real estate home closings (in which fees are collected on a regular basis for very little work) as well as the collection of contingency fees on liability cases (in which lawyers receive a percentage of anything won) (Merry, 1975, p. 1).

Concern has been raised that, with nearly 1 million lawyers, the United States is becoming an overlitigious society that is, one in which too many resources are expended on legal actions. Olson (1991) notes that the United States has 3 times as many lawyers per capita as Great Britain, 10 times the number of lawsuits per capita, 30 to 40 times the number of malpractice suits, and nearly 100 times the number of product claims. The United States is the only society that encourages such lawsuits through our way of financing litigation. Only in the United States must the winning party pay his or her lawyer. Such conduct was even illegal under English law and called "champerty" (lawyers receiving fruits of the successful action) and "barratry" (instigating and maintaining suits and quarrels in courts) (Crovitz, 1991, p. A17).

In 1990, three members of a personal injury law firm in Manhattan were indicted for bribing witnesses to perjure themselves in court and for falsifying evidence in 19 accident cases dating from 1979 (Hevesi, 1990). Fireman's Fund, an insurance company, hired an auditor to examine how their defense attorneys were spending their funds and exposed 20 lawyers, representing plaintiffs and defendants, who had cooperated in manipulating lawsuits and billing up to $100 million in dubious fees to insurance companies (Schmitt, 1992, p. A1). In explaining rising thievery by lawyers, bar association officials, while noting that only a minority are involved, point to tough economic times, the high cost of practicing law, substance abuse, and even glamorized images of lawyers on television (A. D. Marcus, 1990).

Until recently, bar associations published minimum fees and sanctioned attorneys who charged less, even though the Sherman Antitrust Act made no exceptions for professional associations in prohibiting price fixing (Coleman, 1994, p. 61). The concentration of legal talent in the defense of wealthy and corporate violators and the underconcentration in representing their victims (the state and the public) raise questions regarding the ethics of the legal profession itself.

Other Occupations. Examples of crimes against consumers by professionals, merchants, and members of other legitimate occupations are numerous:

- The "greasy thumb on the scale," or short-weighting customers and overcharging for products.
- "Bait and switch" techniques by small merchants, in which the product advertised is unavailable and a more expensive product is pushed on the customer.
- Phony or unnecessary repair work.
- Security violations by stockbrokers, such as misleading clients or insider trading (making use of inside information for personal benefit).
- Abuses in the nursing home industry in which private owners often place profit ahead of the health and safety of elderly residents. Such practices are described by M. A. Mendelson (1975) as "tender loving greed."

Video Link 10.2
View a video of insider trading.

- While not a blanket indictment of the profession as a whole, Mitford's *The American Way of Death* (1963, p. 8) describes illegal or unethical activities of funeral directors, including misuse of the coroner's office in order to secure business, bribery of hospital personnel to "steer" cases, the reuse of coffins, and duplicate billings in welfare cases.
- **"Churning"** by stockbrokers, which involves collecting high commissions by running up sales with unnecessary buy-and-sell orders.
- In "pump and dump" stock scams, online brokers (day traders) buy an inexpensive stock and then hype it to drive up the price, enticing others to buy the stock. Then the stock is sold at the high price after which the stock dives, costing unaware investors plenty. One such price-rigging scam caused $10 million in losses to thousands of customers. Fraud charges were brought by the Manhattan district attorney against three securities firms and 21 brokers. This group manipulated the prices of over-the-counter stocks in certain companies by buying and selling over and over again among themselves ("21 Brokers Charged," 1991).

Insider trading occurs when agents, brokers, or company officials who are aware of pending developments make use of this privileged information to buy or sell stocks before the public learns of these events. Revelations of such wrongdoing led to the collapse and declared bankruptcy of Drexel Burnham Lambert, a major Wall Street investment banking firm. However, in 1989, only 2 months before declaring bankruptcy, the company gave out over $260 million in bonuses to employees, twice the amount of the debt on which Drexel defaulted. A few executives received $10 million each while Drexel, on paper, lost $40 million. While not illegal, such activity certainly fits C. Wright Mills's (1952) theme of the "higher immorality." In 1995, the SEC, in the largest settlement of its kind, had Merrill Lynch and Lazard Freres each agree to pay about $24 million to settle charges that they were involved in a secret fee-splitting scheme with municipal bond underwriters and officers and municipalities ("There's a New Sheriff in Town," 1995, pp. C1, C7). In 2003, major Wall Street brokerage firms including Salomon Smith Barney, Credit Suisse, and Merrill Lynch pleaded nolo contendere and agreed to pay approximately $1.4 billion for knowingly causing investors to lose trillions of dollars in bad investments. Dubious research and insider preference in allocation of new stock shares contributed to huge losses (Morgenson, 2002). In 2007, the backdating of stock options, a practice in which executives improperly change the dates of stock-option grants to increase the value of the grants when cashed in, may be the largest business scandal since the 1980s. Prosecutors charge that backdating is hard-core fraud that hurts earnings and siphons millions from investors (Iwata, 2007). In one case, William McGuire, former chief executive of United Health, agreed to forfeit $418 million in order to settle claims related to backdated stock options. This was in addition to $198 million he had previously agreed to return to his former employer. These represent the first forfeitures exacted by the SEC based on laws put in place after the Enron collapse that forced executives to disgorge ill-gotten gains (Dash, 2007).

Scandals in education are yet another growth industry in the world of crime. The "Coded-Pencil-Caper," which took place in 1996, took advantage of the U.S. time zone difference to assist people in cheating on the Graduate Record Exam (GRE), Test of English as a Foreign Language (TOEFL), and Graduate Management Admission Test (GMAT). Those taking the test on the East Coast would phone the questions and answers to collaborators on the West Coast, who prepared coded pencils with the answers written on them to be used during the tests. Hundreds of prospective test takers paid the American Test Center $6,000 each. The company had advertised a "unique method" for preparing for the exams. The test takers were flown to the West Coast to take the tests and receive the promised "uniquely" high scores before the whole scheme was busted (D. R. Simon & Hagan, 1999, p. 83).

In 1999, a total of 52 educators from 32 New York City public schools were charged with helping students cheat on the standardized reading and math tests. In some cases, teachers actually erased and corrected answers. Many teachers felt pressured by their principals to cheat because success on the tests was tied to school funding (K. Kelly, 1999). In 1995, Steinmetz High School (Chicago) won a statewide academic contest, the Academic Decathlon, by memorizing the answers to a stolen copy of the test. Sponsors of the event became suspicious when they noticed that only 12 students in the country had scored 900 or better on the math quiz and 6 of them were from Steinmetz, a working-class high school.

The title was revoked when the students refused to take a validation test. At a 5-year reunion, some of the students indicated they would do it again, with no guilt, because that is the way the world works (D. Johnson, 2000, p. A6).

Corporate Crime

Video Link 10.3
View a video on corporate corruption.

Organizational crime refers to crime committed on behalf of and for the benefit of a legitimate organization. **Corporate (business) crime** is a type of organizational crime committed in free enterprise economies and thus involves criminal activity on behalf of and for the benefit of a private business or corporation.

Corporate crime takes many forms, including price fixing, kickbacks, commercial bribery, tax violations, fraud against government, and crimes against consumers, to mention a few (Blankenship, 1995). Sutherland's studies of white collar criminality in the 1940s set a tone and sparked other studies during that initial period. Surprisingly, however, with the exception of a few scholarly works, investigative journalistic pieces, and consumer studies (particularly by Ralph Nader and associates), there was a considerable hiatus of research activity in this area until the middle to late seventies. In 1977, Geis and Meier (1977, p. 1), in revising their classic reader on white collar crime originally published 9 years previously, found that they were able to add less than a third new material. With the exception, then, of works by Sutherland (1940, 1941, 1945, 1949, 1956a), Clinard (1946, 1969), Hartung (1950), and Nader and associates (see especially Nader, 1965, 1970, 1973), white collar crime was ripe for the research picking. Friedrichs (2006) adds that it is remarkable that criminology has a "bottomless well of analysis applicable to the delinquency of inner city youths and relatively little to contribute to the crimes of the most powerful adults [crime in high society] in our own society."

A new renaissance in studies of white collar crime took place in the late seventies with publications by Clinard and Yeager: *Illegal Corporate Behavior* (1979) and later *Corporate Crime* (1980). Other than Sutherland's pioneering effort, which was modest by comparison, the research conducted by Clinard and Yeager represents a landmark: the first large-scale, comprehensive investigation of corporate crime. They conducted a systematic analysis of administrative, civil, and criminal actions either filed or completed by 25 federal agencies against 477 of the largest manufacturing corporations in the United States during 1975–1976. In addition, they performed a less comprehensive survey of 105 of the largest wholesale, retail, and service corporations (Clinard & Yeager, 1980, p. 110). Among their findings were the following:

- Sixty percent of the large corporations had at least one action initiated against them during the period.
- The most deviant firms (multiple violators) accounted for 13 percent of those charged (8 percent of all corporations studied) and for 52 percent of all offenses. The average for these corporations was 23.5 violations per firm, while the average for all corporations was 4.2.
- Large corporations were the chief violators, with oil, pharmaceutical, and automobile industries the biggest offenders and the most often cited. These three groups alone accounted for almost half of all the violations.
- The general leniency with which corporate violators are treated, noted over 40 years previously by Sutherland, appeared to persist.

Crimes by Organizations/Corporations Against Individuals (the Public)

Included in the discussion of *crimes by organizations against individuals (the public)* are multinational bribery, corporate fraud, price fixing, manufacturing and sale of faulty or unsafe products, inequitable taxes, and environmental crimes, to mention just a few.

Multinational Bribery

Embarrassed by the public disclosure and international scandal of American-based multinational corporations' expending millions of dollars to bribe foreign officials, the U.S. Congress passed the Foreign Corrupt Practices Act (1977). This law forbids the payment of bribes in order to obtain business contracts. Earlier in this chapter, Transparency International's Corruption Perceptions Index (CPI) was discussed. In 1999 this same organization began producing a Bribe Payers Index (BPI). The questions used in the construction of the index related to leading exporters having to pay bribes to senior public officials. Only 30 exporting countries were analyzed in 2008. A 10 on the index represents negligible bribery, while a zero indicates high levels of bribery. Some select countries and their 2008 bribery scores were (http://www.transparency.org/news_room/latest_news/press_releases/2008/bpi_2008_er):

Belgium	8.8
Canada	8.8
United Kingdom	8.6
Japan	8.6
USA	8.1
Italy	7.4
China	6.5
Russia	5.9

Corporate Fraud

In 1989, an FBI undercover sting operation of commodities traders at the Chicago Board of Trade uncovered traders who overcharged customers, did not pay customers the full proceeds of sales, used their knowledge of customer orders to "inside trade" for their own benefit, and executed orders for fictitious practices (Berg, 1989). Perhaps one of the biggest computer swindles in history, amounting to an estimated $2 billion, came to light in 1973 with the bankruptcy of the Equity Funding Corporation of America. Executives at Equity Funding's life insurance subsidiary used the company computer to create roughly 56,000 phony or "ghost" policies (about 58 percent of all policies the company held). Reinsurers who bought the rights to the dummy policies were out millions of dollars; stockholders alone lost over $100 million. Using computer records rather than hard-copy records, the Equity Funding executives mixed genuine and phony policies in the master tape files; thus, printouts showed that the company had nearly 100,000 policies. When auditors took samples to check against hard copies, they were held off for a day or two during which phony hard-copy records were produced ("Conning by Computer," 1973). The president and 24 other employees and officers were indicted. While the former received an 8-year sentence, the others received shorter terms (Blundell, 1978). Convicted of complicity in the case, outside auditing firms were ordered to pay $39 million to former equity shareholders (Ermann & Lundman, 1982, pp. 43–48).

In 1990, Chrysler Corporation pleaded guilty to selling previously wrecked vehicles as new and disconnecting the odometers on about 60,000 vehicles. Chrysler pleaded no contest and was fined $7.6 million ("Chrysler Fined for Violations," 1990). In 2001, Chrysler was accused of having spent $1.3 billion since 1993 in buying back vehicles with chronic defects (lemons) and then reselling the bulk of these to consumers (Suhr, 2001). Other examples of corporate fraud include a 1985 plea bargain by E. F. Hutton for 2,000 counts of defrauding hundreds of U.S. banks through a check-kiting scheme. Hutton agreed to a record $2 million fine and other settlements (S. Taylor, 1985). In 1992, Sears was accused of overcharging and making unnecessary repairs to customers' vehicles at their auto service centers in California and New Jersey; undercover investigators documented a systematic fraud in California involving overselling 90 percent of the time (Yin, 1992). Stanford University was accused in 1991 of overcharging the federal government for contracted research. One overcharge was for $7,000 for bed sheets for the president of the university (Stout, 1991).

General Electric (GE) was fined $10 million and two executives were sentenced to prison for cheating the government on a contract for battlefield computers in 1990. In 1985, GE paid a fine of roughly $1 million for illegally claiming cost overruns on Minuteman missiles (Stieg, 1990a, p. 2A). An example of "serial fraud," in 1992 GE pleaded guilty to defrauding the federal government in the sale of military engines to Israel and agreed to pay $69 million in a settlement of criminal charges and a civil lawsuit ("GE Pleads Guilty to Fraud," 1992).

The Big Four superbrokers of the Japanese stock market admitted to reimbursing 231 major investors to the tune of $933 million for losses suffered in the 1987 stock market crash. While their actions were not technically illegal, smaller and foreign investors felt they were on the outside of an insiders' game (Ohmar, 1991). In 1999, Cendant Corporation, which owns Days Inn and Ramada hotels, agreed to pay $2.8 billion to stockholders. The company admitted to irregular accounting practices that were used to inflate earnings and permit insiders to sell at a profit ("Cendant to Pay," 1999). In 1994, Prudential Securities, a division of the Prudential Insurance Company of America, paid out over $1 billion in settlements and regulatory fines levied by the SEC and state securities regulators. This is the costliest fraud scandal for any investment in Wall Street's history, exceeding the previous record by Drexel Burnham Lambert of $650 million in 1989 (Eichenwald, 1994). Clients were fraudulently sold risky investments and were lied to and deceived with sales materials. In 1996, Prudential agreed to pay a record fine of more than $20 million and repay policyholders millions more for having "churned" (caused unnecessary sales to gain commissions) customers' accounts. Agents talked customers into trading in paid-up policies in order to finance new, more expensive ones. Some estimate that Prudential may have had to pay between $280 million and $1 billion in order to reimburse cheated customers ("Prudential Fined Millions," 1996).

In 1988, Hertz, the rental car company, admitted to overcharging customers and insurance companies $13 million for accident repairs in which employees forged repair bids ("Hertz Admits," 1988). However, this was minor fraud compared with the operations of defense firms. In 1989, the FBI launched a major investigation into massive fraud, bribery, and bid rigging in defense industry bids on Pentagon contracts. Particularly under attack was "the revolving door," a system in which defense company executives serve stints as Pentagon officials and then return to the industries they previously oversaw as contract officers. Such obvious conflict of interest might be viewed as "deferred bribes" in which cooperative defense contract officers will be later rewarded with defense industry jobs. The losers, of course, are the nation's armed forces and the nation's taxpayers (Waldman & Gilbert, 1989).

The medical and insurance business has been a particular area of fraud. The United States is the only developed country in the world without national health insurance. It pays 50 percent more to run its system, and special interests effectively block any attempt to extend guaranteed health care to all, as is the case in other developed countries. Big profits attract big fraud. In 1997, Blue Shield of California paid $12 million to settle charges for submitting false Medicare claims (Howe, 1997). In 1993 Metropolitan Life was fined $20 million for cheating its customers, and in 1996 Mutual of New York was fined $12.5 million for deceptive sales tactics.

Systematic fraud by Medicare providers is estimated to cost about 10 percent of total Medicare costs in the United States. Some examples of such fraud include the following (Sparrow, 1998):

- In March 1995, the FBI director said intelligence had indicated that cocaine traffickers in Florida and California were switching from drug dealing to the safer and more lucrative health care fraud business.
- A Medicare contractor in 1998 agreed to pay $144 million in civil and criminal penalties for concealing poor performance in reviewing and paying claims of Medicare beneficiaries.
- In an early 1998 scheme, more than $1 billion in phony medical bills using names of unsuspecting patients and doctors had been submitted to private insurers. (p. 2)

In 2008, federal agents raided three medical centers in the Los Angeles area and arrested executives there for alleged schemes involving recruiting homeless people as phony patients and then billing government programs for millions of dollars in unnecessary health service. Some of the skid row recruits went back numerous times (Associated Press, August 2008). In 1995, Caremark International pleaded guilty to paying kickbacks to doctors for steering patients its way. The company agreed to pay $159 million (Burton, 1995). In the largest health care settlement in U.S. history, National Medical Care, Inc., agreed to pay $500 million in civil fines, penalties, and restitution including $101 million in criminal fines for requiring needless tests of Medicare recipients and paying kickbacks for referrals ("Dialysis Chain Agrees," 2000). While there is no annual index of white collar crime comparable to the UCR for street crime, in 2002 the FBI began publishing an annual *Financial Crimes Report to the Public*. While limited to crimes being investigated in a given year by the Financial Crimes Section of the Bureau's Criminal Investigation Division, it serves as one measure of federal activity in investigating corporate fraud, securities fraud, health care fraud, mortgage fraud, and insurance fraud (see Crime File 10.1). The FBI works closely with other regulatory agencies in undertaking these investigations.

Price Fixing

Collusion and price fixing to set artificially high prices had become the norm in the electrical industry, with the firms taking turns (rotational bidding) submitting the lowest bid. This cost the American public untold millions, perhaps billions, of dollars in higher prices.

The Great Electrical Industry Conspiracy. This conspiracy involved price fixing on Tennessee Valley Authority equipment. In February 1961, seven of the highest executives in the electrical industry, from firms such as General Electric and Westinghouse, were given jail sentences of 30 days, an unprecedented benchmark decision that sent a warning to corporate price fixers, bid riggers, and market slicers. In addition, General Electric was fined $437,500 and Westinghouse $372,500. In all, 29 companies and 45 executives were convicted of bid rigging and price fixing estimated at approximately $2 billion (Herling, 1962). The conspirators were well aware of the illegality of their activities: They met under fictitious names in hotel rooms, called their meetings "choir practice," and referred to the list of participants as "the Christmas card list."

Plumbers fix more than leaks. In 1975, the U.S. Department of Justice filed an antitrust suit against three plumbing manufacturers (American Standard, Borg Warner, and Kohler) and three executives for conspiring to fix prices on $1 billion worth of bathroom fixtures. The case actually began in 1966 with 17 corporate and individual coconspirators named. The others pleaded no contest to the charges, got short jail terms, and were fined a total of $370,000 ("U.S. Begins Price-Fixing Prosecution," 1975). The $1 billion stolen by these organizations from the public dwarfs by far the more "mundane" criminal activity that gains so much media attention. For instance, the Boston Brinks robbery netted only $2 million and the largest robbery in U.S. history as of 1980, that of the Lufthansa airport warehouse in New York City, scored only $4 million (Clinard & Yeager, 1980, p. 8). This was superseded in 1983

Crime File 10.1

Financial Crimes: FBI Releases Annual Report to the Public

The arm of the FBI that investigates financial crimes ranging from underground pyramid schemes to institutionalized fraud in the nation's corporate suites has issued its annual report detailing the most prevalent types of schemes investigators tackled in 2006.

The Financial Crimes Report to the Public is prepared each year by the Financial Crimes section of the FBI's Criminal Investigative Division. The report, which covers a 12-month period ending September 30, 2006, explains in detail dozens of fraud schemes, tallies FBI accomplishments combating the crimes, and offers tips the public can use to protect itself. The full report is available on the FBI Web site under "Reports and Publications" (http://www.fbi.gov). The FBI's financial crimes investigations are primarily focused on corporate fraud, health care fraud, mortgage fraud, identity theft, insurance fraud, mass marketing fraud, and money laundering. Within each of those categories are dozens of schemes with a single focus: to illegally beat the system.

Here are some highlights of the report's contents:

Corporate Fraud: The highest priority of the Financial Crimes Section, the FBI was pursuing 490 cases at the end of FY2006, which ended last September, including 19 cases that individually cost investors over $1 billion. Investigations resulted in 171 indictments and 124 convictions, as well as over $1 billion in restitutions, $41 million in recoveries, and $4.62 million in seizures.

Securities Fraud: More than 150 agents were probing 1,655 cases, including "pump and dump" and Ponzi schemes, hedge fund fraud, and late day trading (when mutual funds are illegally traded after the market closes). FBI probes resulted in 302 indictments and 164 convictions last year.

Health Care Fraud: The FBI is the primary investigative agency in the fight against health care fraud, an issue that is expected to grow. Some of the most common schemes include upcoding services (billing for more services than provided), duplicate claims, kickbacks, and providing medically unnecessary services. This section of the report provides tips to protect yourself (review medical bills) and a tip line. More than 2,400 cases investigated last year resulted in 524 convictions, including a doctor whose unnecessary procedures resulted in the deaths of two patients.

Mortgage Fraud: The FBI investigates in two distinct areas: fraud for profit, which often involves insiders inflating a property's value, and fraud for housing, which is typically when borrowers misrepresent their incomes in order to qualify for loans. The report shows a regional analysis of fraud hot spots and lists some indicators of fraud, like requests to sign blank documents or requirements to use an exclusive appraiser. The number of FBI cases has grown steadily in recent years, from 436 in 2003 to 818 last year (2006).

Insurance Fraud: Of the 233 cases investigated last year, 54 resulted in convictions. The number of cases and convictions is expected to rise in the near future as more fraud is uncovered in the wake of Hurricane Katrina, which generated, according to some estimates, more than $34 billion in insurance claims.

The report also provides tips on how to recognize different types of scams and what to do if you are victimized. To that end, you can also read about different financial scams on the FBI's *Common Fraud Schemes* page. The bottom line of the report: Financial schemes, in the end, are designed to game the system and cheat innocent people of their fair share. Be informed to protect yourself.

Source: Financial Crimes Report to the Public, Fiscal Year 2006, FBI, 2006, Washington, DC: Author. Available at http://www.fbi.gov/publications/financial/fcs_report2006/financial_crime_2006.htm.

Web Research Project
Read the Financial Crimes report (http://www.fbi.gov/publications/fimancial/fcs_report2006/financial_crime_2006.htm) and discuss some areas of crime with which you have now become more informed.

by the $11 million robbery of a Sentry Armored Car warehouse in New York City ("Wells Fargo Guard Dopes Boss," 1983). Thus the "great plumbing equipment rip-off," which is far less dramatic and well-known, cost the American public the equivalent of 500 "great Brinks robberies."

The U.S. Justice Department filed suit in 1994 against General Electric Company for fixing prices on industrial diamonds in consort with DeBeers Centenary A.G., which controls 90 percent of the billion-dollar market for synthetic diamonds. This trial constituted the first antitrust case to go to trial in about two decades.

In 1980, the U.S. Department of Energy filed suit against 15 major refining companies and charged them with more than $10 billion in possible pricing violations. As part of some out-of-court settlements, several of the corporations agreed to reimburse overcharged customers; pay the government; give rebates on past charges; cut prices; and accelerate investment in refining, exploration, and production (Lyons, 1980).

The most expensive series of white collar frauds in U.S. history was "The Great Savings and Loan Scandal." Crime File 10.2 provides a brief account.

In 2000, federal investigators claimed that two of the largest auction houses, Sotheby's and Christie's, fixed prices by fixing commissions. The decision was made in secret meetings by the chairpersons of both companies (Frantz, Blumenthal, & Vogel, 2000). While at times it may appear that antitrust enforcement is fruitless, Scott (1989) notes that public exposure of trade conspiracies serves as a deterrent despite weak penalties. Some other examples of price fixing include the case, mentioned earlier in the chapter, of F. Hoffman–La Roche Ltd., which was fined $500 million in 1999 for leading an international conspiracy to fix prices on vitamins (Mokhiber, 1999). BASF Aktiengesellschaft was fined $225 million for a similar venture, while SGL Carbon Aktiengesellschaft was fined $135 million for fixing prices on graphite and carbon products. In 2009, the European Commission fined chipmaker Intel with a record $1.45 billion fine for using illegal monopolistic practices (Acohido, 2009). The commission found that the chipmaker, which controls 77 percent of the global share of chips, was guilty of manipulating rebates that unfairly restricted computers containing the microprocessors of competitors. The fine surpassed a $613 million penalty levied in 2004 on Microsoft for using illegal practices related to its Windows Media Player and Windows operating system.

Also mentioned earlier, in 1996, Archer Daniels Midland was fined $100 million for fixing markets on lysine and citric acid, while in 1997 Haarman and Reiman Corporation and Bayer AG pleaded guilty and paid a $50 million criminal fine for fixing prices on citric acid (Mokhiber, 1999). In 2003, Bristol-Squibb paid $670 million for illegally blocking generic competition for an antianxiety drug, BuSpar, causing consumers to pay more for the drug than they should have (Appleby, 2003, p. 28). Nintendo, the Japanese video game maker, was fined $147 million by the European Commission for colluding with European distributors to fix prices on its products (Meller, 2002). During the G. W. Bush administration, not a single case against big companies for anticompetitive behavior to shut out smaller rivals was brought forth ("Editorial", 2009).

Sale of Unsafe Products

The Ford Pinto Case. In what turned out later to be a bad pun, the advertising slogan for the Ford Pinto was "Pinto leaves you with that warm feeling." In the early sixties, in order to compete with compact foreign imports, the Ford Motor Company rushed the compact Pinto model into production. Since retooling for the assembly line was already a costly investment, the company chose to proceed with production despite the results of its own crash tests, which indicated that the gas tank exploded in rear-end collisions. Choosing profit over human lives, the company continued to avoid, and to lobby against even 8 years later, federal safety standards that would have forced modification of the gas tank (Cullen, 1984; Dowie, 1977).

Crime File 10.2

Audio Link 10.3
Listen to a discussion of the Keating Five.

The Great Savings and Loan Scandal: The Biggest White Collar Crime in U.S. History

The Great Savings and Loan Scandal, with an estimated cost of $500 billion, represents the biggest, costliest series of white collar crimes in U.S. history, far exceeding the Teapot Dome Affair of the 1920s or the **Great Oil Scam** of the 1970s. It represents about 40 years of all other property crime combined, and it also represents the most costly public policy failure in U.S. history. Five hundred billion dollars is more than the cost of all of the bank robberies in the United States since its founding. The U.S. attorney general, the FBI, and the General Accounting Office estimate that at least one third of these losses was due to either regulatory neglect or criminal fraud. The savings and loan (S&L) scandal reflected increased criminal opportunity resulting from an economic crisis and deregulation taken advantage of by greedy insiders who collectively looted financial institutions and sent the bill to the U.S. taxpayers. It represented a criminal justice failure of record-breaking proportions.

A succinct account of the S&L crisis may tend to oversimplify a complex history, but let it suffice to say that the federal government, in order to protect against bank collapses, decided to guarantee bank and savings-and-loan deposits in the 1930s. This $10,000 per deposit would eventually be raised to $100,000 per deposit. In return for this, S&Ls were strictly limited to home mortgages and interest loans/payments. By the 1970s, double-digit inflation wreaked havoc on the industry, which was stuck with 6 percent, 30-year mortgages when inflation was 14 to 16 percent. S&Ls began to collapse and, as a rescue attempt, the federal government in 1982 decided to deregulate them. This included permitting them to charge more competitive interest rates and to invest in other commercial activities and banking services. Despite these measures, over 300 federally insured S&Ls collapsed between 1980 and 1986, and many others were "zombies," technically insolvent with negative net worth (Cranford, 1989; Kane, 1989). The thrift (savings and loan) industry persuaded Congress to continue to postpone inevitable closings, thus raising the final costs. Ignored by Congress and the president, who had their eyes on the next election, S&Ls became unregulated and victimized by congressional incompetence and regulatory ineptitude (Pilzer & Deitz, 1989, p. 126). "Heads I win, tails FSLIC [Federal Savings and Loan Insurance Corporation] loses" became a common phrase in the industry.

Deregulation created a climate of criminal opportunity, a backing of a "junk bond (high-risk) speculative environment" with federal deposit insurance. Wealthy criminals, such as Charles Keating, robbed the S&Ls, and they were aided and abetted by the "best and brightest" professional talent the United States had to offer. More than 80 law firms represented Charles Keating to the tune of $70 million in legal fees. Six of the "Big Eight" accounting firms were charged by federal authorities with illegal conduct. Wall Street brokerage firms unethically took advantage of unsophisticated thrift managers (Mayer, 1990). Pizzo et al. (1989) claim,

> A financial mafia of swindlers, mobsters, greedy Savings and Loan executives, and con men capitalized on regulatory weaknesses created by deregulation and thoroughly fleeced the thrift industry. While it was certainly true that economic factors (like plummeting oil prices in Texas and surrounding states) contributed to the crises, the Savings and Loans would not be in the mess they are today, but for rampant fraud. (p. 289)

Charles Keating had purchased Lincoln S&L with $50 million in junk bonds purchased from Michael Milken, and then used the S&L's money almost exclusively to buy more junk bonds from Milken. Millions in campaign donations and a mistaken overemphasis on constituent service led members of Congress to ignore their oversight function and left the S&Ls as a playground for professional scam artists with the taxpayer as the hapless victim.

Additional Sources: "Fixing the Thrifts," by P. J. Benekos and F. E. Hagan, July 14, 1991, *Journal of Security Administration*, pp. 65–104; "The Great Savings and Loan Scandal," by F. E. Hagan and P. J. Benekos, July 14, 1991, *Journal of Security Administration*, pp. 41–64; "What Charles Keating and 'Murph the Surf' Have in Common: A Symbiosis of Professional and Occupational and Corporate Crime," by F. E. Hagan and P. Benekos, Spring 1992, *Criminal Organizations*, 7, 3–26; *Other People's Money: The Inside Story of the Savings and Loan Mess,* by P. Pilzer and R. Deitz, 1989, New York: Simon & Schuster.

Web Research Project

Locate the online article "Will Charlie Keating Ride Again?" by L. J. Davis. What fears does Davis express?

An estimated 500 persons were burned to death because of the firetrap engineering of the tanks. Once the word spread, Ford withdrew the commercial that the car gave one a "warm feeling." *Mother Jones* (an investigative magazine) collected documents and called to public attention Ford's wrongdoing (Cullen, Makestad, & Cavender, 1987; Dowie, 1977). While the company estimated that it could have made the necessary modifications for about $11 per car, *Mother Jones* estimated the cost at half that. Using the National Highway Traffic Safety Administration estimate of the cost per fatality (assuming lawsuits) of roughly $200,000, Ford had, according to a company memorandum, performed a cost-benefit analysis of the problem. Paying for deaths, injuries, and damages without changing the tanks was guessed to cost about $49.5 million, while the cost of modifying the 12.5 million vehicles would run $137 million. It was "cheaper" to ignore the problem and face the lawsuits.

Photo 10.4

In May 1978, the Department of Transportation recalled all 1971 to 1976 Pintos.

In May of 1978, the U.S. Department of Transportation finally recalled all 1971 to 1976 Pintos and, although it was the biggest auto recall up to that time, the decision amounted to too little too late for the conservatively estimated 500 dead, maimed, or scarred victims (Ermann & Lundman, 1982, p. 18). The **Ford Pinto case** was also a landmark, representing the first time in U.S. history that a corporation was indicted for murder. In 1978, Indiana prosecutors charged Ford with homicide after three people were burned alive in a Pinto (Browning & Gerassi, 1980, p. 406). Even though Ford was acquitted, the trial of a corporation for murder may have served as a signal that the public reaction to corporate crime was changing (Swigert & Farrell, 1980). When asked what fate Lee Iacocca, then president of Ford, deserved, one person sarcastically suggested that someone buy him a Ford Pinto complete with Firestone 500 tires (yet another dangerous product whose manufacturer hid its defects until an unacceptable number of human sacrifices sparked federal action).

The Ford Explorer–Firestone Tires Recall. In 2000, executives of the Ford Motor Company and Bridgestone/Firestone, Inc., both appeared before the U.S. Congress to answer questions regarding possible cover-ups of defects in the Ford Explorer (sport utility vehicle), particularly when equipped with Firestone tires. The tread on Firestone ATX and Wilderness tires would peel off, forcing the Ford Explorer to roll over. By 2000, it became apparent that numerous accidents had taken place due to defects that compounded when the two products were combined. By February 2000, tires were recalled in Thailand and Malaysia. In May, a National Highway Traffic Safety Association (NHTSA) study was begun and Ford recalled 30,000 tires in Venezuela, Ecuador, and Colombia. In August, Firestone recalled 6.5 million tires; Ford CEO Jacques Nasser apologized on television, and Bridgestone/Firestone CEO Masatoshi Ono apologized to the U.S. Congress. By August 31, 2000, NHTSA labeled 1.4 million more tires defective and estimated fatalities at 88 and injuries at 250. Ford had recalled the tires in 16 countries, but had not issued a U.S. recall until later. While both companies pointed the finger at the other for the defects and crashes, in fact neither was as forthcoming as it could have been in identifying problems; it took congressional pressure to force them to act as good corporate citizens.

Reminiscent of C. W. Mills's (1952) "higher immorality" notion, Heilbroner et al. (1973) in their book *In the Name of Profit: Profiles in Corporate Irresponsibility* maintain that the people who run our supercorporations are not merely amoral, but positively immoral. They and other authors cite examples like these:

- B. F. Goodrich plotted to sell defective air brakes to the U.S. Air Force by faking test records and falsifying laboratory reports. National security and the lives of fighter pilots appeared to be of little concern.
- In the early 1970s, a General Dynamics engineer warned his superiors of dangerous defects in DC-10 cargo doors. They ignored this warning. Two years later, a DC-10 crashed in France when the cargo doors opened in flight, killing all 346 passengers (Nader, Green, & Seligman, 1976).
- Theo Colburn in *Our Stolen Future* (1996) claims that toxic chemicals released in the last 50 years mimic natural hormones and may be responsible for human male sperm counts decreasing 50 percent since 1938 and growing infertility, genital deformities, breast and prostate cancer, and neurological disorders.
- In the 1990s, Ford Bronco II's rollover accident rate for rear-wheel drive was double the sport utility vehicle average. In 1992, Ford had spent $113.4 million to settle 334 lawsuits for rollovers.
- In 1994, U.S. Secretary of Transportation Federico Pena made a deal with General Motors (GM) not to order recall of its pickup trucks, whose defects cost 150 lives, in return for a payment of $51 million to support safety programs. The latter were calculated to have the benefit of saving more lives than the calculated 32 more people who would die due to faulty design of the trucks (J. Bennett, 1994).
- In the largest product liability settlement in U.S. history, a federal judge in 1994 granted $4.25 billion in a class action suit against 60 silicone breast implant manufacturers. Dow Corning Corporation agreed to pay the biggest share, $2 billion. The largest previous settlement had been by asbestos manufacturer Manville Corporation for $3 billion. One study showed that Dow knew of the dangers as early as 1975 (Blakeslee, 1994).
- In 1994, the FBI employed a GE engineer to infiltrate a GE jet engine plant to spy on managers who were later charged with compromising the safety of military and commercial aircraft by covering up engine flaws. GE had been involved in more cases of contractor fraud against the Pentagon than any other manufacturer (Frantz & Nasar, 1994). This included 16 criminal convictions and civil judgments.
- In 1995, top executives of seven tobacco companies told a congressional committee under oath that they did not know for certain that tobacco was addictive or caused disease. Attorney General Janet Reno asked the U.S. Justice Department if the companies were guilty of fraud and perjury when it was revealed that Brown and Williamson Tobacco Corporation documents indicated that the company's own research had shown for years that cigarettes were addictive and harmful, and it had covered up such knowledge (Hilts, 1995). A record $368 billion deal between the tobacco companies and 40 states, including punitive damage awards of $50 billion, affected how cigarettes are advertised and sold.

The National Consumer Product Safety Commission estimates that 20 million serious injuries and 30,000 deaths a year are caused by unsafe consumer products (Coleman, 1994, p. 9). When Richardson-Merrell's MER/29, a cholesterol inhibitor, was tested, all the rats died; nevertheless, the company falsified the data and marketed the product. When over 5,000 users suffered serious side effects, it was withdrawn from the market and the company received a minor fine (Coleman, 1994, p. 84). In 1988, the Cordis Corporation pleaded guilty in federal court to concealing defects in thousands of pacemakers (implanted in heart patients to regulate the heartbeat). The company agreed to a fine of $264,000 plus court costs. Executives were charged separately ("4 Indicted Over Pacemaker Scam," 1988).

Defective products continue to plague consumers. More recent examples are defective breast implants that maim, deform, and destroy immune systems; and flawed heart valves (Ingersoll, 1991).

Environmental Crime

In 1962, the publication of Rachel Carson's *Silent Spring* signaled the beginning of the age of environmental awareness. Specifically attacking toxic chemicals and pesticides, Carson's work very dramatically called attention to the irreversible and final genetic and biological harm the poisoning of the environment could bring about. According to Regenstein (1982), "The accuracy and validity of *Silent Spring* was no inhibition to the chemical industry's attacking and attempting to discredit it, a vicious campaign which started even before the work was published and continues today" (p. 132).

Handbook Article Link 10.3
Read an article on
wildlife crime.

Three Mile Island. Ironically, in the film *The China Syndrome* (so-named because of the false belief that a nuclear meltdown in the United States would bore through the earth to the other side—China), a character indicates that a nuclear mishap could render an area the size of Pennsylvania uninhabitable. Almost prophetically, after the release of the film, the worst potential nuclear plant disaster in history occurred at Three Mile Island, Pennsylvania. The accident released radioactivity into the surrounding area and required the temporary evacuation of young children and pregnant women from the immediate vicinity.

On November 7, 1983, a federal grand jury indicted Metropolitan Edison, the owners of the Three Mile Island (TMI) facility, on criminal charges of faking safety test records before the accident. The indictment alleged that the company attempted to conceal from the Nuclear Regulatory Commission the rate of leakage in the main cooling system, in which water passes over the reactor's radioactive core ("Feds Indict TMI," 1983). Allegations had been made that the corporation was eager to have the reactor online by a certain date in order to take advantage of tax benefits.

In April 1984, Metropolitan Edison pleaded guilty to knowingly using inaccurate and meaningless testing methods and agreed to pay a $1 million fine. The company also pleaded no contest to six other criminal counts, including manipulating test results, destroying records, and not filing proper notice of cooling system leaks ("Judge Agrees to TMI Plea Bargain," 1984).

Toxic Criminals. Potential environmental hazards created by new technologies require that corporations and businesses exercise a higher level of ethical behavior than that exhibited in the Ford Pinto incident or other cover-ups and deceptions of the public and of government regulatory agencies. Bhopal (India), Love Canal, Times Beach, Seveso (Italy), and Chernobyl are well-known environmental disasters. Each year, about 300 health care workers die from hepatitis B after exposure on the job (J. Anderson & Van Atta, 1988a). Toxic wastes also expose the public to possible harm. In 1979, the EPA estimated there were 109 very hazardous dumpsites and 32,254 sites where hazardous wastes were buried. That latter figure was subsequently raised to 51,000, with "significant problems" existing at between 1,200 and 34,000 (M. H. Brown, 1982, p. 305). In the late 1980s, beaches in various parts of the United States had to be closed because illegally dumped medical wastes were washing ashore. Blood gushing out of trash compactors and body parts found in trash piles illustrate the ghoulish proportions of such hazards.

Crime File 10.3 reports on the deadliest air pollution disaster in American history, the Donora Fluoride Death Fog.

Alcoa agreed to pay criminal penalties of $7.5 million to the state of New York for illegally holding 33 rail cars full of PCB-tainted soil at a Massena, New York, facility over several months while preparing fake documents to dispose of the material as nonhazardous (Milbank & Allen, 1991).

U.S. v. Allied Chemical. In 1976, Judge Robert Merhige (Richmond, Virginia) fined Allied Chemical $13.2 million after it pleaded nolo contendere to 153 charges of conspiracy to defraud the EPA and Army Corps of Engineers. Allied had polluted the James River and had deceptively blocked the efforts of these agencies to enforce water pollution control laws. In justifying the largest fine ever imposed in a single environmental case, the judge stated, "I don't think that commercial products or the making of profits are as important as the God-given resources of our country" (Beauchamp, 1983, p. 97).

Much of the work social scientists or federal agencies should have been doing in investigating corporate crime has until relatively recently been shouldered by investigative journalists and consumer

Crime File 10.3

**Handbook
Article Link 10.4**
Read an
article on
environmental
crime.

The Donora Fluoride Death Fog: A Secret History of America's Worst Air Pollution Disaster

An environmental horror story similar to fictional works by Michael Crichton visited Donora, Pennsylvania (near Pittsburgh), on Halloween night, 1948. History tells us that a temperature inversion trapped smog in the narrow industrial valley and produced the worst single air pollution disaster in American history, leaving 20 dead and hundreds injured and dying. This incident resulted in the passage of the 1955 Clean Air Act. Bryson (1998) and L. P. Snyder (1994) described the Donora cover-up:

Fluoride emissions from the Donora Zinc works and steel plants owned by U.S. Steel caused these injuries. Philip Sadtler, a chemical consultant who conducted research at the scene of the disaster, concluded that "U.S. Steel conspired with U.S. Public Health Service (PHS) officials to cover up the role that fluoride played in the disaster" (Bryson, 1998). One third of the town's 14,000 residents were affected by the smog, with hundreds evacuated or hospitalized.

While the official PHS report stated "no single substance" was responsible and laid blame on the temperature inversion, Sadtler charged that the PHS report was designed to assist U.S. Steel in escaping liability for the deaths and to prevent controls of toxic fluoride emissions. The national fluoride clean-up would have cost billions.

The PHS was then part of the Federal Security Agency headed by Oscar Ewing, a former top lawyer for Alcoa (third-largest aluminum producer). The latter was facing lawsuits at the time for wartime airborne fluoride emissions. Sadtler, in the December 13, 1948, issue of *Chemical and Engineering News,* reported fluorine blood levels of the dead and ill patients to be 12 to 25 times above normal. Afterward, pressure was brought to bear by manufacturers to prevent the journal from publishing any more articles by Sadtler (Bryson, 1998).

Researching the disaster 50 years later, investigators discovered that important records were missing from the PHS archives and that U.S. Steel records were not open to researchers. Despite the fact that, at the time, the Donora disaster was the largest government environmental investigation ever conducted, almost all the records mysteriously vanished when Dr. Lynne Page Snyder was doing research for her dissertation at the University of Pennsylvania. Snyder's dissertation was titled "The Death-Dealing Smog Over Donora, Pennsylvania: Industrial Air Pollution, Public Health Policy, and the Politics of Expertise, 1948–1949" (L. P. Snyder, 1994). She suspects the archives were determined to be too hot to handle and were gotten rid of.

As a sequel to "The Donora Fluoride Fog," Bryson (2000) describes "Fluoride and the Mohawks" between 1960 and 1975:

Cows crawled around the pasture on their bellies, inching along like giant snails. So crippled by bone disease they could not stand up, this was the only way they could graze. Some died kneeling, after giving birth to stunted calves. Others kept on crawling until, no longer able to chew because their teeth had crumbled down to their nerves, they began to starve.

The cattle belonged to Mohawk Indians on their reservation that straddled the New York–Canadian border; fluoride emissions from nearby aluminum plants devastated their herds and way of life. Crops and trees, birds and bees withered and died. Today, fish caught in the St. Lawrence River by the Mohawks have ulcers and spinal deformities, and Mohawk children also exhibit signs of bone and teeth damage.

In 1980, the Mohawks filed a $150 million lawsuit against two aluminum companies, but after 5 years of legal costs, the bankrupt tribe settled for $650,000 in compensation for damage to their cows.

Web Research Project
Go online and read about the Bhopal disaster in India in 1984. Do you see any parallels with Donora?

advocates, such as Ralph Nader and his associates. A partial list of such studies and their subject matter includes the following:

- Cox, Fellmuth, and Schulz (1969), a report on the FTC
- Esposito and Silverman (1970), *Vanishing Air,* on air pollution regulation
- J. S. Turner (1970), *The Chemical Feast,* on the FDA

- H. Wellford (1972), *Sowing the Wind,* on health and environmental hazards
- M. J. Green et al. (1973), *The Monopoly Makers,* on antitrust activity
- Page and O'Brien (1973), *Bitter Wages,* on occupational safety and health

In addition to these, Nader and his associates have generated numerous other investigations and reports (Nader, 1965; Nader & Green, 1973; Nader, Green, & Seligman, 1976; Nader, Petkas, & Blackwell, 1972). Some success has occurred in the battle against polluters. In 2007, the American Electric Power Company agreed to pay $4.6 billion to settle 8 years of charges that its acid rain–causing chemicals ate away at parks, bays, and the Statue of Liberty (Barrett, 2007). On the subject of environmental and health assaults on consumers, D. R. Simon (1999) describes some cases of "**corporate dumping,**" a practice whereby corporations sell overseas products that have been deemed unsafe in the United States by the EPA, FDA, or other federal agencies. Toxic crime may indeed be the ultimate and most insidious of crimes. Birth defects, long-term genetic damage and mutation, congenital heart defects, and disorders in children—many of these effects may turn up 20 to 30 years later and be difficult to link to the original causative agents or toxic chemicals. In that sense, those who commit environmental crimes may represent the first "intergenerational criminals"—the victimized may not have been born at the time the crime was perpetrated, and the criminal may be deceased by the time the victimization takes place.

Radiation Leaks. In 1988, in the wake of the Chernobyl disaster in the Soviet Union, investigations began to reveal a massive cover-up by the U.S. federal government of the dangers and harm its nuclear facilities and testing program had posed to unwarned workers and neighbors. Fallout from atomic tests in the 1950s and 1960s resulted in little warning by the Atomic Energy Commission of exposure hazards such as birth deformities, cancer, and early death (McGrory, 1988). The Department of Energy runs federally owned nuclear plants that produce the fuel for the nation's nuclear weapons. These obsolete plants have worse safety features than most privately owned plants. The radioactive waste problem at the Energy Department's Hanford nuclear weapons reserve in the state of Washington is unbelievable. It is described as (T. W. Lippman, 1991)

> the most polluted and dangerous nuclear compound in the United States and perhaps the world—the submarine hulks [21 buried radioactive reactor vessels] are little more than a novelty. In fact, they are a stable controllable form of waste in a nightmare world of volatile, explosive, toxic and radioactive junk.

> So great is the mess, so diverse the streams of waste—solids and liquids, above ground and beneath, in the water and in the soil, stationary and migrating—that the most optimistic forecasts say it will take at least 30 years and $30 billion to clean it up. And that's if all goes well, if none of the tanks of lethal liquids explodes and if scientists can figure out what to do with material for which no disposal technology exists. That's if all the waste can be contained before any more of it seeps into the Columbia River. (p. 33)

More and more of our food is processed and packaged by large corporations and, if recent investigations are to be believed, the food processors have not improved much since Upton Sinclair's (1906) expose in *The Jungle.* Despite a 1906 federal Meat Inspection Act and a 1967 Wholesome Meat Act, abuses continue. In a Hormel plant in 1969, a Department of Agriculture inspector was bribed $6,000 annually for overlooking the production of "Number 2" meat (McCaghy, 1976b):

> When the original customers returned the meat to Hormel, they used the following terms to describe it: "moldy liverloaf, sour party hams, leaking bologna, discolored bacon, off-condition hams, and slick and slimy spareribs." Hormel renewed these products with cosmetic measures (reconditioning, trimming, and washing). Spareribs returned for sliminess, discoloration, and stickiness were rejuvenated through curing and smoking, renamed Windsor Loins and sold in ghetto stores for more than fresh pork chops. (p. 216)

Corporate violence. From what has been said so far it should be clear that, while the general public tends to view corporate crime as nonviolent, we might be more persuaded by S. L. Hills, who in

Photo 10.5

Upton Sinclair
became famous for
his novel *The Jungle,*
which brought the
public's attention to
the conditions in the
U.S. meatpacking
industry.

Corporate Violence (1987) describes "'respectable' business executives who impersonally kill and maim many more Americans than street muggers and assailants." He notes that the tools of such violence include

exploding autos, defective medical devices, inadequately tested drugs, and other hazardous products that are manufactured and marketed despite knowledge by corporate officials that such products can injure and kill consumers. There are reports of toxic chemical dumps that have poisoned drinking supplies, caused leukemia in children, and destroyed entire communities; of cover-ups of asbestos-induced cancer, and the gradual suffocation of workers from inhaling cotton dust; of radioactive water leaking from improperly maintained nuclear reactors; of mangled bodies and lives snuffed out in unsafe coal mines and steel mills—and other dangers to our health and safety. (p. vii)

Crimes by Organizations Against Employees

Video Link 10.4
View a video of
the McWane
Corporation,
criticized for
unsafe workplace
practices.

Organizational (corporate) crime against employees (type 6 in Table 10.1) may take many forms; the most insidious relates to purposive violation of health and safety laws that may not only threaten workers' lives, but may also genetically damage their offspring.

During World War II, the large German manufacturing corporation I. G. Farben worked captive workers (slave labor) to death in its factories. More recently, according to Harry Wu, a former Chinese political prisoner, many Chinese exports sold in the United States have been produced in the harsh conditions of Chinese labor camps (Southerland, 1991). While most modern manufacturers do not directly kill their workers, health and safety violations by corporations and organizations against their employees can take many forms (see Frank, 1985). Some occupational exposure to injury and disease may be a necessary part of employment, but unnecessary, preventable hazards and their disregard by employers in the United States are regulated by OSHA and can incur criminal penalties. Terms such as black lung (due to coal exposure), brown lung (due to cotton mill exposure), and white lung (due to asbestos exposure) have become familiar to U.S. workers. The sheer number of new chemicals to which workers are exposed and their long-term impact are enormous. Occupational hazards are not new. In 1812 in Lawrence, Massachusetts, sweatshop conditions in the textile mills produced death in one third of the workers by the age of 25 (Browning & Gerassi, 1980, p. 237).

Just to cite one example of corporate negligence and cover-up, an examination of the asbestos industry is enlightening. Carlson (1979), appearing before the congressional Subcommittee on Compensation, Health, and Safety, indicated the following:

- Examination of corporate memos, letters, and other documents from as early as 1934 showed that senior executives at Johns-Manville and Raybestos-Manhattan (two of the biggest asbestos producers) knew of and covered up company-sponsored research findings that described asbestos-caused diseases.
- Asbestos industry–sponsored research in the 1930s and 1940s also showed asbestos dangers, and researchers were prevented from publishing their results.

- One company, Philip Carey, fired its medical consultant when he warned of possible lawsuits from workers exposed to asbestos.
- Years before the companies acknowledged any awareness of asbestos dangers, documents demonstrated that they had quietly settled injury and death claims from workers who had handled asbestos.
- Johns-Manville purposely did not notify employees of the results of their medical examinations that showed asbestosis, despite executives' knowledge that the disease was progressive and fatal unless treated at an early stage (pp. 25–52).

In 1988, OSHA fined meatpacker John Morrell and Company $4.33 million for having forced hundreds of injured workers in its Sioux Falls, South Dakota, plant to keep working even right after surgery. This was the largest fine against a single employer in the agency's history. Workers in the meatpacking industry chronically suffer from carpal tunnel syndrome and tendonitis, in which joints stiffen because of the erosion of soft tissue ("Meatpackers Hit With Record OSHA Fine," 1988).

In 1990, USX agreed to pay $3.25 million for hundreds of alleged worker safety violations, including what OSHA called fatal, uncorrected hazards (K. Ball, 1990). That same year, a federal jury awarded damages of $26.3 million to a retired insulation worker. Materials once made by Owens Corning Fiberglas Corporation had not been properly labeled as dangerous, causing the worker to develop asbestosis (W. E. Green & Geyelin, 1990).

Larry Agran in "Getting Cancer on the Job" (1982) documents that the cancer epidemic has been primarily fed by many industries' systematic unconcern for workers' health, in which company physicians cover up evidence of unsafe exposure to carcinogenic substances. He concludes that the government regulatory agencies are either too timid to enforce the law or lack staff or resources with which to protect workers.

In 1999, apparel workers and human rights groups filed the largest legal challenge ever against sweatshops on American soil. The suit alleged that major American retailers conspired to place thousands of workers in involuntary servitude and horrible work conditions (Greenhouse, 1999). Poor young women from China, the Philippines, Bangladesh, and Thailand are led to believe that they are going to the United States to work; instead, they are taken to Saipan (Mariana Islands), a U.S. possession, where many work 12 hours a day, 7 days a week, sometimes without pay if they fall behind in their quotas. In some cases, exits are locked, pregnant workers are forced to have abortions, and workers are housed in barracks surrounded by barbed wire. On top of all of this, the clothing labels can read "Made in USA."

In 2007, it was reported that in 2001 nine workers in a microwave popcorn plant developed a rare lung disease due to exposure to an additive that added a buttery taste. OSHA dragged its feet in responding to or enforcing standards even as more workers became ill. The George W. Bush administration vowed to limit cumbersome regulations that it viewed as unnecessary costs on business and consumers (Labaton, 2007).

Such activity represents only the tip of the iceberg in economic globalization, which often represents a "race to the bottom" in a search for the cheapest labor possible, including child labor or, as in China, labor by prisoners. These activities are all in violation of the United Nations' Universal Declaration of Human Rights. In *Nobodies: Modern American Slave Labor and the Dark Side of the New Global Economy,* John Bowe (2007) uses his 6-year field study to describe migrant workers who were murdered as a warning to their peers and workers who are threatened with physical retaliation if they attempt to escape servitude.

If there are any heroes or heroines in the world of corporate crime, they can be found among the ranks of **whistleblowers**—employees who are willing to step forward, usually at great personal sacrifice, to reveal wrongdoing on the part of their employers (see Westin, 1981). "You don't bite the hand that feeds you," states the old adage. The decision to inform on organizational violations has often meant firing, family disruption, and ostracism from friends and former coworkers, as well as the end of one's career, as employers retaliate against the "squealer" or "stool pigeon."

In 1990, jurors ordered Lockheed Corporation to pay $45.3 million in damages to three former employees who had been fired for being whistleblowers regarding safety problems of C-5B military cargo planes. According to the workers, some of these planes with defective mainframes had been used to transport troops to Saudi Arabia ("Lockheed Ordered to Pay," 1990).

Some examples of well-known whistleblowers include

- Frank Serpico, former New York Police Department officer, who informed on fellow officers during the Knapp Commission investigation in the 1960s;
- John Dean, former legal counsel for President Nixon, who cooperated with the government in the Watergate investigation;
- Daniel Ellsberg, who revealed The Pentagon Papers to the press, alleging the government was misleading the public regarding the Vietnam War;
- Engineers at Morton-Thiokol, who had warned NASA regarding unsafe O-rings before the Challenger space shuttle disaster, and who later testified before Congress; and
- Ernie Fitzgerald, a U.S. Defense Department employee, who testified regarding a $2 billion cost overrun in the production of C-5A transports.

In extreme cases, an employer may even threaten an employee's life. While the following horror story is by no means typical, it profiles a true hero in the fight against corporate crime (see Mokhiber, 1988).

The Karen Silkwood Case

Congressional hearings (U.S. Congress, 1976) and Rashke's *The Killing of Karen Silkwood* (1981) describe the Silkwood episode. She was an employee of the Kerr-McGee nuclear plant in Guthrie, Oklahoma. The company used plutonium, one of the most lethal of substances, in its plant. A union activist for stricter safety standards at the company, Silkwood had gathered considerable information documenting the firm's negligence of health and safety measures for employees, as well as dangerous defects in the plutonium compounds being used.

On the evening of November 13, 1974, Silkwood was en route with documents to a meeting with a union official and a reporter from the *New York Times* when her auto crashed into a ditch, killing her. The documents, which had been observed at the scene by state troopers, disappeared. In a subsequent trial investigating her death, Kerr-McGee was found guilty of negligence in health and safety practices, as well as criminally liable in Silkwood's contamination by radiation leaks during her employment. The Atomic Energy Commission found the company to be in violation in the majority of the union complaints, including in the contamination of 73 employees in 17 safety lapses over a 5-year period (Rose, Glazer, & Glazer, 1982, p. 407). The jury also ordered the company to pay Silkwood's estate $10.5 million in damages ("Silkwood Vindicated," 1979, p. 40). The company appealed the case, and in January 1984, the decision was upheld by the U.S. Supreme Court.

In 1986, the U.S. Congress passed additional legislation to protect whistleblowers' jobs, as well as reward them for whistleblowing. They are entitled to as much as 15 percent of what the government collects. Some whistleblowers have collected millions.

Environmentally dangerous occupations include those of chemical and insecticide workers, miners and shipyard workers who deal with asbestos, petrochemical and refinery workers, coal miners, coke-oven workers, textile and lead workers, medical radiation technicians, and those employed in the plastics industry. The exposures and risks are enormous; since most workers cannot easily switch jobs, they are even more dependent on federal regulatory agencies to protect their health and safety. Occupational hazards may be a necessary evil in modern industrial societies, but corporate subterfuge in unnecessarily exposing workers to such threats is not. Weak enforcement of OSHA regulations has resulted in the United States having five times as many work-related deaths per capita as Sweden and three times as many as Japan (J. A. Kinney, 1990).

Crimes by Organizations (Corporations) Against Organizations

Criminal activity by organizations against other organizations (type 9 in Table 10.1) may take many forms, including crimes by private corporations against the state (e.g., wartime trade violations, cheating on government contracts, or income tax violations) and crimes by corporations against corporations (e.g., industrial espionage and illegal competitive practices).

Wartime Trade Violations

Because of their international structure, multinational corporations can sometimes play both sides of the fence in wartime. In *Trading With the Enemy,* Higham (1982) raises eyebrows with the following accusations:

- While gasoline was being rationed in the United States, managers of Standard Oil of New Jersey were shipping fuel through Switzerland to the Nazis.
- Ford trucks were produced for German occupation troops in France with authorization from Ford executives in the United States.
- Chase Manhattan Bank did business with the Nazis during the war.

An early, classic study of "white collar crime" by Marshall Clinard (1969), originally published in 1952, was titled *The Black Market.* Using records of federal regulatory agencies during World War II, Clinard examined wartime trade violations on the part of businesses. He found extensive violations of rationing, price-ceiling offenses, tie-in sales, and lack of quality control. In a study conducted about the same period, Hartung (1950) found many violations of wartime economic regulations in the Detroit wholesale meat industry.

While it is not uncommon for victors to demand that losing countries pay reparations or war debts for damages, it is surprising that the United States paid for damages to U.S. multinational plants that were ruined during Allied bombing of Nazi-occupied Europe. Parenti (1980, p. 76) describes such postwar payments to GM and International Telephone and Telegraph (ITT). ITT had produced Nazi bombers and received $27 million in damages, while GM had produced Nazi trucks and obtained $33 million in compensation. Public furor arose after World War II when it was revealed that many oil companies had collaborated with the Nazis during the war. Although President Truman ordered that the Justice Department investigate and prosecute, the case was finally settled after 15 years of litigation with a minor consent decree (Coleman, 1985, p. 178). In 2007, the U.S. House of Representatives passed new legislation making it easier to convict private contractors who defraud the U.S. government during wartime. Of particular concern were contractors who were charged with overstating the value of goods and services or concealing information or presenting false statements (Flaherty, 2007).

In the 1990s, a renewed effort was undertaken internationally to recover the money of Holocaust victims held in Swiss banks. In addition, survivors of the Holocaust sued German and Japanese companies for damages for slave labor during World War II. Charges were also made that subsidiaries of U.S. auto manufacturers were key elements of Hitler's war machine. Chase National Bank has also been investigated, along with law firms, for collaboration (Hirsh, 1998).

Industrial Espionage

Until recently, **industrial espionage** has been a relatively neglected area of investigation by criminologists. Much of the work in this area has either appeared in trade magazines or has been done by journalists (see Barlay, 1973; Engberg, 1967; P. Hamilton, 1967). Such espionage (literally spying, or the acquiring of information through deceptive or illegal forms) is performed by three different groups: (1) intelligence agencies, (2) competing firms, and (3) disloyal employees. Bergier's highly engaging *Secret Armies* (1975, p. 51) tells the story of an industrial-espionage agent who traveled from office to office of a corporate headquarters with a pushcart telling everyone that he was doing a check on secret documents, which he then proceeded to wheel away. The documents and their collector were never seen again.

Industrial spying goes back at least as far as 3000 B.C., when industrial and commercial secrets relating to silkworms and porcelain were stolen from China by Europeans. In the Middle Ages, it was so widespread that it led to patent laws. Bergier (1975, p. 15) claims that piracy by industrialists and governments was a significant factor in the spreading of the Industrial Revolution. From 1875 to World War I, Japan had the best industrial spies, after which Nazi Germany and the USSR dominated European spying. In the recent Hitachi case, a Japanese corporation attempted to steal state-of-the-art computer secrets from International Business Machines (IBM). Some examples provided by Bergier include the following:

- One large Detroit company found nine television transmitters hidden in the air vents of the main drafting room; these were probably transmitting the company's latest drawings to the competition.
- A telephone tap discovered in Manhattan covered 60,000 phone lines, presumably to pick up useful market tips, blackmail information, and the like.
- Several cases in England involved spies posing as typewriter repairmen and removing typewriters for "repair" in order to peruse used ribbons.
- Cars of important figures are stolen, only to be quickly recovered; the aim is to bug them.

In free societies, about 95 percent of industrial information is available in the trade and popular publications. In fact, a growing area of investigation is called "competitive intelligence," which involves the use of open sources (unclassified documents) to gather information on one's competition. Sources of information on U.S. industry range from legitimate to illegal, as described by the *Wade System of Sources of Information on American Industry* (P. Hamilton, 1967, pp. 222–223). There has been an unexpected wave of foreign espionage with the end of the Cold War. Some examples of such activity include the following:

- A South Korean rival plants a radio transmitter on the target company's fax machine.
- IBM claims to have lost $1 billion because of French and Japanese espionage.
- The French Intelligence Service, the Direction Géneralé de la Sécurité Extérieure (DGSE), has been most brazen, even bugging seats of businesspeople on flights and ransacking their hotel rooms for documents.
- Many companies are canceling plant tours. Americans used to be amused by the number of pictures Japanese business tourists would take when touring their plants. Many of these photographs proved very useful (Hamit, 1991).

In 2000, countries of the European Union alleged that U.S. intelligence agencies (specifically, the Central Intelligence Agency and the National Security Agency [NSA]) were using "Echelon," a worldwide electronic spy network, to benefit U.S. companies in gaining a competitive edge. In effect, the agency had redirected some of its Cold War assets toward economic intelligence. Despite denials, the information is believed to have helped Boeing sell 747s to Saudi Arabia, Raytheon sell a surveillance system to Brazil, and the Hughes Network in contracts for a telecommunications system in Indonesia (Windrem, 2000). The European Parliament alleges that all e-mail and worldwide telephone and fax communications in Europe are intercepted. Whether this is true or not, the NSA appears to have that capability.

A National Institute of Justice survey of the American Society of Industrial Security's list of directors of security in major industries found that 48 percent had experienced the theft of trade secrets (proprietary information) within the past year, and over 90 percent had encountered some

Crime File 10.4

Pirates of the Internet: Criminal Warez Groups

Let's get one thing straight: we're not talking here about kids who make the occasional illegal download of a popular song from the Internet and share it with friends (though that, of course, is wrong). We're talking about big business—professionals who get up in the morning and put in a day of stealing copyrighted music, movies, games, and software from the Internet, processing them and distributing them through peer-to-peer (P2P) or file-sharing networks. How do these "businesses" work? Known as "warez release groups," these syndicates are highly organized:

- Plants in music, film, and software industries supply the newest/hottest items to the groups.
- "Crackers" strip out the embedded source codes and insert new trademarks.
- "Q & A" test the product to make sure it works.
- Distributors transmit the items through networks.

"Executives" not only control these day-to-day operations, they also recruit new members, manage archive sites, and shield their illegal operations from law enforcement with sophisticated encryption.

What's the harm? Economic harm. Online piracy and trading of music, movies, business, and gaming software adds up to lost revenues—enough to put companies out of business, lose jobs, negatively impact the economy, and, in the end, take money out of your pockets as the losses are passed on to you, the consumer, in the form of higher prices.

Not just a U.S. crime problem. These acts of piracy are executed on an international stage—and they need an international law enforcement response. Last month [April, 2004], they got one: OPERATION FASTLINK, the largest global enforcement action ever undertaken against online piracy.

On April 21, 2004, the FBI and our international law enforcement partners conducted some 120 searches in 31 states and 10 countries to dismantle some of the best known and most aggressive online piracy enterprises. We seized over 200 computers and servers, including some that actually housed hundreds of thousands of copies of pirated works. We've identified nearly 100 leaders in these groups and expect that number to go much higher in the days ahead. To report cyber crimes, please contact your local FBI Field Office or file a complaint through the Internet Complaint Center.

Source: Quoted directly from "Pirates of the Internet," *FBI Headline Archives*, 2004. Available at http://www.fbi.gov/page2/may04/051704piracy.htm.

Research Project
Find other examples of the use of the Internet as a tool of crime.

theft within the past 10 years (Mock & Rosenbaum, 1988, p. 18). The major targets were research and development data, new technology, customer lists, program plans, and financial data. Misuse of authority/position was the principal method employed, followed by physical theft, computer penetration, subversion of employees, and false documents/authorization (p. 22). Crime File 10.4 gives an account of the "Pirates of the Internet."

In 1998, textile manufacturer Milliken and Company was charged with hiring consultants to steal customer, supplier, and manufacturing information from nine competitors. They hired a private firm for $500,000 to conduct illegal spying operations (Peterson, 1998). The U.S. Economic Espionage Act (EEA) of 1996 now criminalizes the theft of trade secrets. Two of the earliest cases prosecuted under the act were the Avery Dennison case and the PPG case. In the first example, involving the Avery Dennison Corporation near Cleveland, Ohio, two Taiwanese citizens were charged with stealing millions of dollars worth of trade secrets by bribing an Avery Dennison employee. In the PPG case, the company was informed by a competitor that someone had offered to sell them PPG trade secrets. The suspect (a former PPG employee) had carried secrets out of PPG headquarters in a gym bag. Cooperation of the competitor may have been gained by the fact that under the EEA, they could have been prosecuted as a coconspirator had they not been forthcoming (Nasheri & O'Hearn, 1998).

Warez (pronounced "wares") is derived from the plural form of the word "software" and it means copyrighted material that is illegally traded. It specifically refers to releases by organized groups, a form of commercial profit piracy ("Agents Crack Down on Global Piracy Rings," U.S. Department of State, 2001). Groups in China, Hong Kong, and Russia produce millions of bootleg copies of copyrighted software that are regularly sold on the streets (http://en.wikipedia.org/wiki/warez).

Crimes by private organizations against other private organizations raise problematic areas in jurisprudence, but what if the perpetrator is a country? In the 1990s, China and other countries in Asia tolerated widespread patent and trademark violations within which fake name-brand products were copied and sold at a fraction of their cost. While the United States and other countries continue to threaten trade sanctions over such violations, China seems to make only halfhearted attempts to comply. Calling China "The Pirate Kingdom," Choate (2005) estimates China's bogus goods cost consumers and U.S. manufacturers at least $29 billion annually.

Criminal Careers of Occupational and Organizational Offenders

Occupational and corporate offenders generally do not view their activities as criminal; their violations are usually part of their occupational environment. Such offenders maintain a commitment to conventional society while violating some of its laws, because their activities often are supported and informally approved of by occupational or corporate subcultures or environments (see Frank & Lombness, 1988).

Sutherland (1956a) sees many parallels between the behavior of corporate criminals and that of professional and organized criminals:

1. They are recidivists, committing their crimes on a continual and frequent basis.

2. Violations are widespread, and only relatively few are ever prosecuted.

3. Offenders do not lose status among their peers or associates as a result of their illegal behavior.

4. Like professional thieves, businesspeople reveal contempt for government regulators, officials, and laws that they view as unnecessarily interfering with their behavior (pp. 93–95).

Corporate Environment and Crime

Corporate crime does not occur in a vacuum, but is affected by characteristics of an organization and its market structure. For instance, in an analysis of auto makers, Leonard and Weber (1970) found that price fixing requires two market forces: a few suppliers and inelastic demand (i.e., a steady need or demand for a product irrespective of a rise or fall in cost).

Corporate Concentration

The marketplace in postindustrial or advanced capitalistic societies has moved from competitive capitalism among companies to shared monopolies controlled by huge corporations and conglomerates.

The growing concentration of markets can be demonstrated by the fact that, in 1960, a total of 450 U.S. firms controlled about 50 percent of all manufacturing assets and made 59 percent of all profits. By 1979, as much as 79 percent of the assets and 72 percent of the profits were controlled by these firms (D. R. Simon, 1999).

Overpricing of products is more likely to occur when four or fewer companies control a market. As a result of such shared monopolies, the FTC estimates that prices are 25 percent higher than they should be and that such concentrated market firms enjoy profits that are 50 percent higher than those of less concentrated industries (D. R. Simon, 1999). In size, complexity, assets, and power, these large corporations dwarf most states and most national governments. Their wealth and power in elections, in private foreign policy, and in the international economy make public sector regulation increasingly difficult.

Rationalizations

Having little or no criminal self-concept, offenders view violations as part of their work. Among the rationalizations, or ways of explaining away responsibility, for white collar criminality are those below (Clinard & Yeager, 1980):

- Legal regulations of business are government interference with the free enterprise system.
- Such regulations are unnecessary and reduce profits.
- Such laws are too complex, create too much paperwork, and are incomprehensible.
- Regulatory laws are not needed and govern unnecessary matters.
- There is little deliberate criminal intent (*mens rea*) in corporate violations.
- "Everybody is doing it," and I have to keep up with competitors.
- The damage and loss are spread out among large numbers of consumers, thus little individual loss is suffered.
- If corporate profits do not increase as a result of the violation, there is no wrong.
- Violations are necessary in order to protect consumers (pp. 69–72).

Tax cheats, especially the wealthiest, are regularly beating the federal government out of its fair share of taxes. In 2002, the retiring director of the Internal Revenue Service estimated the losses from partnerships alone at $7 billion, while others estimate those losses alone at $84 billion. Over 82,000 taxpayers used offshore accounts to evade taxes costing as much as $70 billion annually. IRS resources were cut back during the George W. Bush administration (Johnston, 2002).

Societal Reaction

The UCR for the early twenty-first century estimated that property crimes such as robbery, burglary, and larceny cost U.S. society nearly $9 billion. Federal investigators estimate that the federal government is being ripped off by at least $50 billion a year, primarily through fraud. In terms of threat and damage to property, health, theft, and corruption of law enforcement agencies, then, corporate crime is the "big leagues." The cost of the savings and loan scandal of the 1980s was estimated at $500 billion, while the celebrated "Great Brinks Robbery" netted only $2 million. The latter is much better known and has received more publicity than the former, even though 250,000 Brinks robberies would be required to equal the cost of bailing out the savings and loans.

Despite growing public pressure for more severe treatment of higher occupational and corporate offenders, the likelihood of prosecution and conviction remains rare. When offenders are convicted, the penalties remain rather minuscule, considering particularly the economic loss to society. High recidivism rates among such criminals continue. Many are even "deadbeats" in paying assessed fines. The **"big dirty secret"** remains true: Judges and government agencies are "soft" on corporate crime. In 2007, it was reported that enforcement against polluters during the George W. Bush administrations in prosecutions, investigations, and convictions was down by more than one third (Solomon & Eilperin, 2007). EPA civil lawsuits were down 70 percent. The number of investigators at the agency had been cut back.

Returning to the previous example of the "Great Savings and Loan Scandal," the costliest series of white collar crimes in American history, by 1994 the average sentence given for major thrift cases was 36 months compared to 38 months for car thieves and 56 months for burglars. It should be noted that most of the sentences were handed down before more strict federal sentencing guidelines were instituted in 1989 (Pontell, Calavita, & Tillman, 1994).

Why the Leniency in Punishment?

If white collar crimes are economically the most costly crimes to society, why are such acts seldom punished? A number of reasons have been suggested:

- Many acts were not made illegal until the twentieth century. For example, many environmental and occupational health and safety regulations are of post–World War II vintage, and not until the twentieth century were false advertising, fraud, misuse of trademarks and patents, and restraint of trade considered criminal matters.
- American business philosophy has been dominated by beliefs in laissez-faire economics (government noninterference in business) as well as the notion of caveat emptor ("let the buyer beware").
- Public concern with corporate crime is a recent phenomenon. Once this resentment becomes organized, public pressure against white collar crime and pressure for legislation and enforcement can be expected. At least one national survey suggests the general public regards white collar crimes as even more serious than conventional crimes such as burglary, robbery, and the like (Wolfgang, 1980, p. E21). Thus, it seems lenient treatment of elite offenders is not supported by the public.
- In the past, white collar crimes were given less publicity; sometimes the media were owned by businesses that themselves were violators (Snider, 1978). Fear of loss of major advertising revenue may also have an impact.
- White collar criminals and those who make and enforce the laws share the same socioeconomic class and values. They fail to match the public stereotype of the criminal. Vilhelm (1952) suggests that citizens don't oppose such crime because they themselves often violate many of these same laws on a modest scale.
- Political pressure groups often block effective regulation or enforcement. Some of the biggest campaign contributors are also the biggest violators. Funding for such groups may come from previous tax avoidance, laundering, and other shady practices. Since such criminals are seldom prosecuted, many are first offenders and thus are treated with leniency.
- It is easier for politicians and public officials to concentrate on the crimes of the young and lower class, groups that lack political clout.
- The long-term nature of corporate violations and court delays make sanctions difficult.

Black (2004) suggests a main problem with federal regulatory agencies is that they are in desperate need of criminological expertise. No federal, state, or local government agency has a "chief criminologist" position, and criminologists are excluded from policy debates on these issues. We have indicated that asbestos manufacturers had full knowledge that they were killing their workers, as

Johns-Manville did, for instance, when it estimated dust control equipment installation and operation at $17 million and only $1 million for workers compensation payments, and judged that it was cheaper to infect workers (Brodeur, 1974, p. 128).

A 1999 National Public Survey of White Collar Crime conducted by the National White Collar Crime Center (Rebovich & Layne, 1999) revealed that the public regarded many types of white collar crime as serious, or more serious than traditional street crime. For instance, in answering which was more serious, they responded in the following way:

- Someone steals $100 on the street or a contractor cheats someone out of $100: robbery, 41 percent; fraud, 40 percent; equal, 20 percent.
- Someone steals $100 on the street or a bank teller embezzles $100 from his employer: robbery, 27 percent; embezzlement, 54 percent; equal, 18 percent.
- Person robs someone at gunpoint or automaker fails to recall a vehicle with a known defective part: armed robbery, 46 percent; defective product, 40 percent; equal, 14 percent.
- Person robs someone at gunpoint or a store owner sells a shipment of meat he knows is bad: armed robbery, 39 percent; tainted product, 42 percent; equal, 19 percent.

In response to the large number of corporate scandals at the turn of the last century, the U.S. Congress passed the Sarbanes-Oxley Act, which compels the SEC to address weaknesses in corporate oversight. Among the reforms were the following:

- Accountants will no longer be considered independent and objective if they or their auditing partners receive nonaudit (consultant) pay from publicly traded clients.
- Lawyers are required to become whistleblowers and must report legal violations to company officers or the board of directors.
- Mutual fund managers must ensure that shareholder proxies are voted in the best interest of investors and not insiders ("Investor's Guide," 2003).

Theory and Crime

White collar crime is one area where deterrence theory and rational choice theory are most applicable. The basic assumption of such neoclassical theories is that individuals are rational decision makers and would be deterred from committing crime if the penalties exceed the pleasure to be derived. While deterrence theory does not always apply in explaining violent crime or street crime, it does appear relevant with respect to white collar crime. The penalties for white collar crime are often weak, unenforced, and not an effective deterrent in preventing white collar crime.

Institutional anomie theory argues that profit at all cost has come to dominate all institutions in modern society. The recent economic collapse in the United States and the world was brought about by greed on Wall Street and in the insurance and banking industry where individual profits became more important than integrity and professional ethics and government regulation was virtually nonexistent. Using routine activities theory, poor guardianship turned the financial world over to greedy speculators.

Finally, in reading the remaining chapters of this text, keep in mind that if we ended our discussion of crime with this chapter we would have covered the largest, most costly category of crime. All the other forms of criminal behavior together do not equal the costs of occupational and organizational (corporate) crime.

Summary

The formative statement on white collar crime was made by Sutherland in 1939. He defined it as "crime committed by a person of respectability and high social status in the course of his occupation." Despite the much greater cost of widespread corporate violations, the criminal justice system finds it more politically expedient to concentrate on traditional crimes. Related to Sutherland's notion is Ross's notion of criminaloids as "those who prospered by flagitious [shameful] practices which may not yet come under the ban of public opinion" and Mills's concept of the higher immorality or moral insensibility of the power elite. The concept of white collar crime has been criticized as too global in nature, and a variety of other terms have been suggested. Particularly important are the concepts of "occupational," "organizational," and "corporate" crime. *Occupational crime* refers to violations that are committed for self-benefit during the course of a legitimate occupation, while *organizational crime* refers to crimes by businesses or officials on behalf of the employing organization. Organizational crime becomes *corporate crime* when undertaken on behalf of a private business or organization.

The reasons for the lack of research on occupational and corporate crime were detailed in this chapter, indicating that criminologists, because of the lack of readily available data, still rely on many secondary sources. Data and figures from various sources were presented in an attempt to measure the cost of white collar crime. While conservative estimates place the figure at $36 billion, more liberal estimates place the cost for monopolistic practices at over $230 billion. Any of the cost estimates far exceed those for traditional crimes.

Myopia must be avoided, so as not to view the current level of white collar crime as the worst in history; historical analysis suggests that this type of crime may actually have been even more prevalent in the past. Analysis of legal regulation of occupational practice points out that the more developed professions have been granted a mandate for self-governance even though such self-policing has been less than impressive.

While different typologies of white collar crime have been offered (Bloch & Geis; Edelhertz), the author suggests an "occupational/organizational crime grid" as a heuristic device for presentation purposes in this chapter. This results in nine theoretical types based on the criminal (individual, employee, or organization) and the victim (individual, employee, or organization).

Crimes by employees may include a variety of offenses as detailed in Edelhertz's typological examples. Crimes by employees against individuals/the public were portrayed by means of public corruption (the Knapp Commission, **Judgescam,** Watergate, and Abscam), as well as private corruption and sharp practices by auto dealers. Employee vs. employee crime was examined by means of "sweetheart contracts" while crimes by employees against organizations were depicted with descriptions of embezzlement, employee fraud, pilferage, and computer crimes (including the argot of electronic "hackers"). Crimes by individuals (or members of occupations) were delineated by describing crooked practices in the medicine, law, and pharmacy fields as well as in business-related trades and occupations.

Reasons for the dearth of studies of corporate crime were detailed. In the United States, the legal governance of business organizations began in the nineteenth century, particularly with the Sherman Antitrust Act (1890). Much regulation of corporate activity takes place through federal regulatory agencies such as the FCC, ICC, and SEC. These agencies can utilize civil and criminal as well as administrative means of assuring compliance, but they seldom do. Most agencies are "outgunned" by the industries they are supposed to control, and, in fact, they are sometimes controlled by these industries. Gross characterizes this nonenforcement and kid-glove treatment of elite criminals as "the big dirty secret."

Studies by Clinard and Yeager and associates signaled a new renaissance in studies of corporate criminality—the first large-scale, comprehensive study of corporate crime. In an examination of crimes by organizations against individuals/the public, detailed examples were provided, such as multinational bribery, and case examples such as the **Equity Funding Scandal,** the Great Electrical Industry

Conspiracy, and the Great Oil Scam. Other important illustrations presented included the Ford Pinto case, toxic criminals, environmental violations, and corporate dumping of unsafe products.

Crimes by organizations against employees primarily relate to threats to the health and safety of workers, as dramatically illustrated by the tragic **Karen Silkwood case.** Crimes by organizations against organizations were illustrated by examples of wartime trade violations; industrial espionage (such as the Hitachi case); and corporate fraud against government, particularly on the part of defense contractors.

Characteristics of the corporate environment, such as supply and demand, and corporate concentration, such as the number of producers of a product, are predisposing factors in corporate criminality. Societal reaction to higher-level occupational and corporate crime has in the past been characterized by leniency. A number of reasons were provided for such indulgence, including policies of laissez-faire economics and a caveat emptor philosophy prevalent in the past. Recently, public reaction to such crimes has hardened and now rivals or exceeds that for traditional crimes. Recent research suggests some improvement in punishing elite offenders but still not in concomitance with the quantity, prevalence, and cost of such activities. Some retrenchment in regulatory activities may be occurring in response to a more conservative, pro-business political climate. The toleration of white collar criminals and deadbeats raises a major challenge to claims of equitable standards of justice and indirectly fosters crime in the streets through the perpetration of inequality.

The criminal careers of occupational and corporate criminals entail little identification with crime; these offenders enjoy subcultural support and employ rationalizations to explain away responsibility for wrongdoing.

Criminology on the Web

Log on to the Web-based student study site at http://www.sagepub.com/haganintrocrim7e/ for author-created podcasts, e-flashcards, quizzes, and more.

KEY CONCEPTS

Abscam

Big Dirty Secret

Churning

Coerced Crime

Computer Crime

Corporate Crime

Corporate Dumping

Corporate Environment and Crime

Costs of White Collar Crime

Criminaloid

Embezzlement

Equity Funding Scandal

Ford Pinto Case

Great Oil Scam

Higher Immorality

Industrial Espionage

Insider Trading

Judgescam

Karen Silkwood Case

Nolo Contendere

Occupational Crime

Occupational/Organizational Crime Grid

Organizational Crime

Power Elite

Reasons for Lack of Corporate Crime Research

Reasons for Lack Occupational Crime Research

Reasons for Leniency With White Collar Offenders

Revolving Door

Sweetheart Contracts

Types of Computer Crime

Watergate

Whistleblowers

White Collar Crime

White Collar "Deadbeats"

REVIEW QUESTIONS

1. Why, despite its cost, has there been so little research on corporate crime?

2. Who polices corporate crime? Name some of these agencies and their jurisdictions.

3. What have been some criticisms of federal regulatory agencies? Do you see any improvements in these activities?

4. How serious are antitrust violations in the United States? Give some examples.

5. Discuss the "Great Savings and Loan Scandal." How was it possible for what has been described as the "greatest series of white collar crimes in American history" to take place, and why was the American public unaware of this?

6. Discuss Edwin Sutherland's concept of white collar crime. Why was this considered a Copernican revolution or paradigm shift in criminology?

7. Discuss Cressey's theory of embezzlement. Does research support his hypothesis?

8. What were some major public scandals of the Reagan era? What happened as a result of these scandals?

9. What are some different types of computer crime? Give an example of each.

10. Discuss and give examples of crime or unethical practices in the field of medicine.

WEB SOURCES

Antitrust Division Department of Justice
http://www.usdoj.gov/atr/

Corporate Crime Reports
http://www.corporatecrimereporter.com/

Financial Crimes Enforcement Network
http://www.fincen.gov/

Financial Scandals
http://www.ex.ac.uk/~RDavies/arian/scandals/

National White Collar Crime Center
http://www.nw3c.org

Net Scams Online Protection
http://www.netscams.com/scams.jsp

Occupational Safety and Health Administration
http://www.osha.gov/

Public Citizen Corporate Crime Center
http://www.citizen.org/_corpcrime/

Top 100 Corporate Predators
http://www.corporatepredators.org/top100.html

Transparency International
http://www.transparency.org/

WEB EXERCISES

Using this chapter's recommended Web sites, surf the area of white collar crime.

1. What are some of the services offered by the "National White Collar Crime Center"?

2. What is Fincen (Financial Crimes Enforcement Network)?

3. Examine some of the information on the Corporate Crime Reports and Top 100 Corporate Predators. Do these add to your knowledge of white collar crime offending?

4. What is the organization Transparency International, and what does it have to say about the areas of corruption and bribery?

5. Using your Web browser, search the terms "price-fixing" and "accounting scandals."

SELECTED READINGS

Jay Albanese. 1995. *White Collar Crime in America.* New York: Prentice Hall.

A very readable presentation on the issue of organizational offenders by one of America's leading authorities on the subject.

Marshall B. Clinard and Peter Yeager. 1980. *Corporate Crime.* New York: Macmillan.

This is the largest study of corporate crime ever conducted and an excellent update of Sutherland's original investigation.

James Coleman. 2001. *The Criminal Elite* (4th ed.). New York: Worth.

Coleman's text provides in-depth coverage of both occupational and corporate crime. He also explores theoretical explanations for white collar criminality.

Donald Cressey. 1953. *Other People's Money.* New York: The Free Press.

This is the classic case study on embezzlers, which proposes the hypothesis that embezzlers have a "non-shareable problem" that serves to motivate them to embezzle.

Daniel O. Friedrichs. 1995. *Trusted Criminals: White Collar Crime in Contemporary Society.* Belmont, CA: Wadsworth.

Friedrich's text might very well be the most scholarly and thorough contemporary account of white collar crime. It weaves together a mass of literature and is a must-read for anyone who claims to be informed on this subject.

Stephen Pizzo, Mary Fricker, and Paul Muolo. 1989. *Inside Job: The Looting of America's Savings and Loans.* New York: McGraw-Hill.

This is an account by three California journalists of the biggest series of white collar crimes in American history, the "Great Savings and Loan scandal" of the eighties.

David Simon and Frank Hagan. 1999. *White Collar Deviance.* Boston: Allyn & Bacon.

In this work David Simon and the author attempt to reconcile the elite deviance tradition (Simon) with that of white collar crime (Hagan), thus the new concept: "white collar deviance."

James Stewart. 1991. *Den of Thieves.* New York: Simon & Schuster.

Journalist James Stewart examines the world of Wall Street "inside traders" such as Ivan Boesky and Michael Milken.

Neal Shover and John Paul Wright, editors. 2001. *Crimes of Privilege: Readings in White Collar Crime.* New York: Oxford University Press.

The authors supply an excellent selection of 31 articles on the issue of white collar crime. These include classic as well as contemporary selections.

Edwin H. Sutherland. 1949. *White Collar Crime.* New York: Holt, Rinehart and Winston.

Sutherland's classic text is the beginning point for any discussion or analysis of white collar crime.

Ideology

Political Crime: A Definition
  Crime File 11.1 September 11, 2001

Legal Aspects
  The Nuremberg Principle
  The Universal Declaration of Human Rights
  International Law

Crimes by Government
  Secret Police
  Human Rights Violations
  Patriarchal Crime
  Genocide
  Crimes by Police
  Illegal Surveillance, Disruption, and Experiments
  Scandal
  Crime File 11.2 White House Crime and Scandal: From Washington to George W. Bush

Crimes Against Government
  Protest and Dissent
  Social Movements
  Assassination
  Espionage
  Crime File 11.3 The Robert Hanssen Spy Case
  Political "Whistleblowing"
  Terrorism
  Crime File 11.4 The Turner Diaries, ZOG, and the Silent Brotherhood–The Order
  Crime File 11.5 Patterns of Global Terrorism

Criminal Careers of Political Criminals
  The Doctrine of *Raison d'État*
  Terrorism and Social Policy

Societal Reaction

Theory and Crime

Summary

Key Concepts

Review Questions

Web Sources

Web Exercises

Selected Readings

chapter 11

Political Crime and Terrorism

Never believe in anything until it has been officially denied.

—*Otto von Bismarck*

King, there is only one thing left for you to do. You know what it is. You have just 34 days in which to do it. (This exact number has been selected for a specific reason.) It has definite practical significance. You are done. There is but one way out for you.

—*Note sent to Dr. Martin Luther King, Jr., by the FBI 34 days before he was to receive the Nobel Peace Prize. The note allegedly suggested that he should commit suicide (Blackstock, 1976).*

The tree of liberty must be refreshed from time to time with the blood of patriots and tyrants.

—*Thomas Jefferson*

Ideology

An ideology is a distinctive belief system, idea, or abstract ideal. Communism, capitalism, fascism, Islam, Judaism, Christianity, fundamentalism, and the like can serve as ideologies, or can combine to supply their adherents with a guide to societal and individual behavior. "True believers," extremists, or ideologues often are zealots and are absolutely certain of the righteousness of their cause—so much so that they feel justified in forcing their beliefs on others.

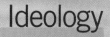

Schafer (1971, 1974) uses the term "convictional criminals" when referring to politically motivated criminals. Such a criminal is "convinced of the truth and justification of his own beliefs" (Schafer, 1974, p. 138). The actual crimes committed by political criminals may be traditional crimes, such as kidnapping, assassination, blackmail, robbery, and the like. It is not the crimes themselves that distinguish political criminals, but their motivations, their views of crime as a necessary means to a higher ideological goal. Some political criminals, particularly human rights advocates, have committed no crime, but have expressed their political views in authoritarian or totalitarian societies that forbid individual expression or criticism of the states, or in free societies where civil disobedience may be viewed with suspicion and considered criminal.

Political Crime: A Definition

Political crime refers to criminal acts committed for ideological purposes. Rather than being motivated by private greed or benefit, these offenders sincerely believe they are following a higher morality that supersedes present society and its laws. Such political criminals may act for social–political reasons (Robin Hood), out of moral–ethical motivations (antiabortion activists), to advance religious causes (Martin Luther), to disseminate scientific beliefs (Copernicus, Galileo), or to publicize political concerns (Nathan Hale, Benedict Arnold) (Schafer, 1976). Such crime may take one of two forms: *crime by government* or *crime against government.*

Crimes by government include violations of human rights, civil liberties, and constitutional privileges, as well as illegal behavior that occurs in the process of enforcing the law or maintaining the status quo. Secret-police violations, human rights abuses, genocide, and crimes by police, as well as illegal surveillance, disruption, and experiments, are just some of the examples of governmental crime to be discussed in this chapter.

Crimes against the government may range from protests, illegal demonstrations, and strikes to espionage, political whistleblowing, assassination, and terrorism. "One person's terrorist is another's patriot" is a common expression that suggests the relative nature of such political crime. In revolutions, the victors' beliefs become the status quo, and the victors inherit the power and privilege by which to brand the acts of their enemies as criminal.

There is a surprising paucity of literature on political crime in criminology (F. E. Hagan, 1986; J. M. Martin, Haran, & Romano, 1988). As of this writing, fewer than 10 works in the field have specifically addressed this issue (see Ingraham, 1979; Kelman & Hamilton, 1988; Kittrie & Wedlock, 1986; Proal, 1973; Roebuck & Weeber, 1978; Schafer, 1974; Schur, 1980; Turk, 1982). More recent additions include F. E. Hagan (1997) and J. Ross (2003). In his classic *Political Crime*, which was originally published in 1898, Louis Proal (1973) notes,

> Political passions have bathed the earth in blood; kings, emperors, aristocracies, democracies, republics, all governments have resorted to murder out of political considerations, these from love of power, those from hatred of royalty and aristocracy, in one case from fear, in another from fanaticism. (p. 28)

While political crimes may be committed by or against the government, seldom do governments or government officials choose to acknowledge their own lawlessness. Sagarin (1973, p. xiv) very aptly points out that political crime includes the tyrant as well as the assassin. The terrorist is the epitome of the political criminal, and no better example of the depths of evil committed in the name of ideology is the mass murder of September 11, 2001. Crime File 11.1 gives an account of that tragic day.

Crime File 11.1

September 11, 2001

September 11, 2001, was a transforming event for a new generation of Americans, similar in effect to the attack on Pearl Harbor and the Kennedy assassinations in previous generations. In the opening salvo of a declared war against this country, simultaneous terrorist attacks in Washington on the Pentagon and New York City at the World Trade Center resulted in the worst loss of American lives on American soil by foreign attack in history. The new enemy in this new world war was not international communism or the Japanese empire, not even a state, but terrorists loosely organized around Osama bin Laden and his organization, al Qaeda. In the name of extremist, fundamentalist Islamic ideology, this group was willing to use any means including weapons of mass destruction to inflict enormous casualties on civilians as well as military targets. The following is the FBI report accounting the activities of 9/11.

On the morning of September 11, 2001, four commercial airliners were hijacked by four coordinated teams of terrorists. The 19 hijackers who carried out the operation were affiliated with al Qaeda, a worldwide terrorist network that had previously attacked U.S. military and diplomatic targets. The hijackers used knives, box cutters, and possibly pepper spray to attack passengers and flight crews and to commandeer the aircraft. After taking control of the aircraft, the hijackers flew toward preselected targets on the U.S. East Coast. Three of the commandeered aircraft reached their destinations, destroying the twin towers of the World Trade Center in New York City and badly damaging the Pentagon in Arlington, Virginia. The fourth aircraft crashed into a remote field in Stony Creek Township, Pennsylvania, as passengers attempted to regain control of the airplane. All of the passengers on each aircraft were killed in the attack, as were more than 2,500 people in the Twin Towers and the Pentagon. In total 2,783 innocent people were murdered along with 19 hijackers, making it the most deadly act of terrorism ever committed.

Photo 11.1

Hijacked United Airlines Flight 175 crashed into the south tower of the World Trade Center and exploded at 9:03 a.m., 17 minutes after hijacked American Airlines Flight 11 crashed into the north tower.

Source: FBI (2002). *Terrorism 2000–2001.* Washington, DC: Government Printing Office, pp. 14–15.

Web Research Project
What progress has been made in the war against al Qaeda?

Legal Aspects

In the United States, various laws are, or have been, intended to protect the government from the clear and present or probable danger of disruption or overthrow. Laws such as the Alien and Sedition Acts, Espionage Act, Voorhis Act, Smith Act, Internal Security Act, and McCarran-Walter Act are examples. The 1940 Voorhis Act requires registration of agents of foreign powers while the Smith Act (1940), which was later struck down by the Supreme Court, outlaws advocating the overthrow of the government. The Internal Security Act (McCarran Act, 1950) calls for registration of communists and communist-front organizations, while the McCarran-Walter Act (1952) provides for deportation of aliens who espouse or have associates who espouse disloyal beliefs (Clinard & Quinney, 1973, p. 155).

Cuba and other authoritarian countries enact criminal laws forbidding propaganda against the state, complaining about social conditions to foreigners, and attempting to publish works not authorized by the state. Many of these laws and their enforcement bear an uncanny resemblance to those in George Orwell's *1984* (1949) and to his descriptions of the Minitrue (Ministry of Truth) in which thought criminals become political criminals or enemies of the state.

Under Anglo-American legal traditions, political crime and political criminals are not recognized as such, and these types of offenders are dealt with under traditional or nonpolitical laws. Anglo-American criminal law considers intent, but not motive. The motive, whether good or bad, has no bearing on guilt. Sagarin (1973) points out, "At one time it was against the law in some parts of this country to preach freedom and abolition of slavery to slaves, or even to free men; it was often against the law to organize into trade unions; at various times political parties have been driven underground and their leaders jailed" (p. ix).

Kittrie and Wedlock in *The Tree of Liberty* (1986) provide historical documents related to elements of political criminality either by the state or by persons accused of such offenses:

- The important Peter Zenger trial of 1735 for false, scandalous, and seditious libel, which established the "freedom of the press" doctrine
- The "crime of being black or Indian," which led to the "Trail of Tears" of the Cherokee Nation and to the outlawing of Abolitionism, the underground railroad, and harboring fugitive slaves
- Subjugation of blacks through private conspiracies and terrorism
- Genocide against American Indians
- Voter registration drives and Freedom Rides during the civil rights struggle
- Imprisonment of Japanese Americans in internment camps during World War II
- Antiwar protest, burning draft cards and records
- Arson, bombing, and other violations against abortion clinics
- "Sanctuary activists" hiding Central Americans, whom they consider to be political refugees

The Nuremberg Principle

After World War II, the victorious Allies held a tribunal and convicted Nazi war criminals. The Nuremberg trials were the first occasion on which defeated war leaders were held responsible in an international legal arena for activities that were legal, even encouraged, by their governments at the time they were committed. Defenses such as "I was just following orders" were rejected and held to be unjustifiable explanations for Nazi atrocities. Kelman and Hamilton (1988) refer to this as a "crime of obedience." Offenses defined by the international tribunal included war crimes and crimes against humanity (Maser, 1979; "Nuremberg Principle," 1970, p. 78; B. F. Smith, 1977).

War Crimes

Violations of law or customs of war include, but are not limited to, murder, ill treatment, or deportation for slave labor—or for any other purpose—of the civilian population of or in occupied territory; murder or ill treatment of prisoners of war or persons on the high seas; killings of hostages; plunder of public or private property; wanton destruction of villages, towns, or cities; or devastation not justified by military necessity.

Crimes Against Humanity

These include murder, extermination, enslavement, deportation, and other inhumane acts committed against any civilian population, before or during the war, or persecutions on political, racial, or religious grounds, whether or not they are in violation of the domestic law of the country where perpetrated.

The Universal Declaration of Human Rights

The concept of human rights is an outgrowth of the period of Enlightenment in Western society and is expressed in such documents as the Magna Carta, the English Bill of Rights, the American Declaration of Independence, the French Declaration of the Rights of Man and of the Citizen, and the United Nations Universal Declaration of Human Rights. All of these documents support the notion of inalienable rights and freedoms that supersede those of government.

International Law

Since much political criminality is international in scope, it theoretically falls under the jurisdiction of international law, the power of which is limited. This covers fairly nonproblematic diplomatic and commercial customs between nations; agreements such as treaties that are drafted in international conventions; and international courts such as the International Court of Justice sponsored by the United Nations. Using precedents (past decisions), customs, and general principles of law, international law is theoretically binding on any signatories to international treaties, although it may also through custom be held to be binding on those who have not ratified the treaties. Stipulations of the Geneva Convention of 1929 regulating wartime conduct serve as an example.

While international bodies past or present, such as the World Court, the League of Nations, and the United Nations, have the facade of law, they lack the crucial power to enforce their decisions, ultimately through force if necessary (Kidder, 1983, p. 34). Essentially, international law lacks teeth—the authority and power to assure compliance. However, the end of the Cold War considerably improved the prospect for international cooperation in enforcing international law and sanctions. UN actions in forcing Iraq to withdraw from Kuwait in 1991 are one such example.

In 2002, the first permanent International Criminal Court opened over U.S. opposition. The United States did not want to have such an international group to have jurisdiction over it. One of the court's mandates was to deal with large-scale human rights abuses. In 2005, it faced its first big case: to investigate genocide in Darfur (western Sudan), where the government is accused of killing roughly 300,000 and displacing another 2 million (Simons, 2005).

Crimes by Government

The first major category of political crime to be discussed is crimes by government. These are crimes or violations of human rights committed for ideological reasons by government officials or their agents. The government political criminal is motivated not by self-interest so much as by a commitment to a particular belief system, a conviction that he or she is defending the status quo or preserving the existing system. Since many such violations are not formally recognized or enforced by the criminal law in most nations, the concept of political crime by government is more a sociological than a political entity.

Secret Police

All countries require some type of **secret police** for clandestine intelligence gathering and internal security. Plate and Darvi (1981) define secret police as

> official or semi-official organs of government. They are units of the internal security police of the state, with the mandate to suppress all serious, threatening political opposition to the government in power and with the mission to control all political activity within (and sometimes even beyond) the borders of the nation-state. (p. 8)

Secret police are often involved in extraordinary illegal surveillance, searches, detention, and arrest; as a matter of practice, they may violate or border on violating human rights.

In totalitarian societies, the effectiveness of secret police in deterring illegitimate violence (crime in the streets) occurs through legitimate violence (crime by the state). The specter and practices of such infamous secret police as Hitler's Gestapo, Stalin's OGPU (later KGB), and Haiti's Tonton Macoutes—midnight raids, tortures, and disappearances—are frightening indeed.

Austin Turk prefers the term "political policing" to refer to secret police operations. In "Organizational Deviance and Political Policing" (1981), Turk provides a number of illustrations:

- Assassination or maiming of political figures
- "Geneva offenses" such as germ warfare, letter bombs, or use of cattle prods to torture political prisoners. The Geneva Convention forbade the mistreatment of sick or wounded soldiers.
- The torture of political detainees, such as those listed by Amnesty International
- Character assassination
- Intervention in conventional politics, such as the FBI campaign against Martin Luther King, Jr.
- Violations of civil or human rights, for example, ranging from illegal surveillance to mental institutionalization of political dissidents in the former USSR
- Economic or political harassment of dissident groups
- Use of "agents provocateurs," informants, and spies in order to manipulate public institutions
- Subversion of economic or other institutions, for example, the overthrow of the Chilean government of Salvador Allende by the U.S. Central Intelligence Agency (pp. 238–239)

If state agents of social control wish to suppress a social movement, the agents have an entire repertoire of actions to choose from (Baylor, 1990). Marx (1979, cited in Baylor) identifies a number of tactics, including the following:

- Litigation against the movement or leaders
- Administrative harassment
- Disinformation campaigns
- Wiretaps and other electronic surveillance methods
- The use of informants and agents provocateurs
- Support of counter or alternative groups
- "Bad jacketing" or "snitch jacketing" (spreading rumors that someone is a snitch)
- Police response, including force (p. 2)

In many of these, the state officially encourages its agents to commit crime (G. Marx, 1990). The criminological study of such crimes by government has been far less a topic of research than similar crimes against the government because of a lack of funding for the former (Longmire, 1988). Chambliss (1988b) calls many of these activities, such as state-sponsored piracy, smuggling, assassinations, murder, and experiments, examples of "state-organized crime," while Barak (1991) simply calls it "state crime."

In 1999, President Clinton issued an apology to the people of Guatemala on behalf of the United States for helping support major atrocities in that country in the past. An independent truth commission had issued a 3,500-page report detailing the Guatemalan government's campaign of terror against its own people during its 36-year civil war (McGrory, 1999). Over 200,000 mainly Mayan Indians were massacred, executed, tortured, or disappeared due to U.S. support for security forces. These poor, uneducated, and voiceless majority native peoples were the hardest hit. Cold War fear of communism had been the motivation. Related to the war on terrorism and war in Iraq, the United States has been accused of using torture at Abu Ghraib prison and at Guantanamo Bay. In addition, concerns have been raised regarding the U.S. using "waterboarding," a practice that simulates drowning in order to gain cooperation of captive suspected terrorists. While opponents charge that such practices are a violation of the Geneva Convention, intelligence agencies claim that the captives are not recognized combatants (so Geneva Convention rules do not apply), and the practices have succeeded in disrupting horrific plans of these terrorists.

Human Rights Violations

Perhaps the most dramatic illustration of crimes by government is the pervasive international violation of **human rights.** Thousands of "political prisoners"—individuals who have committed no

Handbook Article Link 11.1
Read an article
on human trafficking.

crimes other than their espousal of political ideas—are tortured, murdered, or abandoned throughout the world. It is difficult, because of governmental secrecy, to gain an accurate count of such prisoners, although human rights organizations such as Amnesty International provide rough figures. Authoritarian and totalitarian regimes of the political left and right are the least tolerant of dissent and are thus the biggest violators. These countries most resemble what George Orwell described in his novel *1984,* in which the state is preeminent.

While Savak (the Iranian secret police) under the shah was recognized as a brutal force, in the subsequent theocracy created by Khomeini in Iran, as many as 60,000 political prisoners were held and over 25,000 executed as of 1984. In 1990, members of the People's Mojahedin claimed that over 90,000 had been executed and 150,000 tortured (J. Anderson & Van Atta, 1990). According to a report by Amnesty International (M. Lippman, 1987), torture had been routinely practiced on detainees in order to extract confessions. Beatings, floggings, suspension by limbs, and mock executions are common. Thefts were punished by amputations, in accordance with Islamic law. The death penalty is given for acts ranging from adultery or repeated lesbianism to wine drinking (see Elias, 1986).

Beginning in 1975, the Pol Pot regime in Cambodia embarked on a system of mass genocide that destroyed a large portion of that small nation's population. In Latin America, right-wing governments and private government-supported "death squads" kidnap and/or torture and murder individuals who they feel threaten the state. Amnesty International estimates that 98 countries practiced torture during the eighties, primarily as a tool for repression rather than to extract information (Satchell, 1988, p. 38). Amnesty International has expressed concern that, as a result of international attention focused on the plight of political prisoners, many governments may have turned to execution of dissidents (Whitaker et al., 1983, p. 52), assuming that dead people tell no tales. In recent years, both the U.S. State Department and the United Nations Commission on Human Rights in Geneva have identified "rogue nations," ones that refuse to abide by international human rights accords. The list has included Iran, Iraq, and North Korea (labeled by George W. Bush as the "axis of evil"), as well as Burma, China, Somalia, Sudan, Saudi Arabia, and Turkmenistan.

Video Link 11.1
View a video on the Tiananmen Square Massacre.

Perhaps no other event demonstrates the raw, naked power of the state to exercise its political muscle against the popular will than the Chinese government's crackdown on the democracy movement in May of 1989. While the world watched live on television, thousands took to the streets to protest authoritarian rule. Many were massacred by troops. By the time the shooting ended, more than a thousand had been killed or arrested. (Some of the latter were executed after being tried as traitors.) The Chinese government obviously had not forgotten Chairman Mao's axiom, "Power comes from the barrel of a gun."

Slavery still exists throughout the world. Britain's Anti-Slavery International estimates that there are over 100 million slaves worldwide. In some African countries, such as Mauritania and Sudan, the Muslim elite enslave black populations. Many guest workers in Kuwait have been treated as little more than slaves. A large number of young women in India, Bangladesh, and Southeast Asia are sold into prostitution. When demand outstrips supply, women are simply abducted. Most are kept in debt-bondage like that of some Indian workers in Latin America where, no matter how long and hard they work, they still owe the boss. Owners of cane plantations in the Dominican Republic lure poor young Haitians into virtual servitude in which the cost of rent, food, and tools is higher than wages. Despite these and other practices, the United Nations remains reluctant to enforce its own Declaration of Human Rights (Masland, 1992).

Patriarchal Crime

Patriarchal crime refers to crime committed against women and children as part of a system of traditional male dominance and authority. Abuses of such a system include industrial sweatshops, infanticide of female children, sexual mutilation, bride burning, slavery, and human rights abuses.

Ideological justifications for political crime are obviously not limited to political or religious causes, but include preservation of the gender status quo.

In 2000, UNICEF (the United Nations International Children's Fund) declared a global campaign against homicidal violence against women in cultures that sanction such activity (Crossette, 2000). Their focus is on "honor killings," dowry deaths, female infanticide, and acid attacks. Areas of the Mediterranean as well as Pakistan, India, and Bangladesh are particular offenders.

Most of the attacks are technically legal and culturally approved in these countries. UNICEF figures showed that most of the acid attacks in Bangladesh (where men throw acid into the faces of women who reject their request for dates) rose from 47 disfiguring assaults in 1996 to more than 200 in 1998. "Bride burnings" and dowry deaths take place when women are killed because the in-laws consider her dowry (money she brings with her to the marriage) to be inadequate or because the groom was disappointed with the bride. In India in 1997, there were over 6,000 "bride burnings" (Crosette, 2000). The U.S. State Department's Human Rights report estimates 10,000 cases of female infanticide worldwide in 1998. Male dominance, patriarchy, and power predominate in abusive cultures sometimes associated with Islamic fundamentalism, Mediterranean culture, and "machismo" in Latin America. In some Middle Eastern countries, "honor killings" take place whereby female victims of rape are murdered by their own relatives "to preserve the honor of the family." If women stray, the dignity of men can only be restored by killing the women. Marrying without the approval of parents can also provoke murder.

The U.S. State Department (2000) estimates that a million women and children per year, many lured by promises of legitimate employment, are smuggled to other countries and forced into prostitution and a form of slavery. This is common in India, Thailand, Brazil, Ukraine, and Moldova. It is estimated that more than 100,000 people are forced into involuntary servitude in the United States annually, many of them smuggled in from Mexico and Asia. Besides those who are forced into prostitution, some are made to become domestic, migrant, and sweatshop workers.

About 100 million women in Muslim Africa are victims of female genital mutilation (FGM), a procedure which is culturally approved as necessary in order to preserve their virginity. The full butchery of a procedure that amounts to female castration is described by S. McCarthy (1996):

> [This procedure] usually involves the complete removal of the clitoris, and often the removal of some of the inner and outer labia. In its most extreme form—infibulation—almost all the external genitalia are cut away, the remaining flesh from the outer labia is sewn together, or infibulated, and the girl's legs are bound from ankle to waist for several weeks while scar tissue closes up the vagina almost completely. A small hole, typically about the diameter of a pencil, is left for urination and menstruation. (p. 32)

Genocide

Genocide, the mass destruction or annihilation of populations, is the ultimate violent crime by government. The term was coined by jurist Raphael Lemkin (1944), who defined it as the destruction of a nation or of an ethnic group. Genocidal conflicts have a long history, from Roman persecutions, the Crusades, Genghis Khan, and medieval pogroms against European Jews up to the horrors of the present century. In the late 1980s as part of the Iran–Iraq war, Iraq used chemical weapons on civilians as well as on the Iranian military. Such a practice has been outlawed by international conventions (the 1925 Geneva Protocol) since after World War I.

While Germany's genocide of the Jews and other groups during World War II was well documented by the Nuremberg trials, Japan still has not fully acknowledged its atrocities during that war, including the "rape of Nanking," in which thousands of civilians were massacred; the Bataan death march of Allied prisoners of war; and activities of Unit 731, which involved gruesome medical experiments on Allied prisoners (S. H. Harris, 1994).

Video Link 11.2
View a video
on Darfur.

In the early nineties, Brazilian businesspeople were accused of having employed "death squads" to execute poor street children. In 1989 alone, 445 children were murdered as a means of eliminating street crime ("Death Squads," 1991). In another example, despite denials by the Reagan administration, the UN Truth Commission in 1993 concluded that the administration had covered up a massacre of hundreds of innocent civilians at El Mozote, El Salvador, in 1981 by soldiers of that nation's select, U.S.-trained Atlacatl Battalion (Danner, 1995).

Massive genocide did not end with Hitler's attempt to annihilate the Jews. In the late seventies, unbelievable horrors were practiced on Cambodia's own people by Pol Pot's Khmer Rouge regime. If one were to imagine a country ruled by the Charles Manson family, one would be close to picturing the raw terror generated by Angka (the organization of Khmer Rouge) in Cambodia.

In 1948, the United Nations passed a Convention on Genocide, in which it defined genocide as a crime (Kuper, 1981):

> In the present Convention, genocide means any of the following acts committed with intent to destroy, in whole or in part, a national, ethnical, racial or religious group, as such:
> (a) Killing members of the group;
> (b) Causing serious bodily or mental harm to members of the group;
> (c) Deliberately inflicting on the group conditions of life calculated to bring about its physical destruction in whole or in part;
> (d) Imposing measures intended to prevent births within the group;
> (e) Forcibly transferring children of the group to another group. (p. 19)

Despite the concerns expressed in this document, the United Nations has been less than a consistent force in condemning genocide. World inaction found genocide widespread in the Darfur region of Sudan where, despite the United Nations' condemnation, government-sponsored militias called Janjaweed have slaughtered as many as 400,000 and, in a version of ethnic cleansing, have driven 2.5 million from their homes. Rape, torture, murder, and malnutrition plague the helpless refugees (see http://www.savedarfur.org).

Crimes by Police

In democratic societies, the government is expected not only to enforce the law but also, in doing so, to obey the law itself. In the United States, the government is obliged and accountable to certain constitutional guarantees of individual rights, such as freedom of speech, due process, and the right to privacy. Despite this, federal and local law enforcement agencies, being more interested in bureaucratic efficiency than in proper law enforcement, have often ignored and violated these rights in the process of pursuing their mandate.

Prior to the success of the civil rights struggle, local and state officials in the southern United States systematically violated federal law in maintenance of a racist caste system. Murders, lynchings, beatings, and institutionalized denial of constitutional guarantees were all committed in the name of "law and order." It was to the destruction of this *de jure* (by law) discrimination that the civil rights movement was directed; this will be discussed in detail later in this chapter.

Illegal Surveillance, Disruption, and Experiments

In 1967, during the height of dissident activity in the United States, President Johnson directed the Central Intelligence Agency (CIA) to investigate and determine the extent of foreign influence in domestic protest activity. This special operations group, **Operation Chaos,** in surveillance activities of domestic groups, violated the CIA's initial charter, the National Security Act, which clearly excluded its activities from the domestic arena, although pressure to expand activities was ordered by both President Johnson and later Nixon. Operation Chaos and a related Project 2 placed agents in radical groups and collected 13,000 personal files, over half of which were on U.S. citizens. The Rockefeller Commission (1975) investigated the impropriety of the CIA's encroaching on the domestic field of espionage, sabotage, and provocation. In activities related to these operations, the CIA and the FBI in its counterintelligence program, COINTELPRO, committed 238 break-ins (black-bag jobs) and later attempted to destroy records of such activities.

COINTELPRO

The misuse of power by the intelligence agencies was further illustrated in hearings conducted by the U.S. Select Committee to Study Government Operations (1979, cited in C. Thomas & Hepburn, 1983), which revealed that civil rights organizations had been investigated for over 25 years in order to uncover possible communist influences:

> Dr. Martin Luther King, Jr., was harassed by anonymous letters, his telephone was tapped, his speaking engagements were disrupted by false fire alarms—all as a strategy to discredit him and his organization. In addition it is apparent that the FBI and various state police departments used agents provocateurs to infiltrate dissenting groups, radicalize the members, secure the weapons and explosives necessary for violent confrontations, and plan the target of attacks as a means to discredit dissident groups. (Karmen, 1974, cited in C. Thomas & Hepburn, 1983, p. 280)

As part of **COINTELPRO**, the FBI's counterintelligence program to harass and disrupt legitimate political activity such as the Socialist Worker's Party and various black nationalist groups, the FBI employed false letters accusing people of being informants in order to foment internal warfare (Blackstock, 1976, p. 9). The difficulty of separating ideologically motivated actions from personal corruption and vendetta is illustrated by examinations of J. Edgar Hoover's personal files, some of which were released in the 1980s under the Freedom of Information Act. In his nearly 50 years as head of the FBI, Hoover kept personal files replete with gossip and defamatory information on the personal lives of public figures, particularly those whom he happened to dislike either politically or because of his racial bigotry. Eleanor Roosevelt, John and Robert Kennedy, and Martin Luther King, Jr., were just a few of the political figures about whom Hoover gathered revealing information. In addition to surveillance on Dr. King, it is alleged that the FBI sent threatening letters and a tape to Coretta Scott King regarding her husband's sexual trysts. The opening quotation of this chapter suggests an attempt by the FBI to blackmail King into committing suicide (Garrow, 1981).

The Search for the Manchurian Candidate

In 1958, Richard Condon published a novel (later made into a movie) titled *The Manchurian Candidate.* In Condon's very clever and chilling book, which takes place during the Korean War, a character named Raymond Shaw and a U.S. Army squad return after having been missing behind enemy lines. The other members of the squad describe Shaw's heroism in saving them from the enemy; he receives the Congressional Medal of Honor for this. In reality, Shaw and his squad had been "brainwashed" or "hypnotized-programmed" by the Communist Chinese. Asking Shaw to play solitaire until the queen of diamonds appeared would trigger Shaw into zombielike obedience. His own mother (a Communist "mole" or spy) was his operator, and he was able to function as an assassin of the U.S. presidential nominee, thus propelling his stepfather, the vice-presidential nominee, into the Oval Office.

Condon's theme enthralled the Western intelligence establishment, as had the "Moscow show trials" of the period, in which dissidents were paraded before cameras and, as if in a trance, admitted treasonous activities against the state (Scheflin & Opton, 1978, p. 437). How could admissions have been obtained from figures such as Hungary's Cardinal Mindszenty? Cold War propagandist Edward Hunter (1951) coined the term **brainwashing,** which became a household word; however, it is very likely that Hunter popularized the concept as part of his job with the CIA (Scheflin & Opton, 1978, p. 226).

With various code names, such as "Bluebird," "Artichoke," and "MK-ULTRA," the CIA, FBI, and military in the 1950s experimented with various behavioral-control devices and interrogation techniques, including ESP (extrasensory perception), drugs, polygraphs, hypnosis, shock therapy, surgery, and radiation. These projects involved secret testing on private citizens without their permission and, when death or injury took place, a cover-up. In a related example of government agencies using unknowing citizens as guinea pigs, the U.S. Army in the fifties and sixties conducted outdoor tests of poisonous bacteria (serratia), which can cause pneumonia. Due to these bacteriological warfare tests, one hospital reported 12 cases of serratia pneumonia and 1 death (Cousins, 1979; D. R. Simon, 1999, p. 252).

Through various fronts during this period, the CIA, apparently unknown to the recipients, also funded social psychological research by such famous names in the field as the Sherifs, Orne, Rogers, Osgood, and Goffman (Marks, 1979, p. 121) and financed the publication of over 1,000 books, pretending that they were the products of independent scholarship (F. J. Cook, 1984, p. 287).

American Nuclear Guinea Pigs

In 1986, the House Energy and Commerce Subcommittee uncovered the fact that federal agencies had conducted exposure experiments on U.S. citizens, including injecting them with plutonium, radium, and uranium, over a 30-year period beginning in the mid-1940s. The experiments included feeding elderly adults radium or thorium at MIT, inmates receiving x-rays to their testes, open-air fallout tests, and feeding people real fallout (added to their food in powder form) from a Nevada test site (R. Lawrence, 2006). In 1996, the U.S. government agreed to pay 12 victims $4.8 million for injecting the unwitting subjects with plutonium and uranium. Many others remain uncompensated (Dobnik, 1996). Under the shield of national security, major harm was committed by the American nuclear state (Kauzlarich & Kramer, 1998).

Scandal

English historian Edward Gibbon (1737–1794) described history as a record of humanity's crimes, follies, and misfortunes. Nathan Miller in *The Founding Finaglers* (1976) describes corruption in various presidential administrations, which includes activities such as ordinary bribery, conflict of interest, till tapping, and illegal and improper use of government authority for financial gain or political advantage. Presidents themselves may not always be involved in wrongdoing, but as James Madison suggested in the First Congress, a president is "responsible for the conduct of the person he has nominated and appointed" (H. Johnson, 1991, p. 184). Second only to Watergate as the worst public policy scandal in American history was the Iran-Contra conspiracy.

The Iran-Contra Conspiracy

On November 4, 1986, the Lebanese magazine *Al Shiraa* revealed the existence of a secret U.S. arms sale to Iran. This would begin one of the longest (over 5 years) and most expensive probes in the nation's history, as of 1992 costing up to $100 million ("North Freed," 1991, p. A16). Money obtained in the sale of arms to Iran in exchange for American hostages was utilized to secretly fund the Contra rebels opposing the Marxist Sandinista regime in Nicaragua (U.S. House of Representatives, 1987).

It is difficult to succinctly present the tangled web of the privatization of foreign policy that the Iran-Contra affair represented. Marjorie Williams (1991) describes the plot:

> Popular president sells arms to archenemy hostage-taker Iran, violating not one but two U.S. policies (against arming Iran and dealing for hostages), marking up the price of the arms and sending the profit to the Nicaraguan contras in violation of a third policy, the Congressional Boland Amendments forbidding contra aid.

> From there, it was all denouement, a tangled skein of money and guns, middlemen and bank accounts, dates and times and findings and channels. Polls began to show that, as the narrative fragmented, the American people, initially outraged, ceased to follow it. (p. 12)

The Iran-Contra indictment also charged that Marine Corps Lieutenant Colonel Oliver North and retired Air Force General Richard Secord had conspired to divert millions from the sale of U.S. arms to "Enterprise," a secret organization created to privatize foreign policy. Draper (1991) estimated that the Contras may actually have received only about 20 percent of the millions raised (Draper, 1991, p. 114). These activities represented policy disputes between the executive branch and Congress, with Oliver North, having lied to Congress and shredded evidence, the designated scapegoat.

North (1991) was willing to "take the rap" until it became clear that he faced criminal charges without protection from higher-ups. North's boss, General Secord, described how President Reagan was able to truthfully deny knowledge of these activities. Reagan would employ "plausible deniability" by giving general policy guidelines and letting the details be carried out by others without his specific knowledge (Bliven, 1991).

North and former National Security Advisor John Poindexter were convicted in 1989 of various charges, including altering and destroying evidence and obstructing Congress, but these charges were overturned in 1990 and 1991 on the grounds that independent counsel Lawrence Walsh had utilized immunized testimony to Congress to subsequently prosecute them. The Constitutional questions raised by Iran-Contra were more than a partisan policy dispute. Bandow (1991) notes,

> The diversion scheme was a direct assault on our system of constitutional liberty. A small group of men apparently bypassed the president, [perhaps] lied to Congress, and used part of the proceeds of the sale of weapons paid for by taxpayers to implement their own foreign policy. That these people were well-meaning doesn't matter: the Constitution places the power of the purse in Congress, not with a handful of executive appointees. It is for the voters, not the CIA director and a Marine Corps detailee to the NSC, to decide that Congress is "on the other side." (p. A19)

The final chapter in the Iran-Contra affair was written by former President George H. W. Bush, who shortly before leaving office issued full pardons to all who had been convicted or charged with wrongdoing in the affair.

Crime File 11.2, "White House Crime and Scandal: From Washington to George W. Bush," makes the point that scandal did not start with President Clinton.

Crime File 11.2

Audio Link 11.1
Listen to President Clinton deny a relationship with Monica Lewinsky.

White House Crime and Scandal: From Washington to George W. Bush

The impeachment of President Bill Clinton in 1999 and his admission that he did, despite previous denials, have an "inappropriate relationship" with Monica Lewinsky, a 22-year-old intern in the executive mansion, electrified Washington and the nation like nothing since Watergate. How comparable are such scandals and how do they compare with previous executive branch wrongdoing?

This author's review of presidential wrongdoing is problematic due to the fact that recent presidents are subject to more scrutiny and, thus, more reports of scandal, the creation of an independent prosecutor's office in the wake of Watergate, and difficulties in maintaining one's own political objectivity.

The top tier of presidents in terms of public policy scandals is as follows:

- Richard Nixon (Watergate)
- Ronald Reagan (Iran-Contra)
- Warren Harding (Teapot Dome)
- Ulysses S. Grant (Credit Mobilier)
- Lyndon B. Johnson (Tonkin Gulf Resolution)
- George W. Bush (Invasion of Iraq)

Of interest is the observation that, if it is a public policy or an economic scandal, it is usually a Republican (5 of the 6 in this analysis, with Johnson the lone Democrat), while, if it is a sexual scandal, it is usually a Democrat (4 of 5 in this analysis, with Harding the sole Republican).

Watergate was, of course, the benchmark for all other political scandals in the United States. This involved the discovery of the illegal break-in of the offices of the Democratic National Committee located in the Watergate complex in Washington, D.C. Among the charges against and offenses by Nixon and associates were burglary, illegal surveillance, attempted bribery of a federal judge, selling ambassadorships in return for illegal campaign donations, having illegal "slush funds," destruction of evidence, plans for dirty tricks in political campaigns, requests for IRS audits on opponents, perjury, withholding information, altering evidence, and lying to the American public.

Between 1980 and 1988, more than 200 Reaganites came under ethical or criminal investigation, the largest number of scandals in any administration in American history (S. Ross, 1988, p. 1). Major scandals in the Reagan administration included corruption in the Environmental Protection Agency, the Wedtech scandal, the Pentagon procurement scandal, influence peddling at the Department of Housing and Urban Development, and the biggest one—the Iran-Contra conspiracy.

Although Reagan steadfastly refused to admit knowledge of the conspiracy, arms were secretly traded to a terrorist nation (Iran) in return for the release of U.S. hostages. These secret funds were then illegally used to supply arms to Contra rebels in Nicaragua, all against Congressional wishes. Hersh (1990) alleges that senior members of the congressional Iran-Contra committee agreed from the outset that specific evidence of a Presidential "act of commission" would be necessary before Reagan himself would become a target (p. 47). They believed the president lacked the mental ability to fully appreciate what happened (p. 64).

Other public policy scandals included the Credit Mobilier affair of the 1870s affecting the corrupt administration of Ulysses S. Grant. Credit Mobilier was a finance company that bribed members of Congress and inflated profits in a conspiracy of waste, crime, and corruption. Warren Harding's corrupt cabinet was involved in conspiracy, graft, fraud, bribery, and cover-ups related primarily to the "Teapot Dome scandal," the illegal sale for personal profit of U.S. naval oil reserves in Teapot Dome, Wyoming. Lyndon B. Johnson (LBJ) and George W. Bush have both been added to the list for similar reasons. While LBJ was a successful president with respect to his domestic agenda, he was responsible for expanding the costly and ultimately disastrous war in Vietnam, which became his war. The Tonkin Gulf Incident in August 1964 enabled him to persuade Congress to authorize a huge military buildup in Vietnam. The event on which the resolution was based, which probably never took place, involved allegations of a North Vietnamese attack on U.S. destroyers. Unnecessary U.S. escalation as a result produced huge losses in deaths and injured and divided the nation. Similarly, George W. Bush persuaded Congress to give him authorization to invade Iraq, if necessary, in order to force the Iraqis to cooperate in destroying their alleged "weapons of mass destruction." Ultimately, intelligence sources failed to reveal that Iraq had such weapons. The same administration was also implicated in revealing the identity of CIA agent Valerie Plame, apparently in retaliation for her husband revealing that Saddam Hussein had not sought nuclear materials from an African nation as alleged by the George W. Bush administration.

No other presidential administration has suffered the scrutiny that Bill Clinton's presidency experienced, primarily by Special Prosecutor Ken Starr, and under the rubric of "Whitewater," a land deal involving the Clintons in Arkansas. Despite more than 5 years and $100 million in investigations, Starr's office failed to pin policy scandals on Clinton. This would all change with a civil suit for sexual harassment against Clinton for an activity that took place while he was governor of Arkansas.

Sexual Scandal

In contrast with political and economic scandal, a separate list of presidents involved in sexual scandal finds the following as top offenders:

- John Kennedy
- Bill Clinton
- Warren Harding
- Lyndon Johnson
- Grover Cleveland
- Making up the remainder are a host of also-rans such as Jefferson, Garfield, Wilson, and Franklin Roosevelt.

Despite the public image of Camelot, John Kennedy was "the playboy president," the undisputed leader of presidents involved in illicit sexual escapades, having the most active extramarital sex life of any U.S. president. The press at the time followed the custom of not reporting such activity. Kennedy had affairs with movie stars such as Marilyn Monroe; shared a girlfriend with a Chicago mobster; had extramarital sex in the Oval Office; and had two White House "aides," dubbed "Fiddle" and "Faddle" by the Secret Service, who were his regular nude swimming partners in the White House pool.

Although paling in comparison with his idol, John Kennedy, Clinton's political career had been haunted by charges (usually true) from former claimed and real paramours of sexual escapades and indiscretions. Due to the investigations of Special Prosecutor Kenneth Starr, Bill Clinton would become the first elected president in American history to be impeached by a partisan Republican Congress for lying under oath about a sexual affair. Congress refused, however, to remove Clinton from office.

Lyndon Johnson was a major womanizer, and even had a buzzer system installed in the Oval Office so he could be alerted if his wife was approaching, since Lady Bird once caught him having sex with a secretary. He once claimed, "I had more women by accident than Kennedy had on purpose" (Dalleck, 1991, p. 189). Warren Harding was alleged to have had sexual relations with his mistress (a teenager) in the White House coat closet. She had already given birth to his illegitimate daughter. Harding also visited prostitutes. Finally, Grover Cleveland, called "the Beast of Buffalo" by his foes for fathering an illegitimate child, admitted to doing so. When his campaign opponents used the slogan, "Ma, Ma, Where's My Pa," his staff retorted, "Gone to the White House, Ha, Ha, Ha."

None of the worst presidents in public policy or sexual scandal makes the historians' top list of presidents. All of the worst for policy scandal (Reagan not rated) make the worst list by historians, but none of the worst for sexual scandal (except Harding, who was on both lists) makes historians' lists of worst presidents.

Additional Sources: White House Crime and Scandal: From Washington to Clinton. By F. E. Hagan, 1999, paper presented at the American Society of Criminology Meetings, Toronto, Ontario, Canada, November; "White House Scandal: From Washington to George W. Bush," by F. E. Hagan, 2008b, paper presented at the Academy of Criminal Justice Sciences Meetings, Cincinnati, OH, March.

Web Research Project

Choose a presidential scandal since the Carter presidency (e.g., Reagan [Wedtech, Pentagon procurement, HUD, EPA, October Surprise, Debategate] or Clinton [Vince Foster suicide, Travelgate, Filegate, Whitewater, or Billingsgate]) and answer the following: What were the charges? What investigation took place? What was the final resolution?

Crimes Against Government

Protest and Dissent

As previously indicated, crimes against the government may vary from illegal protests, demonstrations, and strikes to treason, sabotage, assassination, and terrorism. At various times in history, social movements that petition for change are viewed as threatening or subversive to the existing society. The American Revolution, the labor movement, the anti–Vietnam War movement, and the struggle for civil rights are examples. Demonstrators for civil rights and other causes may purposely violate laws and be arrested for disorderly conduct, breach of peace, parading without a permit, trespassing, loitering, and the like. They may also be arrested for refusing to pay income taxes, for picketing military bases, for student protests, or for refusing to register for military draft. Many student activists of the sixties viewed their universities as protecting the military, industrial, and racial status quo (Skolnick, 1969, p. xxi).

While dissent and protest activities against the government are usually perceived as "radical" (leftist) in attempting to bring about change in the existing order, they may also represent "reactionary" (rightist) activities aimed at preserving or restoring the old order, institutions, or organizational schemes that are endangered.

Groups express dissent and civil disobedience by employing sit-ins, boycotts, and freedom rides (in order to desegregate facilities) to challenge unjust laws. They consciously decide to violate certain laws to call public attention to their cause and to bring about change in the law. Civil rights leader and director of the Southern Christian Leadership Conference, Dr. Martin Luther King, Jr., a Protestant minister, came under heavy criticism from other clergy for neglecting God's work and becoming too involved in disruptive social activities.

Letter From Birmingham Jail

The Nuremberg principle or precedent supports the view that, when one is faced with the imperative of either obeying unjust laws or following a higher moral conscience, the latter takes precedence; to blindly follow orders when they violate basic human rights and dignity is unacceptable. Martin Luther King, Jr.'s (1963) "Letter From Birmingham Jail" very movingly describes his view that immoral laws must be disobeyed: "I submit that an individual who breaks a law that conscience tells him is unjust and who willingly accepts the penalty of imprisonment in order to arouse the conscience of the community over its injustice is in reality expressing the highest respect for the law."

Photo 11.3

Freedom Riders sit on a bus making a test trip to Mississippi.

Martin Luther King, Jr., and his organization, the Southern Christian Leadership Conference, advocated nonviolent, passive resistance—civil disobedience of the form that was employed so successfully by Mahatma Gandhi in overcoming British rule in India. Gandhi taught that violence on the part of those enforcing unjust laws must be met with nonviolence in order to appeal to the public's sense of justice. Incarcerated members of the Irish Republican Army in Northern Ireland also borrowed a tactic from Gandhi—the "hunger strike." Members of "H-block" (maximum security wing of the prison) starved themselves to death in order to demonstrate their dedication to their cause. Their actions spurred the "Anti H-Block" movement of people who supported the hunger strikers' views.

Social Movements

Illegal protests, demonstrations, and strikes are often associated with social movements that advocate change in the existing order. Members and supporters of such movements are usually deeply committed to altering the status quo. The civil rights battle against racism; the feminist struggle against sexism; the labor and agrarian movements for fair wages; the antiwar movement against the escalation of the Vietnam conflict; the antinuclear, environmental, and anti- or pro-choice movements are all examples. While such groups are intent on altering the status quo and may at times resort to violence, sabotage, and other destructive behavior, most do not resort to treason, assassination, or terrorism. Frequently, political criminals have done nothing more than exist; they suffer attack because of race, gender, ethnicity, or nationality. Expulsion, exile, curfews, confiscations, confinement, restrictions on travel, and controls over associations may all be used to subordinate, enslave, or subject to second-class citizenship subjugated groups.

Conscientious objectors, those who refuse to serve in the military because it violates their personal, religious, or moral principles, may also serve as an example of political offenders. In the eighties and nineties, groups such as the Sanctuary Movement, pro-life and pro-choice groups on the abortion issue, and antinuclear movements participated in various forms of civil disobedience and protest activities. The **Sanctuary Movement** consisted of church and lay workers in the 1980s who ran an "underground railroad" to help keep political refugees (often illegal immigrants) from being deported to their Central American homelands where they often faced political persecution. The U.S. government claimed that such groups were in violation of the immigration laws, that the people they sought to help were economic rather than political refugees, and that the government had a right and responsibility to control the nation's borders (Crittenden, 1988; Tomsho, 1987).

Pro-life (right-to-life) forces are opposed to legalized abortion, viewing it as murder, and seek a reversal of the 1973 *Roe v. Wade* Supreme Court decision, which permits abortion as needed. Besides protests and civil disobedience, more extreme factions have bombed abortion clinics. Opponents (pro-choice activists) argue that such a decision is not the government's to make, but is between a woman and her physician, and that one group's morality should not become public policy for all (Paige, 1985). With over 40,000 arrests for blocking abortion clinics, Operation Rescue, a pro-life campaign to stop abortions, may have become the largest civil disobedience campaign in U.S. history (Lawler, 1991).

Antinuclear forces are convinced that the nuclear industry is unsafe and is sapping funds from more ecologically sane energy policies such as solar energy. Such groups have violated the law in attempts to prevent start-ups of new reactors. One antinuclear activist ran onto a stage in 1992 while former President Reagan was speaking, broke a crystal statue Reagan had received, shoved Reagan aside, and began speaking before being arrested. "ACT-UPpers" are activists who wish to attract public attention and action to fight AIDS. They have interrupted meetings by blowing whistles and using other means of calling attention to their cause.

Another group that became more visible beginning in the late eighties was made up of antivivisectionists—those who oppose using animals in scientific experiments in which maiming, torture, death, or other harm is inflicted. Such groups have protested, raiding laboratories and "liberating" animals (Regan, 1982), and photographed and released to the press some of the more grisly examples. Scientific researchers who use animals claim that such experiments are necessary for discovering medical cures and treatments. Members of the Animal Liberation Front vandalized and set fire to a mink research laboratory at Michigan State University, accusing the professor who ran the lab of killing "thousands of minks in painful and scientifically worthless experiments" ("Animal-Rights Group," 1991). The raid destroyed 30 years of research on the disappearance of minks in the Great Lakes area.

Hate crimes, a growing phenomenon in the United States, are committed for racial, gender, ethnic, or other ideological reasons. Criminals who perpetrate these offenses are convinced of the rightness of their actions (Hamm, 1994) and that their victims deserve their fate. Beginning in 1999 and 2000, the "antiglobalization movement" protested at the World Bank, International Monetary Fund, and World Trade Organization meetings, as well as both the Democratic and Republican national political conventions. The members of various groups associated with the movement charged the World Bank and other groups with imposing crushing debt on poor countries, destroying their natural resources, and exploiting their labor. An unlikely alliance of union members, environmentalists, and religious activists, they have been using mass protests to call attention to their issues.

Assassination

Video Link 11.3
View a video about John Wilkes Booth's plot against President Lincoln.

In 1995, due to assassination attempts and threats against President Bill Clinton, a portion of Pennsylvania Avenue in front of the White House was closed to traffic for the foreseeable future. In 1994, an intoxicated crack user, Frank Corder, crashed a stolen airplane into the south facade of the presidential mansion, killing himself in the process. Francisco Duran, an upholsterer from Colorado, opened fire on the White House with a semiautomatic rifle, claiming he was shooting aliens who were hanging as a mist over the building. On April 4, 1995, Duran was convicted of attempted assassination. Such attacks on public figures unfortunately have a long history in American politics.

In *American Assassins: The Darker Side of Politics,* James Clarke (1982) is highly critical of the popular assumption that all or most assassins suffer from some mental pathology, that they are insane or deranged, and that this causes them to become assassins. Because, according to Clarke, authors incestuously cite each other's works and rely on inaccurate secondary literature, he states that this *pathological myth of assassins* was continually repeated in leading works such as those by Donovan (1952), Hastings (1965), Kirkham (1969), and the Warren Commission (1964). Much observation of assassins' pathological symptoms may result from post hoc error, the false assumption that since one variable or outcome follows another in time, it must be caused by the preceding variable. Clarke believes that most of the major works on assassins simply fail to consider the political context of assassinations (1982, p. 7).

Clarke identifies five types of assassins (the examples have been provided by this writer):

Type 1. *Political assassins* commit their acts (they believe) selflessly, for political reasons. Some examples of such assassins (successful and unsuccessful) are John Wilkes Booth (Lincoln), Leon Czolgosz (McKinley), Oscar Collazo and Griselio Torresola (Truman), and Sirhan Sirhan (Robert Kennedy). Booth committed his crime in support of the Confederacy, Czolgosz's was in support of a class revolt, Collazo and Torresola were Puerto Rican nationalists, and Sirhan felt his act would help the Arab cause.

Type 2. *Egocentric assassins* are "persons with an overwhelming and aggressive egocentric need for acceptance, recognition, and status" (J. W. Clarke, 1982, p. 7). They appreciate the consequences of their acts and do not exhibit cognitive distortion characteristic of delusion or psychoses. Some examples are Lee Harvey Oswald (John F. Kennedy), Samuel Byck (Nixon), Lynette "Squeaky" Fromme (Ford), and Sara Jane Moore (Ford). Such assassins seek attention, which they feel they have been denied, and seek to place a burden on those they feel have denied or rejected them. It would appear that John Hinckley, attempted assassin of Ronald Reagan, would fit this Type 2 description. Oswald and Byck funneled their personal difficulties into political extremism. Oswald wanted to prove himself to the Cuban government and to his wife, neither of whom took him seriously. In February of 1974, Byck died in an attempt to hijack a jetliner, which he planned to crash-dive into the White House in order to kill Nixon. Although "Squeaky" Fromme resembled a Type 1 assassin, her devotion was to a man (Charles Manson) and not a cause (J. W. Clarke, p. 262). Moore wished to demonstrate her commitment to radicals who had rejected her when they discovered she was an FBI informant.

Type 3. *Psychopathic assassins,* unable to relate to others, are emotional cripples who direct their perverse rage at popular political figures. In describing Giuseppi Zangara (Franklin Roosevelt and Chicago Mayor Anton Cermak) and Arthur Bremer (George Wallace), James Clarke (1982) indicates that "their motives were highly personal: they wanted to end their own lives in the most outrageous display of nihilistic contempt possible for a society they hated" (p. 167). Both transferred resentment for their emotional deprivation in childhood to public figures. Bremer targeted Wallace after other presidential candidates he had stalked did not offer the proper opportunity.

Type 4. *Insane assassins* have documented histories of organic psychosis, a type of mental illness caused by physiological factors either environmentally or genetically induced. They exhibit severe emotional and cognitive distortion of reality, such as paranoia, one characteristic of which might be delusions of grandeur. Such psychotic assassins include Richard Lawrence (Jackson), Charles Guiteau (Garfield), and Joseph Schrank (Theodore Roosevelt). Guiteau and Lawrence both believed they had been selected by God to perform his will, while Schrank irrationally believed that he was avenging McKinley's assassination and that Theodore Roosevelt had been the culprit.

Type 5. *Atypical assassins* are those who defy classification, such as Carl Weiss (Huey Long) and James Earl Ray (Martin Luther King, Jr.). Weiss was a successful physician who apparently killed Long because he felt that by so doing he was protecting the lives and political jobs of his relatives. Although racism clearly was behind the King assassination, Ray, an unsuccessful career criminal, appeared to be primarily motivated by an alleged $50,000 payment for the crime (J. W. Clarke, 1982, p. 246).

Clarke concludes his analysis by indicating that, since 1963, Type 2 and 3 assassins have shared a strong desire for media notoriety and that restriction of such coverage could help discourage some assassination attempts. He also feels that there is less need for additional surveillance of suspects than for analysis of information already in the possession of organizations such as the FBI. For instance, the FBI was aware of Byck, Fromme, Hinckley, Moore, and Oswald, and even covered up information after the fact regarding the Oswald and Ray cases. (This, by the way, has led to a variety of conspiracy theories with respect to the King and John Kennedy assassinations.)

Espionage

Espionage, the clandestine theft of information, has been a practice since early recorded history. In the Bible, God commanded Moses to send spies to Canaan, and Joshua sent spies to Jericho. In 1987, archeologists discovered a large collection of 3,700-year-old Mesopotamian clay tablets that described, among other things, the capture and ransom of spies ("Ancient Records Discovered," 1987). Fifth-century B.C. Chinese sage Sun-Tzu in his classic book, *Art of War* (1963), provided a chapter on secret agents and types of spies. While Benedict Arnold, who betrayed the American colonists to the British during the Revolutionary War, is a name that lives in infamy in the United States, a statue of Nathan Hale, an American spy executed by the British, stands outside CIA headquarters in Virginia (F. E. Hagan, 1987b).

Despite images of "cloaks and daggers," Mata Hari and James Bond, "black espionage" or "covert agents" ferreting out secrets, classical forms of spying have for many years has been used less often than "white espionage," which uses space satellites, code breaking, and collection of technical information (Marchetti & Marks, 1974, p. 186; Ranelagh, 1986). The technological revolution in espionage has replaced the "seductive, sable-coated countess traveling first class on the Orient Express" (F. Maclean, 1978, p. 336). "Sub rosa criminals" are spies who steal secrets. One form of spying, treason, is one of the earliest crimes punished by society and the only crime discussed in the U.S. Constitution. Despite lack of attention in the criminological literature, **sub rosa crime** (espionage) is more costly than traditional crime and has altered post–World War II economic and political history.

Defector and former KGB Major Stanislav Levchenko was apparently the first to reveal the acronym MICE for describing the motives of spies (Kneece, 1986): **m**otivation, **i**deology, **c**ompromise, and **e**go. Others have expanded this acronym to **SMICE,** adding sex as a separate motivation. There has been a major shift in the motivations of Eastern and Western spies from the ideological, Cold War fifties to the materialistic/hedonistic eighties and nineties. The ideological motivation has been replaced for the most part by mercenary considerations.

Many previous discussions of types of spies have concentrated on specific role performance or tasks (E. Anderson, 1977; Copeland, 1974; Sun-Tzu, 1963; S. Turner, 1985). This writer proposes a **typology of spies** that includes the following (F. E. Hagan, 1986, 1987b):

Audio Link 11.2
Listen to a discussion of spying.

Mercenary spies trade secrets for personal monetary reward. Andrew Daulton Lee, the "Snowman" described in Robert Lindsey's book *The Falcon and the Snowman* (1979), is an example. Lee was a highly successful drug dealer (hence the "Snowman" title, in reference to his cocaine sales) and began acting as a courier for his friend, Christopher Boyce (the "Falcon") by transporting American military secrets to the Soviets for financial reward. The majority of spy cases since 1980 have been of the mercenary variety.

The *ideological spy* is motivated by strong ideological beliefs. Such spies are political criminals and are often condemned as traitors in one country, while being heralded as heroes in the recipient nation. Julius and Ethel Rosenberg became the first and only native-born Americans to be executed for treason (in March of 1951) for having given the Russians America's atomic secrets (Hyde, 1980). They did so out of devotion to communism. Devotion to communism also motivated the British "establishment spies," Burgess, Maclean, Philby, and Blunt. All four became committed Communists while they were Cambridge University students in the thirties, were recruited by Soviet intelligence, and rose to the highest levels as "moles" (deep cover agents) in British intelligence (Pincher, 1984; N. West, 1982).

The *alienated/egocentric spy* is one who betrays for personal reasons unrelated to monetary or ideological considerations. In 1985, ex-CIA employee Edward Howard Lee, having been fired by the agency, defected to the Soviets and took classified secrets with him.

The *buccaneer* or *sport spy* is one who obtains psychological fulfillment through spying. S. Turner (1985) describes them as "swashbuckling adventurers who spy for kicks." Christopher "The Falcon" Boyce and John Walker are examples, although there are many others. Boyce, the partner of Lee in *The Falcon and the Snowman* (Lindsey, 1979, 1983), was a bored 21-year-old college dropout who gave the Soviets top-secret satellite information in an act of defiance against the CIA. Boyce told a federal marshal, "I guess I'm a pirate at heart. I guess I'm an adventurer" (Lindsey, 1983). John Walker, head of a spy ring that included his son, his brother, and his son's friend, was a former naval officer who passed American cryptographic codes to the Soviets from the 1960s through the mid-1980s. Walker reflected a Walter Mitty—James Bond image of spying, which included props such as umbrella weapons and crossbows. His ring's peddling of American codes to the Soviets cost American pilots' lives in Vietnam and compromised American naval strategy (Kneece, 1986).

Professional spies are agents, careerists, or occupational employees of intelligence bureaucracies. Covert professional agents, such as Richard Sorge and Rudolf Abel, are legends in the history of espionage. Such agents usually operate under diplomatic cover and, when caught, enjoy diplomatic immunity and are dispatched out of the country. Those who lack such cover are usually swapped for other spies at a later date.

Compromised spies are at-first-reluctant traitors who trade secrets either for romantic purposes or because of blackmail and coercion. Many are victims of the SMICE strategy. The most celebrated case was that of U.S. Marine guards working at the U.S. embassy in Moscow, particularly Clayton Lonetree and Arnold Bracey. The guards were allegedly victims of what spy novelist John Le Carré has depicted as "the honey trap." That is, they had been duped by KGB "swallows," or seductive female assistants who trade sex for secrets (Kessler, 1989; Schlachter, 1986). Kessler claims that, although the U.S. government denied it, the KGB had the run of the embassy and its secrets.

The *deceived spy* ("false flag recruit") is one who is led to believe that he or she is working for one organization when, in fact, the work is for another. Edwin Wilson, the subject of Peter Maas's book *Manhunt* (1986), was an ex-CIA employee who recruited assassins, smugglers, technicians, and spies, including high-level moonlighters from the CIA, to work for Libya. He led them to believe it was a "company" (CIA) operation (Epstein, 1983; Goulden, 1984). Industrial spies who believe they are working for a rival company may very well be working for the intelligence agency of a rival power.

The remaining types are the *quasi-agents*—dissenters, such as ex-CIA agent Philip Agee, who released classified information to the public. They resemble whistleblowers. *Escapee spies* are individuals who defect in order to avoid personal problems, while the *miscellaneous category* is for those spies who defy classification. Crime File 11.3 gives a synopsis of the Robert Hanssen spy case. Hanssen was arguably the most effective spy against the United States in FBI history.

Crime File 11.3

The Hanssen Spy Case

On February 18, 2001, FBI Supervisory Special Agent Robert Hanssen was arrested while loading a "dead drop" with secret documents under a footbridge at Foxstone Park for his Russian handlers. The Hanssen case reads better than a spy novel by Le Carré or Ludlum. In fact, he represents the most complex, multidimensional spy of any in U.S. history. He would receive from the Russians an estimated total of $1.4 million for his deceit, including $650,000 in cash, diamonds, and $800,000 in an overseas escrow account (F. E. Hagan, 2002).

Hanssen began spying for the then Soviet Union (now Russian Federation) in 1979 when he walked into Amtorg (a Soviet front company) offices and offered his services. When his wife discovered his treachery in 1980, he promised to quit and to confess his sins to a priest. His confessor, bound to silence by Canon Law, told him to give the proceeds from spying to charity. Hanssen was a self-recruit, a "walk-in" in spy tradecraft jargon. In October 1985, he returned to spying by sending a letter to Victor Cherkashin, a Soviet spymaster, providing secret information and again offering his services. While Hanssen's promotions resulted in higher status within the FBI, his access to information and value to the Soviets (later Russians) increased. In 1990, his brother-in-law, FBI agent Mark Waulk, expressed his suspicions to his superiors that Hanssen was a spy. This was after the discovery by his sister Bonnie (Hanssen's wife) of large sums of unexplained cash at their home. The Bureau did nothing. Perhaps the suspicion was a mild one and difficult to substantiate, or maybe yet another mole higher up

Photo 11.4

Former FBI Supervisory Special Agent Robert Hanssen.

in the organization squelched the investigation (Vise, 2002). Hanssen had worked counterintelligence since the 1980s. As a supervisor and liaison with other intelligence agencies, Hanssen represents the most important FBI agent ever to commit espionage against his country.

Damage by Hanssen

Hanssen may be the most damaging spy in American history or certainly since former CIA turncoat Aldrich Ames. This long-rumored "mole" within the Bureau used various aliases: "B," "Ramon Garcia," and "Jim Baker" among others. Being in counterintelligence, he knew the system and had studied spycraft. A self-trained computer hacker, he continually searched internal FBI systems to make sure that no one was onto him. To protect himself from discovery, he used "dead drops" (secret drop-off points where the participants do not come into physical contact) in parks near his Vienna, Virginia, home. He never personally met any of his Russian handlers.

Hanssen had greater range than previous spies due to his technical and computer expertise and access to intelligence of other federal agencies with whom he served as an FBI liaison. He gave the Soviets the "store," vast amounts of data on top-secret processes, methods, and codes that did grave damage to his country. He provided names of Russian double agents that were confirmed by independent information from other spies such as Aldrich Ames. As a result, at least two of these agents were executed. Other information, the "crown jewels" of U.S. intelligence, included the following (see Affidavit [2001] as well as Vise [2002, Appendix I]):

- A "spy tunnel" beneath the new Soviet embassy in Washington
- U.S. strategy in the event of a nuclear attack
- The National Intelligence Program (Strategic Plan)
- U.S. strategies for recruiting double agents
- The locations of Soviet defectors
- National Security Agency spy and space surveillance operations
- National Security Council secrets
- U.S. intelligence community's private Internet operations

(Continued)

(Continued)

- Information that Soviet defectors had provided
- U.S. recruitment of foreign agents and assets
- FBI and CIA intelligence investigative techniques
- Confirmed information provided by CIA spy Aldrich Ames

For 2 months prior to Hanssen's arrest on February 18, 2001, he had been under 24-hour surveillance by the FBI's Special Surveillance Group. They even occupied a house on Hanssen's street in addition to searching his office, car, and computer files. In the fall of 2001 the FBI obtained the stolen, original KGB file on Hanssen provided by a Russian asset, along with a black garbage bag that Hanssen had used to wrap his secret information before depositing it at dead drops. The fingerprints on the bag matched those of Hanssen, as did odd phrases in his letters to handlers that were immediately recognized by his coworkers. It is possible that the Russians had advance warning of Hanssen's arrest and may have deliberately given him up, since he was due to retire, in order to protect other moles. Ironically, Hanssen included in his last dead drop a letter saying he was giving up the spy business and this was his final contact. It was.

Hanssen pleaded guilty in July 2001 and in return for his cooperation, was not given the death penalty. As part of the deal, his family would receive about half of his yearly pension of $68,000. In 2002, Hanssen received a life sentence without the possibility of a reduced sentence. He had never taken a polygraph examination in his 27 years with the FBI.

Vise (2002) describes Hanssen as a man of contradictions, a man who compartmentalized his life: a family man, loving father, religious person, a spy, and a sexual deviant. He lived modestly as a suburban husband and father of six and was a daily communicant at Roman Catholic Mass. A convert to Catholicism and its conservative wing "Opus Dei," as well as a rampant right-wing anticommunist, he also had a relationship with a stripper, posted sexual fantasies on chat sites, and secretly video recorded his marital sexual activity so that his best friend could observe from another room in their house. He was the son of a disapproving father, a retired Chicago police officer who would humiliate him in public as a child. A quiet loner who lacked social skills in his youth, he became a real fan of spycraft books. His coworkers called him "Dr. Death," "Mr. Mortician," and "Dr. Szell" (the Nazi dentist character in the film *Marathon Man*) due to his dark and somber dress. They described him as unpleasant, unliked, weird, a misfit, a computer nerd, elitist, and arrogant. They did not like him. He was religiously righteous and disapproved of Bureau retirement parties at local strip clubs. He also took exception to female agents being hired by the Bureau. He was suspended for a week for assaulting a female agent who had left a meeting without his approval. Despite his superior intelligence, his lack of social skills and inept demeanor found him continually passed over for promotion and resenting it.

Where does Robert Hanssen fit in the typology of spies? This writer views him as primarily an egocentric spy and to a lesser extent a buccaneer and mercenary spy. Havill (2001, p. 103) describes him as "Walter Mitty squared," after the fictional character who fantasized about exciting lives to make up for his own boring existence. While money was a consideration, "need, not greed" was the motivation. Hanssen was not in search of a lavish lifestyle like Aldrich Ames, but needed money for his kids' parochial school educations. Perhaps wanting to humiliate his "inferior, unappreciative" coworkers, he chose spying as a risk-taking, puppetmaster alternative. His fractured ego seeking recognition would finally "show" the FBI, his corrupt father figure (Vise, 2002). From his point of view, the agency was corrupt because it passed him over for promotion while rewarding those to whom Hanssen felt superior.

Shannon and Blackman (2002) draw an interesting analogy with school shooters. Many are described as loners, victims of derision, unappreciated, and misfits who finally become unglued. They believe the contradictions of the Opus Dei devoutness may have been an attempt to create a "legend" or "cover," as they say in spycraft, or simply an expression of his multiple personality. "Game, not gain" (Vise, 2002, p. 134) motivated Hanssen, who outdid Philby and even the fictional James Bond. Hanssen himself indicates that he did it out of "fear" (of not being able to provide for his family) and "rage" (being passed over for promotions) (Vise, 2002, p. 225). Curiously, he felt that his religious devotion guaranteed his salvation and that espionage, for which he has shown no remorse, was an escape. Not motivated primarily by money, alcohol or drug abuse, ideology, or materialism, Hanssen was an arrogant "gamesman" (Shannon & Blackman, 2002) who wanted to get even.

Additional Sources: Affidavit in Support of Criminal Complaint, Arrest Warrant and Search Warrant in the United States District Court for the Eastern District of Virginia, Alexandria Division, *United States of America v. Robert Hanssen* (a.k.a. "B," "Ramon Garcia," "Jim Baker," "G. Robertson"), February 2001; *The Spy Who Stayed Out in the Cold: The Secret Life of FBI Double Agent Robert Hanssen*, by A. Havill, 2001, New York: St. Martin's Press.

Web Research Project

Search the topics of "spies" and "espionage." Are you able to locate any new spy cases since Hanssen?

Political "Whistleblowing"

Information is usually classified as secret to protect national security; in some instances it is done to misinform the public and shroud questionable activities. It was to protest the latter that Daniel Ellsberg, an employee of the RAND Corporation (a private think tank and research organization), violated his oath of secrecy and turned over secret government documents, The Pentagon Papers, to the press (Gravel, 1971). Ellsberg felt that revealing the government's deceit of the public regarding U.S. involvement in the Vietnam War outweighed his duty to keep government secrets. In an even more controversial case, former CIA agent Philip Agee (1975) wrote personal memoirs of his CIA activities in South America in which he named and, according to some, endangered CIA operatives in those countries. He disagreed ideologically with many covert policies the CIA had been carrying out in that region.

Photo 11.5
Daniel Ellsberg turned over The Pentagon Papers to *The New York Times*, which began publishing excerpts on June 13, 1971.

Terrorism

Viewed outside its political context, international terrorism represents some of the worst examples of mass murder in history.

Definitions and Types of Terrorism

Any definition of **terrorism** is sure to arouse dispute. Definitions by the U.S. Department of Defense, FBI, State Department, Department of Justice, and Vice President's Task Force on Combating Terrorism (1986) include

- The unlawful use of force or violence by revolutionary organizations.
- The intention of coercion or intimidation of governments for political or ideological purposes.
- Premeditated political violence perpetrated against noncombatant targets by subnational groups or clandestine state agents.
- Use of assassination or kidnapping.

Handbook Link 11.2
Read an article on terrorism.

Terrorism may be distinguished from tragic acts of war in its willful and calculated targeting of innocents (Netanyahu, 1986, p. 8). Even during the Nazi occupation of Europe, Allied resistance fighters avoided indiscriminate killing of the families of German soldiers. No such limitations on noncombatants figure into the plans of many current terrorist groups. The Federal Bureau of Investigation (Pomerantz, 1987) defines terrorism as "the unlawful use of force or violence against persons or property to intimidate or coerce a government, the civilian population, or any segment thereof, in furtherance of political or social objectives" (p. 15).

The *Report of the Task Force on Disorders and Terrorism* (National Advisory Committee, 1976b, pp. 3–6) provides the following **typology of terrorism** (Table 11.1): political terrorism, nonpolitical terrorism, quasi-terrorism, limited political terrorism, and official or state terrorism.

The report defines *political terrorism* as "violent criminal behavior designed primarily to generate fear in the community, or a substantial segment of it, for political purposes." *Nonpolitical terrorism* also attempts to elicit fear by means of violence, but is undertaken for private purposes or gain. Examples of this type would include activities of organized crime; the Manson family; or Charles Whitman, "the Texas tower" sniper. *Quasi-terrorism* describes "those activities incidental to the

Journal Article Link 11.1
Examine literature regarding terrorism and Colombia.

TABLE 11.1 ■ Types of Terrorists

Wilkinson's (1976) typology of terrorism:

- Criminal
- Psychotic
- War
- Political
  - Revolutionary
  - Subrevolutionary
  - Repressive

Hacker's (1976) typology of terrorists:

- Crusaders
- Criminals
- Crazies

Schmid and de Graaf's (1982) typology of political terrorism:

- Insurgent terrorism
  - Social Revolutionary
  - Separatist, National, or Ethnic
  - Single Issue
- State or Repressive
- Vigilante

Poland's (1988) typology of political terrorism:

- State or Repressive
- Liberation Theology

- Identity Theology
- Nihilists
- Insurgent
- Revolutionaries
- Separatist/Nationalist
- Single Issue

Sederberg's (1989) typology of terrorism:

- Criminal
- Nihilist
- Nationalist
- Revolutionary
- Establishment
- Vigilante
- Covert Official
- Overt Official
- Genocide

Vice President's Task Force on Disorders and Terrorism's (1976) typology of terrorism:

- Political Terrorism
- Nonpolitical Terrorism
- Quasi-Terrorism
- Limited Political Terrorism
- Official or State Terrorism

commission of crimes of violence that are similar in form and method to true terrorism but which nevertheless lack its essential ingredient." Rather than being ideologically motivated, many skyjackers and hostage takers, although employing terrorist methods, are interested in ransom. *Limited political terrorism* refers to "acts of terrorism which are committed for ideological or political motives, but which are not part of a concerted campaign to capture control of the state." Vendetta-type executions and acts of lone terrorists for essentially private motives are examples. Such terrorism is illustrated by the acts of Theodore Kaczynski, the infamous "Unabomber," apprehended in 1996, who had eluded capture for nearly two decades. An antitechnology, radical environmentalist, he killed 3 and wounded 23 before being captured.

Official or *state terrorism* occurs in "nations whose rule is based upon fear and oppression that reach terroristic proportions" (J. Simpson & Bennett, 1985). J. B. Wolf (1981) differentiates "enforcement terrorism" from "agitational terrorism," the former being used by governments to control populations.

One might add *state-sponsored terrorism,* in which countries support terrorism as "war on the cheap." For example, in 1990, Syria was the home base of the Popular Front for the Liberation of Palestine—General Command (PFLP-GC), which was believed to have been involved in the bombing of the Pan Am jumbo jet over Lockerbie, Scotland, in 1988. Syria also controls the Bekaa Valley, the terrorist training ground in Lebanon. Iran most likely commissioned the bombing of the jet in retaliation for the U. S. Navy's accidental shooting down of an Iranian civilian airliner during a crisis in the Persian Gulf (Wines, 1990). Libyan agents were blamed because the trigger to the device was similar to a type of detonator used by Libyan terrorist bombers. The device was hidden in a Toshiba radio like those used by PFLP-GC terrorists in Germany, who most likely hired the Libyans (Mossberg, 1990). The bombing turned out to be the work of the Libyan secret service, an agent of which was later convicted for the attack.

Brief History

The Assassins of the Middle East were the best-known early terrorist group, although their attacks were confined to officials and authorities. The Jacobin period of the French Revolution and its "reign of terror" provided the name, while the Russian nihilists and "bomb throwers" of the late nineteenth century provided the classic vision of the *terrorist*. Laqueur (1987) notes,

> The popular image of terrorists some 80 years ago was that of a bomb-throwing alien anarchist, disheveled, with a black beard and a satanic (or idiotic) smile, fanatic, immoral, sinister and ridiculous at the same time. (p. 3)

Prior to World War II, most terrorism consisted of political assassination of government officials (G. Martin, 2006). A second, new form of terrorism was inaugurated in Algeria in the late fifties by the FLN (National Liberation Front), which popularized the random attack on enemy civilians. This is depicted well in the classic film *The Battle of Algiers*. A new, third stage of terrorism has become popular since the sixties: "media terrorism"—random attacks on anyone. Palestinian use of suicide bombers inspired al Qaeda and other Muslim extremists to favor this tactic. Concentration on massive civilian casualties has also accelerated. Fears have increased that terrorist groups such as al Qaeda will obtain weapons of mass destruction such as nuclear, chemical, and biological agents that could produce casualties literally in the millions. The prosecution of a jihad (holy war) that is truly international in scope has attracted Muslim religious extremists from throughout the world. The goal shifted from gaining attention to mass destruction for its own sake. Suicidal martyrdom—dying while destroying the enemies of God—dates back hundreds of years, at least as early as the Hashashins in the twelfth and thirteenth centuries. Terrorism has become a divine duty with special rewards in heaven for those who make the ultimate sacrifice. In July 2005, the terrorists who bombed the London subway and a bus were second-generation Muslims who had been born and raised in the United Kingdom.

Indiscriminate terrorism has become widespread only in recent times, with the development of more effective explosives and modern mass media. Such terroristic action is easier to commit than attacks against hardened targets or well-guarded leaders, and since these acts are unlikely to gain political support in one's own country, they are more likely to be committed against foreigners. Most of this terrorism has been directed against democracies, with little against the more totalitarian states. Much of the terrorism in the eighties was "war by proxy" or *state-sponsored terrorism* by countries such as Libya, Syria, and Iran. Terrorism has become a means of waging "war on the cheap."

Frederick Hacker (1976), in *Crusaders, Criminals, Crazies,* points out,

> Contrary to widespread belief, terroristic violence is not always futile and ineffective in transforming reality. If it had not been for IRA terrorist activities, the Republic of Ireland never would have come into being. This is also true of independent Cyprus, Algeria, Tunisia, and possibly Israel. . . . Terrorism often is not confined to outlaws and the dregs of society (riffraff theory); it is supported by responsible citizens and organizations, either openly or in secret. . . . Terrorists are *not* all part of a Leninist-Marxist conspiracy. The IRA, particularly its activist Provisional branch, is actually conservative, patriotic, nationalistic, and rightist, and is denounced by opponents as a bunch of fascists and "crazy drunkards." (p. 69)

Myths Regarding Terrorism

Laqueur (1977) addresses various myths regarding terrorism:

1. Terrorism is new and unprecedented.
2. Since one person's terrorist is another's liberator, the term is "politically loaded" and should be discarded.
3. Terrorism is always "left wing" or revolutionary.
4. Terrorism takes place whenever there are legitimate grievances, and amending such conditions will bring about the cessation of terrorism.

Video Link 11.4
View a video on the trail of a terrorist.

5. Terrorism is always highly effective.

6. Terrorists are idealists.

7. Terrorism is a weapon of the poor.

Laqueur's refutation of each of these myths includes the following:

1. Contrary to popular belief, terrorism is not a new or entirely unprecedented phenomenon. It is at least as old as the Russian Narodnaya Volya, nihilistic bomb-throwers of the nineteenth century.

2. Most terrorism has been directed at democracies or ineffective authoritarian regimes and ignores totalitarian systems such as Nazi Germany, Fascist Italy, and Communist regimes.

3. Revolutionary, intellectual fashions change and slogans should neither be ignored or taken too seriously. Certain right-wing death squads in Latin America or the Ku Klux Klan illustrate terror from the right.

4. The most repressive, unjust societies have been the freest of terrorism.

5. Terrorism is generally effective if it is part of a larger strategy.

6. Terrorists often sacrifice humane behavior for revolutionary goals.

7. Most terrorist leaders come from affluent backgrounds (pp. 219–222).

Terrorists are frequently supported by outside powers such as Russia, Cuba, Libya, and Algeria (see Sederberg, 1989). Terrorism is often an act of desperate revolutionaries, those who lack effective weapons or means of obtaining redress of their grievances through other channels. During the British control of Palestine, Menachim Begin was leader of a group of terrorists, the Irgun, which blew up the King David Hotel, killing innocent victims. Later, as president of the new nation of Israel, Begin sat across the table from Arafat for the Camp David peace talks.

> Attempts to develop profiles of terrorists have not been very successful because the traits of likely extremists are constantly shifting. Islamic radicals in Europe have concentrated recruitment efforts on women, teenagers, white-skinned Europeans and people who have been baptized as Christians. The trait patterns are constantly shifting. (Whitlock, 2007)

Audio File 11.3
Listen to a discussion of crafting a mental profile of a terrorist.

Some terrorists become, to use Sterling's (1981) words, "retail terrorists," a traveling circus of performers such as "Carlos the Jackal," Abu Nidal's group, or Rengo Sekigun (the Japanese Red Army). Some terrorist groups that have been active internationally since the eighties include Basque ETA, Hamas, Hezbollah, Islamic Jihad, M-19, the Palestine Liberation Front, Provisional IRA, Sikh separatists, and Tamil extremists, along with al Qaeda and newer Islamic extremist groups.

The *Basque ETA* attack Spanish targets in their quest for a separate Basque homeland in northern Spain. *Hamas* (in Arabic, "fervor" or "zeal") is an Islamic resistance movement whose primary purpose is to prevent peace between the Israelis and Palestinians. With funds from Iran and Syria, they have made heavy use of young suicide bombers. Rivaling Hamas in fanaticism is **Hezbollah** (the "Party of God") and its action arm, Islamic Jihad. It has been the principal tool by which, since 1979, the theocratic regime in Iran has pressed its jihad (holy war) against the West.

Islamic Jihad (Islamic Holy War) are Shiite fundamentalist extremists. They are responsible for bombings of the U.S. embassy and of the Marines' barracks in Lebanon and for the holding of U.S. hostages. They are backed by Iran. M-19 (April 19 Movement) are leftist guerilla groups in Colombia. In November 1985, they seized the Justice Palace in Bogota, causing the death of 100 people. They are believed to be aligned with Cuba, as well as with "narcoterrorists." The *Palestine Liberation Front* (PLF) is a Palestinian faction headed by Abu el-Abbas, who was blamed for the 1986 hijacking of the Italian cruise ship *Achille Lauro*. The PLF is a subgroup of another breakaway group in the Palestinian movement and is aligned with Palestinian Liberation Organization (PLO) leader Yassir Arafat.

Al Qaeda (Arabic for "the Base") is a group headed by Osama bin Laden, which opposes non-Islamic governments with violence. These veterans of the Afghan war were originally trained and funded by the United States in opposing the Soviet Union. They consider the U.S. military's continued involvement in the Middle East as American occupation of Islamic countries. With cells in more than 60 countries, al Qaeda was implicated in the World Trade Center I and II bombings, as well as the 1997 massacre of tourists in Luxor, Egypt, and is the basis of support for many loosely knit cells throughout the world.

The *Provisional IRA* ("Provos" of the Irish Republican Army) is fighting to unite Northern Ireland (which is part of the United Kingdom) with the Republic of Ireland. They wish to drive the British from Northern Ireland and have ambushed British personnel and bombed British facilities. By the early twenty-first century, the IRA promised to lay down its arms and come to some accommodation with the Ulster (Northern Irish) government.

Sikh extremists seek independence for India's Sikh population in the Punjab. They are responsible for the assassination of Indian Prime Minister Indira Gandhi, the bombing of civilian airlines, and booby-trap bombings throughout India. *Tamil separatists* seek independence for the northern part of the island of Sri Lanka, which is currently dominated by the Sinhalese. Both sides in this controversy have massacred civilians. *Sendero Luminoso* (Shining Path) is a radical Marxist terrorist group in Peru that controls a large chunk of the rural countryside.

Domestic Terrorism, USA

While incidents of international terrorism, particularly with Americans as targets, increased during the eighties, domestic terrorism in the United States in the nineties remained at a relatively low level. The 1993 bomb attack upon the World Trade Center in New York City by Islamic fundamentalists, while dramatic, was atypical. Then came the 1995 Oklahoma City bombing, and in 1996, the possibly terrorist detonation of a pipe bomb at the Atlanta Olympics. The events of 9/11 made all of these pale in comparison.

Prior to the 1980s, most terrorist groups in the United States were either international or strongly identified with separatist or leftist movements. Puerto Rican independence, anti-Castro groups, the Jewish Defense League, and similar groups were active. An explosion of right-wing KKK/neo-Nazi hate groups—such as The Order; Posse Comitatus; American Nazi Party; the Aryan Nations; and the Covenant, Sword, and Arm of the Lord (CSA)—became more prominent in the eighties. Crime File 11.3 discusses some of these groups in detail.

Puerto Rican independence groups have historically been the most active of domestic separatist groups. In 1950, one such group attempted to assassinate President Truman, and in 1954 it shot up the U.S. House of Representatives while it was in session. Such groups want a separate and independent Puerto Rico (which has been a commonwealth of the United States). The two most active groups are the FALN and the Macheteros (Puerto Rican People's Party). FALN (Fuerzas Armadas de Liberacion Nacional—Armed Forces for National Liberation) has been responsible for over 200 bombings in the United States and Puerto Rico. The Macheteros (Machete Swingers) have attacked U.S. military personnel and bases in Puerto Rico (J. W. Harris, 1987). They were also responsible for a Wells Fargo robbery in West Hartford, Connecticut, that netted $7.3 million.

"Single-issue terrorists" are those who use extremist tactics in support of one issue or cause. Examples include animal rights activists, pro-life and pro-choice activists, environmental activists, and others whose zealotry for their cause precipitates extreme tactics.

Radical leftist terrorist groups declined in the eighties. Groups such as the SLA (Symbionese Liberation Army), the SDS (Students for a Democratic Society), Weathermen, and Black Panthers were quite visible in the sixties and seventies. A right-wing faction of interest to law enforcement is the Sheriff's Posse Comitatus, which advocates a tax moratorium and disregard for federal and state authority. The FBI was quite effective in surveillance and deterrence of terrorist acts of such groups by means of "preventive interviews"—that is, interviewing members and letting them know that authorities are well aware of their plans. With the Oklahoma City bombing in 1995, such groups began to be

Video Link 11.5
View a video on domestic terrorism at Fort Hood, Texas.

Crime File 11.4

The Turner Diaries, ZOG, and the Silent Brotherhood—The Order

In the 1930s in beer halls in Munich, Germany, a political criminal and racist misfit, Adolf Hitler, advocated a bizarre future: a world of Wagnerian mysticism, a new order, a Third Reich.

> But we are doing something else which is really more important than our campaign against the System. In the long run, it will be infinitely more important. We are forging the nucleus of a new society, a whole new civilization, which will rise from the ashes of the old. And it is because our new civilization will be based on an entirely different worldview than the present one that it can only replace the others in a revolutionary manner. There is no way a society based on Aryan values and an Aryan outlook can evolve peacefully from a society which has succumbed to Jewish spiritual corruption. (A. MacDonald, 1980, p. 111)

The above statement is not an excerpt from Hitler's *Mein Kampf,* but rather from a book titled *The Turner Diaries* by Andrew MacDonald (1980), the Nazi pen name of the late William Pierce, a leader of a neo-Nazi right-wing extremist group (Holden, 1986; Sapp, 1986; Wiggins, 1986a, 1986b). Groups such as The Order; Aryan Nations; Bruder Schweigen (the Silent Brotherhood); the Covenant, Sword, and Arm of the Lord; and other right-wing extremist groups are linked by "identity theology." This is an anti-Semitic ideology that views Aryans as God's chosen people and Jews as the children of Satan. Some of these groups also practice "survivalism," a belief that they must stock supplies in order to be self-sufficient as the last hold-outs in some final Armageddon.

The Turner Diaries is a thinly disguised blueprint for The Order's battle with ZOG, "Zionist Occupational Government" or "Zombies of Government." The book describes a "white revolution" launched by a terrorist group, "The Organization," to topple the U.S. government (ZOG). They support themselves through bank and armored-car robbery and counterfeiting. They assassinate key leaders and sabotage transportation and power systems. Once in power, they intend to kill Jews, blacks, other minorities, and liberals (Klanwatch, 1985, p. 6). In the eighties, members of these and related groups murdered Jewish talk-show host Alan Berg in Denver (1984), robbed armored cars and banks, killed and had gun battles with police and federal agents, and bombed synagogues and minority-owned businesses—the very acts outlined in *The Turner Diaries.*

J. R. R. Tolkien's *The Lord of the Rings* trilogy is viewed by many as an allegory for the rise of Nazi Germany. Gandalf, the wizard, instructs Frodo, the Hobbit, on the nature of evil: "Always after a defeat and a respite, the Shadow takes another shape and grows again" (Howard, 1990).

By the late eighties, federal authorities had concluded that The Order had been virtually wiped out as a result of FBI and local efforts, although others speculate that the Ku Klux Klan and other neo-Nazi hate groups simply regroup and reappear in new forms under new names. Skinhead white supremacists and "Identity Movement" followers continue their wars of hate.

Web Research Project
What role did *The Turner Diaries* play in the Oklahoma City bombing?

taken far more seriously by federal authorities. This required a shift in thinking. After three decades of watching possible subversion on the left, now authorities must be concerned with thunder on the right from "freemen," militias, and "patriot groups" (see Aho, 1994; K. S. Stern, 1996).

Finally, in a class by itself as the best-known terrorist–hate group is the Ku Klux Klan, whose cross burnings, arson, bombings, vandalism, intimidation, shootings, and assaults continue, although their movement may have gone underground (Klanwatch, 1985) or transformed itself into neo-Nazi or militia groups.

The Oklahoma City bombing. In his *Cycles of American History,* historian Arthur Schlesinger (1986) proposes that the American political mood generally undergoes an ideological shift in every generation. Terrorism in the United States may reflect these ideological cycles. Before World War II,

Photo 11.6

This is the north side of the Alfred P. Murrah Federal Building after Timothy McVeigh's fuel and fertilizer truck bomb detonated in front of the building.

terrorism was perpetrated largely by the right wing; after the war, it shifted to the left; and since the 1970s, it has moved back to the right. On April 19, 1995, Timothy McVeigh filled a rented truck with 4,800 pounds of explosives and detonated the charge at the Oklahoma City federal building, killing 191 men, women, and children. His friend Michael Fortier assisted McVeigh, but testified against him at his trial. Another alleged accomplice, Terry Nichols, belonged to a militia group called the "Patriots," which believed in a federal government conspiracy.

The Oklahoma bombing was, in part, apparently revenge for the deaths of 79 members of David Koresh's Branch Davidian sect in Waco, Texas, who died when their compound was stormed by federal agents. The Oklahoma bombing took place exactly 2 years to the day after the Branch Davidian incident. Federal investigators concluded that, while federal agents were not without fault in managing the Waco incident, most of the casualties were caused by Koresh and his followers, who may have started the fire themselves in a mass suicide as federal agents stormed the compound. Koresh had accumulated a huge illegal arsenal of weapons, which led to confrontation with federal agents. The Waco incident, as well as a federal siege of white separatist Randy Weaver's cabin in Ruby Ridge, Idaho, in which his wife and young son were killed by snipers, became battle cries for right-wing militia movements in the United States.

While the number of domestic antigovernment militia groups and their activities has dropped dramatically since Oklahoma City, the Southern Poverty Law Center indicates that the number of racist and anti-Semitic hate groups has increased (Romano, 2005). Domestic terrorism incidents since Oklahoma City have included the following:

1996—Members of the white supremacist group "Phineas Priests" were convicted of bombings and bank robberies in Spokane, Washington.

1997—Four KKK members were convicted in Texas of robbery and plotting to blow up a natural gas plant.

1997—Members of a militia group headed by Bradley Glover and Michael Dorsett were convicted of plotting attacks against military bases.

Crime File 11.5

Patterns of Global Terrorism

The State Department publishes a report to Congress every year, by April 30, covering developments during the previous year in terrorism. According to the State Department, the report discusses the countries in which acts of terrorism occurred, countries that are state sponsors of terrorism, and countries determined by the Secretary to be of particular interest in the global war on terrorism. The report, renamed "Country Reports on Terrorism" from "Patterns of Global Terrorism" in 2005, also discusses international counterterrorism cooperation.

Strategic Assessment: Trends in 2008

Al-Qa'ida and Associated Trends: Al-Qa-ida (AQ) and associated networks continued to lose ground, both structurally and in the court of world public opinion, but remained the greatest terrorist threat to the United States and its partners in 2008. AQ has reconstituted some of its pre-9/11 operational capabilities through the exploitation of Pakistan's Federally Administered Tribal Areas (FATA), the replacement of captured or killed operational lieutenants, and the restoration of some central control by its top leadership, in particular Ayman al-Zawiri. Worldwide efforts to counter terrorist financing have resulted in AQ appealing for money in its last few messages.

In the years since 9/11, AQ and its extremist allies have moved across the border to the remote areas of the Pakistani frontier, where they have used this terrain as a safe haven to hide, train terrorists, communicate with followers, plot attacks, and send fighters to support the insurgency in Afghanistan. Therefore, FATA provided AQ many of the benefits it once derived from its base across the border in Afghanistan.

The threat from AQ in Iraq continued to diminish. While still dangerous, AQ experienced significant defections, lost key mobilization areas, suffered disruption of support infrastructure and funding, and was forced to change targeting priorities. Indeed, the pace of suicide bombings countrywide, a key indicator of AQ's operational capability, fell significantly during 2008. Initiatives to cooperate with tribal and local leaders in Iraq continued to encourage Sunni tribes and local citizens to reject AQ and its extremist ideology. The sustained growth and improved capabilities of the Iraqi forces increased their effectiveness in rooting out terrorist cells. In Baghdad, Anbar, Diyala Provinces, and elsewhere, local populations turned against AQ and cooperated with the Government of Iraq and Coalition Forces to defeat it.

AQ continued its propaganda efforts seeking to inspire support in Muslim populations, undermine Western confidence, and enhance the perception of a powerful worldwide movement. Terrorists consider information operations a principal part of their effort. Their use of the Internet for propaganda, recruiting and fundraising and increasingly training, has made the Internet a "virtual safe haven." That said, bin Laden and Zawahiri appeared to be in the position of responding to events rather than driving them particularly in the latter half of 2008.

Besides seeking to take advantage of international interventions in Iraq and Afghanistan as tools for radicalization and fundraising, AQ also sought to use the Israeli/Palestinian conflict, but lacked credibility in this regard. The international community has yet to muster a coordinated and effectively resourced program to counter extremist propaganda.

Taliban and other insurgent groups and criminal gangs: The Taliban and other insurgent groups and criminal gangs, some of whom were linked to AQ and terrorist sponsors outside the country, control parts of Afghanistan and Pakistan and threaten the stability of the region. Attacks against our troops, NATO allies, and the Afghan government have risen steadily. Taliban insurgents murdered local leaders and attacked Afghani government outposts in the FATA of Pakistan. Ideological allies of the Taliban conducted frequent attacks into Pakistan's Northwest Frontier Province (NWFP), particularly in the Swat Valley, and have extended operations in to the Punjab and the capital city of Islamabad. Suicide bombers are increasingly used to target Pakistanis, in addition to conducting cross-border raids on ISAF forces.

State Sponsors of Terrorism: State sponsorship of terrorism continued to undermine efforts to reduce terrorism. Iran remained the most significant sponsor of terrorism. Iran has long employed terrorism to advance its key national security and foreign policy interests, which include regime survival, regional dominance and opposition to Arab-Israeli peace, and countering Western influence particularly in the Middle East. Iran continues to rely primarily on its Islamic Revolutionary Guard Corps-Qods Force to clandestinely cultivate and support terrorist and Islamic militant groups abroad, including: Lebanese Hizballah, Palestinian terrorist groups such as Hamas and Palestinian Islamic Jihad, certain Iraqi Shia militant groups, and Islamic militants in Afghanistan, the Balkans, and elsewhere. Iranian weapons transfers to select Taliban members in Afghanistan in 2008 continued to threaten Afghan and NATO troops operating under UN mandate and undermine the stabilization efforts in that country.

Defeating an Agile Terrorist Enemy

The terrorist groups of greatest concern—because of their global reach—share many of the characteristics of a global insurgency: propaganda campaigns, grass roots support, transnational ideology, and political and territorial ambitions. Responding requires a comprehensive approach that focuses on recruiters and their networks, potential recruits, the local population, and the ideology. An holistic approach incorporates efforts aimed at protecting and securing the population; politically and physically marginalizing the insurgents, winning the support and cooperation of at-risk populations by targeted political and development measures; and conducting precise intelligence—led special operations to eliminate critical enemy elements with minimal risk to innocent civilians. Significant achievements in border security, information sharing, transportation security, financial controls, and the killing and capture of numerous terrorist leaders have reduced the threat. But the threat remains, and state sponsorship, improved terrorist propaganda capabilities, the pursuit of weapons of mass destruction by some terrorist groups and state sponsorship of terrorism, and terrorist exploitation of grievances represent ongoing challenges.

Source: Excerpts from the *U.S. Department of State Annual Report on Terrorism,* Washington, DC: Office of the Coordinator for Counterterrorism, April 30, 2009. Available at http://www.state.gov/s/ct/rls/crt/2008/122411.htm.

1997–1998—Members of the North American Militia (Michigan) were convicted of plotting to kill federal agents and bomb various targets.

1998—Members of the New Order (a white supremacist group) were convicted of planning to bomb public buildings, rob banks, poison water supplies, and assassinate federal judges.

2003—Anti-government extremists William Krar and Judith Bruey were convicted of stockpiling an illegal arsenal of weapons, explosives, and bombs (Romano, 2005).

2004—The Earth Liberation Front took credit for a $50 million arson of a housing development in San Diego.

2005—Eric Rudolph pleaded guilty to attacks on abortion clinics and the 1996 Summer Olympics that killed two people.

The weakened state of current domestic terrorist groups is attributed in part to the death or imprisonment of many of their leaders.

Criminal Careers of Political Criminals

For political criminals, crime is instrumental; it is a means of achieving what they perceive as higher moral goals. As Schafer (1976) explains,

The convictional criminal [Schafer's term for the political criminal], with his altruistic moral ideology, places less emphasis upon secrecy and even seeks publicity for his cause. Dramatic publicity, moreover, is almost a necessity for the convictional criminal in order to make the public understand his actions; his crime may serve as an example to would-be followers and generate further convictional crimes. His punishment is not a deterrent and may serve to interest others in the given ideal and to recruit other convictional violators of law. (p. 139)

The only exceptions to this publicity-seeking behavior are government criminals who in most instances prefer secrecy. Political criminals from the left or right tend to be convinced of the rectitude of their cause and their actions. Rather than viewing their behavior as criminal, political criminals either deny the legitimacy of existing laws or view their violation as an essential step in either preserving the existing social order (crime by government) or bringing about change in the existing system (crime against government).

A large proportion of leftist revolutionaries are drawn from educated and middle-class backgrounds rather than from the ranks of the proletariat as Marx had predicted. A similar pattern presents itself with terrorists. Laqueur (1977, p. 207) points out that in West Germany in the late sixties and early seventies, there were more female than male terrorists, and the females were more fanatical than the males. Right-wing groups in the nineties appeared to draw heavily from working-class white males, who often resented minorities and immigrants and blamed them for their economic slippage. Clutterbuck (1975) indicates that "terrorist movements seldom have more than very small minority support from the people . . . [and consist of] earnest young intellectuals increasingly frustrated by their lack of response from the ordinary people" (p. 65). Al Qaeda leadership similarly is represented by highly educated and wealthy individuals.

Political criminals operate within subcultures that define their activities as appropriate or necessary. Whether it be theories of racial supremacy (the Ku Klux Klan), preservation of law and order (illegal police violence), terrorist bombing of innocent victims, the shooting down of civilian airliners (state violence), or nonviolent passive resistance, political criminals feel that they have support of immediate peers. Being convinced of the rightness of their actions, political criminals also assume that others will be impressed with their resolve, "see the light," and eventually agree with their actions. If subcultural support for politically deviant action is not strong, such violators may come to view their actions as illegitimate.

Although some view governmental political criminals as not ideologically committed (H. E. Allen, Friday, & Roebuck, 1981, pp. 201–202), they are in fact ideologically committed to preservation of the status quo, and this convictional devotion may be distinct from the quest to preserve personal power (occupational crime). While governmental political criminals tend to be from more privileged backgrounds, many of their agents (servants of power), such as the police, are not. As previously mentioned, political criminals against the government vary considerably in background, although many leaders of the "new left" in the late sixties and early seventies in the United States and Western Europe were universally educated and drawn from the upper middle class. Even though males dominated numerically, a significant proportion of leaders of radical and terrorist groups during this period were females.

For many terrorists, "the end justifies the means"; believing in the rightness of the cause, their actions are viewed as reactions to repression, injustice, or hostile acts of the enemy. It is the latter who therefore must bear the burden of guilt for their aggression.

The Doctrine of *Raison d'État*

For political crimes, government officials or their agents historically have sought justification in the doctrine of **raison d'état** (reason of state), usually attributed to Italian political philosopher Niccoló Machiavelli (1469–1527). This doctrine holds that some violations of the common law are necessary to serve public utility (Friedrich, 1972, pp. 21–22). This Machiavellian "end justifies the means" is a consistent rationalization of political criminals of all stripes: governmental, religious, or political.

While much of the literature on terrorists plays up their intractability and uncompromising nature, one must also consider the social–structural context in which their activities occur. To take but one example, terrorism by the Provisional wing (Provos) of the Irish Republican Army is in part aimed at uniting Ireland. How much support would the Provos have, however, if a truly successful civil rights movement were to obtain equal jobs, housing, and political influence for Catholics in the North? Similarly, a Palestinian homeland in some form would remove some of the support for Palestinian terrorists.

Terrorism and Social Policy

Terrorist threats of the future promise to be more nuclear, more urban, and to involve wealthier and more skilled terrorists—often as proxy armies for sponsor countries. Terrorism is a problem to be managed rather than solved. Attempts at international cooperation are hindered by the very ideological disputes that often give rise to terrorism. A precedent does exist with respect to international cooperation: Piracy, a historically common practice, has been virtually eliminated through international agreement. At one time, countries hired pirates in a form of "war by proxy;" but for centuries they have been declared *hostis humani generis* (common enemies of mankind), outlaws whose acts fall under the jurisdiction of all states. Perhaps a similar uniform international policy will evolve regarding cross-national terrorism.

The current war on terrorism by the United States and its allies has had tremendous success in Afghanistan and in uncovering al Qaeda cells, but the struggle is a protracted one. The Cold War lasted 45 years and the international war against extremist Islamic terrorists will call for a similar commitment.

Kidder (1983) summarizes counterterrorism measures in terms of nine policies: diplomatic measures, better intelligence gathering, tighter security measures, legal and social measures, more public awareness, military and police action, arms and explosives controls, media self-regulation, and maintenance of public composure. Terrorism by "lunatic minorities" in democratic countries that provide legal recourse (for example, the ballot box) must be condemned as "crime." Sanctions must be imposed on offending regimes (state terrorism and state-sponsored terrorism). This could include withdrawal of financial aid and diplomatic recognition and invocation of strict liability (holding them legally responsible) rather than conducting business as usual (D. W. Martin & Walcott, 1988).

Societal Reaction

The sociological nature of the concept of political crime is illustrated by its relativity with respect to time and place. Ideologically committed spies such as the Rosenbergs, who supplied their country's atomic secrets to a foreign enemy, were traitors in the United States, but heralded as heroes in the recipient country, the Soviet Union. Even one of the most dastardly of terrorist acts, the slaughter of almost the entire Israeli Olympic team in Munich, was applauded in many areas of the Arab world. This very divergence in international ideology explains the relative ineptness of world bodies such as the United Nations to act in unison in condemning global terrorism and atrocities.

Since crimes against the government threaten the status quo of society, societal reaction has been quite strong; however, until recently, public reaction to crimes by the government has been mild. This is partly because, since the government makes and enforces the law, it is hard to imagine it also violating the law. In the United States, public innocence in this regard appears to have matured since revelations of CIA and FBI wrongdoing and the events of Watergate.

The more complex, urban, industrial, and interrelated the world community becomes, the easier it is for a small, fanatical minority of the left or right to disrupt, destroy, or endanger not just their political targets, but all of us. At the level of collective behavior and social change, dynamic societies can continue to be expected to generate new social movements; new demands for change; and, depending on the response, new political criminals either in the form of "bell-ringers" of change or of overzealous guardians at the gates. For further readings on terrorism, see Combs (2003), Kushner (1998), G. Martin (2006), Poland (2005), and J. R. White (2005).

Theory and Crime

Neither greed nor need motivates the political criminal. These individuals are often motivated by a cause, an ideology that excuses all criminal transgressions. Using differential association theory, we find that the political offenders are socialized into their belief system by similarly minded individuals. They often view themselves as patriots, as doing the bidding of higher spiritual forces. Their devotion to a cause or ideology takes precedence over all other commitments and they do not perceive their activities as criminal.

Suicide bombers who are persuaded to commit mass murder of innocent targets are often persuaded by promises of eternal bliss in the arms of Allah in paradise. The moral quagmire of political crime can be explained by the often used phrase that "one man's terrorist is another man's patriot." Much traditional criminological theory does not address itself to political crime even though terrorists, tyrants, assassins, and spies have probably done more harm to society than most traditional criminals.

Summary

Ideology refers to distinctive belief systems, abstract ideals that offer a design for living. Political crime is defined as criminal activity committed for ideological purposes. There are two types of political crime: crimes by government and crimes against government. Crimes by government exclude political corruption, which is an example of occupational crime, and refer instead to violations by secret police, abuses of human rights and constitutional privileges, and genocide, as well as crimes committed by government officials in the act of enforcing the law. Crimes against government range from protests, illegal demonstrations, and strikes to espionage, political whistleblowing, political assassination, and terrorism. The actual definition of political crime is relative to time, place, and the ideological views of those giving the definition.

All governments have criminal laws forbidding activities that threaten the state. In Anglo-American jurisprudence, political criminals are not recognized as such and are dealt with under more traditional, nonpolitical laws. The Nuremberg principle, established by the victorious Allies at the end of World War II, established that individuals faced with the dilemma of obeying orders that involve war crimes and crimes against humanity or following their own consciences, should disobey unjust dictates. Similar documents in the Western political tradition, as well as the UN's *Universal Declaration of Human Rights* (1948), provide customs or standards for international conduct with regard to respecting integrity of persons, basic human needs, and civil liberties. *International law,* however, is handicapped by the lack of a consensual world community as well as by the lack of power of enforcement.

Crime by government is more sociological than political. Secret police (agents of political policing) are units of the internal security police of the state who have a mandate to suppress all serious or threatening political opposition and to control political activity. Their activities often include illegal surveillance, searches, detention, and violations of human rights. *Political prisoners* may include those who have seriously opposed the existing government, but also prisoners of conscience who are tortured, sent into exile, or murdered. Amnesty International finds totalitarian regimes to be the greatest offenders in this area.

Patriarchal crime refers to crime committed against women and children in the name of traditional male dominance. This is illustrated by such practices as female sexual mutilation, bride burning, and honor killings.

Genocide, the mass destruction or annihilation of human populations, is the ultimate violent crime by government; in the modern era, political ideologies have replaced religious justifications for genocide. In 1948, the UN Convention on Genocide defined it as a crime, although this same international body has been less than consistent in condemning such practices. Political crimes by police often involve denial of due process and freedom of speech, as well as invasion of privacy. These and other offenses are committed in the name of "law and order" and preservation of the existing political system.

Other abuses by government agents include illegal surveillance; disruption of democratic processes, including character assassination; and secret experiments on unsuspecting subjects. One such example was Operation Chaos, which among related activities involved illegal surveillance and harassment of domestic dissidents. COINTELPRO was the FBI's counterintelligence program to disrupt legitimate political activity such as that of the Socialist Worker's Party and black nationalist groups. Harassment of Martin Luther King, Jr., has been linked to misuse of intelligence agencies such as the FBI by J. Edgar Hoover. The case of Oliver North and the Iran-Contra debacle was also detailed. Further questionable experiments include "the search for the Manchurian Candidate"— mind-control experiments conducted to develop a secret brainwashing technique. The latter term was coined by Edward Hunter and, according to Scheflin and Opton, was a myth created to justify such experiments on an unsuspecting public. Nuclear exposure experiments were also discussed.

Crimes against government may involve activities of dissent and protest in opposition to the status quo, but may also involve reactionary opposition to changes that have taken place in the existing social or political order. Dissident activities are represented by civil rights, labor, and antiwar groups of the past; reactionary opposition can be found in right-wing "death squads," the Ku Klux Klan, and the American Nazi Party. Excerpts from Martin Luther King, Jr.'s "Letter From Birmingham Jail" provide a very moving defense of civil disobedience and the strategy of the civil rights movement. Social movements advocate change in the existing order and often conflict with responding authorities. Some newer examples include the sanctuary, anti- and pro-choice, antinuclear, and antivivisectionist (animal rights) groups.

Political espionage involves stealing state secrets and is a standard international practice of intelligence agencies. "Sub rosa criminals" are spies who steal secrets. The Robert Hanssen spy case is an example.

The motivation of spies often reveals a SMICE strategy (sex, motivation, ideology, compromise, and ego). A typology of spies includes mercenary, ideological, alienated/egocentric, buccaneer, professional, compromised, deceived, quasi-agent, escapee, and miscellaneous. Treason is the betrayal of one's country out of commitment to either a political ideology or a foreign power. Political "whistle-blowers" such as Daniel Ellsberg violate state secrecy because they believe that the public has a right to know the truth.

Terrorism is the use of cruelty and violence in order to spread fear within a population as an instrument of gaining political power. Types of terrorism include political terrorism, nonpolitical terrorism, quasi-terrorism, limited political terrorism, and official or state terrorism. The last type illustrates the fact that not all terrorism involves crime against the government. While observers such as Laqueur feel the threat of terrorism in the seventies was a media event and an exaggeration, statistics from the early eighties suggest a climbing toll of victims. Examples of both international and domestic terrorist groups were detailed, including the World Trade Center and Oklahoma City bombings. Some possible myths regarding terrorism include beliefs that it is a new phenomenon; an inappropriate, politically loaded term; always leftist in nature due to legitimate grievances; highly effective; idealistic; and a weapon of the poor.

Examination of the criminal careers of political criminals indicates that they view crime as instrumental, a means to ideological ends. Most do not view their activity as criminal and tend to operate within supporting subcultures that reinforce their definitions. For government political criminals, *raison d'état*, national security, and their preservation serve as justification for violations. Some government policies for dealing with terrorism were detailed. Societal reaction to political crime varies, with generally strong disapproval of offenders against the government and mild reaction in the past toward governmental offenders. Divergence in ideology prevents any consistent international reaction to political crime.

KEY CONCEPTS

Brainwashing

Brainwashing Myth

COINTELPRO

Crimes Against Government

Crimes by Government

Enterprise

Espionage

Genocide

Hamas

Hate Crime

Hezbollah

Human Rights

Ideology

Myths Regarding Terrorism

Nuremberg Principle

Operation Chaos

Patriarchal Crime

Political Crime

Project Bluebird

Raison d'État

Sanctuary Movement

Secret Police

SMICE

Sub Rosa Crime

Terrorism

Types of Spies

Types of Terrorism

Universal Declaration of Human Rights

REVIEW QUESTIONS

1. How is the concept of political crime different from other crimes that have been examined in this text? Do you think that political crimes should be treated differently from other crimes?

2. What are secret police and what is their purpose? Give some examples of ways in which they can use political policing.

3. What are some basic human rights recognized by international treaties and the United Nations? Give some examples of their violation.

4. What is the meaning of the phrase "the search for the Manchurian Candidate"? How did this search influence U.S. intelligence agencies to become involved in wrongdoing?

5. Discuss crime and scandal during presidential administrations in the United States. What is the impact of such scandals on a president's standing in history?

6. What were the major offenses in the Iran-Contra conspiracy?

7. Who wrote the "Letter From Birmingham Jail"? To whom was it addressed and what did it say?

8. What is the major issue of the antiglobalization movement and what does it wish to accomplish?

9. What is the "pathological myth of assassins"? What types of assassins were identified by Clarke? Give an example of each.

10. Discuss the various types of spies and give an example of each.

WEB SOURCES

Amnesty International
http://www.amnesty.org

Central Intelligence Agency
http://www.cia.gov//

Crimes of War Project
http://www.crimesofwar.org

Information on Militias
http://www.well.com/user/srhodes/militia.html

FBI Annual Report on Terrorism
http://www.fbi.gov/publications.htm

FBI Freedom of Information Act Declassified Files
http://foia.fbi.gov/spies.htm

U.S. State Department Patterns of Global Terrorism
http://www.state.gov

WEB **EXERCISES**

Using this chapter's recommended Web sites, explore the area of political crime.

1. What tasks are performed by the Southern Poverty Law Center?

2. According to the FBI's Annual Report on Terrorism, what was the major form of domestic terrorism last year?

3. Examine the FBI Freedom of Information Declassified Files and list any three persons or organizations that you were surprised to learn had been under investigation.

4. What activities is Amnesty International currently concerned with?

5. Using your Web browser, search the terms "human rights violations" and "torture."

SELECTED **READINGS**

James Clarke. 1982. *American Assassins: The Darker Side of Politics.* Princeton, NJ: Princeton University Press.

> This is the best work on the subject of assassins. Clarke refutes previous myths and develops an excellent typology of spies.

Richard Condon. 1958. *The Manchurian Candidate.* New York: Random House.

> This novel is a Cold War classic that scared the American intelligence community into undertaking a variety of bizarre experiments in order not to fall behind a perceived Communist mind control gap. This was later made into a movie starring Frank Sinatra and Angela Lansbury.

Congressional Research Service. 1978. *Human Rights Conditions in Selected Countries and the U.S. Response.* Washington, DC: Government Printing Office. Report prepared for the House Committee on International Relations, 95th Congress, 2nd Session.

Theodore Draper. 1991. *A Very Thin Line: The Iran-Contra Affairs.* New York: Hill and Wang.

> An excellent account of the Iran-Contra conspiracy.

Federal Bureau of Investigation. (annually). *Terrorism.* Washington, DC: Government Printing Office (also available free on the World Wide Web).

> This is the FBI's annual report on domestic and international terrorism.

Frank E. Hagan. 1997. *Political Crime: Ideology and Criminality.* Boston: Allyn & Bacon.

> This is the author's attempt to resurrect and develop a successor to Proal's classic work. It provides comprehensive coverage of all areas of political crime, from crime by government to crime against government.

Walter Laqueur. 1987. *The Age of Terrorism.* Boston: Little, Brown.

> This well-written, classic text presents a very scholarly, historical account of terrorism from a European perspective.

Peter Maas. 1986. *Manhunt: The Incredible Pursuit of a CIA Agent Turned Terrorist.* New York: Random House.

> An exciting account of Edwin Wilson and his alleged "false flag" operations in support of Libya.

Nathan Miller. 1976. *The Founding Finaglers.* New York: David McKay.

> An readable, not-too-flattering account of scandal among early presidents and statesmen.

Louis Proal. 1973. *Political Crime.* Montclair, NJ: Patterson Smith. Reprint of an 1898 edition.

> This is the classic on political crime. Proal's work is both eloquent and a benchmark for all other work on this topic.

United Nations. 1948. *Convention of Genocide.* New York: Author.

United States State Department. (annually). *Human Rights Report.* Washington, DC: Government Printing Office (available free on the World Wide Web).

> This annual report to Congress evaluates how well countries throughout the world are abiding by the UN Declaration of Human Rights.

Organized Crime: A Problematic Definition

Sources of Information on Organized Crime

Types of Organized Crime (Generic Definitions)

The Organized Crime Continuum

Street Gangs

International Organized Crime

   Yakuza

   Chinese Triad Societies

   Russian Organized Crime

The Nature of Organized Crime

   Ethnicity and Organized Crime

Money Laundering

Drug Trafficking

   Colombian Cartels

   The Underground Empire

   Mexico's Drug War

   Crime File 12.1 Amado Carrillo Fuentes and Operation Casablanca

Theories of the Nature of Syndicate Crime in the United States

   The Cosa Nostra Theory (The Cressey Model)

   Crime File 12.2 The Origin of the Mafia

   The Patron Theory (The Albini Model)

   The Italian American Syndicate (IAS)

The Classic Pattern of Organized Crime

   Strategic and Tactical Crimes

   Illegal Businesses and Activities

   Crime File 12.3 Snakeheads and Software Mobsters

   Crime File 12.4 Mobsters, Unions, and the Feds

   Big Business and Government

A Brief History of Organized Crime in the United States

   Before 1930

   The Luciano Period

   The Genovese Period

   The Apalachin Meetings

   The Gambino Period

   The Commission Trials

Criminal Careers of Organized Criminals

Public and Legal Reaction

   Drug Control Strategies

   Investigative Procedures

   Laws and Organized Crime

Theory and Crime

Summary

Key Concepts

Review Questions

Web Sources

Web Exercises

Selected Readings

chapter 12

Organized Crime

Cosa Nostra means Our Thing. If you use these words, it means:
I belong to a Mafia family.

—Court testimony by Sicilian Mafia informant Tommaso Buscetta (cited in Alexander, 1988, p. 43)

"Joe, let's stop fooling around. You know I'm here because the Attorney General wants this information. I want to talk about the organization by name, rank, and serial number. What's the name? Is it Mafia?"

"No," Valachi said. "It's not Mafia. That's the expression the outside uses."

"We know a lot more than you think. . . . Now I'll give you the first part. You give me the rest. It's Cosa."

Valachi went pale. For almost a minute he said nothing. Then he rasped back hoarsely, "Cosa Nostra! So you know about it."

—The Valachi Papers (Maas, 1968, pp. 29–30)

Organized Crime: A Problematic Definition

Organized crime has been variously defined and described by the general public, legislatures, law enforcement agencies, social scientists, and syndicate members themselves. Federal agencies such as the FBI and the Department of Justice use the Federal Task Force on Organized Crime's general operational definition, one that best fits the generic type, which will be described shortly:

> Organized crime includes any group of individuals whose primary activity involves violating criminal laws to seek illegal profits and power by engaging in racketeering activities and, when appropriate, engaging in intricate financial manipulations. . . .

Journal Article Link 12.1
Examine literature regarding the organized crime continuum.

Accordingly, the perpetrators of organized crime may include corrupt business executives, members of the professions, public officials, or any occupational group, in addition to the conventional racketeer element. (National Advisory Committee, 1976a, p. 213)

For the purposes of general prosecution and enforcement, most federal and state laws end up including under the definition of organized crime any group crime of a conspiratorial nature that includes types of criminal activity we would more appropriately label as occupational, corporate, political, or even conventional crime (National Advisory Committee, 1976a, pp. 213–215).

Sources of Information on Organized Crime

I remember when Joe was testifying before that Senate committee [McClellan] back in 1963. I was sitting in Raymond Patriarca's office [New England mob boss] . . . and we were watching Joe on television. I remember Raymond saying: "This bastard's crazy. Who the hell is he?" . . . "What the hell's the Cosa Nostra?" Henry asked [Tameleo, the underboss]. "Is he a soldier or a button man?" . . . "I'm a zipper." "I'm a flipper." . . . It was all a big joke to them. In New England we never used names like "soldiers" or "caporegimes." (Teresa, 1973a, pp. 24–25, 28)

Photo 12.1

Joe Valachi (1903–1971) was the first Mafia member to publicly acknowledge the existence of the Mafia.

The above account by Vincent Teresa, author with Thomas Renner of *My Life in the Mafia* (1973b), describes the reaction of a mob boss to the testimony of ex-Mafia member Joe Valachi before a Congressional committee. In *The Valachi Papers* (Maas, 1968), Valachi described the inner workings of something he called "Cosa Nostra" (literally, "this thing of ours"). Other such biographies and autobiographies, although of varying validity, provide rare inside glimpses of organized criminal operations. Pileggi's *Wiseguy* (1985), Pistone and Woodley's *Donnie Brasco:*

My Undercover Life in the Mafia (1987), Bonanno's *A Man of Honor* (1983), and Mustain and Capeci's *Mob Star: The Story of John Gotti* (1988) serve as illustrations. Pileggi's *Wiseguy,* for example, the basis for the film *Goodfellas,* details the life of Henry Hill, a career criminal who literally grew up in the mob. Hill gives an inside account of the Paul Vario organized crime family; the 1983 Lufthansa robbery at Kennedy Airport, which netted $5 million in cash; the Sindona scandal, which nearly collapsed the Vatican bank; and the Boston College basketball point-shaving scandal. Hill followed up with a later autobiography titled *Gangsters and Goodfellas* (Hill & Russo, 2004).

Lupsha (1982) indicates the following sources of information on organized crime: informers, hearings and investigations, court trial transcripts and grand jury depositions, news stories, investigative reporting, wire surveillance transcripts, memoirs/biographies, government reports and releases, law enforcement–assisted research, archives and historical documents, observation, and in-depth interviews. While any source may exhibit varying degrees of validity, far more triangulation (use of multiple methodologies in the same study) is required than has been apparent in past criminological research on organized crime.

Types of Organized Crime (Generic Definitions)

Acknowledging the need for broader (or more generic) definitions of organized crime, like operational policy definitions employed by organizations such as the Federal Bureau of Investigation, Joseph Albini (1971), author of *The American Mafia: Genesis of a Legend,* offers the following definition:

> [A]ny criminal activity involving two or more individuals, specialized or nonspecialized, encompassing some form of social structure, with some form of leadership, utilizing certain modes of operation, in which the ultimate purpose of the organization is found in the enterprises of the particular group. (p. 37)

Albini then identifies four basic types of organized crime: political-social organized, mercenary (predatory), in-group oriented, and syndicate (pp. 38–48).

1. **Political-Social Organized Crime:** This category best fits into the "political criminal" activity discussed earlier. It refers to crime by guerilla and terrorist groups and various militant social movements that use violence, such as the Ku Klux Klan, the Molly Maguires, and the Palestinian Liberation Organization.

2. **Mercenary (Predatory) Organized Crime:** This category refers to crimes committed by groups for direct personal profit, crimes that prey on unwilling victims, such as juvenile and adult criminal gangs who engage in larceny, burglary, and robbery. The Mano Nera (Black Hand) is an example of the last of these. These 1880s extortionist gangs (there was no one Black Hand) in the United States sent threatening notes to fellow Italian immigrants requesting money. The notes usually contained a sinister mark or sign of a black hand. Often erroneously identified as a forerunner of the Mafia, the Black Hand was more a method of crime than an organization. It provided no illicit services and could not assure immunity for its own operators through political corruption.

3. **In-Group-Oriented Organized Crime:** This refers to crimes committed by groups, such as motorcycle gangs and some adolescent gangs, whose major goals are psychological gratification, "kicks," "rep," "highs," "bopping," and "trashing" rather than financial profit. Motorcycle gangs—the post–World War II prototype is Hell's Angels—have branched out since Hollywood portrayals such as Marlon Brando's in The Wild One. These gangs are sometimes used as "muscle" (enforcers) and for low-level jobs by larger syndicate groups (see Abadinsky, 1994, p. 282). The Pagans, begun in Prince George's County, Maryland, in 1959, now have local chapters all along the East Coast, from Connecticut to Florida, with the heaviest membership in the Middle Atlantic states (Pennsylvania Crime Commission, 1980, p. 27). Such groups are involved in narcotics distribution, prostitution, extortion, bribery, contract murders, pornography distribution, and other activities. The Hell's Angels have also moved extensively into drug trafficking, allegedly controlling as much as 90 percent of the "speed" market in northern California ("Hell's Angels," 1979, p. 34). Hopper's (1991) field study of outlaw motorcycle gangs documented their transition from hedonistic hell-raisers to economic entrepreneurs. He also noted that females had lost status in such gangs and now played the dual roles of sex objects and money makers. Perhaps an apt concept to apply to such gangs is that of "semi-organized" crime, since they lack at least one of the key features of our definition of organized (syndicate) crime.

4. **Syndicate Crime:** This is the category of organized crime that is the subject of this chapter and to which most writers refer when speaking of organized crime. Syndicate crime (henceforth a synonym for organized crime) may be defined as having three key features, as suggested by Albini (1971):

   1. A continuing group or organization that participates in illicit activity in any society by the use of force, intimidation, or threats

   2. The structuring of a group or organization whose purpose is to provide illicit services for which there is a strong public demand through the use of secrecy on the part of associates

   3. The assurance of protection and immunity necessary for its operation through political corruption or avoidance of prosecution (pp. 47–48)

In a content analysis of definitions of organized crime provided by various writers and government reports, this author (F. E. Hagan, 1983) discovered that many failed to provide any definition, as such. The following characteristics were identified with some consensus: organized (continuing) hierarchy, rational profit through crime, use of force or threat of force, and corruption to obtain immunity. This analysis has been replicated and updated by Jay Albanese (2004, p. 4), who found that the top six items mentioned included an organized hierarchy; continuing, rational profit through crime; use of force or threat; corruption to maintain immunity; public demand for service; and monopoly over a particular market. Jim Finckenauer (2005) notes the continuing, problematic nature of gaining a consensus in defining organized crime and views violence, illicit services, and immunity as the defining characteristics of organized crime. This content analysis supports a core criminological definition of organized crime that is basically consistent with Albini's (1971, p. 126) definition of syndicate crime. Albini's generic definition of organized crime is actually a definition of "group crime," that is, crime committed by two or more people. Table 12.1 summarizes the concept of "organized crime" from both a general (generic) definitional view and a more specialized (sociological/criminological) definitional view.

Organized crime is used in the most generic sense to refer to group crimes and includes many criminal behavior systems as well as "illicit enterprises" that might more appropriately be labeled professional, occupational, corporate, or even conventional criminal behavior. A more specific criminological definition would refer to groups that (a) utilize violence or threats of violence, (b) provide illicit goods that are in public demand, and (c) assure immunity for their operators through corruption and enforcement. In 2006, this writer did an updated content analysis of definitions of organized crime as presented in textbooks and scholarly books on criminology, criminal justice, and organized crime itself (F. E. Hagan, 2006). Combining these, the most identified traits were that it is an illegal enterprise (vice activities) that uses violence and threats, and that it is self-perpetuating, monopolistic, and relies on corruption. This writer also proposes that "Organized Crime" (capitalized) be used to refer to criminal *organizations* while "organized crime" (lowercase) be used to refer to *activities*. Not all "organized crime" is committed by "Organized Crime" groups. Also, the committing of "organized crime" does not make a group an "Organized Crime" group. While "organized crime" refers to crimes that are organized, "Organized Crime" refers to Organized Crime groups (F. E. Hagan, 2006, p. 134).

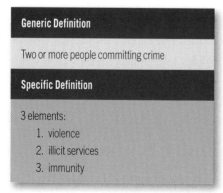

TABLE 12.1 ■ **Generic and Specific Definitions of the Concept of Organized Crime**

Generic Definition
Two or more people committing crime

Specific Definition
3 elements:
1. violence
2. illicit services
3. immunity

The Organized Crime Continuum

A continuum or ordinal model of organized crime has been suggested by others (Albini, 1971, pp. 37–38; Cressey, 1972; McIntosh, 1975; D. C. Smith, 1975, 1978, 1980). In a frequently cited "spectrum-based theory of enterprises," Dwight Smith (1980) proposes that enterprises take place across a spectrum (or continuum) of possible behavior ranging from the legal to illegal, the saintly to the sinful, and that the separation of legitimate business from crime, distinguishing paragons from pariahs from pirates, is an arbitrary point on that range (p. 371).

What all of these models stress is the fact that organized criminal activity is not a simple category. Rather than viewing the concept as a matter of *kind*—that is, is it or is it not—it is far more useful to conceive of it as a matter of *degree*. That is, the concept organized crime is an "ideal type," an abstract generalization that perhaps does not exist in pure form but nevertheless represents a useful, heuristic device for purposes of analysis. Table 12.2 outlines a continuum model of organized (syndicate) crime. Just as medicine may represent the prototype profession, the Cosa Nostra as an ideal type could similarly be a model to which to compare all other groups, although few groups can hope to attain its status or, aside from the Italian American Syndicate, ever have attained it (F. E. Hagan, 1983).

Many profit-oriented or violent criminal groups contain features that may lead us to describe them as examples of semi-organized crime. For example, organizations such as Hell's Angels or the Pagans operate on a fairly highly developed hierarchical structure that uses violence, supplies goods (particularly illicit narcotics) that are in high demand by select segments of the public, and has obtained immunity in outlying geographical areas not through corruption so much as through intimidation of local law enforcement. Thus, Japanese Yakuza, Chinese Triad Societies, and other international criminal organizations to be discussed shortly can be theoretically, if not empirically, placed on the continuum although application of the model may be limited in a non-Western context.

Journal Article Link 12.2
Examine literature regarding the evolution of organized crime in America.

TABLE 12.2 ■ **The Organized Crime Continuum**

Nonorganized Crime (e.g., Intrafamily Assault)	Semi-Organized Crime (e.g., Some Motorcyle Gangs, Narcotics Smuggling Rings)	Organized Crime (e.g., Syndicates, "Cosa Nostra")
No	1. Highly organized	Yes
Not Relevant	A. Hierarchy	Relevant
Absent	B. Restricted membership	Present
Absent	C. Secrecy (codes)	Present
No	2. Violence or threats of	Yes
No	3. Provision of illicit goods in public demand	Yes
No	A. Profit-oriented	Yes
	4. Immunity through	
Unconnected	A. Corruption	Connected
No	B. Enforcement	Yes

Source: An earlier version of this model appeared in: F. E. Hagan, "The Organized Crime Continuum: A Further Specification of a New Conceptual Model," *Criminal Justice Review* 8 (1983): 52–57.

Street Gangs

Handbook Article Link 12.1
Read an article on youth gangs.

Goldstein (1991, pp. 30–32) indicates that the delinquent gang of yesteryear was primarily involved in acts of theft, burglary, and vandalism, with gang fighting ("gang banging") being rare. The 1950s were the era of the rumble, although such skirmishes were exaggerated by the media and by the gangs themselves. M. W. Klein and Maxson (1989) note, "In the 1950s and 1960s, gang members talked much about their fighting episodes, but [homicide] data from several projects revealed their bark to be worse than their bite" (p. 218). Beginning in the 1970s and continuing into the 1990s, gang violence in the United States became worse, reflecting developments on the national scene. The environmental enhancers of this violence are drugs, guns, and territory, although the latter now involves defense of selling (economic) territory and not so much physical turf (Goldstein, 1991, p. 32).

Malcolm Klein (1990) identifies four myths regarding street gangs:

Audio Link 12.1
Listen to a discussion of Hell's Angels.

1. They are highly organized, very cohesive, and have centralized leadership.
2. Street gangs are all violent.
3. Street gangs control drug distribution in our cities.
4. Los Angeles gangs franchise drug distribution to the rest of the country (p. 624).

Klein and associates claim that crack distribution, for example, while involving many individual gang members, was not an organized street gang phenomenon in Los Angeles (M. W. Klein, Maxson, & Cunningham, 1991). A contrary view is suggested by C. Taylor (1990) who, on the basis of field

Photo 12.2

Hell's Angels is perhaps the largest and most notorious outlaw biker gang, with members in countries around the world.

research in Detroit, indicates that the gangs he studied transformed themselves from street punks to drug-dealing entrepreneurs worth millions (see Short, 1990, for a review of recent gang research). Sanchez-Jankowski (1991), in a 10-year participant-observation study of 37 different gangs in New York, Boston, and Los Angeles, was struck by the "defiant individualist character" of many gang members, as well as by their "entrepreneurial spirit." He was stabbed and shot during his research. Sanchez-Jankowski survived the attack, and today is a professor at the University of California, Berkeley.

Mara Salvatrucha (MS-13) is a Salvadoran gang that began with people fleeing to the United States from El Salvador during its civil war. Some members were former guerillas and created the gang with an extensive hierarchy. They traffic in firearms and deal in stolen cars, drugs, murder, and common gang crimes ("Mara Salvatrucha," 2005). In 2005, the FBI arrested 19 suspected members and charged most of them with RICO violations and also with conspiring to intimidate communities through murder, assaults, and kidnappings (K. Johnson, 2005). The **RICO** statute (Racketeer Influenced and Corrupt Organizations) was a feature of the Organized Crime Control Act of 1970 that gave the federal government a powerful legal weapon to prosecute groups for having a pattern of organized crime activity over a period of years. It is estimated that 1 of every 10 federal prisoners is in a gang. These gangs are often based on ethnic/racial ties, such as the Mexican Mafia, the Aryan Brotherhood, or an American group calling itself the Mau Mau.

Some more organized street gangs lie somewhat near the middle of the organized crime continuum, although perhaps they are not as highly developed as some motorcycle gangs. In the 1980s, many tough American street gangs were rapidly converting themselves to ghetto-based drug-trafficking organizations, primarily because of the flood of low-cost cocaine (and crack or rock cocaine) from Colombia. At the onset of the nineties, many of these groups were at about the same place as Italian American groups were in the early 1920s during Prohibition (Morganthau et al., 1988, p. 22). Bloods, Crips (Los Angeles); Montego Bay, Shower, Spangler (Jamaican); Untouchables, 34th Street Players (Miami); or Cobras, Disciples, El Rukns, Latin Kings, and Vice Lords (Chicago) are often big, violent, and increasingly wealthy gangs. In southern California, the majority of street gangs are black or Latino, with Anglos normally joining motorcycle gangs. One such group, the POBOBs ("Pissed Off Bastards Of Bloomington"), emerged to become the Hell's Angels, perhaps the largest and most notorious outlaw biker gang (J. Davis, 1982, p. 42).

Big Hawk 1987 BSVG c 187

The above example of gang graffiti is a marking of gang turf not for bragging rights, but for sales territory and a threat to rivals. The translation is Big Hawk (a member's street name); 1987 (the year); a member of Blood Stone Villain's Gang (BSVG), which is a Bloods set (subgroup of the Bloods gang). The lowercase "c," which is usually X'd out, means that Big Hawk kills Crips, and the number 187 is the section of the California criminal code for murder (Morganthau et al., 1988, p. 23).

For many gang members, self-employment in the underground drug economy provides short-term upward mobility; autonomy; a measure of dignity or self-esteem; and an opportunity to avoid low-level employment under the direction of what is perceived as hostile, outside ethnic or racial groups (Bourgois, 1988, p. 12). Street gangs, despite their penchant for what might appear to be senseless violence, sometimes represent the minor leagues or incubators for future organized criminals and syndicates.

Sanders (1994) notes that there are differences in levels of commitment of gang members (hardcore, affiliate, and fringe). J. M. Hagedorn (1994) notes that, despite high average earnings from drug sales, most gang members would prefer full-time jobs with modest wages and most move in and out of conventional labor markets. Hagedorn identifies four types of gang members: legits, homeboys, dope fiends, and new jacks. "Legits" are those who mature out of the gang, while "homeboys" are a majority of African American and Latino adult gang members who alternate between legitimate jobs and drug sales. While "dope fiends" stayed in the drug business in order to feed their habits, "new jacks" view illegal drug sales as their career (p. 206).

By the 1990s, the FBI began to take gangs very seriously and to apply major resources to their investigation and defeat. By 2007, there were 131 Violent Gang Task Forces and an MS-13 National Gang Task Force. Such actions led to major takedowns of Nuestra Familia, Ruben Castro and the 18th Street Gang, MS-13, the Black P-Stone Bloods, and the Townsend Street Gang (FBI, 2007).

International Organized Crime

Internationally, organized crime is not confined to any single political arena and thrives especially in political climates such as liberal democracies and corrupt dictatorships. Because laws of liberal democracies such as the United States, Canada, post–World War II Japan, and other Western European and former British Commonwealth countries place a priority on individual civil liberties, crime control can suffer; such laws make it difficult to crack down on organized criminals and their political allies. Such nations also emphasize private enterprise, which is not restricted to the legal end of the continuum.

In 2000, the United States and 120 other nations signed the United Nations' Convention Against Transnational Organized Crime (TOC) along with protocols on human smuggling (including that of migrants). The signatories agreed to criminalize TOC actions and cooperate in their investigations. TOC includes, but is not limited to, drug trafficking and abuse, arms trafficking, money laundering, migrant smuggling and other human trafficking, intellectual property theft, and foreign official corruption.

Yakuza

Yakuza, the Japanese term for gangsters (literally, "good-for-nothings") are organized crime syndicates of roughly 90,000 members. Also referred to as *boryokudan* (violent ones), the power of these gangs—the most powerful is the Yamaguchi-gumi—can be illustrated by a recent stock market scandal in Japan in which top firms, such as Nomura and Nikko Securities, allowed affiliates to finance the activities of Yakuza (Kaplan, 1991). Susumi Ishii, the head of the Inagawakai syndicate, received 25 billion yen ($180 million) from these firms. One of Ishii's financial advisors in the United States was a company that employed Prescott Bush, Jr., brother of George H. W. Bush, who was president at the time. While Prescott Bush may not have known with whom he was dealing, the Japanese security firms were aware (Kaplan, 1991).

Organized crime figures in Japan have a curious appearance: crew cuts, elaborate tattoos, and missing tips of the little fingers; they often work as "bouncers" or security guards at corporate conventions, a strategic role that enables them to gather information with which to blackmail corporate officials ("Japan," 1977, p. 40; Rome, 1975).

Representing a traditional part of Japanese society, the Yakuza were originally recruited by right-wing business leaders after World War II to intimidate left-wing opponents. In the eighties, growing concern was expressed regarding Yakuza expansion into the United States (Dubro, 1982). Such groups reportedly owned $100 million in Honolulu real estate, where their restaurants, clubs, and pornography shops catered to Japanese tourists. Active also in California, the groups were involved in smuggling drugs to the United States and guns to Japan as well as recruiting U.S. female "entertainers" as prostitutes in Japan.

Membership in Yakuza groups is claimed to be 20 times larger than membership in the American Mafia. There is considerable acceptance and toleration of such groups by both the public and political powers (Kaplan & Dubro, 1986). They serve a useful function for the right wing in intimidating dissenters, the free press, or any group that may appear critical of the government (CBS Broadcasting, 1989). Yakuza are widely involved in sexual slavery. Thousands of women and children, mainly from developing countries, are forced to work as prostitutes near military bases, to participate in the production of pornography, and to enter into mail-order marriages (Kaplan & Dubro, 1986, p. 201).

One third of the members of Yakuza are Korean and most are from lower-class backgrounds (Kaplan, 1988). Kaplan (1991) explains,

Yakuza gangs occupy a place in Japanese society hard to imagine in the West. Members sport business cards and lapel pins openly identifying their underworld affiliation. Offices proudly display the gang name and insignia, much as if one found the words "Gambino Family, Manhattan Branch" emblazoned on the door of a Mafia concern. (p. 2)

The success of the Japanese police in fighting Yakuza gangs is noted by E. H. Johnson (1990), who reports that their share of the prison population increased from 21 percent in 1975 to 30 percent in 1986. In the early nineties, Yakuza groups had expanded their involvement in coercive resolution of civil disputes stemming from the collection of debts, loan negotiations, bankruptcies, real estate transactions, and other matters (E. H. Johnson, 1990). In 1996, Japanese Prime Minister Ryutara Hashimoto's attempt to bail out leading banking and real estate finance companies met opposition in Parliament due to charges that many of the recipients of the bad loans were companies fronting for organized crime, and some of them had close ties to Hashimoto's party. Bankers fear seizing any real estate linked to the Yakuza for fear of being killed or beaten (WuDunn, 1996).

In the early twenty-first century, some erosion in Yakuza codes and lifestyle began to emerge. In 2001, a turf war between rival factions in Tokyo splintered the discipline within Yakuza clans. Movement into legitimate businesses had eroded some practices such as chopping off the tips of pinkies, bouffant hairstyles, and street-level crimes. A poor Japanese economy and competition with rival ethnic gangs (especially Chinese) have also been identified as precipitating change in Yakuza gangs (French, 2001). Police crackdowns are also reportedly costing Yakuza shrinking membership.

Chinese Triad Societies

Triads are secret Chinese organizations. Referred to as "black societies" by the Chinese, the British called them Triads because of their highly ritualistic use of numerology, a belief in the magical significance of numbers. The number 3 and multiples of 3 were accorded major importance by these groups. The symbol of Triad societies is an equilateral triangle with the three sides representing the three basic Chinese concepts of heaven, earth, and the human being.

Although they are of much more ancient origin and are even more cabalistic, the legends, rituals (such as initiation rites), and early history of Triads bear an uncanny resemblance to the Mafia legend in Sicily (Bresler, 1980; Morgan, 1960). The earliest Triad secret societies were founded in China 2,000 years ago to oppose warlords (Daraul, 1969; Robertson, 1977). The modern Triads are traced to the latter part of the seventeenth century, when members appeared as resistance fighters against the Manchu dynasty, the "barbarian" invaders who defeated the Ming dynasty. Legend dates the founding of the first modern Triad to 128 Buddhist monks at a monastery near Foochow, Fukien province, in 1674. They were well trained in Asian martial arts, including a type they had perfected themselves—kung fu (Bresler, 1980, p. 28; Chin, 1988, p. 7). A Triad called the Fists of Harmony and Justice led the Boxer Rebellion against the European powers, from 1899 to 1901.

Although originating as brotherhoods for freedom (Lyman, 1974), Triads also had elements of banditry and were heavily involved in the control of vice activities. All of the Triad groups shared in common highly ritualized initiation ceremonies, blood oaths, passwords, secret signals, and hierarchical positions. Some positions in a Triad society are described by Bresler in his *Chinese Mafia* (1980):

489. "Shan Chu" (hill chief or head)

438. "Heung Chu" (incense master in charge of ceremonies)

Each cell (branch) had three lower-level offices:

415. The "white paper fan" (financial advisor)

426. The "red pole" (kung fu expert)

432. The "straw sandal" (messenger/liaison with other groups)

49. The ordinary member

The number "4" in all of the titles reflects the ancient Chinese belief that the world was surrounded by four seas.

489 and 438 are said to have been selected because the Chinese characters for 21 (the sum of 4 + 8 + 9) and for 3 and 8, when written together, form the Chinese characters for Hung, the early Ming Emperor in whose name the whole Triad organization began in the first place. 426 is constructed as $4 \times 15 + 4$, which equals 64. This refers to the 64 diagrams of Chinese script invented by a legendary Emperor named Fu the. . . . 432 becomes $4 \times 32 + 4$, giving us 132, which is the actual number of persons (128 monks and 4 others) supposed to have been living in the original Triad monastery near Foochow. Finally, 49 derives from 4×9, which equals 36. This refers to the number of oaths sworn by all new Triad members. (p. 28)

Chin (1988, 1990) claims that many myths similar to early ones of an omnipotent Italian Mafia have been created regarding Triads, and that Chinese small-business owners, not Triads, are responsible for most of the drug trafficking, money laundering, and other criminal activities in U.S. Chinatowns. Care must be taken not to label all Chinese crime groups as Triads. For instance, one Taiwan-based crime group, "the United Bamboo," is not a Triad organization; that is, it has no relationship with mainland Chinese Triad groups (National Central Police, 2005).

With the fall of mainland China to the Communists in 1949, many Triads migrated to Hong Kong. The largest of such groups were the Green Pang (Green Gang), the Chui Chaos (Chiu Chau), and the 14K. Although the Green Pang originally controlled heroin distribution in what was then a British colony, they relied on the Chui Chaos for supplies of Thai morphine and opium (McCoy, 1972, p. 229). The Chui Chaos had important connections and even members within the Hong Kong police; they control much of the drug traffic from the "Golden Triangle" (Northern Burma, Laos, and Thailand) and throughout Southeast Asia. Well-known Triads in Hong Kong in the twenty-first century are Sun Yee On, Wo Shing Wo, and 14K (Finckenauer & Chin, 2004).

Triad groups are nonhierarchical and informal. Each faction is run by an independent boss and is autonomous in planning and executing criminal enterprises. They range from street gangs to sophisticated crime syndicates (Lindberg et al., 1997).

Tongs were Chinese American fraternal and benevolent organizations, the term meaning "town hall" or "large hall." Some of the important Tongs in the United States in the nineteenth century were Bing Kung, Hip Sing, Ying On Ton Su, and Hop Sing. Many of these fraternal organizations relied upon young street gangs to enforce their vice activities. New Tong organizations, formed in the post–World War II period, were more ferocious criminal bands made up of many felons who had fled Hong Kong and the Far East. The Flying Dragons, Ghost Shadows, Gray Shadows, and Black Ghost Shadows were some of these groups. In February 1996, federal law enforcement, after a 1-year sting operation, charged leaders of several Chinese American tongs with drug trafficking and money laundering. Indicted were members and leaders of Hip Sing Tong, the Hung Mung Association, the San Gian Tong, and the Fujian Fellowship Association. In 1994 the Tun On Association in Manhattan was found to be used by a gang for protection of its gambling operations, which later led to expanded operations resulting in 10 murders.

While some observers claim that Tongs, like Chop Suey, were strictly an American invention, organized in the gold fields of California about 1860 (Nash, 1981, p. 337), others see them as branches of Triad societies, mainly the Chee Kung Tong, which generated many feuding rival branches (Bresler, 1980, p. 30). Since the late sixties, members of Triads have emigrated and set up operations in the United States, Canada, and Europe, most notably in the Netherlands and in older established Chinatowns of San Francisco, Vancouver, and Amsterdam (R. Wilson, 1978).

While many modern Triads are respected community organizations, others have developed criminal subgroups. Robertson (1977) claims that, particularly in Western Europe, nearly all Triads are engaged in prostitution, illegal gambling, extortion, and heroin trafficking. They are the major wholesale distributors and processors of opium from the Golden Triangle. "The China White Trail" is a term used by the U.S. Drug Enforcement Administration (DEA) to describe the transportation of heroin from Thailand through the secret societies of Hong Kong and finally the Chinese neighborhoods of New York City (Kerr, 1987). In 1989, the FBI seized 828 pounds of heroin valued at $1 billion (the 1971 "French Connection" bust yielded about 220 pounds). This New York City bust was attributed to the China White Trail. Far more official attention has been paid to the issue of Asian organized crime by U.S. government officials than in the past. The President's Commission on Organized Crime (1984a) issued a report on Asian organized crime, as did the Department of Justice in 1988 (Baridon, 1988) and the Hong Kong Security Forces (Fight Crime Committee, 1986; see also FBI, 1985).

With the return of Hong Kong by the British in 1997 to the People's Republic of China (PRC), most had predicted that the Hong Kong Triads would migrate to the West to escape stricter law enforcement. To the contrary, preliminary signs indicate that at least some Triad groups are thriving and have even extended their operations to areas such as Huizhou, the so-called Palermo of China, where cold cash speaks louder than ideology (Viviano, 1997). Stolen cars from Hong Kong are pervasive, courtesy of the 56,000-member Sun Yee On (New Discipline and Peace) Triad, which also deals in narcotics, money laundering, gambling, and prostitution. Chu (2005) found that from 1997 to 2004, triads increasingly entered the Chinese market, including legitimate businesses.

Russian Organized Crime

The most publicized of organized crime groups in the 1990s were Russian. Some 12 to 15 major "mafiya" groups exist, each with a federation of hundreds of smaller groups. The two largest gangs are the Moscow-based Solntsevo, which includes the U.S.-based group Organizatsiya, and the St. Petersburg group, Tambov, which is less active in the United States. Their prime activities include health care fraud, drug and alien smuggling, prostitution, and financial fraud (Krane, 1999). With the fall of the former Soviet Union, such groups have, in some areas of Russia and the former Soviet republics, challenged the government itself as a source of power. Hundreds of gangs use extortion, fraud, and murder to operate illegal as well as legal businesses. In 1995, they controlled about 400 banks, which explains in part why Moscow, with its exploding crime rate, had few bank robberies at that time (Hockstader, 1995, p. 6). Such groups are well armed and ruthless. While they speak Russian and come from areas of the old Soviet Union or its satellites, numerous groups that are labeled as "Russian organized crime" are from a variety of ethnic backgrounds including Albanian, Armenian, Chechen, Georgian, Jewish, Latvian, Lithuanian, Tatar, and Ukranian (Finckenauer & Waring, 1998, p. 132). Russian crime groups in the United States are fluid, with transient membership in each group varying from 5 to 20 persons. They are loosely structured and often formed on the basis of regional backgrounds or a particular enterprise (Finckenauer & Waring, 1998; Kenney & Finckenauer, 1995).

At the top of such gangs are men such as Vyacheslaw Ivankov, who is one of the *vory v zakone* ("thieves professing the code" or "thieves-in-law"). The Vory had an oath of their own under the Soviet system that shunned accepted society and defied authority. The *vory v zakone* are not members of the same gang, but an honored category of criminals empowered to resolve gang disputes. Predating the Russian revolution, this group's members were recruited in prison and branded with a tattoo of an eagle, usually on their hands. Many gangs, under increasing pressure in Russia, have migrated to Western Europe, particularly Germany, and the United States (Raab, 1994). Russian gangs have set up operations in the United States, particularly in "Little Odessa," the Brighton Beach section of Brooklyn, where they have formed cooperative alliances with traditional Mafia groups. On July 9, 1996, the head of the Odessa Mafiya, Ivankov, and three codefendants were convicted of extorting $3.5 million from owners of an investment company. They had also kidnapped and killed the father of one owner in Moscow. Ivankov's arrest was considered the outcome of growing cooperation between the FBI and Russian police to fight such groups (Kenney & Finckenauer, 1995).

In the United States, Russian groups have been involved in a large jewelry heist, as well as in insurance and Medicare fraud, heroin importation, and control of gasoline distribution in New York City. In the latter alone, they evaded over $5 billion a year in taxes (S. Anderson, 1995, p. 43). According to one source, city police from the 60th and 61st precincts moonlight for them as bagmen, muscle, and chauffeurs, and they even participate in fake accident scams (Friedman, 1994).

Rosner (1995) warns us not to create an overglamorized image of the Russian mafiya in the United States:

> Lastly, the sexy Russian Mafia provides journalists and their readers with a relatively unthreatening, European model of crime—a revisited Marlon Brando world of *consiglieri, caporegima,* and soldiers. At least that is the model which is appealingly seductive, although quite inaccurate. (p. 32)

The Russian **mafiya** is a generic term for a type of criminal (black marketeer, gangster, drug trafficker, and corrupter) who arose out of social, economic, and historical forces in Russia. When all goods were owned by the state, stealing them became a necessity of life, the "Soviet way of crime" (Albini et al., 1995). Such groups often collaborated with state bureaucrats *(nomenklatura)* in what might be called the "gangster industrial complex" (Shelley, 1995). Privatization after the fall of the USSR made Russia what former President Boris Yeltsin has called a "superpower of crime." The Soviet Union itself resembled a criminal racket, and thoroughly corrupt officials were ill prepared for privatization (R. J. Kelly, Schatzberg, & Ryan, 1995).

There have been rising concerns over reports that Russian mobs are recruiting former KGB and former Soviet Special Forces soldiers as members. Many of the groups identified as "Russian" may include others from the former Soviet Union including Armenians, Georgians, Chechens, Ukrainians, and Lithuanians. In addition, there are groups from former Eastern European satellite countries, such as Slovaks, Hungarians, Poles, and Albanians. While it is easy to blame endemic corruption in Russian society for the pervasive organized crime, one must be careful not to replace the Cold War image of Russia as "the evil empire" with one of Russian gangsters. Wedel (1999) points out that in some instances, U.S. policy and institutions have been complicit, either wittingly or unwittingly, in the corruption. Since 1997, the Harvard Institute for International Development has been granted a contract to assist in economic reforms in Russia. The U.S. Justice Department has been investigating the misuse of these development funds in which the privatization of Russian assets have been selectively awarded to insiders through corruption. Members of the Harvard team were criticized by the General Accounting Office for profiting from inside knowledge in these deals. During this time, billions of dollars were being looted from the Russian economy and laundered through U.S. banks such as the Bank of New York (Wedel, 1999).

In 1998, it was revealed that Amy Elliott, a Citibank employee, helped Raúl Salinas, brother of the former president of Mexico, move $100 million into offshore, untraceable accounts through dummy corporations in the Cayman Islands. In the Bank of New York scandal, some $4.2 billion was laundered in over 10,000 separate transactions. The money belonged to Semyon Yukovich Mogilevich, a top Russian boss ("Russian Mob May Have Laundered," 1999). Sukharenko (2004) indicates that Russian groups have yet to cultivate political contacts in the United States to influence the political process. Rather than involving traditional organized crime, their activities tend to include money laundering, tax evasion schemes, insurance fraud, and other white collar offenses.

Countries with corrupt dictatorships, particularly Caribbean vacation spots, became convenient gambling resort areas, especially when organized crime figures such as Meyer Lansky simply "cut in" the authorities, such as Batista in Cuba or Bahamian officials, in return for unencumbered operations.

The Nature of Organized Crime

Given our general definition of organized crime, such groups have existed in varying degrees since, or even before, the advent of modern nation-states. Large, diversified syndicate crime, with control on an extraregional basis over more than a few illegal activities, is primarily a phenomenon of the post–World War I period. While the focus in discussing organized crime is generally on the prototype, what has been called the "Cosa Nostra" (or Mafia), the nature and structure of organized criminal groups are determined by the type of criminal activity they are engaged in as well as by ethnic, subcultural, and cultural values. Criminal gangs, mobs, racketeers, and organized (predatory) criminals share to a lesser degree many of the characteristics of larger syndicates.

Journal Article Link 12.3
Examine literature regarding comparative analysis of perceptions of organized crime.

Ethnicity and Organized Crime

Some believe that organized crime began in the United States as an import, along with mass immigration of Sicilians and Italians in the late nineteenth and early twentieth centuries. Organized crime is not simply a "Mafia transplant" or "alien conspiracy" in the United States; it obviously existed before significant Italian immigration, and it probably will exist long after Italian Americans move out of major involvement in organized crime.

Ianni (1973) proposes an **ethnic succession theory** of organized crime in which organized crime acts as a "queer ladder of mobility" (Bell, 1953, p. 115), an alternative means of upward mobility for ethnic minorities who, because of discrimination or lack of skills, are temporarily lodged at the bottom of the system of reward distribution in a society. Thus, while the last 50 years in the United States have witnessed the period of Sicilian-Italian domination of syndicate crime, this was preceded by Jewish (sometimes facetiously referred to as the "Kosher Nostra") and Irish domination. Prior to these groups, WASPs (White Anglo-Saxon Protestants) controlled organized crime. During these periods, many other ethnic groups—for example, Germans, Lebanese, Greeks, and African Americans—also participated in organized crime. At the present time, with its ethnic base largely middle class, the Italian American Mafia might be described as in its eleventh hour, as black, Latino, Asian, and Russian groups move into positions of power in organized criminal activity with their base of operations in low-income ethnic ghettos, long the wellspring of illegitimate careers (see Kleinknecht, 1996). African American organized crime groups are not an example of ethnic succession in that they have existed for some time, but much of their control has been primarily within black populations. Black Entertainment Television (BET) (2006) channel's production of the documentary *American Gangster* as well as the History Channel's production of *Gangland* give accounts of such organizations. Griffin, in *Black Brothers Inc.: The Violent Rise and Fall of Philadelphia's Black Mafia* (2005), describes a group that controlled drug dealing, loan sharking, the numbers racket, armed robbery, and extortion. Led by Sam Christian, they had close ties with the Nation of Islam. They killed rivals, taxed bookies connected with the Cosa Nostra, and intimidated other gangs. The group's eventual demise as a result of law enforcement led to the emergence of the crack-dealing Junior Black Mafia.

Organized crime is in rapid transition, and much of our image of a dominant Mafia underworld now resembles an old black-and-white gangster movie starring Edward G. Robinson or James Cagney. Former U.S. Attorney General William French Smith, U.S. in remarks before the President's Commission on Organized Crime (1983), stated:

> Even as federal law enforcement agencies have worked hard to catch up on the traditional crime families found in our major cities, new forms of organized crime have emerged throughout the nation. In just the past few years, new groups have organized in pursuit of the lucrative profits that can be made in drug trafficking. Although traditional organized crime is heavily involved in the drug trade also, these new groups do not have places on that family tree. They are distinguishable. They include motorcycle gangs, prison gangs, and foreign-based

organizations. Some of the names of these groups will be familiar, but most are not. They are: Hell's Angels, Outlaws, Pagans, Bandidos, La Nuestra Familia, Mexican Mafia, Aryan Brotherhood, Black Guerilla Family, Japanese Yakuza, Chinese Triad Societies, Israeli Mafia, and many, many more. Of the 425 cases under investigation by the Organized Crime Drug Enforcement Task Force, which this administration established this past year, only a small number involve traditional organized crime. Most involve the new cartels. (pp. 124–125)

This diversity of groups involved in organized crime is certainly well illustrated in the burgeoning international illegal drug business.

The term *transnational crime* refers to criminal activities that take place in more than one country. The United Nations identified these activities as including money laundering, terrorist activities, art and cultural object theft, theft of intellectual property, arms trafficking, aircraft hijacking, sea piracy, land hijacking, insurance fraud, computer crime, environmental crime, human trafficking, trade in human body parts, drug trafficking, fraudulent bankruptcy, infiltration of legitimate business, corruption of public officials, and other businesses committed by organized crime groups (for example, automobile theft). Recognizing the critical relationship between most of these activities and organized crime, the UN drafted the Convention Against Transnational Organized Crime, mentioned earlier in this chapter, which was signed by the 40 countries necessary to make it binding in 2003. The signatories agreed to adopt tougher laws and to cooperate in the enforcement of these laws against organized crime. Three additional protocols were agreed to dealing with human trafficking, the smuggling of immigrants, and arms trafficking (http://www.unodc.org).

As part of the preparation for the Convention Against Organized Crime, the UN commissioned a survey of authorities in 16 countries with respect to 40 organized criminal groups. Some of these groups were familiar, such as Cosa Nostra (Italy), Hell's Angels (Canada), and Yamaguchi-gumi (Japan), while others were simply designated as "Group with no name" (in Germany, the Netherlands, and Russia). Each group was classified regarding its structure, size, activities, level of transborder operations, identity, level of violence, use of corruption, political influence, penetration into the legitimate economy, and cooperation with other organized criminal groups (United Nations Centre for International Crime Prevention [UNCICP], 2000a, 2000b).

The United Nations identified five different types of organized crime groups:

1. Rigid hierarchy—single boss, divisions reporting to the center, with a strong internal system of discipline

2. Devolved hierarchy—regional structures with own leadership hierarchy and autonomy

3. Hierarchical conglomerate—an association of groups with single governing body varying from an umbrella-type body to flexible oversight arrangements

4. Core criminal group—relatively loose group characterized by horizontal rather than vertical arrangements

5. Organized criminal network—shifting alliances that do not regard themselves as an organized crime entity (UNCICP, 2000a, 2000b)

Money Laundering

Money laundering refers to making clean or washing "dirty money" (illegal funds). A classic task of organized crime syndicates has been to somehow convert large amounts of illegally gotten funds into usable money that appears to come from legitimate sources. Drug traffickers are particularly faced with the problem of laundering huge amounts of ill-gotten currency.

Various countries, most notably Switzerland, the Bahamas, Panama, and other "tax havens," have created bank secrecy laws that generally forbid the disclosure of the financial affairs of account holders. This practice apparently began in Switzerland with numbered accounts to protect the finances of those whose holdings were being confiscated in Nazi Germany (T. Clark & Tigue, 1975). Recognizing the seriousness of such activity, the U.S. Congress passed the Money Laundering Act of 1986, which made money laundering a federal crime carrying substantial penalties (Weinstein, 1988). A growing number of countries are passing such laws (Gramckow, 1992).

The easiest way to launder money is to take a suitcase full of it to an unscrupulous bank. This bank may recycle the currency to other countries such as Argentina, where the dollar is used because the local currency is subject to hyperinflation. Other methods may include purchasing luxury goods at inflated prices from a coconspirator who transfers the excess proceeds to the purchaser's account. Proceeds from legitimate businesses such as restaurants can be augmented with illegal funds (Melloan, 1991). The U.S. Department of Justice estimates that drug traffickers launder an estimated $100 billion each year in the United States (Webster & McCampbell, 1992).

Money laundering is a three-step process:

1. Placement—Collect the dirty money and move it into the financial system.

2. Layering—Disguise the money trail by transferring the money into the bank account of phony companies, creating false invoices and enterprises using offshore banks and wire transfers.

3. Integration—The now-clean money can be utilized for investments, political campaign donations, and the infiltration of legitimate enterprise.

Bank personnel are often bribed to accept large deposits without reporting them. These are then wired (transferred) to overseas accounts. The money can also be converted to cashier's checks and money orders or hidden in export items such as cars or televisions. The large-scale interconnections between drug traffickers and money launderers are illustrated by the BCCI (Bank of Credit and Commerce International) scandal, which first unraveled in 1989 (Lohr, 1992). BCCI, which operated in over 70 countries, was controlled by Middle Eastern investors and was heavily involved in the laundering of drug money, worldwide fraud and bribery, and the secret ownership of American banks including First American, whose director was Clark Clifford. Former Panamanian strongman Manuel Noriega, terrorist Abu Nidal, and even the CIA regularly did business with BCCI. Passas (1995) reports,

> BCCI had engaged in a huge Ponzi scheme, defrauding about a million innocent clients around the world. According to the liquidators, $9.5 billion are still unaccounted for but no one knows the precise extent of the loss. Huge amounts disappeared into the Grand Cayman portion of the BCCI Group. Series of complex manipulations and falsification of accounts (e.g., unrecorded deposits and false loans or transactions) hid BCCI's poor financial health and made it virtually impossible to reconstruct the true history of the bank. As investigations intensified and multiplied, revelations were made almost daily, over a long period, about BCCI's banking services to money launderers, drug traffickers, arms dealers, coffee smugglers, tax evaders, political offenders, dictators, and intelligence agencies around the globe. BCCI also conducted some interbank transactions and had a director in common with Banca Nazionale del Lavoro (BNL) whose Atlanta branch extended billions of dollars in illegal loans to Iraq. (p. 382)

Casinos can be an excellent place to launder ill-gotten funds, especially with the cooperation of insiders. In 1998, two employees of the Showboat Casino in Atlantic City were charged with helping an alleged drug dealer launder $100,000 by depositing it in the casino under a phony name. They could exchange the dirty money for checks and avoid a cash transaction report filed with the Internal Revenue Service. Ill-gotten cash was converted to lucky winnings. In a typical scenario, a money launderer buys chips with dirty money, gambles, and then cashes out, obtaining clean money. In this case, the "drug dealer" turned out to be an undercover IRS agent.

Over $500 billion is laundered annually by various global institutions, making money laundering the third-largest industry in the world. U.S. law requires that cash deposits of $10,000 or more be reported to the Internal Revenue Service, but much of the laundered money ends up in secret bank accounts overseas where it can be freely moved.

FINCEN (the Financial Crimes Enforcement Network of the U.S. Department of the Treasury) has uncovered a new twist to money laundering by Colombian drug traffickers in which they use peso brokers and unsuspecting U.S. companies to launder dirty money:

1. Secret stash houses in the United States store large amounts of dollars from street sales. This money is bought from the drug cartels at a 15–25 percent discount by a peso broker with "clean" pesos.

2. Using operatives, or "smurfs," in the United States, they deposit the cash in small increments in U.S. banks.

3. The peso broker finds Colombian businesspersons who need U.S. dollars in order to import goods.

4. In return for the businesspersons' clean money, the broker writes checks from the smurf checking accounts, often exchanging the dollars for pesos at a discount.

5. The orders are welcomed as new sales and the goods are shipped to Colombia (France & Burnett, 1992).

Illustrating the relationships with legitimate society as well as the transnational nature of money laundering operations, a joint international sting operation involving police agencies from Canada, Italy, Spain, the United Kingdom, and the United States called "Operation Dinero" was aimed against the Cali cartel. The Drug Enforcement Agency set up an offshore bank in Anguilla, which had become a favorite money laundering site for the cartel. At one point, they asked the bank to sell art masterpieces (a favorite money laundering investment) for them. Also involved was the Severa crime organization of Italy. The operation netted 88 arrests and seizure of 9 tons of cocaine and $50 million in cash and property (P. Williams, 1997).

Drug Trafficking

While the Italian American Syndicate has been involved in drug trafficking, the business is so large that no one group can hope to control it. Although there are many international sources of illegal drugs, the three primary centers of supply are the Golden Triangle, the Golden Crescent, and Latin America.

The Golden Triangle is the northern border areas of Thailand, Burma (Myanmar), and Laos, which are major heroin-growing areas. Part of this area, called the Shan States, is controlled by an Opium Army made up of the descendants of former Chinese Nationalist troops. The Golden Crescent includes areas of Iran, Afghanistan, Pakistan, and Turkey, which made up the old "French Connection." The latter, which was the basis of a classic movie, involved the smuggling of raw opium to Marseilles, France, for processing into heroin, after which it was sent to the United States to be sold. The third source, Latin America, involves primarily cocaine and marijuana, mainly from Colombia (Abadinsky, 1994; Inciardi, 1992).

Trebach (1984, p. 132) has come up with the notion of the **"Iron Law of Opium Trade"** to describe a situation in which, if one source of supply is closed, another replaces it. In a 6-year observational study of drug smugglers and dealers, P. A. Adler and Adler (1983) found that because of the danger and legal penalties, there were "shifts and oscillations" in drug trafficking careers. Involvement was temporary, but due to the large rewards involved, many successful retirees move in and out of smuggling organizations.

Colombian Cartels

Probably the most powerful international drug trafficking organization in the world was the **Medellin cartel** of Colombia, an organization that used M-19 (the April 19 Movement, a revolutionary group) as protection for their operation (Gugliotta & Leen, 1989). It was the latter terrorist group that gave birth to the cartel. In 1981, M-19 kidnapped the daughter of Fabio Ochoa, the most powerful cocaine boss. In response, the Ochoas formed a cartel of 200 other narcotics trafficking organizations in the city of Medellin and prepared to wage war with M-19. The latter group wisely released Ochoa's daughter and began a hands-off-the-cartel policy in return for a cut of the profits. The cartel used M-19 to storm the country's Palace of Justice in 1985, killing 12 of the 24 Supreme Court justices (J. Anderson & Van Atta, 1988b; Eddy, Sabogal, & Walden, 1988). Narco-terrorist groups also traffic in weapons, launder money, offer mutual assistance, smuggle contraband, and share intelligence. The Medellin cartel was later succeeded by the Cali cartel and other Colombian groups, or baby cartels. By the 1990s, Mexican drug traffickers began to supersede their previous partners, the Colombians, as a base for drug smuggling. In interviews with 34 high-level drug smugglers, Decker and Chapman (2008) found growing decentralization of trafficking.

The Underground Empire

In an investigation of "narcotraficantes" (narcotics traffickers), James Mills in *The Underground Empire: Where Crime and Governments Embrace* (1986) makes some serious charges: "The largest narcotics conspirator in the world is the government of the United States whose intelligence agencies conspire with or ignore the complicity of officials at the highest levels in at least 33 countries" (p. 116). As soon as the DEA closes in on drug *domos* (bosses), the State Department or CIA sabotages their investigations in the name of foreign policy (F. E. Hagan, 1987a). The "underground empire" is a "fourth world" of nations of institutionalized, state-supported crime.

The arrest and conviction of former Panamanian dictator Manuel Noriega, who was also a paid CIA informant, serves as another example. By 2000, law enforcement officials had begun to seize larger and larger shipments of ecstasy pills, a synthetic "psychedelic amphetamine" also known as MDMA. U.S. Customs estimates that it seized 3.5 million pills in the 1999 fiscal year, compared with 750,000 the year before. A pill produced for less than a dollar in Europe sells for up to $40 in the United States ("Ecstasy Trade, Seizures Skyrocket," 2000). Once confined to dance parties called "raves," the drug has spread enormously.

Mexico's Drug War

Crime File 12.1 describes the rise and fall of Amado Carrillo Fuentes, who until his death was the most powerful leader of organized crime/drug traffickers in Mexico. It also examines "Operation Casablanca," the largest sting operation in U.S. history directed at Mexican drug trafficking and money laundering. By 2007, the Zetas, a ruthless Mexican organization that had acted as enforcers for the Gulf drug cartel, superseded their former bosses and moved into the territory of other cartels. Many of its members are from former Mexican elite military units. Their leader, Heriberto Lazcano, is known as El Verdugo, "the Executioner" (Corchado, 2009). The Zetas courted their demise when they began kidnapping wealthy businessmen, attracting the attention of federal law enforcement and the Mexican military.

Crime File 12.1

Amado Carrillo Fuentes and Operation Casablanca

Amado Carrillo Fuentes, who died under mysterious circumstances in July 1997, was the leader of the Juarez plaza (criminal organization), the most powerful drug trafficking–organized crime group in Mexico. In a given year his organization was believed to earn $10 billion dollars. Pem Ex, the Mexican government–owned oil company and the largest corporation in Mexico, earns $7–$8 billion a year. Earnings from cocaine sales alone for all Mexican crime groups are estimated at $30–$32 billion. The story is told that Carrillo was considering closing down his Mexican operations in 1997, and there was fear that the country's economy would collapse.

Carrillo began his apprenticeship in crime growing up in Mexico's drug growing province of Sinaloa, working for his drug trafficking uncle. He was later a member of the Mexican secret police (DFS) for 5 years (Poppa, 1998). His uncle was believed to have been involved in the murder of U.S. DEA agent Enrique Camarena. The Mexican federal police were also involved. Both the Mexican army and secret police were partners with Mexican drug traffickers. Carrillo would later work with Pablo Acosta, one of the top narcotraficantes on the U.S. border. When Acosta was killed, Carrillo took over his "plaza" (criminal organization). He formed relationships with the Colombian Medellin cartel, which was hunting for new trafficking routes as a result of the success of crackdowns in Florida. Later, Carrillo decided to skip the Colombians and began wholesaling cocaine directly from Peru and Bolivia. It was he who consolidated the other cartels and acted as the supplier.

Carrillo had grown to become a legend and, without exaggeration, the single most powerful individual in Mexico. He was a hero to his home province, Sinaloa, where he built a church and assisted local charities (Bowden, 1998). His organization had encryption devices, pagers for border crossings, and spies. License plates of cars entering the DEA's El Paso Intelligence Center were recorded by his agents, and the owners' phones were tapped. The DEA moved its operations to Fort Bliss.

Carrillo was not immortal, however. In July 1977 he visited doctors for plastic surgery. A 9-hour operation involving liposuction and a face lift resulted in his death due to an injection. While it was never confirmed, it was believed to be an assassination. Four months later the bodies of the surgeons involved were discovered. They had been brutally tortured in an attempt to discover how he had died. In the 2 years after his death, a power vacuum existed in Juarez. Over 1,000 drug-related murders had occurred, and the believed successor was his younger brother, Vincente Carrillo Fuentes. Despite the payoffs of high government officials, the police, and the military, Mexican cartels have the same problem as other illegal enterprises—how to launder the money.

Operation Casablanca

The most comprehensive undercover sting operation and criminal investigation in U.S. history was the Customs Bureau's "Operation Casablanca." This was a money laundering sting operation on the Mexican drug cartels. Posing as a U.S. financial organization, U.S. agents met with financial advisors of the Mexican drug cartels and top bankers to set up money laundering operations that they secretly videotaped. The Mexican bankers shared some of their techniques and the names of overseas and American cooperating banks.

The money laundering operation involved smuggling drug money from the U.S. and depositing it in Mexican banks. These banks issued bank drafts in U.S. dollars, which were drawn on the bank's deposits in U.S. banks. The drafts were transported to the U.S. in armored cars and deposited in cooperating U.S. banks, thus making the now "clean" funds available to be withdrawn and used for purchasing legitimate businesses. The operation ended with over 130 arrests and $100 million seized (PBS, 1999). Further investigation revealed as much as $1.5 billion in U.S. bank accounts of the Defense Secretary and "Drug Czar" of Mexico and other high officials, including the president of Mexico. The complicity of such high officials and U.S. banks was not pursued.

The extent of the total corruption of the system can be illustrated by the fact that the Mexican "Drug Czar" lived in an apartment paid for by Carrillo. This same person met with U.S. Drug Czar Barry McCaffrey to share intelligence, after which many U.S. informants in Mexico were found murdered. Despite all of the high-level complicity in drug trafficking, Congress has not seen fit to decertify (decide not to share intelligence and cooperation) with Mexico.

Source: Charles Bowden. 1998. *Juarez: The Laboratory of Our Future.* New York: Aperture Foundation; Terrence Poppa. 1998. *Drug Lord:* A True Story, 2nd edition. New York: Demand Publications; PBS. 1999. "Lords of the Mafia: Mexico" (telecast); U.S. Treasury Department. 1998. "Operation Casablanca Continues Its Sweep." *Office of Public Affairs.* May 20. (http://www.fas.org/irp/news/1998/05/pr2467.htm).

Web Research Project

Do a search on drug trafficking and Mexico. What recent trends have taken place with respect to such activity in that country? What has been the response of the United States to such developments?

In 2008, over 6,200 Mexicans died in drug-related killings. Upon taking office in 2006, President Felipe Calderón deployed 40,000 troops and 8,000 federal police to fight the drug cartels. Reportedly, 90 percent of the weapons seized in the drug wars have been traced to the U.S. The sheer growth of such cartels has spread to the U.S. with gang members in over 200 U.S. cities. The huge U.S. demand for drugs and the fantastic profits to be made has created a monster and the biggest U.S. organized crime problem at the present time. The Mexican cartels are the biggest business in Mexico and their profits are estimated to exceed the U.S. defense budget for the Iraq War. Earning more than the Mexican government, the cartels use their profits to bribe politicians, judges, the police, the military, and other public officials. The Mexican police have historically been poorly paid and encouraged to supplement their salaries with *mordida* ("the little bite"—bribes) (Lacey, 2009, A12).

Mexican cartels control 90 percent of the cocaine market in the U.S. and most of the market in other drugs. Some of the major cartels are: the Sinaloa cartel, the Juarez cartel, the Tijuana cartel, the Gulf cartel, Los Negros, Los Zetas, and LaFamilia. Much of the violence is between these rival groups as well as with the government. The two key contenders are the Tijuana and Gulf cartels versus the Sinaloa and Juarez cartels. The armed wing of the Gulf cartel is Los Zetas, which is countered by Los Negros, the armed wing of the Sinaloa cartel. Both of these consist of former elite paramilitary.

Theories of the Nature of Syndicate Crime in the United States

Jay Albanese (2004) describes three models or paradigms of organized crime that exist in the literature in the field:

1. Conspiracy theory—organized crime as a nationwide conspiracy

2. Organized crime as local, ethnic groups

3. Organized crime as enterprise (p. 96)

The conspiracy theory is what this writer calls "cosa nostra theory," while the local ethnic groups I will call "patron theory." Enterprise theory, as first proposed by Dwight Smith (1975, 1978), argues that organized crime and normal business are similar activities on different ends of a "spectrum of legitimacy." Organized crime represents an extension of the principles of legitimate business in illicit areas (Albanese, 1989, p. 97). Crime File 12.2 examines the legend of the Mafia.

The Cosa Nostra Theory (The Cressey Model)

The Cosa Nostra theory is a theory of the organizational structure of syndicate crime that has the following main proponents:

1. Interpretations of the testimony of informant Joseph Valachi before the McClellan Commission in the sixties, in which the term "La Cosa Nostra" was first officially introduced.

2. The organized crime section of the President's Crime Commission Report of 1967 and theoretical interpretations of its principal consultant, sociologist Donald Cressey.

3. Official although belated policies of federal agencies such as the Federal Bureau of Investigation.

Crime File 12.2

The Origin of the Mafia

In an opening quotation for this chapter, informer and ex-mobster Joseph Valachi, during his testimony before the McClellan Commission, denies that "Mafia" is the name of the organization to which he belonged. As early as 1890, a grand jury investigating the murder of New Orleans police chief David Hennessey concluded that a secret criminal group, the Mafia, was responsible (Albini, 1971, p. 167); the existence of an organization by that name was assumed rather than proven.

Origin of the term *Mafia*

The origin of the term *Mafia* is often assumed but undocumented (that is, without sources referenced). Joseph Albini in his *American Mafia: Genesis of a Legend* (1971, pp. 83–106) notes some of the more commonly cited origins:

Maffia (Tuscan for *misery*)

Mauvias (French for *bad*)

Ma-afir (Arab tribe that settled in Sicily)

MAFIA (**M**azzini **A**utorizza **F**urti **I**ncendi **A**uvelenamenti—Mazzini Authorizes Arson, Thefts, and Poisons)

Mu'afy (Arabic for *protect from death in the night*)

MAFIA (Battle cry during the Legend of Sicilian Vespers—a revolt against the French in 1282: **M**orte **A**lla **F**rancia **I**talia **A**nela— "Death to all French is Italy's cry.")

Mafia (The name of a stone quarry in Sicily)

I Mafiusi di la Vicaria (A popular play by Guiseppe Rizzotto in 1860, *The Heroes of the Penitentiary*)

Of interest, but not mentioned by Albini, is Ma Fia (my daughter) (cited in Talese, 1971, p. 184). On the basis of extensive research on the subject, Albini concludes that the 1860 Rizzotto play is the most likely explanation. The play, which dealt with life among Cammorristi (organized and professional criminals) in a Palermo prison, was very popular; it was later released simply with the title *I Mafiusi,* by then a well-known term. This might explain the fact that the term was not popularly known before 1860, while after this period it became almost a synonym for organized crime. Thus, rather than being the name of an organization, Mafia refers to a method—a syndicate-type organized crime.

Web Research Project
Do an online search using the phrase "Italian Mafia." What new developments have taken place in Italy?

The major elements of "Cosa Nostra theory," as described by Cressey and the Organized Crime Task Force, included the following:

1. A nationwide alliance of at least 24 tightly knit "families" in the United States.
2. Membership that is exclusively of Sicilian or Italian descent, and the organization is referred to as Cosa Nostra particularly by East Coast members. The title of a book by Nicholas Gage (1971) reflects the ethnic exclusivity: *The Mafia Is Not an Equal Opportunity Employer.*
3. The names and criminal activities of approximately 5,000 participants have been assembled and the formal structure (see Figure 12.1) has been pieced together based on Valachi's testimony.

FIGURE 12.1 ▪ Internal Structure of La Cosa Nostra Families

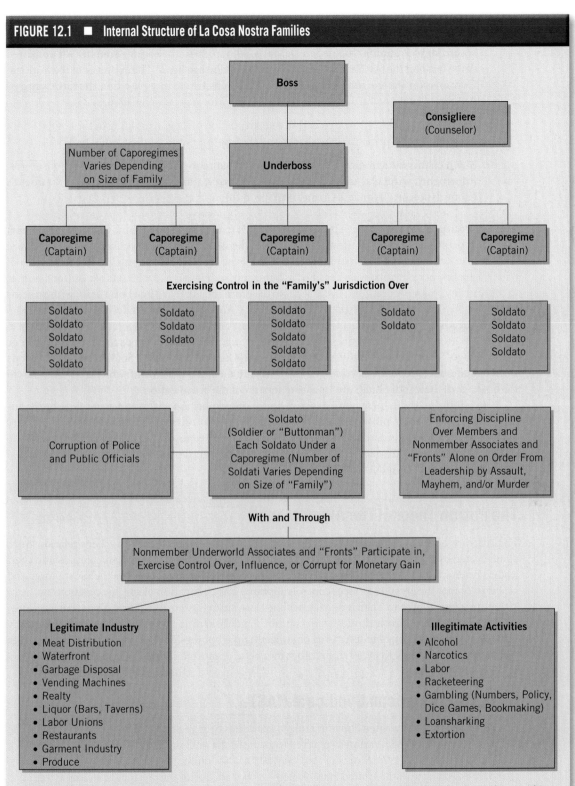

Source: U.S. Senate Permanent Subcommittee on Investigations (1980, April), Committee on Governmental Affairs, Hearings on Organized Crime and Use of Violence, 96th Cong., 2d Session, p. 117.

4. Overseeing the Cosa Nostra is a National Commission made up of the dons (heads) of the most powerful families in the United States. (Originally consisting of 10 to 12 members, according to Fratianno [quoted in DeMaris, 1981, p. 294], the commission in 1981 consisted of the heads of the five New York families plus the Chicago boss.) The existence of the National Commission was corroborated by means of an electronic bug placed in the dashboard of Anthony "Tony Ducks" Corallo's Jaguar (Powell, Emerson, & Kelly, 1986, p. 25).

5. La Cosa Nostra (LCN) controls all but a small portion of illegal gambling in the United States and contains the principal loan sharks and the importers and wholesalers of narcotics.

6. Much of this information is the result of detailed reports of a variety of police observers, informants, wiretaps, and electronic bugs (Cressey, 1969, pp. 99–107, 241–42; President's Commission on Organized Crime, 1967, pp. 6–8).

The description of the internal structure of the LCN in the President's Crime Commission Report was based primarily on Valachi's testimony. Each of the 24 families was described as varying in size from as many as 700 to 1,000 men to as few as 20. Only New York City had more than one family, and it had five. Family organization was described as being rationally designed with sets of positions similar to those in any large corporation. The LCN chain of command is headed by a boss *(don)*, with an advisor *(consigliere)* and underboss (sort of vice president). Answering to the underboss are *caporegimes* (literally, heads of regiments or lieutenants or captains). They are chiefs of operating units or soldiers *(soldati,* "buttons"). "From a business standpoint, the *caporegime* is analogous to plant supervisor or sales manager" (President's Commission on Organized Crime, 1967, p. 451). Soldiers may run various illicit operations on a commission basis or "own" their own operations within which a portion goes to the boss. All of these individuals are "made members" of the organization.

Below and allied with these families are various nonmember associates and employees, individuals who cooperate in and aid organizational operations. Insulation of the boss and other LCN activity is preserved supposedly according to the "oath of omerta": a pledge of loyalty, honor, respect, absolute obedience, manliness, and silence. In the old days, accompanying the initiation was an elaborate ritual in which the novice was inducted into the LCN.

The Patron Theory (The Albini Model)

The "patron theory" views organized crime as consisting of a series of patron–client relationships, as advocated by Albini (1971, 1988). According to this approach, organized crime groups and their leaders resemble a medieval system of shifting warlords in which whoever has the most power and is able to render the greatest services controls support. The occupation of specific positions within a structure is less important than a developmental-association system of peer relations that are informal, flexible, and constantly immersed in conflict. Feudalism rather than the corporate bureaucracy is the appropriate analogy for describing organized crime families, a series of shifting alliances. (See Albanese, 1989, and R. J. Kelly, 1992, for other models of organized crime.)

The Italian American Syndicate (IAS)

Much of what has been written about organized crime has been restricted to an analysis of the Italian American Syndicate (IAS), variously referred to as the Mafia or Cosa Nostra. Critics of this approach (Bell, 1967; Ianni, 1972, 1974; Morris & Hawkins, 1970; D. C. Smith, 1975) have largely made the points that organized crime in the United States is homegrown and is not the product of an imported, alien conspiracy; that no one ethnic group has a monopoly on organized crime; that the picture drawn by the National Advisory Committee on Criminal Justice Standards and Goals (1967), as presented by Donald Cressey (1969) and based predominantly on Joseph Valachi's testimony, is overdrawn; and that it is doubtful that organized crime, or the IAS, which was the most powerful of such syndicates, ever exhibited the extreme bureaucratic, monolithic structure depicted. These critics do not, however, dispute the existence of organized crime and criminal syndicates as distinct phenomena, nor do they dispute that an IAS (Mafia, Cosa Nostra) is more than a creation of moral entrepreneurs or Hollywood.

While writers such as Albini (1988) maintain that "most research has not lent support to the Cressey model [Cosa Nostra theory]" (p. 350), Rogovin and Martens (1989, p. 11) note that evidence produced in recent years, as well as research and informants, has corroborated Cressey many times over. They (Rogovin & Martens, 1992) note,

> The volumes of evidence produced over the past eight years, as well as the books that have been published by researchers, journalists, investigative reporters, private citizens, and members of La Cosa Nostra have corroborated Cressey over and over again. Why ignore Bonanno's (1983) autobiography of his life in LCN? There is a great chapter on the "Commission" and its model of operation. Read Fratianno's account of his life in the LCN (DeMaris and Fratianno, 1981). Seek out the wiretap product which was made public in the Commission trial in New York. . . . Read the transcripts of the Scarfo trials in Philadelphia and critically evaluate the testimony of two LCN members who became state's witnesses (Cooney, 1987; Mallowe, 1988). Obtain the transcripts of the trial of Nick Civella in Kansas City (Turner, 1983, p. 30); the testimony of Cleveland LCN underboss Angelo Lonardo before the Permanent Sub-Committee on Investigations (P.S.I., 1988); or the trial testimony of Gennaro Anguilo in Boston. Surely, the infamous "Pizza Connection I" trial (Alexander, 1988; Blumenthal, 1988); the reporting of Jimmy "the Weasel" Fratianno (DeMaris and Fratianno, 1981) or Joseph Bonanno (1983), Paul Meskil (1976), and Thomas Renner (Teresa and Renner, 1973; Renner and Giancana, 1984) are relevant pieces of research literature. . . . Testimony in these trials admitted the existence of a Mafia (Albanese [1989, p. 66]). (p. 70)

The Classic Pattern of Organized Crime

In their book *The Crime Confederation* (1969), Ralph Salerno and John S. Tompkins describe the "classic pattern of organized crime" as a gradual movement from "strategic and tactical crimes" such as assault, bribery, and extortion to "illegal businesses and activities" to "legitimate businesses" to "big business." Due to their willingness to commit and employ strategic and tactical crimes, organized gangs are thereby able to acquire both the funds and the power to be fairly successful at illegal businesses and activities. They are, of course, not the only types of criminals engaged in these activities, but they are more organized, more persistent in their efforts, and simply better at it. The types of crimes committed under strategic and tactical crimes are for the most part disapproved of by the general public, while the illegal businesses and operations often serve public demand for vices and other illicit activities. Unwittingly the government, by branding much of this activity illegal—such as narcotics and gambling—creates a monopoly for criminal groups that are organized well enough to supply these goods and services.

A major problem of some "unconnected" or nonorganized criminals is laundering of funds obtained in illegal operations. Who brought down Al Capone? Not Elliot Ness with a Tommy gun, but Frank Wilson of the Internal Revenue Service with pencils, ledgers, and a green eyeshade (Marbin, 1989, p. Al). In the earlier history of U.S. organized crime, even "connected" figures such as Al Capone were convicted of income tax violations, and this lesson was not lost on other organized criminals as many began trading in their black shirts and white ties for Brooks Brothers suits. With funds obtained in illegal operations, organized criminals can infiltrate legitimate businesses, an even more fertile field for their activities. Finally, experience in such operations enables movement into even bigger businesses.

Strategic and Tactical Crimes

The strength of organized criminal groups is based on their willingness and ability to use force or threats of force to assure discipline within and outside the organization. Although activities such as assault, coercion, extortion, and murder are not the exclusive property of organized criminals, they seem to be used more frequently by this group than by most other types of criminals.

Assault, coercion, extortion, and murder are bottom-line tools employed by organized crime, weapons of last resort to assure the "rational" pursuit of profit. "Make him an offer he can't refuse" is black humor, but it is an all-too-apt phrase to describe methods used by organized criminals to accomplish their will. Violence and threats of violence; demands for protection money (extortion); and, when necessary as a last resort, murder ("making your bones") is part of the repertoire. Decades of internal wars and assassinations within the ranks of organized crime attest to the fact that outsiders are not the only victims of mob discipline.

Blackmail, bribery, and corruption are ssential strategic tools of organized crime. The American Bar Association (1952) in its *Report on Organized Crime* concluded, "The largest single factor in the breakdown of law enforcement dealing with organized crime is the corruption and connivance of many public officials" (p. 16). Blackmail is more easily achieved by organized crime figures because of their involvement in gambling casinos, pornography, and servicing of vice-related activities. Bribery and corruption of public officials make up the largest operating cost of organized criminal groups, a sort of "underground" tax or license to steal. Organized crime has been so successful in corrupting public officials that cities, counties, and entire states have been "in the bag." Success in strategic and tactical crimes provides the money, "muscle," and "respect" for success at principal illegal businesses and activities. The strategic and tactical crimes are for the most part "rackets," services that lack or do not require public demand. Organized criminal involvement in illicit enterprises or illegal businesses satisfies public demand for services or vice activities that either cannot be or are not met by legal businesses. Thus, the loan shark is a banker of last resort, and the fence is a less expensive shopping center.

Illegal Businesses and Activities

Audio Link 12.3
Listen to a discussion of organized crime and health care.

From the end of the Prohibition Era until recently, gambling has been viewed as the number one moneymaker for organized crime. Although often described as a Depression or ghetto invention, lotteries flourished in the United States as far back as colonial times, used during that period to pay for public works (National Institute of Law Enforcement and Criminal Justice, 1977). In addition to lotteries, syndicates are involved in other gambling activities, such as bookmaking (taking bets on sporting events), illegal gambling devices (slot machines, punchboards, sports polls), and the running of illegal gambling establishments. The **numbers game** is by far the most popular of these activities. In this game, sometimes called policy or "bolita" in the southern United States, the basic strategy is similar to that of legalized state lotteries: The bettor tries to choose a winning three-digit number. Traditionally, number-selection methods differed from those of lotteries: numbers were chosen that matched the numbers of win-place-show horses at a track or Dow Jones averages. State lotteries choose numbers by spinning a wheel or by some other "honest" means.

The advent of legal state lotteries may not have made the dent in the illegal numbers business that many had hoped. Many numbers writers simply use the legal daily number as their own, increasing the nontaxed payoffs and sometimes even laying off bets with the state. In the latter practice, they simply replay the same bet with the state in case it has too big of a run on certain numbers. Illegal casino gambling still exists in many American communities. Dice or craps games, wheels, and high-stakes card games can often be found by simply asking local cabbies, "Where's the action?" Reuter, in his book *Disorganized Crime* (1984b), is quick to point out that the Mafia control of illegal markets is exaggerated and that the numbers game, loan sharking, and bookmaking are very disorganized businesses.

Photo 12.3

Betting on horses is part of the "numbers game," which is the most popular gambling activity.

Related to gambling operations is the **loan shark** or shylock (the latter name is derived from Shakespeare's character Shylock in *The Merchant of Venice*, a money lender who demands his "pound of flesh" as repayment). Loan sharks provide quick loans on the spot to borrowers who are either high risk or in a tight spot. These loans are given at usurious (illegal and exorbitant) rates. Although rates vary, a typical loan might be a "six for five" arrangement. That is, for every five dollars borrowed, six dollars must be repaid (20 percent interest) per week. Sharks are more interested in collecting periodic interest payments (called "juice," "vigorish," or "vig") than in having the loan paid off. Often borrowers give their bodies as collateral since, if they were good risks or could share their problem, they could have gone elsewhere. Gambling is the usual manner in which the successful business person, but poor gambler, is introduced to loan sharks and, if his or her luck does not improve, he or she may have some new business partners. Crime File 12.3 describes organized crime's growing involvement in software piracy.

The Pennsylvania Crime Commission (1992) reports on organized crime infiltration of legitimate, charitable, and Indian reservation bingo games. Racketeers and mobsters were described as practicing fraud, misrepresentation, and diversion of monies from such games and using charitable organizations as fronts for skimming millions of dollars. Ironically, the Pennsylvania Crime Commission itself was dissolved when it began to investigate illegal campaign contributions by gambling operators to the Pennsylvania Attorney General Ernie Preate, who later served time in jail for his activities.

Other successful illegitimate activities of organized criminals involve labor racketeering; narcotics trafficking; prostitution; pornography (although now much of this is legal); stolen property (such as cars, stocks, and bonds); videotape, film, and record piracy; and even illegal sale of alcohol. Organized criminals will involve themselves in any scam (illegal activity) so long as it is relatively safe and profitable (Kwitny, 1979). In 1997 in Russia and Georgia (the country), it is estimated that one in

Crime File 12.3

"Snakeheads" and Software Mobsters

"Snakeheads" (human traffickers), a group of Asian mobsters affiliated with Chinese organized crime groups, have traditionally specialized in illegal immigrant smuggling. They sometimes severely mistreat or even kill their clients by putting them in containers without adequate food or water, or they drop them off too far at sea. Counterfeit software has moved beyond small-time hackers and has become more dangerous with the entry of Asian organized crime. The Business Software Alliance estimates that the U.S. economy loses 130,000 jobs annually to software theft. The international piracy trade is described by Panattieri (1999) as follows:

> First, an organized-crime operation makes millions of dollars pushing drugs and prostitution in Asia. Next, the money is diverted to California, where it is used to purchase hardware, software, and paper goods to produce pirated CDs and user manuals. Third, the counterfeit software is distributed at computer "swap meets" held regularly on college campuses across the nation; shipped abroad; or offered for sale on the Internet. Finally, the proceeds from the pirated software are laundered through real estate purchases in California or illegally wired back to Asia.

Snakeheads, for a $30,000 fee, smuggle illegal aliens into the United States. Some clients become indentured servants, producing illegal software in sweatshops in order to pay off their debt. The multinational nature of such operations produces a real challenge for law enforcement.

In investigating what they call the "software sopranos" or the "digital dons" in California, the Justice Department reports that besides the Snakeheads, the Black Dragons and Wah Ching ("Chinese Youth") gangs are also involved in product counterfeiting. Also operating are Russian groups, as well as a Vietnamese group called "The Company" that was involved in 30 armed robberies of electronics firms (J. Glasser, 2000).

Source: Joseph C. Panattieri. 1999. "The Software Mobsters." *Interactive Week Online.* September 15 (http://www.zdnet.com); Jeff Glasser. 2000. "The Software Sopranos." *U.S. News and World Report.* February 7.

Web Research Project
What has been the controversy related to FAST (the Federation Against Software Theft)?

---

five dwellings produced illegal alcohol. Organized criminal groups, sometimes protected by customs personnel, deliver the products in small ships (Konstantinova, 1997). The mortality rate from the consumption of illegal alcohol in the former Soviet Union is very high.

Labor **racketeering** is the infiltration of unions to use their influence for personal profit. Such operations, which may take the form of bribes, kickbacks, and extortionary threats, permit mobsters to use pension funds and to offer "no-strike insurance" (the guarantee that workers will not strike) and sweetheart contracts (collusion between the employer and union officials at the expense of workers), as well as other operations. Crime File 12.4 gives an account of James Jacobs's book *Mobsters, Unions, and Feds* (2006). Mob infiltration of the American labor movement has not been significantly addressed by scholars in the past.

Four unions with historically substantial organized crime control and influence are the International Longshoremen's Association, the Hotel and Restaurant Employees Union, the Laborers International Union of North America, and the International Brotherhood of Teamsters. Construction costs in New York City are estimated to be 25 percent higher because of the need for organized crime payoffs, and garbage collection is $50 million per year more because of mob control (Powell et al., 1986, p. 28). Mob-controlled Teamster locals 295 and 851 enabled Anthony "Tony Ducks" Corallo to shake down air transport service companies for $1.1 million between

1978 and 1985 (Rowan, 1986, p. 34). In 1990, the federal government charged that for 20 years Nicodemo Scarfo, head of the Philadelphia Mafia, had been running the 22,000-member Local 54 of the Hotel and Restaurant Employees International Union from his cell in prison. The local chapter included Atlantic City casinos (A. Hagedorn & Lambert, 1990, p. B6). "The Outfit" in Chicago charges a "street tax" on all illegal activities. Gamblers, vice operators, even owners of parking lots and legitimate businesses must pay 10 to 50 percent of their gross revenues to the Chicago mob. Crime File 12.4 gives an account of James B. Jacobs's book *Mobsters, Unions, and Feds* (2006; see source note further below). Mob infiltration of the American labor movement has not been significantly addressed by scholars in the past.

Crime File 12.4

Mobsters, Unions, and the Feds

In *Mobsters, Unions, and Feds* (J. B. Jacobs, 2006) and previous works such as *Busting the Mob: The United States v. Cosa Nostra* (J. B. Jacobs, Panarella, & Worthington, 1994) and *Gotham Unbound* (J. B. Jacobs, Friel, & Radick, 1999), James Jacobs tells the tale of labor racketeering as an example of American exceptionalism: "No other country has a history of significant organized crime infiltration of the labor movement, and no other country has an organized crime syndicate with a power base in the labor unions" (p. xi). The mob regularly used union connections to establish and enforce employer cartels that fix prices, control contracts, and suppress competition. Surprisingly, this has not attracted much attention from academic and legal scholars. While there is a subfield of white collar crime, there has not been one devoted to union crime.

The 1957–1959 Senate McClellan Commission Hearings on labor racketeering were the most extensive and largest congressional investigation in U.S. history. Despite this, little attention has been given to the corrosive impact of mob ties on organized labor's dwindling power. Most of the exposure of labor racketeering was left to investigative journalists. While academic scholarship on unions is thin, much can be mined from congressional hearings and reports.

The racketeers' basic modus operandi includes looting union treasuries and pension funds by theft, fraud, and bloated salaries; selling out union members' rights and interests in exchange for employers' bribes and kickbacks; exploiting union power to extort employers; and conspiring with employers to operate cartels that allocate contracts and set prices. Labor racketeering serves the organized crime families as a bridge to the power structure in many American cities (J. B. Jacobs, 2006, p. 2).

For most of the twentieth century, local law enforcement was either too corrupt or incapable of fighting organized crime. The FBI under Hoover denied its existence. When federal law enforcement finally decided to take on labor racketeers, the latter suffered major defeats. Not all unions were highly susceptible to mob takeover. Most vulnerable were those connected to small employers. They were less able to organize against gangsterism and more easily intimidated and subdued. The earliest unions to be infiltrated were the restaurant workers, coach and truck drivers, and construction workers. Large industrial unions were less successful targets for organized crime.

Federal law enforcement efforts during Hoover's FBI were more interested in suppressing communist influence in unions and did not oppose mob influence. In 1986, Reagan's President's Commission on Organized Crime (PCOC) issued its report. One of the volumes titled *The Edge* addressed specifically organized crime's exploitation of labor unions. Organized crime's influence over labor unions provided businesses that were owned, dominated, or favored by organized crime with an edge over competitors (J. B. Jacobs, 2006, p. 42). Organized crime still controls local and regional unions and wields power in several important international unions.

Sources: Jacobs, J. (2006). *Mobsters, Unions, and Feds: The Mafia and the American Labor Movement.* New York: New York University Press; Jacobs, J., Friel, C., & Radick, R. (1999). *Gotham Unbound.* New York: New York University Press; Jacobs, J., Panarella, C., & Worthington, J. (1994). *Busting the Mob: United States v. Cosa Nostra.* New York: New York University Press.

Research Project
Using a Web browser, locate some recent articles on labor racketeering.

The largest illegal business of organized criminals is now drugs. In 1986, the President's Commission on Organized Crime (1986a) estimated that organized crime in the United States took in as much as $106.2 billion, and by far the biggest moneymaker was illegal drug trafficking. Success and money from illegal operations, although welcome, present organized criminals with potential tax problems, further encouraging them to move into legitimate businesses. Such businesses provide many opportunities for organized crime. They provide a source of legal income that can help explain gangsters' high lifestyles. Because of their methods, criminals can monopolize markets and make more money than competitors. Such businesses also yield a "cover," or respectable occupation, as well as a base of operation and a meeting place, particularly for dealing with public officials. They enable the "washing" or "laundering" of funds and provide a diversification of operations.

Favorite businesses of organized criminals include auto sales, bakeries, clothing manufacturing, construction and demolition, import/export, garages, hotels, vending machines, produce, trucking, bars and restaurants, garbage collection, and the like. Businesses such as vending and bars are fertile ground for "skimming" (hiding or not counting money earned, for tax purposes). One hundred dollars skimmed every day from a busy bar would amount to over $30,000 a year, tax free, from just one business. Extortion and monopolization in vending businesses enable organized criminals to force out competitors.

Ianni in *A Family Business* (1972) suggests that the seeds of many American fortunes began with "dirty business" and progressed in a couple of generations to "respectable" business, a natural "ethnic succession" and progression. In discussing federal enforcement, Ianni is concerned that this progression not be entirely blunted. In the main, however, organized criminals in such enterprises often carry over all the same illegal techniques.

With major crackdowns on traditional organized crime activities such as extortion or bid-rigging rackets, Mafia crime groups in New York were shifting some of their focus to health frauds, prepaid phone cards, and Wall Street scams. Raab (1997) indicates,

> The authorities in New Jersey said they uncovered what might be the prototype of the mob's medical care strategy in August, when they arrested 12 men accused of being members of a Genovese crew, or unit. The crew's leaders were charged with siphoning payments from Tri-Con Associates, a New Jersey company that arranged medical, dental, and optical care for more than one million patients in group plans throughout the county. Investigators said that the mobsters set up Tri-Con, investing their own money and using employees as managers, and intimidated some health plan administrators into approving excessive payments to the company. New Jersey authorities said Tri-Con in effect became a broker, linking networks of health-care providers, including physicians, hospitals, and dentists, with group plans for companies and unions. (p. A1)

Prepaid phone cards were grossing $1 billion in the United States in 1996 and provided a new target for organized crime. The Gambino crime family stole over $50 million from companies and phone callers by selling $20 cards that became worthless after only $2 or $3 in calls (Bastone, 1997). Other New York City crime families have infiltrated Wall Street, particularly over-the-counter stocks handled by small brokerage firms. Brokers who are in debt or wish to expand their businesses borrow money from the mob. They are then forced to sell most of the low-priced shares in a company before they are available as initial public offerings. The value is artificially inflated by fake transactions and trading among themselves. At the same time, brokers push the stock on unsuspecting investors. The mobsters then sell, making high profits before the overvalued stock collapses (Bastone, 1997).

Of major concern to law enforcement officials is the burgeoning growth in transnational smuggling of illegal aliens from underdeveloped to developed countries. Tens of thousands from the former Soviet Union, Asia, and Eastern Europe are trafficked each year, often unknowingly into forced prostitution in developed countries. They work as indentured slaves in the global sex industry in bars, massage parlors, and brothels. Most are attracted through deceit and coercion and find themselves without visas or passports in foreign lands. Human rights groups estimate millions of women and children are forced into such lives of criminal exploitation. Such statistics are dubious, however. The U.S. State Department had estimated that 50,000 to 100,000 women and children are forced into the United States each year against their will. These statistics were later seriously questioned and downgraded to less than 17,500 in 2007 (Markon, 2007). In Europe, Albanian clans are middlemen for human smuggling operations of Russian and Chinese organized crime. The lack of strict laws in some countries reflects a patriarchal culture that denigrates women (Fleishmann, 2000).

Incredibly, in something that sounds like a James Bond movie, in 1997 police investigations revealed that Russian organized crime figures in Miami Beach had claimed to Colombian drug cartel members that they could supply them with tactical nuclear weapons, as well as a submarine. Many of the Russian organized crime figures are believed to be ex-KGB members.

Big Business and Government

Success at small business permits mob infiltration of *big business,* the heart of our nation's economy. We have already suggested the impact of mob influence on big labor unions such as the Teamsters and the Longshoremen. Banking, construction, entertainment, insurance, real estate, and even Wall Street are not immune. By the 1990s, the mob had moved from shaking down indebted stockbrokers to stock price manipulation. In a classic "pump and dump" operation, brokers used high-pressure sales tactics to pump up the price of a stock they owned and then the shares were dumped before their worth plummeted.

Given the classic pattern of organized crime, we must ask the question, "What remains?" Only government. Can or does organized crime have the capability of compromising the government itself? In "Operation Mongoose" (Ashman, 1975), the Central Intelligence Agency used syndicate criminals to put a "hit" on Cuban premier Fidel Castro in 1963. Although apparently a scam on the part of the mob in that no serious attempts took place (DeMaris, 1981, p. 267), the deal apparently called for cooperation by the CIA in smuggling prostitutes from Marseilles to staff mob brothels in Las Vegas ("Gangland Enforcer Paid With Life," 1977). One principal figure, John Roselli, who hinted at tie-ins with the Kennedy assassination before the House Assassination Committee, was killed by the Mafia before he could testify further (J. Anderson & Whitten, 1977). The U.S. House of Representatives Select Committee on Assassinations (1979) concluded that there was a conspiracy in the assassination of John Kennedy as well as possible conspiracies in the assassinations of Robert Kennedy and Martin Luther King, Jr. Chief counsels to the committee, G. Robert Blakey and Richard Billing (1981), more specifically point the finger, as the title of their book indicates: *The Plot to Kill the President: Organized Crime Assassinated JFK* (see also Scheim, 1988).

Although it is not proven, it is alleged that the Mafia felt betrayed by John Kennedy. They claimed that they were responsible for getting him elected by stuffing ballot boxes in Illinois. When Kennedy became president and appointed his mob-busting brother, Robert, to the post of Attorney General, a crusade against the mob began, particularly on associates such as Teamster leader Jimmy Hoffa. This campaign did not sit well with his "subordinate," FBI Director J. Edgar Hoover. Although not proven, it has been charged that Hoover avoided any efforts against organized crime because he had been compromised, having received favors from gamblers, and perhaps having been blackmailed regarding an alleged secret sex life. Columnist Jack Anderson and others (Anderson & Whitten, 1977) claimed that JFK was assassinated for two reasons: first, in retaliation for assassination attempts on Castro (Santo Trafficante, one of the would-be assassins, may have defected), and second, as a way of removing Robert Kennedy from power. This is, of course, all speculative, but the day after the assassination of President Kennedy, Hoover, the nation's top law enforcement officer, spent the day at the racetrack.

Concerns have risen regarding organized crime and drug dealers' infiltration of the Medicare system, a $250 billion-a-year business already rife with rip-offs by other crooked operators. Many moved out of the drug trade into the safer and more lucrative medical swindle business. They have set up thousands of fake clinics, medical equipment stores, and laboratories, and they use a maze of bank accounts and offshore accounts to move their money. It is a particular favorite of Russian organized crime groups (Hedges, 1998).

As a testimony to the power of organized crime groups, shortly after the arrest of Cali cartel leader Gilberto Rodrigues Orejuela, Ernesto Samper (1995), then president of Colombia, stated, "In the past decade, Colombia has lost countless lives, including more than 3,000 police officers and soldiers, 23 judges, 63 journalists, and four presidential candidates" (p. A16).

The infiltration of organized criminal groups into large business enterprises is also assisted by activities of legitimate organizations themselves, some operations of which resemble those of organized crime. Bribery and corruption of national and international public officials, violence either indirectly through sale of unsafe products or directly in deposing foreign leaders such as Allende in Chile, and pushing of drugs far in excess of the medicinal needs of consumers, are just a few such examples. A detailing of criminal activities of organized crime syndicates, and particularly the IAS, is not intended to ignore their corporate counterparts, but to recognize the former as having distinct characteristics of their own. Similarly, while the IAS did not invent and does not control all of organized criminal operations, it has been the most powerful of such groups in the United States since the 1930s.

A Brief History of Organized Crime in the United States

Video Link 12.1
View a video on the St. Valentine's Day Massacre.

A detailed account of the history of U.S. organized criminal activity is beyond the scope of this text. Readers are advised to consult sources cited in the reference list, particularly Abadinsky (2006), Talese (1971), Hammer (1975), MacLean (1974), and Gage (1972). However, a brief account will familiarize the reader with some key events in the history of organized crime. Table 12.3 contains a short chronological list.

Before 1930

Organized crime had its beginnings in the New World with colonial pirates, former naval mercenaries working for England in her war against Spain. By the end of the seventeenth century, they were an institutionalized component of international trade, intimately tied up with the business and governmental systems of the time. "The pirates, it is clear, were the racketeers of their day, bribing officials, corrupting entire governments and looting to maintain a vast underworld market in forbidden goods" (Browning & Gerassi, 1980, pp. 71–72). Organized crime appeared to be an intimate component of American cities from their beginnings, with "robber barons" or "industrial pirates" looting the landscape in early capitalism (G. Myers, 1936) and criminals, police, and politicians cooperatively running illicit enterprises in order to satisfy public demand for vice activity. The Irish and Anglo-Saxon street gangs in nineteenth-century New York formed organized criminal groups, just as a later generation of mostly American-born street hoodlums of Italian descent would form the most successful prototype of American syndicates, the IAS.

The view has already been espoused in this chapter that organized crime in the United States existed long before major Italian immigration, being dominated in its early history by small local mobs of WASP origin. Also described was the 1890 "New Orleans Incident" in which a member of a rival Italian American criminal brotherhood killed the city's police chief. The aftermath of that incident was a grand jury report naming a secret criminal group, "Mafia," as responsible. Due to a believed fix in the case, angered citizens stormed the parish prison, executing a number of the gang leaders and almost precipitating a war between the United States and Italy (Albini, 1971, p. 167).

TABLE 12.3 ■ Chronology of Selected Events in the History of Organized Crime in the United States

1700s	Colonial pirates
1800s	WASP, Irish, Jewish gangs
1890	The New Orleans Incident
1920	The Volstead Act (Prohibition)
1930–1931	Castellammarese War in New York City
1931	Would-be "capo di tutti capi" Maranzano murdered
by 1934	Cosa Nostra confederation established
Post-WWII	Mob moves into Las Vegas
1950	Kefauver Committee Hearings
1957	The Apalachin "Gangland" Convention
1963	McClellan Commission Hearing, featuring star witness Joseph Valachi
1970	Organized Crime Control Act passed
1971	Mob boss Joe Colombo shot at Columbus Day rally
1976	Carlo Gambino dies
1979–1980	Mob wars continue (Galante, Bruno, Testa)
1980–1981	FBI round-up of top Cosa Nostra bosses begins
1983–1986	President's Commission on Organized Crime
1983–1987	"The Commission Trials"
1985	Paul Castellano assassinated
1985–1986	"The Pizza Connection"
1985–1986	"The Great Mafia Trial" in Sicily
1992	Sicilian Mafia kill Falcone and Borsellino
1992	John Gotti convicted
1993–1997	Federal prosecutors cripple Cosa Nostra leadership; new ethnic organized crime groups move into power
2000	Remnants of Mafia survive in New York, Chicago, Philadelphia
2001–2006	New mobs become dominant

Up until 1920, most organized crime was confined to relatively small, local mobs whose operations were not particularly sophisticated. Most were controlled by Irish and Jewish gangsters, although a number began to include as "muscle" a growing number of hoodlums of Italian and Sicilian descent. Prohibition was an absolute bonanza for organized crime, the one factor that made it possible for fledgling gangs to become financially successful syndicates. Frank Costello was one of the first Italian American gangsters to make it big.

Photo 12.4

Frank Costello (1891–1973) was one of the first Italian American gangsters to make it big.

George Wolf (1975, pp. 27–29), Costello's lawyer and advisor, described the Prohibition period before 1924 as one in which bootleggers such as Costello organized a virtual naval flotilla with which to smuggle liquor into the United States. The few Coast Guard boats were unable to compete with the much faster skiffs, which raced their contraband from freighters anchored beyond the 3-mile U.S. territorial limit. When the Coast Guard interdicted supplies, newspapers and the public berated them as "pirates." An $8 case of Scotch from the British Isles was sold for $65 on the freighters; $120 at dockside; and, once doctored with three times as much grain alcohol and water, brought around $400. Since many ships carried 20,000 cases, Costello's wealth eclipsed that of his former mentor, Arnold Rothstein. "And along with his ships and yachts Frank introduced a new element into bootlegging, one that was to bring gasps of surprise in the courtroom in 1926: seaplanes for air coverage . . . the air wing of Frank's defense department" (G. Wolf, 1975, p. 29).

Costello, who lived in the Waldorf Astoria in Manhattan, would dine daily with city bigwigs and had so compromised and corrupted city officials that he claimed to possess an authentic, handwritten resignation letter penned by then-Mayor William O'Dwyer, a document that could be turned in any time the later-discredited mayor did not keep his end of the bargain. Naval and air operations by organized crime in the twenties bear an uncanny resemblance to the Colombia runs in Florida that were begun in the late seventies by drug smugglers in what has been called "the Colombian Connection," "the Cocaine Cowboys" (1980), and "Air Ganja" (Plate, 1975, p. 119).

It was Chicago in the twenties, the person of Al "Scarface" Capone, and operations like the St. Valentine's Day massacre that caused a stereotyped picture of the mob to be drawn in the public media. Meanwhile in New York, two old **moustache Petes** or "greasers" (names given to old-time, Italian-born Mafiosi) were involved in a power struggle. The lineup of contending factions reads like a Who's Who of figures who would dominate LCN for decades to come. Initially aligned with the Joe "The Boss" Masseria faction were Luciano, Genovese, Adonis, Anastasia, Costello, Gambino, and, through Masseria's financial assistance, Capone. The rival faction, Castellammarese (named after the home town of most of the members, Castellammare del Golfo, Sicily), was headed by Salvatore Maranzano and included Bonanno, Profaci, Lucchese, Magliocco, Gagliano, and Magaddino. Although the **Castellammarese Wars** were far from the bloodbath erroneously described by chroniclers (Block, 1978), they were important in that the aftermath gave birth to the modern syndicate. Basically, Luciano, Genovese, and others double-crossed Masseria and had him killed.

The victor, Maranzano, often described as "the father of the modern LCN syndicate," was a big fan of Julius Caesar and supposedly modeled his organizational structure after the Roman legions. Unfortunately, Maranzano himself was a victim of overweening ambition, picturing himself as "capo di tutti capi" (boss of all bosses). Six months after taking power, he also was killed by Meyer Lansky and "Bugsy" Siegel, who had been hired by Luciano, Genovese, and company.

The Luciano Period

Charles "Lucky" Luciano took power over the new organization, one that would have a continual alliance (sometimes called the confederation) with other ethnic gangs but which itself would remain exclusively Italian. Avoiding the top-boss role, Luciano supported the autonomy of bosses with a commission for settling disputes. This alliance was apparently further consolidated in the thirties, and a special "hit squad," Murder, Inc., was set up by Louis "Lepke" Buchalter; this group's existence would be later revealed by informant Abe "Kid Twist" Reles. Murder Incorporated's first victim, in 1935, was "Dutch" Schultz, who had unwisely been advertising his plan to kill district attorney Thomas Dewey. The mob's kneeing of a popular district attorney would bring the wrath of the nation and law enforcement upon the mob. In 1934, Vito Genovese fled the country in order to escape a murder charge, and in 1937 Luciano himself was sentenced to a 30- to 50-year term for his prostitution business (so much for insulation from prosecution). It was during this period that Frank Costello acted as boss for the imprisoned Luciano. With the repeal of Prohibition, bookmaking and the numbers racket were now the chief operations.

In 1946, in part for cooperation in "Operation Underworld," a program in which the U.S. Navy had enlisted the help of mobsters to prevent sabotage on the docks during World War II (Gage, 1974; Gosch & Hammer, 1974), Luciano was paroled into permanent exile and, although Costello was still acting boss, the Genovese era was about to begin.

The Genovese Period

In a curious series of events, Vito Genovese, who had voluntarily exiled himself in 1934 in order to escape a murder indictment, was not only decorated with Italy's highest citizen award by Mussolini during World War II, but was also involved in other intrigue. In 1943, he hired Carmine Galante to murder the U.S. editor of an Italian-language newspaper that was critical of Mussolini. Picked up for black-marketing stolen Allied supplies and extradited to the United States for the 1934 murder charge, he was miraculously a free man after the chief corroborative witness against him was poisoned in a Brooklyn jail.

The postwar period found the mob moving into casino building in Las Vegas and also the subject of the live, televised Kefauver Hearings. Only the later Watergate hearings would so captivate the public imagination. After testifying before the **Kefauver Committee,** Costello was a marked man and, barely escaping an assassination attempt, decided to retire. Behind this attempt and other murders (for example, the well-publicized murder by Joey Gallo of Albert Anastasia in the Park Sheraton barbershop in Manhattan) was Genovese, who was now consolidating his power—power that would elude him in what was to have been the "coronation," a mob convention planned for Chicago 3 weeks later.

The Apalachin Meetings

Although a major mob meeting was planned for Chicago, Steve Magaddino (the Buffalo boss), whose ill health prevented long travel, offered the Apalachin, New York, estate of one of his members for the meeting site. Besides the recognition of Genovese as top boss and assurances of peace to Costello, the meeting agenda included confirmation of the mob's antidrug policy and the need for new memberships (Bonanno, 1983; Talese, 1971, p. 213). The last was a sore point because, due to the lack of new blood, the vitality of the

Photo 12.5

Vito Genovese (1897–1969) became leader of the Genovese crime family and mentored many future mob bosses.

syndicate was in peril. Suspicions regarding informers and the lack of discipline among American-born recruits led some families to recruit "greenies" from Sicily's *latifondi* or farming areas (Reid, 1970, p. 72).

Before the session ended, there was a police raid, during which many were temporarily held so that their names could be obtained. This spoiled the anonymity the syndicate valued. Even more catastrophic than the Kefauver Hearings, this evidence of the existence of some type of coordinated, national syndicate was now hard to deny (Bonanno, 1983). In 1958, an informant, Nelson Cantellops, assisted in convicting 24 people, including Genovese, Galante, and Joseph Valachi. Having received "the kiss of death" from Genovese in prison (Vito was sure Valachi was a government informant), and feeling that he was marked for execution, Valachi murdered by mistake a fellow prisoner. To save his own skin, he became the first public "made member" informant from the ranks of organized crime and the star witness at the McClellan Commission Hearings on organized crime.

The Gambino Period

Audio Link 12.4
Listen to a discussion of the charges against the Gambino family.

From the mid-sixties and into the nineties, the IAS, rather than operating as an IBM-type corporation as described in Cosa Nostra theory, resembled instead the "patron model." The Gallo–Profaci Wars, the attempted Colombo assassination, the "Banana Split," and the Gallo–Colombo Wars (Diapoulos & Linakis, 1976; Talese, 1971) suggested continuing internal strife. Perhaps a small, coup-plagued, unstable Latin American country would be a better model than a large corporation in describing the IAS. While Carlo Gambino was consolidating his power during this period, a commentary on the 1969 truce in the Banana Wars (a rift in the Joseph Bonanno family) may have best described the dilemma faced by the IAS: "What the Mafia needed in New York in 1969 was a health clinic, not a gang war" (D. Maclean, 1974, pp. 341–342). Most of the leaders were dying or sick. Many were in their seventies, and their middle-level executives, due to membership moratoriums, were not much younger. Gambino controlled four of the five New York mobs with only the Bonanno family not in the fold. Although forced into involuntary exile in Arizona by the Cosa Nostra Commission, Bonanno still controlled his New York organization through various loyalists. In 1974, his longtime underboss Carmine "the Cigar" Galante was released from a 12-year stretch in federal penitentiaries. Galante announced his return by blasting off the bronze doors on Frank Costello's mausoleum (Costello had died the year before, of a heart attack), warring against blacks and Latinos for the control of narcotics in the South Bronx and Harlem, and finally, according to unnamed mob sources, arranging for the death of Gambino himself by persuading the elderly, coronary-prone man to get a swine flu shot ("After the Don," 1976, p. 32).

With the death of Gambino in 1976, speculation was rife as to who was likely to emerge as the most influential boss. In 1984, boss Paul Castellano of the Gambino family and two of his top lieutenants were indicted for extortion, a pattern of racketeering, and conspiracies to commit murder. Castellano had emerged as the most powerful of the dons and appeared to be in a position to claim the "national crown" when he was gunned down outside a Manhattan restaurant on December 16, 1985. Some members were apparently disenchanted with his leadership, criticizing his favoring and rewarding members who had not "made their bones" (killed others) (R. J. Kelly, 1990, p. 18). The FBI believes that the person responsible for this assassination was an ambitious younger man, John Gotti, who then seized control of the Gambino family.

The Commission Trials

Undoubtedly, the biggest blow ever dealt to the IAS was a series of prosecutions of organized crime figures by the federal government from 1983 to 1987. Indictments in 1985 alone reached almost 5,000, among which were the alleged leaders of 16 of the 24 Mafia families. Albanese (1989, p. 3), in recounting the impact of the trials, pointed out that the existence of both a Mafia and a Commission (with representatives from the major Mafia families) was admitted.

The outcome of the trials crippled the aging upper echelons of the Mafia families. The bosses of the Colombo, Genovese, and Lucchese organized crime families were convicted of being members of "the Commission" established by Luciano in 1931, which settled underworld disputes and authorized gangland killings (Doyle, 1987). The only major survivor of the **"Commission Trials"** was John Gotti, who emerged as the most powerful boss of an organization that had now been weakened by criminal justice and media attention, creating a vacuum to be filled by rival ethnic gangs (Mustain & Capeci, 1988). In referring to Gotti's emergence to head a decimated Mafia, one magazine article was titled "The Last Godfather?" (McKillop, 1989).

Gotti was a "media darling" reminiscent of a Costello or Capone (Capeci, 2003). Gotti, dapper but ruthless, operated out of his headquarters in Queens, the Bergen Hunt and Fish Club. He was called the "Teflon don" because of his ability to escape conviction. Finally convicted in 1992, Gotti was dubbed by the press the "Velcro don." He was convicted, in part, on the basis of the testimony of his former underboss, Sam "Sammy the Bull" Gravano. The American Mafia at the end of the twentieth century represented a dwindling empire with some remaining strongholds, such as New York and the Chicago suburbs. The death of the American Mafia was clear in the late 1990s when the bosses of organized crime families became informants to the federal government. Angelo Lonardo (Cleveland) and Ralph Natale (Philadelphia) both broke the code of omerta and cooperated with government investigations.

Some of the Mafia's competition has included the 6,000-member Herrera family from Mexico, triple the size of the Mafia and developing ties with Colombian groups, particularly in Chicago. Kleinknecht in *The New Ethnic Mobs* (1996) claims that the most important new organized crime gangs are the Chinese, who concentrate on gambling, extortion, alien smuggling, credit card fraud, drug smuggling, and loan sharking. The new Russian mobs are into white collar and financial crime, confidence games, and black market activities. The Arab mob (Christian Iraqis known as Chaldeans) commit grocery coupon fraud. Vietnamese groups in Silicon Valley, California, have staged armed robberies of high-tech firms (Sanoff, 1996). The new "Mafia" is more likely to be multiethnic and global. The U.S. drug czar in the Clinton administration, Barry McCaffrey, indicated that he believed the Russians and Nigerians operated the most threatening criminal organizations based in the United States. (Farah, 1997). The Russian crime groups offer other drug syndicates weaponry previously beyond their reach and have increasingly moved into drug trafficking and money laundering. Nigerian groups have been primarily involved in confidence games and fraud. Outlaw motorcycle gangs, Colombian cocaine crime families, black criminal gangs, and Asian gangs are among the many new mobs contending for this vacated territory.

Black organized crime groups are not examples of emerging groups; they have existed for decades (Abadinsky, 2006; Messick, 1979; Pennsylvania Crime Commission, 1986; Schatzberg & Kelly, 1995). Some of the better-known organizations were those run by Frank Matthews, Charles Lucas, Leroy Barnes, and Jeff Fort (El Rukns), but a variety of other groups are involved in such things as drug trafficking, the numbers racket, extortion, and murder. In Philadelphia in the nineties, a tightly knit criminal organization calling itself "the Family" specialized in drug distribution and murder, as did "the Junior Black Mafia" (Griffin, 2003; Pennsylvania Crime Commission, 1991).

As mentioned earlier, organized crime remains a "queer ladder of mobility" (Bell, 1953) for black mobs in Philadelphia and New York, Colombian and Cuban mobs in Florida, and Chinese and Chicano mobs on the West Coast. Although many of these mobs are not fully developed and structured syndicates as in the organized crime model, they certainly represent an evolving force to be reckoned with.

"The Pizza Connection"

In late 1985 to early 1986, a total of 22 defendants went to trial in New York City on charges that they ran a $1.7 billion drug trafficking organization in the United States, using pizza restaurants as fronts. This group was a "Sicilian mafia," supposedly separate from and with few links to the existing

"American Mafia," according to Tommaso Buscetta, who revealed the group's operations to U.S. authorities (Reuter, 1984a). Federal authorities did, however, find some links with "American Mafia" groups as well as cooperation in money laundering by Swiss and Italian banks and by U.S. brokerage firms E. F. Hutton and Merrill Lynch. Sicilian immigrants, called "zips" because of their rapid speech, staffed these pizza parlors (Alexander, 1988; D. Blumenthal, 1988; Potter, 1989).

The Sicilian Mafia

Audio Link 12.5
Listen to a discussion of the Sicilian Mafia.

Journalist Claire Sterling in her book *Octopus: The Long Reach of the International Sicilian Mafia* (1990) mixes fact with a little fiction to describe the cross-national reach of the post–World War II Italian Mafia. With the closing of the Corsican "French Connection" in the seventies, organized crime figures in Sicily began to supply large amounts of heroin to the United States and in the early eighties began a "civil war" over control of this lucrative trade. Their violence spilled over into the murder of the head of the Italian antimob squad, General Carlo Chiesa, and his pregnant wife, as well as the murder of Judge Caesar Terranova. This enraged the Sicilian public, the church, and public officials, and spelled the doom of Sicilian Mafia groups. In February 1986, over 456 members and associates of organized crime groups were put on trial, including Michele "the Pope" Greco, held by some to be the "boss of bosses." The trial took place in a specially constructed courtroom guarded by 200 crack troops and ended in December 1987 with 338, including Greco, convicted. The key witness in the trial was a former boss, Tommaso Buscetta, who also testified in the **"Pizza Connection"** trials in the United States (Gage, 1988, pp. 36–37).

By 1988, only 112 of those convicted were still in jail, and in April 1991, a total of 28 Sicilian Mafia leaders including Greco were released on a technicality. In March 1992, the Mafia killed a politician in the Christian Democratic Party in Palermo, Salvatore Lima, and in May 1992 assassinated Judge Giovanni Falcone, his wife, and three bodyguards. Falcone was to have enforced stronger anti-Mafia laws. In July 1992, the Mafia killed his successor, Judge Paolo Borsellino (Stille & Robinson, 1992). While the Mafia was sending a message, the Italian government responded by sending the Italian army to Sicily and proposed very strong, perhaps draconian, laws with which to attack the criminal organization (Cowell, 1992).

Criminal Careers of Organized Criminals

Similar to professional crime, but unlike most other types, organized criminal activity is an example of career crime, in which crime is pursued as a livelihood. Organized criminals exhibit varying degrees of the following characteristics, with those who are members of established syndicates expressing these qualities to a greater degree; they identify with crime and criminal activity, possess strong organizational identity, and tend to belong to structured groups that maintain continuance of operation.

Based on our previous description of organized crime, such criminals tend to be bred in low-income, high-crime areas of large central cities, where illegitimate opportunity structures appear more available than legitimate ones. Most begin as conventional criminals, but, rather than retiring as most do in their early twenties, they continue to progress in criminality and in association with organized criminals (Clinard & Quinney, 1973, p. 229).

To varying degrees, organized criminal groups subscribe to a code of secrecy, whether it be the "cosa de hombre" (code of manliness) of the "Nuestra Cosa" (Mexican Mafia), rules of conduct of gangs such as Hell's Angels and the Pagans, or the prototype code of "omerta" described by people like Valachi. Omerta is a Cosa Nostra code of intense loyalty, honor, secrecy, obedience, and "manly" silence—a code that renders loyalty to the organization above loyalty to country, God, or family and whose violation means death.

Secrecy, discipline, corruption, planned violence, and public demand for illicit goods in either compromised or inept political climates provide a continuing good occupational outlook for the next generation of Valachis, whatever their ethnicity. The continuing public demand for illicit goods and services and corrupt relationships with government officials may be more important factors in the persistence of organized groups than the imperviousness of their organizations.

Public and Legal Reaction

In Chapter 13, we will explore the issue of drug abuse from the standpoint of users, but our concern in this chapter has been drug trafficking in which international drug "kingpins" are the new Al Capones and Meyer Lanskys. In the late eighties, the United States began to go after major drug kingpins. In May 1988, Carlos Lehder Rivas, who had been extradited from Colombia, was convicted of being responsible for up to 80 percent of the cocaine smuggled into the United States. In 1989, leaders of Colombia's Medellin cartel were indicted on charges of cocaine trafficking and the slaying of the Colombian justice minister as well as of a U.S. drug informant ("U.S. Indicts Colombian Drug Cartel," 1989). Attempts to put pressure on the "Underground Empire" of drug launderers and officials were illustrated by the capture and imprisonment of Panamanian strongman Manuel Noriega.

Drug Control Strategies

Some drug control strategies or options include legalization, use of diplomacy, interdiction, targeting traffickers, coordination of rival departments, and prevention (J. Adler et al., 1988; M. Moore, 1988). Legalization is viewed by authorities and lawmakers as a last resort, an unnecessary risk, and a questionable moral decision. It would appear to be unwise to overreact to a crack epidemic by legalizing drugs as it might create more demand for drugs at a time when overall drug use is declining. The use of diplomacy or economic and political pressure to halt the drug war being waged against the United States by Colombia, Bolivia, Peru, and Mexico in particular is a supply-side strategy that is not without risk. "Anti-Yanqui" feelings may be fueled, and many countries are dependent on drug money. Interdiction involves stopping the transport and smuggling of drugs into the United States. With many ports of entry and endless borders, some deterrence is possible, but complete interdiction is impossible.

Targeting major traffickers such as the Medellin cartel while enhancing street-level enforcement to totally disrupt street traffic in drugs (Hayeslip, 1989) has possibilities. Departmental coordination and elimination of rivalry are claimed to be aided by the creation of a federal "drug czar" in 1989 to oversee and coordinate agencies involved in the drug war. Finally, a "demand-side" strategy of prevention offers the ultimate hope. Education and rehabilitation programs as well as policy experiments to discover programs that work are greatly needed.

Investigative Procedures

Law enforcement in the eighties and nineties finally became as organized as organized crime and began to effectively apply a variety of investigative procedures including financial analysis, electronic surveillance, use of informants and undercover agents, citizens' commissions (Albanese, 2001, 2004), and computer assistance. Financial analysis involves following paper trails (records of transaction) in order to see if expenditures match earnings. Classic Internal Revenue Service procedures in enforcing tax codes such as analyzing net worth, expenditures, and bank deposits are utilized.

Electronic surveillance (the use of "bugs" and wiretapping in covert eavesdropping) is viewed by many authorities as one of the most effective weapons against organized crime. The use of 150 audio- and videotapes at the "Commission Trials" was very successful. The use of informants (insiders who

Handbook Article Link 12.2
Read an article on drugs and the criminal justice system.

provide information) as well as undercover agents has also been indispensable. Informants such as Jackie Presser (former Teamster president), Angelo Lonardo (former Cleveland don), and Tommaso Buscetta (Sicilian Mafia don) have been devastating to the syndicate. Citizens' commissions such as the Chicago Crime Commission are essential in providing an independent watchdog function in examining organized criminal activity (Albanese, 1989, p. 116).

Finally, computer-assisted investigation has great potential for unraveling complicated transactions and network interrelationships. The FBI uses a sophisticated computer database, the Organized Crime Information System, and is experimenting with artificial intelligence using a supercomputer called Big Floyd.

Laws and Organized Crime

Some specific laws that have been used against organized crime include special laws such as the Hobbs Act; features of the Organized Crime Control Act of 1970, especially RICO; and the Bank Secrecy Act (1970); as well as assets seizure (forfeiture).

Hobbs Act

One effective piece of legislation on the books since the mid-forties is the **Hobbs Act,** an anti-racketeering law that can be interpreted to mean that any interference with interstate commerce to any degree whatsoever is in violation of the act. This statute has been applied, for example, against politicians in Newark, New Jersey, in accepting kickbacks from contractors who had obtained supplies from out of state.

Organized Crime Control Act

The single most effective piece of federal legislation ever passed in the United States to fight organized crime activity is the controversial **Organized Crime Control Act** of 1970, a principal feature of which is the RICO statute. RICO prohibits proceeds from a pattern of racketeering activity from being used in acquiring legitimate businesses that are involved in interstate commerce. Generally, a "pattern of racketeering" involves participation in any two specified crimes, such as murder and extortion, within a 10-year period.

Some of the principal features of the act are the creation of special grand juries to investigate organized criminal activity, and the provision of general immunity for witnesses appearing before the grand jury, in which the protection against self-incrimination is abrogated in return for protection against the use of such compelled information in a criminal proceeding. It provides for the incarceration of witnesses who refuse to testify (recalcitrant witnesses), authorizes a conviction based on irreconcilably inconsistent declarations under oath (perjury), and provides for protected facilities for housing government witnesses and their families. It also authorizes the government to preserve testimony by the use of a deposition (testimony given under oath but outside the courtroom) in a criminal proceeding (a right that previously existed only for the defendant), and prohibits any challenge to the admissibility of evidence based on its being the fruit of an unlawful government act (if such act occurred 5 years or more before the event sought to be proved). The act makes it unlawful to engage in the "illegal gambling business" itself and contains the RICO statute.

The Bank Secrecy Act (1970) is directed at controlling money laundering. It includes features requiring banks to report transactions over $10,000 or file a report if $10,000 or more leaves or enters the country, and citizens to report foreign bank accounts on tax returns (Abadinsky, 1994, p. 430). Forfeiture, the ancient legal practice of government seizure of property used in criminal activity, may prove a particularly useful weapon against illicit narcotics trafficking (Stellwagen, 1985, p. 1).

Assets seizure (forfeiture) has emerged as one of the most powerful tools to break the back of criminal enterprises—"kick them in the assets," so to speak. Imprisonment and fines have been found inadequate in deterring capital organizations, while seizure of assets curtails the financial ability of such groups to continue criminal operations (Bureau of Justice Statistics, 1988, p. 93). Assets may include money, property, businesses, cars, boats, or any item that may have been involved in or is the product of a criminal enterprise (J. E. Jacoby, Gramckow, & Rutledge, 1992; Lombardo, 1990). Forfeiture, the ancient legal practice of government seizure of property used in criminal activity, may prove a particularly useful weapon against illicit narcotics trafficking (Stellwagen, 1985, p. 1).

The RICO Statute

The RICO statute, as mentioned earlier, authorizes the federal government to seize legitimate operations if they have been purchased with illegally gained funds (laundering) or if they are used for criminal purposes. In addition, defendants can be subject to up to 20 years' imprisonment. The law permits prosecutors greater latitude in presenting to the jury a broader picture of patterns of racketeering; this enables them to trace the pattern back to formerly insulated bosses. Because of the broad sweep of the law, lawyers and others are fearful of the application of the law to nonsyndicate crime, such as crimes by legitimate business.

RICO charges offer a unique advantage in targeting an entire enterprise, and civil RICO laws can be used to seize cash and assets. Application of RICO charges to white collar violations, such as insider trading by brokerage firm Drexel Burnham Lambert, raises controversy. Civil RICO permits victims of fraud to bring private civil suits, whether or not the Justice Department files charges ("RICO: An Assault," 1989, p. 18). Threatened companies, it is claimed, are forced to settle or be branded racketeers. While some critics see it as a "statute run amok" (Boucher, 1989), others see it as a powerful tool to control white collar crime in addition to organized crime (Waldman & Gilbert, 1989; see also Safire, 1989). Greek (1990) indicates, "RICO represents a major expansion of the federalization of crime and which for now appears to be quite acceptable, despite those protesting its widespread use, to both the courts and a large segment of the American public as well" (p. 1).

Until 1981, many features of RICO had lain dormant. In a case against IAS boss Frank Tieri, the government alleged that Mafia families themselves constituted illegal enterprises. Los Angeles mobsters were also convicted of racketeering and conspiracy charges (Mitchell, 1981, p. 43).

The civil provisions of RICO permit U.S. attorneys and private citizens to sue for treble damages and the cost of the suit if it can be demonstrated that the plaintiff or his or her business/property was injured as a result of a pattern of racketeering. A Continuing Criminal Enterprise statute is similar to RICO, but targets only illegal drug activity. The statute considers it a crime to commit or conspire to commit a series of felony violations of the 1970 Drug Abuse Prevention and Control Act in concert with five or more other persons (K. Carlson & Finn, 1993). By the mid-nineties, the growth in violent street crime perpetrated by gangs finally received serious federal attention. Federal agencies such as the FBI, DEA, and ATF (Bureau of Alcohol, Tobacco and Firearms) began to team up with local police to target such groups. RICO charges, for instance, were successfully employed against Chicago's Latin Kings, Atlantic City's Abdullahs gang, and Shreveport's Bottoms Boys.

Theory and Crime

Organized crime as a subfield of criminology has developed a number of theories of its own. Most of these relate to the nature, definitions, and characteristics of organized crime. Readers of this chapter have a clear notion by now that not all organized crime is mafia. Organized crime exists internationally where groups use violence or threats of violence in order to profit from supplying illicit goods that are demanded by the public. They also enjoy immunity of operation through intimidation and corruption.

While the ethnicity of the groups involved in organized crime changes, it has represented a "queer ladder of mobility" for various groups. Anomie theory explains that in the pursuit of the American dream of financial success various groups have found themselves at a critical disadvantage and find that legitimate means to success are blocked. Innovators substitute other means (crime) as a ladder to success. According to Cloward and Ohlin's differential opportunity theory, some groups find themselves in neighborhoods where legitimate avenues of upward mobility are blocked; however, illegitimate means may abound. Such mob neighborhoods spawn the next generation of wiseguys. In such subcultures, organized crime is viewed as a good opportunity for being admired and successful. Criminal subcultures may exist which socialize the individual into valuing criminal values and attitudes.

Summary

The subject of much public interest, organized crime has been defined in a variety of ways. In the United States, most federal agency and state statutes use generic definitions, which indicate that organized crime is any criminal activity involving two or more individuals. Utilizing a similar generic (broad or general) definition, Albini identifies four types of organized crime: political-social, mercenary (predatory), in-group, and syndicate. With the exception of the last type, syndicate crime, all of the former refer to other types of criminal activity, such as political, conventional, and professional criminal behavior.

The field of criminology defines organized crime (henceforth synonymous with syndicate crime) as a continuing group or organization

1. That participates in illicit activity in any society by the use of force, intimidation, or threats;

2. That provides illicit services that are in strong public demand; and

3. That assures protection and immunity through corruption.

In this chapter, an organized crime model was proposed as a useful device for avoiding confusion in the process of deciding whether a group's activities represent an example of organized crime. Organized crime as a concept is not a matter of kind, but is rather a matter of degree that is, to what extent this type of crime possesses the characteristics identified in our criminological definition of organized crime. Types of crime may be viewed as distributed along a continuum ranging from nonorganized to organized (syndicate) crime, depending on the degree to which the group exhibits organization, the use or threat of violence, the provision of illicit goods in public demand, and the ability to obtain immunity through corruption and enforcement. Types of organized (syndicate) crime include traditional crime syndicates, nontraditional syndicates, semi-organized crime, local politically controlled organized crime, and national politically controlled organized crime.

In addition to definitional problems, another problem in the study of organized crime is the poor scientific nature of much of the literature, which forces the social scientist to rely on many journalistic and autobiographical accounts.

A variety of street gangs were described, and some were noted to be undergoing transition into ghetto-based drug trafficking organizations.

Internationally, organized crime thrives in two types of political environments: liberal democracies and corrupt dictatorships. Chinese Triad societies, highly ritualized Chinese secret organizations that are often involved in organized crime, were described.

Mafiya is a term used to refer to various Russian organized crime groups, which have grown in power since the collapse of the former Soviet Union. Organized crime, although dominated since the thirties by the Italian American Syndicate (IAS), has participation from a variety of ethnic groups. According to the "theory of ethnic succession," mobs have represented a "queer ladder of mobility" for a variety of minorities.

Money laundering involves making clean or washing "dirty money" (illegal funds). Such operations make use of unscrupulous banks that ask no questions in accepting large deposits of cash.

Primary drug smuggling routes—the Golden Triangle, the Golden Crescent, and Latin America—were discussed, as were the Colombian cartels and the Underground Empire.

Various theories regarding the origin of the term *Mafia* were traced, with the author agreeing with Albini that the most likely source is Risotto's 1860 play, *I Mafiusi de la Vicaria (The Heroes of the Penitentiary)*. Three theories of the nature of syndicate crime in the United States were discussed: (1) Cosa Nostra theory, (2) confederation theory, and (3) patron theory. While the first theory, which has been accepted by federal commissions and agencies, views organized crime as centrally controlled by a formally structured Italian American syndicate, confederation theory views it as controlled by a "combination" of ethnic groups, principally Jewish and Italian. The patron theory views organized crime as a set of shifting alliances, a "client–patron" relationship.

The "Mafia myth" is the belief that organized crime is the product of an alien conspiracy. More moderate expressions of this theme view the IAS as the most powerful of organized crime groups, but not the product of an alien conspiracy. Critics of the Mafia or Cosa Nostra (LCN, La Cosa Nostra) model argue that the terms and descriptions of these organizations are fictitious, the creations of federal law enforcement agencies. A more moderate view admits that many of their criticisms are legitimate, but still argues that the IAS exists, the vision being "skewed, not false."

The classic pattern of organized crime involves a gradual evolutionary development from strategic and tactical crimes, to illegal businesses and activities, to legitimate businesses, to an infiltration of big business and government itself. Some typical operations of organized criminal groups are arson, assault, coercion, extortion, murder, blackmail, bribery, and corruption. Typical illegal businesses include gambling operations, loan sharking, labor racketeering, record and tape piracy, and any number of other activities detailed in the chapter. Infiltration of legitimate businesses such as trucking, construction, and the hotel and restaurant industry provides cover for organized crime operations. Although it is hazardous to guess, estimated gross revenues and untaxed net profits appear to make organized crime wealthier than the nation's largest industrial corporations.

Organized crime infiltration of legitimate business may be viewed on one hand as a natural "ethnic succession" of organized crime, and on the other hand as yet another setting for illegitimate operations. Involvement of organized crime at the highest levels of government is revealed by a **"CIA–Mafia link"** to an attempt to assassinate Fidel Castro.

A brief history of organized crime traces its origins back to colonial times. The New Orleans Grand Jury report of 1890 was the first official recognition of the existence of the Mafia in this country. Other important events in the history of organized crime were traced, such as the Prohibition period, the Castellammarese Wars, the Luciano era, the Genovese era, the Kefauver and McClellan Commission hearings, the Apalachin meetings, the Gambino era, and other, more recent developments in organized crime such as the Commission Trials, the Pizza Connection, and the Great Mafia Trial in Sicily.

Criminal careers of organized criminals were briefly examined. Highlighted was their strong identification with criminal careers, their recruitment, and their relationship with the public. Finally, public and legal reaction to organized crime was discussed. While public reaction to organized crime in the past was characterized as a fascinated apathy and sporadic and unorganized legal reaction, recent events suggest major inroads in the war on organized crime. Application of laws such as the Hobbs Act and the Organized Crime Control Act (1970) represent potent tools. The RICO statutes and "sting" operations by federal agencies represent creative law enforcement efforts in this regard. However, these more aggressive law enforcement efforts have been criticized for threatening civil liberties and for covering up corporate criminality.

Criminology on the Web

Log on to the Web-based student study site at http://www.sagepub.com/haganintrocrim7e/ for author-created podcasts, e-flashcards, quizzes, and more.

KEY CONCEPTS

Activities of Organized Criminals

Apalachin Meetings

Assets Forfeiture

Castellammarese Wars

CIA–Mafia Link

Classic Pattern of Organized Crime

Commission Trials

Continuum Model of Organized Crime

Ethnic Succession Theory

Four Basic Types of Organized Crime (Generic Definitions)

Hobbs Act

Internal Structure of LCN Families

International Political Climates and Organized Crime

Iron Law of Opium Trade

Kefauver Committee

Loan Sharking

Mafiya

Major Elements of La Cosa Nostra (LCN) Theory

Medellin Cartel

"Moustache Petes"

Numbers Game

Organized Crime Control Act

"Pizza Connection"

Racketeering

RICO

Theories of the Nature of U.S. Syndicate Crime

Theories Regarding Origin of Mafia

Three Elements of the Definition of Organized (Syndicate) Crime

Triads

Yakuza

REVIEW QUESTIONS

1. Discuss the various attempts to define organized crime. How does the "organized crime continuum" approach this issue?

2. What are some important sources of information on organized crime?

3. Discuss some features of Triads. Where do they operate and what are their major activities?

4. Discuss the history and present status of Russian organized crime. What are its major criminal operations, and why is it regarded as such an enormous threat?

5. Who was Amado Carrillo Fuentes, and of what importance was he in the history of Mexican organized crime?

6. What is money laundering? How was "Operation Casablanca" successful in investigating and busting a money laundering operation?

7. Why was Joe Valachi so important in the history of the American Mafia?

8. What is the RICO statute? How effective has it been in the war on organized crime?

9. What are some investigative procedures and legal weapons that have been used in the war on organized crime in the United States?

10. Discuss some of the major features of the Organized Crime Control Act of 1970.

WEB SOURCES

Asian Organized Crime
http://www.iaaci.com

Jerry Capeci's Gang Land Web Page
http://www.ganglandnews.com/index.html

Los Mara Salvatrucha (MS-13)
http://www.knowgangs.com/gang_resources/profiles/ms13/

Nathanson Center for the Study of Organized Crime (Canada)
http://www.osgoode.yorku.ca/research/research_centres.html

Organized Crime Registry
http://www.organized-crime.de

Rick Porello's American Mafia.com
http://www.americanMafia.com/index.html

Transnational Crime Center
http://policy-traccc.gmu.edu

United Nation's Convention Against Transnational Crime
http://www.unodc.documents/treaties/UNTOC/Publications/TOC%20Convention/TOCebook-e.pdf

WEB EXERCISES

Using this chapter's recommended Web sites, explore the issue of organized crime.

1. What is Los Mara Salvatrucha 13 (MS-13)? What have been some law enforcement efforts against it?

2. Examine Rick Porello's AmericanMafia.com. What is the meaning of the title of his book, *To Kill the Irishman?*

3. Peruse the GangRule.com Web page . What new developments have taken place in New York City since 2000?

4. What are the latest developments in the "Mexican Drug Wars"?

SELECTED READINGS

Howard Abadinsky. 2000. *Organized Crime* (8th ed.). Belmont, CA: Wadsworth.

This is the bible on organized crime. It is particularly useful and interesting for its historical coverage of the Chicago and New York mobs.

Jay Albanese. 2004. *Organized Crime in Our Times* (4th ed.). Cincinnati: LexisNexis.

This is one of the best written books on the subject. It has excellent coverage of illegal activities as well as the investigation and prosecution of such activities.

Joseph Albini. 1971. *The American Mafia: Genesis of a Legend.* New York: Irvington Press.

This is a classic in the field. Albini uses a variety of sources to trace the legend, myths, and history of the American Mafia.

G. Robert Blakey and Richard Billing. 1981. *The Plot to Kill the President: Organized Crime Assassinated JFK.* New York: New York Times Books.

This controversial book by the chief counsel to the House Assassinations Committee concludes major involvement of organized crime in the assassination of President John Kennedy.

Charles Bowden. 1998. *Juarez: The Laboratory of Our Future.* New York: Aperture Foundation.

This is a biographical account of the life and times of Amado Carrillo Fuentes, who until his death in 1997 was the biggest of the Mexican drug dons.

Robert J. Kelly, editor. 1986. *Organized Crime: A Global Perspective.* Totowa, NJ: Rowman & Littlefield.

Kelly presents a very good selection of readings on transnational organized crime.

Dennis J. Kenney and James O. Finckenauer. 1995. *Organized Crime in America.* Belmont, CA: Wadsworth.

This is an excellent text on organized crime and particularly on new groups such as Russian mafiya groups.

Michael D. Lyman and Gary W. Potter. 2004. *Organized Crime* (3rd ed.). Upper Saddle River, NJ: Prentice Hall.

This is a fresh approach to examining organized crime, presented in a very lucid, readable format for undergraduates.

Pennsylvania Crime Commission. 1990. *Organized Crime in Pennsylvania: A Decade of Change.* Conshohocken: Commonwealth of Pennsylvania.

Until its demise for political reasons by later jailed Attorney General Ernie Preate, who was a subject of its investigation, this organization consistently presented very readable investigations complete with photographs.

Vincent Teresa and Thomas Renner. 1973. *My Life in the Mafia.* Greenwich, CT: Fawcett Publications.

This autobiographical account of a New England Mafia associate provides an inside look at the world of organized crime.

Nuts, Guts, Sluts, and "Preverts"

Broken Windows

Prostitution

   Types of Prostitution

   Massage Parlors

   Johns

   Underaged Prostitutes

Homosexual Behavior

   Crime File 13.1 Laud Humphreys's Tearoom Trade

Sexual Offenses

   Paraphilia

   Nonvictimless Sexual Offenses

   Sexual Predators

   Crime File 13.2 Child Sexual Abuse by Catholic Priests

   Incest

   Characteristics of Sex Offenders

Drug Abuse

   Drugs and History

   Crime File 13.3 Moral Panics and the Strange Career of Captain Richard Hobson—Moral Entrepreneur

   Drug Use in the United States: The Drug Dip?

   Drug Abuse and Crime

   Drunkenness

Special Populations

Societal Reaction

   Overcriminalization

   Decriminalization

Theory and Crime

Summary

Key Concepts

Review Questions

Web Sources

Web Exercises

Selected Readings

chapter 13

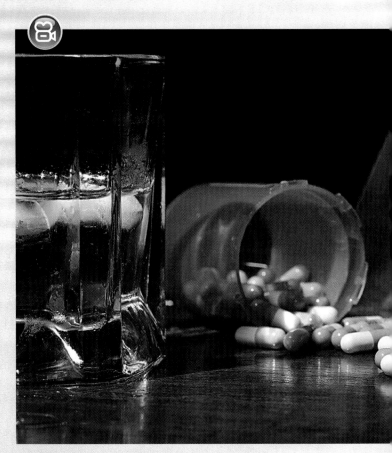

Public Order Crime

Do It Now, Before They Pass a Law Against It

—*Bumper sticker*

It all starts on the streets. What do you see when you walk down a downtown street? Do you see the hooker conning a john? The pool hustler with his permanent pale taking a break between games? The pimp hustling a new woman? The transvestite prostitute with his head in the car window of a potential customer? Do you notice the man on the corner passing baggies to customers or the drunk in the doorway with his brown bag clutched in his hand? When you looked down the alley, did you see the people by the dumpster shooting up? Did you notice the slips of paper being passed between the news vendor and his clients? Were you surprised to see two men having sex when you stepped into the public restroom? . . . These are the players of the deviant street network at work.

—*John H. Lindquist (1988)*

Is vice-related behavior a matter of civil liberties and individual choice in a free society? What is the role of the criminal justice system in enforcing a semblance of public morality and a sense of civic decency? Laws against **public order crime,** sometimes referred to as **crimes without victims** (Schur, 1965) or "legislated morality," refer to a number of activities that are illegal because they offend public morality. Such nonpredatory crime generally includes activities such as prostitution, acts related to homosexuality, alcohol and narcotics abuse, gambling offenses, disorderly conduct, vagrancy, and minor forms of "sexual deviance." These crimes outnumber other recorded crimes and have traditionally represented the bulk of police work.

A number of other concepts have been used to refer to certain categories of public order criminal activity. H. Laurence Ross (1961) coined the term **folk crime** to refer to relatively common violations that occur in part because of the complexity of modern society. Traffic offenses, fish and game law violations, tax offenses, gambling, and sexual deviations all can serve as illustrations. Many, but not all, of the activities to be discussed in this chapter are examples of crimes that are *mala prohibita*—bad because they have been prohibited by law. They violate various conceptions in society as to appropriate moral conduct, but lack the quality of acts *mala in se,* such as murder or rape, in which there is clear and abhorrent victimization of others. Offenses related to prostitution, homosexuality, gambling, and the like serve as examples of "consensual crimes" in that there is free consent on the part of participants. In many "victimless" crimes, the offenders have customers rather than victims (Silberman, 1978, p. 265).

Nuts, Guts, Sluts, and "Preverts"

Alexander Liazos (1972) published what has now become a classic sociological work, an article titled "The Poverty of the Sociology of Deviance: Nuts, Sluts, and 'Preverts.'" Liazos makes the point that sociologists have concentrated too much on the "dramatic" nature of deviance, such as prostitution, homosexuality, and the like, to the neglect of more serious or harmful forms of deviance such as racism, inequitable taxation, and sexism. He states,

> As a result of the fascination with "nuts, sluts, and preverts [*sic*]," and their identities and subcultures, little attention has been paid to the unethical, illegal, and destructive actions of powerful individuals, groups, and institutions in our society. Because these actions are carried out quietly in the normal course of events, the sociology of deviance does not consider them as part of its subject matter. (p. 26)

This chapter will concentrate on behavior that has been labeled criminal or deviant because it is viewed either as different or as immoral or harmful to the individual. Societal attempts to regulate deviance (nuts), sex (sluts), drug and alcohol consumption (guts), and perversion (other activity, "preverts") will be explored.

Broken Windows

Audio Link 13.1
Listen to a discussion of the broken windows theory.

In their classic article "Broken Windows," J. Q. Wilson and Kelling (1982) give a different view of the need to regulate such conduct. Kelling (1988b) explains,

> Just as unrepaired broken windows can signal to people that nobody cares about a building and lead to more serious vandalism, untended disorderly behavior can also signal that nobody cares about the community and lead to more serious disorder and crime. Such signals— untended property, disorderly persons, drunks, obstreperous youth, etc.—both create fear in citizens and attract predators. (p. 2)

Deinstitutionalization of the mentally ill without adequate follow-up or community treatment facilities has added yet another population to an already existing homeless problem. The public wants the police to assist the mentally ill, the public inebriate, and the homeless (Finn & Sullivan, 1988, p. 1). Neighborhood disorder, drunks, panhandlers, youth gangs, and other incivilities unsettle a community; produce fear; and disrupt social, commercial, and political life.

A large decrease in crime in New York City in the nineties was attributed in part to a new policing policy of zero tolerance for previously ignored squeegee men (who try to wash car windows when drivers are stopped at traffic lights), aggressive panhandlers, subway turnstile hoppers, vagrants, and disorderly conduct. By enforcing small things, the police claim to have gotten a better handle on crime in general.

While much of the reduction in crime beginning in the 1990s was attributed to application of **broken windows** theory by its advocates, critics point out that other cities without a broken windows policy also experienced similar decreases during this period.

Prostitution

Prostitution can be defined as the practice of having sexual relations with emotional indifference on a promiscuous and mercenary basis. In some countries and most U.S. states, prostitution itself is not a criminal offense; it is the act of soliciting, selling, or seeking paying customers that is prohibited. Sometimes referred to in jest as the "world's oldest profession," prostitution certainly has been widespread in societies both ancient and modern.

Until the Protestant Reformation in Western society, prostitution was pervasive and tolerated as a "necessary evil." It was often taxed by the church and was a major source of community revenue in the Middle Ages. Public health concerns that arose with the discovery of syphilis and the emergence of the Protestant ethic with its strong emphasis on individual morality were instrumental in prostitution's prohibition. Essentially, concepts of sin were translated into legal notions of crime. Despite its prohibition, prostitution exists internationally, with the exception of some poor and underdeveloped societies where it would be in little demand. While prostitution is generally regarded as a low-status occupation in societies in which it is approved of, in different cultures in the past certain prostitutes have enjoyed high status, such as the *hetaerae* of early Greece, the *lupanaria* in the Roman Empire, the *devadasis* in India, and the *geishas* of Japan (K. Davis, 1961). Such courtesans were often well-educated and trained entertainers or religious performers.

In most states, prostitution is considered a misdemeanor, and laws prohibiting it are generally enforced only when the public insists on it. Typically prostitutes are rounded up, booked, made to pay a small bail or fine, and then are put back on the streets. In order to control the undesirable activities often associated with prostitution and to avoid public complaint, many cities create vice zones or "combat zones"—adult entertainment areas or "red-light districts." Most public order offenders do not regard their behavior as criminal, perhaps in part because of general societal ambivalence toward much of it (Clinard & Quinney, 1973, p. 84). Some of the behavior may reflect personal psychological disability, but much of it reflects either adult consensual relations that are agreeable to both parties and

Photo 13.1

Geishas, among other prostitutes, were considered to have a high-status occupation, and were often well-educated.

Handbook Article Link 13.1
Read an article
on prostitution.

Journal Article Link 13.1
Examine literature
regarding prostitution.

Journal Article Link 13.2
Examine literature
regarding pimp-
controlled prostitution.

harmful to neither or personal choice to participate in activity that the individual desires even though it may be illegal or societally disapproved.

This discussion will concentrate primarily on female prostitution, which appears to persist despite wide variations in economic, political, and social systems. One explanation is that prostitution serves a function in society: it services otherwise unmet sexual needs. There is a strong demand in many societies for no-strings-attached sexual release, particularly in relatively isolated male environments. City leaders—particularly in seaport cities such as Hamburg, Marseilles, and Baltimore—would argue that toleration of prostitution enabled the servicing of armies, the rejected, strangers, and the perverted, thus protecting "decent" females of the community. In addition to the strong demand for such services, prostitution can offer relatively lucrative rewards for females, depending, of course, on the status of customers.

Eleanor Miller, in *Street Woman* (1986), interviewed 64 prostitutes and, like T. Sullivan (1988), found that economic and social problems propelled young women into "hustling" as an alternative to boring, dead-end jobs. For others, it was an escape from abusive or dysfunctional families. Money and survival became key motivations for street hookers (Ritter, 1988).

Armstrong (1983) points out that the role of pimps (procurers) in recruiting women to prostitution is actually minimal. James (1978) notes that this is a direct challenge to the view of prostitutes as victims. James (1977) sees the view of pimps as active panderers (drumming up business and recruits) as based on past behavior of pimps, sensational journalism, and protectionist policies toward women. Lemert (1968) describes the "white slave myth," indicating that "the trauma of forced entry into prostitution inspires sympathy and provides a way to discount responsibility for one's actions" (p. 84). While American explanations of recruitment point to disaffection with family, child abuse, drug use, and the like, McCaghy and Hou (1988) tell us that in Taiwan there is a historical tradition of prostitution that is sanctioned by a patrilineal system that devalues female children. Prostitution often occurs with the families' approval as a means of securing economic well-being during times of family stress.

Types of Prostitution

Prostitution involves a number of types and settings, including the following:

- Brothel prostitutes
- Bar girls
- Streetwalkers
- Massage parlor prostitutes
- Call girls
- Madams
- Other

Video Link 13.1
View a video on the
business of high-end
prostitution.

Like any other occupation, prostitution is stratified, the lowest status and remuneration assigned to brothel or house prostitutes and streetwalkers and the highest prestige and reward attached to expensive call girls, who are able to command higher prices from more exclusive clientele (MacNamara & Sagarin, 1977, p. 99; Perkins & Bennett, 1985).

Brothels—sometimes called whorehouses, cathouses, or bordellos—were widespread in the United States until the post–World War II period. Often clustered together in "red-light districts," brothels were managed by madams with whom prostitutes shared the proceeds from their "tricks" (sexual transactions). The term "red-light district" supposedly had its origin when railroad construction workers in the American West hung their red signal lanterns outside whorehouses they were frequenting in order to keep in contact with their dispatchers (Winick & Kinsie, 1971, p. 132). While some illegal brothels still exist in the United States, most have disappeared, although legal brothels exist in some counties in Nevada.

The following description of a red-light district in Erie, Pennsylvania, in 1907 is illustrative:

> In the three blocks of French Street mentioned, there are roughly sixteen immoral houses. Within these resorts a conservative total of 75 girls have been leading a life of shame. None of these resorts hold a liquor license, but at all of them, drinks of any description can be obtained at any hour of the day . . . by a visitor of almost any age.
>
> Seventy-five percent of the girls of the tenderloin are under 21 years of age. Ninety percent of the visitors are young men under 20. Fully thirty percent are boys of 16 and 17. ("Erie Red Light Resorts," 1988)

Streetwalkers or "hookers" parade and negotiate the sale of their wares on the public streets. The term *hooker* was apparently derived from camp or circuit traveling prostitutes who followed and serviced the Union troops of General Joseph Hooker during the Civil War (Winick & Kinsie, 1971, p. 58). Such "working girls" earn the lowest fees of all prostitutes and are most vulnerable to police interference. Streetwalkers also are most likely to have arrangements with pimps, who play the combined roles of managers, protectors, and pseudo-fathers. In the United States since the sixties, the majority of street pimps have been black. Recent research suggests that pimping is held in less regard than in the past and may be of less importance in the world of prostitution than it was at one time (Winick & Kinsie, 1971, p. 120). While relationships between pimps and their stables of prostitutes vary, many hookers are required by their pimps to earn a certain amount per day or suffer physical harm (Milner & Milner, 1972; Sheehy, 1973; Slim, 1969).

Bar girls, or B-girls, are common in seaport cities and areas serving military populations, such as in combat zones, or adult entertainment sections of some large cities. Such hookers entice customers to buy them drinks, usually nonalcoholic ones, for ridiculous prices, while also arranging for tricks, which may occur on the premises or at nearby "hot sheet" hotels.

Call girls represent the top of the prostitution profession. Such hookers generally are very selective in their clientele and are highly rewarded. Most are from more educated and middle-class backgrounds than streetwalkers or house girls and usually operate on referrals. "Escort services" generally are fronts for prostitution in which clients may pay per hour for the company and "services" of usually attractive young women. In 2008, then New York Governor Elliot Spitzer was forced to resign in disgrace after it was revealed that he paid a call girl over $4,000 for a visit and had done so on numerous occasions.

According to Ruth Rosen (1983), the early twentieth-century campaign against prostitution and the banning of "red-light districts" have made it the lucrative profession it is today. Previously, prostitutes, although exploited, had a certain amount of control over their earnings and working conditions. Their profiteers, madams, landlords, saloon keepers, and other intermediaries were more benign than modern pimps, and the brothel was a rational economic alternative to a sweatshop job at starvation wages.

Photo 13.2

Elliot Spitzer addressed the media with his wife to announce that he would resign from office.

Massage Parlors

"Breslaw Executive Health Spa! 15 Lovely Girls Upstairs and 15 Lovely Girls Downstairs to Serve You!"

Advertisements such as that quoted above appeared on Canadian commercial television (CKCO-TV, Kitchener, Ontario, September 1, 1983, 12:45 a.m. EST). The massage parlor, in which forms of commercial sex are sold under the guise of a health spa or massage service, became quite common in North America in the seventies. Journalist Gay Talese (1979), who recorded extensive participant observation studies of such operations in *Thy Neighbor's Wife,* concluded that for all practical purposes under existing laws such operations constitute legalized prostitution. Since "extras" (prohibited sexual services) must be negotiated and requested by the customer, parlor girls can avoid actual solicitation, and law enforcement agencies must be careful of *entrapment,* or causing illegal activities to occur that would not otherwise have taken place.

Johns

While an extensive literature exists on prostitution, there has been a paucity of information regarding **johns,** or customers, except from interviews with prostitutes. Holzman and Pines (1979) note that much of the literature on johns portrays them as socially, psychologically, or physically inadequate, having to pay for that which others can obtain free as a matter of course (Benjamin & Masters, 1964; A. Ellis, 1959; Gibbens & Silberman, 1960; Laner, 1974; Morris & Hawkins, 1970). The term *trick* is derisively used to refer to the fact that the hooker tricks the john into paying for what he should be able to obtain for free (Milner & Milner, 1972, p. 38). Holzman and Pines indicate that much of this negative evaluation of johns has come from prostitutes and mental health practitioners and that field studies of such customers by social service researchers present a different picture (Armstrong, 1978; M. Simpson & Schill, 1977; M. L. Stein, 1974; Winick, 1962).

M. L. Stein (1974), employing one-way mirrors in order to observe and record hundreds of sessions between prostitutes and clients, was struck by the normal or "straight" quality of the customers. Employing in-depth interviews of a snowball sample of 30 primarily white, middle-class johns (a snowball sample asks the initial interviewee to suggest other subjects), Holzman and Pines (1979) were also unable to support the "pathology-ridden depictions of the clients of prostitutes." All of the subjects indicated current involvements in relationships that involved sex and that they experienced little problem in obtaining sex from nonprostitutes. Some prevailing motivations for visiting prostitutes mentioned by their sample included expectations of mystery and excitement; special "professional" services; and guaranteed, easy, nonentangled sex, which excluded possible rejection. A controversial, although apparently relatively effective, means of cracking down on open solicitation by prostitutes in given urban areas is to prosecute and embarrass johns by publishing their names and addresses in the local newspaper. Modeled after drunk driving and shoplifting programs for first-time offenders, some jurisdictions are experimenting with schools for johns. These remedial classes are designed to make such former customers of prostitutes aware of the seriousness of their offense.

Despite the increased visibility of prostitution in the United States beginning in the seventies, most studies suggest that prostitution has experienced a decline since the pre–World War II period. There appears to be an inverse or negative relationship between sexual permissiveness in a society and the strength of organized prostitution. Prostitution is strongest in countries with traditional concepts of marriage and the double standard–code of sexual behavior that encourages sexual promiscuity on the part of males but discourages such behavior for females. Countries such as France and Italy, for example, discourage divorce and tolerate different expectations of sexual conduct for males and females. Since female sexual expression is discouraged outside marriage, a small proportion of females serve the illicit, erotic desires of the male population. The decline in prostitution can be noted by comparing Kinsey's 1948 and 1953 surveys with more recent ones (Hunt, 1974, p. 144). These show that, while prior to World War II roughly 50 percent of white males had visited prostitutes, in the seventies only 25 percent of college-educated men had visited prostitutes. The erosion of the double standard has eliminated some of the demand for prostitutes' services. Farley (2007) describes Las Vegas as the epicenter of North American prostitution and sex trafficking. She states, "If you peel back the thin, supposedly sexy veneer of the commercial sex trade, you'll quickly see the rotten inside, where females are bought, sold, raped, beaten, shamed and in many, many cases, physically and emotionally wrecked" (pp. 1–2). Mayor Oscar Goodman of Las Vegas wants to legalize prostitution there, as it already is in other parts of Nevada.

Underaged Prostitutes

Since the seventies, concern has been expressed regarding what appears to be growing participation in prostitution by youths, both male and female, under the age of 18. While accurate statistics are difficult to come by, journalistic reports suggest a large recent increase in the sexual exploitation of children, due in part to the youth orientation of society, to growing sexual permissiveness, and to eroding family structures (Booth, 1978, p. 23A). New York City police authorities claim that there are as many as 10,000 "chicken hawks," male prostitutes under 18; another estimate indicates that half of Portland, Oregon's hookers are girls under 18. The so-called Boy Scout sex ring in New Orleans involved 18 Boy Scouts aged 8 to 15; in some cases, the youngsters' mothers were aware of their activities (Booth, 1978).

A large proportion of teenage prostitutes come from "damaged families" and often represent "throwaway children." Many had been raped or sexually abused by surrogate fathers. A Boston ring of homosexual boys being used to sell sex was run by the school bus driver, who peddled the bodies of 8- and 9-year-olds and advertised them through photographs all along the East Coast. When the ring was broken, police arrested a child psychiatrist, a clinical psychologist, a former assistant headmaster, and a teacher at a boys' prep school. The sexual exploitation of children will remain a matter of serious concern, particularly with respect to the long-term psychological impact of such victimizations on the young people involved (A. W. Burgess, 1984; Ritter, 1988; T. Sullivan, 1988; Weisberg, 1985).

One of the biggest trends in prostitution has been a movement from the streets to the Internet. The Internet, pagers, cellular phones, and escort services have all made prostitution less noticeable. Advertisements for sexual services flourish on Web sites.

Video Link 13.2
View a video on fighting child prostitution.

Homosexual Behavior

Homosexuality is the desire for sexual relationships with members of one's own sex. While homosexuality itself is not a crime, certain homosexual activities may be considered criminal, depending on various state or national laws. In certain states, homosexual activity, like some heterosexual activity, may be included under various laws prohibiting adultery, fornication, sodomy, crimes against nature, or lewd and lascivious conduct. *Adultery* is sexual relations of a married person with someone other than her or his spouse. Fornication refers to sexual intercourse between unmarried persons. **Sodomy,** or "crimes against nature," may cover anal intercourse, mouth-genital contact, and even mutual masturbation. The term *sodomy* is derived from the biblical city of Sodom, which (together with Gomorrah) was destroyed by God's wrath because of its rampant eroticism.

Following Judeo-Christian precepts, homosexuality was forbidden and punished in European countries until the French Revolution, after which the laws became more tolerant. Although puritanical America was slower in liberalizing its laws, sodomy statutes generally were not enforced or were selectively enforced for extortion or blackmail purposes.

From the standpoint of law enforcement, there is some criminal activity associated with the homosexual community, primarily on the part of those preying on homosexuals; entrapment, swindling, robbery, blackmail, and sometimes murder may take place. Of primary concern to the criminal justice system are cases in which the behavior is nonconsensual or involves underage minors. Also warranting attention are activities that take place in association with pickups or solicitations in public places, often involving male prostitutes who serve an almost exclusively homosexual clientele.

In 1973, the American Psychiatric Association voted at its national convention that homosexuality no longer be considered a mental illness, although a later survey of members found that a majority still regarded homosexuality as immature and abnormal behavior (Rathus, 1983, p. 395). Up until the sixties, all states forbade homosexual activity.

In 1986, the U.S. Supreme Court upheld a Georgia law that prohibits sodomy (oral or anal sexual relations) between consenting adults. The law carried a punishment of 20 years in prison. That year, 26 states, mostly in the South and West, had existing anti-sodomy laws, and in only 5 states did this refer solely to homosexual activity (Press et al., 1986). Despite decriminalization, homosexual activity is still regulated in most states under sodomy statutes.

Although many variations could be distinguished, those participating in homosexual activity may be simply divided into two types: situational homosexuals and preferential homosexuals. *Situational homosexuals* are those who may prefer heterosexual activity but participate in homosexual activity as a temporary or substitute means of erotic gratification or a means of monetary reward. Preferential homosexuals seek sexual gratification predominantly and continually with members of the same sex. Such individuals tend to develop a homosexual self-concept and to join a gay or homosexual subculture. In actuality, a variety of homosexual roles can be distinguished, including overt or secret, adjusted or maladjusted, true homosexual or situational turnout, and primary or secondary (Clinard & Quinney, 1973, p. 87).

Many individuals participate in homosexual activity but do not identify themselves as homosexual. Much situational homosexuality occurs in isolated sexual environments such as prisons, unisex boarding schools, and military environments. In prisons, for instance, "wolves" exert their masculinity by having fellatio (oral stimulation) performed on them or sodomizing "queens," avowed homosexuals, or "punks," weak males who are forced to perform sexual services (Sykes, 1958, pp. 95–97).

Sexual behavior may reflect opportunity and circumstance in addition to preference. In many traditional Islamic countries, a high premium is placed on virgin brides and there are strong prohibitions against wifely infidelity and even premarital dating. In such a system, heterosexual males use other males for sexual outlets. Sexual assaults on boys are more prevalent than attacks on women in such countries (D. J. West, 1988, p. 183). In 2007, Republican Senator Larry Craig (Idaho) was arrested in a public restroom at the Minneapolis airport that was notorious as a "tearoom" for impersonal homosexual contacts. Interested individuals would use various insider signals such as shoe tapping, hand waving, and body positioning as indicators of interest in such liaisons. Craig's behavior caught the interest of an undercover police officer. Such activity is participated in by consensual adults who are for the most part hidden homosexuals, but since it takes place in public settings it attracts police response. Crime Files 13.1 reports on a study of homosexuality in public restrooms.

Audio Link 13.2
Listen to a discussion of Sen. Larry Craig.

Photo 13.3

Senator Larry Craig was arrested in a public restroom at the Minneapolis airport that was notorious as a "tearoom."

Crime File 13.1

Laud Humphreys's *Tearoom Trade*

The following is a brief account of Laud Humphreys's (1970) controversial and important study of "tearooms," public restrooms where homosexual activity is common. Such solicitations and activities cause most law enforcement attention and arrests for homosexual activity. Despite public fears of child molesters lurking in wait in such places, surprisingly little information existed prior to Humphreys's study of such localities. Using the controversial method of disguised observation, Humphreys posed as a "watch queen," a voyeur who obtains erotic excitement by observing such activities, but also serves the crucial role of lookout for police and other unfriendly strangers who may wander onto the scene and interrupt the homosexual tryst. Humphreys traced the automobile license numbers of participants and showed up at their homes some time later under the guise of performing a mental health survey.

According to Humphreys, tearooms are popular because they provide instant, "no-strings-attached" sex, inexpensive erotic kicks without commitment. They are not gathering places for preferential homosexuals. Such individuals have "come out of the closet," have admitted their sexual preference, and would far more likely be found in gay bars. Tearooms attract a variety of men, only a few of whom are members of the homosexual subculture. The majority have no homosexual self-concept. In fact, over half were married and currently living with their wives.

Humphreys's typology of tearoom participants included: trade, ambisexuals, gay guys, and closet queens. The predominant activity in tearooms is fellatio. Individuals classified as *trade* made up 38 percent of the subjects and were described by Humphreys as "insertors," "fellators," or those who have fellatio performed on them. Most were married, but there was little sex in their marriages. Other sexual outlets, such as affairs, were viewed as too complicated and expensive. Such individuals at the turn of the century would most likely have visited inexpensive red-light districts.

Ambisexuals are more likely to be "insertees"; that is, they perform the oral function. Representing 24 percent of Humphreys's sample, most of these men also were married and indicated that their home sex life was satisfactory. Many enjoyed the adventure, excitement, and risk of such illicit activity. *Gay guys* constituted 14 percent of the participants. Such individuals openly associate in the gay subculture. Humphreys claims that most prefer more permanent, "married" homosexual relationships, which are not to be found in the transitory atmosphere of the tearoom. *Closet queens,* hidden or unavowed homosexuals, made up the remaining 24 percent. Such persons were usually unmarried and fearful of involvement in other areas of the sexual marketplace. Many were particularly interested in young boys, although few of these are to be found in tearooms. Humphreys feels that the unwillingness of "closet queens" to come to terms with their sexual preference and participate in the homosexual subculture makes such individuals potentially dangerous. For most homosexuals the civil consequences of revelation of their activities are a greater personal concern than the threat of criminal penalties.

Web Research Project
Review the topic of "research ethics." Do any of the articles provide guidance in deciding the ethical status of Humphreys's research?

Sexual Offenses

While sexual assault, rape, and adult sexual relations with minors are taken very seriously by the criminal justice system, other acts have been given less attention by authorities in the past three decades and attract a response only when they involve other criminal activity (see Lowman, Jackson, Palys, & Gavigan, 1986; T. Sullivan, 1988). In addition to prostitution and homosexual offenses, some other sexual offenses that have criminal implications include exhibitionism, voyeurism, fetishism, incest, and pedophilia. Related "deviant" sexual activity that may attract criminal predators includes sadism and masochism. **Exhibitionism** usually involves the purposive and unsolicited indecent exposure

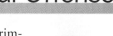

Handbook Article Link 13.2
Read an article
on sexual offenses.

Journal Article Link 13.3
Examine literature regarding arrest histories of sexual offenders.

of sexual parts, usually of the male penis to an unsuspecting female. **Voyeurism** consists of invading the privacy of another by viewing him or her either unclad or in a sexual situation ("peeping Toms"). **Fetishism** involves obtaining erotic excitement through the perception and often collection of objects associated with a desired human sexual object.

Paraphilia

Paraphilia refers to abnormal sexual practices involving sexual interest in nonhuman objects (for example, underwear of the opposite sex), giving or receiving pain, or sexual interest in children who are below the age of consent. Some of these activities, when harmless and committed in private by adults, are of little interest to the criminal justice system (Hickey, 2006b). Fetishism (abnormal interest in objects for sexual arousal) or transvestitism (cross-dressing) would serve as examples. Areas that are of interest to the criminal justice system include asphyxiophilia, or autoerotic asphyxia, such as binding or partial strangulation as a means of sexual gratification; frotteurism (rubbing against people in crowded areas); voyeurism; exhibitionism; sadomasochism; and pedophilia.

Exhibitionism generally involves the purposive public exposure, usually by males, of private sexual parts, in order to elicit shock in unsuspecting victims. While laws prohibiting indecent exposure are equally applicable to both sexes and are usually enforced as a result of public complaint, most "flashers" are male (Forgac & Michaels, 1982). Illustrated by the "dirty old man in a raincoat" who exposes his genitals, exhibitionism may also take the form of adolescent pranksterism consisting of "mooning" (displaying one's bare buttocks to an unsuspecting audience) and "streaking" (running naked through a public gathering). Mooning and streaking are performed for kicks; there appears to be little erotic motivation on the part of the participants. The following annual tradition at one university is illustrative (Landers, 1991):

> Hundreds of Princeton University sophomores shed jeans and down jackets for a traditional event dubbed the "Nude Olympics," held to celebrate the area's first snowfall of the year. About 1,500 spectators cheered them on last week as the students, clad only in boots and hats, lit the torch before doing sit-ups and push-ups in a campus courtyard and ran up and down Nassau Street in Princeton, NJ, reported United Press International. (p. C8)

"Flashers," on the other hand, participate in such activity as a means of sexual arousal and gratification. Most are described as the least harmful of sexual offenders; such exhibitionists are generally immature in their sexual development, wish to evoke fear or shock, and actually would be fearful if the victim acted interested or wanted further contact (Gebhard, Gagnon, Pomeroy, & Christenson, 1965).

Voyeurs attain sexual gratification by viewing others in an unclad state. While legal voyeurism can be practiced in establishments catering to such trade—for example, "topless bars" or adult entertainment districts—illegal voyeurism involves uninvited "peeping" into private homes, parked cars in "lovers' lanes," or other areas. Voyeurs are often called "peeping Toms," a name derived from the fable of the man who stole a peep at Lady Godiva on her naked ride through Coventry. While some patterns of burglary may be associated with voyeurism, in most instances it appears to be pursued as an end in itself. Voyeurism is primarily practiced by juveniles as a means of achieving erotic excitement. Most persistent voyeurs are also at a relatively immature level of psychosexual development and, contrary to the fears of many female victims, do not employ voyeurism as a prelude to sexual attack. In this sense, voyeurs are much like obscene phone callers in that they often fear contact with the opposite sex; otherwise they would avail themselves of readily obtainable erotic outlets in the adult sexual marketplace.

Fetishism involves sexual arousal from the perception of inanimate objects or articles of clothing usually associated with the opposite sex. While some level of fetishism is normal, it becomes abnormal when an individual acquires such items, often through theft, and venerates such articles as a displaced sexual object. There are, for instance, some episodes of shoplifting associated with fetishistic behavior in which the objects are sought because they have significant value for the erotic feelings they arouse.

Of possible concern to law enforcement are sexual practices involving sadism and/or masochism. Sadism involves the attainment of sexual gratification by means of inflicting pain on others. Often unable to achieve sexual arousal or orgasm through any other means, such individuals may harm nonconsenting partners. Masochism refers to the attainment of erotic satisfaction through suffering pain. The masochistic individual must be physically punished in order to gain sexual fulfillment. The leather, whips, and chains school of "kinky sex" is often serviced by prostitutes who specialize in catering to the bizarre needs of their clients.

Not every sexual deviation yields a predictable mode of behavior. While only a very small minority of sexual deviants are potentially dangerous, any type may be associated with more serious criminality in the individual case. The vast majority of persistent offenders exhibit immature psychosexual development and, if their behavior elicits a police response, such offenders are usually treated under civil commitment proceedings. Since most research on sexual offenders relies on official statistics and the majority never come to official attention, far more reliable research is required in this area (Toch, 1979, p. 413).

Nonvictimless Sexual Offenses

Of more serious concern are incest and pedophilia. **Incest** is sexual intercourse between individuals who are legally defined as too closely related to marry. **Pedophilia,** or child molesting, refers to sexual relations between an adult and a child, the latter usually defined as a person under the age of 12 or one who has not yet reached the age of puberty. These two types of offenses clearly are not "victimless" and, as will be described in greater detail shortly, are the most widely condemned and seriously punished of sexual depravations. As defined above, sadism involves the attainment of sexual gratification by means of inflicting cruelty on others, while masochism, sadism's mirror image, involves sexual gratification through suffering physical pain. Unless both parties consent, sadomasochistic activity, sometimes called "S&M," may entail harm or violent victimization.

While many states still have laws prohibiting certain sexual acts between consenting adults, regulations against homosexuality, cohabitation, fornication, adultery, and the like are usually ignored. Societal attempts to regulate obscenity and pornography continue to stir debate.

Sexual Predators

One of the more shocking cases of "serial" child sexual abuse was revealed in an ABC News investigation (ABC, 1992) of former Roman Catholic priest James Porter, who was accused by over 100 former victims of molesting, sodomizing, or raping them when he was their parish priest in the sixties and seventies. A mass silence repressed such memories until one person came forward and organized an investigation into why the Church hierarchy ignored such activity and moved the offender from one parish to another in Massachusetts, New Mexico, and Wisconsin without

Photo 13.4

Former Roman Catholic priest James Porter was accused by over 100 former victims of molesting, sodomizing, and raping them when he was their parish priest in the 1960s and 1970s. He was sentenced to 18–20 years in prison.

Crime File 13.2

Child Sexual Abuse by Catholic Priests

A rising tide of charges against and convictions of Catholic priests for child sexual abuse persuaded the U.S. Conference of Bishops to commission John Jay College of Criminal Justice to undertake an objective national study of the phenomenon. Karen Terry was the principal investigator along with James Levine, who acted as coordinator. Accusations of molestation were found to have been made against 4,392 priests. This represented 4 percent of priests serving between the years 1950 and 2002. The number of people reporting having been abused by priests as children during that time equaled 10,667. The researchers were careful in devising a methodology that would assure confidentiality of both the diocese (which faced possible litigation) and the priests involved. The information was encrypted for each accused priest and then destroyed after analysis. Ninety-seven percent of the 202 dioceses participated by answering the researcher's questionnaire.

While some dioceses reported no cases of abuse, the estimate was as high as 24 percent of priests in one large diocese. Terry suggested that a victimization study of the Catholic population would provide a useful estimate of former abuse. While the study led by Terry probably underestimated the problem, it represented an honest, objective attempt to assess the extent of a previously well-hidden issue.

Sources: "Pulling Back the Veil," by B. Bullag, March 19, 2004, *Chronicle of Higher Education*, pp. A12–13; *The Nature and Scope of Sexual Abuse of Minors by Catholic Priests and Deacons in the United States, 1950–2002*, John Jay College of Criminal Justice, 2004, New York: Author.

Web Research Project
Search for articles that update the issue regarding Catholic priests and pedophilia.

---

warning the new parishes. On December 6, 1993, Porter was sentenced to 18 to 20 years in prison. In 2005, defrocked priest Paul Shanley was convicted of raping and fondling a boy numerous times at his Roman Catholic parish. The activities had begun when the child was 6. Crime File 13.2 summarizes some of the findings of a report commissioned by the U.S. Conference of Catholic Bishops on child sexual abuse by Catholic priests.

Child battering, which can be defined as abuse primarily involving physical assault on children, was discussed in the chapter on violent crime (Chapter 8). While it is difficult to draw clear distinctions between abuse and molestation, this discussion will focus on child molesting, which primarily involves the sexual abuse of children or minors who are past puberty. A child is defined in most states as one who has not yet reached puberty, or age 12 or 14, depending on the state.

Pedophiliacs or *child molesters* are those who have sexual relations with children. For every rape in the United States in a given year, it is estimated that 20 children are molested—roughly 500,000, according to official estimates in 1982 (ABC, 1983a). Many myths exist about child molesters. Some common myths are that molesters are usually strangers, molesters will be caught and jailed, and children quickly get over the emotional harm of having been molested. Most molesters, 85 percent, are known to the child and his or her family. Most are not caught and, if apprehended, are likely to be treated leniently. Fewer than 10 percent of convicted felon child molesters go to prison. One offender claims that psychiatrists will generally release them as long as they do not appear to be "mad dogs" (ABC, 1983a), since such offenders are assumed to be mentally ill, and they are permitted to plead to lesser offenses, even though victims of child molesters report long-term psychological damage as a result of such incidents. Pressure groups such as SLAM—Society for Laws Against Molesters—are lobbying for stricter laws, insisting that repeat offenders receive mandatory prison sentences consisting of a minimum of 4 to 8 years.

The typical act of child molesting involves an adult male and a female victim, usually 11–14 years of age. McCaghy (1976a, p. 87) identified six types of child molesters:

- High-interaction molesters, who have known the children for some time and usually perform or have performed genital fondling
- Incestuous molesters, who take advantage of a child living in the same household
- Asocial molesters, who are involved in illegal careers
- Senile molesters, who are older, poorly educated offenders
- Career molesters, who have persistent offense patterns involving child molestation
- Spontaneous-aggressive molesters, who have had little previous contact with their victims and tend to commit their offenses in a physically forceful and unplanned way

While high-interaction offenders represented only 10 percent of McCaghy's sample, it is likely that they represent the majority of molesters. Since most are well-known to the family and do not employ physical force, they are less likely to be charged with the offense. Although statistics are unreliable, in many cases of sexual child abuse the offenders are family members (Lanning, 2001).

One overlooked aspect of the rising rate of teen pregnancy is the fact that it is often an outcome of child abuse. One study by the Alan Guttmacher Institute found that 66 percent of teen mothers had children by men who were 20 or older. A 1992 Washington state study found 62 percent had been raped or molested before they became pregnant; the offenders' mean age was 27.4 years. "Girls who become pregnant aren't just amoral, premature tarts—they are prey" (J. Klein, 1996). Enforcement of statutory rape laws would be a significant start toward addressing this crime.

There have been other attempts to classify child molesters. Groth, Burgess, Birnbaum, and Gary (1978) describe two types: the "regressed" abuser and the "fixated" abuser. The regressed abuser is one who, having led a fairly normal sexual life, regresses to a sexual interest in children. A previously normal father who suddenly develops sexual interest in children would be an example. Fixated abusers have an early and strongly focused interest in children as sexual objects, often to the exclusion of any other type of adult sexual activity (Crewdson, 1988). While there has been an increase in literature on the topics of sexual molestation, incest, and pedophilia, more such research is needed (Finkelhor, 1986; S. T. Holmes & Holmes, 2008; O'Brien, 1986; Terry, 2005; Vander & Neff, 1986).

In 1990, the "McMartin preschool molestation trial" ended with the acquittal of all the accused. Beginning with accusations that child molesting had taken place in 1983 and continuing through 3 years in court, it was one of the longest and costliest criminal proceedings in U.S. history. The trial may also have represented a "moral panic" and "witch hunt." The jury finally concluded that the state's children's therapist put the child abuse charges into the children's mouths through the therapist's method of questioning the children (Rabinowitz, 1991). The children were believed to be vulnerable to leading questions and wishing to please adults with their answers (A. Hagedorn, 1991). Care must be taken lest we railroad the wrong people in our attempts to fight such abuse. One suggestion that has been made in light of all the charges associated with child care centers is to have a national registry for background checks of child care providers that could be consulted in order to avoid hiring known child abusers.

Incest

Claims of childhood sexual abuse and incest by Roseanne Barr, Oprah Winfrey, La Toya Jackson, and former Miss America Marilyn Van Derbur Atler have opened the door for others to confront long-repressed, painful memories of such abuse.

Incest is related to child molestation, which, although varyingly defined by state or national laws, refers to the universal taboo prohibiting sexual relations or marriage between those who are defined as being too closely related either by blood or marriage. At issue in this discussion is not adult relations, but forms of incest that represent a type of child molesting or sexual victimization in which an adult who is closely related to a child has sexual relations with the child.

The incidence of child sexual abuse by a natural parent is difficult to document, even though the American Humane Association has shown a sharp escalation in such statistics since the mid-seventies. D. Russell (1986) reports that sexual abuse by uncles is more prevalent than father–daughter incest and that incest by stepfathers is seven times more likely than that by biological fathers. "The more 'personal' the relationship between the victim and the offender, the less likely a case of sexual abuse will be reported" (Cardarelli, 1988, p. 9).

Since mother–son incest is rare and brother–sister incest is unlikely to involve as gross an age disparity, father–daughter and father–son incestual relationships are the primary subject of this brief presentation. One study found the average age of female victims to be 10.2 years (Finkelhor, 1979, p. 60). Incestual victimizations may be heterosexual or homosexual. Finkelhor (1979, p. 87) found that, although brother–sister incest was by far the most common, father–daughter incest was most likely to come to the attention of authorities, perhaps because of its more traumatic impact on the family and the child (Goodwin, 1982; Janeway, 1981). Some factors associated with incest include high proportions of stepparent, foster, or adoptive parent relationships; family disorganization (J. D. Herman, 1981); low intelligence; alcoholism; and other types of personality disorganization. While official reports of child battering tend to be more prevalent among low-income families, the American Humane Association Children's Division (1984) reported that sexual abuse and incest are more evenly distributed among social classes. Linda Gordon and Paul O'Keefe (1984), in an analysis of historical records of family violence in the Boston area from 1800 to 1960, did not find that incest offenders were poorer, more alcoholic, or sicker than other assailants. In addition, they questioned the assumption that such violators exhibited pathology or were under external socioeconomic stress.

Characteristics of Sex Offenders

Journal Article Link 13.4
Examine literature regarding recidivism and sexual offenders.

Hans Toch (1979) summarizes much of the research that has been conducted on characteristics of sex offenders:

- Most offenders, far from being "sex fiends," are rather minor offenders.
- Only about 20 percent use force on their victims.
- Untreated, convicted offenders tend to be recidivists in both sexual and nonsexual offenses, but in no greater proportions than nonsexual offenders.
- While few offenders are psychopaths, many suffer from severe neurosis, borderline psychosis, or brain impairment, but most do not fit legal definitions of mental illness.
- Most are emotionally immature and sexually constricted and inhibited, although those involved in rape and incest are more likely to be overimpulsive and oversexed.
- Convicted statutory rapists and those involved in bestiality and incest are more likely to exhibit subnormal intelligence.
- The majority of offenders are young, unmarried, and from poor educational and social-class backgrounds (p. 414; see also R. M. Holmes, 1983, 1991).

Drug Abuse

Video Link 13.3
View a video on America's marijuana industry.

Drugs, chemical substances that alter psychological or physiological functioning, have been used for centuries in various cultures as stimulants or depressants for medical, social, and often religious reasons. Even today in some Middle Eastern countries, alcohol is strictly forbidden by religious law, while the use of other highly addictive substances is tolerated. The fact that drug abuse has moved into the U.S. mainstream can be illustrated by reports of widespread abuse, particularly by professional athletes, entertainers, and other prominent figures. While alcohol is a drug, it is generally not included in most discussions of substances that are abused.

The types of common drugs of abuse include the following:

Cannabis: marijuana, THC, hashish

Depressants: barbiturates, methaqualone, tranquilizers

Stimulants: amphetamines, nicotine, caffeine, methamphetamines

Video Link 13.4
View a video on the meth epidemic.

Hallucinogens: LSD, mescaline, peyote, PCP, psilocybin

Inhalants: nitrous oxide, butyl nitrite, amyl nitrite, and aerosols

Narcotics: opium, morphine, codeine, heroin, and methadone

In the examination of the legal status of drugs and their known harmful effects, the surprising fact that comes to light is that there is often little relationship between the known harmful effects of a particular drug and its legal status in many societies. Substances such as alcohol or nicotine, while possessing mild, short-term, harmful impact, can have lethal, long-term effects and yet they enjoy a legal, sometimes even subsidized status. Drugs such as heroin, which may be lethal in the short run because of overdoses, are not known to be lethal in the long term, but nevertheless are strongly forbidden.

Drugs and History

Opium is believed to have been discovered as early as the Neolithic Age and was used by early physicians such as Hippocrates and Galen (McCoy, 1972, p. 3). Opium, the raw base of other derivatives such as morphine and heroin, was first introduced on a wide scale to the rest of the world by Turkish traders around the eighth or ninth century (Block & Chambliss, 1981, p. 20) and was a trade commodity of European mercantilists as early as the sixteenth century, providing at one point almost half of the revenue of colonial governments. Opium dens controlled by European governments could be found in most Asian cities. When one Chinese emperor objected to such trade, the Opium Wars (1839–1842) were fought, in which the Europeans (the pushers) were the victors. American "China clipper" ships, known as "opium" clippers, had a major piece of this trade (Nash, 1981, p. 166).

In 1805, morphine was derived from opium, but widespread medicinal use of morphine and other derivatives such as codeine brought the onset of serious addiction problems. By 1874, heroin, another opium derivative, was developed and was at first believed to be a nonaddicting miracle drug. Cocaine, which was isolated from coca in 1858, was first thought to be a cure for morphinism and was a popular ingredient of tonics such as Coca-Cola when the first soda fountains were introduced in the 1890s (Inciardi & McElrath, 2001). Its inclusion was outlawed by the Pure Food and Drug Laws of 1906. Backwater patent-medicine peddlers of the late nineteenth century American frontier provided highly addictive remedies such as Dover's Powders, Sydenham's Syrup, and Godfrey's Cordial, which were so widely used that by 1900 an estimated 1 million Americans, mostly women, were opiate users (Brecher, 1972). Brecher described the turn-of-the-twentieth-century United States as a "dope fiend's paradise" (p. 4).

The relatively unregulated distribution of narcotics by physicians and pharmaceutical companies was creating a tremendous drug abuse problem. By 1924, federal authorities estimated that there were 200,000 addicts (McCoy, 1972, p. 5). International concern over growing drug trafficking led to the Hague Convention of 1912, which called for participating nations to crack down on drug distribution. The U.S. response was the **Harrison Act of 1914,** which required a doctor's prescription for narcotics and cocaine. The act required the registration of all legitimate drug handlers, but was not intended to interfere with the legitimate medical treatment of addicts. A vague clause in the law to the effect that physicians could dispense such drugs "only for legitimate medical reasons" led some overzealous federal agents to crack down on offending physicians. By the mid-twenties, an estimated 25,000 physicians had been arrested, with 3,000 serving jail or prison sentences (Goode, 1984, p. 109). The net result of the Harrison Act was that physicians abandoned the treatment of addicts, and the addict as "patient" was replaced by the addict as "criminal," "dope fiend," or outside menace (Duster, 1970; Goode, 1972; Lindesmith, 1965).

Howard Becker (1963) coined the term **moral entrepreneurs** to refer to individuals who personally benefit from convincing the public to label the behavior of others as deviant or criminal. Crime File 13.3 describes two classic moral entrepreneurs: Richard Hobson, "the hero of Santiago Bay," and Harry Anslinger, "the Carrie Nation of marijuana." Thomas Szasz in *Ceremonial Chemistry* (1974),

Crime File 13.3

Moral Panics and the Strange Career of Captain Richmond Hobson—Moral Entrepreneur

"Moral panics" refer to periods in which a previously peripheral issue is pushed onto the social agenda and perceived as a major social menace. Myths associated with such panics may lead to wasteful and dangerous diversion of scarce resources (Jenkins, 1992b; Jenkins & Katkin, 1990). Such panics may be "symbolic crusades" (Gusfield, 1963) in which "moral entrepreneurs" attempt to place their social, moral, or political views at the top of the social agenda or in the forefront of the moral landscape. Even if unenforceable, new laws may make symbolic statements that benefit particular groups (Ben-Yehuda, 1990). Many laws aimed at regulating public morality, though nearly unenforceable, reinforce the values of moral guardians and agencies of social control.

On June 3, 1898, Captain Richmond Hobson, a recent Annapolis graduate and temporary skipper of the *U.S.S. Merrimac*, piloted his vessel into the mouth of Santiago Bay, Cuba. His mission was to sink his ship in the channel, and thus block the Spanish fleet. Premature charges sank the ship before the mission could be accomplished, and Hobson not only failed in his mission but was captured by the Spanish (Epstein, 1977, p. 24).

Ironically, Hobson became the first hero of that short war, lauded in the American press while he was a prisoner in Cuba. When he was released as part of a prisoner exchange, the Navy chose to decorate rather than court-martial him for his incompetence in not accomplishing his mission, and they soon sent him on a cross-country lecture tour. Elected to Congress in 1906, Hobson campaigned first against the "Yellow Peril" (mass immigration of Asian peoples) and later was an organizer of the Women's Christian Temperance Union. A campaigner against the evils of alcohol, he was the highest-paid speaker on the U.S. lecture circuit in 1915. With the demise of the anti-alcohol campaign in the thirties, the undaunted Hobson shifted his crusade to an antiheroin jihad, describing heroin as a vampiric, demonic drug that created the "living dead" and desperadoes. Heroin was viewed as an "enslavement substance" that caused addicts to become criminals.

Hobson's propaganda campaign usefully served the moral entrepreneurship of Harry Anslinger and his efforts to expand the Federal Bureau of Narcotics, which he headed. Anslinger influenced public opinion against marijuana by means of his writing and speeches. One of his articles was titled "Marijuana: Assassin of Youth." In it he portrays a marijuana "addict" who axe-murders his family (Anslinger & Cooper, 1937). Another propaganda feat of the period was a film titled *Reefer Madness*, which similarly displayed marijuana users as rampaging, raving maniacs. Primarily as the result of Anslinger's efforts, the U.S. Congress passed the Marijuana Tax Act of 1937, making marijuana use a criminal matter.

Web Research Project
Examine the issue of "moral panics." In what ways may moral panics be used to marginalize minorities?

perhaps with some exaggeration, compares the drug war with the war on witches and heretics in Europe from 1430 to 1730; the latter cost 300,000 lives and was a reflection of ignorance and superstition. Successive federal legislation from the time of the Harrison Act until the 1970s, such as the Marijuana Tax Act (1937), the Boggs Act (1951), and Narcotics Control Act (1956), were all aimed at controlling drug abuse by means of criminalization and harsher penalties. In a related moral crusade, David Hajdu, in *The Ten Cent Plague: The Great Comic Book Scare and How It Changed America* (2008), describes the "moral panic" surrounding the "comic book panic" of the first half of the twentieth century. The moral entrepreneur that led the onslaught against comic books was Fredric Wertham in his book *The Seduction of the Innocent* (1954). His campaign led to the 1954 televised Senate hearings on the "comic book menace" and delinquency (Kennenberg, 2008).

Drug Use in the United States: The Drug Dip?

Surveys of student drug use had shown declines since the late seventies. Beginning in 1991, however, there was a disturbing reversal of this trend. In the annual Michigan survey of teenagers for the National Institute of Drug Abuse, illicit drug use by 8th graders nearly doubled from 11 to 21 percent, use by 10th graders increased from 20 to 33 percent, and seniors' use grew by about half to about 33 percent. This increase still left the level well below that of peak periods of the 1970s and 1980s (P. Thomas, 1995), and it finally peaked and began to show declines beginning in 1999.

Crack Cocaine

Coke—snow, blow, nose candy, Bolivian marching powder—the drug of Hollywood, of Wall Street, of "sex, drugs, and rock and roll." Cocaine had become the "hip" drug of the last decades of the twentieth century. While at first believed to be nonaddictive, it has emerged as very dangerous, and what was at first thought to be a propaganda film meant to scare people, *Cocaine Fiends,* actually bore a close resemblance to reality (Maranto, 1985). A variation of cocaine, "crack" has considerably raised the level of violence associated with drug trafficking in inner city ghettos. Images of 12- and -13 year-olds carrying Uzi submachine guns and earning more than their parents and teachers are no exaggeration ("Drug Rings Hire Gun-Toting Kids," 1988). The crack cocaine epidemic began in 1986 and peaked by 1992, its declining use reflected in decreasing crime rates.

Photo 13.5

1988 Olympic gold medalist, Ben Johnson, was asked to give up his medals because of steroid abuse.

Another emergent drug is related in part to the fitness craze. "Steroid abuse" came to international attention during the 1988 Olympics when Canadian gold medalist Ben Johnson was asked to give up his medals because of such a drug violation. Bodybuilders and athletes utilize steroids to promote tissue growth and to gain weight and muscle. While outlawed in most athletic organizations, it is used despite increasing research that shows tremendous potential harm. This may include injury to organs, possible increases in aggressive and psychotic behavior, and early death (Weaver, 1988).

A major drug of concern in the twenty-first century is methamphetamine (meth). Meth affects people across the socioeconomic spectrum and seems to be particularly prevalent in white, working-class families in rural areas and small towns. It is considered highly addictive and more powerful than cocaine, but longer lasting in effect. It is produced in simple but dangerous home labs from ingredients found in patented cold medicines such as Sudafed. Crystal meth from Mexico is twice as strong as that produced in home labs.

Drug Abuse and Crime

Some government figures estimating the cost of street crime due to addicts has amounted to statistical overkill. Estimates in the early seventies, such as $18 billion, were several times greater than the total sum of property stolen but unrecovered throughout the entire country using UCR data for the same year (Epstein, 1977, p. 177; Singer, 1971). Despite the attention called to the danger of these "mythical numbers" (Singer, 1971), Reuter (1984b) indicates that in the 13 years since Singer's article, "there is a strong interest in keeping the number high and none in keeping it correct" (p. 136). Estimates both of crime by addicts (J. C. Ball, Rosen, Flueck, & Nurco, 1982; J. M. Chaiken & Chaiken, 1982) and of the estimated size of the illegal drug market remain problematic. Programs such as Arrestee Drug Abuse Monitoring (ADAM), however, have improved the ability to estimate the role of drugs in crime.

The concept of **addiction** is used primarily to describe those who have become dependent on opium and opium derivatives such as morphine, heroin, and various medicines that contain opiates. Addiction involves a physiological dependence commonly referred to as *tolerance,* in which the body requires larger and larger dosages of the substance in order to experience the desired effect. Once this dependence is developed, absence of the required dosage produces the **withdrawal** (or abstinence) **syndrome,** physical discomfort experienced by an addict when deprived of the drug on which he or she has become dependent. Finally, psychological dependence involves mentally connecting the withdrawal syndrome with one's physiological dependence and the decision thereafter to continue to use the substance. The fact that addiction is less a permanent condition than it was once believed to be is illustrated by the statistic that there are few heroin addicts over the age of 35. In addition, a survey of Vietnam veterans found that while about one third used opiates in Vietnam and one fifth were addicted, only 1 percent continued using the drugs on returning to the states (Robins, 1974).

Much of the crime associated with drug addiction is due to the high cost addicts must pay for illegal sources of supply in order to support their need for a "fix." While costs of heroin vary, assuming a hypothetical $50-per-day habit, addicts would have to come up with over $18,000 a year for heroin alone. Unless addicts have occupations that provide either high income or easy access to drugs (such as the medical field), most must steal to support their habit; the major means of support for many is dealing in drugs themselves (Inciardi, 1979, 1981; R. C. Stephens & Ellis, 1975). For others, other types of crime provide the source of funds. Research has suggested that the majority of crimes committed by heroin addicts are crimes against property, such as burglary and shoplifting or prostitution for females, although 47 percent of males had committed robbery (Inciardi, 1981). Gropper (1985) indicates that, contrary to what has been found in past research, heroin-using criminals are as likely as nonusing criminals to kill and rape and are more likely to commit robbery and weapons offenses. In addition, there are a wide variety of different types of drug-involved offenders requiring different types of responses by the criminal justice system (M. R. Chaiken & Johnson, 1988).

B. D. Johnson et al. (1983) did a related in-depth study of 201 New York City heroin abusers. They found most of their subjects were polysubstance abusers. None used only heroin and almost all also used cocaine, alcohol, and other drugs. While most were involved in criminal activity such as shoplifting and burglary, they also supported their habits through being "user-dealers." Nurco, Hanlon, and Kinlock's (1988) study of criminal activity by drug addicts found that, for those previously involved in crime, addiction simply increases an already established criminal lifestyle, while "for those not involved in preaddiction crime, addiction status is associated with a much sharper exacerbation in criminal behavior" (p. 418). In 1988, two thirds of those arrested in New York City, Washington, Chicago, and Los Angeles tested positive for drugs (Anglin & Speckart, 1988; Kurtz, 1989, p. A14). Inciardi (1979), in a study of 356 Miami addicts, found that they had committed 118,134 crimes during a 1-year period.

In examining the relationship between addiction and crime, it is also important to recognize that many addicts were involved in criminal activity prior to their addictions and that many support their habits through funds earned in legitimate occupations. The relationship between addiction and crime is not a necessary one, as is illustrated by the British program of prescribing legal maintenance doses to heroin addicts; in that case, there is little incidental crime associated with addiction. Growth in addiction problems in the United Kingdom has led to some changes (T. Bennett, 1988; "British Clinics," 1985). Much of the crime related to addiction is due to the high price of drugs owing to their prohibition. Erich Goode (1981, pp. 255–256) indicates that there was little crime associated with addiction in the nineteenth century.

Drunkenness

Under English common law, drunkenness itself was not a crime; only when a disturbance of the peace or disorderly conduct occurred was it punished. In the United States, *public drunkenness* is covered by a variety of laws in different jurisdictions, such as public intoxication, breach of peace, disorderly conduct, and inability to care for one's own personal safety. Problem drinking is a primary ingredient in other criminal activity, particularly interpersonal violence. The majority of homicides, aggravated assaults, a large proportion of rapes, and about half of all vehicular deaths are believed to be alcohol-related. Alcoholism or problem drinking remains the number one drug abuse problem in the United States, despite official concern with more esoteric drugs.

The Prohibition Experiment

In a period of missionary zeal, the temperance movement, spearheaded by the WCTU (Women's Christian Temperance Union), pressured the U.S. Congress to pass a Prohibition amendment. This was ratified in 1919 and would, until 1933, constitute what some called "the noble experiment" and others "the great illusion."

Prohibition did not eliminate the alcohol problem. Bootlegging—circumvention of the Volstead Act, the enforcement law of Prohibition—became a national pastime and one of America's largest industries, spawning corruption, organized crime, and public cynicism. As a result of its own failure as a social control policy, as well as of counterpressure from rival urban forces, Prohibition was repealed in 1933. Although alcohol use was decriminalized, it is still regulated by state laws. Just as Brecher (1972) had described the turn-of-the-twentieth-century United States as a "dope fiend's paradise" (p. 4), Rorabaugh (1979) indicates that U.S. consumption of alcohol was much higher during the eighteenth and nineteenth centuries than it is presently. Even today, examination of the problem of drug abuse in the United States finds alcohol abuse still constituting the nation's number one drug problem. "**Problem drinking**" has already been described as a major lethal ingredient in crimes of violence as well as in vehicular homicide (Collins, 1981). The other major alcohol-related problematic area in criminal justice relates to chronic inebriates, who make up over half of U.S. misdemeanor arrestees and county jail inmates.

Problem drinkers consume alcoholic beverages in excess of dietary or social custom to an extent that affects their health and social relationships. Immersed in a drinking culture, some cross the line into problem drinking or alcoholism. While there is no universally accepted medical or psychological model of alcoholism, many accept a "medical model," which describes it as a disease. A useful descriptive model that illustrates this approach is Jellinek's (1960) profile of the stages of alcohol addiction. These are the prealcoholic phase, the intermediate stage, the crucial phase, and the final or bottom phase. The prealcoholic phase involves occasional relief drinking as a means of alleviating tension. After a time, greater amounts of alcohol are needed to generate the desired effect. The intermediate stage occurs when drinking is no longer simply a source of relief, but is sought as a drug. Secretive drinking, occasional blackouts or amnesia, faking alibis, and a compulsion to drink are accompanied by a loss of control. In the crucial phase, the loss of control becomes more complete; the drinker is no longer able to maintain a resolution not to drink. Isolation from others, including family, increases as life becomes alcohol-centered. The final or bottom phase is characterized by the drinker's extensive emotional disorganization. Ethical deterioration, impaired thinking, and obsessive drinking characterize the bottomed-out, chronic alcoholic.

Such chronic inebriates are often handled under what is called the "golden rule disposition" in which, for their own protection, they are picked up by police without formal arrest, jailed overnight, then released in the morning (see Bittner, 1967). Despite frequent arrests or processing, most such individuals do not view themselves as criminals. Arrests for such drunkenness have in fact decreased in many jurisdictions since the early seventies: More police forces have begun to employ strategies of cooperation with local social service agencies that treat such problem drinkers. Typically, local police, when they come across a consistent public inebriate, call such a center. The center's personnel transport the subject to a treatment center where he or she is bathed, "dried out," fed, counseled, and provided the opportunity to break the alcohol-obsessive cycle. In addition to providing more meaningful treatment, such programs relieve the police and jails of an improper burden, freeing up law enforcement resources for more appropriate tasks. The advent of promising drugs such as naltrexone (the "sobriety pill") holds hope for future treatment.

A particularly thorny problem facing campus police has been the problem of binge drinking and alcohol-related arrests on college campuses. In addition to heavier drinking among college students, the upsurge may also reflect greater reporting and enforcement. Alcohol abuse remains a bigger problem on campuses than other drugs. Reporting of such figures is now required by federal law. A survey by the Harvard School of Public Health found about 23 percent of the college student population reporting binge drinking in 1999. This is defined as drinking at least five (men) or four (women) drinks in a row at least three or more times in the 2 weeks before the survey ("Surge in Campus Alcohol Arrests," 2000).

Special Populations

A conservative political climate in the United States in the eighties led to cuts in social programs and a burgeoning homeless population, as well as a deinstitutionalized mentally ill population. The police force became by default social workers of last resort. "They do so because peace officers are unique in providing free, around-the-clock service, mobility, a legal obligation to respond, and legal authority to detain" (Finn & Sullivan, 1988, p. 2).

Isaac and Armat (1990) in *Madness in the Streets* note that the homeless and neglected mentally ill sometimes create the perception of "craziness" on our streets, which is destructive to the social order. The neglectful deinstitutionalization of the mentally ill brings them into oftentimes unpleasant contact with ordinary people, who then have a sense of public disrepair (A. Johnson, 1990). This deinstitutionalization contributes to the overcrowding problem in prisons. Laws related to curfew violations for juveniles also serve as an example of an attempt to regulate public disorder. More kids are arrested in the United States for curfew violations than for any other single category of crime ("Curfew Violations," 1999). Research has so far shown no correlation between such laws and decreases in juvenile crime.

Societal Reaction

The rapid pace of social change and the subsequent cultural lag that it creates have been endemic in the United States in the post–World War II period. We tend to forget that early in the twentieth century cigarette smoking was considered deviant. Similarly, the consumption of alcohol was so dimly viewed that it resulted in a constitutional amendment to forbid its usage.

As previously indicated, Edwin Schur's (1965) concept of "crimes without victims" refers specifically to consensual, adult activities usually conducted in private in which there is no apparent harm to others. The criminalization of such activities, as illustrated by the Prohibition experiment, often involves ineffective overcriminalization, an inappropriate extension of the criminal law into areas of personal conduct and morality. Most public order crime constitutes violations of legislated morality. Prohibition of such crimes represents an effort to control or regulate moral and personal behavior through formal laws, often without attempts to mold public opinion, which is necessary in order to support the legislative and police activity. Since much of the activity is consensual and private, law enforcement efforts often involve invasion of privacy and the use of extraordinary efforts that threaten civil liberties, leading some observers to describe such efforts as not only expensive and ineffective, but also criminogenic (Morris & Hawkins, 1970, p. 2). Sumner's notion, discussed in Chapter 1, that if laws fail to obtain the support of the mores they will tend to be ineffective, suggests that criminalization of these offenses has not markedly decreased their activity. Only a small proportion of offenders are reached by the criminal justice system or deterred by the criminal status of the offense.

In harmful, nonconsensual areas the criminal justice system can have an impact in reducing prohibited activity. For example, in 1988, a U.S. Customs Bureau sting, "Operation Borderline," set up a phony child pornography mail-order house in Toronto and then rounded up many pedophiles who ordered such materials. They were charged under a 1984 Child Protection Act, which outlawed the distribution of all sexually explicit material involving children. Such programs are believed to have reduced considerably the child pornography trade (J. Anderson & Van Atta, 1986; B. Cohn, 1988). In consensual vice activities, however, P. J. Cook (1988) points out, "The criminal law is a cumbersome, costly tool to wield against the harms associated with vice" (p. 1).

In April 1992, the U.S. Supreme Court overturned a conviction and ruled that a Nebraska farmer had been entrapped by postal agents who coaxed him for 2 years to buy mail-order child pornography. "Project Looking Glass," as it was dubbed, resulted in 147 convictions, of which 35 cases showed ongoing or past child abuse, although it also resulted in four suicides by offenders (R. Marcus, 1991).

The history of the regulation of vice has been one of constant symbolic political posturing with little relationship between what is said and what is done. In 1987, President Ronald Reagan not only declared war on drugs but also declared victory, claiming his administration's drug jihad was "an untold American success story." Two years later, national news magazines were sensationally claiming that sections of large cities were so overrun by drug gangs that they were "dead zones" or "Beirut, USA," and that "the drug problem and its accompanying violence has clearly outstripped the resources and capability of local governments, police departments, courts, and prisons to cope with them" (T. Moore et al., 1988, p. 28).

Overcriminalization

Overcriminalization, or extension of the criminal law into inappropriate areas of moral conduct, results in a number of outcomes:

- Many such laws are virtually unenforceable.
- They often lead to corruption of criminal justice personnel and politicians.
- They undermine public respect for the law.
- They create illicit monopolies for organized crime groups.
- They criminalize activities and stigmatize their participants.
- They reflect no consistent, defensible theory of harm (Richards, 1982, p. 194).
- They isolate and embitter offenders.
- Penalties are often ineffective or inappropriate. In the past, for instance, tough drug laws netted many marijuana users and few big drug pushers.
- Such laws tie up law enforcement agencies in thankless tasks that could more appropriately be performed by other social agencies.

Audio Link 13.3
Listen to a
discussion of the
decriminalization
of marijuana.

Decriminalization

Decriminalization refers to the process of lessening the penalties attached to particular offenses. Some arguments in support of decriminalizing many public order crimes include the following:

- Such activities should not be the concern of the state and formal agents of social control, but are more appropriately handled by informal modes of control such as the family, community, church, and the like.
- State interference with much of this behavior often makes matters worse. The criminalization of drug users and view of them as criminals rather than people with medical problems has cut off the legal supply of drugs, created illegal monopolies, and forced many into criminal activity in order to support their habits.
- Such laws tend to accomplish little with those already favorably disposed to such activity. Homosexuality, prostitution, gambling, and the like have been and will continue to be persistent activities in modern society.
- Law enforcement officers' focus on such public order crimes overburdens the criminal justice system with inappropriate tasks, preventing the deployment of resources in combating more serious crimes.

The issue of decriminalization is a matter of degrees of regulation/deregulation rather than of categorical legalization/illegalization. Proposals for decriminalization entail lessening of penalties, but not total abandonment of public or official concern with maintaining some degree of control over such activities. Fears have been raised that decriminalization of such activities constitutes societal approval.

A 1957 British government study, the Wolfenden Report, in examining laws related to homosexuality and prostitution, concluded that private, consensual, adult sexual relations were not the law's business. With respect to homosexuality, one could ask whether individuals really have much choice in or power to change their sexual orientations. Those who oppose decriminalization of homosexual activities fear proselytizing and seduction of the young. Such a **"floodgate theory"** (the assumption that decriminalization of homosexuality will increase such behavior) has not been borne out in England (D. J. West, 1988, pp. 181, 186).

The degree of decriminalization, of course, varies with the type of offense. Few propose decriminalization of predatory or harmful practices such as child molesting or incest, just as few would urge that acts that violate privacy, such as exhibitionism and voyeurism, simply be ignored. Civil commitment proceedings in which psychological treatment is prescribed are an important tool for the protection of society. In public inebriate programs, the police remain involved but maximize the use of community social service agencies. Combined with decriminalization, public media programs can play a role in discouraging undesirable activity. There has been some recent rethinking of the wisdom of decriminalizing public drunkenness. As our earlier discussion of "broken windows" (J. Q. Wilson & Kelling, 1982) suggested, "The presence on the streets of boisterous, obstreperous, and sometimes belligerent drunks contributes to a sense of social disorder" (J. R. Jacobs, 1987, p. 2). A field experiment in Lynn, Massachusetts, demonstrated the efficacy of street-level enforcement (making it difficult for drug dealers to make a sale and for buyers to "score" or purchase) in improving the quality of life in a community (Kleiman, Barnett, Bouza, & Burke, 1988).

The declining number of Americans who smoke tobacco, from 42 percent of adults in 1956 to 33 percent in 1980, shows the effectiveness of a more moderate approach to discouraging the use of harmful substances. Arnold Trebach (1984) in "Peace Without Surrender in the Perpetual Drug War" indicates,

> We did not declare a war on tobacco. We did not make it illegal . . . we did not say that tobacco addicts . . . were evil. We did not seek to disrupt foreign or domestic tobacco supplies. Indeed,

we still subsidize the production of the most dangerous psychoactive drug known to our people. We did not seek to convince our citizens not to smoke through persuasion, objective information and education. . . . Laws do not prohibit smoking entirely, only where and when an addict can take a "fix." In other words, the law did not confront the user head on by absolutely prohibiting this deadly practice. But the law did have a role: it discouraged, it controlled, it curbed, it coaxed. (pp. 136–137)

Erickson (1990) proposed an alternative to the criminalization/decriminalization debate, arguing that a public health approach to demand reduction using social disapproval; informal family, community, and peer group controls; and beefed-up educational efforts could reduce drug usage.

Evidence related to the criminalization/decriminalization debate is uncertain, as can be illustrated by attempts to control heroin or opium abuse (Inciardi, 1990). The pre-1972 British program that medically administered legal doses of heroin to addicts may have meant both a smaller addict population and little crime associated with such addiction (Trebach, 1982). On the other hand, the former British colony of Hong Kong, with the same British program but a different culture, has had an addiction problem much greater than that in the United Kingdom. Others report that international pressure for a consistent drug policy is attempting to force the Dutch government to change its practical and successful drug policy (I. H. Marshall, Anjewierden, & Van Atteveld, 1990). A 5-year experiment in decriminalization of drug use was canceled in Zurich, Switzerland, as the number of addicts and dealers attracted from all over Europe overwhelmed the system (Lynch & Blotner, 1991). Similarly, in the early nineties the Netherlands decided to permit the sale of soft drugs (hashish and marijuana) in order to diminish crime and enable the police to concentrate on hard-drug trafficking. Possession of even small amounts of cocaine, heroin, and other hard drugs was tolerated and viewed as a public health problem to be addressed by treatment centers. However, what began as a successful experiment was flooded by "drug tourists" as European unity collapsed borders. Nevertheless, Dutch policy remains one of rejecting a "war on drugs" model in favor of a harm-reduction model (Leuw & Marshall, 1994).

Arguments for more zealous law enforcement efforts can point to the People's Republic of China, which, through totalitarian policing, appears to have nearly eliminated the problem of drug addiction. Such police powers would be culturally unacceptable in Western democracies and have not been particularly successful in brutal authoritarian regimes such as contemporary Iran. Many of the issues that we have examined in this chapter are complex and laden with heavy moral implications. We have, of course, only scratched the surface of some intense debates on these subjects.

Theory and Crime

The theory that best explains most public order crime is labeling or societal reaction theory. Much of the activity that is covered is deviant, but need not be handled by the criminal law. Sin, amorality, and bothersome behavior becomes criminal when defined so by the criminal law. Moral crusades can result in the addition of activities to be regulated by the criminal justice system, while decriminalization may represent a decision to no longer utilize the criminal justice system as the means of assuring conformity with respect to a particular type of behavior. The war on drugs and the war on alcohol serve as interesting examples. Both have in the past tied law enforcement up with trying to enforce activities that might be better regulated in another manner. Similarly, the law enforcement system must attempt to avoid being used as moral police , as moral busybodies, trying to enforce that which cannot be enforced.

Summary

Public order crime refers to a number of activities that are illegal because they offend public morality. Being primarily nonpredatory, *mala prohibita* acts such as behavior related to prostitution, homosexuality, drug and alcohol abuse, and sexual deviance are sometimes called "crimes without victims," consensual crimes, or folk crimes. Liazos's title "Nuts, Sluts, and 'Preverts'" is an attempt to call attention to the fact that studies of deviance have overconcentrated on the bizarre and kinky at the expense of more serious predatory and elite criminality. Wilson and Kelling's concept of "broken windows" suggests that neglect of public peacekeeping functions encourages disorder.

Prostitution involves sexual relations with emotional indifference on a promiscuous and mercenary basis; the act of soliciting or seeking paying customers is prohibited, not prostitution itself, as defined above. With the exception of some preliterate societies, prostitution exists internationally and has in fact been tolerated throughout most of Western history. Homosexuality, sexual relations with members of one's own sex, may be prosecuted under sodomy statutes that prohibit "crimes against nature," often including mouth-genital and anal intercourse. These same laws, which are generally not enforced, also apply to heterosexual activity. The wide variety of sexual offenses includes exhibitionism (indecent exposure), voyeurism (peeping), and fetishism (unusual veneration of sexual attire or objects). Nonvictimless sexual offenses include incest (intrafamilial sexual intercourse) and pedophilia (child molesting); these two are the most seriously regarded deviations. Sadism (inflicting pain for sexual gratification) and masochism (experiencing pain for sexual gratification) also may entail nonconsensual harm. The legal status of pornography (sexually stimulating media) continues to raise controversy, as does that of gambling and abortion. Drunkenness-related offenses are covered under a variety of state laws such as those prohibiting public drunkenness, breach of peace, disorderly conduct, and so on. The Prohibition experiment, complete criminalization of alcohol usage, was abandoned as a failure. Efforts to control drug abuse, the misuse of chemical substances, has followed much the same pattern as those aimed at controlling alcohol abuse; that is, primary control has been attempted until recently through criminal laws and penalties. There is an inconsistent relationship between drugs' known harmful effects and illegality.

Examination of the criminal careers of most public order offenders finds that most do not view themselves as criminals. Most are participating in either consensual adult relations or—in the case of activities such as exhibitionism—are suffering from some psychological disorder. Prostitution takes a variety of forms, including brothels, bar girls, streetwalkers, massage parlors, and call girls. Johns, prostitutes' customers, do not necessarily fit a pathology-ridden characterization. Underaged prostitution, the sexual exploitation of children, is thought to be growing more prevalent in part because of the erosion of family structure. Homosexuality may be learned as part of the process of socialization; it consists of many types, including preferential and situational patterns. Humphreys's tearoom study of homosexual relations in public restrooms identified several types: ambisexuals, gay guys, closet queens, and trade. Sexual offenders include the more serious child molesters or pedophiles. The latter type is illustrated by cases in which young children have been molested by teachers or other caretakers, showing that many offenders are known and trusted by the victim and his or her family. Incest is another intimate type of victimization. General characteristics of sex offenders were presented.

The history of drugs and drug abuse portrays increased criminalization of drug usage beginning with the Harrison Act of 1914, which resulted in the concept of addiction as a sickness being replaced with that of addiction as criminal. This legal approach to drug policy was viewed as being brought about in part by moral entrepreneurs such as Anslinger and Hobson. Patterns in drug use demonstrate a virtual explosion since the sixties; however, some recent data may signal the first sign of a coming drug dip.

Drug trafficking is highly lucrative and practiced by a large number of groups. While statistics regarding the association between drug abuse and crime have been subject to exaggeration, a tremendous amount of primarily nonviolent property crime is committed by addicts because of the high cost of obtaining illegal drugs. Addiction—which includes physiological dependence (tolerance), psychological dependence, and the abstinence syndrome—is less a permanent condition than is often suggested. Little crime is associated with addiction, however, if legal supplies are available. Problem drinking, is associated with violent crime as well as with the chronic inebriate problem. The law enforcement burden imposed by the latter has been alleviated, in part, through greater utilization of social service agencies.

Societal reaction to public order criminality runs the gamut between overcriminalization and decriminalization. Totalitarian societies simply forbid such activities, while democratic societies must constantly balance the tensions between civil liberties and social morality. Various problems raised by overcriminalization were described, while arguments for decriminalization were also discussed.

Criminology on the Web

Log on to the Web-based student study site at http://www.sagepub.com/haganintrocrim7e/ for author-created podcasts, e-flashcards, quizzes, and more.

KEY CONCEPTS

Addiction

Broken Windows

Closet Queens

Crimes Without Victims

Decriminalization

Exhibitionism

Fetishism

Floodgate Theory

Folk Crime

Harrison Act

Incest

Johns

Moral Entrepreneurs

Overcriminalization

Pedophilia

Problem Drinking

Public Order Crime

Sodomy

Tearoom

Types of Prostitution

Voyeurism

Withdrawal Syndrome

REVIEW QUESTIONS

1. What are "public order crimes"? How do they and their offenders differ from other types?

2. What is the notion of "broken windows"? How has this been applied in modern policing?

3. Why is prostitution such a persistent crime? What have been some recent trends in prostitution and attempts to regulate it?

4. What do you feel should be the role of society in regulating pornography? Defend your views.

5. Discuss the history of attempts to regulate drug abuse. Do you feel that greater criminalization or decriminalization is required to properly deal with this problem?

6. What is a "moral panic" and what are "moral entrepreneurs"? Give some examples.

7. What are paraphilias? Of what concern are these to the criminal justice system?

8. Discuss the problem of sexual abuse of children by Catholic priests. Do you think enough is being done to address this problem?

9. What is intended by the title "Nuts, Sluts, and 'Preverts'"?

10. What was the effect of the Harrison Act of 1914 on drug regulation in the United States?

WEB SOURCES

Against Drug Legalization
http://www.usdoj.gov/dea/demand/speakout

American Civil Liberties Union
http://www.aclu.org

Drug Enforcement Agency
http://www.usdoj.gov/dea

Marijuana Legalization Organization
http://www.mjlegal.org

National Alliance on Mental Illness
http://www.nami.org

National Center for Missing and Exploited Children's CyberTipline
http://www.cybertipline.com

National Institute on Alcohol Abuse and Alcoholism
http://www.niaaa.nih.gov

National Institute on Drug Abuse
http://www.nida.nih.gov

Prostitute's Education Network
http://www.bayswan.org/penet.htm

SexualPredators.com
http://www.einet.net

WEB EXERCISES

Using this chapter's recommended Web sites, investigate the area of public order crime.

1. What did you learn from your visit to Sexual Predators .com?

2. Compare the Marijuana Legalization Organization site with the Against Drug Legalization site.

3. What are some current concerns discussed on the American Civil Liberties Union site?

4. What are some issues raised in the Prostitute's Education Network site?

5. Perform an online search on the "COYOTE" organization as well as on the "meth problem."

SELECTED READINGS

Howard Becker. 1963. *Outsiders: Studies in the Sociology of Deviance.* New York: The Free Press.

  Becker, a major proponent of labeling theory, explores the role of moral entrepreneurs in making deviants into criminals and outsiders.

Georgette Bennett. 1987. *Crimewarps: The Future of Crime in America.* Garden City, NY: Anchor Books, Doubleday.

  Bennett makes a series of predictions regarding future trends in crime as they are affected by technology, population distribution, and crime prevention strategies.

Elliott Currie. 1985. *Controlling Crime.* New York: Pantheon.

  Currie has been a consistent liberal voice in an era of conservative crime control policy who objects to the view of the causes of crime being intractable results of biology, psychology, and personal choice.

Diane Gordon. 1990. *The Justice Juggernaut.* New Brunswick, NJ: Rutgers University Press.

  Gordon argues that reliance on formal modes of control, such as the police, courts, and corrections, while ignoring poverty, unemployment, and urban decay has not worked.

Bertram Gross. 1980. *Friendly Fascism: The New Face of Power in America.* New York: M. Evans and Company.

> Gross claims that the public is not being told the "big dirty secret of crime": that the criminal justice system is soft on white collar crime.

James Inciardi and Karen McElrath, editors. 2001. *The American Drug Scene* (2nd ed.). Los Angeles: Roxbury.

> In perhaps the definitive work on the subject, the authors provide an excellent selection of current and classic articles on the ever-changing drug scene.

Rael J. Isaac and Virginia Armat. 1990. *Madness in the Streets.* New York: The Free Press.

> The subtitle of this book explains its focus—"How Psychiatry and the Law Abandoned the Mentally Ill."

Phillip Jenkins. 1992. *Intimate Enemies: Moral Panics in Contemporary Great Britain.* Hawthorne, NY: Aldine de Gruyter.

> One of the most lucid writers in the field tackles conventional wisdom regarding emergent crime problems, asking whether instead they are examples of moral panics.

Alfred W. McCoy. 1972. *The Politics of Heroin in Southeast Asia.* New York: Harper and Row.

> This controversial work accuses the CIA of encouraging drug trafficking in order to fight the North Vietnamese.

Eleanor Miller. 1986. *Street Woman.* Philadelphia: Temple University Press.

> This is Miller's ethnographic study and interviews with 64 prostitutes. It examines their motivations and escape from abusive domestic settings.

Computer Crime
  Types of Computer Crime
  Types of Attacks on Computer Systems
  Argot of Computer Crime
  Crime File 14.1 Operation Bot Roast: Bot-Herders Charged as Part of Initiative
  Crime File 14.2 Cracking Down on Sexual Predators on the Internet
  Online Predators
  Crime File 14.3 The Bogeyman: Online Sexual Predators
  Crime File 14.4 Protecting Children in Cyberspace: The ICAC Task Force Program
Cyberterrorism
Public and Legal Reaction
Theory and Crime
  Crime File 14.5 Cyberspace Security: Breaking Ground in the New Frontier
  Crime File 14.6 A Fine Point: Mapping Intel Sources
Summary
Key Concepts
Review Questions
Web Sources
Web Exercises
Selected Readings

chapter 14

Computer Crime

A simple reading of history shows that the relationship between crime and technology is by no means new and that the potential for creating harm never seems to be far away from any apparently beneficial technological development.

—David S. Wall (2007, p. 12)

Computer Crime

In July 2009, a major cyberattack took place against U.S. government Web sites by North Korea or some other foreign government. Attacked were sites including the White House and the Pentagon as well as ones in South Korea. The "bot net" assault involved over 100,000 "zombie" computers that were used without the owners' awareness. These third parties were unaware that they were even involved. This is just one illustration of the burgeoning area of computer crime in the twenty-first century.

Beginning in the sixties, computers became essential elements of modern society. The Internet was originally created in the United States for the Department of Defense. Later it was also used in higher education. With the advent of inexpensive personal computers (PCs), it expanded in the 1990s to business and the general public. Other names for computer crime are "high-tech crime" and "cybercrime." Wall (2007, p. 10) tells us that William Gibson (1982) popularized the term *cybercrime* in his 1984 novel *Neuromancer*. Computer or cybercrime refers to crime that is committed using a computer. Major concern about criminal acts committed with computers has led to the passage of much computer crime legislation. Hollinger and Lanza-Kaduce (1990) point out that both the experts at the time (particularly Parker [1979, 1983] and Bequai [1978, 1987]) and legislatures relied heavily on media accounts to encourage more laws. The 1983 movie *WarGames,* in which a fictitious young hacker uses his computer to crack the North American Air Defense Command (NORAD) computer in Wyoming, almost triggering a nuclear war, caused the media to "fixate" (Hollinger & Lanza-Kaduce, 1990, p. 33) on computer crime.

In 1988, in one of the earliest cases of computer forensics, David Copenhefer of Corry, Pennsylvania, had kidnapped and murdered his neighbor's wife. Police were led to him as a suspect when they noticed that a sign in the window of his bookstore had the same distinct script that appeared on the ransom note. Obtaining a search warrant, they examined the files on his computer. Copenhefer did not know

Journal Article Link 14.1
Examine literature regarding cybercrime.

that deleted files could, with painstaking efforts, be rescrambled in part from the hard drive. With hard work, the police were able to reconstruct most of the 22-point kidnap plan, the ransom note, and drafts of the text for phone calls that he thought he had deleted. Their electronic discovery, lifting evidence from hard drives, although routine today, handed Copenhefer the death penalty (Hoppin, 2003). The case well illustrates the use of the computer as a tool for murder and also the fact that computer crime is not a separate type of crime; rather, the computer is a tool that can be used in the commission of a variety of different crimes. The computer could be used as an instrument to commit an offense, it could be the target of the offense, or it could simply be incidental (Grabosky, 2007, p. 11).

Photo 14.1

"Software pirates" cost software manufacturers billions of dollars a year in lost revenue.

Types of Computer Crime

While many traditional crimes such as embezzlement, robbery, and burglary are now facilitated by a new tool—computers—other crimes are more specifically computer related. The illegal production of computer software is very big business in Southeast Asian countries and particularly in Hong Kong, where illegal copies of popular software such as Lotus 1–2–3 are sold for a fraction of their normal cost ("Taking a Byte Out of Crime," 1990). "Software pirates," those who illegally reproduce and use software, cost manufacturers billions of dollars a year in lost revenue. U.S. manufacturers have found hundreds of illegal copies of their software in legitimate firms such as Britain's General Electric or Atari Taiwan Manufacturing. When illegal software is discovered, manufacturers attempt to bring both civil and criminal charges against the offenders and their companies. Part of the problem is the ease of making duplicate copies of computer software. This is comparable to the technological ease of illegally reproducing television movies onto a videocassette recorder or DVD. Some examples of computer crime include insider crime, malicious hacking, activities in support of criminal enterprises, telecommunications fraud, online pedophilia, and high-tech espionage (Spernow, 1995).

It is estimated that about 80 percent of computer crimes are committed by "insiders" or employees. Most of this, however, is not reported to police. Thirty-five percent of the theft of proprietary information is perpetrated by discontented employees; outside hackers steal 28 percent, U.S. companies 18 percent, foreign corporations 11 percent, and foreign governments 8 percent, according to a 1999 survey by Kessler and Associates, a New York security firm (Noack, 2000). Malicious hackers desire forbidden knowledge. Many have an anti-establishment attitude or exhibit a "hacker's ethic" that there should be no restrictions on their right to surf the Net and test systems. Computers can be used to support criminal enterprises. This could entail planning crimes or record keeping, and many such criminal operations are better equipped than law enforcement agencies.

Computers can also be used for gambling, child pornography, security fraud, production of phony documents, counterfeiting, and the like. In 1993, it is estimated that U.S. banks lost $5 billion in fraud and $815 million to fraudulent checks created on PCs. Beginning in 2000, the National White Collar Crime Center's (NW3C) and the FBI's Internet Crime Complaint Center began producing an annual Internet Crime Report (ICR). In 2006, they processed 86,279 complaints of crime to federal, state, and local law enforcement. The total dollar loss of reported complaints was $724 per complainant. In all, they processed over 200,481 complaints. The total loss was $198.44 million. Internet auction fraud was the most reported offense, accounting for 44.9 percent of referred complaints. Nondelivered merchandise or payment accounted for 19 percent. Check fraud made up 4.9 percent, and credit/debit card fraud, computer fraud, confidence fraud, and financial institutions fraud accounted for the remainder

TABLE 14.1 ■ **Types of Computer Crime**

Spernow's (1995) typology of computer crime:
   Insider crime
   Malicious hacking
   Activities in support of criminal enterprise
   Telecommunications fraud
   Online pedophiles
   High-tech espionage

Wall's (2001) typology of cybercrime:
   Cybertrespass—crossing boundaries into other people's property and/or causing damage (e.g., hacking, defacement, viruses)
   Cyberdeceptions and thefts—stealing money, property (e.g., credit card fraud, intellectual property violations, a.k.a. "piracy")
   Cyberpornography—breaching laws on obscenity and decency
   Cyberviolence—doing psychological harm to or inciting physical harm against others, thereby breaching laws relating to the
      protection of the person (e.g., hate speech, stalking; pp. 3–7)

Wall's (2007) three types of online offending:
   Offending related to the integrity of the computer system (e.g., hacking, cracking, vandalism, spying, denial of services, use of
      viruses and Trojans)
   Offending assisted by computers (e.g., used to acquire money, goods, or services dishonestly; Internet frauds; phishing;
      "419" schemes; auction manipulation)
   Offending that focuses on the content of computers (e.g., hate crimes, pornography, incitement; pp. 49–50)

(NW3C, Bureau of Justice Assistance, & FBI, 2007). The highest losses were for Nigerian letter scams ($5,100), check fraud ($3,744), and other investment fraud ($2,695). In California, a Vietnamese Triad gang duplicated payroll checks on PCs. There is even an underground publication called "2600" in which hackers share their findings on compromised numbers. In 1994, Kevin Mitnick (code name "Condor") messed with computer expert Tsutomu Shimomura's home computer files, and Shimomura and others helped the FBI track him down. Mitnick at the time was being hunted by the FBI for various violations, including theft of about 20,000 credit card numbers from computer systems. He was captured in February 1995 after nearly destroying a computer service by reading its Internet subscribers' mail and using their accounts as a way of attacking computers across the Net.

Photo 14.2

Online gambling has become very popular, with online poker revenues increasing dramatically between 2001 and 2005.

Telecommunications fraud can take a variety of forms. A common type is the stealing of pin numbers for making long-distance calls by looking over the user's shoulder or by using scanners. Those stealing cellular phone numbers then use equipment to make the cell phones operate with these stolen codes (C. J. Levy, 1995). Online pedophiles target young "wannabe" hackers and participants in chat rooms, taking advantage of ready access to those with the age, gender, and intellectual and social characteristics they seek in their victims. Those involved in high-tech espionage are often industrial spies who sell their knowledge to rivals or foreign bidders. The FBI has established a computer analysis and recovery team to attempt to keep up with the burgeoning use of cybersleuthing (see also Icove, Seger, & VonStorch, 1995).

Types of Attacks on Computer Systems

Audio Link 14.1
Listen to a
discussion of
spyware.

Coined by hackers in 1985 in order to defend against journalists misusing the term hacker, *crackers* are ill-intended hackers who attempt to crack (break into) computer systems, often in order to do damage. Grabosky (2007, p. 6) explains that in time the term "cracker" fell into disuse and the term "hacker" was used to describe unwanted electronic guests regardless of motive. They might be thought of as cybervandals. The NW3C (2000) identifies the following types of attacks on computer systems:

Denial-of-service attacks

E-mail bombs

Dictionary attacks

Trojan horses

Password phishing

Web spoofing

Worms

Sniffers

Social engineering

Other attacks include

Network scanning

Key loggers

Steganography

Bot nets

Photo 14.3

Kevin Mitnick was the most wanted computer criminal in United States history at the time of his arrest in 1995.

Denial-of-service attacks involve programming a computer to continuously send fake authentication messages to a targeted server, keeping it constantly busy and forcing out legitimate users. On February 9, 2000, an orchestrated attack of bogus traffic shut down Amazon, eBay, CNN, Buy.com, and Yahoo. Such a denial-of-service or synchronized (SYN-flood) attack floods the computer with requests; when the computer attempts to reply, it is unable to connect to the phony addresses. This type of attack is now routine, and many site operators claim they fend off 9 or 10 similar attacks per week. The suspected culprits of the February 9 attacks were two teenage boys (NW3C, 2000).

E-mail bombs involve overwhelming an e-mail system with an enormous amount of mail. A Monmouth University (New Jersey) student crashed the university's e-mail system by sending e-mail bombs to two university administrators, consisting of 24,000 random text messages that damaged the system (NW3C, 2000). In a related "ping flood" attack, a "smurf" (malicious Internet user) fooled hundreds of computer systems into sending traffic into one location, flooding the location with "pings" (hits).

A dictionary attack consists of guessing numerous common passwords in order to log on to a computer system. Software programs such as Cracker are used to run through potential passwords until the correct one is found.

Trojan horses are subprograms (hidden in a program) that contain a virus, bomb, or other harmful feature. These often masquerade as inviting attachments that offer harmless software upgrades, help files, screen savers, or pornography. When users open the attachment, a secret program steals their password and mails it back to the cracker, and then that outgoing message is deleted from the victim's e-mail outbox (NW3C, 2000).

Password phishing entails a cracker stealing password, account, or credit card information. An America Online (AOL) subscriber could not understand why his dial screen told him he was inactive. AOL had canceled his account because 500 people had received a Trojan horse from his e-mail. Crackers had stolen his password and used his account to spread a Trojan horse designed to steal more numbers.

Web spoofing creates a false (shadow) version of a Web site that is controlled by the attacker. All network traffic between the victim's browser and the shadow Web is funneled through the attacker's machine. This enables interception and alteration of information and the acquisition of passwords, credit card numbers, and account numbers. In 1999, clients of the second-largest bank in the Netherlands attempted to log on to the bank's Web site to access their accounts and received an error message the first time they entered their password and user name, but were able to get on the second time. The first attempt was actually on a shadow site set up by a cracker who used the stolen information to take five guilders (about $2.35) from each account (a "salami slice"), a sum not immediately noticed (Regan, 1999).

Worms are similar to viruses that reproduce themselves and subvert computer systems. While a virus must be carried from system to system, worms can spread with no assistance. The term "worm" originated in a science fiction novel, *The Shockwave Rider,* by John Brunner (1975), where freedom fighters attack a totalitarian government's computer with a program called Tapeworm, which shuts down the network (NW3C, 2000).

Sniffer programs are used to gain passwords to access accounts. This can be used to access software, impersonate the owner, or gain access to other accounts.

Social engineering is the manipulation of people in order to obtain critical information about a computer/network system. A typical example involves the target receiving a telephone call from someone identifying him- or herself as being with tech support and claiming the server is being reset with new passwords and that the caller needs the old password to get the account running again. Users of AOL or MSN, for example, might receive fake messages stating there is an account billing problem and requesting that the user enter his or her name and password. An old, low-tech standby is "dumpster diving" in which crackers seek out old manuals, memos, program printouts, internal phone books, and other items from which to retrieve user information.

Network scanning programs are used to find vulnerable networked computers, while key loggers are secret programs that record a user's keystrokes. Steganography hides text or information within other information or images. Bot nets are computer robots—programs that run automatically. These are often used to transmit "spam," or unsolicited bulk mail messages. Crime File 14.1 reports on the operations of a "bot herder."

Argot of Computer Crime

The world of computer offenses has produced its own argot, a fraction of which is defined below.

- **Cyberpunks** are computer hackers who develop harmful programs.
- **Salami techniques** are used to steal small amounts (slices) from assets of many sources and transfer them to the thief's account. For example, a fraction of a percent of thousands of savings accounts could be retained for the thief's account.
- **Time bombs** (logic bombs) are computer programs that perform a task, such as printing a message or destroying data, on a certain date.
- **Vaccines** are computer programs that seek out and destroy viruses.
- **Viruses** are rogue programs that copy themselves onto other programs or disks (Markhoff, 1988).

Crime File 14.1

Operation Bot Roast: Bot-Herders Charged as Part of Initiative

They're called "bot-herders": hackers who install malicious software on computers through the Internet without the owner's knowledge. Once the software is loaded, they can control the computer remotely. And once they've compromised enough computers, they have a robot network to botnet. Some botnets are huge: tens of thousands of infected computers. Or more. As a result of Operation Bot Roast, an ongoing and coordinated initiative to disrupt and dismantle these bot-herders, we've identified about 1 million computers across the country that have been compromised.

- The FBI has also charged numerous individuals with cyber crimes. Robert Alan Soloway of Seattle, Washington, is accused of using botnets to send tens of millions of spam messages touting his website;
- James C. Brewer of Arlington, Texas, is accused of infecting tens of thousands of computers worldwide, including some at Chicago-area hospitals; and
- James Michael Downey of Covington, Kentucky, is charged with botnets to disable other systems.

As the investigations continue to unfold, it is possible we will uncover more victims. Here are some important things to remember:

First, if you believe your computer has been compromised, do not call the FBI directly. We are not in a position to provide technical assistance. Please see Ways To Protect Your Computer.

Second, if you determine you are a victim, then we encourage you to file a complaint online through our Internet Crime Complaint Center.

Third, the FBI will not contact you online and request your personal information: Be wary of fraud schemes that request this type of information, especially via unsolicited e-mails.

Operation Bot Roast was launched because the national security implications of the growing botnet threat are broad. The hackers may use the computers themselves, or they may rent out their botnets to the highest bidder. The more computers they control, the more they can charge their clients.

A bot-herder can do a lot with compromised computers:

- Steal the computer's identity;
- Launch massive spam campaigns;
- Engage in click-fraud—schemes which artificially inflate the number of visitors to a website; and
- Launch denial of service attacks that can cripple web servers and crash sites.

One of the difficulties in fighting this type of cyber crime is that it is difficult for computer owners to know their machines have been infected. There is no easy way to tell, unfortunately. It may be running slowly, your outbox may be full of mail you did not send, and you may get mail stating you've sent spam. "The majority of victims are not even aware that their computers have been compromised or their personal information exploited," said FBI Assistant Director James Finch, who heads our Cyber Division. That's why we urge every computer owner to implement the security precautions that are available. Prevention is always better than reaction.

Resources:

Botnet Consumer Alert

Onguard Botnet Guide

Home Computer Security Guidelines

Botnet Tips

FBI Internet Crime Complaint Center

Source: "Operation: Bot Roast," *FBI Headline Archives,* June 2007. Available at http://www.fbi.gov/page2/june07/botnet061307.htm.

Journal Article Link 14.2
Examine literature regarding identity theft.

Computer crimes come in a wide variety of forms:

- Sunbelt Software discovered a "massive identity theft ring" that used "keystroke loggers" (or "keylogging"—a method of capturing and recording user keystrokes) to obtain information that could be used to make fake online identities. They could then log in information connected to bank accounts.
- In 2007, Robert Soloway, the "Spam King," was arrested and charged with fraud, money laundering, and identity theft. He had hijacked computers from unsuspecting users and used them to send spam advertising his fraudulent marketing schemes. This was the first time that federal prosecutors used identity theft laws to prosecute a spam case ("Longtime Spam King Charged," 2007).
- In 2005, Sven Jaschan, creator of the Sasser worm and Netsky virus, was sentenced in Germany to 21 months' probation and 30 hours of community service. The Sasser worm attacked PCs throughout the world while the Netsky e-mail virus accounted for 25 percent of all viruses for the first 6 months of 2005 (NW3C et al., 2007).
- In 2007, it was disclosed that hackers stole information from 45.7 million credit and debit cards from retailers such as TJ Maxx and Marshalls in what was believed to be the largest breach of consumer information in history.
- Volkswagen lost nearly $260 million because of an insider computer scam involving fake currency exchanges.
- In Fort Worth, Texas, a former insurance company employee deleted more than 160,000 records from the company's computer.
- "Demon dialers," automatic speed dialers used by computer hackers, were used to dial 800 numbers until the codes were broken. The hackers then used the ID codes to steal $12 million in phone service from NASA (Rogers, 1992).
- In August of 1995, the managers at the Naval Command, Control and Ocean Surveillance Center in San Diego discovered in their computer system several innocuous-looking, unauthorized files with names like "sni 256" and "test." When they opened them up, they found a "sniffer" program. As noted above, this is a program that secretly copies vital information such as the passwords given by legitimate users when they log into the system. Investigators tracked the intruder to Telecom Argentina and Julio Ardita ("Hacker Traced," 1996).
- Online stock traders and day traders utilize "pump and dump" techniques to personal advantage. Traders pump up their stocks in newsgroups and chat rooms to boost sales, then dump the stocks at a profit, causing the value of the stock to plummet.
- Of critical concern is information warfare by foreign intelligence services and terrorists.
- A bank employee hacked a computer and placed an order for Brinks to deliver 44 kilograms of gold to a remote site, collected it, and vanished.

Part of the problem with attempting to crack down on juvenile computer hackers is the "hacker's ethic," noted above—a view that such illegal and potentially dangerous experimentation is a necessary part of the computer training for a creative next generation of hackers. Apple Computer cofounder Steve Wozniak claims that a little mischief is essential in the quest for knowledge. He points out that his experience in building illegal "blue boxes" for "phone-phreaking" (devices for making free phone calls) helped him develop his later computer hardware skills (J. Schwartz, 1990, p. 37).

Handbook Article Link 14.1
Read an article
on identity theft.

Crime File 14.2

Cracking Down on Sexual Predators on the Internet

In 1993, a 10-year-old boy disappeared from his Brentwood, Maryland, neighborhood. Within weeks, the investigation would uncover two pedophiles and a larger ring of online child pornographers. Within two years, it would spawn a national initiative that is now the centerpiece of the FBI's efforts to protect children from predatory pedophiles in cyberspace.

Here's How the Events Unfolded. When FBI Agents and Prince George's County police detectives went door-to-door to talk with neighbors following the boy's disappearance in 1993, they encountered a pair of suspicious men who had been "befriending" local children, showering them with gifts and even taking them on vacation. Evidence followed that the men had been sexually abusing children for a quarter century. More recently, they had moved online, setting up a private computer bulletin board service not only to "chat" with boys and set up meetings with them but also to share illicit images of child pornography. That, in turn, led investigators to a large ring of computer pedophiles. When a similar case with national reach turned up the following year, the FBI realized it was onto an alarming new trend: sexual exploitation of children via the Internet.

A Program Is Born. In 1995, the FBI created its Innocent Images National Initiative (IINI). Its goal: to break up networks of online pedophiles, to stop sexual predators from using the Internet to lure children from their families, and to rescue victims. Today, 28 of the FBI's 56 field offices have undercover Innocent Images operations. More than 200 FBI agents work these cases. Some pose as teenagers or pre-teens in chat rooms to identify "travelers" who seek to meet and abuse children. Others focus on dismantling major child exploitation enterprises. Since 1995, the FBI has opened more than 10,000 total cases and helped secure nearly 3,000 convictions.

Keeping Safe. To report child pornography and/or potential cases involving the sexual exploitation of children, please contact the Crimes Against Children Coordinator at your local FBI Field Office. You can also file an online report at the National Center for Missing and Exploited Children's CyberTipline at http://www.cybertipline.com; these reports are forwarded to the appropriate law enforcement authorities.

Source: Quoted directly from "10 Years of Protecting Our Children: Cracking Down on Sexual Predators on the Internet," FBI Headline Archives, 2003. Available at http://www.fbi.gov/page2/dec03/online120203.htm; also check out the FBI's "Innocent Images" Web site, particularly the "Guide to Internet Safety."

Research Project
Find articles on "pedophiles on the Internet." What have been some recent developments in this regard?

Online Predators

Video Link 14.1
View a video of how the Internet is transforming the childhood experience.

Online pedophiles represent an electronic version of letting perverted bogeymen into a child's bedroom. The bedrooms in this case are hooked up to the Internet and chat rooms and Web sites that entice naive youngsters. One example was the "Wonderland Club," an international pedophile and pornography network. Members had to be approved by at least three other club members, and prospects had to demonstrate that they had access to 10,000 images of child pornography. The images they traded were of children as young as 2 being raped and tied up, and having sex with animals, adults, or other children. Police in 12 countries synchronized simultaneous raids and arrested 100 suspects. In the operation, police were unsuccessful in infiltrating the group but traced about 200 members through wiretaps, online transmissions, and agents in chat rooms (Goodspeed, 1998). In 2007, undercover officers in Britain, Canada, and the U.S. busted a pedophile ring. Its chat room featured images, including live videos of children, some as young as 1 month old, being subject to sexual abuse such as rape and brutality. Fifteen children were rescued in the U.K. and seven in Canada. Crime File 14.2 reports on attempts to crack down on sexual predators.

Crime File 14.3

The Bogeyman: Online Sexual Predators

"Evil does exist and so does the Bogeyman. I know, I met him."
—Alicia Kozakiewicz, kidnapped by an online sexual predator at the age of 13

On New Year's Day 2002, Alicia Kozakiewicz, then 13 years old, disappeared from her suburban Pittsburgh home. "I walked out my door and found out the bogeyman was real and he lives on the Web," Alicia explains. Family and investigators had suspicions that the teen had been lured by someone she had met in an Internet chat room. Acting on a tip from the predator's online acquaintance to whom the predator had been bragging that he was holding the young lady captive, authorities were able to track Alicia to a townhouse in Herndon, Virginia, where they found her physically restrained. FBI agents arrested a 38-year-old computer programmer, who is currently serving a sentence of almost 20 years.

Source: Presentations by Alicia and her mother, Mary Kozakiewicz, at the Mercyhurst College Criminal Justice Conference on "Cyber Crimes in Cyber Times," October 11, 2007.

Photo 14.4

Today co-anchor Meredith Vieira interviews Chris Hansen, host of *To Catch a Predator*, about keeping children safe online.

Crime File 14.3 reports on an incident of online sexual predation while Crime File 14.4 describes the FBI program to catch online predators.

In 1999 the FBI put online its National Sex Offender Registry, a computerized database of convicted pedophiles. This provides instantaneous background checks to law enforcement agencies. The assumption of such a registry is that, once predatory pedophiles are convicted, they lose forever their right to hide.

Crime File 14.4

Protecting Children in Cyberspace: The ICAC Task Force Program

Children have embraced the Internet with remarkable alacrity, as they go online to learn, play, and communicate with their friends. The Internet clearly influences how a growing number of children discover and interact with the world around them. Unfortunately, cyberspace is not always a safe place for youngsters to visit. Some sex offenders use the privacy and anonymity of the Internet to prey on vulnerable children and teenagers, whose access is often unsupervised. In exchanging child pornography or seeking victims online, sex offenders may face little risk of interdiction of their criminal activities. To combat computer-related sex crimes, the Office of Juvenile Justice and Delinquency Prevention funds the Internet Crimes Against Children (ICAC) Task Force Program, which protects children in cyberspace. Although apprehending sex offenders who use the Internet to facilitate their crimes presents significant challenges, the ICAC Program can help state and local law enforcement agencies to develop an effective response to online enticement and child pornography cases. Included in this response are community education, forensic investigation, and victim service components.

Large numbers of young people are encountering unwanted sexual solicitations that, in the most serious cases, involve being targeted by offenders seeking children for sex. Research conducted by the University of New Hampshire and the National Center for Missing and Exploited Children (NCMEC) disclosed that 1 in 5 children ages 10–17 received a sexual solicitation over the Internet in the past year (Finkelhor, Mitchell, & Wolak, 2000). One in 33 received an aggressive solicitation—that is, the solicitor asked to meet them somewhere, called them on the telephone, or sent them regular mail, money, or gifts. Cloaked in the anonymity of cyberspace, sex offenders can capitalize on the natural curiosity of children, seeking victims with little risk of interdiction. These offenders no longer need to lurk in parks and malls. Instead, they roam the chat room looking for vulnerable, susceptible children. Today's Internet is also becoming the marketplace for offenders seeking to acquire material for their child pornography collections. More insidious than the exchange of sexually explicit material among adults, child pornography often depicts the sexual assault of a child and is often used by child molesters to recruit, seduce, and control their victims. Although not all molesters collect pornography and not all child pornography collectors molest children, significant consensus exists among law enforcement officials about the role pornography plays in the recruiting and controlling of new victims.

The debate about the role child pornography plays in triggering actual victimization continues. Many in the law enforcement community believe that the validation and nearly constant stimulation afforded to sex offenders by the Internet put minors at great risk for sexual exploitation. The purpose of the ICAC Program is to help state and local law enforcement agencies develop an effective response to cyberenticement and child pornography cases that encompass forensic and investigative components, training and technical assistance, victim services, and community education.

ICAC Task Force Program in Action

- A citizen contacts NCMEC's CyberTipline [http://www.cybertipline.com] to report that an adult male is using the Internet to locate minors for sex. CyberTipline analysts forward the information to the ICAC Task Force officers, who begin an inquiry into the subject's online activities. The subject forwards nude photographs as he schedules a meeting with the undercover officers. Upon arresting the offender, the officers learn that he had previously victimized four children who were ages 8–11.

- A middle-age male sends child pornography photos and a video to an undercover officer posing as a 13-year-old boy. He invites the "boy" to his house to watch some other child pornographic videos and later admits his sexual intent. A search of his home reveals more than 1,000 pornographic videotapes (many of them containing child pornography), hundreds of photographs of child pornography, and numerous magazines and calendars depicting children in sexually explicit positions.

- Task Force members execute a search warrant at the residence of a registered sex offender and seize his computer. An examination of the computer locates three additional victims living more than 2,500 miles away whom the suspect had abused for years.

- A Task Force undercover operation results in the arrest of a middle-aged man after he drives nearly 1,000 miles to meet a 14-year-old girl for sex. Following his arrest, officers seize an axe handle, shovel, and several gallons of gasoline from the trunk of his car. The suspect's house includes a dungeon outfitted with torture instruments and an extensive collection of serial killer videotapes.

- Parents notify Task Force investigators of their concerns about a chat room relationship their 14-year-old son has developed with a stranger. Officers assume the boy's online identity and, within 2 days, the suspect makes arrangements for a sexual encounter with the boy.

Conclusion

If children are to thrive in today's world and compete in tomorrow's workplace, they must be able to safely play, learn, and grow in cyberspace. Challenges exist, but the work of the ICAC's Task Force Program and its federal agency partners is a vital step in protecting children in the Information Age.

References

D. Finkelhor, K. Mitchell, and J. Wolak. (2000). *Online Victimization: A Report on the Nation's Youth.* Arlington, VA: The National Center for Missing and Exploited Children.

Source: Quoted directly from "Protecting Children in Cyberspace," by M. Medaris and C. Girouard, January 2002, *OJJDP Juvenile Justice Bulletin* 191213. Available at http://www.ncjrs.gov/pdffiles1/ojjdp/191213.pdf.

> **Web Research Project**
> Visit the FBI's Innocent Images National Initiative Web site (http://www.fbi.gov/contact/fo/balt/major.htm) and report on some of its services and activities.

The television public was mesmerized in 2006 by *Dateline NBC*'s "To Catch a Predator " series in which online predators were lured on Internet chat rooms to show up at the homes of minors whose parents were supposedly not home. They were confronted instead by the *Dateline* host, who proceeded to interrogate them on camera before they were arrested by the police. While performing a public service in alerting parents to dangers on the Web, the impression may have been given that the major threat exists in cyberspace rather than the fact that the majority of sex crimes against minors are from someone known to the child, often a parent or guardian (Levy, 2006, p. 20).

With the advent of the Internet, many sexual predators moved from the schoolyard to the Internet in search of young victims. Such stalking online is usually of adolescents (over 12 years of age) and not young children. The greatest number of victims have been 14 to 17 years old, thus the perpetrators were technically not pedophiles who prey on children. Many predators utilize romance and seduction in order to corner their victims, and a surprising number are known to the youth in person (Finkelhor, Mitchell, & Wolak, 2001). Online predators troll the Internet, actively seeking recruits. Finding an ASL (A Sex Location), they groom (brainwash) and build up the trust of their victims and then move toward actual contact. The Youth Internet Safety Survey (Finkelhor et al., 2001) found that 6 percent of the young people they surveyed reported having been the victims of reputed harassment (threats, rumors, or other offensive behavior) during the past year. Two percent reported disturbing harassment. Few of the victims of solicitations, exposure to pornography, or cyberbullying reported this to authorities.

Experts suggest that parents should play a more proactive role in monitoring their child's use of the Internet and urge them to report inappropriate behavior to authorities. Online reporting resources include the National Center for Missing and Exploited Children's (NCMEC) CyberTipline (http://www.cybertipline.com). One mother of a victim of abduction by an online predator simply states, "Our children are not supposed to be our friend. We have a right to read their e-mail. Who lets their children spend time with perverts? Who has computers in their bedroom?" (Kozakiewicz, 2007).

Cyberterrorism

Video Link 14.2
View a video of
cyberterrorism.

Video Link 14.3
View a video on
fighting terrorism
with technology.

If savvy young hackers can do major damage over the Internet, one can only begin to imagine the possibilities of a hostile group or foreign power intent on doing harm to another developed country via computer. This could include, but is not limited to, critical infrastructure attacks, and use for propaganda purposes, fund-raising, fraud, recruitment, and covert communications. Virus attacks can cripple Internet operations, and competitive intelligence of open source data by hostile groups could supply information that would be very helpful in planning and executing terrorist attacks. (Competitive intelligence involves investigating and analyzing business operations, while open source data are data that are freely available either in hard copy or on the Internet.) Terrorists, transnational criminals, and intelligence services are using exploitation tools such as computer viruses. Trojan horses, worms, logic bombs, and eavesdropping sniffers can all be used to compromise the integrity of or deny access to data. In 2007, European security officials were successful in identifying the attackers who hit computers in the offices of the head of a European government: units of the Chinese military, the People's Liberation Army, in Shanghai and Beijing. The Chinese government denied such allegations (Hosenball, 2007). Crime File 14.5 looks at the latest in cybersecurity from the FBI.

Public and Legal Reaction

Audio Link 14.2
Listen to a
discussion of
prosecuting
cybercrime.

This chapter began with a description of the Copenhefer case in which a PC was used to commit a kidnap-murder and the contents of this same computer were utilized to solve the case. The advances in law enforcement since that time in the use of computers are described in Crime File 14.6.

The twenty-first century is still in its infancy, and one can only wonder what unbelievable technological breakthroughs await us. Sociologist William Ogburn (1922) proposes the concept of "cultural lag" in order to explain the influence of technology on social institutions. That is, he views change as taking place unevenly—technological changes take place far more quickly than the social aspects of a culture such as its laws and customs. Technological change is often a neutral force that occurs very quickly, and often the social values and institutions of the society are left behind, unable to adjust to the effects of such rapid technological change. Robert Merton (1968) views social change—including technological change—as often producing a two-edged sword. The **manifest functions** of social change are the anticipated or intended consequences of, for instance, new technology. The **latent functions** are the unanticipated, unintended, hidden functions of this same change. Modern postindustrial societies will continue to accelerate the pace of change, and certainly cybernetics (replacement of human mental processes by machines), the second industrial revolution, will continue to lead the way.

Theory and Crime

Computer crime is so relatively new that most traditional theories do not specifically address it. One classic sociological theory that provides some explanation is Ogburn's theory of cultural lag (1922). As previously indicated, he argues that social change takes place unevenly and that technological change occurs more quickly than the social aspects of a culture. Computers can change communications, commerce, and even dating behavior far more quickly than the capability of societal values and norms to keep pace.

Crime File 14.5

Cyberspace Security: Breaking Ground in the New Frontier

Bots. Worms. Spoofing. Phishing. Cyber terrorism and espionage. Malware. Hacking. Virtual copyright and identity theft. Online child exploitation. Internet scams and spam.

Crime has a new frontier—the vast, digitized realm of wired and wireless communications. And for the FBI, it's meant building a new set of technological and investigative capabilities and partnerships—so we're as comfortable chasing outlaws in cyberspace as we are down back alleys and across continents.

Last week, Director Robert Mueller and FBI Cyber exec Steve Martinez traveled to Silicon Valley to talk about these growing capabilities and partnerships. The venue: the annual RSA Security Conference, which brings together thousands of representatives from technology companies and their customers to discuss cyber security issues. In his keynote address at a town hall meeting sponsored by the Business Software Alliance, the Director spelled out key capacities and new initiatives that are helping us to address cyber threats:

- A new Cyber Division at FBI Headquarters "to address cyber crime in a coordinated and cohesive manner";
- Specially trained Cyber Squads at FBI Headquarters and in each of our 56 field offices, staffed with "agents and analysts who protect against and investigate computer intrusions, theft of intellectual property and personal information, child pornography and exploitation, and online fraud";
- New Cyber Action Teams that "travel around the world on a moment's notice to assist in computer intrusion cases" and that "gather vital intelligence that helps us identify the cyber crimes that are most dangerous to our national security and to our economy";
- Our 93 Computer Crimes Task Forces nationwide that "combine state-of-the-art technology and the resources of our federal, state, and local counterparts";
- A growing partnership with the U.S. Secret Service—[with which] we share federal jurisdiction for fighting cyber crime—through its nationwide network of electronic crimes task forces.

As a result, the Director said the FBI continues to "break new ground in the investigation and prosecution of cyber criminals." He cited some recent examples, including:

- The arrest of 20-year-old hacker Jeanson James Ancheta, a well-known member of the botmaster underground, who pled guilty last month to "seizing control of hundreds of Internet-connected computers and renting the network to people who mounted attacks on websites";
- Within two weeks' time, the capture of the cyber criminals responsible for creating and spreading the Mytob and Zotob worms, thanks to the help of law enforcement officials from Turkey and Morocco and to the intelligence provided by private senior sector partners.

The Director's final message to cyber professionals: report your security breaches and alert us to pending threats so we can work together to head off cyber attacks. The Director said he realized some companies fear possible negative publicity and loss of marketplace competitiveness if they do so, but he indicated that the FBI will minimize disruptions to their businesses and guard proprietary or confidential information.

Source: Quoted directly from "Cyberspace Security: Breaking Ground in the New Frontier," *FBI Headline Archives,* February 2006. Available at http://www.fbi.gov/page2/feb06/rsamueller022106.htm.

Web Research Project
Visit the FBI Headlines Archives (http://www.fbi.gov/page2/page2archive.htm) and find an updated story on FBI investigations of computer crime.

Crime File 14.6

A Fine Point: Mapping Intel Sources

When investigators arrived at a bar in Philadelphia after a shooting in July that left four people dead, their questions elicited an all-too-common response: nobody saw anything. "It was a quadruple homicide in a neighborhood bar and nobody's talking," said Special Agent Bill Schute of our Philadelphia [FBI] field office. "It happens all the time. Witness intimidation is a very real factor. So what you have to do as law enforcement is go out and get information. Nobody's going to hand it to you."

But where exactly is the information? The answer, Shute says, can usually be found within a 400-yard radius of a crime scene, especially when it's within an inner city. Shute knows this because that's one of the first things he does when police detectives come up empty at a crime scene. He draws a big circle around it on a digital map and then watches as potential leads—informants, sex predators, probation violators, bail jumpers, community leaders—appear as color-coded icons on the grid. The map also plots locations of past homicides and non-lethal shootings. Shute's three-year-program, called Project Pinpoint, integrates existing court, police, and FBI records into an off-the-shelf street map program that gives investigators visual leads on whose doors to knock on first for valuable information on current cases or even emerging threats. "What Project Pinpoint does is take a street-level approach to gathering intelligence," Shute said.

For example, a probation violator living near a crime scene might be motivated by a police visit to tell what he knows about a crime. Likewise, a witness might share information with an informant or confide in a community leader rather than go to law enforcement authorities. The reluctance to share information with law enforcement—particularly in urban settings—is a driving force behind the program. Once confronted privately, however, most witnesses tend to cooperate with authorities and give good information. On occasion, witnesses become trusted informants; the program has tripled the informant base for the Violent Crimes Task Force in Philadelphia since 2004, Shute said.

Shute says working with local detectives and warrant officers has given him valuable insight into street-level law enforcement. "The incredible amount of criminal intelligence that exists on the street is the very thing that I wanted to see us capture," he said. "This can only be accomplished through the data provided by local law enforcement and is one of the biggest reasons the FBI needs to remain strongly involved in the violent crime program." The program has been adopted by the violent crime task forces in Detroit and Denver and is slated for use in at least a half-dozen other major metropolitan areas. Agents in most of our 56 field offices have used the program as well. In fact, its use has spread beyond the violent crimes investigative areas [and is now being used to address] public corruption, gangs, and counterintelligence.

In Philadelphia, two of Pinpoint's biggest successes were in 2005. The program led agents to arrests in the separate slayings of a city police officer and a nine-year-old boy. In days after the multiple shootings at the Southwest Philadelphia bar in July, it helped identify potential witnesses and assisted with the recovery of the murder weapon. Shute says Pinpoint's beauty is its simplicity. "You don't need a Master's Degree in computer science to use this program. We can teach it to anyone in ten minutes. We use it every day."

Source: Quoted directly from "A Fine Point: Mapping Intel Sources," *FBI Headline Archives*, October 2007. Available at http://www.fbi.gov/page2/oct07/pinpoint100507.htm.

Web Research Project
Using a Web browser, locate and review some recent articles on new technology and policing.

The difficulty of applying specific theories to computer crime is that computer crime is not a separate type of crime. It represents a tool or means of committing other types of crime. It makes possible more sophisticated means of committing vandalism, robbery, embezzlement, and espionage. It certainly adds a dimension to be reckoned with in the twenty-first century.

Summary

Computer crime (cybercrime) promises to be the major technologically fueled crime of the twenty-first century. The Copenhefer case illustrated the use of a computer to commit a kidnap-murder and to solve the crime. Computer crime is not a separate type of crime; rather, the computer is a tool or means of committing many different crimes. It can be used to commit an offense, be the target of the offense, or simply be incidental to the offense. The types of computer crime include insider crime, malicious hacking, activities in support of criminal enterprises, telecommunications fraud, online pedophilia, and high-tech espionage. Malicious hackers seek forbidden knowledge and often support the "hacker's ethic," which is that there should be no restrictions on a hacker's right to surf the Net. Types of attacks on computers have generated their own *argot* (specialized language), which includes Trojan horses, phishing, spoofing, worms, bot nets, key loggers, salami slices, time bombs, and viruses. Recent law enforcement efforts in beefing up cybersecurity are described in this chapter, as well as "Project Pinpoint," which illustrates the potential of using computer software in order to solve crimes. "Cultural lag" refers to a gap created when technology changes faster than other aspects of a culture. *Manifest functions* are intended or planned consequences of social change, while latent functions are unintended, negative, or hidden consequences.

Criminology on the Web

Log on to the Web-based student study site at http://www.sagepub.com/haganintrocrim7e/ for author-created podcasts, e-flashcards, quizzes, and more.

KEY CONCEPTS

Argot of Computer Crime	Hackers	Spoofing
Bot Nets	Latent Functions	Trojan Horses
Computer Crime	Manifest Functions	Types of Crackers
Cultural Lag	Phishing	Virus
Hacker's Ethic	Salami Techniques	Worm

REVIEW QUESTIONS

1. Discuss the various types of computer crime.

2. What have we learned about online sexual predators?

3. Discuss the concept of the "hacker's ethic."

4. What have been some recent law enforcement efforts in cybersecurity and the use of the computer in criminal investigations?

5. Discuss some recent examples of attacks on computer systems.

6. Discuss some of the terminology of computer crime that you learned in this chapter.

7. What are virtual crimes? Give some examples?

WEB SOURCES

Botnets, Cybercrime and Cyberterrorism
http://www.fas.org/sgp/crs/terror/RL32114.pdf

Computer Crime and Intellectual Property Theft
http://www.cybercrime.gov

Computerworld Security
http://www.computerworld.com

Cyber Criminals Most Wanted
http://www.ccmostwanted.com

Department of Defense Cyber Crime Center
http://www.dc3.mil/

FBI Cyber Investigations
http://www.fbi.gov/cyberinvest/cyberhome.htm

Infosyssec: Information System Security Professionals
http://www.infosyssec.com/index.shtml

National Institute of Justice Electronic Crime Program
http://www.ojp.usdoj.gov/nij/topics/technology/electronic-crime/welcome.htm

U.S. Department of Justice—Searching and Seizing Computer Evidence
http://www.cybercrime.gov/ssmanual/index.html

WEB EXERCISES

Using this chapter's recommended Web sites, explore the issue of computer crime.

1. What are some of the types of issues with respect to computer crime that are currently of concern to the FBI in its cyber investigations?

2. What are some concerns in searching and seizing computer evidence?

3. What are some features of the National Institute of Justice Electronic Crime Program?

4. Using your browser, search the term "bot net" and explain and give examples of such attacks.

Peter Grabosky. 2006. *Electronic Crime.* Upper Saddle River, NJ: Pearson Prentice Hall.

This short, engaging book provides the reader with a very useful introduction to computer crime.

Michael Knetzger and Jeremy Muraski. 2008. *Investigating High-Tech Crime.* Upper Saddle River, NJ: Pearson Prentice Hall.

This is an excellent detailed presentation on the latest in computer crime investigations.

Jesus Mena. 2003. *Investigative Data Mining for Security and Criminal Detection.* Boston: Butterworth Heinemann.

This handbook describes in detail the many computer-based intelligence programs that are available to fight computer crime.

Deborah Osborne. 2006. *Out of Bounds: Innovation and Change in Law Enforcement Intelligence Analysis.* Washington, DC: Center for Strategic Intelligence Research.

Osborne addresses the changing nature and role of analysis in policing. Her case studies make excellent use of computer analysis in crime solving.

Robert W. Taylor, et.al. 2006. *Digital Crime and Digital Terrorism.* Upper Saddle River, NJ: Pearson Prentice Hall.

This is a detailed textbook on computer crime, particularly as it applies to terrorism. It provides excellent coverage and examples for law enforcement investigation.

D. S. Wall. 2007. *Cybercrimes*: *The Transformation of Crime in the Information Age.* Manchester, UK: Replika.

Wall's book is a must for obtaining a historical as well as current perspective on computer crime.

Majid Yar. *Cybercrime and Society.* London: Sage.

This is a very informative account of computer crime and provides excellent examples and case studies.

The Future of Crime
  Predicting the Future of Crime: Methods
  Other Crime Predictions
  Crimewarps
  The Future of Digital Crime
  Other Predictions
  British Home Office Predictions
  Crime File 15.1 Anticipating Future Trends in Crime and Disorder Audits
  Crime File 15.2 Hot Products: Understanding, Anticipating, and Reducing Demand for Stolen Goods
Summary
Key Concepts
Review Questions
Web Sources
Web Exercises
Selected Readings

chapter 15

Epilogue

The Future of Crime

A century ago the famous philosopher Santayana observed that those who forget the past are condemned to repeat it. Today those who ignore the future are in for a rude shock when it arrives.

—*Peter Grabosky (2007, p. 107)*

Predicting the Future of Crime: Methods

In describing the difficulty of performing accurate, long-term forecasts of the nature and future of crime, Schneider (2002) points out,

> Although all uncertainty cannot be removed, it is still possible to systematically formulate a range of possibilities using established methods and analytical tools. The tools of the disciplined futurists according to Cole (1995) are a sound methodology, a sense of history and theory, knowledge of key factual data, and the ability to examine in the contexts of broader social, political, technological, and economic trends. The data sources and methods used to guide forecasting include crime statistics; surveys of experts, practitioners, and the general public; literature reviews; scenario writing; and statistical (time series) models that extrapolate crime trends into the future. (p. 1)

Journal Article Link 15.1
Examine literature regarding crime waves, law creation, and crime prevention.

Mathematical models use quantitative time series to forecast crime trends. Environmental scanning attempts to identify future developments (trends or events that might influence crime). These may include economic conditions, demographic shifts, governmental policies, enforcement resources, international events, social attitudes, and technological advances. Surveys and the judgment of experts are also used. Delphi techniques, named for the Greek oracle, are procedures that ask a panel of experts to predict the future. Schneider (2002) cites the 2000 Canadian Foresight

Program that adopted the Delphi technique in using a questionnaire and surveying 80 experts in law enforcement, insurance, loss prevention, academia, science, and the computer industry to assess future developments affecting criminal activity. Scenario writing is yet another method. It attempts to describe how present developments might appear in the future. Usually, a number of possible scenarios are presented at the same time. Great Britain and its Home Office have done the most work on crime forecasting.

Police Futurist International is an organization that exists for the purpose of predicting the future of policing. Founded by former FBI Special Agent William Tafoya, the organization seeks to act as a forum for those interested in crime forecasting. The three primary goals of futures research are to form perceptions of the future (the possible), study likely alternatives (the probable), and make choices to bring about desirable outcomes (the preferable).

Criminologists lack a crystal ball or legerdemain with which to look into the future and forecast its likely direction. However, along with other social scientists and futurists, they can demarcate some likely directions, though even these are affected by a myriad of variables whose trends may not be fully appreciated (G. Stephens, 1982; Tafoya, 1992).

The cost, quantity, and international scope of criminality are likely to grow and will continue to play a major role in future street crimes. Having the largest civilian armed population in the world, the United States is likely to continue to lead developed nations in criminal homicide. Urban robbery rates are likely to remain high, depending on a number of factors, including employment opportunities for young, minority males. Crime rates in central cities are likely to remain high as long as federal policies use them as dumping grounds for national problems such as racism and inequality. Official statistics on violence in the family, spouse and child abuse, and rape are likely to continue to increase in response to better reporting and more supportive social programs for victims. Most areas of public order criminality are likely to continue to experience decriminalization combined with better regulation and social, psychiatric, and medical support systems. Common property crimes are likely to decrease, being most responsive to the aging demographic profile. In the area of professional criminal behavior, frauds related to securities, credit cards, and computer records as well as "knockoffs" and related counterfeit products will remain problematic. The illegal marketing of human body parts to physicians and researchers will likely be a growth area.

Organized criminal activity will persist, although its dominance by Italian American syndicates will continue to wane while a multitude of groups capitalize on the criminal seed money of the new Prohibition era (drug prohibition)—catering to the demand for illicit drugs. Transnational crime will continue to expand in response to global interdependence. Political criminality will continue, particularly in the international arena, with terrorism employed as war on the cheap. Radical Islamic extremists such as al Qaeda are likely to continue to plague the international community. It is hazardous to make even these predictions, however; criminologists and others do not agree in their forecasts.

Other Crime Predictions

Some other crime predictions include the following:

- Electronic tagging, DNA analysis, and recognition systems based on retinas or fingerprints could all be used to tackle crime in the twenty-first century. In 2007, the FBI was building the world's largest computer database of people's physical characteristics such as digital images of faces, fingerprints, and palm patterns. Biometric information will include iris patterns, face shapes, scars, and the way people walk and talk, and will be used to solve crimes and identify criminals (Nakashima, 2007). Called "next generation identification," this technology also opens the door to "biometric spoofing." That is, in the future, criminals may be able to use this same technology to copy faces, irises, voices, fingerprints, and DNA in order to commit crimes.

Handbook Article Link 15.1
Read an article on crime prevention.

Journal Article Link 15.2
Examine literature regarding lessons learned from Enron.

- Australian Federal Police Commissioner Mick Keelty predicts that robotics and cloning may be future challenges for law enforcement ("Top Cop Predicts Robot Crimewave," 2007).
- There will be an increase in scams related to virtual worlds such as Second Life, where people spend real money via credit cards to purchase products including virtual real estate, roles, or gifts ("Top Cop," 2007).

Crimewarps

In *Crimewarps,* Georgette Bennett (1987) made a number of interesting predictions, which over two decades later we can better judge:

Photo 15.1

Second Life is a virtual world that launched in 2003 and has been growing ever since. Users can attend classes, look for jobs, and purchase/sell goods and services.

- The computer will be the biggest generator of crime in the future.
- The concentration of crime in the United States will continue to shift to the Sunbelt and the West.
- Low birth rates and high work rates will leave a plethora of unguarded homes ripe for daytime burglary.
- The growing service economy will create many part-time jobs that, combined with fewer student dropouts, will mean less crime.
- More abusive families will emerge as the number of young, single, poor, and undereducated mothers grows.
- Industries with older workers will experience less theft.
- The growth in the elderly population will increase medical quackery and insurance fraud.
- Fear of AIDS will reduce the demand for streetwalkers.

The Future of Digital Crime

R. Taylor, Caeti, Loper, Fritsch, and Liederbach, in *Digital Crime and Digital Terrorism* (2006), provide eight forecasts for the future of digital crime:

1. Crimes involving computers will increase substantially, requiring major changes in resource allocation and training, and the creation of new police specialties.
2. Internet fraud, including identity theft, will be the largest computer problem.
3. Virtual crimes against persons will increase at a faster pace, requiring new laws and new types of crime investigation and prevention.
4. Some hacker groups will evolve into networked criminal enterprises.
5. Organized crime groups will increasingly use the computer as a criminal instrument.
6. Terrorist groups will increasingly use the Internet for communications as well as for cyberterrorism.
7. Espionage will continue to evolve into information warfare, economic espionage, and intellectual property theft.
8. Criminals, terrorists, and anarchists will accelerate the use of technology to steal data, destroy communications, and in other ways do harm (pp. 354–379).

Journal Article Link 15.3
Examine literature regarding the new war on terrorism.

Audio Link 15.1
Listen to a discussion of cyberdefense.

Video Link 15.1
View a video of white-collar crime trends.

Other Predictions

- Employee computer crime will increase as computer literacy increases.
- Financial crimes and white collar crime in general will increase.
- More harmful computer viruses will be generated and spread.
- Future technologies will produce new vices and crimes such as pleasure robots, sensic addiction, and neural simulation, which will become legally regulated.
- Internet threats will continue to evolve in their malice and technology.
- Public observation via closed-circuit television will increase, as will privacy invasion.

British Home Office Predictions

Crime File 15.1 reports on some of the predictions and methods of the British Home Office.

Crime File 15.1

Anticipating Future Trends in Crime and Disorder Audits

Modeling trends in crime, developing models to explain changes in crime and examining the way trends might develop in the future is complex and resource intensive. It may well be beyond the capabilities of most local [crime reduction] partnerships. However, there are a number of ways in which local partnerships can usefully look at future trends in crime and disorder when undertaking crime and disorder audits.

- Look at local crime trend analysis
- Look at macro-level research predicting crime trends
- Think about new technology

A distinction should be made between predicting crime trends and forecasting future crime levels:

- A forecast of future crime levels would require a model that took account of all relevant social, demographic and economic factors that are causes of crime, as well as the likely effect of current and future policy initiatives likely to impact upon crime. As such, it is conceptually and practically impossible to provide forecasts with any degree of accuracy.
- Predictions in future crime trends indicate the likely effect of specific factors that have been shown to have a relationship with crime, assuming that no other factors come into play. They do not take into account the impact of other factors that have a relationship with crime.

Local Crime Trend Analysis

Much of the data analysis described above could be used to make simple predictions about the way in which local crime trends are likely to fluctuate in the near future. For instance, a temporal analysis of crime might indicate that crime rates fluctuate seasonally according to seasonally changing leisure habits (for instance, more town centre drinking during the summer) or seasonal changes in populations (for instance, an influx of tourists during the summer or seasonal agricultural workers in the summer or autumn). Such trends should be used to shape the development of strategies. Developing formal models to predict crime trends would be a more complicated and resource-intensive exercise.

Macro-Level Research Predicting Crime Trends

There is a wide range of research that examines the causes and trends in crime at the macro level, including predictions of future crime trends. It is not suggested that local partnerships should attempt to replicate that work. However, it is suggested that partnerships should be aware of that work, and that in some cases it might be profitable to replicate aspects of that work at the regional level.

Economics and Demographics

Home Office research over the last ten years has attempted to model historical trends in the level of recorded property crime in England and Wales. The main economic findings of a 1990 study were that:

- In the short term the economic factor associated with rates of crime is "per capita real personal consumption." This is the amount that each person in the country spends, on average, in any year. When personal consumption increases, property crime tends to grow more slowly or falls. When personal consumption grows more slowly or falls, property crime grows more rapidly. This is thought to be because upturns in economic growth have the most immediate effect on those who are economically marginalized and provide an increased capacity for the lawful acquisition of goods, thereby reducing the temptation of unlawful acquisition through theft.
- In the longer term, trends in property crime rise with rises in personal consumption so that trends in theft and burglary are linked to the stock of crime opportunities as measured by the stock of consumer goods. For every one percent increase in this stock, burglary and theft increase by around two percent (http://www.homeoffice.gov.uk/rds/pdfs/hors198.pdf).

The main demographic finding was that:

- Trends in thefts and burglaries were associated with the number of young males. For every one percent increase in the number of young males aged 15 to 20, burglary and theft increases by about one percent (http://www.homeoffice.gov.uk/rds/pdfs/hors198.pdf).

This research has been used as the basis for predicting future trends in property crime. The models used for modeling historical trends in crime were updated and modified to try to project how trends in burglary and theft might change. The projections made were not forecasts because they only examined the likely effect of the specific economic and demographic trends referred to above. The Association of British Insurers (2000; http://www.insurance.org.uk?ResearchInfo/) suggests several implications:

- The increase in the elderly population may produce an increase in vulnerability to criminal attack.
- An older population may mean that there will be fewer people in the peak offending group (young males). However, the reduction in the relative size of the peak offending group may be outweighed by an increase in the "vulnerable" population most likely to offend.

Source: Quoted directly from *Anticipating Future Trends in Crime and Disorder Audits*, Home Office, 2002, London: Author. Available at http://www.crimereduction.homeoffice.gov.uk/toolkits/p0317.htm.

Web Research Project
Review the Crime Reduction Toolkits of the Home Office at the Internet address above. What have you learned from this review?

Some of the British Home Office predictions are based on social and lifestyle trends and consider the makeup of communities, changes in lifestyles, the interaction between technology and society, and developments in popular culture that impact crime. Other such predictions include the following:

- The development of new synthetic drugs like ecstasy poses future challenges for crime control.
- Virtual-reality addiction may increase.
- Alcopops, designer alcoholic drinks, will continue to increase in popularity with underage drinkers.
- Increases in telecommuting, more women in the workforce, and increased leisure activity outside the home will create new opportunities for criminals.

Photo 15.2

Increases in telecommuting will create new opportunities for criminals.

- Central city regeneration based around alcohol and leisure may increase violent crime and disorder.
- The introduction of the 24-hour city could create more opportunities for crimes such as burglary and shoplifting.
- Decreases in the use of cash will continue to result in greater theft involving credit cards and checks.
- As skills in information communications technology become a prerequisite for work, those not computer literate will be more excluded from work.

Photo 15.3

Greater theft involving credit cards will be part of the future of digital crime.

- Identifying hot products may assist in understanding, anticipating, and reducing demand for stolen goods. Examples of hot products are mobile phones, digital television, and portable DVD players (Home Office, 2002).

Looking back at the tremendous change unleashed in the twentieth century by technological advances such as automobiles, aircraft, television, telephones, and computers, one can only begin to imagine the fantastic changes awaiting us later in this century. Such changes, of course, will most likely be a mixed blessing containing the seeds of crime as well as providing the tools for the resolution of crime.

Crime File 15.2

Hot Products: Understanding, Anticipating, and Reducing Demand for Stolen Goods

Summary

This report focuses attention on the so-called 'hot products' that are most likely to be taken by thieves. A better understanding of which products are 'hot' and why, could help businesses protect themselves from theft and also the police in advising them how to do this. Earlier research into hot spots of crime and repeat victimization has led to new thinking in the field of crime reduction, and there is every reason to expect the same from research focusing on hot products. Some strategies will center on police work; others relate to the wider field of action covered by the Home Office Crime Reduction Program—in particular, on product design. This report is the first to review comprehensively what is known about hot products and what further research is needed to assist policy.

Main Findings

- **Theft is concentrated upon relatively few products.** These products share a number of common attributes in that they are generally concealable, removable, available, valuable, enjoyable and disposable. While each of these elements may be important in explaining which products are stolen, how much they are stolen may depend critically on just one attribute—the ease of disposal.
- **For each kind of theft, specific items are chosen by thieves.** In residential burglaries, for example, thieves are most likely to choose jewelry, video players, cash, stereos and televisions. The British Crime Survey shows that for thefts involving personal possessions, cash is more frequently stolen than anything else—followed in order by vehicle parts (even when car stereos are excluded), clothing, and tools.
- **Certain items are at risk of being shoplifted wherever they are sold.** These include music, cassettes, cigarettes, alcoholic drinks, and certain brands of fashion items. What these goods have in common is that they are enjoyable things to own and consume.
- **Which cars are most likely to be stolen depends on the purpose of the theft.** An American study found, for example, that joy riders prefer sporty models. Thieves looking for cars to sell prefer expensive luxury models. Those seeking components to sell prefer models with easily-removable, good quality, radios. The cars most frequently stolen both here (UK) and in the USA are popular cars that are several years old. It appears that these are stolen to supply demand in the used car parts market.
- **Though more research is needed, relatively few hot products may account for a large proportion of thefts.** For example, research in the USA found that theft insurance claims for new cars in 1993–95 were twenty times higher for models with the worst theft record than those with the best.

Points for Action

This report is very much a first step on the road to providing valuable information that will help in tackling acquisitive crime using the Market Reduction Approach. The Home Office is currently funding two projects under the Crime Reduction Program to look at the cost-effectiveness of reducing crime by tackling stolen goods markets.

For Designers

- The report should help designers to anticipate which products under development are likely to become 'hot,' and therefore which ones need enhanced crime resistance designed in.

For the Police

- Analysis of crime patterns can help to identify new hot products. This information can be used to inform operational decision making and crime prevention advice.
- A further police role here is, for example, to help identify those products that are actually or potentially 'hot' and to feed information to designers on the products' vulnerability and the MOs used by offenders to overcome existing crime resistance features.

(Continued)

(Continued)

For Research

- The Home Office Crime Reduction Program is funding two projects to explore the potential benefits of raising awareness among designers, manufacturers and retailers of the need to avoid making and selling products likely to be easy and tempting targets for thieves.
- It would be helpful to develop our knowledge in the following areas:
  - How to disrupt theft markets, especially those serving particular hot products
  - The amount of theft accounted for by hot products
  - To determine when particular products are most at risk and who will bear the cost of the theft.
  - To develop techniques to assist in anticipating technological developments and new technology that could result in new hot products—or in new ways of preventing their theft.

Source: Ronald V. Clarke. (1999) *Hot Products: Understanding, Anticipating and Reducing Demand for Stolen Goods,* London: Home Office, Police Research Series Paper 112, July.

Web Research Project
Review the Home Office's other papers in their "Police Research Series" at http://www.homeoffice.gov.uk/rds/policerspubs1 .html. What other topics are addressed?

Summary

Journal Article Link 15.4
Examine literature regarding guns, crime, and social order.

Futurists use a number of methods in order to predict the future of crime. These include *mathematical models, environmental scanning, surveys, Delphi techniques,* and *scenario writing.* The chapter features a large number of predictions from Police Futurists International, the British Home Office, Georgette Bennett's *Crimewarps,* Taylor et al.'s *Digital Crime and Digital Terrorism,* and a variety of other efforts.

Criminology on the Web
Log on to the Web-based student study site at http://www.sagepub.com/haganintrocrim7e/ for author-created podcasts, e-flashcards, quizzes, and more.

KEY CONCEPTS

Delphi Techniques

Future of Crime

Hot Products

REVIEW QUESTIONS

1. Discuss some of the predictions of the future of crime in this chapter. Which ones do you view as the biggest threats in the future?

2. Discuss some of the methods used by futurists in attempting to predict the future.

3. Of what use is the concept "hot products"?

4. What are virtual crimes? Give some examples.

WEB SOURCES

Future Crime: Technology and the Future of Crime
http://futurecrime.wordpress.com

IBM and the Future of Crime
http://odeo.com/episodes/22029211-IBM-and-the-Future-of-Crime

The Future of Crime
http://futuresworkinggroup.cos.ucf.edu/publications/futureofcrimeApril%202007.doc

The Future of Crime Management Tools
http://www.crimescenesoftware.com

Police Futurists International
http://www.policefuturists.org

WEB EXERCISES

Using this chapter's recommended Web sites, explore the area of the future of crime.

1. What is the perceived impact of technology on the future of crime?

2. What predictions does IBM make with respect to the future of crime?

3. What types of future forensic and crime investigation tools are discussed by crime futurists?

4. Using your browser, search the terms "scenarios" and " Delphi techniques."

SELECTED READINGS

John Arquilla and David Ronfeldt. 2009. *Networks and Netwars: The Future of Terror, Crime and Militancy.* Santa Monica, CA: Rand.

The notion of netwar examines the networked organizational structure of its practitioners (many leaderless) and their quickness in coming together in swarming attacks.

Dangrsmind. April 29, 2008. "Microsoft Ships Future Crime Fighting Tool and Fights Cybercrime." http://futurecrime.word press.com.

Microsoft has developed a tool which will enable forensic investigators to easily gather digital evidence after a crime has been committed. The COFEE is a USB device that supports 150 commands that dramatically cuts the time it takes to gather digital evidence including decrypting passwords and analyzing Internet activity. Prior to COFEE the equivalent work would require a computer forensics expert to enter 150 complex commands manually. This process would take 3 to 4 hours. COFEE takes 20 minutes.

Joseph A. Schafer, editor. 2007. *Policing 2020: Exploring the Future of Crime, Communities and Policing.* Quantico, VA: Futures 2007 Working Group.

This series of papers generated from the conference provides some leads in the latest thinking by police futurists.

Glossary

Abscam   (Arab Scam or Abdul Scam) an FBI sting operation in which agents posed as rich oil sheiks and bribed members of Congress.

Addiction   an extreme physical and psychological dependence on drugs.

Adversary system   a practice in which the prosecution and defense take sides and vigorously represent their client.

Age-crime debate   a disagreement among criminologists as to whether all "mature out of crime" or whether some remain career criminals.

Androcentric bias   the charge that criminology has reflected a male-centered bias.

Anomie   a moral confusion or breakdown in mores or a gap between goals and means in society.

Antisocial potential   Farrington's notion that bad life events increase one's antisocial disposition.

Apalachin Meetings   a gangland convention in 1957 in the small hamlet of Apalachin, NY, that was raided by police.

Argot   the distinctive or specialized language of a group.

Arson   the purposeful setting of fires.

Assault   threatening to do bodily harm to a person or placing him or her in fear of such harm.

Asset forfeiture   laws that require that property obtained through criminal activity is surrendered to the state.

Astrological theory   the pseudoscientific belief that astrological signs such as configurations of the planets or stars influence human behavior.

Atavism   the belief that criminals represent genetic "throwbacks to the ape" or earlier, more primitive humanity.

Badger game   a scam that preys on naïve elderly victims. For example, an elderly man might be falsely accused of sexual relations with a young female con artist.

Bank examiner's scam   swindlers pretend to be bank examiners and ask to borrow "buy money" to catch a dishonest teller.

Battery   (aggravated assault) offensive, unconsented to, unprivileged, and unjustified bodily contact.

Behavioral modification   the use of a system of rewards and punishments in order to modify or engineer behavior.

Big cons   more sophisticated confidence games involving complicated schemes, prominent clients, and large sums of money.

Big dirty secret   the criminal justice system is soft on white collar crime.

Biological positivism   theories that propose that crime is caused by inherited genetic and other biological causes.

Boojo   dishonest gypsy fortune-telling scams.

Booster   professional shoplifters.

Bourgeoisie   Marx's term for the capitalists or owners of industry.

Box man   a professional safecracker.

Brainwashing   a form of drastic resocialization of personality.

Brainwashing myth   the belief that the term "brainwashing" was created by intelligence agencies in order to justify bizarre mind control experiments.

Broken windows   an approach to crime control that advocates that police not ignore small disturbances lest they lead to bigger crimes.

Burglars, professional   specialists at burglary who concentrate on big, well-planned operations.

Burglary   the unlawful entry of a structure to commit a felony or theft.

Bust out   an organized crime operation in which criminals take over a legitimate company, loot it, burn it down, and collect the insurance.

Campbell Collaboration   an international organization that conducts meta-analyses in order to discover what works in criminology.

Cannon   a name for a professional pickpocket.

Career criminals   criminals who have a life-long involvement in crime.

Castellammarese Wars   Mafia wars in New York City in 1929–1931 that are believed to have resulted in the formation of the Cosa Nostra.

Catharsis hypothesis   the belief that the observation of media violence serves as an emotional release and lessens violence.

Cheater theory   theory that holds that males have a greater interest in mating and little interest in child rearing and use illegitimate means to maximize their offspring.

Chicago school   a school of sociology in the 1920s and 1930s that produced many urban ecological and ethnographic studies of Chicago.

Child abuse   excessive mistreatment, either physically or emotionally, of children beyond any reasonable explanation.

Churning   a dishonest practice by stockbrokers of unnecessarily causing many buys and sells of stock in order to collect commissions.

CIA-Mafia link   a secret deal by the CIA to enlist the support of the Mafia to assassinate Cuban Premier Fidel Castro.

Classical theory   contained in the writings of Beccaria and Bentham, these theories assume that criminals are rational actors who weigh the pleasure and pain of an activity.

Classic experimental design   a research design that has equivalence, an experimental and a control group, and pre- and post-test.

Closet queens   hidden homosexuals.

Code of ethics   requirements that researchers behave ethically in conducting research.

Coerced crime   a term used to describe a practice in which auto dealer franchises are required to force accessories and unnecessary sales in order to keep their franchise.

Cohen's "lower-class reaction" theory   delinquency is a lower-class reaction to unfulfillable middle class values.

COINTELPRO   a secret FBI program, Counter Intelligence Program, to discredit legitimate social movements.

Commission trials   trials in New York City, 1983–1987, that convicted Cosa Nostra dons of being members of a commission.

Community policing   policing that emphasizes working with citizens to solve crime problems.

Computer crime   crimes that use the computer as a tool in crime commission.

Computer virus   rogue programs that copy themselves onto other computer programs.

Confidence game   "con games" that win the confidence of victims in order to take advantage of them.

Confidentiality   the requirement in research that the researcher protect the identity of his or her subjects.

Conflict model of law   the belief that criminal law reflects the conflicts of interest of groups and that the more powerful groups define the law.

Consensus model of law   the belief that criminal law originates in the will of the majority.

Containment theory   Reckless's theory that crime takes place when pressures are high and containments (protections) are low.

Continuum model of organized crime organized criminal groups exist along a continuum from nonorganized crime to highly organized crime groups.

Conventional criminals semiprofessional criminals who are generally unsuccessful at their trades of larceny and burglary.

Copycat crime crimes in which criminals imitate crimes that were previously publicized.

Corporate crime crime within a legitimate occupation on behalf of one's employer.

Corporate dumping the selling of products that are prohibited in this country overseas where there is less regulation.

Cosa Nostra "This thing of ours." Believed to be the real name of the American Mafia.

Cost of crime the cost of crime includes financial and other costs, such as psychological and health costs. Estimates have been as high as $1.7 trillion.

Crime violations of criminal law.

Crime against government political crimes against the government, from protests to terrorism.

Crime by government political crimes by the government, from human rights violations to genocide.

Crime clocks a very inaccurate graphic device that reports crimes per time unit such as minute or hour.

Crime control model the purpose of the criminal justice system is to be efficient in enforcing law and order.

Crime dip the decrease in recorded crime in the U.S. beginning in 1992.

Crime index a measure of crime (Part I crimes) that results in the calculation of the crime rate.

Crime rate the number of index crimes divided by the population × 100,000.

Crime trends in the post–World War II period, crime had declined until the mid-sixties, when it exploded. The crime dip began in the U.S. in 1992.

Crimes without victims crimes that are described as "legislated morality" in which there is no identifiable victim other than the person himself or herself.

Criminal behavior systems a typology of crimes that looks at identification with crime, societal reaction, and group involvement.

Criminal law violations of law than are enforced by the state in order to protect victims.

Criminal saturation, law of Ferri's theory that crime expands to fit the amount of law enforcement machinery.

Criminal typologies attempts to identify types of crime.

Criminaloid criminal-like behavior. A term first used by A. E. Ross (1907) to refer to flagitious (wicked) practices that had not yet been condemned by criminal law.

Criminology the scientific study of crime and criminal behavior.

Critical criminology consists of a variety of perspectives that challenge basic assumptions of mainstream criminology.

Cultural lag change takes place unevenly—technological changes take place far more quickly than the social aspects of a culture such as its laws and custom.

Culture of violence ways of life in societies that approve of violence as a means of resolving disputes.

Daisy chain scam sellers buy and sell items or property in order to artificially pump up its price.

Dark figure of crime unmeasured or undiscovered crime.

Decriminalization the lessening of penalties attached to a particular behavior.

Delinquency and drift Matza's theory that delinquents exist in a limbo wherein they drift back and forth between delinquency and conventionality.

Demonological theory assumes that supernatuaral forces cause and control crime commission.

Desistance in life course criminality theory, the quitting or cessation of criminal activity.

Developmental/life course theory the belief that criminal activity changes over an individual's lifetime from onset to persistence to desistance.

Deviance   behavior that is outside the limits of societal toleration.

Differential association theory   Sutherland's theory that crime is learned due to exposure to an excess of contacts that advocate criminal behavior.

Differential opportunity theory   Cloward and Ohlin's theory that crime takes place due to a lack of legitimate opportunity and is also due to the availability of illegitimate opportunities.

Dirty secrets of crime   the criminal justice system can do very little about street crime since it is due to poverty and discrimination. Also, the criminal justice system is soft on white collar crime.

Drug court   special courts organized to rehabilitate drug offenders.

Due process model   the criminal justice system must protect the rights of the accused in the process of enforcing the law.

DUI   driving under the influence.

Durham decision   individuals are not guilty if their acts are the product of mental disease or defect.

Ecological fallacy   a problem in which group rates are used in order to describe individual behavior.

Ecological theory   theories that posit that crime is caused by environmental or geographic forces.

Economic theory   is influenced by the writings of Karl Marx and views inequality and capitalism are the causes of crime.

Embezzlement   theft from an employer by an employee who has reached a level of trust.

Enterprise, the   the secret organization within the Reagan White House that was responsible for the Iran-Contra affair.

Equity funding scandal   a corporate scam in which fake insurance policies (ghosts) were created in order to take advantage of reinsurers.

Espionage   spying, the stealing of secrets.

Ethnic succession theory   the belief that organized crime has been a ladder of mobility for a succession of ethnic groups.

Evidence-based research   research findings that are based on replicated, experimental research.

Exhibitionism   individuals who gain sexual excitement by sexually exposing themselves in order to shock their victims, i.e., flashers.

Experiments   research that involves variations of the classic experimental design employing equivalence, pre- and post-tests, and experimental and control groups.

Fallacy of autonomy   the erroneous belief that failings of the family are separate and independent of inequality, racism, and discrimination.

Feeblemindedness   the belief by Goring that criminals were mentally defective.

Felony   a more serious crime generally punished by at least a year in jail.

Felony-murder doctrine   if a person is murdered during the course of a crime, any accomplice is equally guilty of murder.

Feminization of poverty   poverty increasingly takes place within female-headed households.

Fence   a dealer in stolen property.

Fetishism   when individuals have a sexual fixation on objects, attire, or parts, usually associated with the opposite sex, e.g., feet, lingerie.

Fix, the   the act of forestalling legal actions through bribery and corruption.

Floodgate theory   the belief that decriminalization of some illegal behavior will open the gates to a flood of illegal activity.

Folk crime   crimes that everyone commits, e.g., traffic violations.

Folkways   nice customs, traditions, or less serious norms.

Forcible rape   nonconsensual sexual relations where force or intimidation is used.

Ford Pinto case   in the early 1960s, firetrap engineering of gas tanks on Ford Pintos resulted in the death and injury of hundreds of victims.

Freudian theory   the belief that human behavior is motivated by unconscious sexual desires.

Functional necessity of crime   Durkheim's theory that society defines itself by reacting to crime and wrongdoing.

Gemeinschaft   a communal or folk society.

Genocide   the eradication of a group.

Gesellschaft   an associational or heterogeneous society.

Global fallacy   the error of attempting to have a specific theory explain all crime.

Great oil scam   in the 1970s, the major oil companies were accused by the Department of Energy of conspiring to raise prices.

Gun control   a heated debate exists between those who view gun ownership as a constitutional right and those who see it as responsible for the high U.S. homicide rate.

Hacker's ethic   there should be no restrictions on their right to surf the Net and test systems.

Hackers   unwanted electronic guests regardless of motive.

Hamas   in Arabic, "fervor" or "zeal"; a Middle East, Islamic resistance movement.

Harrison Act   an antidrug act passed in 1914 that required a doctor's prescription for narcotics and cocaine.

Hate crime   a bias crime that is motivated by dislike of a race, religion, disability, sexual orientation, or ethnicity/national origin.

Hedonism   a pleasure-seeking philosophy.

Heel   sneak thieves who operate in stores and offices.

Hezbollah   an Islamic, Iranian-backed terrorist group whose name means "party of God."

Higher immorality   a term used by C. Wright Mills to describe the moral insensibility of the Power Elite.

Hobbs Act   a 1940s antiracketeering act that holds that any interference with interstate commerce is a violation.

Human ecology   the study of the interrelationship between human organisms and the physical environment.

Human rights   basic human activities that are guaranteed in democratic societies such as the right to life, liberty, and the pursuit of happiness.

Identity theology   a neo-Nazi racist belief that Aryans are God's people.

Ideology   a distinctive belief system.

Incest   sexual activity between individuals who are viewed as too closely biologically related.

Index crimes   Part I crimes in the Uniform Crime Reports that are used to construct the violent crime and property crime indexes.

Industrial espionage   spying in the commercial sector.

Insider trading   using prohibited, confidential information in order to trade for one's advantage.

Integrative theory   theories that combine features of other theories.

Iron law of opium trade   Arnold Trebach's thesis that as soon as one source of heroin supply is shut off, another will emerge to meet the demand.

Johns   customers of prostitutes.

Judgescam   an FBI sting operation of the Cook County, Illinois, judiciary (also called "Operation Greylord").

Karen Silkwood case   a case in the 1970s of an employee of the Kerr McGee nuclear plant who died in an auto accident on her way to report on company wrongdoing.

Kefauver Committee   Congressional hearings in the 1950s that examined labor racketeering by organized crime.

Labeling theory   crime is a label attached to wrongdoing and often the label becomes a stigma that increases criminality (also called societal reaction theory).

Larceny-theft   miscellaneous property theft.

Latent functions   unexplained, unanticipated (hidden) consequences of social activity.

Laws   codified (written) rules that are more serious norms and contain sanctions.

Left realism   a theoretical attempt to translate radical ideas into realistic social policy.

Legalistic style of policing a formal style of policing that emphasizes enforcing the letter of the law.

Life course/developmental theory a theoretical approach that looks at crime over the life course consisting of onset, persistence, and desistance.

Life history method a qualitative, case study method of examining a subject's life history.

Loan sharking the illegal lending of money at usurious interest rates.

Lobotomy the surgical removal of the frontal lobes of the brain.

Logic bomb computer programs that print messages or destroy data on a certain date.

Looking-glass self Cooley's theory of personality as a perceived perception of the reaction of others.

Lower-class reaction theory Cohen's theory that delinquency involves a lower-class reaction to unachievable middle-class values.

Mafiya a term for describing the Russian Mafia.

Mala in se acts that are bad in themselves.

Mala prohibita acts that are bad due to being prohibited.

Manifest functions intended or planned consequences of social arrangements.

Manslaughter homicide without malice aforethought.

Mass murder the killing of four or more victims at one location on a single occasion.

M'Naughten rule not guilty by reason of insanity (NGRI).

Medellin cartel a Colombian drug trafficking organization.

Methodological narcissism the belief that one's favorite method is best.

Methodology tbe attempt to gather data that are accurate.

Merton's anomie theory crime occurs when there is a gap between societal goals and the means provided for their achievement.

Miller's "focal concerns" theory crime reflects the overemphasis on lower-class values.

Minneapolis Domestic Violence Experiment a domestic violence experiment that concluded that arrest works best in deterring repeat domestic violence.

Misdemeanor less serious crimes that result in less than 1 year in jail.

MKUltra mind control ultra, a secret government mind control program.

Modes of personality adaptation this is part of Merton's theory of anomie that results in personality adaptations: conformist, innovator, ritualist, retreatist, and rebel.

Monozygotic concordance similar behavior in identical twins.

Moral entrepreneurs those who benefit by labeling activity as criminal.

Mores more serious informal social controls.

Moustache Pete elderly Mafioso from the old country.

Murder, first degree a deliberate, premeditated intent to murder with malice aforethought.

Murder, second degree intent to murder with malice aforethought, but without deliberation or premeditation.

Naive check forgers those who pass bad checks in order to resolve a temporary financial crisis.

NCVS (National Crime Victimization Survey) a survey of the general public to measure claimed crime victimization.

NIBRS (National Incident-Based Reporting System) a system for recording far more detail on crime incidents that is intended to replace the UCR.

Nature-nurture controversy an attempt to examine whether crime is inherited (nature) or learned (nurture).

Natural areas according to the Chicago school, these are subcommunities that emerge to serve specific, specialized functions.

Neoclassical theory new classical theories that view crime as influenced by criminal opportunities to commit crime.

NGRI (not guilty by reason of insanity) the earliest insanity ruling, sometimes called the right/ wrong test.

Nolo contendere   a plea of no contest to charges.

Norms   prescribed rules of conduct.

Numbers game   an illegal game of betting of numbers.

Nuremberg principle   the precedent established at the Nuremberg trials that says that, if a person has a conflict between morality and immoral orders, one is to obey morality.

Objectivity   researchers should be neutral or unbiased in examining their subject matter.

Occasional property crime   crimes committed by avocational criminals who do not identify with criminal careers.

Occupational crime   crime committed during the course of a legitimate occupation for one's own benefit.

Occupational/organizational crime grid   a model that looks at white collar crimes in terms of whether the perpetrator and victim are individuals, employees, or groups.

Operationalization   describing how a concept is being measured.

Operation Chaos   a secret CIA program to investigate activities of domestic groups.

Organizational crime   crime committed during the course of a legitimate occupation for the benefit of the employee.

Organized crime   refers to organizations that use violence, provide illegal services, and have immunity of operation.

Organized Crime Control Act   the most powerful legislation ever passed to attack organized crime. It was passed in 1970.

Overcriminalization   the overuse of the criminal law as an attempt to control deviant activity.

Palmistry   a pseudoscience that claims to be able to read personality characteristics on the basis of lines on the palm of the hand.

Paper hangers   professional counterfeiters.

Paradigm shift   a dramatic change in the basic view or model of reality.

Parole   a supervisory period on release from prison.

Part I crimes   the index crimes or crimes that are used in order to calculate the crime rate.

Part II crimes   the nonindex crimes that are not used in the calculation of the crime rate.

Participant observation   a methodology that involves the observation of a group by participating in varying degrees in its activities.

Patriarchal crime   crime that is committed in the name of male dominance.

Peacemaking   a theoretical approach that advocates peace and justice as the solution to the crime problem.

Pedophilia   sexual child molestation.

Pennyweighting   thieves who specialize in jewelry stores.

Phrenology   the pseudoscience that claims to determine personality and intelligence on the basis of the size and shape of a person's skull.

Physical stigmata   Lombroso's theory that criminals could be denoted on the basis of their facial appearance.

Physiognomy   a pseudoscience that measures facial and other body characteristics and their relationship to personality.

Phishing   entails a cracker stealing password, account, or credit card information.

Plea bargaining   an agreement that is made between the prosecuting and defense attorneys and the judge that the defendant will be granted a lesser penalty if he or she agrees to plead guilty.

Pigeon drop   a con game in which a victim is duped into putting up a good-faith deposit to collect "lost money."

Pizza Connection   a Sicilian mafia group (called "the Zips") that was convicted in the late 1980s of running a drug trafficking organization from pizza shops in the U.S.

Political crime   crime that is committed for ideological reasons.

Ponzi scheme   a con game (named after Charles "Get Rich Quick" Ponzi) that involves paying early investors high investment returns with money from later investors in a nonexistent enterprise.

Positivism   a scientific or quantitative approach to criminology that searches for pathology, uses the scientific method, and suggests therapy.

Post hoc error   "after-the-fact" error. If a person exhibits a characteristic after the fact, it is mistakenly assumed to be the cause of the behavior.

Postmodernism   a theoretical approach that attacks modernity or scientific rationality.

Power control theory   in egalitarian households, both boys and girls have more similar delinquency levels.

Power elite   C. Wright Mills's theory that argues that a small group of corporate, military, and political elite run the United States.

Praxis (practical critical action)   in Marxist theory, such action is more important than theory.

Precipitation hypothesis   the belief that the observation of media violence increases the propensity to violence.

Primary deviance   in labeling theory, this refers to the initial criminal act itself.

Probation   supervision in lieu of prison.

Problem drinking   also known as alcoholism; a dependency on or addiction to alcohol.

Profession   an occupation that exhibits esoteric knowledge and a service orientation and achieves autonomy of operation.

Professional crime   a type of career crime that involves skill, specialization, progression in knowledge, full-time status, success, and high status.

Professional crime, model of   professionalism in crime is a matter of degree. The more a criminal exhibits each characteristic of professional crime such as skill, high status and the like, the more professional they are regarded.

Progression of knowledge   Comte's theory that knowledge has historically progressed from theological to metaphysical to scientific.

Project Bluebird   secret CIA mind control research.

Proletariat   refers to the working class in Marx's writings.

Psychological positivism   a group of theoretical approaches that look to the personality as the cause of criminal activity.

Psychometry   attempts to measure personality.

Psychopath (sociopath, antisocial personality)   a personality disorder in which, due to inadequate childhood socialization, an individual never develops full range of adult personality.

Public order crime   vice crimes or crimes without victims.

Pyramid schemes   a financial scam relying on the continual recruitment of investors in a nonexistent product.

Racketeering   dishonest practices, particularly in the labor sector.

Radical Marxism   a theory that blames capitalism for crime and advocates violent revolution as a means of its eradication.

Raison d'état (reason of state)   Machiavelli's notion that any means are necessary in order to protect the state.

"Real rape"   the mistaken belief that it is not a rape unless the victim physically resists to the point of experiencing physical harm.

Reciprocity   a system of mutual obligation between researcher and subject.

Reliability   the consistency or stability of measurement.

Revolving door   prisoners who are constantly in and out of prison are described as being in a revolving door.

RICO   Racketeer influence in corrupt organizations statute of the Organized Crime Control Act of 1970. It prohibits the use of the proceeds from criminal operations from being used to acquire legitimate operations.

Routine activities approach   Felson's theory that crime can be predicted by three elements: likely criminals, likely targets, and guardianship.

Salami techniques   a computer crime in which small slices (amounts of money) are taken from many accounts and transferred to that of the thief.

Sampling   selecting a group for study from a larger population.

Sanctuary Movement   a U.S. social movement in the 1980s that attempted to provide refuge for illegal political refugees from Latin America.

Scams   illegal games to swindle people out of their money.

Secondary deviance   deviance that ensues as a result of a person being labeled and stigmatized as a criminal.

Secret police   domestic security police whose purpose it is to protect the existing regime from violent overthrow.

Self-reports of crime   self-admission surveys of admitted crime.

Serial murder   a series of murders over a period of time.

Shaming   offenders are made to feel humiliation due to their transgressions.

Shoplifting   the stealing of merchandise from stores.

Simulation   games that mimic reality.

SMICE   an acronym for the method of recruiting spies: sex, motivation, ideology, commitment, and ego.

Snitch   in Mary Cameron Owen's study, an amateur shoplifter.

Social bond theory   people become criminal when their stakes in society are broken.

Social class   socioeconomic status is measured using income, education, and occupation.

Social control theory   views crime as taking place when social control or bonds to society breaks down.

Social Darwinism   the belief that there is a survival of the fittest in society.

Social disorganization theory   Shaw and McKay's theory that crime is due to social disorganization and social breakdown of an area.

Sodomy   laws forbidding particular types of sexual activity.

Soft determinism   theories that indicate that certain forces have an influence, but do not determine behavior.

Somatotypes   body types as described by Sheldon: endomorphs, mesomorphs, and ectomorphs.

Spree murder   murder at two or more locations with almost no time between the murders.

Spoofing   creates a false (shadow) version of a Web site that is controlled by the attacker.

Spouse abuse   domestic battering by a spouse or lover.

Statutory law   legislative-made law.

Strain theory   Agnew's theory that anomie may include more general gaps such as interpersonal problems or disrespect.

Strike forces   federal task forces consisting of members from various departments to attack organized crime.

Subcultural theories   theories that view the type of crime as due to various forms of delinquent subcultures.

Sub rosa crime   another name for espionage (literally, "under the rose").

Subterranean values   underground values that exist alongside conventional values.

Surveys   various ways of gathering data that include: mail questionnaires, interviews, and telephone and Internet surveys.

Sweetheart contracts   secret deals between labor negotiators and management to the disadvantage of the workers.

Systematic check forgers   professional criminals who specialize in passing bad checks (paperhangers).

Tearoom   a slang term for public restrooms that are used for homosexual liaisons.

Techniques of neutralization   rationalizations (excuses) used by juveniles to explain away responsibility for their actions.

Terrorism the purposeful targeting of innocents for political purposes in order to spread fear and intimidation.

Theoretical range the different types of crime that a theory is intended to cover.

Theory plausible explanations of reality.

Thermic law of crime Quetelet's theory that violent crime increases toward the equator.

Triads Chinese organized crime groups.

Triangulation the use of multiple research methods.

Trojan horse a secret, malicious computer program hidden within other programs.

Twinkie defense a legal defense that argues that a defendant is not responsible for his or her actions due to having consumed too much sugar.

UCR (Uniform Crime Reports) official police reports on crime maintained by the FBI.

Undercriminalization the underuse of the criminal law to control deviant activity.

Unfounded crimes crimes that the police decide never took place.

Universal Declaration of Human Rights a United Nations declaration that calls for the respect of human rights.

Unobtrusive measures nonreactive methods of data gathering in which the subjects are unaware that they are being studied.

Urban/rural differences in crime crime is highest in urban areas and lowest in rural areas.

Validity the accuracy of measurement.

Variables concepts that vary or that can take on different numerical values.

Victim precipitation the victim contributes to or brings on their victimization.

Victim surveys surveys of the general public to measure the rate and circumstances of victimization.

Voyeurism ("peeping Toms") those who gain sexual excitement by secretly observing unknowing victims.

Watergate the burglary of the Democratic National Headquarters in the Watergate complex by Nixon's "plumbers group."

Whistleblowers those who inform on wrongdoing within their organization.

White collar crime crime committed by those within legitimate occupations or organizations.

White collar "deadbeats" convicted white collar offenders often do not pay penalties.

Withdrawal syndrome uncomfortable feelings experienced when an addict who has developed tolerance to a substance is deprived of his or her drug.

Workplace violence violence that occurs within a place of work.

Worm similar to viruses that reproduce themselves and subvert computer systems but can spread with no assistance.

XYY syndrome the theory that males who possess an extra Y chromosome are more prone to violence.

Yakuza Japanese organized crime groups (literally, "good for nothings").

References

A confession ends career of robbery, 56-bank toll. (1994, July 14). *New York Times*, p. A8.

Abadinsky, H. (1983). *The criminal elite: Professional and organized crime*. Westport, CT: Greenwood.

Abadinsky, H. (1994). *Organized crime* (4th ed.). Chicago: Nelson-Hall.

Abadinsky, H. (2006). *Organized crime* (8th ed.). Belmont, CA: Wadsworth.

Abrahamsen, D. (1960). *The psychology of crime*. New York: Columbia University Press.

Abram, S. (1998, February 24). Insurer sues 45 people in auto accident. *Los Angeles Times*, p. A1.

Academy of Criminal Justice Sciences. (1998). *Code of ethics*. Arlington, VA: Author.

AccountingWEB. (2004). *Long-serving male executives most likely to commit fraud*. Retrieved April 3, 2007, from http://www.accountingweb.com/cgi-bin/item.cgi?id=99078

Acohido, B. (2009, May 14). Intel hit with $1.45 billion antitrust fine. *USAToday*, p. B1.

Adams, V. (1976). *Crime*. New York: Time-Life.

Adler, F. (1975). *Sisters in crime*. New York: McGraw-Hill.

Adler, F. (1983). *Nations not obsessed by crime*. Littleton, CO: Fred B. Rothman.

Adler, F., & Simon, R. J. (Eds.). (1979). *The criminology of deviant women*. Boston: Houghton Mifflin.

Adler, J., et al. (1988, November 28). Getting tough on cocaine. *Newsweek*, pp. 76–79.

Adler, P. A., & Adler, P. (1983). Shifts and oscillations in deviant careers: The case of upper-level drug dealers and smugglers. *Social Problems, 31,* 195–207.

After the don: A donnybrook. (1976, November 1). *Newsweek*, p. 32.

Agee, P. (1975). *Inside the company: CIA diary*. New York: Stonehill.

Agnew, R. (1985). Social control theory and delinquency: A longitudinal test. *Criminology, 23,* 47–61.

Agnew, R. (1992). Foundation for a general strain theory of crime and delinquency. *Criminology, 30,* 47–87.

Agnew, R. (1994). The contributions of social psychological strain theory to the explanation of crime and delinquency. In F. Adler & W. S. Laufer (Eds.), *Advances in criminological theory* (Vol. 6). New York: Transaction Books.

Agnew, R. (1995). Strain and subcultural theories of criminality. In J. F. Sheley (Ed.), *Criminology: A contemporary handbook* (2nd ed., pp. 305–327). Belmont, CA: Wadsworth.

Agnew, R. (1997). The nature and determinants of strain: Another look at Durkheim and Merton. In N. Passas & R. Agnew (Eds.), *The future of anomie theory* (pp. 27–51). Boston: Northeastern University Press.

Agnew, R., Cullen, F., Burton, V., Evans, T., & Dunaway, R. (1996). A new test of classic strain theory. *Justice Quarterly, 13,* 681–704.

Agnew, R., & White, H. R. (1992). An empirical test of general strain theory. *Criminology, 30,* 475–499.

Agran, L. (1982). Getting cancer on the job. In J. H. Skolnick & E. Currie (Eds.), *Crisis in American institutions* (5th ed., pp. 408–419). Boston: Little, Brown.

Aho, J. (1994). *This thing of darkness: A sociology of the enemy*. Seattle: University of Washington Press.

Akers, R. (1967). Problems in the sociology of deviance: Social definitions and behavior. *Social Forces, 46,* 455–465.

Akers, R. (1980). Further critical thoughts on Marxist criminology: Comments on Turk, Toby and Klockars. In J. A. Inciardi (Ed.), *Radical criminology: The coming crisis* (pp. 133–38). Beverly Hills, CA: Sage.

Akers, R. (1994). *Criminological theories: Introduction and evaluation*. Los Angeles: Roxbury.

Akers, R., & Sellers, C. (2004). *Criminological theories* (4th ed.). Los Angeles: Roxbury.

Albanese, J. (1989). *Organized crime in America*. Cincinnati, OH: Anderson.

Albanese, J. (1995). *White-collar crime in America*. Upper Saddle River, NJ: Prentice Hall.

Albanese, J. (2001). The prediction and control of organized crime. *Trends in Organized Crime, 6*(3, 4), 4–29.

Albanese, J. (2004). *Organized crime in our times* (4th ed.). Cincinnati, OH: Anderson.

Albini, J. L. (1971). *The American Mafia: Genesis of a legend*. New York: AppletonCentury-Crofts.

Albini, J. L. (1986, March). *The Guardian Angels: Vigilantes or protectors of the community?* Paper presented at the Academy of Criminal Justice Sciences Meeting, Orlando, FL.

Albini, J. L. (1988). Donald Cressey's contribution to the study of organized crime: An evaluation. *Crime and Delinquency, 34,* 338–354.

Albini, J. L., Kutushev, V., Rogers, R. E., Moiseev, V., Shabalin, V., & Anderson, J. (1995). Russian organized crime: Its history, structure and function. *Journal of Contemporary Criminal Justice, 11,* 213–243.

Alexander, S. (1988). *The pizza connection.* New York: Weidenfeld and Nicholson.

Allen, H. E., Friday, P. C., & Roebuck, J. B. (1981). *Crime and punishment: An introduction to criminology.* New York: Free Press.

Allen, J. (1977). *Assault with a deadly weapon: The autobiography of a street criminal.* New York: McGraw-Hill.

American Bar Association. (1952). *Report on organized crime.* New York: Author.

American Bar Association. (1976). *Final report of the committee on economic offenses.* Washington, DC: Author.

American Broadcasting Company. (1983a, February 3). Child molesters [Television series episode]. *20/20.* New York: Author.

American Broadcasting Company. (1983b, January 6). Scientology fraud [Television series episode]. *20/20.* New York: Author.

American Broadcasting Company. (1983c, February 24). *The media and violence* [Television broadcast]. New York: Author.

American Broadcasting Company. (1987, August 17). *Nightly News* [Television broadcast]. New York: Author.

American Broadcasting Company. (1991, November 21). Men of God [Television series episode]. *Primetime Live.* New York: Author.

American Broadcasting Company. (1992, July 19). Child molesting [Television series episode]. *Primetime Live.* New York: Author.

American Humane Association Children's Division. (1984). *Trends in officially reported child neglect and abuse in the United States.* Denver, CO: Author.

American Society of Criminology. (1998). *Code of Ethics.* Retrieved on April 3, 2007, from http://www.asc41.com/ethicspg.html.

American Society of Criminology and Academy of Criminal Justice Sciences. (2004). *Oral history project* [Videotapes, 3 volumes]. Belmont, CA: Wadsworth.

American Society of Criminology task force report to Attorney General Janet Reno. (1995, November/December). [Special issue]. *The Criminologist, 20*(6).

Amir, M. (1971). *Patterns in forcible rape.* Chicago: University of Chicago Press.

Ancient records discovered. (1987, December 29). *Erie Morning News,* p. 4A.

Anderson, D. A. (1999). The aggregate burden of crime. *Journal of Law and Economics, 42,* 47.

Anderson, E. (1977, January and March). A study of industrial espionage: Parts I and II. *Security Management.*

Anderson, E. (1981). *A place on the corner: Identity and rank among black streetcorner men.* Chicago: University of Chicago Press.

Anderson, E. (1990). *Streetwise: Race, class and change in an urban community.* Chicago: University of Chicago Press.

Anderson, J., & Van Atta, D. (1986, July 13). The amount of child pornography declines. *Erie Times News,* p. 5B.

Anderson, J., & Van Atta, D. (1987, July 19). Danger lurks in counterfeit bolts. *Erie Times News,* p. 3B.

Anderson, J., & Van Atta, D. (1988a, October 16). Medical waste poses health threat. *Erie Times News,* p.3B.

Anderson, J., & Van Atta, D. (1988b, August 28). Terrorists act as hired guns for drug cartel. *Erie Times News,* p. 3B.

Anderson, J., & Van Atta, D. (1990, March 24). Human rights—Iranian prisons remain a horror. *Erie Times News,* p. 3B.

Anderson, J., & Whitten, L. (1977, March 24). Mafia chieftain. *Erie Times News,* p. 3B.

Anderson, S. (1995, December). Looking for Mr. Yaponchik: The rise and fall of a Russian mobster in America. *Harper's,* pp. 40–51.

Anglin, M. D., & Speckart, G. (1988). Narcotics use and crime: A multisample, multimethod analysis. *Criminology, 16,* 197–233.

Animal-rights group destroys mink research. (1991, March 11). *Chronicle of Higher Education,* p. A5.

Annin, P., & Rhine, J. B. (1999, September 6). The gang that loves glitter. *Newsweek,* p. 32.

Anslinger, H., & Cooper, C. R. (1937). Marijuana: Assassin of youth. *American Magazine, 74,* 19, 50.

Appleby, J. (2003, January 8). Bristol to pay $670 million to settle antitrust suits. *USA Today,* p. 28.

Archer, D., & Gartner, R. (1984). *Violence and crime in cross-national perspective.* New Haven, CT: Yale University Press.

Ardrey, R. (1963). *African genesis.* New York: Atheneum.

Armstrong, E. G. (1978). Massage parlors and their customers. *Archives of Social Behavior, 7,* 117.

Armstrong, E. G. (1983). Pondering pandering. *Deviant Behavior, 40,* 203–217.

Armstrong, E. G. (1991, March). *Music and violence.* Paper presented at the meeting of the Academy of Criminal Justice Sciences, Nashville, TN.

Arson for hate and profit. (1977, October 31). *Time,* pp. 22–25.

Ashman, C. (1975). *The CIA–Mafia link.* New York: Manor Books.

Associated Press. (1994, October 23). Criminologists' file found.

Associated Press. (2007, August 18). Orphans granted settlement for monster study.

Associated Press. (2008, August 7). Hospital CEO charged in healthcare scheme. *Erie Times News,* p. 15A.

Atkinson, A. B. (1975). *The economics of inequality.* Oxford, UK: Clarendon.

Auerbach, A. H. (1998). *Ransom: The untold story of international kidnapping.* New York: Henry Holt.

August, O. (1997, February 17). Car theft rivals drugs in world crime earnings. *The Times.*

Austin, T. (1986). Book review of James Q. Wilson and Richard J. Herrnstein's *Crime and human nature. Criminal Justice Policy Review, 1,* 241–242.

Auto thieves. (1992, June 13). *Primetime Live* [Television broadcast]. New York: American Broadcasting Company.

Badillo, H., & Haynes, M. (1972). *A bill of no rights: Attica and the American prison system.* New York: Outerbridge and Lazard.

Bailey, W. C. (1971). Correctional outcome: An evaluation of 100 reports. In L. Radzinowicz & M. E. Wolfgang (Eds.), *Crime and Justice: Vol. 3.* New York: Basic Books.

Bakan, D. (1975). *The slaughter of the innocents.* San Francisco: Jossey-Bass.

Ball, J. C., Rosen, L., Flueck, J. A., & Nurco, D. N. (1982). Lifetime criminality of heroin addicts in the United States. *Journal of Drug Issues, 1,* 1–2.

Ball, K. (1990, December 21). USX Corp. to pay OSHA record $3.25 million fine. *Erie Morning News,* p. A1.

Ball, R. (1980). An empirical evaluation of neutralization theory. *Criminologica, 4,* 22–32.

Balrig, F. (1988). *The snow-white image: The hidden reality of crime in Switzerland.* Oslo: Norwegian University Press.

Bandow, D. (1991, September 12). Robert Gates: A case worth investigating. *Wall Street Journal,* p. A19.

Bandura, A. (1973). *Aggression: A social learning approach.* Englewood Cliffs, NJ: Prentice Hall.

Barak, G. (1991). *Crimes by the capitalist state: An introduction to state criminality.* Albany: SUNY Press.

Baridon, P. (1988). *Report on Asian organized crime* (U.S. Department of Justice, Criminal Division). Washington, DC: Government Printing Office.

Barker, T., & Carter, D. L. (1986). *Police deviance.* Cincinnati, OH: Anderson.

Barlay, S. (1973). *The secrets business.* New York: Thomas Y. Crowell.

Barovick, H. (1999, December 27). Bad to the bone. *Time,* 130–131.

Barrett, D. (2007, October 10). Major acid rain lawsuit settled. *Pittsburgh Post Gazette,* pp. A1–A3.

Barry, V. (1983). *Philosophy: A text with readings.* Belmont, CA: Wadsworth.

Bartol, C., & Bartol, A. M. (1986). *Criminal behavior: A psychosocial approach* (2nd ed.). Englewood Cliffs, NJ: Prentice-Hall.

Bastone, W. (1997). Mob Bell, they're all connected at Gotti Jr.'s telephone card company. *Village Voice.*

Baumrind, D. (1978). Parental disciplinary patterns and social competence in children. *Youth and Society, 9,* 239–276.

Bayer, R. (1981). Crime, punishment and the decline of liberal optimism. *Crime and Delinquency, 27,* 169–190.

Bayley, D. H. (1978, March). *Comment: perspectives on criminal justice research.* Speech delivered to the Academy of Criminal Justice Sciences, New Orleans, LA. Reprinted in *Journal of Criminal Justice, 6,* 287–289.

Baylor, T. (1990, August). *Informants/agent provocateurs—violators of trust: One element in the tactical repertoire of social control agents.* Paper presented at the meeting of the American Sociological Association, Washington, DC.

Beauchamp, T. L. (Ed.). (1983). *Case studies in business, society and ethics.* Englewood Cliffs, NJ: Prentice Hall.

Beccaria, C. (1963). *On crimes and punishments* (H. Paolucci, Trans.). Indianapolis, IN: Bobbs-Merrill. (Original work published 1764)

Beck, M., & Cowley, G. (1990, March 26). Beyond lobotomies. *Newsweek,* 44.

Becker, G. (1968). Crime and punishment: An economic approach. *Journal of Political Economy, 76,* 169–217.

Becker, H. (1950). *Through values to social interpretation.* Durham, NC: Duke University Press.

Becker, H. (1954). Anthropology and sociology. In J. Gillin (Ed.), *For a science of social man* (pp. 102–159). New York: Macmillan.

Becker, H. (1963). *Outsiders: Studies in the sociology of deviance.* New York: Free Press.

Becker, H. (Ed.). (1964). *The other side: Perspectives on deviance.* New York: Free Press

Beekman, M. E., & Daly, M. R. (1990, September). Motor vehicle theft investigations. *FBI Law Enforcement Bulletin,* pp. 14–17.

Beirne, P. (1987). Adolphe Quetelet and the origins of positivist criminology. *American Journal of Sociology, 92,* 1140–1169.

Beirne, P. (1991, November). Inventing criminology: The "science of man" in Cesare Beccaria's "Dei Delitti e Delle Pene (1764)". *Criminology, 29,* 777–820.

Beirne, P., & Messerschmidt, J. (2000). *Criminology* (3rd ed.). Boulder, CO: Westview.

Belknap, J. (1990, October). Review of "Fraternity gang rape: Sex, brotherhood and privilege on campus" by Peggy Reeves Sanday. *Criminal Justice Policy Review, 4,* 285–287.

Belknap, J. (1996). *The invisible woman: Gender, crime and justice.* Belmont, CA: Wadsworth.

Bell, D. (1953). Crime as an American way of life. *Antioch Review, 13,* 131–154.

Bell, D. (1967). *The end of ideology.* Glencoe, IL: Free Press.

Belson, W. A. (1978). *Television violence and the adolescent boy.* Farmborough, UK: Saxon House.

Benekos, P. (1995). Women as victims and perpetrators of murder. In A. V. Merlo & J. Pollock (Eds.), *Women, law, and social control* (pp. 219–237). Boston: Allyn & Bacon.

Benekos, P. J., & Hagan, F. E. (1991a, July 14). Fixing the thrifts. *Journal of Security Administration,* 65–104.

Benekos, P. J., & Hagan, F. E. (1991b, July 14). The Great Savings and Loan Scandal. *Journal of Security Administration,* 41–64.

Benjamin, H., & Masters, R. E. L. (1964). *Prostitution and morality.* New York: Julian Press.

Bennett, G. (1987). *Crimewarps: The future of crime in America.* New York: Anchor Books.

Bennett, J. (1994, December 5). Cost of saving lives. *New York Times,* p. A16.

Bennett, R. R., & Lynch, J. P. (1990, February). Does a difference make a difference? Comparing cross-national indicators. *Criminology, 28,* 153–181.

Bennett, T. (1988). The British experience with heroin regulation. *Law and Contemporary Problems, 51,* 299–314.

Bennett, W. J., DiIulio, J. J., & Walters, J. P. (1996). *Body count: Moral poverty and how to win America's war on crime and drugs.* New York: Simon & Schuster.

Bensinger, G. (1987, November 14). *Operation Greylord and its aftermath.* Paper presented at the meeting of the American Society of Criminology, Montreal, Quebec, Canada.

Bentham, J. (1823). *Introduction to the principles of morals and legislation.* Oxford: Oxford University Press.

Ben-Yehuda, N. (1990). *The politics and morality of deviance: Moral panics, drug abuse, deviant science and reversed stigmatization.* Albany: SUNY Press.

Bequai, A. (1978). *Computer crime.* Lexington, MA: Lexington Books.

Bequai, A. (1987). *Technocrimes.* Lexington, MA: Lexington Books.

Berdie, R. (1947, January). Playing the dozens. *Journal of Abnormal and Social Psychology, 42,* 102–121.

Berg, E. N. (1989, January 30). FBI commodities "sting": Fast money, secret lives. *New York Times,* p. A1.

Berger, V. (1988). Review essay: Not so simple rape. *Criminal Justice Ethics, 7,* 69–81.

Bergier, J. (1975). *Secret armies: The growth of corporate and industrial espionage* (H. Salemson, Trans.). Indianapolis, IN: Bobbs-Merrill.

Berk, R. A., & Newton, P. J. (1985). Does arrest really deter wife battery? An effort to replicate the findings of the Minneapolis Spouse Abuse Experiment. *American Sociological Review, 50,* 253–262.

Bernard, T. J. (1987). Structure and control: Reconsidering Hirschi's concept of commitment. *Justice Quarterly, 4,* 409–424.

Berton, L. (1991, May 15). Malleable money men. *Wall Street Journal,* p. A12.

Best, J., & Luckenbill, D. F. (1982). *Organizing deviance.* Englewood Cliffs, NJ: Prentice Hall.

Biderman, A. D., et al. (1967). *Report on a pilot study in the District of Columbia on victimization and attitudes toward law enforcement, field surveys I.* Commission on Law Enforcement and Administration of Justice. Washington, DC: Government Printing Office.

Big Belfast heist began with a cruel lie. (2004, December 23). Associated Press.

Binder, A., & Meeker, J. W. (1988). Experiments as reforms. *Journal of Criminal Justice, 16*, 347–358.

Bittner, E. (1967). The police on skid row: A study of peace keeping. *American Sociological Review, 32*, 699–715.

Bjorklund, D., & Pellegrini, A. (2002). *The origins of human nature: Evolutionary developmental psychology.* Washington, DC: American Psychological Association.

Black, D. J. (1970). Production of crime rates. *American Sociological Review, 35*, 733–748.

Black, D. (1999). *Bad boys, bad men: Confronting antisocial personality.* New York: Oxford University Press.

Black, W. K. (2004). Why doesn't the SEC have a chief criminologist? *The Criminologist, 29*(6), 1–4.

Blackman, A., & Simmons, A. (1995, September 18). Bury my heart in committee. *Time*, pp. 48–51.

Blackstock, N. (1976). *Cointelpro: The FBI's secret war on political freedom.* New York: Random House.

Blakeslee, S. (1994, April 7). Lawyers say Dow study saw implant danger. *New York Times*, pp. A1, A11.

Blakey, G. R., & Billing, R. (1981). *The plot to kill the president: Organized crime assassinated JFK.* New York: New York Times Books.

Blakey, G. R., & Goldsmith, M. (1976). Criminal redistribution of stolen property: The need for law reform. *Michigan Law Review, 74*, 1518–1545.

Blankenship, M. B. (Ed.). (1995). *Understanding corporate criminality.* New York: Garland.

Blankenship, M. B., & Brown, S. E. (1993). Paradigm or perspective: A note to the discourse community. *Journal of Crime and Justice, 16, 1*, 167–175.

Blaum, P. A. (1991, October 17). Crime is a male domain: Researchers explain why women are less likely to commit crimes. *Intercom: Focus on Research*, p. 1.

Bliven, N. (1991, June 17). Books: All the president's men II. *The New Yorker*, pp. 113–116.

Block, A. A. (1978). The history and study of organized crime. *Urban Life, 6*, 455–474.

Block, A. A., & Chambliss, W. J. (1981). *Organizing crime.* New York: Elsevier.

Block, C. R., & Block, R. (1988, November). *Is violent crime seasonal? Victimization and visibility.* Paper presented at the meeting of the American Society of Criminology, Chicago, IL.

Bloom, M. T. (1957). *Money of their own: The great counterfeiters.* New York: Scribner.

Blumberg, A. S. (1967). The practice of law as a confidence game: Organizational cooptation of a profession. *Law and Society Review, 1*, 15–39.

Blumenthal, D. (Ed.). (1988). *The last days of the Sicilians.* New York: Times Books.

Blumenthal, R. (1994, December 15). The maddening mysteries of the greatest art theft ever. *New York Times*, pp. B1, B6.

Blumstein, A. (1994). *Youth, violence, guns, and the illicit-drug industry.* Pittsburgh, PA: Carnegie Mellon University.

Blumstein, A. (1995, August). Violence by young people: Why the deadly nexus? *National Institute of Justice Journal*, pp. 2–9.

Blumstein, A., & Cohen, J. (1987, August 28). Characterizing criminal careers. *Science, 237*, 985–991.

Blumstein, A., Cohen, J., & Farrington, D. P. (1988a). Criminal career research: Its value for criminology. *Criminology, 26*, 1–35.

Blumstein, A., Cohen, J., & Farrington, D. P. (1988b). Longitudinal and criminal career research: Further clarifications. *Criminology, 26*, 57–74.

Blumstein, A., Cohen, J., Roth, J., & Visher, C. A. (Eds.). (1986). *Criminal careers and career criminals.* Washington DC: National Academy Press.

Blundell, W. E. (1978). I did it for jollies. In J. M. Johnson & J. Douglas (Eds.), *Crime at the top* (pp. 153–185). Philadelphia: Lippincott.

Boccella, K. (1994, May 31). Grocery store workers steal more food, money than customers do, survey finds. *Erie Times News*, p. 9S.

Bogert, C. (1990, June 4). On reform: Prime time for crime. *Newsweek*, p. 25.

Bohm, R. M. (1982). Radical criminology: An explication. *Criminology, 19*, 565–589.

Bohm, R. M. (1997). *A primer on crime and delinquency.* Belmont, CA: Wadsworth.

Bonanno, J. (1983). *A man of honor: The autobiography of Joseph Bonanno.* New York: Simon & Schuster.

Bonger, W. A. (1969). *Criminality and economic conditions.* Bloomington: Indiana University Press.

Bonn, R. L. (1987). Review of James D. Wright and Peter H. Rossi's "Armed and Considered Dangerous." *Justice Quarterly, 4*, 133–136.

Booth, C. (1978, March 12). Prostitutes in America: Teens, chicken hawks, networks abound. *Erie Times News*, p. 23A.

Bordua, D. J. (1961). Delinquent subcultures: Sociological interpretations of gang delinquency. *Annals of the American Academy of Political and Social Science, 338,* 119–136.

Bordua, D. J. (1962). Delinquency and opportunity: Analysis of a theory. *Sociology and Social Research, 46,* 167–175.

Bottomley, A. K. (1979). *Criminology in focus: Past trends and future prospects.* New York: Barnes and Noble.

Boucher, R. (1989, March 12). Trying to fix a statute run amok. *New York Times,* p. 4E.

Boudreau, J. F., Kwan, Q. Y., Faragher, W. E., & Denault, G. C. (1977). *Arson and arson investigations: Survey and assessment.* Washington, DC: U.S. Department of Justice.

Bourgois, P. (1988, November). *Fear and loathing in el barrio: Ideology and upward mobility in the underground economy of the inner city.* Paper presented at the meeting of the American Anthropological Association, Phoenix, AZ.

Bourgois, P. (1995). *In search of respect: Selling crack in el barrio.* New York: Cambridge University Press.

Bowe, J. (2007). *Nobodies: Modern American slave labor and the dark side of the new global economy.* New York: Random House.

Bowers, W. B., & Pierce, G. (1975). The illusion of deterrence in Isaac Ehrlich's research on capital punishment. *Yale Law Journal, 85,* 164–227.

Bowker, L. (Ed.). (1998). *Masculinities and violence.* Thousand Oaks, CA: Sage.

Braithwaite, J. (1976). Functional and conflict theories of crime. In W. J. Chambliss and M.Mankoff (Eds.), *Whose law, what order?* New York:Wiley.

Braithwaite, J. (1989a). *Crime, shame and reintegration.* Cambridge, UK: Cambridge University Press.

Braithwaite, J. (1989b). Criminological theory and organizational crime. *Justice Quarterly, 6,* 333–358.

Brantingham, P. J., & Brantingham, P. L. (1984). *Patterns in crime.* New York: Macmillan.

Brecher, E. (1972). *Licit and illicit drugs.* Boston: Little, Brown.

Brenner, M. H. (1978). Economic crises and crime. In L. Savitz & N. Johnston (Eds.), *Crime and society* (2nd ed., pp. 555–572). New York: Wiley.

Bresler, F. (1980). *The Chinese Mafia.* New York: Stein and Day.

Briar, S., & Piliavin, I. (1965). Delinquency, situational inducements and commitment to conformity. *Social Problems, 13,* 35–45.

British clinics are struggling as heroin addiction climbs. (1985, April 11). *New York Times,* p. 15A.

Brodeur, P. (1974). *Expendable Americans.* New York: Viking.

Brooke, J. (1995, April). Kidnappings soar in Latin America, threatening region's stability. *New York Times,* p. A7.

Brown, B. S., Wienckowski, L. A., & Bivins, L. W. (1973). *Psychosurgery: Perspectives on a current problem.* Washington, DC: Government Printing Office.

Brown, C. (1964). *Manchild in the promised land.* New York: Macmillan.

Brown, M. H. (1982). Love Canal and the poisoning of America. In J. H. Skolnick & E. Currie (Eds.), *Crisis in American institutions* (5th ed., pp. 297–316). Boston: Little, Brown.

Brown, R. M. (1969). Historical patterns of violence in America. In H. D. Graham & T. R. Gurr (Eds.), *Violence in America, a staff report to the National Commission on the Causes and Prevention of Violence* (pp. 43–80). New York: New American Library.

Browning, F., & Gerassi, J. (1980). *The American way of crime.* New York: Putnam.

Brownmiller, S. (1975). *Against our will: Men, women and rape.* New York: Simon & Schuster.

Brunner, J. (1975). *The shockwave rider.* New York: Harper & Row.

Brush, P. (1999, May 9). Honor student, teachers charged in shoplift ring. APBnews.com.

Bryson, C. (1998, Fall). The Donora fluoride fog. *Earth Island Journal.*

Bryson, C. (2000). *Fluoride and the Mohawks.* Retrieved November 18, 2009, from www.fluoridation.com/donora.htm

Bureau of Justice Statistics. (1983). *Report to the nation on crime and justice: The data.* Washington, DC: Government Printing Office.

Bureau of Justice Statistics. (1988, March). *Report to the nation on crime and justice* (2nd ed.). Washington, DC: Government Printing Office.

Bureau of Justice Statistics. (1994, October). National crime victimization survey redesign. *Bureau of Justice Statistics Fact Sheet.* Washington, DC: Government Printing Office.

Bureau of Justice Statistics. (2003). *National crime victimization survey.* Washington, DC: Government Printing Office.

Bureau of Justice Statistics. (2004, November 29). *The nation's school crime: Rates continue to decline* [Press release] (NCJ 205290). Washington, DC: Author.

Burgess, A. W. (1984). *Child pornography and sex rings.* Lexington, MA: D. C. Heath.

Burgess, A. W., Hazelwood, R. R., Rokous, F. E., Hartman, C. R., & Burgess, A. G. (1987). Serial rapists and their victims: Reenactment and repetition. In R. R. Hazelwood & A. W. Burgess (Eds.), *Practical aspects of rape investigation.* New York: Elsevier.

Burgess, E. W. (1925). The growth of the city. In R. E. Park, E. W. Burgess, & R. D. McKenzie (Eds.), *The city* (pp. 47–62). Chicago: University of Chicago Press.

Burgess, R. L., & Akers, R. L. (1966). A differential association-reinforcement theory of criminal behavior. *Social Problems, 14,* 128–147.

Burnett, C. (1986). Review essay. *Criminology, 24,* 203–211.

Bursik, R. J., Jr. (1988). Social disorganization and theories of crime and delinquency: Problems and prospects. *Criminology, 26,* 519–551.

Burton, T. M. (1995, June 19). Caremark paid physicians to obtain patients, government documents say. *Wall Street Journal,* p. B8.

Byrne, J., & Sampson, R. (1986). Cities and crime: The ecological/non-ecological debate reconsidered. In J. Byrne & R. Sampson (Eds.), *The social ecology of crime.* New York: Springer-Verlag.

Cameron, M. O. (1964). *The booster and the snitch: Department store shoplifting.* New York: Free Press.

Campbell, D. T., & Stanley, J. C. (1963). *Experimental and quasi-experimental designs for research.* Chicago: Rand McNally.

Campo-Flores, A. (2005, August 22). Sex offenders: Do you live next to one? *Newsweek,* p. 11.

Cantor, D., & Land, K. C. (1985). Unemployment and crime rate in the post–World War II United States: A theoretical and empirical analysis. *American Sociological Review, 50,* 317–332.

Capeci, J. (2003). *Gangland.* New York: Alpha.

Cardarelli, A. P. (1988, May/June). Child sexual abuse: Factors in family reporting. *NIJ Reports, 209,* 9–12.

Carey, B. (2008, September 25). Two big drug manufacturers begin disclosing payments they make to doctors, *New York Times,* p. A16.

Carey, J. (1988, December). From revival tent to mainstream. *U.S. News and World Report,* pp. 52–61.

Carey, J. T. (1972). Problems of access and risk in observing drug scenes. In J. D. Douglas (Ed.), *Research on deviance* (pp. 71–92). New York: Random House.

Carey, J. T. (1975). *Sociology and public affairs: The Chicago school.* Beverly Hills, CA: Sage.

Carley, W. M. (1991, March 19). Glory of Rome: An ancient treasure leads to dark intrigue in the world of art. *Wall Street Journal,* pp. A1, A4.

Carlisle, C. (1998, October 6). As Latin American art prices rise, so do forgeries. *New York Times,* p. C1.

Carll, E. (2007, June 29). Violent video games: Rehearsing aggression. *Chronicle of Higher Education,* p. B20.

Carlson, K. W. (1979). Statement before the Congressional Committee on Education and Labor, Subcommittee on Compensation, Health, and Safety, hearings on asbestos-related occupational diseases, 95th Congress, Second session (pp. 25–52). Washington, DC: Government Printing Office.

Carlson, K., & Finn, P. (1993, November). *Prosecuting criminal enterprises* (Bureau of Justice Statistics special report). Washington, DC: Bureau of Justice Statistics.

Carson, R. (1962). *Silent spring.* New York: Houghton Mifflin.

The case of the doctored transcripts. (1974, July 22). *Time.* Available online at http://www.time.com/time/magazine/article/0,9171,942925,00.html

Catalano, R. F., & Hawkins, J. D. (1996). The social developmental model: A theory of antisocial behavior. In J. D. Hawkins (Ed.), *Delinquency and crime.* Cambridge: Cambridge University Press.

Caudill, W. A. (1958). *The psychiatric hospital as a small society.* Cambridge, MA: Harvard University Press.

CBS Broadcasting, Inc. (1989, November 20). Japan [Television series episode]. In S. Zirinsky (Producer), *48 Hours.* New York: Author.

Cendant to pay $2.8 billion fraud settlement. (1999, December 8). APBnews.com.

Chaiken, J. M., & Chaiken, M. R. (1982). *Varieties of criminal behavior.* Report prepared for the National Institute of Justice, U.S. Department of Justice. Santa Monica, CA: The Rand Corporation.

Chaiken, M. R., & Johnson, B. D. (1988). Characteristics of different types of drug-involved offenders. *Issues and practices in criminal justice.* Washington, DC: National Institute of Justice.

Chambliss, W. J. (1964). A sociological analysis of the law of vagrancy. *Social Problems, 12,* 67–77.

Chambliss, W. J. (1975a). *Box man: A professional thief's journal, by Harry King.* New York: Harper.

Chambliss, W. J. (1975b). Toward a political economy of crime. *Theory and Society, 2,* 152–153.

Chambliss, W. J. (1988a). *On the take: From petty crooks to presidents* (2nd ed.). Bloomington: Indiana University Press.

Chambliss, W. J. (1988b). State-organized crime. *Criminology, 27,* 183–208.

Chambliss, W. J., & Seidman, R. B. (1971). *Law and order and power.* Reading, MA: Addison-Wesley.

Chamlin, M. B., & Cochran, J. K. (1995, August). Assessing Messner and Rosenfeld's institutional anomie theory: A partial test. *Criminology, 33,* 411–429.

Chappell, D., & Fogarty, F. (1978). *Forcible rape: A literature review and annotated bibliography.* Washington, DC: National Institute on Law Enforcement and Criminal Justice.

Cheatwood, D. (1988). Is there a season for homicide? *Criminology, 26,* 287–306.

Cheesman, B. (1999, December 30). The Thai dying you won't find in a brochure. *World Financial Review,* pp. 1–4.

Chen, D. W. (2007, September 7). 11 New Jersey officials are arrested in corruption investigation. *New York Times,* p. A26.

Chesney-Lind, M. (1989). Girls' crime and woman's place: Toward a feminist model of female delinquency. *Crime and Delinquency, 35,* 5–30.

Chesney-Lind, M., & Shelden, R. G. (1998). *Girls, delinquency, and juvenile justice* (2nd ed.). Belmont, CA: Wadsworth.

Chesterton, G. K. (1935). *Avowals and denials: A book of essays.* London: Methuen.

Chin, K.-L. (1988, November). *Chinese organized crime: Myth and fact.* Paper presented at the meeting of the American Society of Criminology, Chicago.

Chin, K.-L. (1990). *Chinese subculture and criminality.* Westport, CT: Greenwood.

Choate, P. (2005, May 12). The pirate kingdom. *New York Times,* p. A27.

Christensen, H. T., & Gregg, C. F. (1970). Changing sex norms in America and Scandinavia. *Journal of Marriage and Family, 32,* 616–627.

Christiansen, K. (1968). Threshold of tolerance in various population groups illustrated by results from a Danish criminological twin study. In A. V. S. de Reuck (Ed.), *The mentally abnormal offender.* Boston: Little, Brown.

Chrysler fined for violations. (1990, August 11). *Erie Morning News,* p. 2A.

Chu, Y. K. (2005). Hong Kong Triads after 1997. *Trends in Organized Crime, 8*(5), 5–12.

City inspectors extorted hundreds of thousands from N.Y. restaurants. (1988, March 25). *Erie Morning News,* p. 3A.

Clark, A. L., & Gibbs, J. P. (1965). Social control: A reformation. *Social Problems, 13,* 399–415.

Clark, T., & Tigue, J. J., Jr. (1975). *Dirty money: Swiss banks, the Mafia, money laundering and white-collar crime.* New York: Simon & Schuster.

Clarke, J. W. (1982). *American assassins: The darker side of politics.* Princeton, NJ: Princeton University Press.

Clarke, S. H. (1974). Getting 'em out of circulation: Does incarceration of juvenile offenders reduce crime? *Journal of Law and Criminology, 65,* 528–535.

Clinard, M. B. (1946, June). Criminological theories of violation of wartime regulations. *American Sociological Review, 11,* 258–270.

Clinard, M. B. (1969). *The black market: A study of white-collar crime.* Montclair, NJ: Patterson Smith. (Original work published 1952)

Clinard, M. B. (1978). *Cities with little crime: The case of Switzerland.* Cambridge, UK: Cambridge University Press.

Clinard, M. B., & Abbott, D. J. (1973). *Crime in developing countries: A comparative perspective.* New York: Wiley.

Clinard, M. B., & Quinney, R. (1973). *Criminal behavior systems: A typology.* New York: Holt, Rinehart and Winston.

Clinard, M. B., & Quinney, R. (1986). *Criminal behavior systems: A typology* (2nd ed.). New York: Holt, Rinehart and Winston.

Clinard, M. B., & Yeager, P. C. (1978). Corporate crime: Issues in research. *Criminology, 16,* 255–272.

Clinard, M. B., & Yeager, P. C. (1979). *Illegal corporate behavior.* Washington, DC: Law Enforcement Assistance Administration.

Clinard, M. B., & Yeager, P. C. (1980). *Corporate crime.* New York: Macmillan.

Cloward, R., & Ohlin, L. (1960). *Delinquency and opportunity: A theory of delinquent gangs.* New York: Free Press.

Clutterbuck, R. (1975). *Living with terrorism.* New Rochelle, NY: Arlington House.

Cocaine cowboys running rampant in Florida. (1980, May 6). *Erie Morning News.*

Cohen, A. K. (1955). *Delinquent boys.* New York: Free Press.

Cohen, D. (1979). *Mysteries of the world.* Garden City, NJ: Doubleday.

Cohen, L. E., & Felson, M. (1979). Social change and crime rate trends: A routine activities approach. *American Sociological Review, 44,* 588–608.

Cohen, L. E., & Land, K. (1987). Age and crime: Symmetry vs. asymmetry and the projection of crime rates through the 1990s. *American Sociological Review, 52,* 170–183.

Cohen, L. E., & Stark, R. (1974). Discriminatory labeling and the five-finger discount: An empirical analysis of differential shoplifting dispositions. *Journal of Research on Crime and Delinquency, 11,* 25–35.

Cohn, B. (1988, March 14). A fresh assault on an ugly crime. *Newsweek,* pp. 64–65.

Cohn, E. G. (1990, Winter). Weather and crime. *British Journal of Criminology, 30,* 51–63.

Colburn, T. (1996). *Our stolen future.* New York: Dutton.

Cole, D. (1999). *No equal justice: Race and class in the American criminal justice system.* New York: New Press.

Coleman, J. W. (1985). *The criminal elite.* New York: St. Martin's Press.

Coleman, J. W. (1994). *The criminal elite* (3rd ed.). New York: St. Martin's Press.

Collins, J. J., Jr. (Ed.). (1981). *Drinking and crime: Perspectives on the relationships between alcohol consumption and criminal behavior.* New York: Guilford Press.

Colvin, M., & Pauly, J. (1983). A critique of criminology. *American Journal of Sociology, 89,* 513–551.

Combs, C. C. (2003). *Terrorism in the twenty-first century* (3rd ed.). Upper Saddle River, NJ: Prentice Hall.

Comstock, G. A. (1975). The effects of television on children and adolescents: The evidence so far. *Journal of Communication, 25,* 25–34.

Comte, A. (1877). *A system of positive polity.* London: Longmans. (Original work published 1851)

Con game nets $3,600 from widow. (1972, November 7). *Erie Morning News,* p. 11A.

Conklin, J. E. (1972). *Robbery and the criminal justice system.* New York: Lippincott.

Conklin, J. E. (1981). *Criminology.* New York: Macmillan.

Conning by computer. (1973, April 23). *Newsweek,* p. 26.

Cook, F. J. (1984). *Maverick: Fifty years of investigative reporting.* New York: Putnam.

Cook, P. J. (1983, June). Robbery. *Research in Brief.* Washington, DC: U.S. Department of Justice, National Institute of Justice.

Cook, P. J. (Ed.). (1988). Vice. *Law and Contemporary Problems, 51* (entire issue).

Cooley, C. H. (1902). *Human nature and social order* (1964 ed.). New York: Schocken Books.

Copeland, M. (1974). *Beyond cloak and dagger: Inside the CIA.* New York: Pinnacle Books.

Corchado, A. (2009). *Los Zetas: Inside Mexico's dangerous gang.* Retrieved November 18, 2009, from www .airpower.au.af.mil/apjinternational/apj-s/2009/3tri09/brandseng.htm

Cornish, D. B., & Clarke, R. V. (Eds.). (1986). *The reasoning criminal: Rational choice perspectives on offending.* New York: Springer-Verlag.

Cortés, J., with Gatti, F. M. (1972). *Delinquency and crime.* New York: Seminar Press.

Coser, L. (1956). *The functions of social conflict.* New York: Free Press.

Courtright, K. E., & Mutchnick, R. J. (1999). *The cartographic school of criminology.* Unpublished paper.

Cousins, N. (1979, November 10). How the US used its citizens as guinea pigs. *Saturday Review,* p. 10.

Cowell, A. (1992, June 21). Inquiry into Sicilian slaying looks for Mafia link to Colombia drug cartel. *New York Times,* p. 3.

Cowell, A. (1994, March 22). Where Juliet pined, youths now kill. *New York Times,* p. A4.

Cowley, G. (1992, May 25). Fueling the fire over Halcion. *Newsweek,* 84.

Cox, E. R., Fellmuth, R. C., & Schulz, J. E. (1969). *Nader's Raiders: Report on the Federal Trade Commission.* New York: Grove Press.

Cranford, J. (1989, August 12). Congress OKs sweeping bill to save thrift industry. *Congressional Quarterly, 47*(32), 2113–2151.

Craven, D. (1996, December). *Female victims of violent crime* (NCJ162602). Washington, DC: Bureau of Justice Statistics.

Cressey, D. (1953). *Other people's money.* New York: Free Press.

Cressey, D. (1960). Epidemiology and individual conduct: A case from criminology. *Pacific Sociological Review, 3,* 47–58.

Cressey, D. (1969). *The theft of the nation: The structure and operations of organized crime in America.* New York: Harper & Row.

Cressey, D. (1972). *Criminal organization.* New York: Harper & Row.

Crewdson, J. (1988). *By silence betrayed: Sexual abuse of children in America.* Boston: Little, Brown.

Crichton, R. (1959). *The great imposter.* New York: Permabooks.

Criminologists' file found. (1994, October 23). Associated Press.

Crittenden, A. (1988). *Sanctuary: A story of American conscience and law in collision.* New York: Weidenfeld and Nicolson.

Crockett, A. (Ed.). (1991). *Spree killers.* New York: Pinnacle.

Cromwell, P. (Ed.). (1996). *In their own words: Field research on crime and criminals—an anthology.* Los Angeles: Roxbury.

Crossette, B. (2000, March 9). UNICEF opens a global drive on violence against women. *Wall Street Journal,* p. A6.

Crovitz, L. G. (1991, April 17). The more lawsuits the better and other American notions. *Wall Street Journal,* p. A17.

Crowe, R. R. (1974). An adoption study of antisocial personality. *Archives of General Psychiatry, 31,* 785–791.

Cullen, F. T. (1984). The Ford Pinto case and beyond. In E. C. Hochstedler (Ed.), *Corporations as criminals.* Beverly Hills, CA: Sage Publications.

Cullen, F. T., & Agnew, R. (Eds.). (1999). *Criminological theory: Past to present: Essential readings.* Los Angeles: Roxbury.

Cullen, F. T., & Agnew, R. (Eds.). (2003). *Criminological theory: Past to present: Essential readings* (2nd ed.). Los Angeles: Roxbury.

Cullen, F. T., & Gilbert, K. (1982). *Reaffirming rehabilitation.* Cincinnati, OH: Anderson.

Cullen, F. T., Link, B. G., & Polanzi, C. W. (1982). The seriousness of crimes revisited: Have attitudes toward white-collar crime changed? *Criminology, 20,* 83–102.

Cullen, F. T., Makestad, W., & Cavender, G. (1987). *Corporate crime under attack: The Ford Pinto case and beyond.* Cincinnati, OH: Anderson.

Cullen, F. T., et al. (1991, November). *Testing the general theory of crime: Self-control, age and lawbreaking.* Paper presented at the American Society of Criminology Meetings, San Francisco, CA.

Curfew violations—A useless statutory crime. (1999, September 15). *Washington Post,* p. A24.

Curran, D., & Renzetti, C. M. (2001). *Theories of crime* (2nd ed.). Boston: Allyn & Bacon.

Currie, E. (1985). *Confronting crime: Why there is so much crime in America and what we can do about it.* New York: Pantheon.

Dabney, D. A., Hollinger, R. C., & Dugan, L. (2004). Who actually steals? A study of covertly observed shoplifters. *Justice Quarterly, 21*(4), 693–728.

Dahrendorf, R. (1959). *Class and class conflict in industrial society.* Stanford, CA: Stanford University Press.

Dalgard, O. S., & Kringlen, E. (1975). A Norwegian twin study of criminality. *British Journal of Criminology, 16,* 213–232.

Dalleck, R. (1991). *Lone star rising: Lyndon Johnson and his times, 1908–1960.* New York: Oxford University Press.

Dalton, K. (1961). Menstruation and crime. *British Medical Journal, 2,* 1752–1753.

Daly, K. (1989, November). Gender varieties of white-collar crime. *Criminology, 27,* 769–793.

Daly, K., & Chesney-Lind, M. (1988). Feminism and criminology. *Justice Quarterly, 5,* 497–533.

Daly, K., & Tonry, M. (1997). Gender, race and sentencing. In M. Tonry (Ed.), *Crime and justice: A review of research, Vol. 22* (pp. 201–252). Chicago: University of Chicago Press.

Dammer, H., with Fairchild, E., & Albanese, J. (2006). *Comparative criminal justice systems,* 3rd ed. Belmont, CA: Wadsworth.

Danner, M. (1989). Socialist feminism: A brief introduction. *Critical Criminologist, 1,* 1–2.

Danner, M. (1995). *The massacre at El Mozote: A parable of the Cold War.* New York: Random House.

Daraul, A. (1969). *A history of secret societies.* New York: Pocket Books.

Darwin, C. (1859). *Origin of species.* London: John Murray.

Darwin, C. (1871). *The descent of man.* London: John Murray.

Dash, E. (2007, December 7). Former chief will forfeit $418 million. *New York Times,* pp. C1, C5.

Davis, H., & Gurr, T. (1969). *Violence in America: Historical and comparative perspectives.* New York: The New American Library.

Davis, J. (1982). *Street gangs: Youth, biker and prison gangs.* Dubuque, IA: Kendall/Hunt.

Davis, K. (1961). Prostitution. In R. K. Merton & R. A. Nisbet (Eds.), *Contemporary social problems* (pp. 275–276). New York: Harcourt, Brace and World.

Death squads prey on Rio street children. (1991, April-May). *CJ The Americas, 4,* 12.

Decker, S., & Chapman, M. (2008). *Drug smugglers on drug smuggling.* Philadelphia: Temple University Press.

DeFleur, M. L., & Quinney, R. (1966, January). A reformulation of Sutherland's differential association theory and a strategy for empirical verification. *The Journal of Research in Crime and Delinquency, 3,* 1–22.

DeKeseredy, W. (1988). The left realist approach to law and order. *Justice Quarterly, 5,* 635–640.

DeKeseredy, W., & MacLean, B. D. (1993, Fall). Critical criminological pedagogy in Canada: Strengths, limitations, and recommendations for improvements. *Journal of Criminal Justice Education, 4,* 361–376.

DeKeseredy, W. S., Saunders, D. G., Schwartz, M. D., & Alvi, S. (1997). The meanings and motives for women's use of violence in Canadian college dating relationships: Results from a national survey. *Sociological Spectrum, 17,* 199–222.

Del Piano, A. J. (1993). The fine art of forgery, theft and fraud. *Criminal Justice, 8*(2), 16–20, 56–57.

DeMaris, O. (1981). *The last mafioso: The treacherous world of Jimmy Fratianno.* New York: Times Books.

Denfield, D. (1974). *Streetwise criminology.* Cambridge, MA: Schenkman.

Denno, D. (1985). Sociological and human developmental explanations of crime: Conflict and consensus? *Criminology, 23,* 711–742.

Denno, D. (1990). *Biology and violence: From birth to adulthood.* New York: Cambridge University Press.

Department of Corporations. (1999, September 7). *Department of Corporations issues warning on "Affinity Fraud."* California news release 99–13. Available at http://www.corp.ca.gov/press/news/1999/nr9913.asp.

Dershowitz, A. (1994). *Abuse excuse: Cop-outs, sob stories and other evasions of responsibility.* Boston: Little, Brown.

Detlinger, C., with Prugh, J. (1983). *The list.* Atlanta: Philmay Enterprises.

Detroit's former chief convicted. (1992, May 8). *Erie Morning News,* p. 16B.

Dialysis chain agrees to $500 million fraud settlement. (2000). APBnews.com.

Diapoulos, P., & Linakis, S. (1976). *The sixth family: The true inside story of the execution of a Mafia chief.* New York: Bantam.

Dickey, C. (1989, May 29). Missing masterpieces. *Newsweek,* pp. 65–68.

Dionne, E. J., Jr. (1990, May 21–27). The death penalty: Getting mad and getting even. *Washington Post National Weekly Edition,* p. 37.

Dobash, R. E., Dobash, R. P., & Noaks, L. (1995). *Gender and crime.* Cardiff: University of Wales Press.

Dobnik, V. (1996, November 20). Never again! U.S. government to pay $4.8 million. *Erie Morning News,* pp. 1A, 2A.

Dodge, K. (1991). The structure and function of reactive and proactive aggression. In D. J. Pepler & K. H. Rubin (Eds.), *The development and treatment of childhood aggression* (pp. 54–61). Hillsdale, NJ: Erlbaum.

Doerner, W. G. (1988). The impact of medical resources on criminally induced lethality: A further examination. *Criminology, 26,* 171–179.

Doerner, W. G., & Speir, J. C. (1986). Stitch and sew: The impact of medical resources upon criminally induced lethality. *Criminology, 24,* 319–330.

Dolive, L. L. (1999, November). *When criminals rule: Corruption and politics.* Paper presented at the meeting of the American Society of Criminology, Toronto, ON, Canada.

Donovan, R. (1952). *The assassins.* New York: Harper Brothers.

Dornfeld, M., & Kruttschnitt, C. (1991, November). *Is there a weaker sex? Mapping gender-specific outcomes and risk factors.* Paper presented at the meeting of the American Society of Criminology, San Francisco.

Douglas, J., & Olshaker, M. (1995). *Mindhunter.* New York: Pocket Star Books.

Douglas, J., & Olshaker, M. (1997). *Journey into darkness.* New York: Scribner.

Dowie, M. (1977, September). Pinto madness. *Mother Jones, 2,* 18–32. Reprinted in J. H. Skolnick & E. Currie (Eds.), *Crisis in American Institutions* (1982). Boston: Little, Brown.

Doyle, J. M. (1987, January 14). Aging Mafia bosses get century in jail. *Erie Morning News,* p. 2A.

Draper, T. (1991). *A very thin line: The Iran-Contra affairs.* New York: Hill & Wang.

Drug raid conducted at Pa. prison. (1995, October 24). *Erie Morning News,* p. 2A.

Drug rings hire gun-toting kids. (1988, June 5). *Erie Times News,* p. 4A.

Du Bois, W. E. B. (1901). The spawn of slavery: The convict-lease system in the South. *The Missionary View of the World, 14,* 737–745.

Du Bois, W. E. B. (1973). *The Philadelphia Negro: A social study.* Millwood, NY: Kraus-Thomson. (Original work published 1899)

Dubro, A. (Writer). (1982, May 27). Yakuza [Television series episode]. *20/20*. New York: American Broadcasting Company.

Duckworth, M. (1991, October 28). Counterfeiting credit cards is a science only one step behind bona fide plastic. *Wall Street Journal*, p. B5.

Dugan, L., Nagin, D., & Rosenfeld, R. (2000). *The declining rate of intimate partner homicide*. National Criminal Justice Reference Service, NIJ Videotape (NCJ 180212).

Dugdale, R. (1877). *The Jukes: A study in crime, pauperism and heredity*. New York: Putnam.

Duhart, D. (2001, December). *Violence in the workplace: 1993–1999* (Bureau of Justice Statistics special report, NCJ 190076). Washington, DC: Bureau of Justice Statistics.

Durkheim, E. (1950). *The rules of the sociological method*. Glencoe, IL: Free Press. (Original work published 1895)

Durkheim, E. (1951). *Suicide*. New York: Free Press. (Original work published 1897)

Durkheim, E. (1964). *The division of labor in society*. New York: Free Press. (Original work published 1893)

Duster, T. (1970). *The legislation of morality: Laws, drugs and moral judgment*. New York: Free Press.

Dutton, D. (Ed.). (1983). *The forger's art: Forgery and the philosophy of art*. Berkeley: University of California Press.

Ecstasy trade, seizures skyrocket. (2000, April 3). APBnews.com.

Eddy, P., Sabogal, H., & Walden, S. (1988). *The cocaine wars*. New York: Norton.

Edelhertz, H. (1970). *The nature, impact, and prosecution of white collar crime*. National Institute of Law Enforcement and Criminal Justice. Washington, DC: Government Printing Office.

Editorial: Return of the trustbusters. (2009, May 13). *New York Times*, p. A26.

Egger, S. A. (1984). A working definition of serial murder. *Journal of Police Science, 12*, 348–357.

Ehrhart, J. K., & Sandler, B. R. (1985). *Campus gang rape: Party games?* Project on the Status and Education of Women. Washington, DC: Association of American Colleges.

Eichenwald, K. (1994, July 13). Prudential's fraud costs to exceed $1.1 billion. *New York Times*, pp. C1, C13.

Eichenwald, K. (2006, July 6). Enron founder, awaiting prison, dies in Colorado. *New York Times*, p. A1.

Elder abuse. (2008). *Medline Plus* (U.S. National Library of Medicine). Available at http://www.nlm.nih.gov/medlineplus/elderabuse.html

Elias, R. (1986). *The politics of victimization: Victims, victimology, and human rights*. New York: Oxford University Press.

Elliott, D. S., & Ageton, S. S. (1980, February). Reconciling race and class differences in self-reported and official estimates of delinquency. *American Sociological Review, 45*, 95–110.

Elliott, D. S., Ageton, S. S., & Cantor, R. J. (1979). An integrated theoretical perspective on delinquent behavior. *Journal of Research on Crime and Delinquency, 16*, 3–27.

Ellis, A. (1959). Why married men visit prostitutes. *Sexology, 25*, 344.

Ellis, L. (1982, May). Genetics and criminal behavior. *Criminology, 20*, 43–56.

Ellis, L. (1985). Religiosity and criminality: Evidence and explanation surrounding complex relationships. *Sociological Perspective, 28*, 501–520.

Ellis, L. (1996). Arousal theory and the religiosity-criminality relationship. In P. Cordello & L. Siegel (Eds.), *Readings in contemporary criminological theory* (pp. 65–83). Boston: Northeastern University Press.

Ellis, L., & Walsh, A. (1997). Gene-based evolutionary theories in criminology. *Criminology, 36*(2), 229–276.

Ellis, L., & Walsh, A. (2000). *Criminology: A global perspective*. Boston: Allyn & Bacon.

Engberg, E. (1967). *The spy in the corporate structure*. Cleveland: World Publishing.

Ennis, P. H. (1967). *Criminal victimization in the United States: A report of a national survey*. Field Surveys II, The President's Commission on Law Enforcement and Administration of Justice, Washington, DC: Government Printing Office.

Epstein, E. J. (1977). *Agency of fear*. New York: Putnam.

Epstein, E. J. (1983, September 18). Edwin Wilson and the CIA: How badly one man hurt our nation. *Parade*, 22–24.

Erickson, P. G. (1990). A public health approach to demand reduction. *The Journal of Drug Issues, 20*(4), 563–575.

Erie red light resorts pay more than $50,000 annually in rentals. (1988, January 22). *Erie Morning News*.

Erikson, E. H. (1950). *Childhood and society*. New York: Norton.

Erlanger, H. S. (1974, March). The empirical status of the subculture of violence thesis. *Social Problems, 22*, 280–292.

Ermann, M. D., & Lundman, R. J. (1982). *Corporate deviance.* New York: Holt, Rinehart and Winston.

Esposito, J. C., & Silverman, L. J. (1970). *Vanishing air: Ralph Nader's study group report on air pollution.* New York: Grossman.

Estrich, S. (1987). *Real rape: How the legal system victimizes women who say no.* Cambridge, MA: Harvard University Press.

Eysenck, H. (1977). *Crime and personality* (3rd ed.). London: Routledge/Kegan Paul.

Fagan, J. (1990). Intoxication and aggression. In M. Tonry & J. Q. Wilson (Eds.), *Drugs and crime* (pp. 241–320). Chicago: University of Chicago Press.

Fagan, J., Piper, E., & Moore, M. (1986). Violent delinquents and urban youths. *Criminology, 24,* 439–471.

Fagan, J., & Wexler, S. (1987). Family origins of violent delinquents. *Criminology, 25,* 643–669.

Fancher, R. E. (1985). *The intelligence men.* New York: W. W. Norton.

Farah, D. (1997, October 6). A red alert on drugs. *Washington Post National Weekly Edition,* p. 17.

Farley, M. (2007). *Prostitution and trafficking in Nevada: Making the connection.* San Francisco: Prostitution Research and Education.

Farrington, D. P. (1986). Age and crime. In M. Tonry & D. Farrington (Eds.), *Crime and justice* (Vol. 7, pp. 189–250). Chicago: University of Chicago Press.

Farrington, D. P. (2003). Developmental and life-course criminology: Key theoretical and empirical issues—The 2002 Sutherland Award address. *Criminology, 41*(2), 221–265.

Farrington, D. P., Ohlin, L. E., & Wilson, J. Q. (Eds.). (1986). *Understanding and controlling crime.* New York: Springer-Verlag.

Farnworth, M., & Leiber, M. J. (1989). Strain theory revisited. *American Sociological Review, 54,* 263–274.

FBI gambling sting nets 23 Cleveland police officers. (1991, May 31). *Erie Morning News,* p. 1A.

FBI Uniform Crime Reports statistics. (2007). *Crime in the United States, 2006.* Washington, DC: Government Printing Office.

Federal Bureau of Investigation. (1985). *Oriental organized crime.* Washington, DC: Government Printing Office.

Federal Bureau of Investigation. (2004). *Crime in the United States, 2003. Uniform Crime Reports.* Washington, DC: Government Printing Office.

Federal Bureau of Investigation. (2005). *Crime in the United States, 2004. Uniform Crime Reports.* Washington, DC: Government Printing Office.

Federal Bureau of Investigation. (2007, March 30). *Over 100 arrests in effort to disrupt neighborhood gangs and their criminal activity* [Press release]. Available at http://www.fbi.gov/pressrel/pressrel07/safestreet033007.htm.

Feds indict TMI operator over reports. (1983, November). *Erie Morning News.*

Feds: Philly guards gave inmates drugs, phones. (2007, September 22). *Erie Times-News,* p. 2B.

Feeney, F., & Weir, A. (1975, May). The prevention and control of robbery. *Criminology, 13,* 102–105.

Feldman, R. E. (1968). Response to compatriot and foreigner who seeks assistance. *Journal of Personality and Social Psychology, 10,* 202–214.

Felson, M. (1983). Ecology of crime. In S. H. Kadish (Ed.), *The encyclopedia of crime and justice.* New York: Macmillan.

Felson, M. (1987). Routine activities and crime prevention in the developing metropolis. *Criminology, 25,* 911–931.

Ferdinand, T. N. (1987). The methods of delinquency theory. *Criminology, 25*(4), 841–862.

Ferdinand, T. N. (1991, Spring). The theft/violence ratio in antebellum Boston. *Criminal Justice Review, 16,* 42–58.

Ferguson, C. (2007, June 22). Video games: The latest scapegoat for violence. *Chronicle of Higher Education,* p. B20.

Ferracuti, F. (1968). European migration and crime. In M. E. Wolfgang (Ed.), *Crime and culture: Essays in honor of Thorsten Sellin.* New York: Wiley.

Ferrell, J., Hamm, M. S., & Adler, P. (1998). *Ethnography at the edge: Crime deviance and field research.* Boston: Northeastern University Press.

Ferri, E. (1917). *Criminal sociology* (J. I. Kelley & John Lisle, Trans.). Boston: Little, Brown. (Original work published 1878)

Fight Crime Committee. (1986, April). *A discussion document on options for changes in the law and in the administration of the law to counter the Triad problem.* Hong Kong: Government Security Office.

Finckenauer, J. O. (2005). Problems of definition: What is organized crime? *Trends in Organized Crime, 8*(5), 63–83.

Finckenauer, J., & Chin, K. (2004, November). *Asian transnational organized crime and its impact on the United States: Developing a transnational crime research agenda* (National Institute of Justice special report). Washington, DC: U.S. Department of Justice. Available at http://www.ncjrs.gov/pdffiles/1nij/grants/213310.pdf.

Finckenauer, J. O., & Waring, E. J. (1998). *The Russian Mafia in America.* Boston: Northeastern University Press.

Finestone, H. (1976). *Victims of change.* Westport, CT: Greenwood.

Finkelhor, D. (1979). *Sexually victimized children.* New York: Free Press.

Finkelhor, D. (1986). *A sourcebook on child sexual abuse.* Beverly Hills, CA: Sage Publications.

Finkelhor, D., Mitchell, K., & Wolak, J. (2001, March). *Highlights of the Youth Internet Safety Survey* OJJDP Fact Sheet #4. Washington, DC: U.S. Department of Justice, Office of Juvenile Justice and Delinquency Prevention.

Finn, P. E., & Sullivan, M. (1988). Police respond to special populations: Handling the mentally ill, public inebriate, and the homeless. *NIJ Reports, 209,* 2–8.

Firestone, D. (1994, August 3). 22 are arrested in burglaries of Kennedy Airport cargo. *New York Times,* p. A1.

Fishbein, D. H. (1990, February). Biological perspectives in criminology. *Criminology, 28,* 27–72.

Fishbein, D. H. (2001). *Behavioral perspectives in criminology.* Belmont, CA: Wadsworth.

Fishbein, D. H. (2002). Biocriminology. In D. Levinson (Ed.) *Encyclopedia of crime and punishment* (Vol. I, pp. 109–118). Thousand Oaks, CA: Sage Publications.

Flaherty, A. (2007, October 10). House OKs bill to curb wartime contractor fraud. *The Cleveland Plain Dealer,* p. A6.

Fleishmann, J. (2000, May 12). Fleeing poverty, finding slavery. *Philadelphia Inquirer,* p. A1.

Fletcher, R. (1991). *Science, ideology and the media: The Cyril Burt affair.* New Brunswick, NJ: Transaction Publishers.

Flowers, R. B. (1988). *Minorities and crime.* Westport, CT: Greenwood Press.

Fontana, V. J. (1973). *Somewhere a child is crying.* New York: Macmillan.

Forgac, G. E., & Michaels, E. J. (1982). Personality characteristics of two types of male exhibitionists. *Journal of Abnormal Psychology, 91*(4), 287–295.

Former Ohio State president criticizes fans in job interview, softens stance. (2007, August 30). Associated Press. Available at http://sports.espn.go.com/ncaa/news/story?id=2997669

Forsthoffer, D. (1999, December 24). Robbery suspect wants cultural insanity test, *Erie Morning News,* p. 8B.

4 indicted over pacemaker scam. (1988, September 1). *Erie Morning News,* p. 1A.

Fox, J. A., & Levin, J. (1985). *Mass murder: America's growing menace.* New York: Plenum.

Fox, R. G. (1971). The XYY offender: A modern myth? *Journal of Criminal Law, Criminology, and Police Science, 62,* 59–73.

Fox, V. (1976). *Introduction to criminology.* Englewood Cliffs, NJ: Prentice-Hall.

Fox, V. (1985). *Introduction to criminology* (2nd ed.). Englewood Cliffs, NJ: Prentice Hall.

France, M., & Burnett, V. (1992, December 1). Corporate America's Colombian connection. *Business Week,* pp. 168–170.

Frank, N. (1985). *Crimes against health and society.* New York: Harrow and Heston.

Frank, N., & Lombness, M. (1988). *Controlling corporate illegality.* Cincinnati, OH: Anderson.

Frantz, D., Blumenthal, R., & Vogel, C. (2000, October 8). Ex-leaders of 2 auction grants are said to initiate price-fixing. *New York Times,* p. A11.

Frantz, D., & Nasar, S. (1994, July 18). FBI inquiry on jet engines new jolt to company image. *New York Times,* p. A1.

Freedman, M. H. (1976). Advertising and soliciting: The case of ambulance chasing. In R. Nader & M. Green (Eds.), *Verdicts on lawyers.* New York: Thomas Y. Cromwell.

French, H. W. (2001, October 10). Even in Ginza: Honor among thieves crumbles. *USA Today,* p. A4.

French, L. (1989, November) *Post-traumatic stress disorder and violence: three forensic cases.* Paper presented at the American Society of Criminology Meetings, Reno, NV.

Freud, S. (1930). *Civilization and its discontents* (J. Riviere, Trans.). Garden City, NJ: Doubleday.

Friedlander, K. (1947). *The psychoanalytic approach to juvenile delinquency.* New York: International Universities Press.

Friedman, R. (1994, November 7). The Organizatsiya. *New York Magazine*, pp. 52–58.

Friedrich, C. J. (1972). *The pathology of politics*. New York: Harper & Row.

Friedrichs, D. O. (1980a). Carl Klockars vs. the "heavy hitters": A preliminary critique. In J. A. Inciardi (Ed.), *Radical criminology: The coming crisis* (pp. 149–160). Beverly Hills, CA: Sage.

Friedrichs, D. O. (1980b). Radical criminology in the United States: An interpretive understanding? In J. A. Inciardi (Ed.). *Radical criminology: The coming crisis* (pp. 35–60). Beverly Hills, CA: Sage.

Friedrichs, D. O. (1995). *Trusted criminals: White-collar crime in contemporary society*. Belmont, CA: Wadsworth.

Friedson, E. (1970). *The profession of medicine*. New York: Dodd, Mead.

Fry, F., Jr. (1986, September 21). Consumer bag: Airport theft. *Erie Times News*, p. 7B.

Gabbidon, S. L. (1999, January). W. E. B. DuBois on crime: American conflict criminologist. *The Criminologist, 24,* 1, 3, 20.

Gabbidon, S., & Greene, H. (2005). *Race and crime*. Thousand Oaks, CA: Sage Publications.

Gabor, T., & Normandeau, A. (1989). Armed robbery: Highlights of a Canadian study. *Canadian Police College Journal, 13*(4), 273–282.

Gage, N. (1971). *The mafia is not an equal opportunity employer*. New York: McGraw-Hill.

Gage, N. (1972). *Mafia, U.S.A.* Chicago: Playboy Press.

Gage, N. (1974, December 17). Questions are raised on the Lucky Luciano book. *New York Times*, pp. 1, 8.

Gage, N. (1988, January 18). A tale of two Mafias. *U.S. News and World Report*, pp. 36–37.

Gamble, T., & Eisert, A. (2004). Delinquency theory: Emerging explanations from the biopsychological perspective. In P. J. Benekos & A. Merlo (Eds.), *Controversies in juvenile justice and delinquency*. Cincinnati, OH: LexisNexis.

Gangland enforcer paid with life for rifts with mob over C.I.A. (1977, March 31). *United Press International*.

Gans, H. (1962). *The urban villagers*. New York: The Free Press.

Garner, J., & Visher, C. A. (1988, September/October). Policy experiments come of age. *NIJ Reports, 201,* 2–8.

Garofalo, R. (1914). *Criminology* (R. W. Millar, Trans.). Boston: Little, Brown. (Original work published 1884)

Garrow, D. (1981). *The FBI and Martin Luther King*. New York: Norton.

GE pleads guilty to fraud charges. (1992, July 23). *Erie Morning News*, p. 2A.

Gebhard, P., Gagnon, J. H., Pomeroy, W. B., & Christenson, C. V. (1965). *Sex offenders: An analysis of types*. New York: Harper & Row.

Geis, G. (1974a). Avocational crime. In D. Glaser (Ed.), *Handbook of criminology* (pp. 273–298). Chicago: Rand McNally.

Geis, G. (1974b). Upperworld crime. In A. S. Blumberg (Ed.), *Current perspectives on criminal behavior: Original essays on criminology* (pp. 179–198). New York: Knopf.

Geis, G. (2007). *White collar and corporate crime*. Upper Saddle River, NJ: Prentice Hall.

Geis, G., & Meier, R. F. (Eds.). (1977). *White-collar crime: Offenses in business, politics, and the professions* (Rev. ed.). New York: Free Press.

Geis, G., & Meier, R. F. (1979). Looking backward and forward: Criminologists on criminology as a career. In E. Sagarin (Ed.), *Criminology: New concerns* (pp. 173–188). Beverly Hills, CA: Sage.

Gelles, R. J. (1977). *Etiology of violence: Overcoming fallacious reasoning in understanding family violence and child abuse*. Manuscript available through the National Criminal Justice Reference Service, Rockville, MD.

Gelles, R. J. (1978, October). Violence toward children in the United States. *American Journal of Orthopsychiatry, 48,* 580–592.

Gelles, R. J., & Straus, M. (1979, March). Violence in the American family. *Journal of Social Issues, 35,* 15–39.

Georges-Abeyie, D. (Ed.). (1984). *The criminal justice system and blacks*. New York: Clark Boardman.

Georges-Abeyie, D. (1989, Summer). Review of William Wilbanks: The myth of a racist criminal justice system. *The Critical Criminologist, 1,* 5–6.

Gerber, J. (1991, March). Heidi and Imelda: The changing image of crime in Switzerland. *Criminology, 8,* 121–128.

Gerlin, A. (1999, October 21). Cure or curse. *Erie Morning News*, p. 1A.

Germany's "punks" descend on Hanover for repeat performance of "chaos days." (1995, August 5). *Erie Morning News*, p. 1A.

Gibbens, T. C. N., & Silberman, M. (1960). The clients of prostitutes. *British Journal of Venereal Disease, 36*, 113.

Gibbon, E. (1776–1789). *The decline and fall of the Roman empire* (6 volumes). London: Meuthen.

Gibbons, D. C. (1977). *Society, crime and criminal careers* (3rd ed.). Englewood Cliffs, NJ: Prentice Hall.

Gibbons, D. C. (1979). *The criminological enterprise: Theories and perspectives.* Englewood Cliffs, NJ: Prentice Hall.

Gibbons, D. C. (1982). *Society, crime and criminal behavior* (4th ed.). Englewood Cliffs, NJ: Prentice Hall.

Gibbons, D. C. (1992). Talking about crime: Observations on the prospects for causal theory in criminology. *Criminal Justice Research Bulletin, 7* [entire issue].

Gibbons, D. C., & Garabedian, P. (1974). Conservative, liberal and radical criminology: Some trends and observations. In C. E. Reasons (Ed.), *The criminologist: Crime and the criminal* (51–56). Pacific Palisades: Goodyear.

Gibbs, J. B. (1985). Review essay of crime and human nature by James Q. Wilson and Richard Herrnstein. *Criminology, 23*, 381–388.

Gibson, W. (1982). *Neuromancer.* London: HarperCollins.

Gilham, J. R. (1992). *Preventing residential burglary: Toward more effective community programs.* New York: Springer-Verlag.

Gladwell, M. (1990, July 9–15). Hot time, bummer in the city. *Washington Post National Weekly Edition*, p. 39.

Glaser, D. (1978). *Crime in our changing society.* New York: Holt, Rinehart and Winston.

Glaser, D. (1990). Science and politics as criminologists vocations. *Criminal Justice Research Bulletin, 5*,(6), 1–6.

Glaser, D. (1994). What works and why it is important: A response to Logan and Gates. *Justice Quarterly, 11*, 711–724.

Glasser, J. (2000, February 7). The software sopranos. *U.S. News and World Report.*

Glasser, W. (1965). *Reality therapy.* New York: Harper & Row.

Glick, S. (1994). Good news, bad news about FBI crime data. *Erie Morning News*, p. 7A.

Glueck, S., & Glueck, E. (1950). *Unraveling juvenile delinquency.* Cambridge, MA: Harvard University Press.

Glueck, S., & Glueck, E. (1956). *Physique and delinquency.* New York: Harper and Row.

Goddard, H. H. (1912). *The Kallikak family.* New York: Macmillan.

Goff, C., & Reasons, C. (1986). Organizational crimes against employees, consumers, and the public. In B. MacLean, *The political economy of crime* (pp. 204–231). Toronto, ON: Prentice Hall of Canada.

Goldkamp, J. S. (1987, December 30). Rational choice and determinism. In M. B. Gottfredson & T. Hirschi (Eds.), *Positive criminology* (pp. 125–137). Beverly Hills, CA: Sage Publications.

Goldstein, M. J. (1991). *Delinquent gangs: A psychological perspective.* Champaign, IL: Research Press.

Goode, E. (1972). *Drugs in American society.* New York: Knopf.

Goode, E. (1981). Drugs and crime. In A. Blumberg (Ed.), *Current perspectives on criminal behavior* (pp. 227–272). New York: Knopf.

Goode, E. (1984). *Deviant behavior* (2nd ed.). Englewood Cliffs, NJ: Prentice Hall.

Goode, E. (2000, March 14). Human behavior: Born or made? *New York Times*, pp. D1, D9.

Goodspeed, P. (1998, September 6). Cracking the secret code of an elite porn "club." *Toronto Sunday Star*, p. A1.

Goodwin, J. (1982). *Sexual abuse: Incest victims and their families.* Boston: John Wright.

Gordon, D. M. (1973, April). Capitalism, class and crime in America. *Crime and Delinquency, 19*, 163–186.

Gordon, G. R. (1991, November). *Economic crime: An international perspective.* Paper presented at the meeting of the American Society of Criminology, San Francisco.

Gordon, L., & O'Keefe, P. (1984, February). Incest as a form of family violence: Evidence from historical case records. *Journal of Marriage and the Family, 46*, 27–34.

Gordon, R. A. (1987). SES versus IQ in the race-IQ-delinquency model. *International Journal of Sociology and Social Policy, 7*(3), 30–96.

Goring, C. (1913). *The English convict.* London: His Majesty's Stationery Office.

Gosch, M. A., & Hammer, R. (1974). *The last legacy of Lucky Luciano.* New York: Dell.

Gottfredson, M. R., & Hindelang, M. J. (1977, Fall). A consideration of telescoping and memory decay biases in victimization surveys. *Journal of Criminal Justice, 5*, 205–216.

Gottfredson, M. R., & Hirschi, T. (1986). The true value of lambda would appear to be zero: An essay on career criminals, criminal careers, selective incapacitation, cohort studies and related topics. *Criminology, 24*, 213–234.

Gottfredson, M. R., & Hirschi, T. (1987). The methodological adequacy of longitudinal research on crime. *Criminology, 25,* 581–614.

Gottfredson, M. R., & Hirschi, T. (1990). *A general theory of crime.* Stanford, CA: Stanford University Press.

Gould, L. C. (1969, September). The changing structure of property crime in an affluent society. *Social Forces, 48,* 50–59.

Gould, S. J. (1981). *The mismeasure of man.* New York: Norton.

Goulden, J. C. (1984). *The death merchant: The rise and fall of Edwin P. Wilson.* New York: Simon & Schuster.

Grabosky, P. (2007). *Electronic crime.* Upper Saddle River, NJ: Prentice Hall.

Gramckow, H. (1992, October). New money laundering law. *German American Legal Journal, 2,* 12–13.

Gravel, M. (Ed.). (1971). *The Pentagon Papers* (4 vols.). Boston: Beacon Press.

Greek, C. (1990, March 17). *Is this the end of RICO?* Paper presented at the Academy of Criminal Justice Sciences meeting, Denver, CO.

Green, G. (1990). *Occupational crime.* Chicago: Nelson-Hall.

Green, H. T. (1979). *A comprehensive bibliography of criminology and criminal justice literature by black authors from 1895–1978.* Hyattsville, MD: Ummah Publications.

Green, M. J. (Ed.). (1973). *The monopoly makers: Ralph Nader's study group report on regulation and competition.* New York: Grossman.

Green, W. E., & Geyelin, M. (1990, November 27). Asbestos case jury awards $26.3 million. *Wall Street Journal,* p. B6.

Greenberg, D. (1975, Summer). The incapacitative effects of imprisonment: Some estimates. *Law and Society Review, 9,* 541–580.

Greenberg, D. (1981). *Crime and capitalism.* Palo Alto, CA: Mayfield.

Greenberg, D. (Ed.). (1993). *Crime and capitalism: Readings in Marxist criminology.* Philadelphia: Temple University Press.

Greenfield, L. A. (1996, March). Child victimizers: Violent offenders and their victims. *Bureau of Justice Statistics Executive Summary* (NCJ 158625). Washington, DC: Bureau of Justice Statistics.

Greenfield, L. A., & Smith, S. K. (1999, February). *American Indians and crime* (NCJ 173386). Washington, DC: Bureau of Justice Statistics.

Greenhouse, S. (1999, January 14). 18 major retailers and apparel makers are accused of using sweatshops. *New York Times,* p. A9.

Griffin, S. P. (2003). *Philadelphia's black Mafia.* Dordrecht, Netherlands: Kluwer.

Griffin, S. P. (2005). *Black brothers, Inc.: The violent rise and fall of Philadelphia's Black Mafia.* Preston, UK: Milo Books.

Gropper, B. A. (1985, February). Probing the links between drugs and crime. *Research in Brief.* Washington, DC: U.S. Department of Justice, National Institute of Justice.

Gross, B. (1980). *Friendly fascism: The new face of power in America.* New York: M. Evans.

Groth, N., & Birnbaum, H. J. (1979). *Men who rape: The psychology of the offender.* New York: Plenum Press.

Groth, N. A., Burgess, A. W., Birnbaum, H. J., & Gary, T. S. (1978). A study of the child molester: Myths and realities. *LAE (Journal of the American Criminal Justice Association), 41,* 17–22.

Guerry, A. M. (1833). *An essay on moral statistics.* Paris: Crochard.

Gugliotta, G., & Leen, J. (1989). *Kings of cocaine.* New York: Simon & Schuster.

Gusfield, J. (1963). *Symbolic crusades.* Urbana: University of Illinois Press.

Haas, S. (1985, December). Bad seeds and social policy: Two histories. *Psychology Today,* pp. 73–74.

Hacker traced to Argentina. (1996, March 30). *Erie Morning News,* pp. A1–A2.

Hacker, F. J. (1976). *Crusaders, criminals, crazies: Terror and terrorism in our time.* New York: Norton.

Hagan, F. E. (1975). *Comparative professionalism in an occupational arena: The case of rehabilitation.* Unpublished doctoral dissertation, Case Western Reserve University.

Hagan, F. E. (1983, Fall). The organized crime continuum: A further specification of a new conceptual model. *Criminal Justice Review, 8,* 52–57.

Hagan, F. E. (1986). Sub rosa criminals: Spies as neglected criminal types. *Clandestine Tactics and Technology* (A technical and background data service, International Chiefs of Police), *11* [entire issue].

Hagan, F. E. (1987a). Book review: James Mills' The underground empire: Where crime and government embrace. *American Journal of Criminal Justice, 11,* 128–130.

Hagan, F. E. (1987b). The global fallacy and theoretical range in criminological theory. *Journal of Justice Issues, 2,* 19–31.

Hagan, F. E. (1987c). The global fallacy and theoretical range in criminological theory. *Journal of Justice Issues, 2*, 19–31.

Hagan, F. E. (1991, March). *The professional criminal in the nineties.* Paper presented at the meeting of the Academy of Criminal Justice Sciences, Nashville, TN.

Hagan, F. E. (1996). Panopticon. In M. D. McShane & F. P. Williams (Eds.), *Encyclopedia of American prisons* (pp. 341–342). New York: Garland.

Hagan, F. E. (1997). *Political crime: Ideology and criminality.* Boston: Allyn & Bacon.

Hagan, F. E. (1999, November). *White House crime and scandal: From Washington to Clinton.* Paper presented at the meeting of the American Society of Criminology, Toronto, ON, Canada.

Hagan, F. E. (2000, May). *The ghost of Chic Conwell: Professional crime and fraud in the twenty-first century.* Paper presented at the National White-Collar Crime Summit, Austin, TX.

Hagan, F. E. (2002, March 7). *The Robert Hanssen spy case.* Paper presented at the American Society of Criminology Meetings, Anaheim, CA.

Hagan, F. E. (2005). International corruption and extortion. *International Journal of Comparative Criminology, 4*(2), 250–264.

Hagan, F. E. (2006). Organized crime and "organized crime." *Trends in Organized Crime, 9*(4), 127–137.

Hagan, F. E. (2008a). *Introduction to criminology: Theories, methods, and criminal behavior.* Thousand Oaks, CA: Sage Publications.

Hagan, F. E. (2008b, March). *White House scandal: From Washington to George W. Bush.* Paper presented at the Academy of Criminal Justice Sciences Meeting, Cincinnati, OH.

Hagan, F. E. (2010). *Research methods in criminal justice and criminology* (8th ed.). Upper Saddle River, NJ: Prentice Hall.

Hagan, F. E., & Benekos, P. J. (1992). What Charles Keating and "Murph the Surf" have in common: A symbiosis of professional and occupational and corporate crime. *Criminal Organizations, 7*, 3–27.

Hagan, F. E., & Sussman, M. B. (Eds.). (1988). *Deviance and family.* New York: Haworth Press.

Hagan, F. E., & Tontodonato, P. (2004). Classical and sociological theories of delinquency. In P. Benekos & A. Merlo (Eds.), *Corrections: Dilemmas and directions.* Cincinnati, OH: Anderson.

Hagan, J. (1987). Review essay: A great truth in the study of crime. *Criminology, 25*, 421–428.

Hagan, J. (1989). *Structural criminology.* New Brunswick, NJ: Rutgers University Press.

Hagan, J. (1993, November). The social embeddedness of crime and unemployment. *Criminology, 31*, 465–491.

Hagan, J. (1994). *Crime and disrepute.* Thousand Oaks, CA: Pine Forge Press.

Hagan, J., Gillis, A. R., & Simpson, J. (1985). The class structure of gender and delinquency: Toward a power-control theory of common delinquent behavior. *American Journal of Sociology, 90*, 1151–1178.

Hagan, J., Gillis, A. R., & Simpson, J. (1987). Class in the household: A power-control theory of gender and delinquency. *American Journal of Sociology, 92*, 788–816.

Hagedorn, A. (1991, April 15). Prosecution of child-molestation cases grows more wary in wake of acquittals. *Wall Street Journal*, p. B1.

Hagedorn, A., & Lambert, W. (1990, December 20). U.S. alleges mob runs big casino union. *Wall Street Journal*, p. B6.

Hagedorn, J. M. (1994, May). Homeboys, dope fiends, legits, and new jacks. *Criminology, 32*, 197–219.

Hajdu, D. (2008). *The ten-cent plague: The great comic book scare and how it changed America.* New York: Farrar, Straus and Giroux.

Hall, J. (1952). *Theft, law and society* (Rev. ed.). Indianapolis, IN: Bobbs-Merrill.

Hall, K. (1992, July 19). Truckers set up for wrecks in deadly insurance scam. *Erie Times News*, p. 7E.

Hamilton, A. (1987, April 17). Gang leader's manual gave youths pointers on shoplifting at malls. *Erie Morning News*, p. 4A.

Hamilton, P. (1967). *Espionage and subversion in an industrial society.* London: Hutchinson.

Hamit, F. (1991, October). Taking on corporate counterintelligence. *Security Management*, 35–38.

Hamlin, J. E. (1988). The misplaced role of rational choice in neutralization theory. *Criminology, 26*, 425–38.

Hamm, M. (Ed.). (1994). *Hate crime: International perspectives on causes and control.* Cincinnati, OH: Anderson.

Hamm, M. S., & Ferrell, J. (1994, May/June). Raps, cops and crime: Clarifying the "cop killer" controversy. *ACJS Today, 1*(3), 29.

Hammer, R. (Ed.). (1975). *Playboy's illustrated history of organized crime.* Chicago: Playboy Press.

Haran, J. F. (1982). *The loser's game: A sociological profile of 500 armed robbers.* Unpublished doctoral dissertation, Fordham University.

Hardt, R. H., & Hardt, S. P. (1977, July). On determining the quality of the delinquency self-report method. *Journal of Research in Crime and Delinquency, 14,* 247–261.

Harmer, R. M. (1975). *American medical avarice.* New York: Abelard-Schuman.

Harrington, A. (1972). *Psychopaths.* New York: Simon and Schuster.

Harris, J. W., Jr. (1987, October). Domestic terrorism in the 1980s. *FBI Law Enforcement Bulletin,* pp. 5–13.

Harris, P., & Clarke, R. V. (1991, July). Car chopping, parts making, and the Motor Vehicle Theft Law Enforcement Act of 1984. *Social Science Review, 75,* 228–237.

Harris, S. H. (1994). *Factories of death: Japanese biological warfare, 1932–1945, and the American cover-up.* New York: Routledge.

Harrist, R. (1995, August 28). Notorious bank robber met his match in Mississippi. *Erie Morning News,* pp. A1–A2.

Hartung, F. E. (1950, July). White-collar offenses in the wholesale meat industry in Detroit. *American Journal of Sociology, 56,* 25–32.

Haskell, M. R., & Yablonksy, L. (1983). *Crime and delinquency* (3rd ed.). Chicago: Rand McNally.

Hastings, D. W. (1965). The psychiatry of presidential assassination. *The Journal Lancet, 85,* 93–100, 157–162, 189–192, 294–301.

Hawkins, D. F. (Ed.). (1986a). *Homicide among black Americans.* Lanham, MD: University Press of America.

Hawkins, D. F. (1986b). Race, crime type and imprisonment. *Justice Quarterly, 3,* 253–269.

Hawkins, D. F. (1987). Beyond anomalies: Rethinking the conflict perspective on race and punishment. *Social Forces, 65,* 719–745.

Hawkins, J. D., & Lishner, D. M. (1987). Schooling and delinquency. In E. H. Johnson (Ed.), *Handbook of crime and delinquency prevention* (pp. 179–221). Westport, CT: Greenwood.

Hayeslip, D. W., Jr. (1989, March/April). Local-level drug enforcement: New strategies. *NIJ Reports,* pp. 2–7.

Haywood, I. (1987). *Faking it: Art and the politics of forgery.* New York: St. Martin's Press.

Hazelwood, R. R., & Burgess, A. W. (1987). An introduction to the serial rapist: Research by the FBI. *FBI Law Enforcement Bulletin, 58,* 16–24.

Hazelwood, R. R., & Warren, J. (1989). The serial rapist: His characteristics and victim (Part I). *FBI Law Enforcement Bulletin, 60,* 10–17.

Healy, W. (1915). *The individual delinquent: A textbook and prognosis for all concerned in understanding offenders.* Boston: Little, Brown.

Heath, J. (1963). *Eighteenth century penal theory.* New York: Oxford University Press.

Hedges, S. J. (1998, February 2). The new face of Medicare. *U.S. News and World Report,* pp. 46–53.

Heilbroner, R. L., et al. (1973). *In the name of profit: Profiles in corporate irresponsibility.* New York: Warner Paperback Library.

Heise, L. (1991, December 16–22). Assaulted first by the rapist. Then by the societal response. *Washington Post National Weekly Edition,* p. 23.

Hell's Angels: Some wheelers may be dealers. (1979, July 2). *Time,* p. 34.

Hellman, P. (1970, March 15). One in ten shoppers is a shoplifter. *The New York Times Magazine,* p. 34.

Henderson, J. H., & Simon, D. R. (1994). *Crimes of the criminal justice system.* Cincinnati, OH: Anderson.

Henry, S., & Milanovic, D. (1993). Back to basics: A postmodern redefinition of crime. *The Critical Criminologist, 5,* 1–2, 12.

Henry, S., & Milanovic, D. (1996). *Constructive criminology: Beyond postmodernism.* London: Sage.

Hepburn, J. R. (1984). Occasional property crime. In R. F. Meier (Ed.), *Major forms of crime* (pp. 73–94). Beverly Hills, CA: Sage Publications.

Herling, J. (1962). *The great price conspiracy: The story of the antitrust violations in the electrical industry.* Washington, DC: Luce.

Herman, J. D. (1981). *Father-daughter incest.* Cambridge, MA: Harvard University Press.

Herrnstein, R. J. (1983). Some criminogenic traits of offenders. In J. Q. Wilson (Ed.), *Crime and public policy* (pp. 31–49). San Francisco: Institute for Contemporary Studies.

Herrnstein, R. J., & Murray, C. (1994). *The bell curve: The reshaping of American life by differences in intelligence.* New York: Free Press.

Hersh, S. (1990, April 29). The Iran-Contra committees: Did they protect Reagan? *New York Times Magazine,* pp. 47–49.

Herskovits, M. J. (1930). *The anthropometry of the American Negro.* New York: Columbia University Press.

Hertz admits to driving over the line. (1988, February 8). *Newsweek,* p. 48.

Hevesi, D. (1990, January 12). 8 at law firm accused of bribing witnesses and faking evidence. *New York Times,* p. 28.

Hickey, E. W. (2006a). *Serial murderers and their victims* (4th ed.). Belmont, CA: Wadsworth.

Hickey, E. W. (2006b). *Sex crimes and paraphilia.* Upper Saddle River, NJ: Prentice Hall.

Hidden Cameras Project. (1978). Seattle, Washington.

Higham, C. (1982). *Trading with the enemy: An exposé of the Nazi-American money plot, 1933–1949.* New York: Delacorte Press.

Hill, H., & Russo, G. (2004). *Gangsters and goodfellas.* New York: MJF Books.

Hills, S. L. (Ed.). (1987). *Corporate violence: Injury and death for profit.* Totowa, NJ: Rowman & Littlefield.

Hilts, P. J. (1995, July 26). U.S. turning to grand juries to scrutinize tobacco industry. *New York Times,* pp. A1, C19.

Hindelang, M. (1970, Spring). The commitment of delinquents to their misdeeds: Do delinquents drift. *Social Problems, 17,* 509.

Hindelang, M. (1971, January). Extroversion, neuroticisms and self-reported delinquent involvement. *Journal of Research in Crime and Delinquency, 8,* 23–31.

Hindelang, M. J. (1973, Spring). Causes of delinquency: A partial replication and extension. *Social Problems, 21,* 471–487.

Hindelang, M. J., et al. (1977). Correlates of self-reported victimization and perceptions of neighbourhood safety. In L. Hewitt and D. Brusegard (Eds.), *Selected papers from the social indicators conference* (1975). Edmonton: Alberta Bureau of Statistics.

Hindelang, M. J., Hirschi, T., & Weis, J. (1979). Correlates of delinquency: The illusion of discrepancy between self-report and official data. *American Sociological Review, 44,* 95–110.

Hirschi, T. (1969). *Causes of delinquency.* Berkeley: University of California Press.

Hirschi, T. (1983). Crime and the family. In J. Q. Wilson (Ed.), *Crime and public policy* (pp. 53–68). San Francisco: Institute for Contemporary Social Studies.

Hirschi, T., & Gottfredson, M. R. (1987). Causes of white-collar crime. *Criminology, 25*(4), 949–974.

Hirschi, T., & Gottfredson, M. R. (1989). The significance of white-collar crime for a general theory of crime. *Criminology, 27*(2), 359–371.

Hirschi, T., & Hindelang, M. J. (1977, August). Intelligence and delinquency: A revisionist review. *American Sociological Review, 42,* 571–587.

Hirschi, T., & Gottfredson, M. R. (1990, October). Substantive positivism and the idea of crime. *Rationality and Society, 2,* 412–428.

Hirsh, M. (1998, December 14). The hunt hits home. *Newsweek,* p. 48.

Hockstader, L. (1995, March 20–26). Crime atop chaos: In post-communist Russia, the strong arm of the Mafiya is everywhere. *Washington Post National Weekly Edition,* pp. 6–7.

Holden, R. (1986, March). *The road to fundamentalist and identity movements.* Paper presented at the Academy of Criminal Justice Sciences Meeting, St. Louis, MO.

Hollin, C. (1989). *Psychology and crime: An introduction to criminological psychology.* London: Routledge.

Hollinger, R. C., & Davis, J. L. (2002). *2002 National Retail Security Survey, final report.* Gainesville: University of Florida.

Hollinger, R. C., & Lanza-Kaduce, L. (1990). The process of criminalization: The case of computer crime laws. In D. H. Kelly (Ed.), *Criminal behavior* (pp. 29–43). New York: St. Martin's Press.

Holloway, M. (1999, January). Intelligence scores are rising, James R. Flynn discovered—but he remains very sure we're not getting any smarter," *Scientific American,* pp. 4–7.

Holmes, R. M. (1983). *The sex offender and the criminal justice system.* Springfield, IL: Charles C Thomas.

Holmes, R. M. (1989). *Profiling violent crimes: An investigative tool.* Newbury Park, CA: Sage Publications.

Holmes, R. M. (1991). *Sex crimes.* Newbury Park, CA: Sage Publications.

Holmes, R. M., & DeBurger, J. (1988). *Serial murder.* Beverly Hills, CA: Sage Publications.

Holmes, S. T., & Holmes, R. M. (2008). *Sex crimes.* Thousand Oaks, CA: Sage Publications.

Holt, J. (1994, October 19). Anti-social science? *New York Times,* p. A15.

Holzman, H. R., & Pines, S. (1979, November). *Buying sex: The phenomenology of being a "John."* Paper presented at the meeting of the American Society of Criminology, Philadelphia.

Home Office. (2002). Anticipating future trends in crime and disorder audits. *Crime reduction toolkits partnership working.* London: Author. Available at http://www.crimereduction.homeoffice.gov.uk/toolkits/p0317.htm.

Home repair scam draws jail term. (1986, May 1). *Erie Morning News,* p. 22A.

Hood, R., & Sparks, R. (1971). *Key issues in criminology.* New York: McGraw-Hill.

Hooton, E. (1939). *Crime and the man.* Westport, CT: Greenwood.

Hoover, E. (2002, December). Reading and rioting. *Chronicle of Higher Education, 13,* 40–41.

Hopper, C. B. (1991, November). *The changing role of women in outlaw motorcycle gangs: From partner to sexual property.* Paper presented at the meeting of the American Society of Criminology, San Francisco.

Hoppin, J. (2003, January 16). Discoveries future is electronic. *The Recorder.* Available at http://www.law.com/jsp/article.jsp?id=1039054556627.

Hosenball, M. (2007, October 15). Security whacking hackers. *Newsweek,* p. 10.

Hot spots for stolen cars. (2004, November 30). *CNN Money.* Retrieved April 2, 2007, from http://money.cnn.com/2004/11/16/pf/hot_spots_for_hot_cars.

Howard, P. (1990, January 6). The shadow grows. *Erie Morning News,* p. 2B.

Howe, K. (1997, May 3). Blue Shield pays fine for fraud. *San Francisco Chronicle,* p. B1.

Huang, C. (1999, November 24). Study: Shoplifters prefer small and expensive. APBnews.com.

Huang, C. (2000, January 7). Merrill Lynch probes $40 million theft. APBnews.com.

Hubbard, L. R. (1963). *Dianetics.* New York: Paperback Library.

Heussenstamm, F. K. (1971). Bumper stickers and the cops. *Trans-Action, 8,* 32–33.

Huff, C. R. (1990). Historical explanations of crime: From demons to politics. In D. H. Kelly (Ed.), *Criminal behavior: Texts and readings in criminology* (pp. 161–176). New York: St. Martin's Press.

Huff, D. (1966). *How to lie with statistics.* New York: Wiley.

Humphreys, L. (1970). *Tearoom trade: Impersonal sex in public places.* Chicago: Aldine.

Huizinga, D., Wylie Weiher, A., Espiritu, R., & Esbensen, F. (2003). Delinquency and crime: Some highlights from the Denver Youth Survey. In T. P. Thornberry & M. D. Krohn (Eds.), *Taking stock of delinquency: An overview of findings from contemporary longitudinal studies.* New York: Kluwer/Plenum.

Hunt, M. (1974). *Sexual behavior in the 1970s.* New York: Dell Books.

Hunter, E. (1951). *Brain-washing in red China.* New York: Vanguard Press.

Hutchings, B., & Mednick, Sarnoff A. (1977). Criminality in adoptees and their adoptive and biological parents: A pilot study. In S. A. Mednick & K. Christiansen (Eds.), *Biosocial bases in criminal behavior* (pp. 127–142). New York: Gardner Press.

Hyde, H. M. (1980). *The atom bomb spies.* New York: Ballantine.

Ianni, F. A. J. (1972). *A family business: Kinship and social control in organized crime.* New York: Russell Sage.

Ianni, F. A. J. (1973). *Ethnic succession in organized crime.* Washington, DC: Government Printing Office.

Ianni, F. A. J. (1974). *Black Mafia: Ethnic succession in organized crime.* New York: Simon & Schuster.

Icove, D., Seger, K., & VonStorch, W. (1995). *Computer crime: A crime fighter's handbook.* Sebestopol, CA: O'Reilly and Associates.

Ignatius, D. (2000, January 5). Buccaneers of the 21st century. *Washington Post,* p. A21.

Inciardi, J. A. (1970, August). The adult firesetter: A typology. *Criminology, 8,* 145–155.

Inciardi, J. A. (1975). *Careers in crime.* Chicago: Rand McNally.

Inciardi, J. A. (1977). In search of the class cannon: A field study of professional pickpockets. In R. S. Weppner (Ed.), *Street ethnography* (pp. 55–78). Beverly Hills, CA: Sage Publications.

Inciardi, J. A. (1979). Heroin use and street crime. *Crime and Delinquency, 25,* 335–346.

Inciardi, J. A. (1980). *Radical criminology: The coming crisis.* Beverly Hills, CA: Sage.

Inciardi, J. A. (Ed.). (1981). *The drugs–crime connection.* Beverly Hills, CA: Sage Publications.

Inciardi, J. A. (1983). On grift at the Superbowl: Professional pickpockets and the NFL. In G. Waldo (Ed.), *Career criminals* (pp. 31–41). Beverly Hills, CA: Sage Publications.

Inciardi, J. A. (1984). Professional theft. In R. F. Meier (Ed.), *Major forms of crime* (pp. 221–243). Beverly Hills, CA: Sage Publications.

Inciardi, J. A. (1990). *The drug legalization debate.* Newbury Park, CA: Sage Publications.

Inciardi, J. A. (1992). *The war on drugs II.* Mountain View, CA: Mayfield.

Inciardi, J. A., & McElrath, K. (Eds.). (2001). *The American drug scene* (2nd ed.). Los Angeles: Roxbury.

Ingersoll, B. (1991, September 13). FDA panel charges Pfizer unit sought to prolong sale of flawed heart valve. *Wall Street Journal,* p. A2.

Ingraham, B. L. (1979). *Political crime in Europe: A comparative study of France, Germany and England.* Berkeley: University of California Press.

Investor's guide to new SEC rules. (2003, January 24). *USA Today,* p. 3B.

Irvine, M. (2000, February 1). Governor orders execution moratorium. *Erie Morning News,* p. 5A.

Isaac, R. J., & Armat, V. C. (1990). *Madness in the streets: How psychiatry and the law abandoned the mentally ill.* New York: Free Press.

Iwata, E. (2007, March 9). Justice, SEC questioned on pace of probes. *USA Today,* p. 4B.

Jacks, I., & Cox, S. G. (Eds.). (1984). *Psychological approaches to crime and its correction: Theory, research, practice.* Chicago: Nelson-Hall.

Jackson, B. (1972). *In the life: Versions of the criminal experience.* New York: New American Library.

Jackson, J. (1994, January). Fraud masters: Professional credit card offenders and crime. *Criminal Justice Review, 19,* 34–58.

Jacobs, J. B. (2002). *Can gun control work?* New York: Oxford University Press.

Jacobs, J. B. (2006). *Mobsters, unions, and feds: The Mafia and the American labor movement.* New York: New York University Press.

Jacobs, J. B., Friel, C., & Radick, R. (1999). *Gotham unbound.* New York: New York University Press.

Jacobs, J. B., Panarella, C., & Worthington, J. (1994). *Busting the mob: United States v. Cosa Nostra.* New York: New York University Press.

Jacobs, J. R. (1987). Drinking and crime. *Crime File* (NCJ 100737). Washington, DC: U.S. Department of Justice, National Institute of Justice.

Jacobs, P. A., Brunton, M., Melville, M. M., Brittain, R. P., & McClemont, W. F. (1965). Aggressive behaviour, mental sub-normality, and the XYY male. *Nature, 208,* 1351–1352.

Jacoby, J. E. (Ed.). (2004). *Classics of criminology* (3rd ed.). Oak Park, IL: Moore Publishing.

Jacoby, J. E., Gramckow, H. P., & Rutledge, E. C. (1992, February). *Asset forfeiture programs.* Washington, DC: Jefferson Institute for Justice Studies.

Jacoby, T. (1988, October 24). A web of crime behind bars. *Newsweek,* pp. 76–81.

James, J. (1977). Prostitutes and prostitution. In E. Sagarin & F. Montanino (Eds.), *Deviants: Voluntary actors in a hostile world* (pp. 365–429). New York: General Learning.

James, J. (1978). The prostitute as victim. In R. Chapman & M. Gates (Eds.), *The victimization of women* (pp. 175–201). Beverly Hills, CA: Sage Publications.

Jamieson, K. M. (1995). *The organization of corporate crime: An inquiry into the dynamics of antitrust violation.* Newbury Park, CA: Sage Publications.

Janeway, E. (1981, November). Incest: A rational look at the oldest taboo. *Ms.,* pp. 61–64, 78, 81, 109.

Japan: Putting the Mafia to shame. (1977, October 17). *Time,* pp. 40, 46.

Japanese TV: Prime-time violence and mayhem. (1992, June 20). *Erie Times News,* p. 3A.

Jaroff, L. (1988, June 13). Fighting against flimflam. *Time,* 72.

Jeffrey, C. R. (1978, August). Criminology as an interdisciplinary behavioral science. *Criminology, 16,* 153–156.

Jellinek, E. M. (1960). *The disease concept of alcoholism.* New Brunswick, NJ: College and University Press.

Jenkins, P. (1982, March). *The long resistance: The enemies of positivism 1890–1945.* Paper presented at the Academy of Criminal Justice Sciences Meetings, Louisville, KY.

Jenkins, P. (1984). *Crime and justice: Issues and ideas.* Monterey, CA: Brooks/Cole.

Jenkins, P. (1988). Myth and murder: The serial killer panic of 1983–1985. *Criminal Justice Research Bulletin, 3,* 1–7.

Jenkins, P. (1992a, June). *African-American and serial homicide.* Paper presented at the meeting of the Northeastern Academy of Criminal Justice Sciences, Newport, RI.

Jenkins, P. (1992b). *Intimate enemies: Moral panics in contemporary Britain.* Hawthorne, NY: Aldine de Gruyter.

Jenkins, P., & Katkin, D. (1990, September). *Occult criminality: Myth and reality in contemporary moral panic.* Paper presented at the American Society of Criminology Meeting, Baltimore.

Jesilow, P. D., Pontell, H. N., & Geis, G. (1985). Medical criminals: Physicians and white-collar offenses. *Justice Quarterly, 2,* 151–165.

Jindal delivers GOP response to Obama's speech. (2009, February 24). *The Online NewsHour.* Retrieved November 18, 2009, from www.pbs.org/newshour/bb/politics/jan-june09/gopresponse_02-24.html

Johnson, B. D., Goldstein, P., Preble, D., Schmeidler, J., Lipton, D. S., Spunt, B., et al. (1983). *Economic behavior of street opiate users.* New York: Narcotic and Drug Research, Inc.

Johnson, D. (2000, May 16). Cheaters' final response: So what? *New York Times,* p. A6.

Johnson, E. A., & Monkkonen, E. H. (1996). *The civilization of crime: Violence in town and country since the Middle Ages.* Champaign: University of Illinois Press.

Johnson, E. H. (1990, March). *Yakuza (Criminal Gangs) in Japan: Characteristics and management in prison.* Paper presented at the meeting of the Academy of Criminal Justice Sciences, Denver, CO.

Johnson, H. (1991). Sleepwalking through history: America in the Reagan years. New York: W. W. Norton.

Johnson, K. (2005, August 26). 19 indicted as authorities target MS-13. *USA Today,* p. 5A.

Johnson, K. (2007, August 31). Cities study victims' criminal past. *USA Today,* p. 3A.

Johnson, M. P. (1995). Patriarchal terrorism and common couple violence: Two forms of violence against women. *Journal of Marriage and the Family, 57,* 283–294.

Johnson, R. E. (1979). *Juvenile delinquency and its origins.* Cambridge, MA: Cambridge University Press.

Johnson, R. E. (1986). Family structure and delinquency: General patterns and gender differences. *Criminology, 24,* 65–84.

Johnson, T. (1987, April 6). A little house of horrors: Murder in Philadelphia. *Newsweek,* p. 29.

Johnston, D. (2002). Departing chief says the IRS is losing its war on tax cheats. *New York Times,* November 5, pp. A1, C2.

Jones, D. (1997, April 3). 48% of workers admit to unethical or illegal acts. *USA Today,* pp. 1–4.

Judge agrees to TMI plea bargain. (1984, March 1). *Erie Morning News,* p. 3A.

Judge, A. V. (1930). *The Elizabethan underworld.* London: Routledge.

Kalven, H., Jr., & Zeisel, H. (1966). *The American jury.* Boston: Little, Brown.

Kane, E. (1989). *The S&L insurance mess: How did it happen?* Washington, DC: The Urban Institute.

Kandel, E., & Mednick, S. A. (1991, August). Perinatal complications predict violent offending. *Criminology, 29,* 519–529.

Kaplan, D. (Writer). (1988, November 20). Godfather of the ginza [Television series episode]. *60 Minutes.* New York: CBS Television.

Kaplan, D. E. (1991, Fall). Japanese mob. *Criminal Organizations, 6,* 1–3.

Kaplan, D. E., & Dubro, A. (1986). *Yakuza: The explosive account of Japan's criminal underworld.* Reading, MA: Addison-Wesley.

Kappeler, V. E., Sluder, R. D., & Alpert, G. P. (1994). *Forces of deviance: Understanding the darker side of policing.* Prospect Heights, IL: Waveland Press.

Karchmer, C. (1977, October 4). The underworld turns fire into profit. *Firehouse Magazine.* Read into the Congressional Record, U.S. Senate, p. S-16263.

Katz, J. (1988). *Seductions of crime: Moral and sensual attractions in doing evil.* New York: Basic Books.

Kauzlarich, D., & Kramer, R. C. (1998). *Crimes of the American nuclear state.* Boston: Northeastern University Press.

Keil, T. J., & Vito, G. F. (1989, August). Race, homicide severity, and application of the death penalty: A consideration of the Barnett scale. *Criminology, 27,* 511–531.

Kelling, G. A. (1988a, November). *Eliminating graffiti from New York subway trains.* Paper presented at the meeting of the American Society of Criminology, Chicago.

Kelling, G. A. (1988b, June). Police and communities: The quiet revolution. *Perspectives in Policing.* Washington, DC: U.S. Department of Justice, National Institute of Justice.

Kelly, K. (1999, December). A New York shell game: Cheating 101. *U.S. News and World Report,* p. 57.

Kelly, R. J. (1990). Succession by murder: Reflections on Paul Castellano's funeral and the rise of John Gotti. *Criminal Organizations, 5*(1), 16–18.

Kelly, R. J. (1992, February). Trapped in the folds of discourse: Theorizing about the underworld. *Journal of Contemporary Criminal Justice, 8,* 11–35.

Kelly, R. J., Schatzberg, R., & Ryan, P. J. (1995). Primitive capitalist accumulation: Russia as a racket. *Journal of Contemporary Criminal Justice, 11,* 257–275.

Kelman, H. C., & Hamilton, V. L. (1988). *Crimes of obedience: Toward a social psychology of authority and responsibility.* New Haven, CT: Yale University Press.

Kempe, R. S., & Kempe, C. H. (1978). *Child abuse.* Cambridge, MA: Harvard University Press.

Kennenburg, G. (2008, May 23). The not-so-untold story of the great comic book scare. *The Chronicle Review,* pp. 1–3.

Kenney, D. J., & Finckenauer, J. O. (1995). *Organized crime in America.* Belmont, CA: Wadsworth.

Kerner, O. (1968). *The Report of the National Advisory Commission on civil disorders: The riot commission report.* New York: Government Printing Office.

Kerr, P. (1987, August 11). Chasing the heroin from plush hotel to mean streets. *New York Times,* p. 1B.

Kerr, P. (1993, September 30). National Medical to pay $125 million in accord. *New York Times,* p. 1B.

Kessler, R. (1989). *Moscow station: How the KGB penetrated the American embassy.* New York: Scribner.

Kidder, R. L. (1983). *Connecting law and society.* Englewood Cliffs, NJ: Prentice Hall.

King, H., & Chambliss, W. J. (1984). *Harry King: A professional thief's journal.* New York: Wiley.

King, M. L., Jr. (1963). Letter from Birmingham Jail. In M. L. King, *Why We Can't Wait.* New York: Harper & Row.

Kinney, D. (1998, December 1). Fourteen students charged in riot no longer attend Penn State. *Erie Morning News,* p. 6C.

Kinney, J. A. (1990, September 10). Why did Paul die? *Newsweek,* p. 11.

Kinsey, A. (1948). *Sexual behavior in the human male.* Philadelphia: W. B. Saunders.

Kinsey, A. (1953). *Sexual behavior in the human female.* Philadelphia: W. B. Saunders.

Kinsey, R., Lea, J., & Young, J. (1986). *Losing the fight against crime.* London: Blackwell.

Kirkham, J. F. (1969). *Assassination and political violence.* Washington, DC: Government Printing Office.

Kirkpatrick, C., & Kanin, E. J. (1957, February). Male sex aggression on a university campus. *American Sociological Review, 22,* 52–58.

Kitsuse, J. L., & Dietrick, D. C. (1970). Delinquent boys: A critique. In H. L. Voss (Ed.), *Society, delinquency, and delinquent behavior* (pp. 238–245). Boston: Little, Brown.

Kittrie, N. N., & Wedlock, E. D., Jr. (Eds.). (1986). *The tree of liberty: A documentary history of rebellion and political crime in America.* Baltimore: Johns Hopkins University Press.

Klanwatch. (1985, February). Domestic terrorists: The KKK in the "Fifth Era." *Klanwatch Intelligence Report,* pp. 5–10.

Klaus, P. (2004). Carjacking, 1993–2002. *Research in Brief.* Washington, DC: Bureau of Justice Statistics (document 205123).

Kleiman, M. A. R., Barnett, A., Bouza, A. V., & Burke, K. M. (1988, August). *Street-level drug enforcement: Examining the issues.* Washington, DC: U.S. Department of Justice, National Institute of Justice.

Klein, J. (1996, April 29). The predator problem. *Newsweek,* p. 32.

Klein, J. F. (1974, April). Professional theft: The utility of a concept. *Canadian Journal of Criminology and Corrections, 16,* 133–143.

Klein, J. F., & Montague, A. F. (1977). *Check forgers.* Lexington, MA: Lexington Books.

Klein, M. W. (1990, November). *Having an investment in violence: Some thoughts about the American street gang.* Presentation at the meeting of the American Society of Criminology, Baltimore.

Klein, M. W., & Maxson, C. L. (1989). Street gang violence. In N. A. Weiner & M. W. Wolfgang (Eds.), *Violent crime, violent criminals* (pp. 198–234). Newbury Park, CA: Sage Publications.

Klein, M. W., Maxson, C. L., & Cunningham, L. C. (1991, November). "Crack" street gangs and violence. *Criminology, 29,* 623–650.

Klein, S. P., Turner, S., & Petersilia, J. (1988). *Does race make a difference in sentencing?* Santa Monica, CA: Rand Corporation.

Kleinknecht, W. (1996). *The new ethnic mobs.* New York: Free Press.

Klemke, L. W. (1992). *The sociology of shoplifting: Boosters and snitches today.* Westport, CT: Praeger.

Klockars, C. (1974). *The professional fence.* New York: Free Press.

Klockars, C. (1979, Fall). The contemporary crisis of Marxist criminology. *Criminology, 16,* 477–515.

Kneece, J. (1986). *Family treason: The Walker spy ring case.* New York: Stein and Day.

Koenig, D. J. (1991). Conventional crime. In R. Linden (Ed.), *Criminology: A Canadian perspective* (2nd ed.). Toronto, ON, Canada: Holt, Rinehart and Winston.

Kohut, J. (1997). Kidnap Corp. *Asia, Inc.* Available at http://www.asiainc.com.

Kolata, G. (1994, December 15). Bodies of patients newly dead used for practice by hospitals. *New York Times,* p. A14.

Kolbert, E. (1994, December 14). Television gets a closer look as a factor in real violence. *New York Times,* pp. A1, A13.

Konstantinova, N. (1997). South of Russia may become the centre of illegal import. *The Independent* (Moscow), *8,* p. 16.

Kornhauser, R. (1978). *Social sources of delinquency and its origins.* Chicago: University of Chicago Press.

Koski, P. R. (1988). Family violence and nonfamily deviance: Taking stock of the literature. In F. E. Hagan & M. B. Sussman (Eds.), *Deviance and the family* (pp. 23–46). New York: Haworth Press.

Kozakiewicz, M. (2007, October 11). *The Surviving Parent's Coalition.* Presentation at the Cyber Crimes in Cyber Times Conference, Mercyhurst College, Erie, PA.

Krahn, H., Hartnagel, T., & Gartrell, J. (1986). Income inequality, and homicide rates: Cross-national data and criminological theories. *Criminology, 24,* 269–295.

Krane, J. (1999, March 7). Russian mobsters kick down world doors. APBnews.com.

Krane, J. (2000, February 7). Databases revolutionize pawnshop policing. APBnews.com.

Kratcoski, P. C. (1988). Families who kill. In F. E. Hagan & M. B. Sussman (Eds.), *Deviance and the family* (pp. 47–70). New York: Haworth.

Kretschmer, E. (1926). *Physique and character* (W. J. H. Sprott, Trans.). New York: Harcourt, Brace.

Krisberg, B. (1975). *Crime and privilege: Toward a new criminology.* Englewood Cliffs, NJ: Prentice-Hall.

Krzycki, L. (1994, September/October). It's not that simple. *ACJS Today, 1*(3), 28.

Kuhl, A. F. (1985). Battered women who murder: Victims or offenders? In I. L. Moyer (Ed.), *The changing roles of women in the criminal justice system* (pp. 197–216). Prospect Heights, IL: Waveland Press.

Kuhn, T. S. (1962). *The structures of scientific revolutions.* Chicago: University of Chicago Press.

Kuper, L. (1981). *Genocide: Its political use in the twentieth century.* New Haven, CT: Yale University Press.

Kurtz, H. (1989, April 4). Across the nation, rising outrage. *Washington Post,* pp. 1A, 14A, 16A.

Kushner, H. W. (Ed.). (1998). *The future of terrorism.* Thousand Oaks, CA: Sage Publications.

Kutchins, H. (1988, November). *Making criminals crazy: The impact of new psychiatric diagnosis on the criminal justice system.* Paper presented at the American Society of Criminology, Chicago.

Kwitny, J. (1979). *Vicious circles: The Mafia in the marketplace.* New York: Norton.

Lab, S. P., & Hirschel, J. D. (1988a). Climatological conditions and crime: The Forecast is . . . ? *Justice Quarterly, 5,* 281–299.

Lab, S. P., & Hirschel, J. D. (1998b). Clouding the issues: The failure to recognize methodological problems. *Justice Quarterly, 5,* 281–299.

Labaton, S. (2007, April, 25). OSHA leaves worker safety largely in the hands of the industry. *New York Times,* pp. A1, A20.

Lacey, M. (2009, March 30). In drug war, Mexico fights the cartels and itself. *New York Times,* pp. A1, A12.

LaFraniere, S. (1999, September 13). A money trail that leads to Yeltsin. *Washington Post,* p. 16.

Landers, A. (1991, October 6). Bare facts shocking to mother of son bound for Princeton. *Erie Times News,* p. C8.

Laner, M. R. (1974). Prostitution as an illegal vocation: A sociological overview. In C. Bryant (Ed.), *Deviant behavior: Occupational and organizational bases* (pp. 406–418). Chicago: Rand McNally.

Langan, P. A., & Farrington, D. P. (1998, October). *Crime and justice in the United States and in England and Wales, 1981–96* (Executive Summary, NCJ 173402). Washington, DC: U.S. Department of Justice, Bureau of Justice Statistics.

Lange, J. (1931). *Crime as destiny: A study of criminal twins* (C. Haldane, Trans.). London: Allen & Unwin.

Langway, L., & Smith, S. (1975, October 6), Warning! Someone may try to steal your money. *Newsweek,* p. 67.

Lanning, K. V. (2001, September). *Child molesters: A behavioral analysis* (4th ed.). Washington, DC: National Center for Missing and Exploited Children.

Laqueur, W. (1977). *Terrorism.* Boston: Little, Brown.

Laqueur, W. (1987). *The age of terrorism.* Boston: Little, Brown.

Larimer, T. (1996, March 8). A freer Vietnam brings rise in violent crime. *New York Times,* p. A7.

Larzelere, R. E., & Patterson, G. R. (1990, May). Parental management: Mediator of the effect of socioeconomic status on early delinquency. *Criminology, 28,* 301–323.

Laub, J. H., & Sampson, R. J. (1988). Unraveling families and delinquency: A reanalysis of the Gluecks' data. *Criminology, 26,* 355–380.

Laub, J. H., & Sampson, R. J. (2003). *Shared beginnings, divergent lives: Delinquent boys to age 70.* Boston: Harvard University Press.

Launer, H. M., & Palenski, J. E. (Eds.). (1988). *Crime and the new immigrants.* Springfield, IL: Charles C Thomas.

Lauter, D. (1988, June 30). Children must testify face-to-face in abuse cases. *Erie Morning News,* p. 1.

Lawler, P. F. (1991, August 29). An issue this paper can't sidestep. *Wall Street Journal,* p. A11.

Lawrence, R. (2006). *School crime and justice* (2nd ed.). New York: Oxford University Press.

Lea, J., & Young, J. (1984). *What is to be done about law and order.* Harmondsworth, UK: Penguin.

LeBeau, J. L. (1988). Comment—weather and crime: Trying to make social sense of a physical process. *Justice Quarterly, 5,* 301–309.

LeBeau, J. L., & Langworthy, R. H. (1986). The linkages between routine activities, weather, and calls for police service. *Journal of Police Science and Administration, 14,* 137–145.

LeBlanc, M. (1996). Changing patterns in the perpetration of offenses over time. *Studies on Crime and Crime Prevention, 5,* 151–165.

Lefer, D. (1999, March 14). Pyramids of the greedy: Worldwide bank note scam bilks investors of millions. *New York Daily News.*

Lemert, E. M. (1951). *Social pathology.* New York: McGraw-Hill.

Lemert, E. M. (1953, September). An isolation and closure theory of naïve check forgery. *Journal of Criminal Law, Criminology, and Police Science, 44,* 296–307.

Lemert, E. M. (1958, Fall). The behavior of the systematic check forger. *Social Problems, 6,* 141–149.

Lemert, E. M. (1967). *Human deviance, social problems, and social control.* New York: Prentice Hall.

Lemert, E. M. (1968). Prostitution. In E. Sagarin & D. E. J. MacNamara (Eds.), *Problems in sex behavior.* New York: Crowell.

Lemkin, R. (1944). *Axis rule in occupied Europe.* Washington, DC: Carnegie Endowment for International Peace.

Leonard, W. N., & Weber, M. G. (1970, February). Automakers and dealers: A study of criminogenic market forces. *Law and Society Review, 4,* 407–424.

Lessons turn orphans into outcasts. (2001, May 20). *Erie Times News,* 1A.

Letkemann, P. (1973). *Crime as work.* Englewood Cliffs, NJ: Prentice Hall.

Leuw, E., & Marshall, I. H. (1994). *Between prohibition and legalization: The Dutch experiment in drug policy.* Amsterdam: Kugler.

Levathes, L. E. (1985, August). The land where the Murray flows. *National Geographic,* pp. 252–278.

Levy, C. J. (1995, September 12). A bulletin board is virtual but hacker arrests are real. *New York Times,* pp. A1, A14.

Levy, S. (2006, July 3-July 10). All predators, all the time? Maybe not. *Newsweek,* p. 20.

Levy, S., & Stone, B. (2005, July 4). Grand theft identity, *Newsweek,* pp. 38–47.

Lewin, T. (1995, December 7). Parents poll finds child abuse to be more common. *New York Times,* p. A17.

Leyton, E. (1986). *Compulsive killers: The story of modern multiple murders.* New York: Washington News Book.

Liazos, A. (1972, Summer). The poverty of the sociology of deviance: Nuts, sluts and "preverts." *Social Problems, 20,* 103–120. Reprinted in *Readings in social problems, 79/80* (pp. 22–31). Guilford, CT: Dushkin.

Lieber, A. L., & Sherin, C. R. (1972, July). Homicides and the lunar influence in human emotional disturbance. *American Journal of Psychiatry, 129,* 69–74.

Liebert, R. M., & Baron, R. A. (1972). Some immediate effects of televised violence on children's behavior. *Developmental Psychology, 6,* 469–475.

Liebow, E. (1967). *Tally's corner: A study of Negro streetcorner men.* Boston: Little, Brown.

Light, I., & Bonacich, E. (1988). *Immigrant entrepreneurs: Koreans in Los Angeles, 1965–1982.* Berkeley: University of California Press.

Lilly, J. R., Cullen, F. T., & Ball, R. A. (2007). *Criminological theory: Context and consequences* (4th ed.). Thousand Oaks, CA: Sage Publications.

Lindberg, K., et al. (1997, November). The changing faces of organized crime. *Crime and Justice International, 13,* 5–6, 20–33.

Lindesmith, A. R. (1965). *The addict and the law.* Bloomington: Indiana University Press.

Lindesmith, A. R., & Levin, Y. (1937, March). The Lombrosian myth in criminology. *American Journal of Sociology, 42,* 653–671.

Lindquist, J. H. (1988). *Misdemeanor crime: Trivial criminal pursuit.* Newbury Park, CA: Sage Publications.

Lindsey, R. (1979). *The falcon and the snowman.* New York: Simon & Schuster.

Lindsey, R. (1983). *The flight of the falcon.* New York: Simon & Schuster.

Lippman, M. (1987). Iran: A question of justice? *C. J. International, 3,* 5–6.

Lippman, T. W. (1991, December 9–15). At Hanford's nuclear graveyard, a nightmare of endless potential. *Washington Post National Weekly Edition,* p. 33.

Lockheed ordered to pay $45 million. (1990, November 16). *Erie Morning News,* p. 12A.

Loeber, R., & Stouthamer-Loeber, M. (1986). Models and metaanalysis of the relationship between family variables and juvenile conduct problems and delinquency. In N. Morris & M. Tonry (Eds.), *Crime and justice: An annual review of research* (Vol. 7, pp. 29–149). Chicago: University of Chicago Press.

Loeber, R., Wung, P., Keenan, K., Giroux, B., Stouthamer-Loeber, M., Van Kammen, B. & Maughan, B. (1993). Developmental pathways in disruptive behavior. *Development and Psychopathology, 4,* 12–48.

Lohr, S. (1992, August 2). Trial of a bank scandal leads on and on. *New York Times,* pp. B1, B3.

Lombardo, R. M. (1990). Civil forfeiture: A powerful tool against commercial gambling. *Criminal Organizations, 5,* 3–5.

Lombroso, C. (1911). Introduction. In G. Lombroso-Ferrero (Ed.), *Criminal man according to the classification of Cesare Lombroso.* New York: Putnam.

Lombroso-Ferrero, G. (Ed.). (1972). *Criminal man according to the classification of Cesare Lombroso* [reissue of 1911 work]. Montclair, NJ: Patterson Smith.

Londer, R. (1987, August 3). Can bad air make bad things happen? *Parade,* 6–7.

Longmire, D. R. (1988, November). *Crimes of power and opulence: Criminological researcher's experience with taboos.* Paper presented at the meeting of the American Society of Criminology, Chicago.

Longtime spam king charged with fraud. (2007, June 1). *Washington Post,* p. 1.

Lorenz, K. (1966). *On aggression.* New York: Harcourt, Brace, Jovanovich.

Lowenthal, M. (1950). *The Federal Bureau of Investigation.* New York: William Sloane Associates.

Lowman, J., Jackson, M. A., Palys, T. S., & Gavigan, S. (Eds.). (1986). *Regulating sex: An anthology of commentaries on the Badgley and Fraser reports.* Burnaby, BC, Canada: Simon Fraser University.

Luckenbill, D. F. (1991). Criminal homicide as a situated transaction. *Social Problems, 25,* 176–186.

Luckenbill, D. F., & Doyle, D. P. (1989, August). Structural position and violence: Developing a cultural explanation. *Criminology, 27,* 419–436.

Lupsha, P. A. (1982, November). *Networks vs. networking: An analysis of organized criminal groups.* Paper presented at the meeting of the American Society of Criminology, Toronto, ON, Canada.

Lynam, D. R., Moffitt, T. E., & Stouthamer-Loeber, M. (1993). Explaining the relationship between I.Q. and delinquency. *Journal of Abnormal Psychology, 102,* 187–196.

Lyman, S. M. (1974). *Chinese Americans.* New York: Random House.

Lynch, G. W., & Blotner, R. (1991, April 15). Failed Zurich test a fatal blow to case for decriminalization. *Law Enforcement News,* p. 8.

Lyons, R. (1980, June 23). Reports on U.S. oil companies. *Erie Morning News,* p. 4.

Maas, P. (1968). *The Valachi papers.* New York: Bantam.

Maas, P. (1975). *King of the gypsies.* New York: Bantam.

Maas, P. (1986). *Manhunt: The incredible pursuit of a CIA agent turned terrorist.* New York: Random House.

MacDonald, A. (1980). *The Turner diaries.* Arlington VA: National Vanguard Books.

MacDonald, J. M. (1971). *Rapists and their victims.* Springfield, IL: Charles C Thomas.

Mack, J. A. (1972, January). The able criminal. *British Journal of Criminology, 12,* 44–54.

MacLean, D. (1974). *Pictorial history of the Mafia.* New York: Galahad Books.

Maclean, F. (1978). *Take nine spies.* New York: Atheneum.

MacNamara, D., & Sagarin, E. (1977). *Sex crime and the law.* New York: Free Press.

Mankoff, M. (1980). A tower of Babel: Marxist criminologists and their critics. In J. A. Inciardi (Ed.), *Radical criminology: The coming crisis* (pp. 139–148). Beverly Hills, CA: Sage.

Mann, C. M. (1989, Summer). Random thoughts on the ongoing Wilbanks-Mann discourse. *The Critical Criminologist, 1,* 3–4.

Mann, C. M. (1993). *Unequal justice: A creation of color.* Bloomington: Indiana University Press.

Mann, C. R. (1984). *Female crime and delinquency.* University: University of Alabama Press.

Mannheim, H. (1965). *Comparative criminology.* Boston: Houghton Mifflin.

Manning, P. K. (1975, January). Deviance and dogma. *British Journal of Sociology, 15,* 1–20.

Mara Salvatrucha: MS-13. (2005). Retrieved April 2, 2007, from http://www.knowgangs.com/gang_resources/profiles/ms13/index.php.

Maranto, G. (1985, March). Coke: The random killer. *Discover.*

Marbin, C. (1989, April 17). IRS "giant killers" fighting drug war. *Palm Beach Post,* pp. 1A, 6A.

Marchetti, V., & Marks, J. D. (1974). *The CIA and the cult of intelligence.* New York: Dell.

Marcus, A. D. (1990, November 26). Thievery by lawyers is on the increase, with duped clients losing bigger sums. *Wall Street Journal,* p. B1.

Marcus, R. (1991, December 16–22). When does the government make criminals out of honest citizens? *Washington Post National Weekly Edition,* p. 34.

Mark, V., & Ervin, F. (1970). *Violence and the brain.* New York: Harper and Row.

Markhoff, J. (1988, September 4). "Virus" in military computers disrupts systems nationwide. *New York Times,* p. 1.

Markon, J. (2007, October 11). Rights groups sue over crackdown on illegal immigration. *Washington Post,* B6.

Marks, J. (1979). The search for the Manchurian candidate: The CIA and mind control. New York: Times Books.

Marsh, F. H., & Katz, J. (Eds.). (1985). Biology, crime and ethics: A study of biological explanations for criminal behavior. Cincinnati, OH: Anderson.

Marshall, I. H., Anjewierden, O., & Van Atteveld, H. (1990, June). Toward an "Americanization" of Dutch drug policy? *Justice Quarterly, 7,* 393–420.

Marshall, J. (1992, May 21–31). Targeting the drugs, wounding the cities. *Washington Post National Weekly Edition,* p. 23.

Martin, D. W., & Walcott, J. (1988). Best laid plans: The inside story of America's war against terrorism. New York: Harper & Row.

Martin, G. (2006). *Understanding terrorism: Challenges, perspectives and issues* (2nd ed.). Thousand Oaks, CA: Sage Publications.

Martin, J. M., Haran, J. F., & Romano, A. T. (1988, April). *Espionage: A challenge to criminology and criminal justice.* Paper presented at the meeting of the Academy of Criminal Justice Sciences, San Francisco.

Martin, R., Mutchnick, R. J., & Austin, W. T. (1990). *Criminological thought: Pioneers past and present.* New York: Macmillan.

Martinson attacks his own earlier work. (1978, December). *Criminal Justice Newsletter, 9,* 4.

Martinson, R. (1974, Spring). What works? Questions and answers about prison reform. *The Public Interest, 35,* 22–54.

Martinson, R. (1979). New findings, new views: A note of caution regarding sentencing reform. *Hofstra Law Review, 7,* 242–258.

Marquart, J. W. (1986, March). Doing research in prison: The strengths and weaknesses of full participation as a guard. *Justice Quarterly, 3,* 15–32.

Marx, G. (1990). *Undercover: Police surveillance in America.* Berkeley: University of California Press.

Marx, K. (1848). The Communist manifesto. In *Karl Marx and Frederick Engels: Selected works* (Vol. 1, pp. 108–137). Moscow: Progress Publishers.

Marx, K. (1967). *Das Kapital* (3 volumes). New York: International Publishers. (Original work published 1868)

Maser, W. (1979). *Nuremberg: A nation on trial* (R. Barry, Trans.). New York: Scribner.

Masland, T. (1992, May 4). Slavery. *Newsweek,* 30–39.

Matsueda, R. L. (1988). The current state of differential association theory. *Crime and Delinquency, 34,* 277–306.

Matza, D. (1964). *Delinquency and drift.* New York: Wiley.

Maurer, D. W. (1940). *The big con.* Indianapolis, IN: Bobbs-Merrill.

Maurer, D. W. (1964). *Whiz mob.* New Haven, CT: College and University Press.

Mayer, M. (1990). *The greatest-ever bank robbery: The collapse of the savings and loan industry.* New York: Scribner.

Mayhew, H. (1862). *London's underworld.* London: Spring Books.

McCaghy, C. (1976a). Child molesters: A study of their careers as deviants. In M. B. Clinard & R. Quinney (Eds.), *Criminal behavior systems* (pp. 75–88). New York: Holt, Rinehart and Winston.

McCaghy, C. (1976b). *Deviant behavior.* New York: Macmillan.

McCaghy, C. (1980). *Crime in American society.* New York: Macmillan.

McCaghy, C., Giordano, P., & Henson, T. K. (1977). Auto theft. *Criminology, 15*(3), 367–381.

McCaghy, C., & Hou, C. (1988, November). *Cultural factors and career contingencies of prostitution: The case of Taiwan.* Paper presented at the meeting of the American Society of Criminology, Chicago.

McCall, N. (1994). *Makes me wanna holler: A young black man in America.* New York: Random House.

McCandless, B. R., Persons, W. S., & Roberts, A. (1972, April). Perceived opportunity, delinquency, race and body build among delinquent youth. *Journal of Consulting and Clinical Psychology, 38,* 281–287.

McCarthy, B. (1995, November). Not just "for the thrill of it": An instrumentalist elaboration of Katz's explanation of sneaky thrill property crimes. *Criminology, 33,* 519–538.

McCarthy, E. D., Langner, T. S., Gersten, J. C., Eisenberg, J. G., & Orzeck, L. (1975). The effects of television on children and adolescents: Violence and behavior disorders. *Journal of Communications, 25,* 71–85.

McCarthy, S. (1996, July/August). Fleeing mutilation, fighting for asylum. *Ms.*

McCord, W., & McCord, J. (1958). The effects of parental role model on criminality. *Journal of Social Issues, 14,* 66–75.

McCoy, A. W. (1972). *The politics of heroin in Southeast Asia.* New York: Harper & Row.

McDermott, M. J. (1979). *Rape victimization in 26 American cities.* Washington, DC: Government Printing Office.

McFadden, R. D. (1994, September 9). The sting, FBI gets hot cars, great deals and 30 suspects. *New York Times,* p. A13.

McGrory, M. (1988, March 27). Owning up to atomic injustice. *Washington Post,* p. 1.

McGrory, M. (1999, March 22). An overdue apology: President Clinton has finally expressed the nation's regret for its role in Guatemala. *Washington Post National Weekly Edition,* p. 23.

McIntosh, M. (1975). *The organization of crime.* London: Macmillan.

McKillop, P. (1989, February 6). The last godfather? Going after Gotti. *Newsweek,* p. 25.

McKinley, J. Jr., (2009, April 15), U.S. stymied as guns flow to Mexican cartels. *New York Times,* pp. A1, A14.

McKinney, J. C. (1966). *Constructive typology and social theory.* New York: Appleton-Century-Crofts.

McMullan, J. L. (1984). *The Canting crew: London's criminal underworld, 1550–1700.* New Brunswick, NJ: Rutgers University Press.

McNamara, R. P. (Ed.). (1995). *Sex, scams and street life: The sociology of New York City's Times Square.* Westport, CT: Praeger.

Mead, G. H. (1934). *Mind, self, and society.* Chicago: University of Chicago Press.

Meatpackers hit with record OSHA fine. (1988, October 29). *Erie Morning News,* p. 2A.

Mednick, S., & Volavka, J. (1980). Biology and crime. In N. Morris & M. Tonry (Eds.), *Crime and justice: An annual review of research* (Vol. 1, pp. 85–159). Chicago: University of Chicago Press.

Meller, P. (2002, October 31). Europe fines Nintendo $147 million for price fixing. *New York Times,* pp. W1, W7.

Mellgren, D. (1994, October 19). Norwegian girl's slaying spurs TV violence debate. *Erie Morning News,* p. 1A.

Melloan, G. (1991, April 29). It's hard to make money laundries come clean. *Wall Street Journal,* p. A11.

Mellow, J. (1996, March/April). Measuring race: Historical and contemporary issues. *The Criminologist,* pp. 6–7.

Menard, S. (1987). Short-term trends in crime and delinquency: A comparison of the UCR, NCS and self-report data. *Justice Quarterly, 4,* 455–474.

Mendelson, B. (1963, June). The origin and doctrine of victimology. *Excerpta Criminologica, 3,* 239–244.

Mendelson, M. A. (1975). *Tender loving greed.* New York: Vintage.

Merry, R. W. (1975, November 1). The law is on trial. *The National Observer,* pp. 1–3.

Merton, R. K. (1938, October). Social structure and anomie. *American Sociological Review, 3,* 672–682.

Merton, R. K. (1957). *Social theory and social structure* (Rev. ed.). New York: Free Press. (Original work published 1949)

Merton, R. K. (1961). Social problems and sociological theory. In R. K. Merton & R. Nisbet (Eds.), *Contemporary social problems.* New York: Harcourt, Brace and World.

Merton, R. K. (1968). *Social theory and social structure* (Rev. ed.). New York: Free Press.

Messerschmidt, J. W. (1986). *Capitalism, patriarchy, and crime: Toward a socialist feminist criminology.* Totowa, NJ: Rowman & Littlefield.

Messerschmidt, J. W. (1993). *Masculinities and crime.* Lanham, MD: Rowman & Littlefield.

Messerschmidt, J. W. (1997). *Crime as structured action.* Thousand Oaks, CA: Sage.

Messick, H. (1979). *Of grass and snow: The secret criminal elite.* Englewood Cliffs, NJ: Prentice Hall.

Messner, S. F. (1980, January). Income inequality and murder rates: Some cross-national findings. *Comparative Social Research, 3,* 185–198.

Messner, S. F., Krohn, M. D., & Liska, A. E. (Eds.). (1989). *Theoretical integration in the study of deviance and crime: Problems and prospects.* Albany: State University of New York Press.

Messner, S. F., & Rosenfeld, R. (1994). *Crime and the American Dream.* Belmont, CA: Wadsworth.

Messner, S. F., & Tardiff, K. (1985). The social ecology of urban homicide: An application of the "routine activities" approach. *Criminology, 23,* 241–267.

Miami police scandal called worst in U.S. since Prohibition. (1987, November 8). *Erie Times News,* p. 1C.

Michalowski, R. J. (1993). (De)construction, postmodernism, and social problems: Facts, fiction and fantasies at the "end of history." In J. A. Holstein & G. Miller (Eds.), *Reconsidering social constructionism: Debates in social problems theory.* New York: Aldine De Gruyter.

Mihalic, S., Fagan, A., Irwin, K., Ballard, D., & Elliott, D. (2004, July). Blueprints for violence prevention report. *Office of Juvenile Justice and Delinquency Prevention.* NCJ 204274.

Milgram, S. (1974). *Obedience to authority: An experimental view.* New York: Harper.

Milbank, D., & Allen, M. (1991, July 12). Alcoa pleads guilty in toxic waste case. *Wall Street Journal,* p. B2.

Miller, E. (1986). *Street woman.* Philadelphia: Temple University Press.

Miller, J. (1989, November). *The relationship between child abuse and the commission of violent crime by adolescents and adults: An overview.* Paper presented at the meeting of the American Society of Criminology, Reno, NV.

Miller, J. M., & Tewksbury, R. (2000). *Extreme methods: Innovative approaches to social science research.* Needham: Allyn & Bacon.

Miller, S. L. (Ed.). (1998). *Crime control and women.* Thousand Oaks, CA: Sage.

Miller, W. (1958, May). Lower class culture as a generating milieu of gang delinquency. *Journal of Social Issues, 14,* 5–19.

Mills, C. W. (1952). A diagnosis of moral uneasiness. In I. L. Horowitz (Ed.), *Power, politics and people* (pp. 330–339). New York: Ballantine.

Mills, J. (1986). The underground empire: Where crime and governments embrace. New York: Doubleday.

Milner, C., & Milner, R. (1972). *Black players: The secret world of black pimps.* Boston: Little, Brown.

Milanovic, D. (1992). *Postmodern law and disorder: Psychoanalytic semiotics, chaos, and juridic exegeses.* Liverpool: Deborah Charles.

Milovanovic, D. (1996). Postmodern criminology. *Justice Quarterly, 13,* 567–610.

Minor, W. W. (1981). Techniques of neutralization: A reconceptualization an empirical examination. *Journal of Research in Crime and Delinquency, 18,* 295–318.

Minor, W. W. (1984). Neutralization as a hardening process. *Social Forces, 62,* 995–1019.

Mitchell, G. (1981, May). The trouble with RICO. *Police Magazine,* pp. 39–44.

Mitford, J. (1963). *The American way of death.* New York: Paperback Library.

Mock, L. F., & Rosenbaum, D. (1988, May). *A study of trade secret theft in high-technology industries.* Washington, DC: National Institute of Justice.

Moffit, T. E. (1999). Adolescence limited and life-course persistent antisocial behavior: A developmental theory. In F. R. Scarpitti & A. L. Nielsen (Eds.), *Crime and criminals* (pp. 206–231). Los Angeles: Roxbury.

Moffit, T. E., & Walsh, A. (2003). The adolescence-limited/life-course persistent theory and antisocial behavior: What have we learned? In A. Walsh & L. Ellis (Eds.), *Biosocial criminology: Challenging environmentalis's supremacy* (pp. 123–144). Hauppage, NY: Nova Science.

Mokhiber, R. (1988). *Corporate crime and violence.* San Francisco: Sierra Club Books.

Mokhiber, R. (1999). *Top 100 corporate criminals of the 1990s.* Retrieved April 4, 2007, from www.corporatepredators.org/top100.html

Monahan, J., & Splane, S. (1980). Psychological approaches to criminal behavior. In E. Bittner & S. L. Messenger (Eds.), *Criminology review yearbook, vol. 2* (pp. 17–47). Beverly Hills, CA: Sage.

Monmaney, T., & Robins, K. (1988, May 23). The insanity of steroid abuse. *Newsweek,* 75.

Monroe, R. R. (1978). *Brain dysfunction in aggressive criminals.* Lexington, MA: D.C. Heath.

Moore, M. (1988). Drug trafficking. *Crime File* (NCJ 097224). Washington, DC: U.S. Department of Justice, National Institute of Justice.

Moore, T., et al. (1988, April 10). Dead zones. *U.S. News and World Report,* pp. 20–33.

Morash, M., & Chesney-Lind, M. (1991, September). A reformulation and partial test of the power control theory of delinquency. *Justice Quarterly, 8,* 347–377.

Morgan, W. P. (1960). *Triad societies in Hong Kong.* Hong Kong: Government Press.

Morgantheau, T., et al. (1988, October 31). The drug gangs. *Newsweek,* pp. 26–29.

Morgenson, G. (2002, December 20). Accord highlights Wall Street failures. *New York Times,* pp. C1, C5.

Morris, N. (1987). Insanity defense. *Crime File* (NCJ 097226). Washington, DC: U.S. Department of Justice, National Institute of Justice.

Morris, N., & Hawkins, G. (1970). *The honest politician's guide to crime control.* Chicago: University of Chicago Press.

Mossberg, W. S. (1990, October 11). Pan Am bombing probe takes new turn as device points to Libyan-based agents. *Wall Street Journal,* p. A11.

Most stolen cars: Is yours on the list? (2004, October 19). *CNN Money.* Retrieved April 2, 2007, from http://money.cnn.com/2004/10/19/pf/autos/stolen_vehicles.

Mounties, U.S. agents nab Canada counterfeiters. (1998, December 11). Reuters.

Moushey, B. (1998, November 22). Win at all costs [10-part series]. *Pittsburgh Post Gazette.*

Moyer, I. L. (Ed.). (1990). *The changing roles of women in the criminal justice system* (2nd ed.). Prospect Heights, IL: Waveland Press.

Moyer, K. E. (1976). *The psychology of aggression.* New York: Harper & Row.

Moynihan, D. P. (1986). *Family and nation.* New York: Harcourt, Brace, Jovanovich.

Mui, Y. (2007, July 12). Updates to retail theft laws urged. *Washington Post,* p. 4A.

Muraskin, R. (2000). *It's a crime: Women and justice.* Upper Saddle River, NJ: Prentice Hall.

Murr, A., & Rogers, A. (1995, December 11). Violence, reel to real. *Newsweek,* pp. 45–47.

Murray, C. A. (1976). *The link between learning disability and juve-nile delinquency.* Washington, DC: National Institute of Juvenile Justice and Delinquency Prevention.

Museum jewel robbery. (1964, November 6). *Time,* p. 23.

Mustain, G., & Capeci, J. (1988). *Mob star: The story of John Gotti.* New York: Dell.

Myers, G. (1936). *The history of great American fortunes.* New York: Modern American Library.

Myers, M., & Talarico, S. (1987). *The social contexts of criminal sentencing.* New York: Springer-Verlag.

Nader, R. (1965). *Unsafe at any speed.* New York: Grossman.

Nader, R. (1970). Foreword to *The vanishing air,* by J. C. Esposito. New York: Grossman.

Nader, R. (1973). *The consumer and corporate accountability.* New York: Harcourt, Brace, Jovanovich.

Nader, R., & Green, M. J. (Eds.). (1973). *Corporate power in America.* New York: Grossman.

Nader, R., Green, M. J., & Seligman, J. (1976). *Taming the giant corporation.* New York: Norton.

Nader, R., Petkas, P. J., & Blackwell, K. (Eds.). (1972). *Whistle blowing: The report of the conference on professional responsibility.* New York: Viking Penguin.

Nagel, I. H., & Hagan, J. (1983). Gender and crime: Offense patterns and criminal court sanctions. In M. Tonry & N. Morris (Eds.), *Crime and justice: An annual review of research* (Vol. 4, pp. 91–144). Chicago: University of Chicago Press.

Nakashima, E. (2007, December 23). FBI prepares vast data base of biometrics. *Washington Post,* p. B1.

Naked vandalism replaced nude run. (1994, May 5). *Erie Morning News,* p. 3A.

Nash, J. R. (1975). *Bloodletters and badmen* (3 vols.). New York: Warner.

Nash, J. R. (1981). *Almanac of world crime.* Garden City, NY: Doubleday.

Nasheri, H., & O'Hearn, T. J. (1998). Crime and technology: New rules in a new world. *Information and Communications Technology Law, 7*(2), 145–157.

National Advisory Committee on Criminal Justice Standards and Goals. (1976a). *Organized crime: Report of the Task Force on Organized Crime.* Washington, DC: Law Enforcement Assistance Administration.

National Advisory Committee on Criminal Justice Standards and Goals. (1976b). *Report of the task force on disorders and terrorism.* Washington, DC: Government Printing Office.

National Association of Elementary School Principals. (1980). *The most significant minority: One-parent children in the schools.* New York: Charles F. Kettering Foundation.

National Broadcasting Company. (1983, April 26). Crime and insanity [Television series episode]. *NBC white-paper.* New York: Author.

National Central Police. (2005). Organized crime gangs in Taiwan. *Trends in Organized Crime, 8*(3), 13–23.

National Commission on the Causes and Prevention of Violence. (1969). *To establish justice, to insure domestic tranquility.* New York: Award Books.

National Council on Crime and Delinquency. (2000). *And justice for some.* San Francisco: National Council on Crime and Delinquency.

National Criminal Justice Reference Service. (1979). *We are all victims of arson.* Washington, DC: Government Printing Office.

National Criminal Justice Reference Service. (2003, May/June). Females in the criminal justice system. *NCJRS Catalog, 70,* 14.

National Institute of Justice. (2003). *ADAM: 2002 annual report on adult and juvenile arrestees.* Washington, DC: Author.

National Institute on Law Enforcement and Criminal Justice. (1977). *The development of the law and gambling, 1776–1976.* Washington, DC: Government Printing Office.

Neergaard, L. (1994, May 14). Consumer group claims U.S. doctors performing unnecessary C-sections. *Erie Morning News,* p. 8C.

Nelan, B. W. (1997, March 17). The Ponzi revolution. *Time,* p. 32.

Nelkin, D. (1995, September 28). Biology is not destiny. *New York Times,* p. A1.

Netanyahu, B. (Ed.). (1986). *Terrorism: How the West can win.* New York: Farrar, Straus, and Giroux.

Nettler, G. (1974, January). Embezzlement without problems. *British Journal of Criminology, 14,* 70–77.

Nettler, G. (1978). *Explaining crime* (2nd ed.). New York: McGraw-Hill.

Nettler, G. (1982). *Criminal careers* (4 vols.). Cincinnati, OH: Anderson.

Nettler, G. (1989). *Criminology careers* (4 vols.). Cincinnati, OH: Anderson.

Newman, A. (1990, November 12). Sensormate Electronics' investors expect anti-theft devices to be a hit in recession. *Wall Street Journal,* p. C6.

Newman, G. (1979). *Understanding violence.* New York: Lippincott.

Newman, G. (Ed.). (1999). *Global report on crime and justice.* New York: Oxford University Press.

Newman, G., & Marongiu, P. (1990, May). Penological reform and the myth of Beccaria. *Criminology, 28,* 325–346.

Nicolova, R. (1999, January 25). Global crime puts business people at risk. *Kansas City Business Journal,* pp. 1–4.

Nielson, M. (1998, January). A comparison of Canadian youth justice commission and Navajo peace makers. *Journal of Contemporary Criminal Justice, 14,* 6–25.

Nielson, M., Fulton, D., & Tsosie, I. (2000, March). *Recent trends in community-based strategies for dealing with juvenile crime on the Navajo nation.* Paper presented at the Academy of Criminal Justice Annual Meeting, New Orleans, LA.

Nisbet, L. (2001). *The gun control debate: You decide* (2nd ed.). New York: Prometheus Books.

Noack, D. (2000). Employees, not hackers, greatest computer threat. APBnews.com. Available at http://www .investigation.com/articles/library/2000articles/articles18.htm

Nordstrom, C. (2007). *Global outlaws.* Berkeley: University of California Press.

North freed. (1991, September 17). *Wall Street Journal,* p. A16.

North, O., with Novak, W. (1991). *Under fire: An American story.* New York: HarperCollins.

Nurco, D. N., Hanlon, T. E., & Kinlock, T. W. (1988). Differential criminal patterns of narcotic addicts over an addition career. *Criminology, 26,* 407–423.

Nuremberg principle. (1970, January 26). *The Nation,* p. 78.

NW3C. (2007). *"True" computer crime: Computer as target.* Available at http://www.nw3c.org/papers.

NW3C, Bureau of Justice Assistance, & FBI. (2007). *Internet crime report.* Internet Complaint Center. Available at http://www.nw3c.org/research/site_files .cfm?mode=r.

NYC's Mollen Commission paints grim corruption picture. (1993, November 30). *Law Enforcement News, 19,* 11.

Nye, F. I., Short, J. F., & Olson, V. J. (1958, January). Socioeconomic status and delinquent behavior. *American Journal of Sociology, 63,* 381–389.

O'Brien, S. (1986). *Why they did it: Stories of eight convicted child molesters.* Springfield, IL: Charles C Thomas.

O'Connor, T. (1987, February 2). The misfortune tellers. *Woman's World,* p. 41.

Ogburn, W. (1922). *Social change.* New York: Heubsch.

Ohmar, K. (1991, August 6). The scandal behind Japan's financial scandals. *Wall Street Journal,* p. B1.

Olson, W. K. (1991). *The litigation explosion: What happened when America unleashed the lawsuit?* New York: Dutton.

Orcutt, J. D. (1987). Differential association and marijuana use: A closer look at Sutherland (with a little help from Becker). *Criminology, 25,* 341–358.

Orwell, G. (1949). *1984.* New York: Harcourt, Brace.

Packer, H. (1968). *The limits of criminal sanction.* Stanford, CA: Stanford University Press.

Page, J., & O'Brien, M. W. (1973). *Bitter wages: Ralph Nader's study group report on disease and injury on the job.* New York: Grossman.

Paige, C. (1985). *The right to lifers: Who they are, how they operate, where they get their money.* New York: Summit Books.

Palmer, L. (1990, July 16). Coming home from Nam. *Erie Morning News,* p. 4A.

Panattieri, J. C. (1999, September 15). The software mobsters. *Interactive Week Online* (http://www.zdnet .com).

Panel for the Evaluation of Crime Surveys. (1976). *Surveying crime.* Washington, DC: National Academy of Sciences.

Parenti, M. (1980). *Democracy for the few* (3rd ed.). New York: St. Martin's Press.

Park, R. E. (1952). *Human communities.* Glencoe, IL: Free Press.

Parker, D. (1979). *Computer crime: Criminal justice resource manual.* Washington, DC: Government Printing Office.

Parker, D. (1983). *Fighting computer crime.* New York: Scribner.

Partridge, B. (1999). *EU commissioner worried by corruption in candidate states.* Retrieved April 4, 2007, from http://www.rferl.org/features/1999/03/f.ru.990322131614.asp

Passas, N. (1995, June). The mirror of global evils: A review essay on the BCCI affair. *Justice Quarterly, 12,* 377–406.

Paternoster, R., & Bachman, R. (2001). The structure and relevance of theory in criminology. In R. Paternoster & R. Bachman (Eds.), *Explaining criminals and crime: Essays in contemporary criminological theory* (pp. 1–10). Los Angeles: Roxbury.

Paternoster, R., & Mazerolle, P. (1999). General strain theory and delinquency: A replication and extension. In F. R. Scarpitti & A. L. Nielsen (Eds.), *Crime and criminals* (pp. 274–283). Los Angeles: Roxbury.

Patterson, G. R. (1982). *Coercive family process.* Eugene, OR: Castalia.

Patterson, G. R., & Dishion, T. J. (1985). Contributions of families and peers to delinquency. *Criminology, 23,* 63–79.

Pauly, D., Friday, C., & Foote, J. (1987, July 27). A scourge of video pirates. *Newsweek,* pp. 40–41.

Pearson, G. (1982). *Hooligans: A history of respectable fears.* New York: Schocken.

Pelfrey, W. V. (1980). *The evolution of criminology.* Cincinnati, OH: Anderson.

Pennsylvania Crime Commission. (1980). *A decade of organized crime, 1980 report.* St. Davids, PA: Author.

Pennsylvania Crime Commission. (1986). *The changing face of organized crime, 1986 report.* Conshohocken, PA: Author.

Pennsylvania Crime Commission. (1991). *Organized crime in Pennsylvania: A decade of change, 1990 report.* Conshohocken, PA: Author.

Pennsylvania Crime Commission. (1992). *Racketeering and organized crime in the bingo industry.* Conshohocken, PA: Author.

Pepinsky, H. E., & Quinney, R. (Eds.). (1991). *Criminology as peace making.* Bloomington: Indiana University Press.

Perkins, R., & Bennett, G. (1985). *Being a prostitute: Prostitute women and prostitute men.* Sydney, Australia: Allen & Unwin.

Perls, F. (1970). Four lectures. In J. Fagan & I. L. Shepherd (Eds.), *Gestalt therapy now* (pp. 14–38). New York: Harper & Row.

Perry, S. W. (2004, December). *American Indians and crime.* Washington, DC: Bureau of Justice Statistics (NCJ 203097).

Peter, L. J. (Ed.). (1977). *Peter's quotations.* New York: William Morrow.

Petersilia, J. (1983). *Racial disparities in the criminal justice system.* Santa Monica, CA: Rand Corporation.

Petersilia, J., Greenwood, P. W., & Lavin, M. (1977). *Criminal careers of habitual felons.* Santa Monica, CA: Rand Corporation.

Peterson, M. (1998). *Assessing criminal organizations through their management of profits.* Trenton: New Jersey Division of Criminal Justice (unpublished paper).

Petrosino, A., et al. (2003, June). Toward evidence-based criminology and criminal justice: Systematic reviews and the Campbell Collaboration, and the crime and justice group. *International Journal of Comparative Criminology,* 42–61.

Pfohl, S. J. (1985). *Images of deviance and social control.* New York: McGraw-Hill.

Pfohl, S. J. (1993). Twilight of the parasites: Ultramodern capital and the new world order. *Social Problems, 40,* 125–151.

Phar-Mor discloses "financial swindle" of $350 million. (1992, August 5). *Erie Morning News,* pp. 1A, 2A.

Philadelphia crime statistics questioned. (1998, November 2). Associated Press.

Pileggi, N. (1985). *Wiseguy: Life in a Mafia family.* New York: Simon & Schuster.

Piliavin, I., Gartner, R., Thornton, C., & Matsueda, R. L. (1986). Crime, deterrence, and rational choice. *American Sociological Review, 51,* 101–119.

Pincher, C. (1984). *Too secret, too long.* New York: St. Martin's Press.

Pinter, R. (1923). Intelligence testing: Methods and results. New York: Holt.

Piquero, A., Oster, R. P., Mazerolle, P., Brame, R., & Dean, C. W. (1999). Offense age and offense specialization. *Journal of Research on Crime and Delinquency, 36,* 275–299.

Pistone, J., & Woodley, R. (1987). *Donnie Brasco: My undercover life in the Mafia.* New York: New American Library.

Piven, F. F. (1981, June). Deviant behavior and the remaking of the world. *Social Problems, 28,* 489–508.

Pizzo, S., Fricker, M., & Muolo, P. (1989). *Inside job: The looting of America's savings and loans.* New York: McGraw-Hill.

Plagens, P., Starr, M., & Robins, K. (1990, April 2). To catch an art thief. *Newsweek,* pp. 52–53.

Plate, T. (1975). *Crime pays: An inside look at burglars, car thieves, loan sharks, hit men, fences and other professional criminals.* New York: Ballantine.

Plate, T., & Darvi, A. (1981). *Secret police: The inside story of a network of terror.* Garden City, NY: Doubleday.

Platt, T. (1985). Criminology in the 1980s: Progressive alternatives to law and order. *Crime and Social Justice, 22–22,* 191–199.

Pokorny, A. D., & Jachimczyk, J. (1974, June). The questionable relationship between homicides and lunar cycle. *American Journal of Psychiatry, 131,* 827–829.

Poggio, E. C. et al. (1985, May). *Blueprint for the future of the uniform crime reporting program: final report of the UCR study.* Washington, DC: U.S. Department of Justice.

Poland, J. (1988). *Understanding terrorism: Groups, strategies, and responses.* Englewood Cliffs, NJ: Prentice Hall.

Poland, J. (2005). *Understanding terrorism: Groups, strategies and responses* (2nd ed.). Upper Saddle River, NJ: Prentice Hall.

Police Foundation. (1977). *Domestic violence and the police: Studies in Detroit and Kansas City.* Washington, DC: Author.

Pollock, J. (1999). *Criminal women.* Cincinnati: Anderson.

Polsky, N. (1967). *Hustlers, beats and others.* Chicago: Aldine.

Pomerantz, S. L. (1987, October). The FBI and terrorism. *FBI Law Enforcement Bulletin,* pp. 14–17.

Pontell, H., Calavita, K., & Tillman, R. (1994). *Fraud in the savings and loan industry: White-collar crime and government response.* Washington, DC: U.S. Department of Justice, National Institute of Justice.

Pope, C. (1980). Patterns in burglary: An empirical examination of offense and offender characteristics. *Journal of Criminal Justice, 8*(1), 39–51.

Post, T., & Field, C. (1992, April 13). The strict rules of revenge. *Newsweek,* p. 45.

Postal violence said to be a myth. (2000, September 1). Associated Press. Retrieved July 1, 2008, from http://www .arizona.edu/papers/94/10/01_20_m.html.

Potter, G. W. (1989). Book review: Zips, pizza and smack. *Criminal Organizations, 4,* 17–18.

Powell, S., et al. (1986, February 3). Busting the mob. *U.S. News and World Report,* pp. 24–32.

President's Commission on Law Enforcement and the Administration of Justice. (1967). *The challenge of crime in a free society.* Washington, DC: Government Printing Office.

President's Commission on Organized Crime. (1983). *Organized crime: Federal law enforcement perspective.* Record of Hearing I. Washington, DC: Government Printing Office.

President's Commission on Organized Crime. (1984a). *Organized crime of Asian origin.* Washington, DC: Government Printing Office.

Press, A., et al. (1986, July 14). A government in the bedroom. *Newsweek,* pp. 36–38.

Pressley, S. A. (1998, December 9). Spiderman scaling condos to rob from rich. *Erie Times News,* p. A1.

Preston, D. J. (1986). *Dinosaurs in the attic: An excursion into the American Museum of Natural History.* New York: St. Martin's Press.

Proal, L. (1973). *Political crime.* Montclair, NJ: Patterson Smith. (Original work published 1989)

Prudential fined millions. (1996, July 6). *Erie Morning News,* p. A1.

Prus, R., & Sharper, C. R. D. (1977). *Road hustler.* Toronto: Gage.

Quetelet, L. A. J. (1969). *A treatise on man and the development of his faculties.* Gainesville, FL: Scholar's Facsimiles and Reprints. (Original work published 1842)

Quetelet, L. A. J. (1984). *Research on the propensity for crime at different ages* (S. F. Sylvester, Trans.). Cincinnati, OH: Anderson.

Quinney, R. C. (1963, Fall). Occupational structure and criminal behavior: Prescription violations by retail pharmacists. *Social Problems, 11,* 179–185.

Quinney, R. C. (1970). *The social reality of crime.* Boston: Little, Brown.

Quinney, R. C. (1974a). *Criminal justice in America: A critical understanding.* Boston: Little, Brown.

Quinney, R. C. (1974b). *Critique of legal order: Crime control in capitalist society.* Boston: Little, Brown.

Quinney, R. C. (1974c). *Criminology: Analysis and critique of crime in the United States.* Boston: Little, Brown.

Quinney, R. C. (1977). *Class, state and crime: On the theory and practice of criminal justice.* New York: David McKay.

Quinney, R. C. (1980). *Providence.* New York: Longman.

Quinney, R. C. (1988, Winter). Crime, suffering, service: Toward a criminology of peacemaking. *The Quest,* pp. 102–116.

Quinney, R. C. (1991). The way of peace: On crime, suffering, and service. In H. E. Pepinsky & R. C. Quinney (Eds.), *Criminology as peacemaking.* Bloomington, IN: Indiana University Press.

Raab, S. (1994, August 23). Top echelon of mobsters pose threat. *New York Times,* p. A7.

Raab, S. (1997, February 10). Officials say mob is shifting crimes to new industries. *New York Times,* p. A1.

Rabinowitz, D. (1991, May 6). Parents and children on trial. *Wall Street Journal,* p. A14.

Radzinowicz, L., & King, J. (1977). *The growth of crime: The international experience.* New York: Basic Books.

Raffali, H. C. (1970). The battered child. *Crime and Delinquency, 16,* 139–150.

Rafter, N. (2008). *The criminal brain: Understanding biological theories of crime.* New York: New York University Press.

Rafter, N. H., & Maher, L. (Eds.). (1995). *International feminist perspectives in criminology.* New York: Open University Press.

Ragavan, C., & Kaplan, D. E. (1999, June 14). Why auto theft is going global. *U.S. News and World Report,* 16–20.

Randi, J. (1988). *The faith healers.* New York: Prometheus.

Radzinowicz, L. (1966). *Ideology and crime.* New York: Columbia University Press.

Ranelagh, J. (1986). *The agency: The rise and decline of the CIA: From Wild Bill Donovan to William Casey.* New York: Simon & Schuster.

Rashke, R. (1981). *The killing of Karen Silkwood.* Boston: Houghton Mifflin.

Rathus, S. (1983). *Human sexuality.* New York: Holt, Rinehart and Winston.

Rebovich, D., & Layne, J. (1999). The 1999 National Public Survey of White-Collar Crime completed. *NCWCCR Focus* (National Consortium for White-Collar Crime Research), *3*(1), 1.

Reckless, W. C. (1967). *The crime problem* (4th ed.). New York: Appleton-Century-Crofts.

Reed, G. E., & Yeager, P. C. (1991, November). *Organizational offending and neoclassical criminology: A challenge to Gottfredson and Hirschi's general theory of crime.* Paper presented at the meeting of the American Society of Criminology, San Francisco.

Reed, G. E., & Yeager, P. C. (1996, August). Organizational offending and neoclassical criminology: Challenging the reach of a general theory of crime. *Criminology, 34,* 357–382.

Regan, T. (1982). *All that dwell therein: Animal rights and environmental ethics.* Berkeley: University of California Press.

Regan, T. (1999, July 8). Logging on to cyber-crime. *Christian Science Monitor,* p. 1.

Regenstein, L. (1982). *America the poisoned.* Washington, DC: Acropolis Books.

Reibstein, L., & Drew, L. (1988, September 12). Clean credit for sale. *Newsweek,* p. 49.

Reichstein, K. J. (1965). Ambulance chasing: A case study of deviation and control within the legal profession. *Social Problems, 13,* 3–17.

Reid, E. (1970). *The grim reapers: The anatomy of organized crime in America, city by city.* New York: Bantam.

Reckless, W. C. (1961). *The crime problem* (3rd ed.). New York: Appleton-Century-Crofts.

Reckless, W. C., & Dinitz, S. (1967, December). Pioneering with self-concept as a vulnerability factor in delinquency. *Journal of Criminal Law, Criminology and Police Science, 58,* 515–523.

Reckless, W. C., Dinitz, S., & Kay, B. (1957, October). The self-component in potential delinquency and potential nondelinquency. *American Sociological Review, 22,* 566–570.

Reckless, W. C., Dinitz, S., & Murray, E. (1957, May). The "good boy" in a high delinquency area. *Journal of Criminal Law, Criminology, and Police Science, 48,* 18–25.

Reidel, M., & Zahn, M. (1985). *The nature and pattern of American homicide.* Washington, DC: Government Printing Office.

Reiman, J. H. (1984). *The rich get richer and the poor get prison* (2nd ed.). New York: Wiley.

Reiman, J. H. (1998). *The rich get richer and the poor get prison* (5th ed.). New York: Macmillan.

Reiman, J. (2000). *The rich get richer and the poor get prison* (6th ed.). Boston: Allyn & Bacon.

Reiss, A. J., Jr. (1967). *Studies in crime and law enforcement in major metropolitan areas.* Field Surveys III, President's Commission on Law Enforcement and the Administration of Justice. Washington, DC: Government Printing Office.

Rengert, G., & Wasilchick, J. (1985). *Suburban burglary: A time and place for everything.* Springfield, IL: Thomas.

Rennison, C. M., & Welchans, S. (2000, May). *Intimate partner violence* (Bureau of Justice Statistics Special Report, NCJ 178247). Washington, DC: Bureau of Justice Statistics.

Renzetti, C. M. (1993). On the margins of the malestream (or, they still don't get it, do they?): Feminist analyses in criminal justice education. *Journal of Criminal Justice Education, 4,* 219–234.

Repetto, T. A. (1974). *Residential crime.* Cambridge, MA: Ballinger Press.

Reuter, P. (1984a, Spring). The continued vitality of mythical numbers. *The Public Interest, 75,* 135–147.

Reuter, P. (1984b). *Disorganized crime.* Cambridge, MA: MIT Press.

Revell, O. B. (1988, January). *Terrorism: A law enforcement perspective.* Federal Bureau of Investigation.

Richards, D. (1982). *Sex, drugs and the law: An essay on human rights and overcriminalization.* Totowa, NJ: Rowman & Littlefield.

RICO: Assault with a deadly weapon. (1989, January 30). *New York Times,* p. 18.

Regoli, R., & Poole, E. (1978). The commitment of delinquents to their misdeeds: A reexamination. *Journal of Criminal Justice, 6,* 261–269.

Rittenhouse, C. A. (1991, August). The emergence of premenstrual syndrome as a social problem. *Social Problems, 38,* 412–423.

Ritter, B. (1988). *Sometimes God has a kid's face: The story of America's exploited street kids.* New York: Covenant House.

Robertson, F. (1977). *Triangle of death: The inside story of the Triads.* London: Routledge/Kegan Paul.

Robins, L. N. (1974). *The Vietnam drug user returns.* Monograph, series A, number 2. Rockville, MD: National Institute on Drug Abuse.

Robinson, W. S. (1950, June). Ecological correlations and the behavior of individuals. *American Sociological Review, 15,* 351–357.

Rockefeller Commission. (1975). *The Rockefeller report to the president by the Commission on CIA activities.* Washington, DC: Government Printing Office.

Roebuck, J., & Weeber, S. G. (1978). *Political crime in the United States: Analyzing crime by and against the government.* New York: Praeger.

Roebuck, J., & Windham, G. O. (1983). Professional theft. In G. B. Waldo (Ed.), *Career criminals* (pp. 13–29). Beverly Hills, CA: Sage Publications.

Rogers, T. (1992, June 8). Outgunned "computer posse" tracks high-tech criminals. *Erie Morning News,* p. A4.

Rogovin, C. H., & Martens, F. T. (1989). Albini on Cressey. *Criminal Organizations, 4*(4), 11–14.

Rogovin, C. H., & Martens, F. T. (1992, February). The evil that men do. *Journal of Contemporary Criminal Justice, 8,* 62–79.

Rohrlick, J. (2007, June 5). Shoplifting boosting everyone's cost of living. *Minyanville Financial Infotainment.* Available at http: www.minyanville.com/articles/Retail-Bush-Claude+Allen-shoplifting/index/a/13022.

Romano, L. (2005, April 25–May 1). Where have all the militias gone? *Washington Post National Weekly Edition,* p. 30.

Rome, F. (1975). *The tattooed men: An American woman reports on the Japanese criminal underworld.* New York: Delacorte Press.

Rorabaugh, W. J. (1979). *The alcoholic republic: An American tradition.* New York: Oxford University Press.

Rosanoff, A. J., Handy, L. M., & Plesset, I. R. (1934, May). Criminality and delinquency in twins. *Journal of Criminal Law and Criminology, 24,* 923–934.

Rosberg, R. R. (1980). *Game of thieves.* New York: Everest House.

Rose, P. I., Glazer, M., & Glazer, P. M. (1982). *Sociology: Inquiry into society* (2nd ed.). New York: St. Martin's Press.

Rosen, L., & Neilson, K. (1978). The broken home and delinquency. In L. Savitz & N. Johnson (Eds.), *Crime in society* (2nd ed., pp. 406–415). New York: Wiley.

Rosen, R. (1983). *The lost sisterhood: Prostitution in America, 1900–1918.* Baltimore: Johns Hopkins University Press.

Rosenbaum, J. (1987). Social control, gender and delinquency: An analysis of drug, property and violent offenders. *Justice Quarterly, 4,* 117–132.

Rosenbaum, J. (1989a, January). Family dysfunction and female delinquency. *Crime and Delinquency,* pp. 31–44.

Rosenbaum, J. (1989b). Women and crime [Special issue]. *Crime and Delinquency, 35* [entire issue].

Rosenberg, T. (1991). *Children of Cain: Violence and the violent in Latin America.* New York: Morrow.

Rosner, L. S. (1995, Fall). The sexy Russian Mafia. *Criminal Organizations,* pp. 28–32.

Ross, A. E. (1907, January). The criminaloid. *Atlantic Monthly, 99,* 44–50.

Ross, H. L. (1961, Winter). Traffic law violation: A folk crime. *Social Problems, 9,* 231–241.

Ross, J. (2003). *The dynamics of political crime.* Thousand Oaks, CA: Sage Publications.

Ross, L., & Edwards, W. (1998). Publishing among African American criminologists: A devaluing experience. *Journal of Criminal Justice, 26*(1), 29–40.

Ross, L. A., & McMurray, H. L. (1996). Dual realities and structural challenges of African-American criminologists. *ACJS Today, 15*(1), 3, 9.

Ross, S. (1988). *Fall from grace: Sex, scandal, and corruption in American politics from 1702 to the present.* New York: Ballantine.

Rossi, P., Waite, E., Bose, C. E., & Berk, R. E. (1974, April). The seriousness of crimes: Normative structure and individual differences. *American Sociological Review, 39,* 224–237.

Rounds, D. (Ed.). (2000). *International criminal justice: Issues in a global perspective.* Boston: Allyn & Bacon.

Rovetch, E. L., Poggio, E. C., & Rossman, H. H. (1984). *A listing and classification of identified issues regarding the Uniform Crime Reporting program of the FBI.* Cambridge, MA: ABT Associates.

Rowan, R. (1986, November 10). The biggest Mafia bosses. *Fortune,* pp. 24–38.

Rowe, D. (2002). *Biology and crime.* Los Angeles, CA: Roxbury.

Rupe, R. A. (1980, March). Formula for loss prevention. *Retail Control,* pp. 2–15.

Russell, D. (1986). *The secret trauma: Incest in the lives of girls and women.* New York: Basic Books.

Russell, K. K. (1992). Development of a black criminology and the role of the black criminologist. *Justice Quarterly, 9*(4), 667–683.

Russian mob may have laundered billions at Bank of New York. (1999, August 23). *Russia Today.* Available at http://www.russiatoday.com.

Safire, W. (1989, January 30). The end of RICO. *New York Times,* p. 19.

Sagarin, E. (1973). Introduction to *Political Crime,* by Louis Proal. Montclair, NJ: Patterson Smith.

Sagarin, E. (Ed.). (1980). *Taboos in criminology.* Beverly Hills, CA: Sage Publications.

Salerno, R., & Tompkins, J. S. (1969). *The crime confederation.* Garden City, NY: Doubleday.

Salwen, K. G. (1991, May 17). SEC charges firm with defrauding 40,000 investors. *Wall Street Journal,* p. B8.

Samper, E. (1995, June 30). Colombia's war on drugs. *Wall Street Journal,* p. A16.

Sampson, R. J. (1985). Structural sources of variation in race and age specific rates of offending across major U.S. cities. *Criminology, 23,* 647–673.

Sampson, R. J., & Laub, J. H. (2003). Life-course desisters? Trajectories of crime among delinquent boys followed to age 70. *Criminology, 41,* 555–572.

Sanchez-Jankowski, M. (1991). *Islands in the street: Gangs and American urban society.* Berkeley: University of California Press.

Sanday, P. R. (1990). *Fraternity gang rape: Sex, brotherhood, and privilege on campus.* New York: New York University Press.

Sanders, W. B. (Ed.). (1976). *The sociologist as detective* (2nd ed.). New York: Praeger.

Sanders, W. B. (1994). *Gangbangs and drivebys: Grounded culture and juvenile gang violence.* New York: Aldine de Gruyter.

Sanoff, A. P. (1996, September 30). The hottest import crime. *U.S. News and World Report,* p. 4.

Sapp, A. (1986, March). *Organized linkages of right wing extremist groups.* Paper presented at the Academy of Criminal Justice Sciences Meeting, St. Louis, MO.

Sarbin, T. R., & Miller, J. E. (1970, Summer). Demonism revisited: The XYY chromosome anomaly. *Issues in Criminology, 5,* 195–207.

Satchell, M. (1988, December 19). The just war that never ends. *U.S. News and World Report,* pp. 31–38.

Savage, G. (1976). *Forgeries, fakes and reproductions.* London: White Lion Publishing.

Savitz, D. (1959, July). Automobile theft. *Journal of Criminal Law, Criminology, and Police Science, 50,* 132–143.

Savitz, L. D. (1978). Official police statistics and their limitations. In L. D. Savitz & N. Johnston (Eds.), *Crime and society* (pp. 69–81). New York: Wiley.

Savitz, L. D., Kumar, K. S., & Zahn, M. (1991). Quantifying Luckenbill. *Deviant Behavior, 12,* 19–29.

Scared Straight found ineffective again. (1979, September). *Criminal Justice Newsletter, 10,* 7.

Scarpitti, F., Murray, E., Dinitz, S., & Reckless, W. (1960, August). The good boy in a high delinquency area: Four years later. *American Sociological Review, 23,* 555–558.

Scarr, H. A. (1973). *Patterns of burglary.* Washington, DC: Government Printing Office.

Schafer, S. (1969). *Theories in criminology: Past and present philosophies of the crime problem.* New York: Random House.

Schafer, S. (1971, Spring). The concept of the political criminal. *Journal of Criminal Law, Criminology and Police Science, 62,* 380–387.

Schafer, S. (1974). *The political criminal.* New York: Free Press.

Schafer, S. (1976). *Introduction to criminology.* Reston, VA.: Reston Publishing Company.

Schatzberg, R., & Kelly, R. J. (1995). *African-American organized crime.* New York: Garland.

Schauss, A.(1980). *Diet, crime and delinquency.* Berkeley, CA.: Parker House.

Scheflin, A. W., & Opton, E. M., Jr. (1978). *The mind manipulators.* New York: Paddington Press.

Scheim, D. E. (1988). *Contract on America: The Mafia murder of President John F. Kennedy.* New York: Shapolski.

Schellhardt, T. D. (1990, October 15). Anti-shoplifting statutes. *Wall Street Journal,* p. B2.

Schichor, D. (1982, March). An analysis of citations in introductory criminology textbooks. *Journal of Criminal Justice, 10,* 231–237.

Schlachter, B. (1986, June 15). Women find success in espionage. *Buffalo News,* p. A14.

Schlegel, K., & Weisburd, D. (Eds.). (1994). *White-collar crime reconsidered.* Boston: Northeastern University Press.

Schloss, B., & Giesbrecht, N. A. (1972). *Murder in Canada: A report on capital and non-capital murder statistics 1961–1970.* Toronto: Centre of Criminology, University of Toronto.

Schlesinger, A. (1986). *Cycles of American history.* Boston: Houghton Mifflin.

Schmalleger, F. (1990, January-February). A call for caution on capital punishment policy. *The Criminologist, 15,* 4.

Schmid, A., & de Graaf, J. (1982). *Violence as communication: Insurgent terrorism and the Western news media.* Beverly Hills, CA: Sage Publications.

Schmitt, R. B. (1992, July 21). An insurer's sleuth sniffs out lawyers inflating their bills. *Wall Street Journal,* pp. A1, A5.

Schneider, S. (2002). Summary and analysis: Methodological design. *Predicting crime: The review of research.* Ottawa: Canadian Department of Criminal Justice. Accessed July 1, 2007, at http://section15.gc.ca/en/ps/rs/rep/2002/rr2002–7/rr2002–7_007.html.

Schrag, C. (1962, January). Delinquency and opportunity: Analysis of a theory. *Sociology and Social Research, 46,* 165–175.

Schrag, C. (1971). *Crime and justice: American style.* Washington, DC: Government Printing Office.

Schrager, L. S., & Short, J. F. (1978, April). Toward a sociology of organized crime. *Social Problems, 25,* 407–419.

Schuerman, L. A., & Kobrin, S. (1986). Community careers in crime. In A. J. Reiss, Jr., & M. Tonry (Eds.), *Communities and crime.* Chicago: University of Chicago Press.

Schuessler, K. F. (1952). The deterrent influence of the death penalty. *Annals of the Academy of Political and Social Sciences, 284,* 54–62.

Schuessler, K. F. (1954). Review. *American Journal of Sociology, 49,* 604.

Schuessler, K. F., & Cressey, D. R. (1953). Personality characteristics of criminals. *American Journal of Sociology, 55,* 166–176.

Schulsinger, F. (1972, January). Psychopathy, heredity and environment. *International Journal of Mental Health, 1,* 190–206.

Schur, E. M. (1965). *Crimes without victims: Deviant behavior and public policy.* Englewood Cliffs, NJ: Prentice Hall.

Schur, E. M. (1969, November). Reactions to deviance: A critical assessment. *American Journal of Sociology, 75,* 309–322.

Schur, E. M. (1971). *Labeling deviant behavior.* New York: Harper and Row.

Schur, E. M. (1980). *The politics of deviance.* Englewood Cliffs, NJ: Prentice Hall.

Schwartz, J. (1990, July 2). Hackers of the world, unite! *Newsweek,* pp. 36–37.

Schwartz, M. D., & Friedrichs, D. O. (1994, May). Postmodern thought and criminological discontent: New metaphors for understanding violence. *Criminology, 32,* 221–246.

Schwartz, M. D., & Tangri, S. (1965). A note on self-concept as an insulator against delinquency. *American Sociological Review, 30,* 922–926.

Schweinhart, L. J., & Weikart, D. P. (1980). *Young children grow up: The effects of the Perry preschool program on youths through age 15.* Ypsilanti, MI: High/Scope.

Scott, D. W. (1989). Policing corporate collusion. *Criminology, 27,* 559–588.

Sederberg, P. C. (1989). *Terrorist myths: Illusion, rhetoric, and reality.* Englewood Cliffs, NJ: Prentice Hall.

Sellin, T. (1938). Culture conflict and crime. *Social Science Research Council Bulletin, 41,* 1–7.

Sellin, T. (1959). *The death penalty.* Philadelphia: American Law Institute.

$700 million Ponzi schemer draws record prison term. (2000, April 29). *Bloomberg News,* p. 1.

Senate Permanent Subcommittee on Investigations. Committee on Governmental Affairs. (1979, December 7, 11–14). 96th Congress, first session.

Shah, S. A., & Roth, L. H. (1974). Biological and psychophysiological factors in criminality. In D. Glaser (Ed.), *Handbook of criminology* (pp. 101–173). Chicago: Rand McNally.

Shannon, E., & Blackman, A. (2002). *The spy next door.* Boston: Little, Brown.

Shapiro, J., & Wright, A. R. (1995, August 14). Sins of the father. *U.S. News and World Report,* pp. 52–53.

Shapiro, M. (1992, March 9–15). A freedom mugged by crime. *Washington Post National Weekly Edition,* p. 9.

Shaw, C. R. (1929). *Delinquency areas: A study of the geographic distribution of school truants, juvenile delinquents, and adult offenders in Chicago.* Chicago: University of Chicago Press.

Shaw, C. R. (1930). *The jack roller.* Chicago: University of Chicago Press.

Shaw, C. R., & McKay, H. D. (1942). *Juvenile delinquency and urban areas.* Chicago: University of Chicago Press.

Shaw, C. R., McKay, H. D., & McDonald, J. F. (1938). *Brothers in crime.* Chicago: University of Chicago Press.

Shaw, G. B. (1941). Preface to *The doctor's dilemma,* by George Bernard Shaw. New York: Dodd, Mead and Company.

Sheehy, G. (1973). *Hustling: Prostitution in our wide open society.* New York: Delacorte Press.

Sheldon, W. H. (1940). *The varieties of human physique.* New York: Harper & Row.

Sheley, J. F., & Wright, J. D. (1995). *In the line of fire: Youth, guns, and violence in urban America.* Hawthorne, NY: Aldine de Gruyter.

Shelley, L. I. (1995, December). Privatization and crime: The post-Soviet experience. *Journal of Contemporary Criminal Justice, 11,* 244–256.

Sherif, M., & Sherif, C. (1966). *Groups in harmony and tension.* New York: Octagon.

Sherman, L., & Berk, R. (1984). The specific deterrent effects of arrest for domestic assault. *American Sociological Review, 49,* 261–272.

Sherman, L. W., Shaw, J. W., & Rogan, D. P. (1995, January). The Kansas City gun experiment. *Research in Brief.* Washington, DC: U.S. Department of Justice, National Institute of Justice.

Shoemaker, D. L., & Williams, J. S. (1987). The subculture of violence and ethnicity. *Journal of Criminal Justice, 15,* 461–472.

Shoplifting costs U.S. retailers $40.5 billion according to ADT-sponsored Survey. (2007, December 4). *PR Newswire.*

Short, J. F. (1990). Gangs, neighborhoods, and youth crime. *Criminal Justice Research Bulletin, 5*(4), 1–11.

Short, J. F., Jr., & Nye, F. I. (1958). Extent of unrecorded delinquency: Tentative conclusions. *Journal of Criminal Law, Criminology and Police Science, 49,* 296–302.

Shover, N. (1973, December). The social organization of burglary. *Social Problems, 20,* 499–514.

Shover, N. (1983, December). The later stages of ordinary property offender careers. *Social Problems, 31,* 208–218.

Siconolfi, M., & Johnson, R. (1991, August 29). Broker grandmother accused of losing clients' cash at baccarat. *Wall Street Journal,* pp. C1, C11.

Sigler, R. T., & Haygood, D. (1988). The criminalization of forced marital intercourse. In F. E. Hagan & M. B. Sussman (Eds.), *Deviance and the family* (pp. 71–85). New York: Haworth Press.

Silberman, C. E. (1978). *Criminal violence, criminal justice.* New York: Random House.

Silkwood vindicated. (1979, May 28). *Newsweek,* p. 40.

Simcha-Fagan, O., & Schwartz, J. E. (1986). Neighborhood and delinquency: An assessment of contextual effects. *Criminology, 24,* 667–703.

Simmel, G. (1955). *Conflict and the web of group affiliations* (K. H. Wolff & R. Bendix, Trans.). New York: Free Press. (Original work published 1908)

Simon, D. R. (1996). *Elite deviance* (5th ed.). Boston: Allyn & Bacon.

Simon, D. R. (1999). *Elite deviance* (6th ed.). Boston: Allyn & Bacon.

Simon, D. R., & Hagan, F. E. (1999). *White-collar deviance.* Boston: Allyn & Bacon.

Simon, D. R., & Swart, S. L. (1984, January). The Justice Department focuses on white-collar crime: Promises and pitfalls. *Crime and Delinquency, 30,* 107–119.

Simon, R. J. (1975). *Women and crime.* Lexington, MA: Lexington Books.

Simon, R. J. (1990). Women and crime revisited. *Criminal Justice Research Bulletin, 5*(5), 1–8.

Simon, R. J., & Aaronson, D. E. (1998). *The insanity defense: A critical assessment of law and policy in the post-Hinckley era.* New York: Praeger.

Simons, M. (2005, April 29). Sudan poses first big trial for the World Court. *New York Times,* p. A10.

Simpson, J., & Bennett, J. (1985). *The disappeared and the mothers of the plaza.* New York: St. Martin's Press.

Simpson, M., & Schill, T. (1977). Patrons of massage parlors: Some facts and figures. *Archives of Sexual Behavior, 6,* 521.

Simpson, S. (2003, July/August). The criminological enterprise and corporate crime, *The Criminologist, 28*(4), 1, 3–5.

Simpson, S. S., & Ellis, L. (1995). Doing gender: Sorting out the caste and crime conundrum. *Criminology, 33,* 47–81.

Sinclair, U. (1906). *The jungle.* New York: Doubleday and Page.

Singer, M. (1971, Spring). The vitality of mythical numbers. *The Public Interest, 23,* 3–9.

Singer, M., & Levine, M. (1988). Power control theory, gender, and delinquency: A partial replication with additional evidence on the effects of peers. *Criminology, 26,* 627–647.

Skinner, B. F. (1953). *Science and human behavior.* New York: Macmillan.

Skinner, B. F. (1971). *Beyond freedom and dignity.* New York: Knopf.

Skolnick, J. H. (1969). *The politics of protest: The Skolnick Report to the National Commission on the Causes and Prevention of Violence.* New York: Ballantine.

Skolnick, J. H., & Currie, E. (Eds.). (1982). *Crisis in American institutions* (5th ed., p. 390). Boston: Little, Brown.

Skolnick, J. H., & Currie, E. (Eds.). (1988). *Crisis in American institutions.* Boston: Little, Brown.

Slade, M. (1994, May 26). At the bar. *New York Times,* p. B12.

Slim, I. (1969). *Pimp: The story of my life.* Los Angeles: Holloway House.

Snodgrass, J. D. (1972). *The American criminological tradition: Portraits of the men and ideology in a discipline.* Unpublished doctoral dissertation, University of Pennsylvania.

Snodgrass, J. D. (1982). *The jack roller at seventy: A fifty-year follow up.* Lexington, MA: D.C. Heath.

Smigel, E. O., & Ross, H. L. (Eds). (1970). *Crimes against bureaucracy.* New York: Van Nostrand Reinhold.

Smith, A. (1953). *The wealth of nations.* Cambridge, MA: Harvard University Press. (Original work published 1776)

Smith, B. F. (1977). *Reaching judgment at Nuremberg.* New York: Basic Books.

Smith, D. C., Jr. (1975). *The Mafia mystique.* New York: Basic Books.

Smith, D. C., Jr. (1978, May). Organized crime and entrepreneurship. *International Journal of Criminology and Penology, 6,* 161–177.

Smith, D. C., Jr. (1980, July). Paragons, pariahs and pirates: A spectrum-based theory of enterprise. *Crime and Delinquency, 26,* 358–386.

Smith, H. (1989, February 6). *The power game* [Television broadcast]. Arlington, VA: Public Broadcasting Service.

Smith, J. D. (1985). *Minds made feeble: The myth and legacy of the Kallikaks.* Aspen, CO: Aspen.

Snider, D. L. (1978, April). Corporate crime in Canada. *Canadian Journal of Criminology, 20,* 142–168.

Snyder, H., & Sickmund, M. (2005). *Juvenile offenders and victims.* Pittsburgh, PA: National Center for Juvenile Justice.

Snyder, L. P. (1994). *The death-dealing smog over Donora, Pennsylvania: Industrial air pollution, public health policy and the politics of expertise, 1948–1949.* Doctoral dissertation, University of Pennsylvania.

Solomon, J., & Eilperin, J. (2007, October 8–14). Dwindling pursuit of polluters. *USA Today,* pp. 34–35.

Solzhenitsyn, A. I. (1975). *The GULAG Archipelago.* New York: Harper and Row.

Somers, C. H. (1994). *Who stole feminism?* New York: Simon & Schuster.

Southerland, D. (1991, October 7–13). A witness against China's export practices. *Washington Post National Weekly Edition.*

Soviet crime rate up. (1989, February 14). *Los Angeles Times,* p. 1A.

Sparks, R. F. (1980). A critique of Marxist criminology. In N. Morris & M. Tonry (Eds.), *Crime and justice, vol. 2* (pp. 159–208). Chicago: University of Chicago Press.

Sparrow, M. (1998, December). Fraud control in the health care industry. *Research in Brief.* Washington, DC: U.S. Department of Justice, National Institute of Justice.

Spencer, C. (1966, September). A typology of violent offenders. *Administrative Abstracts, 23.*

Spernow, B. (1995, November). Videoconference presented by the National White-Collar Crime Center, Morgantown, WV.

Spitzer, R. J. (2003). *The politics of gun control.* New York: CQ Press.

Spitzer, S. (1975, September). Toward a Marxian theory of deviance. *Social Problems, 22,* 638–651.

Spohn, C., & Cederblom, J. (1991, September). Race and disparities in sentencing: A test of the liberation hypothesis. *Justice Quarterly, 8,* 305–327.

Springer, K. (1991, July 22). A slippery pyramid: To some Nu Skin may be nu scam. *Newsweek,* p. 39.

Staats, G. R. (1977, May). Changing conceptualizations of professional criminals: Implications for criminology theory. *Criminology, 15,* 53–63.

Staba, D. (2007, August 15). Killer of 3 women in Buffalo area is given a life term. *New York Times,* p. A26.

Stark, R. (1987). Deviant places: A theory of the ecology of crime. *Criminology, 25,* 893–909.

State v. Rideout. (1969). 450 P.2d 452 (Wyo. 1969).

Stecklow, S. (1996, June 27). Trustee for new era is suing Prudential. *Wall Street Journal,* p. A3.

Steffens, L. (1904). *The shame of the cities.* New York: McClure, Phillips.

Steffensmeier, D. (1978). Crime and the contemporary woman: Analysis of changing levels of female property crime, 1960–1975. *Social Forces, 57,* 566–584.

Steffensmeier, D. (1986). *The fence: In the shadow of two worlds.* Totowa, NJ: Rowman & Littlefield.

Steffensmeier, D. (1989a). Age and the distribution of crime. *American Journal of Sociology, 94,* 803–831.

Steffensmeier, D. (1989b, May). On the causes of "white-collar" crime: An assessment of Hirschi and Gottfredson's claims. *Criminology, 27,* 345–358.

Steffensmeier, D., & Allan, E. A. (1988). Sex disparities in arrests by residence, race and age: An assessment of the gender convergence/crime hypothesis. *Justice Quarterly, 5,* 53–80.

Steffensmeier, D., & Allan, E. A. (1990, August). *Physical fitness, age, and crime: The significance of biological aging vs. social aging in explaining the rapid decline in offending in the late teens.* Paper presented at the meeting of the American Sociological Association, Washington, DC.

Steffensmeier, D., & Kramer, J. F. (1990, November). *Race differences in sentencing: Research continuities and further developments.* Paper presented at the meeting of the American Society of Criminology, Baltimore.

Steffensmeier, D., & Ulmer, J. (2006). *Confessions of a dying thief.* Somerset, NJ: Transaction Publishing.

Stein, M. L. (1974). *Lovers, friends, slaves . . . The nine male sexual types.* Berkeley, CA: Berkeley Publishing Corporation.

Stein, M. R. (1964). *The eclipse of community: An interpretation of American studies.* New York: Harper & Row.

Steinmetz, S. K., & Straus, M. (1978). The family as a cradle of violence. In P. Wickman & P. Whitten (Eds.), *Readings in criminology* (pp. 59–65). Lexington, MA: D. C. Heath.

Stellwagen, L. D. (1985, July). The use of forfeiture sanctions in drug cases. *Research in Brief.* Washington, DC: U.S. Department of Justice, National Institute of Justice.

Stephens, G. (Ed.). (1982). *The future of criminal justice.* Cincinnati, OH: LexisNexis, Anderson.

Stephens, R. C., & Ellis, R. D. (1975). Narcotics addiction and crime: An analysis of recent trends. *Criminology, 12,* 474–487.

Sterling, C. (1981). *The terror network: The secret war of international terrorism.* New York: Holt, Rinehart and Winston.

Sterling, C. (1990). *Octopus: The long reach of the international Sicilian Mafia.* New York: Norton.

Stern, K. S. (1996). *A force upon the plain: The American militia movement and the politics of hate.* New York: Simon & Schuster.

Stern, P. (2008). *America's most stolen cars* [Editorial]. Available at http://editorial.autos.msn.com/article.aspx?cp-documentid=434545

Stevens, M. (1991). *The big six: The selling out of America's top accounting firms.* New York: Simon & Schuster.

Stewart, D. W., & Spille, H. A. (1988). *Diploma mills: Degrees of fraud.* New York: Macmillan.

Stewart, J. E., & Cannon, D. A. (1977). Effects of perpetrator status and bystander commitment on response to a simulated crime. *Journal of Police Science and Administration, 5,* 318–323.

Stieg, B. (1990a, July 27). Judge fines GE $10 million; sentences 2 employees to prison. *Erie Morning News,* p. 2A.

Stieg, B. (1990b, November 4). A Philly favorite: Faking injuries. *Erie Times News,* p. 16A.

Stille, A., & Robinson, L. (1992, August 3). Guns drawn, the Mafia turns against the rest of Italy. *U.S. News and World Report,* p. 42.

The sting. (1976, March 15). *Newsweek,* p. 35.

Stirling, N. (1974). *Your money or your life.* Indianapolis, IN: Bobbs-Merrill.

Stout, H. (1991, March 14). Stanford accused of overcharging U.S. for research. *Wall Street Journal,* p. C15.

Straus, M. A. (1994). *Beating the devil out of them: Corporal punishment in American families.* San Francisco: Lexington Books.

Straus, M. A. (1999, December 15). *The benefits of avoiding corporal punishment: New and more definitive evidence.* Unpublished paper, University of Maryland.

Straus, M. A., & Gelles, R. J. (1986). Societal change and change in family violence from 1975 to 1985 as revealed by two national surveys. *Journal of Marriage and the Family, 48,* 465–479.

Straus, M. A., Gelles, R. J., & Steinmetz, S. (1980). The marriage license as a hitting license. In J. H. Skolnick & E. Currie (Eds.), *Crisis in American institutions* (5th ed., pp. 273–287). Boston: Little, Brown.

Suhr, J. (2001, March). Chrysler accused of reselling "lemons" to dealers. Associated Press, p. 20.

Sukharenko, A. (2004). The use of corruption by "Russian" organized crime. *Trends in Organized Crime, 8*(2), 118–129.

Sullivan, M. (1989). *Getting paid: Youth crime and work in the inner city.* Ithaca, NY: Cornell University Press.

Sullivan, T. (1988). Juvenile prostitution: A critical perspective. In F. E. Hagan & M. B. Sussman (Eds.), *Deviance and the family* (pp. 113–134). New York: Haworth Press.

Sumner, W. G. (1906). *Folkways.* New York: Dover.

Sun-Tzu. (1963). *The art of war* (S. B. Griffith, Trans.). New York: Oxford University Press.

Super sleuths, Erie Insurance Group. (1991, Summer). *In Sync,* pp. 2–5.

Surge in campus alcohol arrests. (2000, June 4). Associated Press.

Sutherland, E. H. (1937). *The professional thief.* Chicago: University of Chicago Press (Reissued 1956. Chicago: Phoenix)

Sutherland, E. H. (1940, February). White-collar criminality. *American Sociological Review, 5,* 1–12.

Sutherland, E. H. (1941, September). Crime and business. *Annals of the American Academy of Political and Social Science, 217,* 112–118.

Sutherland, E. H. (1945, April). Is "white-collar crime" crime? *American Sociological Review, 10,* 132–139.

Sutherland, E. H. (1947). *Principles of criminology* (4th ed.). Philadelphia: Lippincott.

Sutherland, E. H. (1949). *White-collar crime.* New York: Holt, Rinehart and Winston.

Sutherland, E. H. (1956a). Crime of corporations. In A. Cohen, A. Lindesmith, & K. Schuessler (Eds.), *The Sutherland Papers* (pp. 78–96). Bloomington: Indiana University Press.

Sutherland, E. H. (1956b). The development of the theory. In A. Cohen, A. Lindesmith, & K. Schuessler (Eds.), *The Sutherland Papers* (pp. 13–29). Bloomington: Indiana University Press.

Sutherland, E. H., & Cressey, D. C. (1960). *Criminology.* Philadelphia: Lippincott.

Sutherland, E. H., & Cressey, D. C. (1974). *Principles of criminology* (9th ed.). Philadelphia: Lippincott.

Sutherland, E. H., & Cressey, D. C. (1978). *Principles of criminology* (10th ed.). Philadelphia: Lippincott.

Swigert, V., & Farrell, R. (1980, Autumn). Corporate homicide: Definitional processes in the creation of deviance. *Law and Society Review, 15,* 161–182.

Swisher, K. (1994, May 31). Office violence is on the rise, and firms aren't ready. *Erie Times News,* p. 11S.

Sykes, G. M. (1958). *The society of captives: A study of maximum security prison.* Princeton, NJ: Princeton University Press.

Sykes, G. M., & Matza, D. (1957, December). Techniques of neutralization: A theory of delinquency. *American Sociological Review, 22,* 664–670.

Szasz, T. (1974). *Ceremonial chemistry: The ritual persecution of drugs, addicts and pushers.* New York: Anchor Books.

Tafoya, W. (1992, March). *Law enforcement in the year 2000.* Paper presented at the Academy of Criminal Justice Sciences Meeting, Pittsburgh, PA.

Taking a byte out of crime. (1990, Winter). *Lotus Quarterly,* pp. 27–28.

Talese, G. (1971). *Honor thy father.* Greenwich, CT: Fawcett.

Talese, G. (1979). *Thy neighbor's wife.* Greenwich, CT: Fawcett.

Tannenbaum, F. (1938). *Crime and the community.* Boston: Ginn.

Tappan, P. (1960). *Crime, justice and corrections.* New York: McGraw-Hill.

Tarde, G. (1912). *Penal philosophy.* Boston: Little, Brown.

Taylor, C. (1990). *Dangerous society.* East Lansing: Michigan State University Press.

Taylor, I., Walton, P., & Young, J. (1973). *The new criminology: For a social theory of deviance.* New York: Harper & Row.

Taylor, I., Walton, P., & Young, J. (1975). *Critical criminology.* London: Routledge and Kegan Paul.

Taylor, L. (1984). *In the underworld.* Oxford, UK: Basil Blackwell.

Taylor, R., Caeti, T., Loper, K., Fritsch, E., & Liederbach, J. (2006). *Digital crime and digital terrorism.* Upper Saddle River, NJ: Prentice Hall.

Taylor, R. B., & Harrell, A. V. (1996, May). *Physical environment and crime.* Washington, DC: National Institute of Justice.

Taylor, S., Jr. (1985, May 16). U.S. defends disputed Hutton decision. *New York Times,* p. D5.

Tennenbaum, D. (1977, January). Research studies of personality and criminality. *Journal of Criminal Justice, 5,* 1–19.

Teresa, V. (1973a, February). A mafioso cases the Mafia craze. *Saturday Review,* pp. 23–29.

Teresa, V., with Renner, T. (1973b). *My life in the Mafia.* Greenwich, CT: Fawcett.

Terrill, R. (2007). *World criminal justice systems: A survey* (6th ed.). Cincinnati, OH: Anderson.

Terry, K. (2005). *Sexual offenses and offenders.* Belmont, CA: Thomson Wadsworth.

There's a new sheriff in town. (1995, November 10). *New York Times,* pp. C1, C7.

Thibault, E. A. (1992, June). *The violent woman at home.* Paper presented at the Northwestern Academy of Criminal Justice Sciences Meeting, Newport, RI.

13th victim of cult discovered. (1989, April 14). *Erie Morning News,* p. 1A.

Thomas, C., & Hepburn, J. R. (1983). *Crime, criminal law and criminology.* Dubuque, IA: William C. Brown.

Thomas, P. (1995, December 25–31). Using more, worrying less. *Washington Post National Weekly Edition,* p. 32.

Thomas, W. I., & Swaine, D. (1928). *The child in America.* New York: Knopf.

Thornberry, T. (1987). Toward an interactional theory of delinquency. *Criminology, 25,* 863–891.

Thornberry, T. P., Lizotte, A. J., Krohn, M. D., Farnworth, M., & Jang, S. J. (1991). Testing interactional theory. *Journal of Criminal Law and Criminology, 82,* 3–35.

Thornberry, T. P., & Krohn, M. (2001). The development of delinquency: An interactional perspective. In S. O. White (Ed.), *Handbook of youth and justice.* New York: Plenum.

Thornberry, T., Smith, C., Rivera, C., Huizinga, D., & Stouthamer-Loeber, M. (1999, September). *OJJDP Juvenile Justice Bulletin.* Rockville, MD: Juvenile Justice Clearinghouse.

Tittle, C. R. (1988). Two empirical regularities (maybe) in search of an explanation: Commentary on the age/crime debate. *Criminology, 26,* 75–85.

Tittle, C. R., Villemez, W., & Smith, D. (1978). The myth of social class and criminality: An empirical assessment of the empirical evidence. *American Sociological Review, 43,* 643–656.

Tjaden, P. (1997, November). Summary of a presentation entitled "The crime of stalking: How big is the problem?" *National Institute of Justice Research Preview.*

Toby, J. (1980). The new criminology is the old baloney. In J. A. Inciardi (Ed.), *Radical criminology: The coming crisis* (pp. 124–132). Beverly Hills, CA: Sage.

Toch, H. (1979). *Psychology of crime and criminal justice.* New York: Holt, Rinehart and Winston.

Toch, H., & Adams, K. (1991). *The disturbed violent offender.* New Haven: Yale University Press.

Toennies, F. (1957). *Community and society.* East Lansing: Michigan State University Press.

Tolson, J. (2002, July 1). The age of excess. *U.S. News and World Report,* 24–27.

Tomsho, R. (1987). *The American sanctuary movement.* Austin, TX: Monthly Press.

Top cop predicts robot crimewave. (2007, July 6). Available at http://www.theage.com.au/.

Trasler, G. (1962). *The explanation of criminality.* London: Routledge and Kegan Paul.

Trebach, A. S. (1982). *The heroin solution.* New Haven, CT: Yale University Press.

Trebach, A. (1984, March). Peace without surrender in the perpetual drug war. *Justice Quarterly, 1,* 125–144.

Triplett, R., & Jarjoura, R. (1994). Theoretical and empirical specification of informal labeling. *Journal of Quantitative Criminology, 10,* 241–276.

Truzzi, M. (1976). Sherlock Holmes: Applied social psychologist. In W. B. Sanders (Ed.), *The sociologist as detective* (2nd ed., pp. 50–86). New York: Praeger.

Tunnell, K. (1991). *Choosing crime: The criminal calculus of property offenders.* Chicago: Nelson-Hall.

Turk, A. T. (1969a). *Criminality and the legal order.* Chicago: Rand McNally.

Turk, A. T. (1969b). Introduction. In W. Bonger, *Criminality and economic conditions.* Bloomington: Indiana University Press.

Turk, A. T. (1972). *Legal sanctioning and social control.* Washington, DC: Government Printing Office.

Turk, A. T. (1980). Analyzing official deviance: For nonpartisan conflict analyses in criminology. In J. A. Inciardi (Ed.), *Radical criminology: The coming crisis* (pp. 78–91). Beverly Hills, CA: Sage.

Turk, A. T. (1981, August). Organization deviance and political policing. *Criminology, 19,* 231–250.

Turk, A. T. (1982). *Political criminality: The defiance and defense of authority.* Beverly Hills, CA: Sage Publications.

Turner, J. S. (1970). *The chemical feast: Nader's raiders study group report on the Food and Drug Administration.* New York: Grossman.

Turner, J. H. (1974). *The structure of sociological theory.* Homewood, IL: Dorsey.

Turner, S. (1985). *Secrecy and democracy: The CIA in transition.* New York: Houghton Mifflin.

Twain, M. (1899). *Following the equator: A journey around the world.* New York: Harper.

21 brokers charged in price-rigging scheme. (1991, January 24). *Erie Morning News,* p. 5A.

United Nations Centre for International Crime Prevention. (2000a). Assessing transnational organized crime: Results of a pilot survey of 40 selected organized criminal groups in 16 countries. *Trends in Organized Crime, 6*(2), 44–92.

United Nations Centre for International Crime Prevention. (2000b). Appendix—overview of the 40 criminal groups surveyed. *Trends in Organized Crime, 6*(2), 93–140.

U.S. begins price-fixing prosecution. (1975, September 9). *Erie Times News,* p. 10.

U.S. Congress. (1976, April 26; May 20). Hearings Before the Subcommittee on Energy and Environment of the Committee on Small Business. 94th Congress, second session.

U.S. Department of Health and Human Services. (1982). *Television and behavior.* Washington, DC: Government Printing Office.

U.S. Department of Justice. (1974). *Crime in eight American cities.* National Criminal Justice Information and Statistics Service. Washington, DC: Government Printing Office.

U.S. Department of Justice. (1975a). *Criminal victimization surveys in the nation's five largest cities.* National Criminal Justice Information and Statistics Service. Washington, DC: Government Printing Office.

U.S. Department of Justice. (1975b). *Criminal victimization in thirteen American cities.* National Criminal Justice Information and Statistics Service. Washington, DC: Government Printing Office.

U.S. Department of Justice. (1976). *Criminal victimization in the United States.* National Criminal Justice Information and Statistics Service.Washington, DC: Government Printing Office.

U.S. Department of Justice. (1978). *Myths and realities about crime.* Washington, DC: Government Printing Office.

U.S. Department of Justice. (1988, April). *Bureau of Justice Statistics annual report, fiscal 1987.* Washington, DC: Government Printing Office.

U.S. Department of State. (2000). *Human rights report, 1999.* Washington, DC: Government Printing Office.

U.S. Department of State. (2001, December 11). *Agents crack down on global piracy rings* [Press release].

U.S. House of Representatives. (1979). *Select committee on assassinations hearings.* 95th Congress, Second session.

U.S. House of Representatives. Select Committee to Investigate Covert Arms Transactions with Iran. (1987). *Report of the congressional committees investigating the Iran-Contra affair: With supplemental, minority, and additional views.* Washington, DC: U.S. Government Printing Office.

U.S. indicts Colombian drug cartel. (1989, May 23). *Erie Morning News,* p. 1A.

U.S. Surgeon General's Scientific Advisory Committee on Television and Social Behavior. (1972). *Television and growing up: The impact of televised violence.* Washington, DC: Government Printing Office.

Van den Berghe, P. (1974, December). Bringing beasts back in: Toward a biosocial theory of aggression. *American Sociological Review, 39,* 777–788.

Van den Haag, E. (1966, January). No excuse for crime. *Annals of the American Academy of Political and Social Sciences, 423,* 133–241.

Van den Haag, E., & Conrad, J. P. (1983). *The death penalty: A debate.* New York: Plenum.

Van Dijk, J. (2008). *The world of crime: Breaking the silence on problems of security, justice, and development across the world.* Thousand Oaks, CA: Sage Publications.

Van Dijk, J., & Kangaspunta, K. (2000, January). Comparing crime across countries. *National Institute of Justice Journal,* pp. 35–41.

Van Voorhis, P., Cullen, F. T., Mathers, R. A., & Garner, C. C. (1988). The impact of family structure and quality on delinquency: A comparative assessment of structural and functional factors. *Criminology, 26,* 235–261.

Vander, B. J., & Neff, R. J. (1986). *Incest as child abuse: Research and applications.* New York: Praeger.

Vetter, H. J., & Silverman, I. J. (1978) *The nature of crime.* Philadelphia: W. B. Saunders.

Vice President's Task Force. (1986). *Report on the vice president's task force on combating terrorism.* Washington, DC: Government Printing Office.

Vilhelm, A. (1952, November). White-collar crime and social structure. *American Journal of Sociology, 58*(3), 263.

Vise, D. (2002). *The Bureau and the mole.* New York: The Atlantic Monthly Press.

Viviano, F. (1997, May 28). Hong Kong Triad's new frontier. *San Francisco Chronicle,* p. A1.

Vold, G. B. (1958). *Theoretical criminology.* New York: Oxford University Press.

Vold, G. B. (1979). *Theoretical criminology* (3rd ed.). New York: Oxford University Press.

Vold, G. B., with T. J. Bernard (1979). *Theoretical criminology* (2nd ed.). New York: Oxford University Press.

Vold, G. B., with T. J. Bernard (1986). *Theoretical criminology* (3rd ed.). New York: Oxford University Press.

Vold, G. B., Bernard, T. J., & Snipes, J. (2002). *Theoretical criminology* (5th ed.). New York: Oxford University Press.

Volk, K. (1977). Criminological problems of white-collar crime. In *International summaries* (Vol. 4, pp. 13–21). Rockville, MD: National Criminal Justice Reference Service.

Von Hentig, H. (1948). *The criminal and his victim.* New Haven, CT: Yale University Press.

Voss, H. (1963, September). Ethnic differentials in delinquency in Honolulu. *Journal of Criminal Law, Criminology and Police Science, 54,* 322–327.

Wade, A. L. (1967). Social processes in the act of juvenile vandalism. In M. B. Clinard & R. Quinney (Eds.), *Criminal behavior systems: A typology* (pp. 94–109). New York: Holt, Rinehart and Winston.

Wade, N. (1976). IQ and heredity: Suspicion of fraud beclouds classic experiment. *Science, 194,* 916–919.

Wade, W. C. (1987). *The fiery cross: The Ku Klux Klan in America.* New York: Simon & Schuster.

Waldman, M., & Gilbert, P. (1989, March 12). RICO goes to Congress. *New York Times,* p. 4E.

Waldo, G., & Dinitz, S. (1967, July). Personality attributes of the criminal: An analysis of research studies, 1950–1965. *Journal of Research in Crime and Delinquency, 4,* 185–201.

Walker, S., Spohn, C., & DeLone, M. (1995). *The color of justice: Race and crime in America.* Belmont, CA: Wadsworth.

Wall, D. (2001). Cybercrime and the Internet. In D. Wall (Ed.), *Crime and the Internet.* London: Routledge.

Wall, D. (2007). *Cybercrime.* Manchester, UK: Replika Press.

Wallace, B. (1998, September 16). Credit and counterfeiting on the rise. *San Francisco Chronicle,* p. A1.

Wallerstein, J. S., & Wyle, C. J. (1947, April). Our law-abiding law breakers. *Probation,* 107–118.

Walsh, A. (2002). *Biosocial criminology 7: Introduction and integration.* Cincinnati, OH: Anderson.

Walsh, A. (2009). *Biology and crime: the biosocial synthesis.* New York: Routledge.

Walsh, A., & Beaver, K. M. (2008). *Biosocial criminology: A primer.* New York: Routledge.

Walsh, A., & Hemmens, B. (2008). *Introduction to criminology: A text/reader.* Thousand Oaks, CA: Sage.

Walsh, M. E. (1977). *The fence.* Westport, CT: Greenwood Press.

Walters, G. D., & White, T. W. (1989a, August). Heredity and crime: Bad genes or bad research? *Criminology, 27,* 455–486.

Walters, G. D., & White, T. (1989b). The thinking criminal: a cognitive mode of lifestyle criminality. *Criminal Justice Research Bulletin.* Sam Houston State University.

Warchol, G. (1998, July). *Workplace violence 1992–1996* (Bureau of Justice Statistics Special Report, NCJ 168634). Washington, DC: U.S. Department of Justice, Office of Justice Programs.

Warner, B. D., & Pierce, G. L. (1991, November). *Testing social disorganization theory using calls to police.* Paper presented at the American Society of Criminology Meeting, San Francisco.

Warr, M., & Stafford, M. (1991, November). The influence of delinquent peers: What they think and what they do. *Criminology, 29,* 851–866.

Warren Commission. (1964). *Report of the President's Commission on the Assassination of President Kennedy.* Washington, DC: Government Printing Office.

Wartzman, R. (1992, July 2). Counterfeit bills confound detectors at the Fed, sleuths at the Secret Service. *Wall Street Journal,* p. A8.

Weaver, W., Jr. (1988, December 11). Justice Department detects a gain in drive on steroid abuse. *New York Times,* p. 2A.

Webb, E. J., Campbell, D. T, & Schwartz, R. D. (1981). *Nonreactive measures in the social sciences.* New York: Houghton Mifflin.

Weber, M. (1949). *The methodology of social sciences* (E. A. Shils and H. A. Finch, Trans.). New York: Free Press.

Webster, B., & McCampbell, M. S. (1992, September). International money laundering: Research and investigation join forces. *Research in Brief.* Washington, DC: U.S. Department of Justice, National Institute of Justice.

Wedel, J. R. (1999, September 14). Harvard's complicity in laundering billions in Western aid for Russia. *Erie Times News,* p. 11A.

Weinstein, A. K. (1988). Prosecuting attorneys for money laundering: A new and questionable weapon in the war on crime. *Crime and Contemporary Problems, 51,* 369–386.

Weisberg, D. K. (1985). *Children of the night: A study of adolescent prostitution.* Lexington, MA: Lexington Books.

Weisel, D. T. (2004, August). *Problem-oriented guides for police* (Problem-specific guide series). Center for Problem-Oriented Policing, Guide 9.

Weiser, B. (1999, October 28). U.S. charges 8 inspectors in kickback scheme at Bronx produce market. *New York Times,* p. A21.

Wellford, C. (1975, February). Labelling theory and criminology: An assessment. *Social Problems, 22,* 332–345.

Wellford, H. (1972). *Sowing the wind: A report from Ralph Nader's Center for Study of Responsive Law.* New York: Grossman.

Wells Fargo guard dopes boss, robs armored car. (1983, September 14). *Erie Morning News,* p. 5A.

Wertham, F. (1954). *Seduction of the innocent.* New York: Rinehart.

Werthman, C. (1967). The function of social definitions in the development of delinquency career. *Task force report: Juvenile delinquency and youth crime.* The President's Commission on Law Enforcement and Administration of Justice. Washington, DC: Government Printing Office.

West, D. J. (1988). Homosexuality and social policy: The case for a more informal approach. *Law and Contemporary Problems, 51,* 181–199.

West, D. J., & Farrington, D. P. (1977). *The delinquent way of life.* London: Heinemann.

West, N. (1982). *The circus: MI5 operations, 1945–1972.* New York: Stein and Day.

Westin, A. L. (1981). *Whistle-blowing! Loyalty and dissent in the corporation.* New York: McGraw-Hill.

Wheeler, D. L. (1995, October 6). Protesters disrupt meeting on possible genetic basis of criminal behavior. *Chronicle of Higher Education,* p. 1.

White, J. (1990, March). Neighborhood permeability and burglary rates. *Justice Quarterly, 7,* 59–67.

Whitaker, M., et al. (1983, February 14). The forgotten: World's prisoners of conscience. *Newsweek,* pp. 40–55.

White, J. (1990, March). Neighborhood permeability and burglary rates. *Justice Quarterly, 7,* 59–67.

White, J. L., et al. (1991, November). *Preliminary results from a study of the relationship between impulsivity and delinquency.* Paper presented at the meeting of the American Society of Criminology, San Francisco.

White, J. R. (2005). *Terrorism: An introduction* (4th ed.). Upper Saddle River, NJ: Prentice Hall.

White-collar crime: Second annual survey of laws. (1981). *American Criminal Law Review, 19,* 173–520.

Whitelaw, K. (1999, March 22). Your money or your life. *U.S. News and World Report,* pp. 34–41.

Whitlock, C. (2007, March 12). Terrorists proving harder to profile. *Washington Post,* p. A1.

Whitman, D. (1987, April 27). The numbers game: When more is less. *U.S. News and World Report,* pp. 39–40.

Whyte, W. F. (1955). *Streetcorner society.* Chicago: University of Chicago Press. (Original work published 1943)

Widom, C. S. (1989, May). Child abuse, neglect and violent criminal behavior. *Criminology, 27,* 251–271.

Widom, C. S. (1992, October). The cycle of violence. *Research in Brief.* Washington, DC: U.S. Department of Justice, National Institute of Justice.

Wiggins, P. (1986a, March). *An extremist right-wing group and domestic terrorism.* Paper presented at the Academy of Criminal Justice Sciences Meeting, St. Louis, MO.

Wiggins, P. (1986b, March). *The Turner diaries: Blueprint for right-wing extremist violence.* Paper presented at the meeting of the Academy of Criminal Justice Sciences, St. Louis, MO.

Wilbanks, W. (1987). *The myth of a racist criminal justice system.* Monterey, CA: Brooks/Cole.

Wilbanks, W., & Kim, K. H. (1984). *Elderly criminals.* Lanham, MD: University Press of America.

Wilkinson, P. (1976). *Political terrorism.* New York: Wiley.

Williams, F., & McShane, M. (1988). *Criminological theory.* Englewood Cliffs, NJ: Prentice Hall.

Williams, F., & McShane, M. (1994). *Criminological theory* (2nd ed.). Englewood Cliffs, NJ: Prentice Hall.

Williams, J. (1991, October 13). Japan: The price of safe streets. *Washington Post,* pp. C1, C4.

Williams, M. (1991, September 30–October 6). Getting a grip (or losing it) on Iran-Contra. *Washington Post National Weekly Edition,* pp. 11–12.

Williams, P. (1997, Spring). Money laundering. *Criminal Organizations, 10,* 18–27.

Wirth, L. (1938, July). Urbanism as a way of life. *American Journal of Sociology, 44,* 8–20.

Wilson, E. O. (1975). *Sociobiology.* Cambridge, MA: Harvard University Press

Wilson, J. Q. (Ed.). (1983a). *Crime and public policy.* San Francisco: Institute for Contemporary Studies.

Wilson, J. Q. (1983b). *Thinking about crime* (revised edition). New York: Basic Books.

Wilson, J. Q., & Herrnstein, R. J. (1985). *Crime and human nature.* New York: Simon & Schuster.

Wilson, J. Q., & Kelling, G. L. (1982, March). Broken windows: The police and neighborhood safety. *Atlantic Monthly,* pp. 27–28.

Wilson, J. Q., & Lowry, G. C. (Eds.). (1987). *From children to citizens: Vol. 3, families, schools and delinquency prevention.* New York: Springer-Verlag.

Wilson, R. (1978, July). Chinatown: No longer a cozy assignment. *Police Magazine, 1,* 19–29.

Wilson, W. J. (1987). *The truly disadvantaged: The inner city, the underclass, and public policy.* Chicago: University of Chicago Press.

Windrem, R. (2000, April 19). *U.S. spying paying off for business.* Retrieved July 1, 2008, from http://www.msnbc.com

Wines, M. (1990, December 21). Ex-C.I.A. official says U.S. ignores Syrian terror. *New York Times,* p. A7.

Winick, C. (1962). Prostitutes' clients perceptions of prostitutes and of themselves. *International Journal of Psychiatry, 8,* 289.

Winick, C., & Kinsie, P. M. (1971). *The lively commerce: Prostitution in the United States.* Chicago: Quadrangle Books.

Winslow, R. (1970). *Society in transition: A social approach to deviancy.* New York: Free Press.

Witkin, G. (1988, November 23). Making war on handguns. *U.S. News and World Report,* p. 28.

Witkin, H. A., Mednick, S. A., Schulsinger, F., Bakkestrom, E., Christiansen, K. O., & Goodenough, D. R. (1976, August 13). Criminality in XYY and XXY men. *Science, 193,* 547–555.

Wolf, G. (1975). *Frank Costello: Prime minister of the underworld.* New York: Bantam.

Wolf, J. B. (1981). Enforcement terrorism. *Police Studies, 3,* 45–54.

Wolf, R. (2007, November 14). Probe uncovers 30,000 Medicaid providers cheating IRS. *USA Today,* p. 6A.

Wolfgang, M. E. (1958). *Patterns in criminal homicide.* Philadelphia: University of Pennsylvania Press.

Wolfgang, M. E. (1960). Cesare Lombroso. In H. Mannheim (Ed.), *Pioneers in criminology* (pp. 168–227). Chicago: Quadrangle Books.

Wolfgang, M. E. (1980, March 2). Crime and punishment. *New York Times,* p. E21.

Wolfgang, M. E. (1987). *From boy to man, from delinquency to crime.* Chicago: University of Chicago Press.

Wolfgang, M. E., & Ferracuti, F. (1967). *The subculture of violence: Towards an integrated theory in criminology.* London: Tavistock.

Wolfgang, M. E., Figlio, R. M., & Sellin, T. (1978). *Delinquency in a birth cohort.* Chicago: University of Chicago Press.

Wolfgang, M. E., Figlio, R. M., & Thornberry, T. B. (1978). *Evaluating criminology.* New York: Elsevier.

Wright, J. D., & Rossi, P. H. (1986). *Armed and considered dangerous: A survey of felons and their firearms.* Hawthorne, NY: Aldine.

Wright, K. N., & Wright, K. E. (1995, August). *Family life, delinquency, and crime: A policymaker's guide: Research summary.* Washington, DC: Office of Juvenile Justice and Delinquency Prevention.

Wright, R., & Decker, S. (1996). *Burglars on the job.* Boston: Northeastern University Press.

Wright, R., & Decker, S. (1997). *Armed robbers in action: Stickups and street culture.* Boston: Northeastern University Press.

WuDunn, S. (1996, February 14). Uproar over a debt crisis. *New York Times,* p. C1.

Wunderlich, R. (1978). Neuroallergy as a contributing factor to social misfits. In L. Hippchen (Ed.), *Ecologic-biochemical approaches to treatment of delinquents and criminals* (pp. 229–253). New York:Van Nostrand Reinhold.

Yablonsky, L. (1965). *Synanon: The tunnel back.* Baltimore, MD: Penguin.

Yablonsky, L. (1967). *The violent gang.* New York: Penguin.

Yin, T. (1992, June 12). Sears is accused of billing fraud at auto centers. *Wall Street Journal,* p. B1.

Yochelson, S., & Samenow, S. E. (1976). *The criminal personality* (Vols. 1 and 2). New York: Jason Aronson.

Young, V., & Sulton, A. T. (1991). Excluded: The current status of African-American scholars in the field of criminology and criminal justice. *Journal of Research in Crime and Delinquency, 28*(1), 101–116.

Zalba, S. R. (1971, July). Battered children. *Transaction, 8,* 58–61.

Zeisel, H. (1957). *Say it with figures* (4th ed.). New York: Harper.

Zeitz, D. (1991). *Women who embezzle or defraud: A study of convicted felons.* New York: Praeger.

Zimbardo, P. (1972). Pathology of imprisonment, *Society, 9,* 4–6.

Zimbardo, P. (1973, March). On the ethics of intervention in human psychological research: With special reference to the Stanford prison study. *Cognition, 22,* 243–246.

Zimbardo, P. (1974). The psychology of imprisonment: Privation, power and pathology. In Z. Rubin (Ed.), *Doing unto others* (pp. 61–73). Englewood Cliffs, NJ: Prentice Hall.

Zimbardo, P. (2007a). *The Lucifer effect: Understanding how good people turn evil.* New York: Random House.

Zimbardo, P. (2007b, March 30). Revisiting the Stanford prison experiment: A lesson in the power of persuasion. *Chronicle Review,* pp. B6–B7.

Zimring, F. E., & Hawkins, G. (1992). The citizen's guide to gun control. New York: Macmillan.

Photo Credits

Chapter 1

Chapter 1 opener: Jupiter Images.
1.1: Hulton Archive/Getty Images.
1.2: Jupiter Images.
1.3: Used with permission of Criminals Hall of Fame, Niagara Falls, Canada.
1.4: Robert Nickelsberg/Getty Images News/Getty Images.
1.5: Mike Nelson/AFP/Getty Images.

Chapter 2

Chapter 2 opener: iStockphoto.com.
2.1: Associated Press/Paul Sakuma.
2.2: Associated Press.
2.3: Robert Nickelsberg/Getty Images News/Getty Images.
2.4: Associated Press/Craig Schreiner.
2.5: Jupiter Images.

Chapter 3

Chapter 3 opener: iStockphoto.com/Andrey Prokhorov.
3.1: David Einsel/Getty Images News/Getty Images.
3.2: Popperfoto/Getty Images.
3.3: Tim Zielenbach/AFP/Getty Images.
3.4: John Moore/Getty Images News/Getty Images.
3.5: Jupiter Images.
3.6: Janette Beckman/Premium Archive/Getty Images.

Chapter 4

Chapter 4 opener: Kean Collection/Hulton Archive/Getty Images.
4.1: Bettmann/Corbis.
4.2: Bettmann/Corbis.
4.3: Jupiter Images.
4.4: Library of Congress.
4.5: Jupiter Images.

Chapter 5

Chapter 5 opener: Hans Casparius/Hulton Archive/Getty Images.
5.1: Gramercy Pictures/Photofest.
5.2: Bettmann/Corbis.
5.3: Bettmann/Corbis.
5.4: Bettmann/Corbis.
5.5: John Storey/San Francisco Chronicle/Corbis.

Chapter 6

Chapter 6 opener: © Bettmann/CORBIS.
6.1: Mark Ralston/AFP/Getty Images.
6.2: Toby Canham/Getty Images Entertainment/Getty Images.
6.3: Bettmann/Corbis.
6.4: Andrew Lichtenstein/Corbis.
6.5: Steve Starr/Corbis.

Chapter 7

Chapter 7 opener: iStockphoto.com/Andrejs Zemdega.
7.1: © iStockphoto.com/Ernst Daniel Scheffler.
7.2: Herbert Orth/Time & Life Pictures/Getty Images.
7.3: AP Photo/Rogelio Solis.
7.4: AP Photo/ Dee Marvin.
7.5: © iStockphoto.com/Jerry Koch.
7.6: Don Emmert/AFP/Getty Images.

Chapter 8

Chapter 8 opener: iStockphoto.com/Nuno Silva.
8.1: William F. Campbell/Time & Life Pictures/Getty Images.
8.2: Michael Ochs Archives/Getty Images.
8.3: © HO/Reuters/Corbis.
8.4a: AP Photo/Steve Helber.
8.4b: AP Photo/Mike Morones.
8.4c: AP Photo/Jeff Tuttle.
8.4d: AP Photo/Denis Paquin.
8.5: © Ted Soqui/Sygma/Corbis.

Chapter 9

Chapter 9 opener: Jupiter Images.
9.1: Steve Grayson/WireImage/Getty Images.
9.2: Jupiter Images.
9.3: Tim Chapman/Getty Images News/Getty Images.
9.4: © Bettmann/CORBIS.
9.5: AP Photo/David Crews.

Chapter 10

Chapter 10 opener: iStockphoto.com/stephanie phillips.
10.1: Time & Life Pictures/Getty Images.
10.2: Stuart Ramson/Getty Images News/Getty Images.
10.3: Rolls Press/Popperfoto/Getty Images.
10.4: Associated Press.
10.5: Murray Garrett/Hulton Archive/Getty Images.

Chapter 11

Chapter 11 opener: iStockphoto.com.
11.1: Spencer Platt/Getty Images News/Getty Images.
11.2: Popperfoto/Getty Images.
11.3: Paul Schutzer/Time & Life Pictures/Getty Images.
11.4: FBI/Getty Images News/Getty Images.
11.5: Pictorial Parade/Hulton Archive/Getty Images.
11.6: Bob Daemmrich/AFP/Getty Images.

Chapter 12

Chapter 12 opener: iStockphoto.com/kostas koutsoukos.
12.1: Washington Bureau/Hulton Archive/Getty Images.
12.2: Chris Jackson/Getty Images News/Getty Images.
12.3: Jupiter Images.
12.4: George Tames/Hulton Archive/Getty Images.
12.5: Katherine Young/Hulton Archive/Getty Images.

Chapter 13

Chapter 13 opener: iStockphoto.com/Gord Horne.
13.1: iStockphoto.com/Nuno Silva.
13.2: Timothy A. Clary/AFP/Getty Images.
13.3: Alex Wong/Getty Images News/Getty Images.
13.4: AP Photo/Keith Nordstrom.
13.5: Ron Kuntz/AFP/Getty Images.

Chapter 14

Chapter 14 opener: Source: Jupiter Images.
14.1: iStockphoto.com/Pali Rao.
14.2: Karen Bleier/AFP/Getty Images.
14.3: Dan Callister/Getty Images Entertainment/Getty Images.
14.4: AP Photo/Lisa Berg.

Chapter 15

Chapter 15 opener: iStockphoto.com/savas keskiner.
15.1: Robin Utrecht/AFP/Getty Images.
15.2: Source: Jupiter Images.
15.3: iStockphoto.com/Marie Fields.

Index

ABSCAM (Arab Scam or Abdul Scam), 310–311
Abuse excuse, 136–138
Abused child syndrome, 235–236
Acquaintance rape, 130
Act-uppers, 359
Addict robber, 231–232
Addiction, 364–365
Administrative law, 12
Adoption studies, 124–125
Affinity group fraud, 274
African Americans, as serial murderers, 211–212
Age and crime, 65–68
    recidivism, 66
Agent provocateurs, 349
Agnew, Robert, 151–152
Akers, Ronald, 160
Al Qaeda, 345, 372–373
Alcohol:
    role in crime, 443–444
Alcohol addiction, 443–444
Alcoholic robbers, 231–234
Alienated/egocentric spy, 362
Alternative data gathering strategies, 34
Ames, Aldrich, 362–363
Amir v. Brownmiller, 229
Androcentric bias, 70
Anger rape, 229–230
Anomie, 146–147
    See also Anomie theory
Anomie theory, 149–153
    Agnew's general strain theory, 150–153
    Cloward and Ohlin's differential opportunity
        theory, 152–153
    Cohen's lower class reaction theory, 151–152
    delinquent subcultures, 152–153
    Durkheim and anomie, 147
    institutional, 151
    Merton's, 148–150
    modes of personality adaptation in, 148–149
    subcultural theory, 151–152
Antiglobalization Movement, 352–353
Antinuclear movement, 352–363
Antisocial personality, 138
Antisocial potential (AP), 168–169
Antivivisectionists, 352–353
Apalachin meetings, 413
Armed robbery, 231–234
Arsonist, 259–260
    definition of, 259
    excitement, 259–260
    professional, 259–260
    profit-motivated, 259–260
    revenge, 259–260

Art forgery, 275–276
Art theft, 275–276
Assassins:
    atypical, 361
    Clarke's typology of, 360
    definition of, 360
    in twentieth-century U.S., 360–361
    political assassins, 360–361
Assault, 216–218
    aggravated, 217
    definition of, 216–217
Astrology, 107
Atavism, 117–119
Atypical assassins, 361
Auto theft, 255
Auto theft rings, 255
Avocational crime, 264

Badger game, 267
Bait and switch, 314
Bandura, Albert, 133
Bank examiner's scam, 267
Bank Secrecy Act, 418
Basque ETA, 366
Battered child syndrome, 235
Battery (aggravated assault), 216–218
BCCI (Bank of Credit and Commerce
        International), 395
Beccaria, Cesare, 3, 97–99
Behavioral genetics, 128
Behavioral modification, 133
Bell curve, 135
Bentham, Jeremy, 99–100
Big cons, 269–277
Bimodal theory of distribution of crime
        commission, 70–71
Bin Laden, Usama, 365–366
Biological determinism, 117–122
Biological theories, 115–130
    early, 115–118
    neobiological theories, 122–130
Black organized crime, 393–394
Body types, 120–121
Bonger, Willem, 109–110
Boojo, 268
Boosters, 250–251
Bourgeoisie, 108–109
Box man (safecracker), 264–265
Brain disorders, 123–124
Brainwashing, 353–354
Brainwashing myth, 353–354
Braithwaite, John, 9, 180–181
Brawner test, 136

Broken windows, 425
Bribe payers' index, 317
BTK killer, 214–215
Buccaneer/sport spy, 362
Burgess, Ernest, 154–155
Burglar, professional, 256–258
Burglary:
    breaking or entering, 256
    definition of, 256
    dip in, 259
    fencing operations, 257–258
    rates of, 255
    types of burglars, 256
Buscetta, Tommaso, 415
Bush, George W., 357–358

Cable piracy, 283
Campbell Collaboration, 37–38
Campus gang rape, 227–228
Cannons (pickpockets), 264–265
Car cloning, 285–286
Career criminals, 240–242
    definition of, 240–242
    intensives, 240–242
    intermittents, 240–242
    typology of, 242
    See also Criminal careers
Carjacking, 230
Castellammarese Wars, 412
Catharsis hypothesis, 83–84
Chambliss, William, 182–183,191–192
Character assassination, 349
Character contests, 240–241
Characteristics of criminal law, 11
Cheater theory, 127
Check forgery, 253–255
Chicago school, 154–155
social process theory, 153–163
Child abuse, 235–236
Child molesters, 436
Chinese triad societies, 3389–391
Churning, 314–315
CIA-Mafia link, 409
Clarke's typology of assassins, 359–360
Classical theory, 92–104
Cleary law, 130
Cleveland, Grover, 357–358
Clinton, Bill, 357–358
Closure, 311–313
Cloward, Richard, 151–152
Cocaine, 397
Coded-Pencil-Caper, 314
Coerced crime, 310
Cohen, Albert, 151–152
COINTELPRO, 353
Colombian cartels, 397
Colombia, kidnappings in, 2
Commission Trials, 414
Compromised spy, 362–363
Computer crime:
    argot of, 456–460
    cyberspace security, 3458
    cyberterrorism, 383–390
    online predators, 381

Operation Bot Roast, 464
    protecting children in cyberspace, 460–463
    public and legal reaction, 464–465
    types of attacks, 454–456,458–460
Computer virus, 456–458
Concentric zone theory, 154–156
Confidence game (con), 267–270
Conflict criminology, 181–183
    Chambliss and Quinney on, 182
    Du Bois on, 182–183
    Reiman on, 183
    Turk on, 181–182
Conflict model of law, 8
Conflict vs. Marxist criminology, 192
Conklin's typology of robbers, 233–234
Consensus model of law, 8
Consigliere, 399–403
Containment theory, 164–165
Continuum model of organized crime, 301
Convention against Transnational Organized Crime
    (TOC), 387–388
Convention on Genocide, 350–352
Conventional criminals, 255–260
Conventional property crime, 255–260
    burglary, 256–257
    fencing operations, 257–258
    international burglary rates, 257
    larceny/theft, 258–260
    stings, 258–260
Conventional property offender, 255–260
    occasional property offender vs., 255
    societal reaction to, 261–262
Convictional criminals, 344
Copycat crime, 84–85
Corporate crime, 316–334
    auto industry, 321–322
    corporate concentration, 334
    corporate fraud, 317–318
    corporate violence, 328–329
    criminal careers of organizational offenders, 347–348
    Donora Fluoride Death Fog, 326–327
    Edelhertz's typology, 305–306
    environmental crime, 325–328
    Ford Explorer-Firestone tire recall, 323–324
    Ford Pinto case, 322–323
    Great Electrical Industry Conspiracy, 319
    history of, 299–302
    industrial espionage, 321–323
    insider training, 314
    internet pirates, 333
    Karen Silkwood case, 330–331
    leniency in punishment, 336–337
    measurement/cost, 298–299
    medical and insurance industries, 312–314
    military industries, 331
    multinational bribery, 317
    oil pricing, 316–317
    price-fixing, 319–320
    radiation leaks, 327–328
    rationalizations, 334
    sale of unsafe products, 3211–322
    savings and loan scandal, 322–323
    societal reaction, 335–336
    stockbrokers, 314–315

Three Mile Island, 325
toxic criminals, 325
U.S. vs. Allied Chemical, 325–326
Warez groups, 333–334
wartime trade violations, 331–332
See also Occupational crime; White collar crime
Corporate dumping, 327
Corporate environment and crime, 335
Corporate fraud, 339
Corporate violence, 328
Corruption:
    police, 308–309
    private, 311
    public, 306–311
Corruption Perception Index, 307
Cosa Nostra:
    Cressey model of, 399–400
    definition of, 399
    internal structure of, 399–402
Cost of crime, 14–15
Costello, Frank, 412–413
Counterfeiting, 281–282
Crack cocaine, 397
Crashers/cappers, 275
Credit doctors, 281
Credit Mobilier scandal, 357–358
Cressey, Donald, 399–400
Crime against government, 358–374
Crime against humanity, 349–351
Crime by government, 348–358
Crime by police, 348–354
Crime dip, 32–33
Crime index, 27–28
Crime profiling, 131–132
Crime/criminals, general characteristics of, 28–53, 65–85
    age and crime, 65–68
    copycat crimes, 83–84
    cost of, 14–15
    economy and crime, 82–83
    education and crime, 80
    family and crime, 78–80
    functional necessity of, 146–147
    gender differences in criminality, 68–70
    institutions and crime, 78–85
    international variations in crime, 56–58
    mass media and crime, 83–85
    minorities and crime, 73–76
    problem of crime, 14
    race and crime, 71–73
    regional variation in crime, 77
    religion and crime, 81–82
    social class and crime, 70–71
    trends in crime, 60–61
    trends in crime, against persons, 64–65
    trends in crime, against property, 64–65
    urban/rural differences, 77
    war and crime, 82
    See also individual type of crime
Crimes of the Century, 9
Crimes without victims, 425
Criminal behavior systems, 198–199
Criminal personality, 133–134

Criminal saturation, law of, 118–119
Criminal typologies, 195–196
    Abrahamsens's types of criminals, 197
    criminal behavior systems, 198–199
    Glaser's types of crime, 197
    Lombroso's types of criminals, 197
Criminalistics, 2
Criminaloid, 296
Criminology, introduction to, 1–19
    characteristics of criminal law, 11–12
    consensus vs. conflict model of law, 8
    cost of crime, 14–15
    crime and criminal law, 11
    crime and deviance, 3–4
    crimes of twentieth century, 9
    criminology, 2–5
    criteria of criminal law, 11
    defining crime, 11–12
    defining criminology, 1–2
    emergence of criminology, 3–4
    fads and fashions in crime, 2
    felonies, 13
    mala in se, 6
    mala prohibita, 6
    misdemeanors, 13
    overcriminalization, 7
    scientific research in, 5
    social change and, 7
    sociological definitions of crime, 7
    Sumner's norm types, 6
    undercriminalization, 7
Critical criminology, 176–177
Critical feminism, 183–186
CSA (Covenant Sword and Arm of Lord), 369–372
Cult slaying, 211
Cultural insanity defense, 136–138
Cultural values, 7
Culturally violent offender, 238–239
Culture of violence, 238–239
Cutpurses, 264–265

Dahrendorf, Ralf, 176–177
Dark figure of crime, 38
Darwin, Charles, 108, 117
Date rape, 227–228
D.C. snipers, 214
Deceived (false flag recruit) spy, 362–363
Decriminalization, 446
Delinquency and drift, 162
Delinquent subculture, 152–153
Demonological theory, 96–97
Designing out crime, 102
Desistance, 168–170
Developmental/life course (DLC) theory, 167–169
    assumptions of, 167
    crime vulnerability assessment, 168
    Farrington's antisocial potential (AP) theory, 168
    Sampson/Laub's life course criminality, 168–169
Deviance:
    definition of, 3–7
    primary, 178–179
    secondary, 178–179
Deviant places, 157–158
Differential association theory, 157–160

contacts in, 159
critique of, 160
propositions of, 159
Differential opportunity theory, 152–153
Dizygotic twins, 123–124
Domestic violence, 234–235
child abuse, 235
elder abuse, 238
kidnapping, 235
spouse abuse, 236–237
Donora Fluoride Death Fog, 323–324
Dowry deaths, 350–351
Dramatization of evil, 178–179
Drug abuse, 438–442
crime and, 441–442
drug dip, 441
history of, 439
moral entrepreneurs, 178–179
Drug abuse violations, 441–442
Drugs, 438–439
cocaine, 438–439
crack cocaine, 438–439
marijuana, 438–439
morphine, 438–439
types of, 438–439
Drunkenness, 443
Du Bois, W. E. B., 182–183
Dualistic fallacy, 122–123
Durham decision, 136
Durkheim, Émile, 146–147

Early/classical criminological theory, 92–110
astrology, 96–97
Beccaria and, 97–100
Bentham and, 99–100
Bonger and, 109–110
classical theory, 97–98
demonological theory, 96–97
ecological theory, 104–105
economic theory, 108–110
forerunners of modern criminological thought,
108–109
Guerry and, 105–106
Marx and, 108–109
neoclassical theory, 100–101
other geographical theory, 107–108
palmistry, 116–117
phrenology, 116–117
physiognomy, 116–117
Quetelet and, 105–106
Echelon, 332
Ecological fallacy, 155
Ecological theory, 104–108
Ecology, 104
Economic Espionage Act, 334
Economic theory, 108–110
Economy and crime, 82–83
Ectomorphs, 120–121
Edelhertz's typology, 266–267
Education and crime, 80–81
Egocentric assassins, 360–362
Elder abuse, 236–237
Electrical industry conspiracy, 339–340
Elliot, Delbert, 194–195

E-mail bombs, 456–458, 281–282
Embezzlement, 311–313
Endocrine imbalance, 125–126
Endomorph, 120–121
Enterprise, 354–355
Environmental crime, 322–323
Equity funding scandal, 318–319
Error:
in self report surveys, 41–42
in victim surveys, 40
post hoc, 130–131
Escapee spy, 362–363
Espionage, 361–364
definition of, 361
white vs. black, 361–362
Ethics in research, 22–24
Ethnic succession theory, 393–394
Ethnicity, and organized crime, 393–394
Eugenics, 120
Evidence-based research, 37–38
Evolutionary psychology, 128
Excitement arson, 260
Exhibitionism, 434–436
Experiments, 34–38
Candid Camera, 35
classic design, 35
control group for, 35
definition of, 35–36
Scared Straight, 35
Eysenck, Hans, 131–132

Fads in crime, 2–3
Fallacy of autonomy, 79
FALN (Fuerzas Armadas de Liberacion Nacional),
370–371
False flag recruits, 361–363
Family and crime, 78–80
Family violence. *See* Domestic violence
Farrington, David, 168–169
FBI's Ten Most Wanted Fugitives, 4–6
Federal Bureau of Investigation (FBI):
behavior control experiments, 353–354
COINTELPRO, 353–354
Fee splitting, 313–314
Feeblemindedness, 118–120
Felony, 13
Felony-murder doctrine, definition of, 217
Female genital mutilation (FGM), 350–351
Female infanticide, 350–351
Feminist criminology theory, 183–185
generalizability problem and, 185
liberal feminism, 184–185
radical feminism, 184–185
victimization issues, 184–185
Feminization of poverty, 79–80
Fence, 257–258
Ferri, Enrico, 118–119
Fetishism, 431–436
FINCEN (Financial Crimes Enforcement Network),
395–396
Financial crime, 313–314, 320–321
Flashers, 434–436
FLN, 365–368
Floodgate theory, 425

Focal concerns theory, 160–161
Folk crime, 424
Folkways, 6
Forcible rape, 225–230
Ford Explorer-Firestone tire recall, 323–324
Ford Pinto scandal, 322–323
Foreign Corrupt Practices Act, 317
Forerunners of Criminological Thought, 108–110
Forgery, 59, 253–254
   juveniles and, 253–254
   naïve, 253–254
   systematic, 281–282
Freud, Sigmund, 130–131
Freudian theory, 130–131
Functional necessity of crime, 7–8
Funeral directors, 314–315
Future of crime, 471–479
   anticipating the, 477–479
   British Home Office predictions, 474–475
   crimewarps, 473
   demand for stolen goods, 477–479
   future of digital crime, 473
   hot products, 477–478
   predicting, 471–473

Gambino, Carlo, 413–414
Gambling, 404–405
   organized crime and, 404–405
Gang graffiti, 386–388
Gang rape, 228–229
Gangs, 386–388
Garofalo, Raffaelo, 118–119
Gemeinschaft, 7
Gender and crime, 68–70
   See also Gender differences, in criminology
Gender differences, in criminology, 183–185
General theory of crime, 1666–167
Genocide, 350–351
Genovese, Vito, 412–413
Gesellschaft, 7
Global fallacy, 199
Goddard, Henry, 119–120
Goring, Charles, 119
Gottfredson, Michael, 166–167
Gotti, John, 415
Graffiti, 252
Grant, Ulysses S., 357–358
Great Electrical Industry Conspiracy, 399–340
Grifters, 268
Group conformer rapist, 230
Guerry, A. M., 105–107
Guilty, but mentally ill, 137
Gun control, 223–224
Guns, 223–224

Hackers, 458–460
Hackers' ethic, 458–460
Hagan, John, 166–167
Hamas, 368–369
Hanging paper, 282–284
Hanssen, Robert, 363–364
Harding, Warren, 358–358
Harrison Act of 1914, 440
Hate crime, 359

Health fraud, 313
Hedonism, 99–100
Heel, 276–278
Heroin, 438–439
Hezbollah, 368–369
Higher immorality, 201
Hinckley case, 136–138
Hirschi, Travis, 165–167
Hobbs Act, 418
Hobson, Richmond, 439–440
Homicide. *See* Murder
Homosexual behavior, 431
Honor killing, 350–351
Hookers, 430
Hooligans, 252
Hooton, Ernest, 120
Hotel prowling, 264
Human ecology, 154–155
Human rights, 349–351
Human trafficking, 349–350

IAS (Italian-American Syndicate), 402–403
Identity theft, 270–271
Identity theology, 365–366
Ideological spy, 362–363
Ideology, 345
Incest, 433–435
Index crimes, 27–28
Industrial espionage, 332
In-group oriented organized crime, 381–383
Insane assassins, 361
Insanity defense, 136–138
Insider trading, 225
Institutional anomie theory, 151
Institutions and crime, 78–85
Integrated theory, 193–195
   Elliot's integrative theory of juvenile delinquency, 194–195
   Thornberry's interactional theory of delinquency, 195
Intellectual property theft, 382–384
Intelligence and crime, 134
Intensive career criminal, 240–242
Interactional theory, 195
Interactionist model of law, 8
Intergenerational criminals, 322–323
Internal Security Act, 342–343
International crime statistics, 56–58
   auto theft, 257(table), 57–58
   rape, 57–58
International human trafficking, 349–350
International organized crime. *See* Organized crime, international
International public corruption, 306–311
International variations in crime, 56–59
International violent crime, 56–59
Inventory shrinkage, 248–249
Investigator scam, 267
IQ and crime, 134–135
Iran-Contra conspiracy, 354–355
Islamic Jihad, 368–369

Johns, 430
Johnson, Lyndon, 357–358

Joyriding, 253
Judgescam, 309
Jukes family, 119–120
Junkie burglars, 256
Justice, community/restorative, 187–189
Justifications for punishment, 103–104
Juvenile subcultures:
    conflict subculture, 151–152
    criminal subculture, 151–152
    retreatist subculture, 151–152

Kallikaks, 119–120
Karen Silkwood case, 330
Kefauver Commission, 414
Kennedy, John F., 357–358
Kerner (Riot) Commission, 73–76
Kickbacks, 306–311
King, Martin Luther, Jr. 353
Kleptomania, 249
Knowledge. *See* Progression of knowledge
Ku Klux Klan, 370–371

Labeling theory, 177–181
    assumptions of, 177–178
    Braithwaite's shaming theory, 180–181
    critique of, 181
Labor racketeering, 406–407
Laissez-faire, 335–336
Larceny/theft, 255–259
    definition of, 255
Latent functions, 7, 464
Laub, John, 168–169
Law:
    conflict model of, 8–9
    consensus model of, 8–9
    social change and, 7
    statutory, 7
Law of criminal saturation, 118–119
Left realism, 185
Legend of Sicilian Vespers, 400
Lemert's secondary deviance, 178–179
"Letter from a Birmingham Jail," 358
Lex talionis, 103–104
Liberation hypothesis, 69–70, 73
Life course criminality, 168–170
Life course/developmental theory, 168–170
    definition of, 168
    See also Developmental/life course
        (DLC) theory
Loan sharking, 414
Lobotomy, 123
Lombroso, Cesare, 117–119
Lower class reaction theory, 151–152
Luciano, Charles "Lucky," 412–413
Lunar cycles, 106

Macheteros, 369
Machismo, 240–241
Madoff, Bernie, 273–274
Mafia:
    origin of, 400
Mainstream vs. critical criminology, 175–176
Mala in se, 61
Mala prohibita, 61

Malice aforethought, 211–213, 217
Manchurian Candidate, 353–355
Manifest functions, 7, 464
Manslaughter, 211–213, 217
Manson, Charles, 209
Mapping intel sources, 466–467
Marijuana, 438–439
Marital rape, 130
Marx, Karl, 108–109
Marxist criminology, 108–109
Masochism, 435
Mass media and crime, 83–84
    copycat crimes, 83–84
    movie violence, 83–84
    print violence, 83–84
    television violence, 83–84
Mass murder, 211–213
Massage parlors, 430
Master fence, 257–258
Matza, David, 162–163
Maurer's big con, 269
Mazzini Organization, 400
McCarran-Walter Act, 344
McClelland Commission, 414
McKay, Henry, 154–155
M'Naghten rule, 136
Meaning stretcher rapist, 129–130
Medellin cartel, 397
Medicare fraud, 313
Megan's law, 229
Mens rea, 121, 211–213
Mercenary (predatory) organized crime, 381–383
Mercenary spy, 362–363
Merton, Robert, 146–147
Merton's anomie theory, 148–152
Mesomorph, 120–121
Messner, Steven, 151
Methodological narcissism, 34, 47–49
MICE (SMICE) strategy, 362
Milken, Michael, 322–323
Miller, Walter, 160
Miller's focal concerns theory, 160
Minority status and crime, 73–76
Misdemeanor, 13
MKULTRA, 353
Model of professional crime, 265–266
Modes of personality adaptation, 148–150
    conformist, 149
    definition of, 148
    innovator, 149
    rebel, 149
    retreatist, 149
    ritualist, 149
Mollen Commission, 308–309
Money laundering, 395–396
Moniz, Antonio, 123
Monozygotic concordance, 123–124
Moral entrepreneurs, 178–179, 439–440
Moral panics, 440
Mores, 6
Motor vehicle theft, 253–254
Moustache Petes, 412–413
Muggers, 216–218
Multifactor approach, 192

Multinational bribery, 317
Murder (homicide):
    first degree, 211–213
    mass, 211–213
    second degree, 211–213
    serial, 211–212
    spree, 213–214
    types of, 210–211
Murph the Surf, 263
Music and crime, 83–84

Nacirema, 93–95
Native check forgers, 253–254
Naive grasper rapist, 229
Narcoterrorists, 366
Natural areas, 154–155
Nature-nurture controversy, 122
NCVS, 38
Neighborhood fence, 257–258
Neobiological positivism, 122–130
Neoclassical theory, 100–101
Neurosciences, 128
New critical criminology, 185–186
    left realism, 186–187
    peacemaking, 186–187
NIBRS (National Incident-Based Reporting System), 31
NGRI (Not Guilty by Reason of Insanity), 136
Nigerian letter scams, 276–277
Nixon, Richard, 357–358
Nolo contendere, 302–303
Norms, 6–7
Numbers game, 404
Nuremberg principle, 347–348
Nursing home industry, 314

Occasional property crime, 247–254
    booster, 249
    check forgery, 254
    motor vehicle theft, 253
    shoplifting, 249–251
    snitches, 249
    vandalism, 251–252
Occasional property offender, 247–262
    conventional property offender vs., 1248
    societal reaction to, 261–262
Occupational crime, 305–316
    ABSCAM, 310
    bait and switch, 314
    crimes by individuals, 315
    Edelhertz's typology, 305–316
    embezzlement, 311
    funeral directors, 315
    greasy thumb on the scale, 314
    in education field, 315
    in financial field, 313
    in law field, 314
    in medical field, 313
    in nursing home industry, 314
    insider trading, 314
    Judgescam, 309
    legal regulation of, 302–303
    measurement/cost of, 298
    police corruption, 308–310
    private corruption, 310

    public corruption, 305–306
    reasons for lack of research, 298–299
    repair work, 314
    stockbrokers, 314
    Watergate, 309
    See also Corporate crime; White collar crime
Occupational/Organizational Crime Grid, 301
Ohlin, Lloyd, 151–152
Oklahoma City bombing, 370–371
Olson, Frank, 353
Omerta, 400
Online sexual predators, 456–460
Operation Bot Roast, 455
Operation Chaos, 353
Operation Dinero, 397
Operation Fastlink, 333
Operation Greylord, 309
Operation Mongoose, 409
Operation Rescue, 359
Operation Road Spill, 257
Operation Underworld, 412–413
Opium Wars, 439
Opportunist robber, 231
Oppositional culture, 84–85
Oppositional defiant disorder (ODD), 132–135
    See also Corporate crime
Organized crime, 380–419
    big business/government, 409
    brief history in U.S., 410–414
    classic pattern of, 403
    Cosa Nostra theory, 399–403
    criminal careers of organized crime criminals,
        416–419
    definition issues, 381–383
    definition of, 381–383
    drug control strategies, 417
    drug trafficking, 395–400
    ethnicity and, 493–494
    Hobbs Act, 418
    illegal businesses/activities, 404
    information sources on, 382
    international organized crime, 388–389
    investigative procedures, 417–418
    Italian-American Syndicate, 403–404
    Mafia, origins of, 400
    money laundering, 395–396
    nature of, 400–401
    organized crime continuum, 385–386
    Organized Crime Control Act, 418
    Patron theory (Albini model), 401–402
    pizza connection, 415–416
    public and legal reaction, 417–418
    RICO statute, 419–420
    snakeheads, 406
    software mobsters, 406
    street gangs, 386–387
    types of, generic, 383–384
    underground empire, 397
    See also Organized crime, brief history in U.S.;
        Organized crime, international
Organized crime, brief history in U.S., 410–414
    Apalachin meetings, 413
    before 1930, 410
    commission trials, 414–415

Gambino period, 414–415
Genovese period, 413
Luciano period, 412–413
selected events, 411–412
Organized crime, international, 388–397
Chinese triad societies, 306–389–391
Colombian cartels, 397
Mexican, 397–400
Russian organized crime, 391–393
Sicilian Mafia, 416
Yakuza, 388–389
Organized crime continuum, 385–386
Organized Crime Control Act, 418–419
Outlet fence, 257–258
Overcriminalization, 6–7,445

Palmistry, 7, 116
Paper hangers, 282–284
Paradigm shift, 425
Paraphilia, 434–436
Park, Robert, 154–155
Part I and Part II crimes, 27–28
Participant observation, 42–44
Password phishing, 454–460
Pathologically violent offender, 216
Patriarchal crime, 350–351
Patron theory (Albini model), 402–403
Peacemaking, 187–189
Pedophilia, definition of, 230
Pedophiliac/child molester, 230
Pentagon procurement scandal, 357–358
Phrenology, 116
Physical stigmata, 117–118
Physiognomy, 116
Pigeon drop, 267
Pirates of the internet, 332
Pizza connection, 415
Pleasure principle, 99–100
PMS (Premenstrual Syndrome) defense, 127
Pol Pot regime, 349
Police:
corruption of, 308–309
overview of, 217
secret, 353
Political assassins, 361
Political crime, 342–376
American nuclear guinea pigs, 354
assassination, 360–362
COINTELPRO, 353
crimes against government, 358–374
crimes against humanity, 344–345
crimes by government, 344–358
crimes by police, 352–354
definition of, 344
espionage, 361–364
genocide, 351–352
human rights violations, 349–350
ideology, 344
illegal surveillance/disruption/experiments,
352–354
international law and, 348
Iran-Contra conspiracy, 354–355
legal aspects of, 348
letter from Birmingham jail, 358

Nuremberg principle, 346–347
Oklahoma City bombing, 371
patriarchal crime, 350–351
political assassins, 361
protest/dissent, 358
raison d'état, 376
scandal, 354–355
secret police, 348–350
social movements, 359–360
terrorism, 365–373
Tiananmen Square, 350
Universal Declaration of Human Rights, 347–348
war crimes, 347–348
whistleblowing, 365
White House crime/scandal, 356–357
Political prisoners, 353
Political-social organized crime, 400–401
Ponzi scheme, 269–270
Positivism, 115–118
precursors of, 116–117
See also Positivist theory, biological; Positivist
theory, psychological
Positivist theory, biological, 115–126
adoption studies, 124–125
biological theories, critique of early, 121–122
biological positivism, early, 121–122
body types, 120–121
brain disorders, 123–124
endocrine imbalance, 125–126
Goring and, 119
Hooton and, 120
Jukes/Kallikaks, 119–120
Lombroso and, 117–119
neobiological theories, 122–123
neurological studies, 123
twin studies, 123–124
XYY Syndrome, 124–126
See also Positivist theory, psychological
Positivist theory, psychological, 130–139
crime profiling, 132
Eysenck and, 131–132
Freudian theory and, 130–131
insanity defense, 136–139
intelligence and crime, 134–135
psychometry, 131
Skinner and, 133
Yochelson/Samenow and, 133–134
See also Positivist theory, biological
Post hoc error, 131
Postmodernism, 190–191
Power control theory, 166–167
Power elite, 296
Power rape, 229
Precipitation hypothesis, 83–84
Price-fixing, 339–340
Primary deviance, 178–179
Prison gangs, 395
Private corruption, 311
Pro-choice movement, 359, 369
Profession, definition of, 263
Professional auto theft rings, 285–286
Professional burglars, 279–280
Professional crime, model of, 265–266
Professional criminal, characteristics of, 264

Professional fence, 280
Professional killers, 286–287
Professional property crime, 262–289
  argot, 264
  arsonists, 285
  auto theft rings, 285–286
  boosters, 276–277
  box man, 280
  burglars, 279–280
  cable piracy, 283
  cannons (pickpockets), 276–277
  car cloning, 287
  characteristics of, 263–265
  con, big, 267–269
  con, small, 267–269
  concept of, 263–264
  concept of professional crime, 263–264
  criminal careers of professionals, 287–288
  definition of, 262–263
  Edelhertz's typology of white color crime, 266–276
  emerging patterns of, 275–276
  fence, 279–280
  heels, 276–277
  identity theft, 269–271
  intellectual property theft, 1282
  killers, 286–287
  Maurer's big con, 269–270
  Nigerian letter scams, 276
  paper hangers, 281–282
  pennyweighting, 264
  Ponzi schemes, 270–272
  PTL scandal, 274–275
  pyramid schemes, 270–271
  religious cons, 273–274
  robbers, 284–285
  scams, 267–270
  societal reaction to, 289
Professional robber, 284–285
Professional spy, 362–363
Progression of knowledge, 3
  metaphysical stage, 3
  scientific stage, 3
  theological stage, 3
Prohibition, 443
Pro-life movement, 352–353
Property crime, 247–289
  arson, 259–260
  vandalism, 250–252
  See also Conventional property crime; Occasional
      property crime; Professional property crime
Prostitution, 427–431
  brothels, 429
  decriminalization of, 430
  johns, 430
  massage parlors, 430
  pimps, 430
  types of, 428–429
  underage prostitutes, 431
Protestant ethic, 427
Protest/dissent, as political crime, 357–359
Psychiatry, 130
Psychoanalysis, 130
Psychological theories, critique of, 130–140
  definition of, 130

Psychometry, 131
  definition of, 131
  positivist, 131
Psychopath, 138
Psychopathic assassins, 361
Psychosurgery, 123, 140
PTL scandal, 274–275
Public corruption, 306–311
Public-order crime, 424–447
  broken windows, 426
  crimes without victims, 425
  decriminalization, 446–447
  drunkenness, 443
  folk crime, 425–426
  homosexual behavior, 431–434
  legislated morality, 425–426
  moral panics, 440
  overcriminalization, 445
  Prohibition experiment, 443–444
  prostitution, 428–431
  sexual offenses, 433–436
  special populations, 444
PTL scandal, 274–275
Public-order criminal behavior, 425–447
Pyramid scheme, 291–292
Pyromaniac, 260

Quasi-agent spy, 362–363
Queer ladder of mobility, 393–394
Quetelet, Adolphe, 105–107
Quinney, Richard, 182, 190–191

Racial profiling, 131–132
Racketeering, 406–407
Radiation leaks, 327
Radical criminology, 190–192
Radical feminism, 184
Radical Marxist criminology, 190–192
  Chambliss on, 191–192
  Quinney on, 190–191
Raison d'état, 374–375
Rap music, violence in, 84
Rape, 225–231
  acquaintance rape, 228
  Amir v. Brownmiller, 229
  anger rape, 229
  factors in, 226–227
  gang rape, 227–228
  honor killings and, 350–351
  marital rape, 230
  power rape, 229
  real, 225
  simple, 225
  statutory, 217
Rape drugs, 229
Rational choice theory, 101
Rationalizations, 162
Reagan, Ronald, 357–358
Real rape, 225
Reckless, Walter, 164–165
Red Lake massacre, 214–215
Red light district, 430
Region and crime, 77
Regulatory agencies, 302–303

Reiman, Jeffrey, 182–183
Religion and crime, 81–82
Religious cons, 273–274
Research methods, 20–53
  alternative data gathering strategies, 33–34
  Candid Camera, 35
  confidentiality, 22
  content analysis, 45–46
  controlling for error in self reports, 42
  controlling for error in victim surveys, 41–42
  crime dip, 32–33
  crime indexes, 27
  crime rate, 31
  crime statistics, sources of, 33–34
  data gathering strategies, 33–34
  ethics and, 22–23
  evidence-based research, 37
  experiments in criminology, 34–35
  FBI reading room, 48
  issues/cautions in UCR data, 29–30
  issues/cautions in victim surveys, 40
  life history/case study, 44–45
  methodological narcissism, 47–48
  National Crime Victimization Survey, 41
  National Incident-Based Reporting System, 31
  objectivity, 22
  participant observation, 43–44
  physical trace analysis, 45–46
  posttests, 34–35
  pretests, 34–35
  qualitative strategies, 33–34
  quantitative strategies, 33–34
  reciprocity, 22–23
  reliability, 47–48
  Scared Straight experiment, 35
  secondary analysis, 45–46
  self reports, 41–42
  simulation, 45–46
  surveys, 38
  Uniform Crime Reports, 25–26
  unobtrusive measures, 45–46
  useful sources for research, 46
  validity/reliability/triangulation, 47–48
  victim surveys, 38–39
Restitution, 187–189
Restorative justice, 187–189
RICO statute, 387
r/K continuum theory of crime, 127
Robber barons, 299
Robbers, 231–234
  addict, 231
  alcoholic, 231
  mugger, 232
  opportunists, 231
  professional, 231
  typology of, 231
Robbery, 231–234
  armed, 231–234
  domestic rates, 231–234
  international rates, 231–234
Rock cocaine (crack), 438–439
Rockefeller Commission, 353
Rosenfeld, Richard, 151

Routine activities approach, 156–157
Rule of thumb, 237
Russian organized crime, 391–395

Sadism, 435
Sadistic rape, 229–230
Salami techniques, 456–460
Samenow, Stanton, 133–134
Sampling, 40
Sampson, Robert, 168–169
Sanctuary movement, 359
Savings and loan scandal, 322–323
Scams, 267–270
  badger, 267
  bank examiner, 267–268
  boojo, 269
  circus grifter, 268
  identity theft, 269–270
  investigator, 267–268
  Nigerian fee fraud, 276
  pigeon drop, 267
  postal fraud, 267
  three-card monte, 268
Scared Straight, 35
School violence, 222–223
Scientific research in criminology, 3
Scientific stage, features of, 3
Secondary deviance, 178–179
Secret police, 347–348
Secret Service study of school shooters,
    222–223
Self report surveys, 41–42
  class/criminality relationship, 41
  controlling for errors, 42
  delinquency items, 42
Serial murder, 211–213
  African Americans and, 213
Serial rapist, 229–230
Sex looter rapist, 229–230
Sexual assault, 225–231
  acquaintance rape, 227–228
  Amir v. Brownmiller, 229
  anger rape, 229–230
  group conformer rapist, 229–230
  marital rape, 230
  meaning stretcher rapist, 229
  naive grasper rapist, 229
  power rape, 229
  sadistic rape, 229
  serial rapist, 229
  sex looter rapist, 229
  sexual predators, 230
  stalking, 230–231
Sexual offenses, 433–435
  exhibitionism, 434
  fetishism, 434
  incest, 435, 437
  masochism, 435
  paraphilia, 434
  sadism, 435
  sexual predator, 435
  voyeurism, 434
Sexual predators, 435–439

characteristics of, 438
online, 360, 460–464
pedophiliac/child molester, 436
registration of, 461
U.S. Catholic priests, 436
Sexual scandal, in White House, 357–358
Shake down, 405–406
Shaming theory, 180–181
Shaw, Clifford, 154–155
Shining Path, 368
Shoplifting, 248–249
boosters, 249
definition of, 248–249
snitches, 249
Sicilian Mafia, 416
Sikh extremists, 368
Silent Brotherhood, 368
Silent Spring (Carson), 322–323
Silkwood, Karen, 323
Simple assault, 220
Skinner, B. F., 133
Skyjacking, 3
SMICE, 362
Smith Act, 344
Snakeheads, 406
Sniffer program, 455
Snitch, 250–251
Social bond theory, 165–166
Social class and crime, 70–71
Social control theory, 164–166
Gottfredson and Hirschi's general theory of
crime, 168
Hirschi's social bond theory, 164–166
Reckless's containment theory, 163–165
Social Darwinism, 118
Social disorganization theory, 154–155
Social movements, 349–350, 352–353
Social process theory, 152–163
Chicago school, 153–156
contacts in differential association, 150
designing out crime, 158–159
human ecology, 154–155
Matza's delinquency and drift theory,
162–163
Miller's focal concerns, 160–161
routine activities approach, 156–157
Shaw/McKay's social disorganization theory,
150–156
Sutherland's theory of differential association,
157–159
Sociological critical/integrated theory,
175–193
conflict criminology, 181–185
labeling theory, 177–189
new critical criminology, 185–191
radical Marxist criminology, 190–191
Sociological mainstream theories, 145–169
anomie, 146–153
developmental and life course, 167–169
social control, 163–167
social process, 153–163
Sociopath, 138
Sodomy, 431

Software piracy, 406–407
Somatotypes, 120–121
Spies, 361–365
definition of, 361
Hanssen spy case, 363–364
motives of, 362
typology of, 362–363
Spouse abuse, 236–237
Spree murder, 213–214
Stalking, 230–231
Statutory rape, 217
Sting operations, 258–259
Strain theory, 150–152
general, 150–152
Merton and, 147–150
Streaking, 434–436
Street gangs, 386–388
Sub rosa crime, 361
Subcultural theories, 151–153
Subculture of violence, 240–241
Subterranean values, 162
Sumner, William Graham, 6
Sumner's norm types, 6
Surveys, 38–42
Sutherland, Edwin H., 157–160
Sutton, Willie, 5
Sweetheart contracts, 312
Symbolic interaction, 8–9
Systematic check forgers, 282–284

Tax havens, 395–396
Teapot Dome scandal, 357–358
Techniques of neutralization, 162
Telecommunications fraud, 274, 455
Televangelists and crime, 274
Television violence, 83–84
Ten Most Wanted, 4
Terrorism, 365–373
brief history of, 365–368
definition of, 365–366
domestic, U.S., 369–372
limited political, 366
myths regarding, 367–368
nonpolitical, 366
official/state terrorism, 366
political, 366
quasi-terrorism, 366
retail terrorist, 366
separatists, 366
single issue terrorist, 366
social policy and, 375–376
state-sponsored, 366
types of, 365–366
Theoretical range, 198–199
Theory, in criminology, 21, 92
definition of, 21, 92
global fallacy and, 199–200
theoretical range, 198–199
See also Early/classical criminological theory;
Individual theory
Thermic law of crime, 106
Three-card monte, 268
Time bomb (computer program), 456–460

Tongs, 398–391
Transnational crime, 394
   See also International crime
Transparency international index, 307
Triads, 389–391
Trojan horse, 456–458
Truman, Harry, 357–358
Turk, Austin, 182–183
Turner diaries, 369
Twinkie defense, 125, 137
Twin studies, 123–124

UCR (Uniform Crime Reports), 25–33
   assault statistics, 27
   calculating crime rate, 27–28, 31–32
   definition of, 25
   index crimes, 27–28
   Part I crimes, 27–28
   Part II crimes, 27–28
   redesign of, 31
   sources of statistics, 25–26
Undercriminalization, 6–7
Underground Empire, 398
Universal Declaration of Human Rights,
   347–348
Unobtrusive research measures, 45–47
   content analysis, 45–46
   definition of, 45
   observation, 46–47
   physical trace analysis, 47
   secondary analysis, 47
   simulation, 46–47
Unsafe products, 322
Urban/rural differences in crime, 77

Valachi, Joe, 382
Validity, 49–50
   controlling for error, 50
   external, 49
   internal, 49
Vandalism, 250–252
   predatory, 251
   tagging, 251
   types of, 250–252
   vindictive, 251
   wanton, 251
Victim precipitation, 213–214
Victims:
   typology of, 214–215
Victim surveys, 38–41
   controlling for error in, 40
   definition of, 38
   international, 38
   issues/cautions, 40–41
Video piracy, 283
Violent crime, 207–242
   acquaintance rape, 228–229
   armed robbers on job, 234–236
   assault, 216–217
   BTK killer, 214–215
   child abuse, 235–237
   Conklin's typology of robbers, 233–234
   D.C. snipers, 214–215

   domestic violence, 234–238
   forcible rape, 228–231
   guns and, 222–223
   history in U.S., 207–208
   international, 210–211
   legal aspects, 216–217
   murder, 208–213
   patterns/trends in, 219–220
   Red Lake massacre, 214–215
   robbery, 231–234
   school violence, 222–223
   Secret Service study of school shooters, 223
   sexual assault, 225–226
   victim precipitation, 214–215
   Virginia Tech massacre, 212–213
   workplace violence, 220–221
   See also Violent offenders
Violent offenders:
   criminally, 216–217
   culturally, 216–217
   culture of violence, 239–240
   pathologically, 216–217
   situationally, 216–217
   subculture of violence and, 239–240
   types of, 216–217
Virginia Tech massacre, 212
Virus (computer program), 456
Vold, George, 181
Voluntary manslaughter, 211–213
Vory v zakone, 391–395
Voyeurism (Peeping Toms), 434–436

Walker spy ring, 365–366
War and crime, 82
War crimes, 347
Warez, 333
Warren Commission, 360
Wartime trade violations, 331
Watergate, 357–358
Wedtech scandal, 357–358
What works?, 36
Whistleblowers, 329–330
White collar crime:
   classic statement on, 295
   cost of, 297–298
   Occupational/Organizational Crime Grid, 301
   reasons for leniency, 336
   related concepts, 296
   See also Corporate crime; Occupational crime
White House crime/scandal, 354–355
Whitewater, 357–358
Williams, Wayne, 213
Workplace violence, 220–221
Worm, computer, 456–458

XYY syndrome, 124–125

Yakuza, 387–389
Yochelson, Samuel, 133

Zips, 415
ZOG, 369
Zone of transition, 154–156

About the Author

Frank E. Hagan is a native of the North Side of Pittsburgh and has earned degrees at Gannon, Maryland and Case Western Reserve. He is the Director of the James V. Kinnane Graduate Program in Administration of Justice and is the author of eight books. These are *Deviance and the Family* (with Marvin B. Sussman), *Introduction to Criminology* (7th edition), *Crime Types and Criminals, Research Methods in Criminal Justice and Criminology* (8th edition), *Essentials of Research Methods in Criminal Justice, Political Crime, White Collar Deviance* (with David Simon), and *The Language of Research* (with Pamela Tontodonato).

He is also the author or coauthor of many journal articles and articles in edited volumes. A recipient of the Academy of Criminal Justice Sciences Fellow Award (2000), he was also awarded the Teacher's Excellence Award by Mercyhurst College in 2006. His major interests are research methods, criminology and organized crime, white collar crime, and political crime and terrorism.